ISBN 978-1-330-19992-3
PIBN 10051001

FB &c Ltd, Dalton House, 60 Windsor Avenue, London, SW19 2RR.
Company number 08720141. Registered in England and Wales.

LONGMANS' CIVIL ENGINEERING SERIES

CIVIL ENGINEERING

LONGMANS' CIVIL ENGINEERING SERIES

CIVIL ENGINEERING

AS APPLIED IN

CONSTRUCTION

BY

LEVESON FRANCIS VERNON-HARCOURT

M.A., M.INST.C.E.

AUTHOR OF "SANITARY ENGINEERING," "RIVERS AND CANALS," "HARBOURS AND DOCKS," AND "ACHIEVEMENTS IN ENGINEERING"

SECOND EDITION

REVISED BY

HENRY FIDLER

M.INST.C.E., F.G.S.

AUTHOR OF THE ARTICLE ON "DOCKYARDS" IN THE "ENCYCLOPÆDIA BRITANNICA," AND "NOTES ON CONSTRUCTION IN MILD STEEL"

WITH NUMEROUS ILLUSTRATIONS OF WORKS IN THE TEXT

LONGMANS, GREEN, AND CO.

39 PATERNOSTER ROW, LONDON

NEW YORK, BOMBAY, AND CALCUTTA

1910

PREFACE

CIVIL engineering, even in its more restricted signification as applied to works of construction, covers such a wide range that it might reasonably appear somewhat presumptuous for any civil engineer to endeavour to deal with so far-reaching a subject, more especially within the limits of a single volume; and my sole excuse for making the attempt is that nearly ten years ago I unexpectedly received a request from Messrs. Longmans to undertake this onerous task. Possibly the publication of "Achievements in Engineering" in 1891, may have led to the request being addressed to me; but that book, giving descriptions, in a popular form, of some of the most notable engineering works, and only touching incidentally upon the principles involved in their construction for the sake of engineering students, was totally different in scope to this book, which treats primarily of the principles involved in the various branches of engineering construction, and refers to a great variety of works chiefly with the view of illustrating the methods by which these principles receive their practical application. Descriptions of a few important engineering works, in popular language, are comparatively easy of accomplishment; and the facility with which the earlier book was carried out, led me to underrate the difficulties and labour inseparable from the aims of the present book, which, had I appreciated them at the time, would probably have decided me to decline the proposal. Moreover, the leisure at my disposal proved much less than I had anticipated; and, owing to the various professional demands on my time, the preparation of the book has, to my regret, been greatly delayed, so that at times I almost despaired of bringing it to a conclusion; and it was only the courteous consideration accorded me by Messrs. Longmans, and their strongly expressed wish that I should not relinquish the undertaking, that have at last led to the completion of the book according to the scheme originally laid down by me. The long delay, however, has not been devoid of some compensating advantages, for it has enabled me to summon to my aid an extended experience of various engineering works and the conditions for their application, gained in the interval, not merely in the course of my ordinary professional avocations in England and Ireland, but also on the Continent and in distant countries. Thus a professional visit to India afforded me the opportunity of passing through the Suez Canal, of inspecting Port Said Harbour, of spending several days on the river Húgli and its estuary and viewing the inlet from the Ganges of its principal feeder, and of studying the changes the river has undergone by a comparison of the various charts. The British Association Meeting at Toronto, which I attended as a member of the Council, gave me an occasion for going up the St. Lawrence to Montreal, of seeing the various bridges across the Niagara Rapids below the Falls, and of traversing the Canadian Pacific Railway from Toronto to Vancouver, and back to Montreal. Attendances, also, at four International Navigation Congresses, with their accompanying visits to works, within the last ten years, in France, Holland, and Belgium, have enabled me to visit several of the principal navigation and maritime works in those countries, under the most favourable conditions, in company with the engineers in charge of the works, one congress alone, held in Paris, having led to visits to various inland navigation works, the weirs on the Lower Seine, the ports of Calais and Havre, the Furens Reservoir Dam, the Lyons Cable Railway, the Mulatière weir on the Saône, the regulation works along the Rhone from Lyons nearly to its outlet, the St. Louis Canal forming the navigable outlet from the Rhone, Marseilles Harbour, and the Marseilles Cliff Railway. A short visit to Chamonix and Switzerland enabled me to see the Chamonix Railway then in course of construction, and specimens of the Swiss rack and cable railways; whilst my duties in 1900, as the British Member of the International Jury for Civil Engineering at the Paris Exhibition, led to my receiving particulars about several large works in progress abroad, including the actual condition of the Panama Canal Works. The experience thus gained, and the knowledge thereby acquired, together with the information derived from earlier visits to works abroad, have enabled me to deal with many foreign works in the light of the results of personal observation and intercourse with foreign engineers.

In the preparation of this book, besides utilizing the above sources of information, together with a varied professional practice which has extended over a period of thirty-six years, particulars have been gathered, especially as regards details of works, from the most reliable sources available; and in this respect, I am specially indebted to the *Proceedings of the Institution of Civil Engineers* and the excellent technical library of that Institution. Every endeavour has been made to acknowledge in the notes the source from which any information or illustration may have been derived; and in many instances, owing to the necessarily concise character of the descriptions or details of works, references have been given where fuller accounts or information can be found, more with the view of assisting the reader, than because any acknowledgment is due. On account of the number of subjects which have had to be dealt with, and the restricted space available, the descriptions or details of the works selected as examples have been limited to what is necessary for the elucidation of the subject under consideration, and indications of the practice followed in actual execution. The classification of a subject under various headings, has often led to a subdivision of the descriptions of works in illustration amongst the several headings to which their different parts relate, as specially noticeable in the chapter on "Ship-Canals"; and this necessary subdivision has involved the preparation of a full index, so that details of any particular work, scattered over some pages in the book, may be collected together under the heading of the work in the index, with a concise indication of the purport of each reference. Moreover, by giving the purport, as well as the page of each reference in the index, the reader is saved the trouble of a wearisome search, possibly through several pages, before the required reference is found.

Care also has been taken to draw the illustrations, as far as practicable, to simple fractional scales, easily comparable with each other, so that the relative sizes of different works may be readily perceived; and where several illustrations are given of works of a similar class, such, for instance, as arched, suspension, girder, and cantilever bridges respectively, and cross sections of tunnels, movable weirs, canals, dock and quay walls, breakwaters, and masonry reservoir dams, the typical examples given of each class of works are grouped as much as possible together, and drawn to the same scale for greater facility of comparison, except where unusual divergencies of size, as in the cases of the Brooklyn and Forth bridges, precluded this arrangement. The numerous illustrations distributed throughout the book will, I trust, materially aid the elucidation of the principles involved, and add to the intelligibility and interest of the descriptions of works; and I desire to take this opportunity of acknowledging the care my assistant, Mr. Edward Blundell, has bestowed on the preparation of these illustrations, under my direction.

I now venture to submit this book, the product of much time and thought spread over some years, to the generous consideration of my professional brethren, both British and foreign, with the earnest hope that, in spite of many deficiencies, of which no one can be more conscious than its author, it may prove of interest, and perhaps of some service to them, in view of the concise grouping together in a single volume of the various branches of constructive civil engineering, and their illustration by numerous references to works, and may also to some extent assist the progress of engineering science, and the advance of the profession to which all engineers are so proud to belong; and in that case, I shall feel that my time and labour have not been spent in vain. I trust, moreover, that the book may be of considerable value to engineering students, in directing their attention to the principles forming the basis of design and construction in civil engineering, and in indicating the different ways in which these principles have been applied to actual practice. The book also, I venture to hope, may prove of use to many persons who, though not engineers, are concerned in some way or other with engineering undertakings, and desire to gain some insight into engineering practice, or take an interest in engineering progress.

L. F. VERNON-HARCOURT.

6, QUEEN ANNE'S GATE, WESTMINSTER, S.W.,
12th December, 1901.

PREFACE TO THE SECOND EDITION

A NEW edition of this work having been called for since the decease of its distinguished Author, it has been necessary to determine to what extent revision of the text was desirable.

The descriptions of engineering works which, at the date of the publication of the first edition, were in course of construction or in an early stage of initiation, were naturally subject to such revision as the progress of events had rendered necessary; while it was evident that works of magnitude which, during the period which has elapsed since the issue of the first edition, have been commenced, and, for the most part, brought to a successful conclusion, must receive such brief notice as the scope of this work would permit.

Subject to these considerations the Author's text has been, as far as possible, left intact, while on the other hand it is hoped that the attempt which has been made to include such examples as would best illustrate the progress of Civil Engineering within the last eight or ten years will not be without advantage to the student of construction.

HENRY FIDLER.

January, 1910.

CONTENTS

PART I.

MATERIALS, PRELIMINARY WORKS, FOUNDATIONS, AND ROADS.

PART II.

RAILWAY, BRIDGE, AND TUNNEL ENGINEERING.

PART III.

RIVER AND CANAL ENGINEERING, AND IRRIGATION WORKS.

PART IV.

DOCK WORKS; AND MARITIME ENGINEERING.

PART V.

SANITARY ENGINEERING.

LIST OF ILLUSTRATIONS

LIST OF TABLES

PART I.

MATERIALS, PRELIMINARY WORKS, FOUNDATIONS, AND ROADS.

CHAPTER I.

INTRODUCTION.

Objects of civil engineering—Scope of the book—Relation of civil engineering to science—Mathematics as applied to civil engineering—Physics as applied to civil engineering—Chemistry in relation to civil engineering—Geology in relation to civil engineering—Meteorology in relation to civil engineering—Remarks on the scientific requirements of civil engineers.

CIVIL engineering was defined by Thomas Tredgold, in 1828, as "the art of directing the great sources of power in nature for the use and convenience of man." The epithet *civil* has been applied to this great science of construction in order to distinguish it from military engineering, to which its main objects present a very striking contrast. Military engineering, indeed, is concerned with provisions for the attack and defence of fortresses, and the defeat of opposing armies; whereas civil engineering is directed to the extension of the means of communication and commerce, and to the promotion of the well-being and prosperity of mankind.

The civil engineer, by extending railways through undeveloped or barbarous countries, becomes the pioneer of progress and civilization; and he is largely instrumental in developing the intercourse and trade of communities and nations, by improving rivers, and constructing roads, railways, canals, harbours, and docks. Moreover, the civil engineer, by carrying out irrigation works in hot, dry countries, converts arid regions into fertile plains, thereby greatly increasing the productiveness of these districts, and averting the famines resulting from a deficiency in rainfall. Furthermore, by providing an ample supply of good water, and efficient sanitary arrangements, he is enabled to preserve from ravaging epidemics and untimely death the vast populations which, in recent years, have been crowding more and more into cities and towns.

Scope of the Book.—Civil engineering, in its widest signification, comprises a great range of subjects, embracing not merely works for developing means of communication, for facilitating sea-going trade, and for securing water-supplies and efficient drainage for towns, but also mining and metallurgy; the lighting of towns with gas and electricity; telegraphic and telephonic inter-communication; the construction of steam-engines, other forms of motors, and machinery of all kinds; ship-building, including ironclads and swift cruisers; and the manufacture of heavy ordnance, projectiles, and torpedoes. Such a great variety of subjects, however, could not be properly dealt with by a single author, or condensed into a single book. Moreover, the vast and increasing development of engineering science, and its numerous ramifications have necessarily led to the subdivision of engineering into different branches, such as mechanical engineering, mining engineering, electrical engineering, and naval architecture, as well as civil engineering, which is more and more being employed to denote the special constructive branch of engineering as distinguished from the other four. It is, accordingly, proposed to deal in this book with civil engineering in its modern, restricted sense; and even in this limited application of the term, it comprises some distinct branches which are often regarded as more or less subdivisions of the science, such as railway, canal, river, maritime, hydraulic, and sanitary engineering, though much less distinctly separated from one another than the five branches of engineering enumerated above.

Relation of Civil Engineering to Science.—In order to be able to direct rightly the forces of nature for the benefit of mankind, it is essential to possess some knowledge of the principles of these forces, which have been long studied by scientific workers, the recorded results of whose labours are termed "natural science." Civil engineering is, accordingly, based primarily upon natural science; and it may, indeed, be regarded as the practical application of the discoveries of science for industrial purposes and the general well-being of the human race. It is evident, therefore, that a scientific education should constitute an indispensable part of the preliminary training of every person who aspires to embrace the profession of a civil engineer. Some branches of science, however, are more intimately connected with the problems involved in civil engineering than others; and their relative importance depends upon the special branch of the subject to which a civil engineer may have occasion to devote his attention. Mathematics, physics, chemistry, geology, and meteorology find practical applications in the designs and works of civil engineers; and a knowledge of the general principles of some of these sciences is an invaluable assistance in the succcessful practice of civil engineering.[1] The aims, however, and the requisite knowledge of the man of science, and of the civil engineer, are essentially different. The man of science generally devotes himself to the minute investigation of some special branch of one of these sciences, with which long study has rendered him intimately familiar. The civil engineer, on the contrary, merely needs an adequate general acquaintance with these sciences to be enabled to select those portions of the discoveries of the great body of scientific investigators, which are capable of useful application to the requirements of his profession.

[1] *Report of the British Association*, 1895. Ipswich Meeting, President's Address to the Mechanical Science Section, pp. 782-788.

Mathematics as applied to Civil Engineering.—Trigonometry constitutes the basis of surveying and tacheometry, which are employed for the preliminary examination of sites before works are undertaken; and it is also made use of in setting out the lines of works about to be carried out. Logarithms, moreover, are useful in simplifying the calculations required in triangulation, and the ranging of railway curves.

Statics supply the groundwork for the design of bridges and other structures, by determining the direction and amount of the strains at different parts of the structure, which would result from the loads liable to be imposed on the structures after their erection. Graphic statics, however, in which the strains are represented in magnitude and direction by lines on a diagram, have to a great extent advantageously superseded analytical methods. Hydrostatics serve for the calculation of the pressures to which reservoir dams, lock-gates, and weirs are exposed when supporting given heads of water; and hydrodynamics deal with the general laws of fluid motion, which, however, require to be supplemented by experiment in order to determine with accuracy the flow of water in pipes and open channels, points of great importance in hydraulic engineering.

Geometrical optics are utilized in determining the forms to be given to the lenses for concentrating the rays of the lamps of a lighthouse into a single beam of parallel rays, directed to the required quarter of the sea for giving due warning to vessels. Astronomical observations have also been employed to enable surveyors to determine directions when traversing unexplored regions. Mathematics, indeed, enter so largely into the solution of problems presented in the various branches of civil engineering, that no engineer would be justified in dispensing with their aid.

Physics as applied to Civil Engineering.—The researches of physics have such an intimate connection with civil engineering, that they in reality occupy as important a position in relation to it as mathematics. The physical properties of matter, indeed, cannot be disregarded by the civil engineer. He has to provide, in his designs of large structures, for expansion by heat and contraction by cold. The feasibility of constructing long tunnels under high mountains, depends upon the heat which is liable to be encountered at great depths below the surface; whilst the adoption of compressed air in the advanced headings of such tunnels has greatly facilitated their construction, by serving to drive the perforators for drilling holes in the hard rock, and by the improved ventilation from the supply of air thereby afforded. Compressed air has also been employed for many years past in executing subaqueous foundations with the same security as on dry land; and this system has been more recently applied to the driving of tubular tunnels through water-bearing strata. Congelation of the soil is another process which has been used for sinking wells through soils charged with water and running sand.

The properties of light as regards its visibility at considerable distances, and the relative penetration of different kinds of light through fog, have to be considered by the civil engineer with reference to light-house illumination; and investigations in acoustics have to be made for determining the best kind of sonorous signal for giving warning of danger in foggy weather. A just appreciation, indeed, of general physical considerations, in relation to civil engineering, appears to be indispensable for the satisfactory prosecution of large undertakings.

Chemistry in Relation to Civil Engineering.—The chemical constitution of various materials employed in construction by civil engineers, furnishes in some instances important evidence as to their soundness. The elements of arsenic, silicon, sulphur, phosphorus, and manganese in iron and steel each exert an important influence on the strength and other qualities of the metal, while the percentage of carbon largely determines the quality and grade of the material. The strength and durability of mortar and concrete depend greatly upon the chemical composition of the lime or cement with which they are made, more especially when employed in structures in the sea. The powerful explosives, also, used in carrying tunnels through hard rock, and in blasting rocky shoals under water, are chemical compounds, and are comparatively recent discoveries as compared with gunpowder.

Chemical processes have been employed in the numerous efforts to utilize sewage economically for agricultural purposes; and chemical analysis has to be resorted to for determining to what extent the effluent water, from land irrigated with sewage, has been rendered innocuous. Recourse, moreover, is had to chemical analysis for deciding as to the adequate purity of any proposed source of supply, and in the periodical testing of the state and due filtration of the potable waters supplied to towns.

Geology in Relation to Civil Engineering.—The nature of the ground to be excavated in ordinary cuttings for railways or canals, is generally adequately indicated by borings or trial pits; but where the cuttings to be executed are very deep and extensive, and rock or treacherous strata are liable to be met with at some depth below the surface, as happens occasionally in railway works, and more commonly in the construction of large ship-canals, some knowledge of the general geology of the district is valuable. Such knowledge becomes of much greater importance in driving long tunnels at a considerable depth below the surface, where the difficulties to be encountered, the possible influx of large volumes of water during construction, and the cost and period required for the work greatly depend upon the nature and dip of the strata to be traversed. Geological considerations are also important in judging as to the impermeability of a site proposed for a reservoir, and the adequate stability of the foundation for a reservoir dam; and indications of the thickness of water-bearing strata, their depth below the surface, their dip, and the extent and position of their outcrop, as well as the prospect of fissures, and the possibility of faults, are of the utmost importance in determining the position where a deep well for water-supply should be sunk, the depth to which it would require to be carried, and the probability of obtaining a suitable supply of water.

Meteorology in Relation to Civil Engineering.—Gales exert a considerable pressure against high walls, bridges, and roofs; and therefore a knowledge of the maximum force of the wind in different localities is important in designing structures, so as to ensure their stability during exceptional storms. The maritime engineer has to ascertain the direction and prevalence of the strongest gales to which his works on the sea-coast are liable to be exposed, so as to provide shelter from the worst quarters by works of requisite stability; and he should also know the general direction and average force of the wind at different periods of the year, in order to select the calmest period of the year for the prosecution of his works. In localities visited periodically by cyclones or earthquakes, special precautions have to be adopted to prevent the structures erected from succumbing to these calamitous visitations.

Observations of the amount and distribution of rainfall are very valuable for hydraulic engineers, to enable them to determine the minimum flow available in rivers for navigation or irrigation in the summer or the dry season, and the maximum discharge for which an adequate channel has to be provided to prevent inundations. The varying influences of evaporation and percolation in reducing the actual amount of rainfall available, and the equalizing effect of forests and vegetation on the flow of mountain streams, are of considerable interest to engineers concerned in water-supply and river works; and meteorological considerations are essential for forming a reliable estimate of the average amount of water that can be collected for distribution from a given catchment area.

Remarks on the Scientific Requirements of Civil Engineers.—The foregoing observations indicate that engineers have numerous opportunities for utilizing the researches of science in the practice of their profession, and show how intimately science is bound up with civil engineering, which is its practical exponent. It is, moreover, evident that a fair knowledge of mathematics and physics, in their practical aspects, should constitute a recognized necessary portion of the professional equipment of every civil engineer; whilst some acquaintance with chemistry, geology, and meteorology is desirable, and in some branches essential, for the intelligent and systematic carrying out of civil engineering works. An engineer cannot devote sufficient time to become very proficient in one or more of the sciences enumerated, except under exceptionally favourable conditions in his preliminary education; but a general scientific training furnishes the most valuable preparation for a profession which deals with the practical applications of science; and, other conditions being equal, those civil engineers will be the best qualified for the successful pursuit of their profession, and the advancement of the science of civil engineering, who have been most thoroughly trained in the principles of science, and have become imbued with scientific methods of observation and inquiry. In complicated or abstruse cases, a civil engineer should necessarily seek

the assistance of a scientific expert; but it is necessary that
understand the principles of the particular science involved
to appreciate properly the views of his adviser; and an en
be less liable to errors, and better able to utilize fully the
scientific investigations, in proportion as he has a clear insig
principles of the sciences which he is called upon to apply for t
of mankind.

CHAPTER II.

MATERIALS EMPLOYED IN CONSTRUCTION.

Choice of materials—Materials used in works—Standardization—**Timber**: nature, protection, uses, strength of various kinds—**Fascines**: uses, construction, mattresses—**Building-stone**: principal kinds ; from primitive rocks, special uses ; sandstones, nature and uses ; limestones, forms and qualities ; strength of different kinds—**Bricks**: conditions of use and advantages, composition and manufacture, form and qualities, strength—**Limes**: ordinary, sources and uses ; hydraulic, composition and advantages—**Cements**: natural, sources ; artificial, introduction ; slag cement, composition ; Portland cement, composition, fineness, strength, tests—**Mortars**: proportions of ingredients, manufacture, strength, remarks—**Concrete**: importance, composition, proportions in blocks, bags, and mass ; forms and uses, mixing and depositing, advantages of, in mass, effect of sea-water—**Iron and Steel**: differences in composition ; cast iron, uses and strength, disadvantages ; wrought iron, uses and advantages ; steel, manufacture, advantages, strength, uses—**Safe Strains on Materials**: limit of elasticity ; dead and moving loads ; stresses allowed on structures ; wind-pressure and snow.

THE duty of a civil engineer is to carry out the works entrusted to him with the most suitable materials, and in as economical a manner as practicable, consistently with efficiency and stability. The materials employed should therefore depend, not merely on the nature of the work, but also on the conditions of the site, and the materials most readily available in the locality.

Materials used in Works.—The materials commonly employed by civil engineers are timber, fascines, stone, bricks, lime, cement, mortar, concrete, iron, and steel, as well as various substances in their natural condition, such as sand, gravel, shingle, rubble stone of different kinds, chalk, and clay. In some cases, inspection by an experienced person is sufficient to determine whether the materials are suitable for the particular purpose. Mechanical tests, however, and in some instances chemical tests also, are required to secure that materials employed in construction are up to the required standard of strength and durability, as in the case of cement, iron, and steel; whilst where high pressures are liable to be reached, and the slightest failure might lead to disastrous results, as for instance in high reservoir dams, the stone, bricks, mortar, and concrete have to be subjected to tests.

The important question of the standardization of the tests of materials employed in Engineering Construction has received much attention in recent years, and the student will do well to carefully study the reports and specifications issued and revised from time to time by the Engineering Standards Committee, more especially those dealing with Portland

Cement, Structural Steel, Railway and Tramway Rails, the Properties of British Standard Sections, Test Pieces, Nuts and Bolt Heads, etc.

TIMBER.

The timber usually employed in permanent, as well as temporary, works is fir, obtained from the large forests of Northern Europe and America, as it is generally the cheapest wood of adequate size, and is more easily worked than hard wood. Where, however, hard-wood trees abound, or in positions where the timber is specially exposed to wear and tear, or to other sources of injury or decay, hard woods are adopted. Thus fenders put along jetties, piers, and quays, against which vessels rub, and the wedges used for fixing rails in chairs, are made of hard wood; and greenheart, from British Guiana, has been extensively employed for dock-gates and jetties exposed to the ravages of the teredo, which is found in salt water, and soon honeycombs most sorts of timber when immersed in sea-water.

Nature of Timber.—Timber should be compact, free from cracks, shakes, hollows, knots, and other defects, and should be procured from trees cut down in their prime, and at the period of the year when most devoid of sap. Most varieties of timber, and particularly soft woods, tend gradually to decay when exposed to the weather, more especially in positions where they are alternately wet and dry, as at the surface of the ground or when standing in water varying in level; whereas timber buried in the ground, or always under water, remains sound for a long period. Sap in timber is liable to cause decay by its decomposition; and the seasoning of timber by the removal of its moisture, can be effected either naturally by long exposure to a current of air under shelter, or artificially by subjecting it to a current of hot air. Teak, oak, greenheart, ironbark, and jarrah are some of the most durable and strongest woods; whilst larch and pitch pine, though decidedly less durable and less strong, are superior in these respects to red pine and spruce, but more difficult to work.

Protection of Timber.—Timber must be thoroughly dried before being protected externally by paint or tar, otherwise the imprisoned moisture rots the wood. A more effective method, however, of protecting timber exposed to the weather or damp, consists in thoroughly filling up the pores of the wood with some preservative liquid, such as creosote, mercuric or zinc chloride, or copper sulphate, thereby precluding the entrance of moisture. Zinc chloride or copper sulphate is sometimes injected on the spot into recently felled trees, thus expelling the sap directly; but by the ordinary process, the liquid is forced under pressure into the seasoned timber, for which purpose creosote is very extensively employed.

Uses of Timber.—Timber is largely used for jetties, landing-places, floors, roofs, and sleepers, and also for bridges and viaducts where economy is a paramount consideration, as in the extension of railways through undeveloped countries where wood is abundant; but though numerous timber bridges were erected in the first instance on the Great Western Railway, timber is only now used for such structures under exceptional conditions. Owing, also, to the attacks of the white ant, wood has been abandoned for sleepers in some tropical countries. Moreover, though greenheart is found to resist successfully the attacks of the teredo in temperate seas, it is liable to be injured by them in tropical waters, as, for instance, in the River Hugli at Calcutta; and the immunity of jarrah under such conditions, which has been suggested as a substitute, has not been proved. In France, wooden dock-gates, exposed to sea-water, have been protected by studding them all over with large-headed nails.

Strength of Timber.—Timber possesses about double the strength in tension that it has in compression; and most of the Australian hard woods are about twice as strong in tension as red pine and spruce, the former possessing an average tensile strength of over 9 tons per square inch, and the latter about 4½ tons.[1] The strength of timber in compression, being less than in tension, and influenced by the length of the specimen, is an important factor in timber construction, together with the transverse and shearing resistances. The crushing strength of white pine is between ½ ton and 1 ton per square inch; of Danzig fir and yellow pine, ¾ ton to 2 tons; of English oak, 1½ tons; of pitch pine, about 2 tons; of Norway spruce, nearly 2½ tons; of Oregon pine, nearly 3 tons; of Australian hard woods, from 2 to 4 tons;[2] and of greenheart, about 4 tons per square inch. The shearing strength along the fibres has been found to range, in American timbers, from between 253 and 374 lbs. per square inch for spruce, up to between 726 and 999 lbs. for red oak; and in Australian timbers, from between 700 and 1400 lbs. per square inch for gum, up to between 1100 and 1400 lbs. for ironbark. The coefficient of the bending strength, or modulus of rupture of timber beams, is about 2 tons per square inch for Danzig fir, Baltic red pine, and American spruce and white pine; 3 tons per square inch for Swedish and Russian pine, and American yellow pine; about 4⅓ tons for English oak; and it ranges, in the hard woods of Australia, from about 3 tons for the red gum of Victoria, up to 8 tons per square inch for ironbark and 9 tons in the salmon gum of West Australia.[3]

FASCINES.

Uses of Fascines.—When a road or railway has to be carried across a soft marsh or a bog, or when protection works, training works, or dams have to be carried out in rivers at places where stone is difficult to procure, faggots of wood or fascines are employed. For instance, the Liverpool and Manchester Railway was carried across Chat Moss by laying a number of bundles of brushwood across the bog along the line of the railway, which served to support the railway in crossing the soft marshy ground;[4] and fascines are extensively used in river works in Holland and the United States.

[1] "Engineering Construction in Steel and Timber," W. H. Warren, p. 60.
[2] *Ibid.*, p. 73. [3] *Ibid.*, p. 73.
[4] "Lives of the Engineers," Samuel Smiles, vol. iii. p. 224.

Construction of Fascines and Mattresses.—Fascines are ordinarily formed of a bundle of five or six sticks or pieces of brushwood, about 7 to 11 feet long, bound firmly together near each end by an osier. Recently-cut willow furnishes the best material; but the brushwood of other trees, such as oak, hazel, ash, and alder, is also employed. The fascines are fixed in position by pegs or stakes, and are weighted with stones, bricks, or clay. For large works in rivers, the fascines are laid together in rows for three or more layers, connected together by strong intertwined bands of fascines to form a large mattress, the layers being fastened together by stakes driven down through the interstices of the upper network and round the edges of the mattress, and joined together by bands of interwoven sticks. The mattress, on completion, is launched from the bank, towed into position, and sunk by being loaded with rubble stone thrown out from boats alongside. In constructing dams, training walls, or jetties, several rows of mattresses are laid one over the other till the desired height is reached, the mound thus formed being capped with stone pitching or concrete blocks. The silt deposited by the river in the interstices of the mattresses, and over the surface of the mound, protects the fascines from decay.

BUILDING-STONE.

The employment of stone in engineering works depends on the accessibility and nature of the stone in the neighbourhood of the works. If stone is readily available, is durable, and fairly easily dressed, it is naturally used extensively; but when the stone is very hard, or has to be procured from a distance, in the first case it is only used for special purposes, or as rough walling or backing, and in the second case brickwork or concrete generally proves preferable. Granite, sandstone, and limestone are the principal kinds of stone used in construction.

Stones from Primitive Rocks.—Granite, though very valuable for certain classes of work where great hardness and durability are essential, as for the sills and hollow quoins of large locks, the coping of docks, the voussoirs of large arched bridges, and the bed stones of large girders or columns, the difficulty of working it and its consequent cost render it unsuitable for ordinary masonry. Other igneous rocks, such as gneiss, whinstone, trap, and basalt, are used to some extent for masonry in the districts where they are found; but gneiss is more specially suited for flagstones, and the others for paving and metalling roads, on account of their hardness and durability, and the small blocks in which they are generally found. The igneous rocks are particularly durable, owing to their compactness and consequent imperviousness to water; and they are specially adapted for bearing high pressures, heavy weights, and great wear and tear.

Sandstones.—Sandstones consist of grains of quartz cemented together into a solid mass by a substance of variable composition, upon the durability of which the value of the stone depends. When the cementing substance consists mainly of silica, mica, or hard felspar, the sandstone is hard and durable; but when it is composed of alumina or oxides of iron, the sandstone is softer and more liable to disintegration. When calcium carbonate forms the cementing material, the sandstone can be easily sawn and worked, but it is somewhat readily injured by the weather.[1] Sandstones, accordingly, vary considerably in their qualities, some of the best for building being the millstone grit, the sandstones of the coal-measures, and the red sandstones. They are largely used for facework and ashlar masonry, on account of the facility with which they are dressed.

Limestones.—Calcium carbonate forms the main ingredient of limestones, which vary considerably both in composition and quality, the hardest forms being marble, which is an almost pure crystallized calcium carbonate, and dolomite, which is a double carbonate of calcium and magnesium. Oolitic limestones vary greatly in durability, some forms being subject to rapid decay; whereas Portland stone is durable, and also some varieties of Bath stone, which are easily worked and harden by exposure to the air, though others readily disintegrate. There are also shelly limestones, of which Purbeck limestone is an instance, being largely composed of small shells. As a rule, limestones are softer, absorb more water, and are more subject to injury from the weather than sandstones.

Strength of Building-stones.—In practice, masonry, brickwork, and concrete are almost always wholly subjected to compressive strains, for mortar is incapable of undergoing tensional strains of any importance; and therefore only the crushing strength of stone has to be ascertained, as well as its relative impermeability to the absorption of water, which furnishes some measure of its durability, though this and its rate of wear are matters which are largely based on experience. Granite and Italian white marble possess a very similar compressive strength of about 1400 tons per square foot, whilst in basalt and slate it averages about 1200 tons; whereas the compressive strength of sandstones ranges from about 700 to 260 tons, and of the better class of limestones, from about 600 to 200 tons per square foot.[2] The percentage of water absorbed, which is under 1 per cent. for granite, whinstone, slate, and marble, amounts to between $3\frac{1}{2}$ and 10 per cent. for sandstones, and between $5\frac{1}{4}$ and $12\frac{3}{4}$ per cent. for limestones.

BRICKS.

When a district is devoid of stone, and clay is obtainable, brickwork is commonly employed in place of masonry for constructive purposes. Brickwork possesses the advantages of being carried out with less skilled labour, and with lighter staging, termed scaffolding, than masonry, as no dressing is required, and no heavy weights have to be lifted and put in position. On the other hand, there are many more

[1] "Chemistry for Engineers and Manufacturers," B. Blount and A. G. Bloxam, vol. i. p. 2.

[2] "Notes on Building Construction," Part iii.

joints in brickwork than in masonry, necessitating the use of more mortar, and affording much less bond in the work, which is sometimes compensated for by the insertion of long strips of iron at intervals along some of the joints, an arrangement known as hoop-iron bond.

Composition of Bricks.—Clay containing some sand, or pure clay mixed with about a fourth of its volume of sand, is used for the manufacture of bricks, the usual composition of the material being chiefly silicate of alumina and silica in the form of sand, combined generally with lime, potash, and magnesia, together with some ferric oxide, which, when present in a fair proportion, gives a red colour to the bricks. The bricks are manufactured by making the mixture of clay and sand into a paste with water, which is then moulded into bricks; and the bricks are burnt in clamps or stacks, or in kilns. There are various kinds of bricks, as their colour, porosity, weight, and strength depend upon the composition of clay, loam, or marl from which they are manufactured. Gault clay, containing 25 per cent. of calcium carbonate, forms a close, light-coloured brick, suitable for facework; and pressed bricks are also used for the same purpose. The hardest and strongest bricks are made from a highly ferruginous clay, and are generally known as Staffordshire blue bricks, which are much less porous than ordinary bricks.

Form and Qualities of Bricks.—The dimensions of ordinary bricks, including the thickness of the joints, are 9 inches long, 4½ inches wide, and 3 inches high. They are generally laid in alternate rows of headers and stretchers on the outside in large engineering structures, for the sake of bond; and the best bricks, or sometimes special bricks, are laid along the face. Hand-made bricks are generally formed with a hollow in the top, which keys the bricks into the mortar.

Bricks when knocked together should give a sort of metallic ring, and when broken across should exhibit a compact uniform texture, quite free from iron pyrites, and any lumps of lime, which in slaking would crack the brick. Bricks weigh from about 6 to 10 lbs., according to their quality, some being so porous as to absorb a fifth of their weight of water; whilst others, such as blue bricks, are so compact that they absorb hardly any water. Ordinary bricks in hot, dry weather should be wetted just before being laid, as this promotes the adherence of the mortar. Smooth, compact, non-porous bricks, like blue bricks, should be laid in cement mortar or neat cement, as ordinary lime mortar does not adhere firmly enough to their smooth, impervious surfaces.

Strength of Bricks.—London stock bricks have a crushing strength when tested singly of from 80 to 180 tons per square foot; gault bricks from 100 to 180 tons; Leicester red, 380 tons; Staffordshire blue from 400 to 700 tons; but the compressive resistance of brickwork in the form of piers is much below the above figures, being influenced by the nature of the mortar and other conditions.

LIMES.

Limes employed with sand for making mortar, may be divided into two classes, namely, ordinary lime obtained by calcining pure limestones, and hydraulic lime made from limestones containing clay or other substances, such as are found in the Lias formation.

Ordinary Lime.—Common lime is obtained by calcining limestone composed almost wholly of calcium carbonate, such as chalk, marble, and some oolites, resulting in the expulsion of the carbonic acid and water. The quicklime thus formed slakes when mixed with water, producing calcium hydrate, which gradually hardens in the air by drying and the absorption of carbonic acid, a change termed setting. This class of lime, however, does not set at all under water, and is therefore only suitable for structures out of the reach of water. It is largely used for ordinary building operations, owing to its furnishing the cheapest form of mortar, such as greystone lime mortar in London; and the addition of sand to the lime in making mortar, by exposing a much larger surface of lime, facilitates the absorption of carbonic acid from the air, and consequently hastens the setting.

Hydraulic Lime.—Lias lime in England, Aberthaw lime in Wales, Theil lime in France, and other limes of similar character, containing 10 to 30 per cent. of clay, other silicates, and sometimes certain other substances, in addition to calcium carbonate, are able to set under water, and have therefore been employed for hydraulic works. These limes, accordingly, are more generally serviceable than ordinary limes, but they are also usually somewhat more costly.

Cements.

When water-tightness, rapid setting, or special strength are requisite, cements are employed instead of lime. Some cements are obtained from limestones containing between 20 and 40 per cent. of clay; whilst other cements are manufactured by mixing the suitable ingredients before burning, in the proportions which experience has shown to be satisfactory. There are, therefore, natural and artificial cements.

Natural Cements.—The materials from which natural hydraulic cements are obtained, consist of nodules found in the London clay and other beds, and thin strata interspersed amongst hydraulic limestone beds.

Roman cement, the best known of these natural cements, was discovered in 1796, and is formed by calcining nodules found on the island of Sheppey, at Harwich, and other places, containing about 66 per cent. of calcium carbonate, 25 per cent. of clay, and some ferrous oxide. It sets quickly, and therefore is very useful as a protection for the face-joints of tidal work in exposed situations; but its ultimate strength is only between one-half and one-third that of Portland cement.[1] The cement should be used soon after its manufacture, for exposure to the air soon reduces its strength.

Medina cement, another quick-setting hydraulic cement, obtained from septaria procured from the Isle of Wight, the bed of the Solent, and Hampshire, has been used for tidal work. It is very similar to

[1] *Proceedings Inst. C.E.*, vol. xxxii. pp. 280 and 281.

Roman cement, but contains rather more lime; and though it apparently reaches its maximum strength somewhat sooner than Roman cement, it is liable to deteriorate in course of time, which is a fatal objection to its use in permanent works exposed to the sea.

Certain volcanic earths, consisting very largely of clay which has been subjected at some period to heat, possess the property of converting ordinary limes into a sort of hydraulic cement when mixed with them; such, for instance, as pozzolana, found at Pozzuoli near Vesuvius, and trass, obtained on the sites of extinct volcanoes.

Artificial Cements.—In 1824 it appears to have been first discovered that a cement could be manufactured from a mixture of chalk and clay, which was called Portland, from the supposed resemblance of the blocks made with it to Portland stone. Portland cement, however, was not extensively used for many years after its discovery, being only gradually adopted for dock and harbour works on the south coast of England and on the continent, till numerous systematic tests, commenced in 1858, led to its use for the main drainage works of London; and these trials having resulted in great improvements in the quality of the cement, and increased reliance in its soundness, its employment has been extended to most classes of works where Roman cement and hydraulic limes were formerly used.

Slag cement is another artificial cement which has occasionally been used. It is made by granulating slag from the blast furnaces by chilling it in water, and then grinding it with lime, to which it imparts hydraulic properties. Slag cement contains rather less than 50 per cent. of lime, in place of about 60 per cent. contained in Portland cement; but it contains rather more silica, and considerably more alumina and oxide of iron than Portland cement. Slag cement, however, though it has been used for concrete deposited under water in some harbour works,[1] sets slowly as compared with Portland cement, and does not increase in strength as quickly. More uncertainty, moreover, has been experienced hitherto as to the quality of the cement produced by this process, than in the now well-understood manufacture of Portland cement.

Portland Cement.—Great attention has been paid to the manufacture of Portland cement for many years past, and to the tests requisite to ensure that any unsoundness or deficiency in strength of the cement shall be detected before it is used. Portland cement is commonly made by an intimate mixture of clay and chalk, varying in proportion from about 3 of clay and 7 of chalk, to 1 of clay and 4 of chalk, according to the differences in constitution of these substances, which is afterwards burnt in a kiln, whereby the water and carbonic acid are driven off, and the calcium carbonate is converted into quicklime, which, decomposing the clay, forms calcium silicate and calcium aluminate, of which the clinkered cement should chiefly consist.[2] Clay consists almost wholly of silicate of alumina, and chalk of calcium

[1] *Proceedings Inst. C.E.*, vol. cv. pp. 234 and 235.
[2] *Ibid.*, vol. cvii. p. 32.

carbonate, which together, after burning, form Portland cement, which should be composed of about 62 per cent. of lime, 22 per cent. of silica, 9 per cent. of alumina, 2 to 3 per cent. of oxide of iron, and not more than 2 to 3 per cent. of magnesia. Similar cements can also be made from various other substances, provided they contain the necessary ingredients in the requisite proportions when mixed together.[1] The presence of free lime or magnesia is liable in time to injure works, by the swelling of these substances in combining with water, though free lime in a cement increases its strength in the first instance.

The adequate burning of the cement is tested by its specific gravity, which should not be less than 3·1, as a lower specific gravity shows that the cement is underburnt. The value of any cement largely depends upon the fineness to which the clinker is ground; for the extent to which the cement can coat over, and consequently cement together a certain quantity of sand of definite size, depends upon its fineness. Samples of the cement supplied are, accordingly, passed through a very fine sieve, having 76 meshes of wire, 0·0044 inch diameter, or 5776 meshes per square inch, leaving not more than 3 per cent. residue, and through a finer sieve of 32,400 meshes per square inch, with a residue of not more than 18 per cent. The coarser cement left on the sieve has little cementing value, and should be treated as so much sand in determining the proportion of cement to be used.

The strength of Portland cement is almost always tested in tension, as the most convenient method, owing to its compressive strength averaging about nine times its tensile strength. The cement is formed into a paste with water, and moulded into briquettes contracted in the middle to a sectional area of $2\frac{1}{4}$ square inches,[2] or more recently 1 square inch, immersed in water when one day old, and taken out and tested seven days after manufacture. The standard tensile breaking strain of the cement under these conditions has been gradually raised, with improvements in manufacture, to not less than 400 lbs. per square inch at seven days. A much lower breaking strain indicates that the cement is comparatively weak; and a much higher breaking strain points to the probable presence of free lime, which is liable to be a source of subsequent injury, especially in sea-water. The test after seven days is the regulation one, in order to avoid delay in using the cement; but later tests are expedient to prove a due increase in the strength of the cement with age; and a breaking strain of 500 lbs. on the square inch 28 days after the making of the briquettes, has been included as a standard qualification.[3] Tests, however, of cement in tension do not furnish quite exact indications of the strength of a substance subjected in practice to compressive strains, as the strength in compression does not always bear the same relation to the tensile strength in different samples of Portland cement.

The soundness, or prospect of durability, of cement is tested by forming small pats of cement with water on a glass plate, and then, when set, immersing them in water for seven days, at the end of which

[1] *Proceedings Inst. C.E.*, vol. cvii. p. 118. [2] *Ibid.*, vol. xxv. plates 3 and 5.
[3] See *British Standard Specification for Port and Cement.*

time reliable cement should show no signs of disintegration or cracks. The tests for the strength of cement mortar prepared with standard sand, for the expansion of the cement by the Le Chatelier method, and for the determination of the setting period by means of the "needle," are all included among the standard qualifications.[1]

Mortars.

Mortar is made by mixing lime or cement with a definite proportion of clean, sharp sand, which is determined according to the quality of the lime or cement, and the position in which the mortar is to be used. The choice of proportions should also be guided by the quality of the sand available for making mortar.

Proportions of Sand in Mortars.—Under favourable conditions, the highest proportions of sand generally used for mortar, are 3 parts of sand to 1 part of lime, and 5 parts of sand to 1 part of Portland cement; whilst 2 of sand to 1 of lime, and 3 of sand to 1 of Portland cement, are proportions often used under ordinary conditions. Roman cement, however, is so much weakened by admixture with sand, that equal proportions of sand and cement are commonly adopted in this case. When the work in which the mortar is to be used is liable to be exposed to water, the proportions should be reduced to equal quantities of sand and hydraulic lime, and 2 of sand to 1 of Portland cement; and where the work is actually exposed to water, 1 of sand to 1 of Portland cement may become necessary, the face joints being protected by a coating of neat Portland or Roman cement when the work is subject to the wash of the sea shortly after construction.

In important sea-works, the actual proportion of sand chosen for the mortar should be regulated by the coarseness or fineness of the sand to be employed; for it has been found by experiment that fine sand requires to be mixed with about twice as much cement as coarse sand to attain the same strength, owing to the much larger surface needing coating with cement in the case of fine sand, than of coarse sand, to cement the particles together.[2] On the other hand, mortar made with very coarse sand, though needing less cement, is not so compact as mortar made with finer sand, and consequently is not so suitable for a work subject to a pressure of water; and the admixture of some fine sand with the coarse becomes necessary to ensure the impermeability of the work.

Manufacture of Mortars.—In making lime mortars, the lime has to be slaked with water before it is mixed with sand; whereas finely-ground Portland cement can be mixed at once with sand, and made into mortar by the addition of water and turning the materials over sufficiently to ensure thorough admixture. Ordinary lime mortar is mixed in large quantities in a mortar-mill, the measured ingredients being placed in the revolving pan, and guided by iron scrapers under

[1] *British Standard Specification for Portland Cement.*

[2] "International Maritime Congress, London, 1893, Proceedings of Section I," p. 91.

a pair of heavy rollers, which grind any lumps of lime, as well as assist in mixing the ingredients. With finely-ground hydraulic lime or cement, the mixing machine is only required to ensure the complete incorporation of the materials. Where fine sand is used for the mortar, a more thorough mixing, as well as a larger proportion of lime or cement, is required than with coarse sand, owing to the far greater number of particles of sand which have to be coated with cement.

The amount of water used should be just sufficient to make the materials into a paste, and must necessarily vary according to the condition of the materials and the state of the weather; whilst an excess of water washes fine lime or cement away from the sand, and delays the setting of hydraulic limes and cements. Ordinary lime mortar can be safely used some hours after manufacture, on account of its slowness in setting; but hydraulic lime mortars, and especially quick-setting cement mortars, should be used as soon as possible after mixing, for their strength is considerably impaired by disturbance after setting has commenced.

Strength of Mortars.—The strength of mortars depends, not merely upon the strength and soundness of the lime and cement used, but also upon the quality of the sand and the thoroughness of the mixing. Accordingly, in important works, tests are made of cement mortars as well as of neat cement. Except in the case of certain substances, such as the volcanic earths already alluded to, the materials mixed with the lime or cement are simply inert substances, added for the sake of economy, and reducing the strength of the lime or cement. One of the advantages of Portland cement is that, owing to its strength and fineness, it can form an adequately strong mortar with a considerably larger proportion of sand than lime can, and still more than Roman cement. In this respect, some experiments seem to indicate that the best slag cement is even superior to Portland cement, on account of its still greater fineness;[1] but a greater reliability in the manufacture of slag cement, and much more extended experience of its qualities and durability will be needed, before it can have any prospect of competing successfully with Portland cement.

To make comparable tests of Portland cement mortars, sand of similar quality and size must be used in the experiments. It has been found that the tensile strength of Portland cement mortar at the end of a year, when mixed with equal quantities of sand and cement, is about three-fourths that of neat cement; and with larger proportions of sand, from 2 of sand to 1 of cement up to 4 of sand to 1 of cement, the tensile strength of the mortar varies approximately according to the ratio of cement to sand as compared with neat cement, having about one-half the strength of neat cement in the first case, and one-fourth in the latter case.[2] Beyond this limit, however, the strength appears to diminish more rapidly with an increase in the proportion of sand, since mortar with 5 of sand to 1 of cement has only one-sixth the tensile strength of neat cement.

[1] *Proceedings Inst. C.E.*, vol. cv. p. 225. [2] *Ibid.*, vol. xxv. pp. 77 and 88.

The strength of Portland cement mortars in compression, though much greater than in tension, diminishes more rapidly with an increase in the proportion of sand; for the crushing strength, at the end of a year, of a mortar made with equal quantities of sand and cement, was found to be rather more than half the crushing strength of neat cement, a 2 to 1 mortar one-third, a 3 to 1 mortar one-fifth, a 4 to 1 mortar one-eighth, and a 5 to 1 mortar only one-thirteenth that of neat cement.[1]

The strength of mortars is greatly affected by the nature, as well as the size of the sand, the mortar made with clean, sharp, or angular sand being much stronger than mortar made with dirty sand with rounded grains. The presence of loam, silt, or clay in sand can be detected by taking up a handful of moist sand, and closing one's hand on it with a little pressure, as dirty sand will bind together, instead of falling to pieces, when the grasp is relaxed. Dirty sand must be washed by a stream of water before being used for mortar.

The increase in the strength of cement mortar is slower and more prolonged than that of neat cement; and the increase, which is fairly rapid during the first two or three years, gradually decreases in amount, till at last, after the lapse of several years, the ultimate strength is reached, and a slight tendency to deteriorate is exhibited by the mortar.[2]

Remarks on Mortars.—The strength and permanence of any structure depends so largely upon the strength and durability of the mortar which binds the stones or bricks together, that the utmost care is needed in selecting the most suitable materials for the mortar, in mixing it, and in laying it in the work. As frost injures newly laid mortar, building has to be discontinued during frosty weather. The proportion of cement to sand must be regulated by the required strength and the exposure of the work, as well as by the strength and fineness of the cementing material; and where the best class of sand is not procurable at a reasonable cost, compensation must be made for the poorer quality of sand available, by an increased proportion of lime or cement. Portland cement has been so much improved in quality in recent years, and cheapened in cost, that it is very generally used now for mortar in important works; and it is regarded by most engineers as almost indispensable in works exposed to water or the sea, not merely on account of its strength and quick setting under water, but also owing to the preservation of its strength during storage on a long voyage if not exposed to damp, and the maintenance of its qualities when mixed with salt water or made into mortar with sea-sand.

CONCRETE.

Concrete has gradually become one of the most important materials of construction at the disposal of the civil engineer, on account of its cheapness, the facility with which materials for it are often procured where ordinary building materials are scarce, the ease with which it

[1] *Proceedings Inst. C.E.*, vol. xxxii. pp. 288–291. [2] *Ibid.*, vol. cvii. p. 388.

is manufactured, its adaptability to various situations, and its good strength and durability when carefully made.

Composition of Concrete.—Concrete is generally formed of sand mixed with gravel, small rubble stone, broken bricks, slag, or other inert hard materials cemented together into a solid mass by lime or cement; and it may be regarded as composed of lime or cement mortar filling up the spaces between, and connecting together the larger hard materials, which take the place of stone or bricks in ordinary constructions. The materials should be so proportioned according to their size, that the whole, when thoroughly mixed with water, may harden gradually into a solid, impervious mass. A large proportion of the big materials expedites the work, whilst strengthening it by causing it to approximate to rubble masonry; it involves less labour in mixing; and it reduces the amount of mortar required, and the quantity of the cementing material, which is the most costly item. Large materials, however, must only be introduced in the proportions, and in the class of work, in which an adequate thickness of mortar can be ensured between every large piece. The proportion of sand to the lime or cement, and its quality, must be determined by the same considerations as the formation of any strong, impervious mortar referred to above. The smallest proportion of cement is allowable in the manufacture of concrete blocks, which can be left to harden for some time before being used; and a specially large proportion of cement is necessary when concrete is deposited under water, and more particularly where it is likely to be partially exposed, before becoming set, to a current of water or the wash of the sea.

Proportions employed for Concrete.—The actual proportions of the materials used in making concrete vary considerably according to the quality and accessibility of the materials, the form in which the concrete is placed in the work, and the conditions of the site. Thus concrete made into blocks has been formed in proportions ranging between 3½ of sand to 1 of Theil lime at Port Said harbour,[1] and 2 of sand, 6 of broken stone, and 1 of Portland cement in sloping and horizontal blocks in various recent breakwaters. The proportions adopted for concrete in bags deposited under water, were 3 of sand, 4 of shingle, and 1 of Portland cement at the north breakwater of Aberdeen harbour, and 2 of sand, 5 of shingle, and 1 of Portland cement at Newhaven breakwater, or 7 of materials to 1 of cement in both cases. The concrete-in-mass deposited below low water, was composed of 7 of gravel and sand to 1 of Portland cement at Wicklow harbour,[2] 2 of sand, 4 of stone, and 1 of Portland cement for Babbacombe fishery pier, and 4 of sand and stone to 1 of Portland cement at Buckie harbour; whilst the concrete-in-mass above low water, was formed of 3 of sand, 4 of shingle, and 1 of Portland cement at the north pier of Aberdeen harbour, about 11 of sand and stone to 1 of Portland cement at Buckie, and 2 of sand, 5 of shingle, and 1 of

[1] "Harbours and Docks," L. F. Vernon-Harcourt, p. 642.
[2] *Proceedings Inst. C.E.*, vol. lxxxvii. p. 118.

Portland cement at Newhaven. The proportions of concrete-in-mass in recent important dockworks have been as follows: Dock floors and walls below low water, 1 Portland cement, 1½ sand, 5 broken stone; dock and basin walls between high and low water, 1 Portland cement, 2 sand, 6 broken stone; above high water, 1 Portland cement, 3 sand, 8 broken stone.[1]

The proportion of stone in concrete-in-mass above low water, or in very large blocks, can be advantageously increased by embedding large blocks of rubble stone at intervals in the concrete, provided these blocks are kept away from the face; but it is not safe to introduce such blocks in concrete deposited under water, as their proper bedding in the concrete, apart from one another, cannot be secured.

Cement and mortar naturally adhere more firmly to angular, rough, clean surfaces than to smooth or dirty surfaces; and therefore, just as a stronger mortar is formed by sharp sand than by rounded sand, so angular or broken stones make a stronger concrete than smooth shingle or boulders; and in concrete, as in mortar, all the materials composing it should be perfectly clean.

Forms and Uses of Concrete.—One of the earliest applications of concrete to engineering works was in the form of blocks, where stone was scarce or difficult to dress, and where close-fitting work was required. Concrete blocks are made within timber frames to the required size; and when the concrete has set sufficiently, the frames are taken off to form other blocks, and the blocks are left to harden in the open air for a month or more before they are placed in the work, holes having been left in forming the blocks, by the insertion of pieces of wood in suitable positions inside the frames, for the introduction of long iron bolts when the blocks have to be lifted. These blocks are very commonly used in the construction of breakwaters under water, where the blocks have to be laid as close together as possible, in a position in which the joints cannot be filled with mortar. The advantages of this form of concrete are, that the blocks being made beforehand, and given ample time to set hard, the work can be conducted rapidly, and with certainty as to the condition of the concrete when immersed in water, and that a smaller proportion of cement can be safely used than in concrete work executed under less favourable conditions.

Concrete has been employed enclosed in very large bags for the portions of some breakwaters below low water, as the yielding nature of the bag-work enables the bottom bags to adapt themselves to the irregularities of a hard chalk or rocky bottom; whilst the covering provided by the bag prevents the concrete being injured by the sea when being deposited, and after it has been put in position. Moreover, a sufficient portion of the finest part of the concrete oozes through the jute sacking forming the bag, under pressure, to unite the heap of bags into a solid mass. This system, though involving the use of large plant of a special character for the deposit of the bags, as in the case where large concrete blocks have to be handled, has the advantages of

[1] *Proceedings Inst. C.E.*, vol. clxxii.

obviating the great cost of levelling a rocky bottom under water, and of requiring only a moderate amount of cement in making the concrete, owing to the protection afforded by the bag. Concrete in bags has also been used, like concrete blocks, as an apron for protecting the upper exposed surface of the rubble mound of a breakwater, near the sea-face of the superstructure resting on the mound, and also in filling up hollows in foundations exposed to a current of water.

Concrete-in-mass has been very extensively used in foundations, dock walls, retaining walls, and the portions of breakwaters above low water, and occasionally for reservoir dams, lighthouse towers, and arched bridges. The reliance, moreover, felt of late years in Portland cement, has led to the use of concrete-in-mass for breakwaters and other work under water, deposited within framing lined with jute canvas, by closed skips with movable bottoms. The mixing of the concrete deposited in mass or into large bags, has to be very efficiently performed, either by hand for small quantities, or by mixing machines for large amounts. In making concrete by hand, the measured materials are turned twice over when dry, on a clean floor, and then again after admixture of water. Concrete-mixing machines consist of revolving boxes or cylinders worked by steam-power, in which the materials are turned over several times, and thoroughly mixed, before being discharged ready for use. The concrete should not, if possible, be deposited from a height, or through a long shoot, for the stones, descending faster than the rest, tend to form a heap by themselves; and the deposited concrete needs somewhat mixing again. The greatest care is required in depositing concrete under water, to prevent the concrete losing a portion of its cement by falling through the water; and the concrete must not be disturbed when once deposited. Concrete has sometimes been allowed to set partially before being deposited under water, with the view of preventing the washing out of the cement from the concrete in its passage through the water; but it is probable that the concrete loses as much strength by the disturbance of its setting when in what is termed the plastic state, as by the loss of cement under ordinary conditions. Concrete deposited under water must be given an ample proportion of cement; it must be protected from any wash or current; and it must be lowered in a closed skip as near the bottom as practicable, before being deposited. When concrete-in-mass is deposited in large amounts in successive layers, the top surface of the layer of the previous day's work must be cleaned from any coating of silt before the next layer is commenced; and in the case of concrete out of water, the surface should be moistened, and roughened when necessary, to ensure the adherence of the succeeding layer.

Concrete-in-mass possesses several advantages. It furnishes a cheap method for forming broad foundations for heavy structures, and readily fills up hollows or other irregularities in its excavated foundations; and it can easily be finished off to a level surface at the top, for the structure to be erected upon it. Moreover, though requiring constant care and supervision in its mixing and deposit, concrete-in-mass can be carried out with very little plant, and without skilled labour. This system of construction has been resorted to in dock walls and graving docks from economical considerations. For breakwaters, however, concrete-in-mass possesses the additional advantage that it forms the structure into a huge monolith, not liable to disturbance by the sea; though its deposit under water in such conditions needs very special precautions. On the other hand, porous concrete, subject to the percolation of sea-water under pressure, has in some instances become disintegrated, in consequence, apparently, of the substitution of magnesia from sea-water for the lime in the cement, which has led to the expression of doubts as to the permanence of structures exposed to sea-water, which have been formed with Portland cement.[1] The durability, however, of many breakwaters, docks, and basin walls of Portland cement concrete, and exposed to the action of sea-water, indicates that Portland cement in concrete properly proportioned and mixed so as to form an impervious mass, and devoid of free lime and magnesia, is not liable to decomposition by the sea-water, which merely comes in contact with the face of the work.

IRON AND STEEL.

Iron is used in three forms in construction, namely, as cast iron, wrought iron, and steel. Cast iron contains between 90 and 95 per cent. of pure iron, the remaining constituents being from 2 to 4 per cent. of carbon, mainly in the form of graphite, together with small amounts of silicon, manganese, phosphorus, and sulphur, in variable proportions according to the nature of the ore from which the iron is obtained. Wrought iron is the purer, malleable metal obtained from molten cast iron in the puddling process, by which the greater portions of the impurities are removed, so that their combined residues form less than 1 per cent. of the mass, the rest consisting of pure iron. Steel, which is of various grades of tensile strength and hardness according to its chemical composition, contains proportions of carbon ranging from about the same minute percentage in mild steel as in wrought iron, up to between 1 and 2 per cent. in hard-tool steel; whilst the quantities of silicon, sulphur, and phosphorus, are decidedly less in steel than in wrought iron, and the amount of manganese is greater.

Cast Iron.—Towards the close of the eighteenth century, and in the earlier half of the nineteenth century, cast iron was the form of metal almost wholly employed for bridges, the use of wrought iron having been confined to suspension bridges, as, for instance, the Menai Suspension Bridge, where the strains are entirely tensile. Cast iron was, indeed, very naturally adopted for arches, as the strains in these structures are wholly compressive, which cast iron is specially suited to sustain, as its crushing strength is about 40 to 45 tons per square inch, whereas its tensile strength only amounts to between 7 and 10 tons per square inch. Accordingly, cast iron has been successfully employed for arched bridges of considerable span, as, for example,

[1] *Proceedings Inst. C.E.*, vol. cvii. pp. 74-77; and "Report on the Causes of Damage to the Aberdeen Graving Dock," 1887, P. J. Messent.

Southwark Bridge over the Thames in London, erected in 1819, with a central span of 240 feet; but it was ill adapted for the small girder bridges constructed with it in the early days of railway enterprise, on account of the tensile strains on the bottom flange, which had to be made considerably larger in area than the top flange subject only to compression. It was not, however, till after the successful completion, in 1850, of the Britannia Tubular Bridge across the Menai Straits, made of wrought iron, that cast iron was abandoned in favour of wrought iron for girder bridges.

The use of cast iron has been gradually relinquished, even in arched bridges, for railways, for cast iron is liable to be fractured by sudden, severe shocks; and serious flaws have been discovered in some of the castings forming the ribs of old railway bridges, when taken down owing to their proving unsafe under the increased strains due to the heavier locomotives and higher speeds of the present day. Cast iron, however, is still very largely employed for columns, cylinders, tubes, and pipes, bed-plates, chairs for rails, and various other purposes.

Wrought Iron.—The use of wrought iron in construction, which was at first mainly employed for suspension bridges, in the form of link chains, suspending rods, and bolts, and later on in wire cables, and also for rails, was extended to large-span bridges in the case of the Britannia and Conway tubular bridges, and the Saltash Bridge over the Tamar, after it had been proved by experiments that wrought-iron plates and angle-irons riveted together could form beams capable of sustaining heavy loads over wide spans. Wrought iron was subsequently adopted for girders of various forms for many years, and has occasionally been used for arched bridges; but it has, in its turn, been gradually superseded by steel, owing to the greatly cheapened production of this stronger metal, especially for rails and bridges of large span. Wrought iron possesses the advantage over cast iron that its strength in compression is not much less than its strength in tension, the crushing strength of wrought iron being between one-half and one-third that of cast iron, and its tensile strength being about three times that of cast iron. Moreover, wrought iron is not liable to fracture or to have flaws like cast iron; and whereas the joining of several castings together by bolts only forms a thoroughly satisfactory connection when the strains on the structure are wholly compressive, the riveting together of plates and angle-irons, forming a wrought-iron girder, rigidly connects the different parts for resisting tensile as well as compressive strains.

Steel.—The old method of making steel from wrought iron manufactured from the purest Swedish or other ores, by combining it with carbon in the cementation process, rendered steel a costly product. The introduction, however, of the Bessemer process, in which mild steel, approximating in composition to wrought iron, can be manufactured more directly in large quantities from impure ores, resulting in a very great reduction in cost, opened out a wide field for steel as a material for engineering construction.

The original aim of the Bessemer process was to arrest the removal of carbon from cast iron at the intermediate stage, when steel might be assumed to have been produced, instead of first removing almost all the carbon so as to form wrought iron, and then adding carbon, in a second process, to wrought iron in order to make steel. Owing, however, to the presence of phosphorus, silicon, and sometimes sulphur in the iron obtained from impure ores, it proved impracticable, except in the case of the purest iron, to stop the blast of air through the molten metal, in the Bessemer converter, at the stage when the carbon originally existing in the cast iron has been reduced by the blast to the proportion forming steel; and the carbon has to be restored to the molten metal by the addition of the requisite proportion of the pure iron ore, rich in carbon, known as spiegeleisen, after the other impurities have been removed by the prolongation of the blast. The Thomas-Gilchrist process introduces a basic lining into the converter, which, entering into combination with the impurities in the molten metal, especially phosphorus, removes them from the steel, which is thereby procured in an adequately pure condition. In the Siemens-Martin process, a bath of molten pig-iron mixed with more or less wrought iron, steel, or similar iron products with the addition of manganese or spiegeleisen is converted into steel by exposure to the direct action of the flame in a regenerative gas furnace, the operation being so conducted that the final product is entirely fluid. Various modifications of the open hearth process, which may be acid or basic, are in operation.

Steel manufactured by the above processes has been fused, and more effectually freed from impurities than wrought iron, and steel formed from puddled iron, which are never wholly freed from the impurities of the slag in which they have been immersed. The tensile strength of steel varies with the proportion of carbon in its composition, ranging from that of wrought iron in very mild steel, up to a maximum with about 1·5 per cent. of carbon, when it attains more than double that strength, its breaking strain sometimes reaching about 60 tons on the square inch. The strength, moreover, of steel is very similar in tension and compression. Cast steel, much used for special purposes, is generally considered to have three times the strength of cast iron; but owing to its much higher fusing point, its contraction in cooling is nearly double that of cast iron.

Mild steel, owing to its greater strength and elasticity, has practically superseded wrought iron not only in bridge work but in most other branches of iron construction, but is, like its predecessor, open to the destructive effects of corrosion, unless carefully and efficiently protected against rusting.[1] Steel has also been very advantageously adopted for rails, which are now exposed to increased strains and wear by the augmented weight of locomotives and the use of continuous brakes; and as rails must be both tough and hard, to sustain the shocks of heavy trains travelling at a high speed and to support the wear of the brakes, they are tested for toughness by the impact of a falling weight, and for hardness by their breaking strain.

[1] "Notes on Construction in Mild Steel," Longmans' Civil Engineering Series.

SAFE STRAINS ON MATERIALS.

In all structures, it is of the utmost importance that none of the materials of which they are formed should ever be strained to an extent liable to produce any permanent injury. In certain materials, such, for instance, as wrought iron and steel, there is a limit, varying with the nature of the material, up to which changes in form are proportionate to the load imposed; and the original form is gradually regained on the removal of the load. When, however, this *limit of elasticity* is exceeded, the change in form becomes greater than in proportion to the increase of load; and the material no longer returns to its original condition after the withdrawal of the load, but acquires a *permanent set.* It would be unsafe to strain materials close up to this limit; for repeated applications of such a load, causing *fatigue* of the metal, is liable to lower the elastic limit, and may eventually even produce fracture. Factors of safety have, accordingly, been deduced from the results of experiments for various materials, representing the relation between the breaking load and the stresses to which the different materials may be safely subjected, which are considerably below the elastic limit.

Dead and Moving Loads.—The loads which some structures have to bear, such as bridges, floors, and roofs, are of two kinds, namely, the dead or permanent load, and the live or moving load. The dead load consists of the roadway or other permanent weight borne by the structure, together with the weight of the supporting structure itself between the points of support. The moving or variable load comprises trains, vehicles, and foot-passengers crossing over bridges, and wind, and in some countries snow, especially on roofs. The moving load imposes a more severe strain upon a structure than the dead load, owing to its frequently rapid and unequal application, the shocks with which it is often accompanied in the case of a train going at a high speed, and the alteration sometimes produced in the nature of the stress, putting certain portions into tension and compression alternately, a change which is much more trying to materials than similar stresses of unvarying kind. The greater strain, however, produced by the moving load is to some extent compensated for, in actual practice, by the structure being designed to bear the maximum moving load possible, which may, indeed, be imposed in the testing of the strength of the structure before its opening for traffic, but which is rarely, if ever, attained in actual working. The ordinary allowances for the moving loads on bridges are 1 to 1½ tons per lineal foot per line of way for railway bridges, 1 cwt. per square foot for roadways, and 70 lbs. per square foot for footways.

Stresses allowable on Structures.—Certain general rules have been framed for ensuring the stability of structures, some of which are enforced by the Board of Trade with regard to railway bridges in the United Kingdom. In timber structures, the dead load, together with twice the maximum moving load, should not produce a stress in any part of the structure exceeding one-fifth of the breaking stress of the timber employed. Masonry and brickwork should only be weighted to the extent of about one-eighth of their crushing strength on the average. Cast iron should be only used in compression; and the stress due to the dead load, together with twice the maximum moving load, should not exceed one-third to one-fourth of the breaking stress in any part of a cast-iron structure.

The standard maximum working stresses on metal structures allowed by the Board of Trade, are 5 tons per square inch in tension, and 4 tons per square inch in compression for wrought iron, and $6\frac{1}{2}$ tons per square inch for steel; whereas the elastic limit has been found to average about 12 tons per square inch for wrought iron, and 18 tons per square inch for mild steel. In the above limits of safe stresses, no distinction is made between the dead and moving loads; and the limits are somewhat low for the large members of a girder of large span, and high for the small members subjected to shocks and to alternate tensile and compressive stresses. Moreover, the load which can be safely borne by a steel structure, varies considerably according to the nature of the steel employed.

Wind-pressure, and Snow.—In addition to the dead and moving loads, allowance has to be made, in designing railway bridges, for a wind-pressure against the exposed surfaces, of 56 lbs. per square foot—a condition imposed by the Board of Trade after the overthrow of the Tay Bridge during a gale in December, 1879. The pressure of wind has also to be considered in the case of roofs, and the occasional additional load due to snow in cold countries. The wind-pressure, however, is greater in proportion to the pitch of the roof; whereas the depth of snow which may accumulate, increases in proportion to the flatness of the roof; but the pressure imposed on a roof by these two influences, constitutes a considerable portion of the stresses which have to be borne by a roof. The weight or compactness of the snow is in inverse proportion to the size of the flakes, and may be assumed to amount to from 5 to 10 lbs. per cubic foot; but the weight of snow on bridges in North America, has been estimated to attain 30 lbs. per square foot under unfavourable conditions.

The subject of wind-pressure and its influence on the design of roofs and bridges has received much attention in recent years, and the experimental investigations which have been carried out on the influence of the pitch of roofs, negative wind-pressure, the results of wind-pressure on lattice girders, and the sheltering influence of the windward girder should be carefully studied.[1]

[1] *Proceedings Inst. C.E.*, vols. cxviii., clvi.

CHAPTER III.

PRELIMINARY ARRANGEMENTS FOR CARRYING OUT WORKS.

Surveys—Investigations of physical conditions—Borings, and trial pits; methods and objects—Parliamentary plans and sections; objects, and general conditions—Working or contract drawings—Methods of carrying out works; with, and without a contractor—Specification; general description, and forms appended—Tenders; contents and stipulations—Lump-sum contract; based on schedule of prices, extras—Progress of the works; conditions, inspection of materials, supervision of works, stipulations as to plant and materials—Payments for works; arrangements, retentions—Completion of the works; stipulations, variable competency of contractors—Remarks on the carrying out of works; relative efficiency and cost of methods, importance of supervision.

BEFORE any work is designed, a careful survey of the site has to be made, with a longitudinal section along the line of the work, and often cross-sections as well, so that the levels of the ground to be traversed or built upon may be accurately known, the best position for the work may be selected, and the amount of earthwork to be carried out may be ascertained. Reference should be made to any previous survey of the locality, to facilitate the execution of the more detailed, and generally larger survey required, on which the general plan of the proposed work is laid down. Moreover, if a geological map of the district has been made, it is very desirable to consult it, in order to obtain a general idea of the strata through which excavations will have to be carried, or on which foundations for structures will have to be laid. More particular information as to the nature of the soil in which the works are to be carried out, should be procured by borings and trial pits; and some data as to the rainfall of the locality, and the height of the floods of the streams and rivers to be traversed or dealt with by the works, should be obtained. In fact, as intimate a knowledge as practicable of the physical characteristics of the district in which the works are to be carried out is most desirable, for enabling engineers to prepare suitable designs and reliable estimates of cost, the importance of special branches of information varying with the nature and object of the work. Thus the nature and dip of the strata, and the existence of underground springs, exercise an important influence on the design and cost of deep cuttings and tunnels; the rainfall, the catchment area, and the presence of faults or fissures, are additional essential factors in the provision of water-supplies for towns; the maximum and minimum discharge of rivers, the nature of their bed, their fall, and the amount of detritus they bring down, determine to a great extent their capability for improvement, and the works that should be carried out; whilst the direction of the strongest and prevailing winds, the drift of sand or shingle along the coast, the rise of tide, and the slope of the sea-bottom, are of paramount importance in the design of works for the formation of harbours, the improvement of river outlets, and the protection of coasts.

Borings, and Trial Pits.—Ordinary borings furnish the cheapest and most expeditious means of ascertaining the nature of the strata below the surface for a considerable depth. The bore-hole is excavated through soft soil by turning round in it a hollow cylindrical auger, somewhat pointed at the lower end, which collects and brings up the material when lifted; and as the bore-hole is carried down, it is lined by thin wrought-iron pipes. When rock or other hard material is encountered, it has to be broken up by the blows of a jumper before it can be raised by the auger. The nature, thickness, and position of the several strata are thus readily indicated; and by the aid of a series of borings, a sort of geological section of the ground can be drawn out. The process of sinking the boring, however, so disintegrates the more compact materials, such as indurated silt, clay, or rock, that it is impossible to form an accurate opinion of the condition of the materials in position; and in this respect, diamond boring machines, which bring up a complete core of the strata traversed, afford much more satisfactory indications of the actual condition of the materials, but the method is considerably more expensive than the ordinary process.

Trial pits enable a far more correct idea to be formed of the exact condition of the soil at the various depths, and are not subject, like borings, to the error of mistaking a chance solitary boulder for a stratum of rock; but these pits occupy a much longer time in excavating, and are considerably more costly; and, moreover, their excavation, even at moderate depths below the surface, is liable to be impeded by the influx of water directly the line of saturation of the soil is reached. Accordingly, the number and depth of the trial pits have generally to be strictly limited, and are confined to the most important spots; whilst extended investigations of the nature of the ground are commonly effected by borings.

These preliminary investigations of the strata are not merely needed for determining the proper side slopes for the cuttings, the nature, and consequently the cost, of the excavations, and the depth to which the foundations of bridges, walls, and buildings, will have to be carried, but they are also very valuable in indicating to what extent the materials obtained from the excavations may be suitable for the purposes of the works, such, for instance, as masonry, pitching, ballast, or concrete, which has an important bearing on the cost of the works.

Parliamentary Plans and Sections.—Except where the works are kept within the property of a single landowner, it is generally necessary, in the United Kingdom, to obtain parliamentary sanction for the works, so as to secure the compulsory purchase of the land required for the undertaking. In the case of large works, this is accomplished by means of a private Bill in Parliament; and often, for small works, the cost is notably reduced by applying for a "Provisional Order," with the sanction and aid of the Board of Trade.

In return for the privileges conferred on the promoters of the scheme, when the Bill, after passing through both Houses of Parliament, is converted into an Act on receiving the Royal assent, certain preliminary conditions have to be complied with, in order to safeguard the interests of the public, and to give due notice to the owners and occupiers of the lands proposed to be taken for the works. Plans and sections of the proposed works, drawn up in accordance with the "Standing Orders" of Parliament published each year, have to be deposited before the 30th of November, and estimates of cost furnished before the end of the year, for all schemes for which an Act is to be sought in the following session. The plans show the general lines of the works; and each separate enclosure within the "Limits of Deviation" is indicated by a special number in each parish, corresponding with similar numbers in the "Book of Reference," in which a description of the properties is given, with the names of the owners and occupiers, to whom notice of the application has to be sent, and who have access to a copy of the plans relating to their parish. The sections show the levels of the proposed works, and consequently indicate how they will affect the adjacent land. The works, when sanctioned, have to be carried out within the limits of deviation shown by dotted lines on the deposited plans; and the levels generally can only be raised or lowered 5 feet from the levels shown upon the deposited sections. A period is named in the Act within which the purchase of land can be effected under it; and a longer limit is also stated for the completion of the authorized works. Any relaxation of these restrictions has to be sought by application for a fresh Act.

Working or Contract Drawings.—As soon as it has been decided to carry out any work, and the preliminary surveys and investigations have been completed, the working or contract drawings are prepared, showing in adequate detail the various works proposed, to enable operations to be commenced, or a contract to be obtained for their execution, and from which a more detailed and exact estimate of cost can be prepared. These drawings are supplemented, as the work proceeds, by numerous detailed drawings of the smaller and more intricate portions of the works, to larger scales for the guidance of the workmen or the contractor; and drawings showing the foundations and works as actually executed, should be prepared by the engineering staff as the work progresses, so that an exact record may be preserved for future reference, in the event of repairs or extensions being required.

Methods of carrying out Works.—The execution of the works is generally entrusted to a contractor, who enters into a contract to carry out the works according to specified conditions, under the supervision and direction of the engineer, for a definite sum, or in accordance with a fixed schedule of prices for the different classes of work. Occasionally, however, the authorities or companies undertake the works themselves, entrusting the execution to their executive engineer and other officials; and this course is more particularly resorted to in dredging operations for deepening a river or the approaches to a port, where the amount of work cannot be easily defined, and may vary in character, and also for the erection of breakwaters in the sea, for which a contractor is liable to demand an excessive price for the risks of sea-works. In most cases the contract system is preferred, on account of the limit that it places on the cost of the works, and the freedom it affords from liability for damages, accidents, and other unforeseen contingencies; but generally additional works raise the actual expenditure above the contract sum, especially in large undertakings; and occasionally the bankruptcy or death of the contractor entails delays in the completion of the works, and an increase in their cost.

Specification.—Every contract for the execution of engineering works is based on the "Contract Drawings," which indicate the general character and extent of the works, their construction, and the materials of which they are composed; and the "Specification," which explains in detail the nature of the works, the quality and proportions of the materials to be employed, and provisions and stipulations as to plant, temporary works, progress, liability for accidents and damages, the setting out and carrying out of the works, supervision, measurement, payment, time of completion, and all other matters incidental to the contract. The contractor, in fact, is expected to bind himself, in consideration of a stipulated sum or a defined rate of payment, to carry out, and maintain for a certain time after completion, usually six months or a year, the works indicated in the contract drawings, under the conditions and in the manner laid down in the specification, to the satisfaction of the engineer. To the specification is appended a "Form of Tender," a "Schedule of Prices," giving a list of the various kinds of work to be carried out; a "Form of Bond" to be executed by the sureties for the due performance of the contract, proposed by the contractor and approved by the company; and a "Form of Contract" to be signed and sealed by the company and the contractor.

Tenders.—As soon as the contract drawings and the specification are ready, contractors are invited by advertisement to tender, or in some cases the invitation is confined to a selected list of contractors; and each contractor desiring to tender is given access to the contract drawings, the results of borings, and any other particulars of importance; he is sometimes furnished with the quantities of the different classes of work taken out by the engineer, which he is expected to check; and he is provided with a copy of the specification with its appended forms. A date is named on or before which the tenders have to be sent in, with the form of tender filled in, giving the contract sum and the names of the proposed sureties, and a price affixed opposite each item in the schedule of prices. It is invariably stated that the company do not bind themselves to accept the lowest or any tender; but generally, provided the contractor is supposed to be competent, and his sureties are considered capable of paying up the penalty named in the event of the non-performance of the contract, the lowest tender is accepted.

Lump-sum Contract.—The most satisfactory form of contract is where a total or lump sum is named, which is based upon a schedule of prices, for in this manner the expenditure of the company upon the proposed works is clearly defined at the outset; and if changes in the quantities of the work are introduced during the progress of the works, they are regulated by the schedule of prices. Properly, in this form of contract, the total quantities of the different works, as indicated by the drawings, multiplied by the respective prices in the schedule, should correspond to the contract sum; and if less or more of any classes of work are actually executed than shown on the drawings, a corresponding reduction in, or addition to, the contract sum should be made, reckoned at the prices given in the schedule. In practice, it is impossible, and indeed often undesirable, to adhere strictly to the contract drawings; and the above adjustment of cost is fair for both sides; but a contractor is much more ready to agree to additions to the sum he is to receive, than to any reductions from it on account of work omitted as unnecessary. It is, however, most important to include, as far as possible, in the specification and schedule of prices, all classes of work likely to be required; for any deviations from the specified works, and any kinds of work not put down in the schedule of prices, afford the contractor an opportunity of claiming a higher proportionate rate, which it is difficult to adjust fairly; and these modifications and additions make up the claim for "extras," which not unfrequently constitute a considerable and unwelcome augmentation of the original contract sum. A specification carelessly drawn up, or ambiguously worded, and an inadequate list of classes of work in the schedule of prices, may eventually involve a company in considerable additional expense; for a contractor is only bound to carry out the works in the manner, and according to the stipulations contained in the specification; and the cost of alterations in the works is only regulated by the schedule of prices so far as they are applicable.

Progress of the Works.—As soon as the contractor is ready to commence the works he has undertaken to carry out, the company must put him in possession of the required land, so that he may be unable to claim compensation for delay or hindrance on this account. In most specifications, the contractor is made responsible for the correct setting out of the works; but without relieving the contractor of his legal liability, this work is generally best performed by a competent resident engineer, leaving the contractor free to check the correctness of the lines and levels if he thinks fit.

The stability of many classes of work essentially depends upon the strength and good quality of the materials employed, and the manner in which the work is executed. It is the engineer's duty to see that samples of all materials used in the works are subjected to adequate tests or inspection, as the case may be, and to ensure thorough supervision of the execution of all important portions of the work. Thus, whilst large quantities of earthwork may be carried out in excavations and embankments with little looking after, supervision is required in all kinds of construction work, such as brickwork, masonry, and ironwork, and more particularly in works which depend for their soundness on their manner of execution, and those which are subsequently hidden from view, such as foundations, concrete, especially under water, reservoir dams, retaining and dock walls, tunnels, sewers, and conduits. In fact, a thorough and efficient system of testing, inspection, and supervision, is essential for the satisfactory carrying out of works.

All the plant and materials required for the execution of the works are usually provided by the contractor; and a stipulation is generally inserted in the specification that, on being brought on to the works, they shall become the property of the company, with the object of their not being liable to be seized in the event of the contractor becoming bankrupt, and in order that they may be retained for the prosecution of the works if the contractor fails to carry out his contract. The plant and unused materials only revert to the contractor on the completion of the works.

Payments for Works.—Provision is made in the specification for making payments on account to the contractor, at intervals of one or two months, for work done, according to measurements made by the resident engineer, calculated at the rates given in the schedule of prices. In practice, the most convenient method of ascertaining the value of the work done during the specified interval, is to measure the total amount of work executed, and, after arriving at its value, to deduct the total value of work done at the preceding measurement. A deduction of ten per cent. is usually made from each certificate, thus made out, of the value of the work done during the several intervals, which is retained by the company till the completion of the works, when half of the total amount thus retained is paid over to the contractor, and the other half is only paid at the expiration of the term of maintenance, on the engineer certifying that the works have been completed and maintained to his satisfaction, in accordance with the provisions of the contract.

Completion of the Works.—As the capital expended on works is generally unremunerative till the works have been finished, and the benefits to be derived from the works remain in abeyance, a period is commonly named in the specification within which the works are to be completed. Often, however, unforeseen contingencies, accidents, and additional works delay the final completion of the works; and though sometimes a penalty is named in the specification, proportionate to the time by which the time occupied in the performance of the contract exceeds the stipulated period, its enforcement upon the contractor is surrounded by difficulties. Generally the best security for the speedy completion of works, is the interest the contractor has in releasing his plant and being paid his retentions. Contractors and their agents naturally vary considerably in their capacity for organization, and the power of carrying out works with efficiency and rapidity; and the command of a large capital is a great assistance in the prosecution of works. A contractor with limited capital will naturally try to do the most paying work first, regardless of the general progress of the work, and will place as little plant as possible on the ground; whereas a contractor with ample means is able to provide an adequate supply of plant at the outset, and concerns himself more about the final outcome from the works on their completion, than about an immediate profit. The works, indeed, are necessarily far more likely to be efficiently and expeditiously carried out under the latter conditions than under the former; but, though the engineer would naturally prefer to entrust the work to the most competent of the contractors tendering for it, the lowness of a tender is generally allowed to override all other considerations.

Remarks on the Carrying out of Works.—Probably the best work is accomplished when entrusted to a competent executive engineer, without the intervention of a contractor; but the ultimate cost, under these conditions, is necessarily more uncertain, and, except in the case of dredging and certain classes of sea-works, is liable to be larger than with a contract. Works are executed under the most unsatisfactory conditions when the contractor provides the funds for the undertaking, or finances it, as it is said; for then the contractor can, in a great measure, dictate his own terms, and secure a predominating influence in the management, so that the engineer is hampered in the performance of his duty of rigorous supervision, which furnishes the main safeguard for the efficient carrying out of the works.

Thorough inspection and supervision are second only in importance to correctness of design for the successful prosecution of works, especially those carried out by contract; for the contractor naturally pays most attention to the economical conduct of the works, and generally relies on the engineer and his subordinates seeing to their efficiency and conformity with the contract. Excellency of workmanship cannot, indeed, make up for deficiencies in the design; but a good design may be seriously impaired by inferior work; and sound work must be combined with a suitable design, to secure a successful result.

CHAPTER IV.

EXCAVATIONS, DREDGING, PILE-DRIVING, AND COFFERDAMS.

Objects of operations—**Excavation:** ordinary excavation, removal of earthwork, executed by piecework; excavators and steam-navvies, methods of working; excavation in hard rock; formation of embankments; methods of removing excavations—**Dredging:** objects; types of dredgers; bag and spoon; bucket-ladder dredgers, stationary and hopper; dipper bucket dredger; grab; sand-pump—Remarks on dredging machines—Methods of breaking up rock under water, blasting, floating diving-bell, in New York harbour, rock-breaking plant; conveyance of dredged materials, by chain of buckets and waggons, by waggons on barges, through long shoots, through floating tubes; remarks, comparison of excavating and dredging, cost of dredging—**Pile-driving:** piles, cap and shoes; ordinary pile-driving; steam pile-drivers; gunpowder pile-drivers; water-jet; bearing piles; sheet piling—**Cofferdams:** objects and construction; single-sheeted cofferdam; double row of piles with puddle wall; iron or steel caissons, instances; earthwork dams; closure of dams, timber, and embankments.

CERTAIN operations will be considered in this chapter which are very largely employed in carrying out civil engineering works. Excavation, in some form or other, can very rarely be dispensed with in the execution of works; and it often constitutes a very important branch of the undertaking, and involves a large proportion of the expenditure, as, for instance, in the construction of railways, canals, and docks. Dredging, though not so extensively applicable to the execution of works as excavation, is essentially necessary for certain classes of works, such as the deepening of rivers and canals, the improvement of the approach channel to harbours, and the maintenance of the depth in docks and harbours. Pile-driving is resorted to, in many cases, for the formation of cofferdams to exclude water from foundations, and for providing stable foundations on soft soil by the insertion of bearing piles, as well as in the construction of quays, wharves, and jetties; whilst cofferdams are often indispensable preliminary works in the formation of docks, and for laying the foundations of piers for bridges across rivers.

EXCAVATION.

Excavation and dredging consist essentially of two operations, namely, the digging out of the material from the site which it occupies in a more or less compact mass, and its removal to a suitable place of deposit; but whilst excavation relates to digging operations conducted in the open air, dredging consists in removing the material from under water.

Ordinary Excavation.—All excavations are commenced by manual labour, the soil being dug out by a gang of labourers with spades or shovels, and loaded into barrows, which are then wheeled by another gang along lines of planks to the place of deposit, or tipped into waggons. The earth has often to be loosened by pickaxes from the face of the excavation before it can be shovelled up; and when a gully or trench has been carried into a cutting, bars are commonly driven down at intervals along the top on each side, in a line with the face and a short distance back, eventually forming a longitudinal crack, thereby releasing a long block of material from the mass, so that it can be thrown down into the trench to be removed by shovels, or tipped direct into waggons standing below on a line of rails laid along the trench. In India, baskets are used by the natives instead of wheelbarrows, which they carry to the place of deposit. When the excavations are considerably below the level of the depositing ground, as in dock works, steep inclined planes are made with planks along the side slopes, up which the men ascend holding their wheelbarrows with straight arms, being hauled up by means of a chain attached to the front of the wheelbarrow, which passes over a pulley on a staging at the top of the incline, and is drawn along by a horse. Skips also, raised by a steam-crane standing on a wall, are employed for removing excavated material from the bottom.

Waggons drawn by horses along a line of rails are the ordinary means of conveying the earthwork from the cutting, for a moderate distance, to the place of deposit. When, however, the "lead," or the distance to be traversed, is long, or the excavation is proceeding rapidly, so that a train of waggons is quickly filled, locomotives are employed for conveying the waggons to their destination.

The powerful men who generally carry out extensive excavations on large public works are called navvies. Such work is usually executed as piecework, the men being paid in proportion to the amount of work done, which leaves them more freedom of action as to the periods of their work, and obviates any necessity for supervision, beyond seeing that the excavation is performed at the required places, and to the proper lines and levels.

Excavators and Steam-navvies.—Where large quantities of excavation have to be carried out, as, for instance, for docks, ship-canals, and large railway cuttings, and especially where the climate is unhealthy, labour is scarce, or strikes are liable to occur, mechanical appliances have been resorted to for saving manual labour and expediting the work. These excavators and steam-navvies are made according to two distinct types, the excavators, worked by steam, being furnished with a chain of buckets, like an ordinary bucket-ladder dredger, and the steam-navvies working with a single large bucket.

An excavator generally works along the top of a cutting, with the ladder carrying the chain of buckets extending down the slope, and planes off a slice of the slope by the revolution of its buckets and its gradual onward motion along a line of rails laid at the top of the slope, as shown in Fig. 1; and therefore it is very serviceable for the enlargement of cuttings. Sometimes, indeed, with a modified arrangement, this form of excavator is employed at the bottom of a cutting for extending it, by placing the ladder in a horizontal position, and projecting it in advance of the truck which carries it, so that the chain of buckets in revolving may excavate the face of the cutting.[1] This type of excavator, however, though tried at the Panama Canal works, excavates somewhat at a disadvantage in the bottom of a cutting as compared with a steam-navvy; whilst the ordinary type working from

BUCKET-LADDER EXCAVATOR.

Fig. 1.

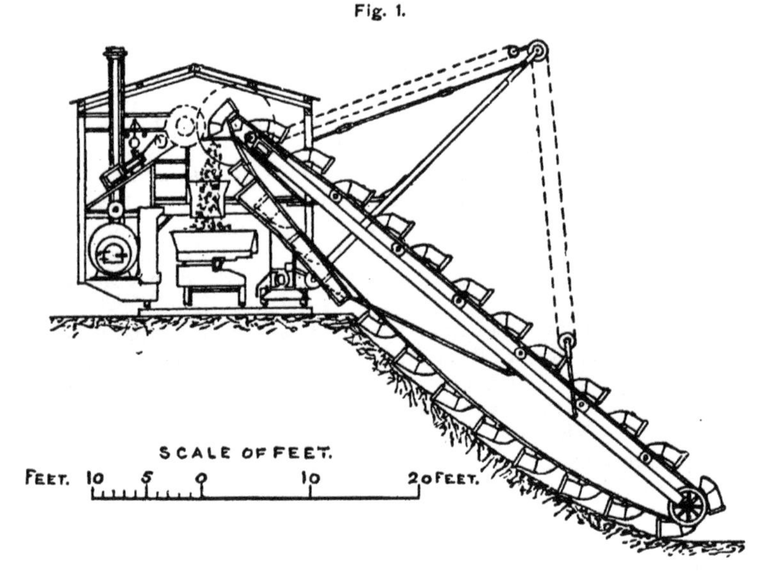

the top has been very extensively used, having been first introduced on the Suez Canal works, and subsequently adopted on the Danube regulation works, the enlargement of the Ghent-Terneuzen Canal, the formation of the Tancarville Canal, the Panama Canal works, and the Manchester Ship-Canal.[2] These excavators possess the advantage of raising the material to the top of the cutting, thereby reducing the cost of haulage; but the track upon which they travel has to be laid very solidly, with heavy steel rails and stout sleepers, to carry the heavy weight without settlement; and where the ground is soft, precautions have to be taken to prevent the machine tipping over into the cutting.

[1] *Le Génie Civil*, vol. v. p. 393 and plate 42.

[2] *Proceedings of the Institution of Mechanical Engineers*, Liverpool Meeting, July, 1891, pp. 419-424.

Moreover, these excavators, which work very economically and quickly in loose, soft soils, such as sand, gravel, loam, and light clay, are unable to cope efficiently with stiff clay, boulders, and soft rock.

The steam-navvy excavating with a single bucket, which is sometimes given a capacity of 2½ cubic yards, is a more powerful machine than the steam-excavator, and able to attack any kind of material up to soft rock; and it is very suitable for excavating in the bottom of a cutting and extending it. In the large machines, the bucket is carried by a stout, long beam, moving in a strongly-framed revolving jib by means of chains, affording the bucket a considerable range of work in front and round to the sides (Fig. 2). In the smaller machines, the bucket is hinged to the lower part of the jib of a revolving steam-crane, and worked by chains and a pulley hanging from the top of the jib, which form of machine possesses the advantage of being lighter, and therefore

STEAM-NAVVY.

Fig. 2.

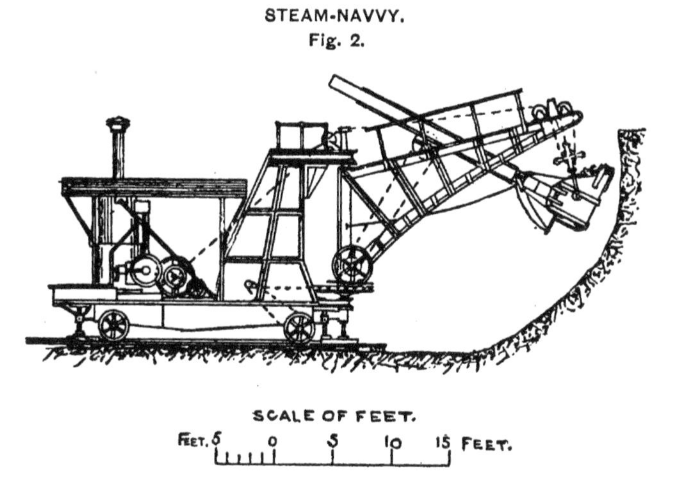

more easily shifted from place to place, and less damaging to the roads, and is capable of being used as a steam-crane when not required for excavating. The bucket of these machines is provided with a movable bottom turning on a hinge, so that when a catch is released, the flap bottom opens downwards, releasing the contents of the bucket into a waggon below without the necessity of turning the bucket over. Steam-navvies are now ordinarily used in most large, deep excavations, such as those required for docks and ship-canals.

Excavating in Hard Rock.—Where rock has to be excavated which is too hard and compact to be loosened by a pick or attacked by a steam-navvy, holes have to be drilled in it, either by hand or by drilling machines worked by steam or compressed air. Charges of blasting powder, dynamite, rackarock, tonite, or other powerful explosive, are then inserted in the holes, formed at suitable distances apart; and by firing these mines, the rock is shattered. The use of percussion and rotary drilling machines, first introduced at the headings of the Mont

Cenis tunnel, have greatly expedited excavation in rock; and the adoption of compressed air for driving them, facilitates the ventilation of the long headings in forming tunnels under high mountains.

Formation of Embankments.—In the construction of a railway, the formation of embankments across valleys constitutes as important a part of the work as excavating cuttings through ridges. One operation, indeed, is the complement of the other, as the materials removed from the cuttings serve to form the embankments. The earthwork from a cutting is loaded into tip-waggons, with a hinged-movable side in front, retained in place by an iron hook; and a train of loaded waggons is drawn by horses, or a locomotive, near to the end of the tip of the adjacent embankment in course of formation. The waggons are then uncoupled, to be drawn singly in succession to tip their contents over the extremity of the embankment. Some old sleepers are placed across the termination of the line of rails at the tip-end of the embankment; and the body of the tip-waggon is so connected in front to the frame carrying the wheels, and unconnected at the back, as to be capable of tipping up. Each waggon is drawn separately by a horse, at an increasing speed, towards the end of the tip; and on approaching the extremity, the horse is pulled aside by his driver, who unhooks the hauling chain; the man at the tip-end knocks up the hook which kept up the movable side of the waggon in front, as the waggon runs past him; and, the waggon being suddenly arrested in its course by the sleepers across its road, the acquired momentum causes the free body of the waggon to rise at the back and shoot out its contents down the slope of the tip. The embankment is thus gradually pushed forward, and is at the same time consolidated by the passage of the waggons over it on the top, and by the shock of the tipping.

Various Methods of removing Excavations.—In mountainous districts, quarries, bridge, dock, and other engineering works, the transport of spoil and materials by means of cable-ways and skips is now commonly employed. This system as applied to the removal of excavated material at the Chicago Drainage Canal is sketched in Fig. 3, the steel cables with a span of 700 feet being supported from two movable towers.[1] In dock-work the cable-way has been used at Keyham, Gibraltar, and Malta Dockyard Extensions, the cable-ways in the first example having a span of 1520 feet, the load, 8 tons, being lifted at 150 feet per minute and traversed to tip at 1000 feet per minute. The main cables were steel-wire r e 9 inches in circumference with a breaking stress of 190 tons.[2] The system is also to be applied at the Gatun Locks in the Panama Canal. In bridge-work cable-ways have been used at the building of Vauxhall, Newcastle-on-Tyne, and Victoria Falls Bridges; while the Famatina cable-way in the Argentine Andes is a remarkable example of the system applied to the transport of ore from the mines to the smelting works.

[1] "The Travelling Cableway and some other Devices employed by Contractors on the Chicago Main Drainage Canal," Lidgerwood Manufacturing Company, New York, pp. 7–29.

[2] *Proceedings Inst. C.E.*, vol. clxxii.

Another method of conveying the excavated materials from the bottom of a cutting to a high spoil-bank at some distance from the side,

CABLE-WAY.

Fig. 3.—Chicago Drainage Canal.

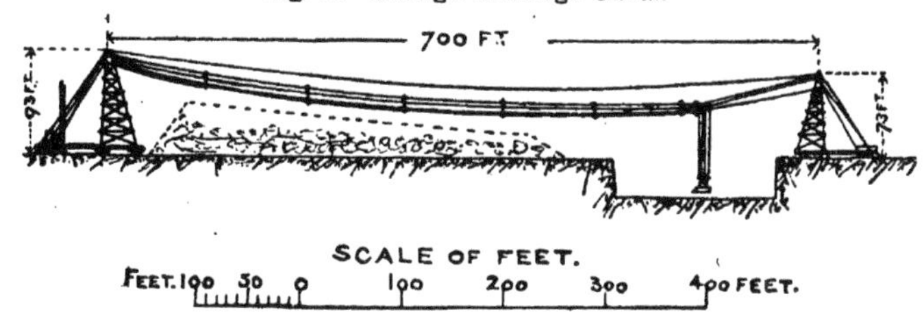

was adopted on the works of the Chicago Drainage Canal, by the help of a cantilever crane, or hoist, travelling along one edge of the cutting. This crane consisted of a straight, steel truss, 353 feet long, resting on two central supports, and braced above. The truss dipped downwards over the cutting to the further side, and rose at the opposite end over the spoil-bank at the back (Fig. 4). Loaded skips were conveyed along this truss, and their contents were deposited on the mound, which

CANTILEVER CRANE.

Fig. 4.—Chicago Drainage Canal.

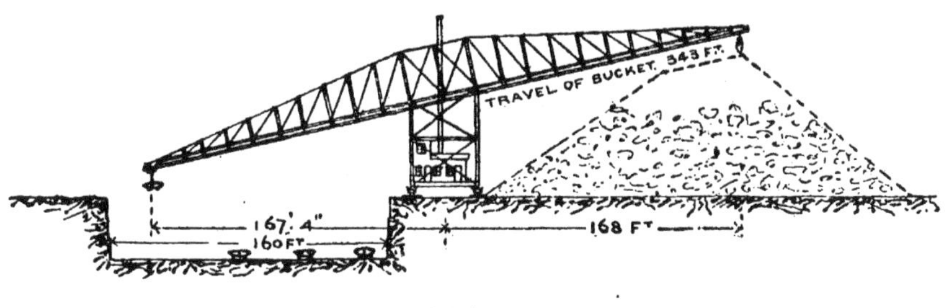

could be thus raised to a height of 90 feet from the surface of the ground; and each of these cranes conveyed and deposited about 600 cubic yards of material per day.[1]

Endless travelling belts are well-known methods of conveyance, especially in grain warehouses; and a machine was designed for the Chicago Drainage Canal works, for conveying the excavations from the bottom of the canal to the spoil-banks on each side, worked on this principle. This cantilever steel-belt conveyor consisted of a bridge spanning the cutting, with cantilever arms at the two ends rising above

[1] "The Chicago Drainage Canal," Ingersoll Sergeant Drill Company, New York, pp. 6 and 19.

the spoil-banks on each side, having lengths of 172 and 148 feet respectively, the total length of the machine between the extreme ends of the cantilevers being 640 feet. The belt, formed of steel pans linked together, dipped down into the cutting, where the pans were loaded by ploughs worked by steam, and passed over drums at the ends of the two cantilever arms; the material raised from the cutting being deposited on the bank from a height of 90 feet.[1] This machine, which was wrecked by a gale in 1895, and had to be rebuilt, was able to carry and deposit about 1000 cubic yards in the day.

Bridges with open floors, spanning the spoil-banks at the sides of the Chicago Drainage Canal, and travelling sideways, were also used for the deposit of the excavations, the material being brought on to the bridge either by waggons drawn up an incline from the cutting, or by a rubber-belt conveyor fed with crushed material by a granulator supplied by a steam-navvy.[2]

DREDGING.

Dredging consists in excavating under water, and raising the excavated materials; but just as in ordinary excavation on land, the removal of the dredged materials to a suitable site forms an essential portion of such operations. Moreover, when the material to be removed consists of rock too hard to be broken up by the steel teeth of a bucket or grab, blasting, as in ordinary excavation, though under more difficult conditions, or the breaking up of the rock by blows, constitutes an indispensable preliminary operation to dredging.

Four distinct types of machines are used for dredging, namely, the bucket-ladder, the dipper bucket, the grab, and the sand-pump, the choice depending upon the nature of the work and the conditions under which it has to be carried out; whilst the primitive contrivance termed the bag and spoon, is employed for the removal of small quantities of loose material.

Bag and Spoon.—A leather or canvas bag is encircled at its mouth by an iron ring fastened to the end of a long pole, constituting the spoon; and it is worked by two men in a barge, to which it is loosely attached, one man guiding the upper end of the pole, whilst the other drags the bag along the bottom of the river by means of a rope or chain fastened to the lower end of the pole, which is wound up by a crab at the far end of the barge (Fig. 5). The material is thus scooped up by the flattened ring at the bottom, and, entering the bag, is drawn up by the chain and deposited in the barge. This machine is only suitable for deepening small channels, removing small isolated shoals, or procuring gravel and sand from a river for building purposes.

In the aquamotrice,[3] an iron scoop or bucket is substituted for the

[1] "Contractors' Methods employed on the Chicago Drainage Canal," Ligerwood Manufacturing Company, New York, p. 34.

[2] *Ibid.*, pp. 30–33 and 36–39.

[3] *Annales des Ponts et Chaussées*, 1874 (2), p. 188.

bag, so hinged to the handle that it can be turned over to release its contents; and the barge being moored in position, the chain, dragging the scoop or bucket along the bottom, and eventually raising it to the surface, is wound up by paddlewheels turned by the current at the sides of the barge. This machine raised gravel and shingle from the bed of the Garonne, at a total cost of 1½*d.* per cubic yard.

BAG AND SPOON.
Fig. 5.

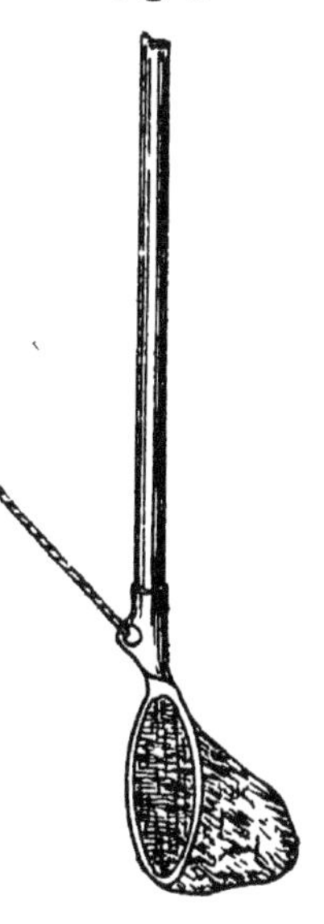

Bucket-ladder Dredger.—This dredger consists of a vessel of small draught, carrying a long girder or beam, one end of which is held up by a staging on the deck of the vessel, and the other end can be lowered as required for dredging; and a continuous row or ladder of iron or steel buckets, connected by a pair of endless chains, revolves round the girder by means of a tumbler at each end, the motion being imparted by turning the top tumbler by steam-power (Fig. 6). The buckets, provided with strong steel lips, and occasionally with claws to disintegrate stiff material, passing in succession horizontally along the bottom, scoop up the material; and, then assuming a more upright position, they raise their contents out of water, and turning bottom upwards after going round the top tumbler, discharge their load into a shoot, which leads the dredged material into a central tank in the vessel, or into a barge alongside. The vessel, which is anchored, is slowly moved forwards or sideways by hauling on its moorings to extend the deepening. The ladder of buckets is generally placed centrally in the vessel, a long well being provided in the central line of the vessel, through which the ladder is lowered for dredging; but sometimes the ladder is situated on one side of the vessel, and occasionally a ladder of buckets has been put on each side of a vessel. By placing the ladder-well in the fore part of the dredger, extending out to the bows, and adopting traversing gear by which the ladder can be carried forward when required, the dredger can cut its own channel in advance; but, on the other hand, the opening in the bows somewhat impairs the speed and seaworthiness of the vessel. The depth to which a bucket-ladder dredger can excavate depends on the length of the ladder, and therefore in a great measure on the size of the vessel, for the ladder-well must be long enough to enclose the raised ladder when the vessel is steaming along; and the size of the buckets must be regulated according to the strength of the machinery employed for driving them, and may be larger for soft soil than for compact strata.

There are two distinct forms of bucket-ladder dredgers, known respectively as stationary, and hopper dredgers. Stationary dredgers, though often provided with propelling machinery for shifting their position, remain dredging as continuously as practicable at the required site, and discharge the dredged materials into barges towed by tugs to the place of deposit, or into steam hopper barges which discharge their contents in deep water by opening flap doors at the bottom of the hopper or central well.

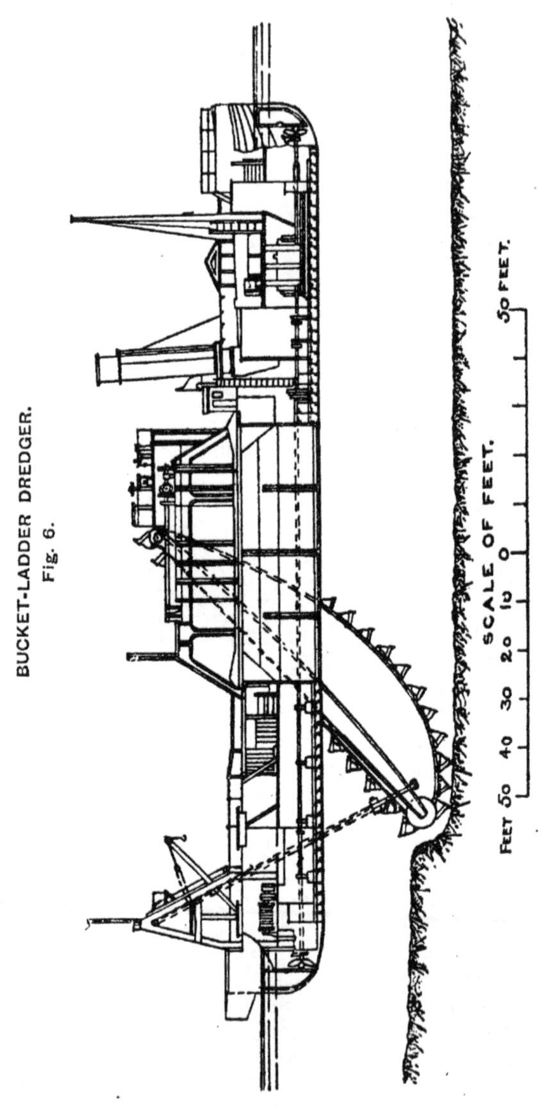

BUCKET-LADDER DREDGER.
Fig. 6.

Hopper dredgers combine the dredging machinery and the central well with flap doors at the bottom, or hopper, in the same vessel, so that these dredgers load themselves, and convey the dredged material to the place of deposit. Their dredging is, accordingly, intermittent, being only carried on till their hopper is full; after which some time is occupied in going to and from the depositing ground.

Stationary dredgers more fully utilize their dredging machinery, and have a smaller draught than hopper dredgers with their load; but they necessitate a larger plant, and take up a larger width in the channel with their attendant barges. Stationary dredgers, accordingly, are the best when the amount of dredging is large and concentrated, the available depth small, the width of the navigable channel ample, and the place of deposit distant; whilst hopper dredgers are preferable where the dredging is moderate and scattered, the money available for dredging plant limited, the channel narrow, and the depositing ground within a moderate distance.

Dipper Bucket Dredger.—This type of dredger, which may be regarded in principle as a greatly magnified form of the bag and spoon worked by steam, is practically the same machine as the steam-navvy, with the only difference that the huge bucket with its beam and supporting derrick are carried by a barge instead of a truck; and the machine is transported on water and excavates in it, in place of on land and in dry earth (see Fig. 2, p. 39). As, however, the weight of the machine is of little consequence when floating on water, as compared with running on land along lines of way which can sometimes with difficulty be maintained on soft ground, dipper bucket dredgers can be provided with larger buckets, worked by more powerful machinery, and having a greater range of work than steam-navvies.

Grab Bucket Dredger.—The grab, as this type of dredger is often designated, consists of an iron or steel semi-cylindrical or

GRAB BUCKETS.

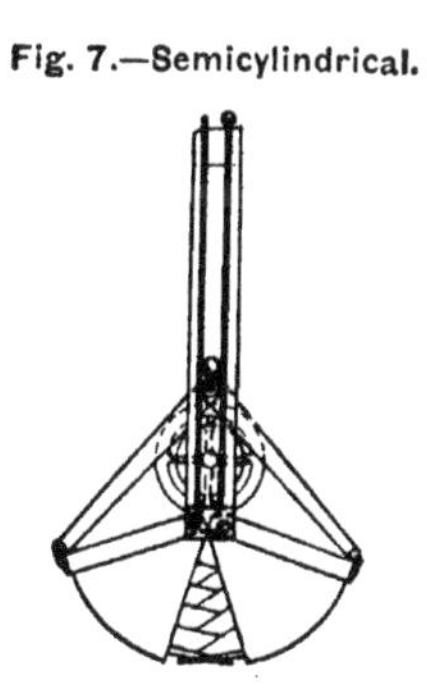

Fig. 7.—Semicylindrical.

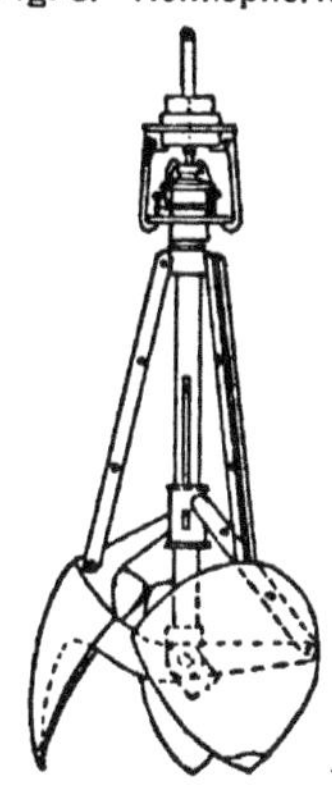

Fig. 8.—Hemispherical.

Scale $\frac{1}{75}$.

hemispherical bucket, separating when opened from the bottom into two, three, or four sections, with sharp edges or claws at their lower ends for penetrating and disintegrating the ground on to which it is lowered, or for gripping boulders, piles, trunks of trees, or other obstacles to be raised (Figs. 7, 8, and 9). The grab is hung by chains from a crane, or attached to the end of a spear pushed down from a derrick or beam; and the penetration is effected by the impact of the open bucket, or the push of the spear, aided by the weight of the bucket. The jaws, which have been held open by chains in lowering the bucket, tend to close by their weight when released in the act of raising, the closing being aided by levers; and the sections, in coming together, enclose the material penetrated, or grip the obstruction, which is then raised and removed. The inability of the grab to penetrate adequately into compact or adhesive ground, unless forced down by hydraulic power (Fig. 9), limits its general applicability to loose soils and rough obstacles; but it can raise suitable materials from any depth required in practice, and its operations are not stopped by small waves.

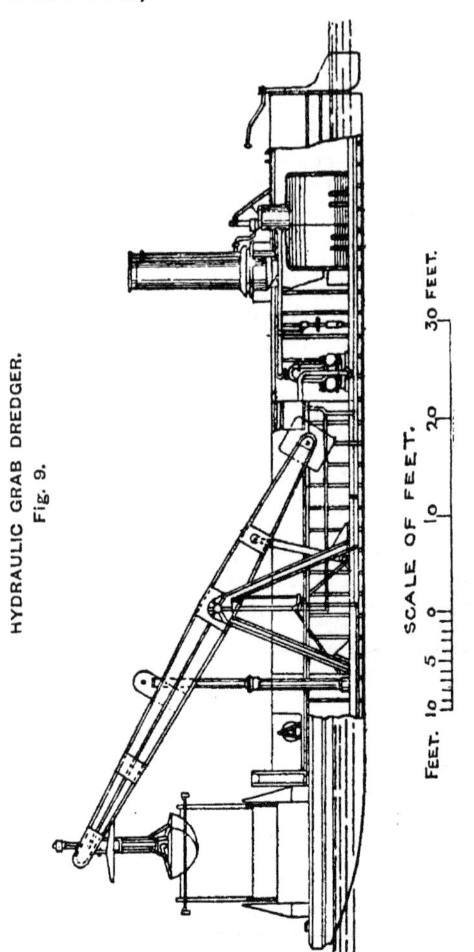

HYDRAULIC GRAB DREDGER.
Fig. 9.

Sand-pump Dredger.—The raising of pure sand from the bed of a channel is most readily and economically effected by a sand-pump or suction dredger, which consists of a long pipe lowered from a vessel and dipping slightly into the sand to be dredged, up which a stream of water and sand is drawn by a centrifugal pump on the vessel, and discharged into the hopper of the dredger or into a barge alongside, when the sand settles in the hopper or barge, and the water eventually escapes overboard (Fig. 10). Silt can also be raised in the same manner, but as it settles less readily than sand, a much larger proportion flows away with the water; whilst silt mixed with the sand greatly reduces the quantity of material raised in a given time, by its being much less readily drawn up in suspension, owing to the adherence of the

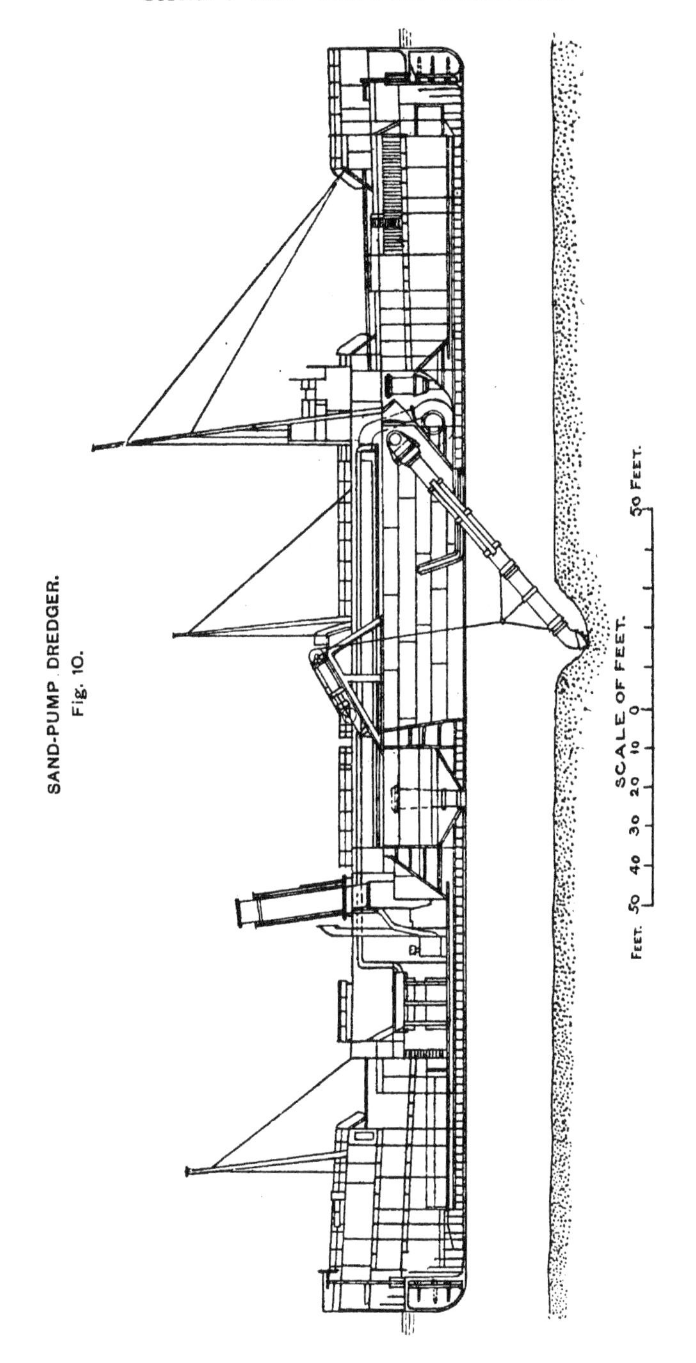

SAND-PUMP DREDGER.
Fig. 10.

particles together. The employment, however, of rapidly revolving cutters encircling the bottom of the pipe, and thus disintegrating and putting into suspension the material to be dredged, has been found, not only to increase largely the proportion of pure sand raised with the stream of water drawn up the pipe, but also to render the raising of a mixture of silt and sand, and even clay, by suction comparatively easy (Fig. 11). By fitting a flexible piece, or a telescopic joint at the end of

APPLIANCES FOR SUCTION DREDGERS.

Fig. 11.

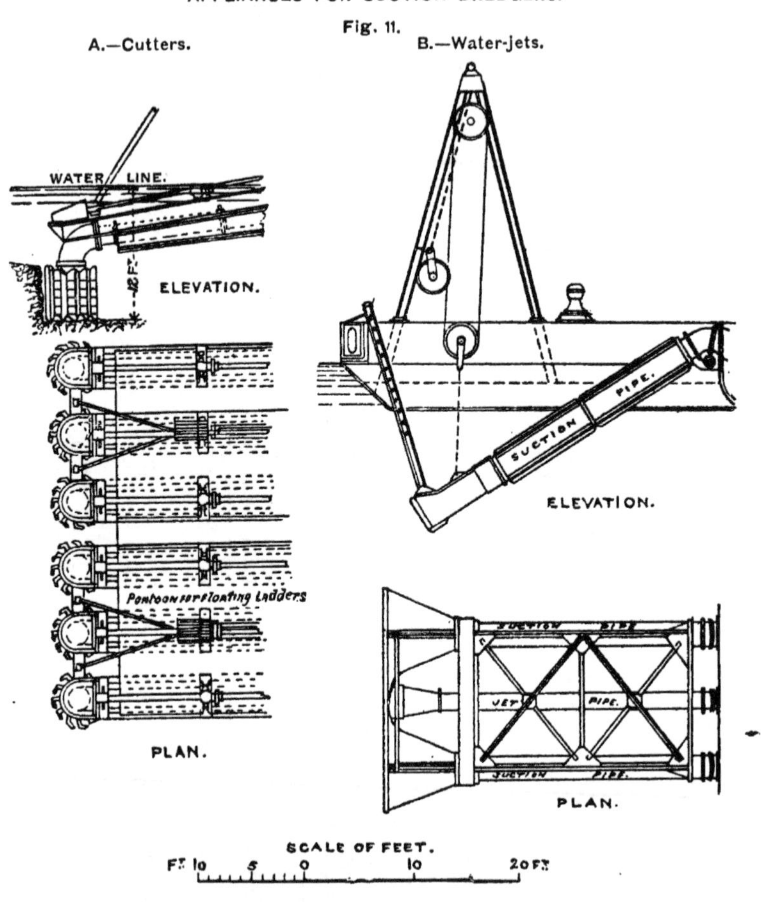

the pipe, the suction dredging can be continued in exposed places when the waves do not exceed three or four feet in height. Powerful sand-pump dredgers have been used since 1890 on the sand bar at the mouth of the Mersey in Liverpool Bay, the latest of these, the "Leviathan," being capable of dredging 70,000 tons of sand in 24 hours, and

300,000 tons in one week of $5\frac{1}{2}$ days;[1] whilst on the Mississippi below Cairo, a powerful suction dredger, provided with revolving cutters surrounding the suction-pipe, has proved able to raise nearly 5000 cubic yards of sand per hour from the shifting bars encumbering the river.[2] Still more recently, however, comparative trials, made in 1898, with two precisely similar suction dredgers purposely built for dredging on the Mississippi bars, one furnished with water-jets and the other with cutters, for stirring up the sand and silt close to the mouth of the suction-pipe, conclusively proved that the water-jets render the dredging much more effective than the cutters, for the materials forming the bars of the Mississippi; and the second vessel, accordingly, has had water-jets substituted for the cutters.[3]

Remarks on Dredging Machines.—The selection of the type of dredger to be employed must be determined by the conditions of the site, the extent and position of the shoals to be removed, and the nature of the material to be dredged. The bag-and-spoon dredger is only suitable for a very limited amount of work, in soft or loose soil. The bucket-ladder dredger is best adapted for removing large quantities of materials, ranging in quality from silt to boulder-clay and soft laminated rock, in fairly sheltered sites; and the systematic, extensive deepening of rivers, harbours, and canals, has been mainly carried out by dredgers of this type.

The dipper bucket dredger has been largely used in the United States for the improvement of navigation on large rivers; and it is very serviceable for extensive deepening, and for rough work in compact strata; but it is not so well suited for continuous deepening to a moderate extent over a long distance, and for maintaining the depth, as the bucket-ladder dredger with its smaller and numerous buckets.

The grab dredger is a comparatively cheap machine, and is very useful in removing silt and loose soil from docks, foundations, the inside of cylinders, and other confined situations, and in raising rough obstructions, boulders, and large pieces of blasted rock, from considerable depths in exposed places; but unless furnished with specially powerful appliances, it is unable to penetrate effectively any compact strata. Moreover, though it may be advantageously employed for removing detached shoals of loose material of limited extent, a grab is unsuited for rapidly and economically carrying out extensive works of deepening.

The sand-pump dredger has proved the most economical and efficient machine for lowering sandy shoals; and these qualities, and the power of working in exposed situations, have enabled this type of dredger to open out and maintain, at a reasonable cost, an adequately deep, navigable, approach channel across a flat sandy foreshore on the seacoast, to the ports of Calais, Dunkirk, and Ostend, and other ports on sandy coasts, and more recently to form a low-water channel across

[1] "The Sand Dredger Brancker," A. Blechynden, International Maritime Congress, London, 1893, and Annual Report, 1909.

[2] "Report of the Chief of Engineers, U.S. Army, for the year 1896," part vi. p. 3423.

[3] *Ibid.*, 1898, part v. p. 3138 and plates 6-11.

the Mersey bar. The sand-pump, moreover, when provided with water-jets or revolving cutters, is rendered capable of dealing efficiently with silty and clayey shoals.

Breaking up Rock under Water.—Sometimes rocky reefs, stretching across rivers and harbours, present barriers to improvements of the navigable channel, which require to be broken up before deepening by dredging can be accomplished; and the enlargement of the Suez Canal where it traverses rocky strata, has also necessitated the breaking up of the rock before the widening and deepening could be effected. Ordinarily a barge is moored over the site of the reef, from which holes are drilled in the rock, in which dynamite cartridges are inserted; and the rock is shattered by the explosion of the charges, and can then be raised by a dredger. An improved modification of this system, adopted for the removal of rocks in the St. Lawrence and at the "Iron Gates" of the Danube, consists in boring holes by steam Ingersoll drills inside iron tubes resting on the rock, to protect the drills from the current, clearing out the holes by a jet of water, and exploding simultaneously a series of dynamite charges in the holes by electricity.[1] Props are lowered down from the moored drilling barge, by aid of which the barge is raised slightly out of water to secure it from wave motion during the drilling.

ROCK-BLASTING DIVING-BELL.
Fig. 12.

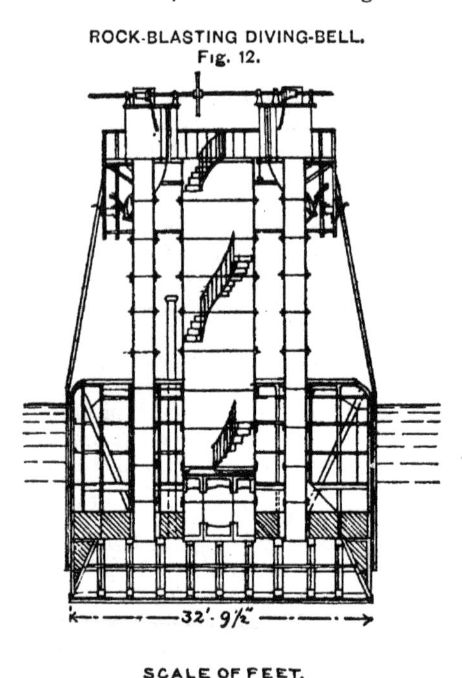

Sometimes the drilling of the holes for blasting isolated, submerged rocks in harbours, has been effected by divers, as, for instance, in Holyhead and Alderney harbours; and occasionally diving-bells have been resorted to, thereby giving the miners more freedom for their work, as in blasting some small detached rocks impeding the approach-channel to New York harbour.

A special form of diving-bell, supplied with compressed air, has been used for lowering a rock in Brest harbour, and for deepening the rocky channel of access to the naval dockyard at Cherbourg, having a bottomless working chamber, 6½ feet high, below, and an air compartment above, ballasted so as to float upright, with a central shaft rising

[1] "La Régularisation des Portes de Fer," Béla de Gonda, Congrès international de Navigation intérieure, Paris, 1892, pp. 75–78.

above the highest water-level, and affording communication, by the aid of air-locks, between a raised platform at the top and the working chamber at the bottom[1] (Fig. 12). This diving-bell is sunk on to the submerged rock by letting water into the air compartment, above the working chamber, through valves in the side; and the men mine the rock forming the floor of the working chamber, by the help of compressed air. When a charge has to be fired, the men seek shelter in the large air-locks at the base of the shaft on the roof of the working chamber; and the diving-bell can be readily floated off the rock again when required, by admitting compressed air into the air compartment through valves in the roof of the working chamber, which ejects the water previously introduced to sink the apparatus.

In the approach to New York harbour from Long Island Sound, two extensive rocky shoals presented serious dangers to navigation, namely, Hallett's Reef projecting out from the shore, and Middle Reef rising up in mid-channel, 3 acres and 9 acres in area respectively, above the 26-foot contour of depth. As the ordinary methods of blasting rock under water would have been too slow and costly under such conditions, a shaft was sunk in each case, and a network of galleries was formed under the whole area, leaving pillars of rock supporting the roof.[2] Holes were then bored in the pillars and roof, and blasting charges inserted, which, after the admission of water for tamping, were exploded simultaneously by electricity from the shore, dynamite being mainly used in the first case, and rackarock composed of potassium chlorate and nitro-benzol, together with the explosion of most of the charges by the concussion resulting from the firing of the primary exploders, in the second case. Most of the rock thus shattered had to be removed by grab dredgers, in order to obtain the requisite navigable depth of 26 feet over the site of the reefs.

ROCK-BREAKING RAMS.
Fig. 13.

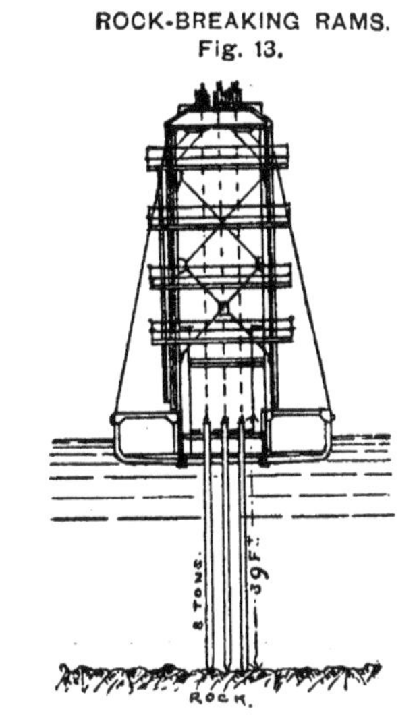

A method of shattering submerged rock without the use of explosives, consisting in breaking up the rock by repeated blows caused by the fall of a chisel-pointed heavy weight, has been adopted with success in widening the Suez Canal through the rocky cuttings, in the Danube, the Manchester Canal, at Malta, Blyth and elsewhere. The long steel

[1] *Mémoires de la Société des Ingénieurs Civils*, 1880 (2), p. 455.
[2] *Proceedings Inst. C.E.*, vol. lxxxv. p. 264, and plate 6.

rock cutters, with hardened, detachable points, weighing from 6 to 15 tons, have a fall of 6 to 10 feet, guided by staging, and are raised again by steam winches. These cutters, delivering from 35 to 150 blows per hour, will break up about 2 cubic feet per blow, the effect varying with the nature of the rock. The distance between the points on the rock where the blows are delivered is from 3 to $4\frac{1}{2}$ feet (Fig. 13).[1]

Removal and Deposit of Dredged Materials.—Besides the system of hopper dredgers or hopper barges depositing their loads in deep water, so commonly employed for the disposal of dredged materials, other methods have sometimes been resorted to. When dredgings discharged into barges have to be deposited on land, the unloading of the barges has been facilitated by using a chain of buckets, projecting and worked from a jetty extended out from the bank, to remove the soft material from the barge, and discharge it through a shoot into waggons which convey it to the spoil-bank on the land.[2] Sometimes, in order to avoid this duplicated scooping up and discharge of the material, the waggons are run on to the barges from a specially-constructed landing-

LONG SHOOT FOR DISCHARGING DREDGINGS.

Fig. 14.—Suez Canal.

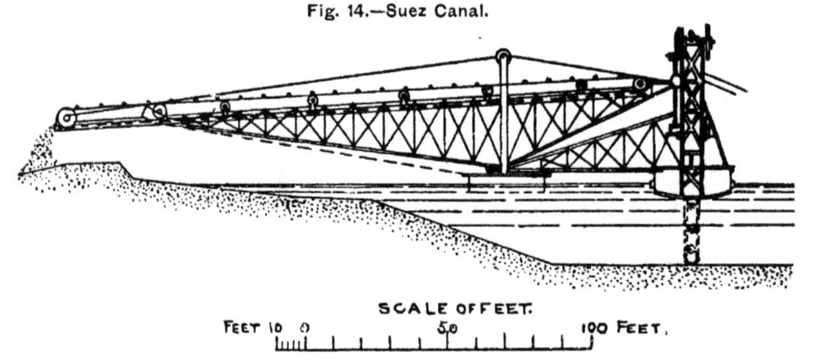

stage, so that, being brought alongside the dredger, they can be loaded directly with the material falling from its buckets; and, being run again on to the land at the stage, they convey their load straight to its destination.[3]

Where large quantities of material have to be conveyed only a short distance to the side, as in dredging ship-canals through lakes or low ground, and occasionally in deepening rivers, long shoots or floating tubes from the dredger to the bank have been adopted with advantage. Thus in constructing the Suez Canal, shoots 5 feet wide, and having a maximum length of 230 feet, conveyed the materials from the dredgers to form the side banks (Fig. 14),[4] the passage of mud or sand along the shoot, having a minimum inclination of 1 in 25 to 1 in 20, being aided by a stream of water, and in the case of clay or other

[1] *Proceedings Inst. C.E.*, vol. xcvii., clxx., clxxiv.

[2] *Annales des Ponts et Chaussées*, 1875 (1), p. 405.

[3] *Ibid.*, 1880 (1), p. 29, and plate 2, figs. 12 and 13.

[4] *Mémoires de la Société des Ingénieurs Civils*, Paris, 1866, p. 494, and plate 70.

stiff material, by travelling endless chains as well; and similar shoots have been used on the Maas, the Ghent-Terneuzen and Panama canals, and other works.

Long wooden tubes, 15 inches in diameter, connected by leathern joints and buoyed up by floats, were employed during the formation of the Amsterdam Ship-Canal through the lakes by dredging, for conveying the dredged materials a maximum distance of 900 feet to the side banks, the dredgings, mixed with water, being forced through the tubes by a centrifugal pump capable of discharging the materials 8 feet above the water-level.[1] By joining on a centrifugal pump about halfway along the floating tube, the sand and silt dredged for the enlargement of the Ghent-Terneuzen Canal could be discharged as much as 28 feet above the water-level, and conveyed considerably further off along a shoot; and by adjusting another pump further along the tube, the materials could be deposited more than half a mile away from the dredger.[2]

Remarks on Dredging.—Dredging possesses advantages over excavating in the facility with which dredgers of the largest size and capacity for work can be moved from place to place, in comparison with the care and labour required in shifting large excavators and steam-navvies, and also in the cheap transport by water of the dredgings to the place of deposit. On the other hand, excavation can be executed with greater precision than dredging; and whilst the excavated materials are utilized on railways in forming embankments, and on dock works in forming quays on low-lying lands, dredgings, except when discharged on to the side banks, are generally merely got rid of in the cheapest available manner. On the whole, however, dredging is, in the majority of cases, the cheaper method of executing earthwork; and under conditions when either system can be used, the preference is generally given to dredging, of which portions of the Suez Canal works were a notable instance.

The cost of dredging depends upon the nature of the material, the concentration of the work, and the distance of the place of deposit, even more than the cost of excavation does; and though often merely the actual working expenses and cost of repairs are adopted as the basis for ascertaining the cost of dredging, interest on the capital expended in the dredging plant, and an allowance for the depreciation of the plant, constitute essential items in a proper estimate of cost. Extensive dredging operations have been carried out with bucket-ladder dredgers on various rivers, at prices ranging between 7*d.* and 3*d.* per cubic yard,[3] though in some instances the expenditure has been notably larger; whilst on the Manchester Ship-Canal, grabs, steam-navvies, and excavators removed earthwork at a cost of from only 3*d.* to 1½*d.* per cubic yard,[4] which cannot, however, be regarded as an ordinary price for

[1] *Proceedings Inst. C.E.*, vol. lxii. p. 6.

[2] "Le Canal de Terneuzen-Gand, et ses Installations Maritimes," O. Bruneel and E. Braum, p. 21.

[3] "Rivers and Canals," L. F. Vernon-Harcourt, 2nd Edit., vol. i. pp. 79 and 80.

[4] *Proceedings of the Institution of Mechanical Engineers*, Liverpool Meeting, July, 1891, p. 424.

excavation. The cost of dredging in the extensive deepening of channels has been very materially reduced of late years, by the introduction of larger and more efficient machines; and this reduction has been specially noticeable in dredging with sand-pumps. Thus whereas the first price, in 1876, for removing sand from the approach-channel to Dunkirk harbour, across the sandy foreshore, by sand-pumps, amounted to 1*s.* 9*d.*, it was gradually reduced during subsequent years down to about 2*d.*; and the raising of large quantities of sand from the Mersey bar by powerful suction dredgers, has been accomplished at a cost of about 1¾*d.* per cubic yard.[1]

PILE-DRIVING.

The driving of piles into the ground constitutes an important part of many temporary and permanent works, where the ground is soft, the influx of water has to be guarded against, or works have to be carried out in water.

Piles.—Small, straight trunks of trees, stripped of their branches, and driven down with their smaller, upper end downwards, are sometimes used as bearing piles for supporting structures on an alluvial foundation, as, for instance, in Holland. Usually, however, piles are formed from long, squared, fir timber, encircled with an iron ring shrunk on at the head to prevent the pile splitting under the blows of the ram, and protected at the pointed end by an iron shoe fastened to the sides of the pile by straps and spikes. Single piles, or the main piles of a cofferdam, are furnished with a pointed shoe to facilitate the penetration of the pile into the ground; but in driving a row of piles close together to form sheet-piling for cofferdams, or other purposes, the intermediate piles are sometimes provided with a wedge-shaped cast-iron shoe, with its bottom edge in line with the piling, and slanting slightly downwards towards the adjacent pile previously driven, so as to make the pile in driving come close up to its neighbour. Concrete piles, plain or reinforced, have been successfully used. In the former case, hollow forms or moulds are first driven and concrete filled in, the mould being withdrawn. In the latter, the reinforcement consists of steel bars bedded in the concrete pile which is moulded vertically and driven in the usual way.

Ordinary Pile-driving.—Piles are driven into the ground by successive blows of a heavy cast-iron weight falling on their head. This weight, called a monkey or ram, weighs generally from half a ton to over a ton; it is raised by a rope or chain passing over a pulley at the top of a wooden framework, termed a pile-engine (Fig. 15), which, by means of a groove in its nearly vertical front face, in which a projection from the ram slides, serves to guide the weight in its descent. The fall given to the ram varies from a few feet up to a maximum of about 40 feet, according to circumstances; and the

[1] *British Association Report*, Liverpool Meeting, 1896, p. 556.

catch at the end of the lifting tackle which holds the ram in its ascent, on encountering an obstacle on the pile-engine, placed at the desired height, on being pulled by a rope, releases the ram, which drops upon the head of the pile below. Though the ram is sometimes raised by manual labour by the help of a winch, an endless chain, worked by steam, provided with catches at intervals to raise the ram, enables more rapid blows to be dealt with ease and regularity.

PILE-ENGINE.
Fig. 15.

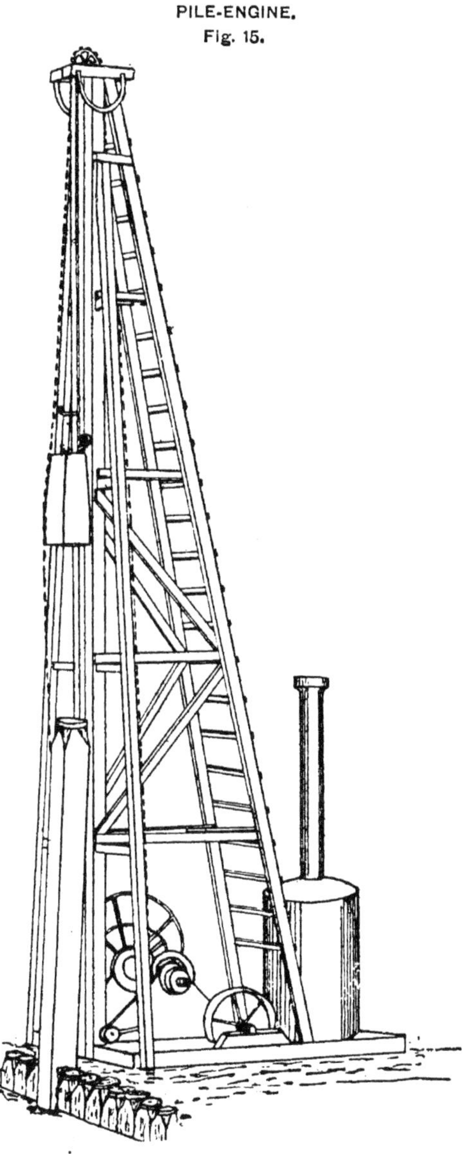

The pile must be of sound timber free from cracks or knots, otherwise it is liable to be injured in the process of driving. As the driving proceeds, the descent of the pile at each blow diminishes, owing partly to the increasing area of pile exposed to friction against the ground, and partly to the greater compactness of the soil generally as greater depths below the surface are reached, so that the fall of the ram has to be increased; and often at last, three or four blows are needed to drive the pile down an inch. When this resistance to driving has been reached, the driving should be discontinued, as such a slow descent is liable to be merely the result of the brushing up of the pointed end of the pile against the iron shoe under the blows of the ram. The efficiency of the blow is greatly affected by the solidity of the pile-head, for when fibres are brushed up, a considerable portion of the force of the blow is dissipated in bending the elastic fibres;[1] and a notable reduction of power is experienced when, owing to the low level of the pile, a solid piece of timber has to be interposed between the pile and the ram.[2]

Steam Pile-drivers.—Nasmyth's pile-driver is the oldest and best-known form of these machines.[3] It consists of a heavy iron frame resting on the pile, which forms the cylinder, with a hammer or ram constituting the piston inside it, which is rapidly raised and lowered three or four feet by steam-power, dealing blows on the head of the pile at the rate of 60 to 80 per minute (Fig. 16). Though the hammer in this machine has been made considerably heavier than ordinary rams, reaching 4 tons in weight, its small fall, in spite of the aid of steam, renders its impact much less than that of a ram with a large fall; but, nevertheless, the rapidity of its blows makes it much more efficient in sandy or silty soils, for the pile is kept in almost

NASMYTH'S PILE-DRIVER.
Fig. 16.

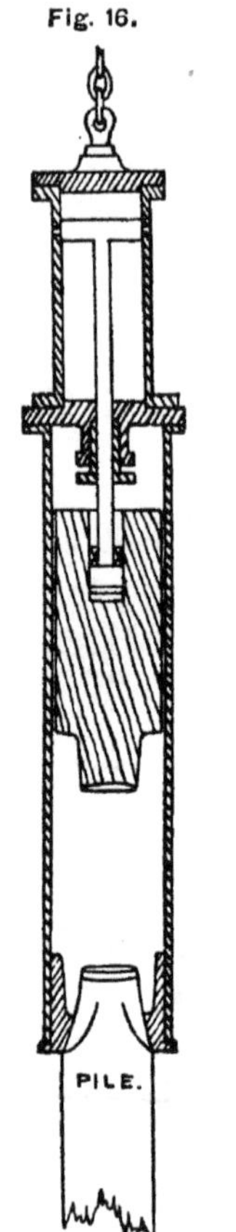

STEAM PILE-DRIVER.
Fig. 17.

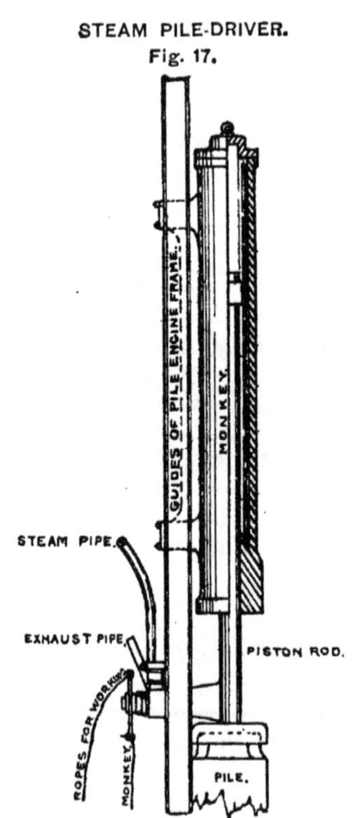

[1] *Transactions of the American Society of Civil Engineers*, vol. xii. p. 441.
[2] *Proceedings Inst. C.E.*, vol. xxvii. p. 277.
[3] "James Nasmyth, an Autobiography," pp. 274–275.

constant motion, so that the soil has not time to settle round the pile between the blows, as occurs in ordinary pile-driving. Moreover, the jumping up of a pile after a blow, which is a common occurrence in driving through quicksand, is prevented by the weight of the frame and hammer, amounting to 7 tons, which also assists the descent of the pile, as well as by the quick succession of the blows. The driving of a pile with this machine is effected in a few minutes, which would occupy over an hour or more by the ordinary method. In other forms of steam pile-drivers, the ram itself forms the cylinder in which a piston is fitted. In one type, the piston-rod rests upon the head of the pile (Fig. 17); and the ram is raised and lowered by admitting steam between the piston-head and the top of the ram, and then letting it escape when the ram has reached its highest point, causing the fall of the ram. In another type, the piercing of the bottom of the ram for the exit of the piston-rod is obviated by providing a frame supported on the pile, from the top of which the ram is hung by its piston, and is raised, as in the former case, by admitting steam between the top of the ram and the piston-head, which, however, in this instance, is at the bottom of the piston-rod instead of at the top.

Gunpowder Pile-driver.—By fitting a steel cap on the top of a pile, with a central hollow in it for receiving a cartridge, piles can be driven by the explosion of gunpowder.[1] The ram is provided at the bottom with a piston which fits the cylindrical hollow in the mortar or cap; and the ram having been raised and a cartridge inserted in the mortar, on releasing the ram the piston enters the barrel of the mortar, and developing heat by the compression of the air, explodes the cartridge, causing the projection of the ram upwards again, being controlled in its course by guides, and driving down the pile by the reaction resulting from the explosion. The ram can be arrested at any desired point in its ascent; and the best rate of working has been found to be about twelve blows per minute—a convenient rate for the introduction of the cartridges, which, if supplied more rapidly, are liable to be exploded at once by the undue heating of the mortar or the ignited remains of the preceding cartridge, involving the waste of the cartridge and the drawing up of the ram. The charge of the cartridge is varied according to the nature of the ground, to avoid injuring the pile by an undue strain; and the system has proved efficient, rapid, and economical.

The Water-jet in Aid of Pile-driving.—By placing two flexible pipes, or hoses, down opposite sides of a pile, with their nozzles extending below the pile and directed centrally under its point, from each of which a powerful jet of water under pressure is discharged, the soil under the pile is greatly loosened, and to some extent scoured out, so that the pile descends much more freely, and far less driving is required. The system is most efficient in pure sand; and the driving of piles and panels of sheeting in this material at the Calais Harbour Works, was effected by the water-jet in about one-eighth the time occupied by the ordinary method.[2] The water-jet is also advantageous

[1] *Deutsche Bauzeitung*, 1875, p. 433.
[2] *Annales des Ponts et Chaussées*, 1878 (1), p. 74.

in mud and soft clay; but where gravel predominates, or in hard clay or peat, or where driftwood or other large obstacles are encountered, the jet is of little value. As the force of the blows for making the piles penetrate the soil, can be greatly reduced by the diminution in compactness and adherence produced by the water-jet, it is possible to drive timber by its aid, which would be too fragile to be used under ordinary conditions.

Bearing Piles.—Piles driven into a sandy or silty stratum for supporting the piers of bridges, dock walls, or other structures, bear the weight imposed upon them, either by reaching firm ground below the soft stratum on which they rest, or by the mere adherence or compression of the sand or silt encompassing them, varying with the compactness of the material pressing against their sides and the depth to which they are driven. Bearing piles, however, should, if practicable, be driven down to a solid stratum, as their supporting power is thereby much increased, having been found in some instances to amount to five times that due to adherence alone. Nevertheless, where the sand or silt extends to a considerable depth, a long pile derives a fair supporting power from the enveloping material when time is allowed for it to consolidate round the pile after driving.

Sheet-Piling.—Piles are often driven close together in a row on works, to exclude water from excavations, to prevent a run of water under walls or the sills of a lock, to facilitate the excavation of foundations, or to retain slopes of earth. The main piles of this sheet-piling are driven well down a few feet apart; and the intermediate sheeting of piles, half-balks, planks, or panels, are driven close together between them, forming a continuous sheeting of timber, the joints being sometimes caulked, where exposed, to make the structure more water-tight.

Iron Sheet-Piling.—Sheet-piling to stop a run of water, of cast-iron or mild steel, has been frequently made use of. A modern example of the former is found in the Nile Barrage at Asyût,[1] while in the Hodbarrow Outer Barrier, Millom, Cumberland,[2] mild steel piling, consisting of riveted combinations of rolled steel sections and plates, was successfully employed. The total amount of metal used at Asyût was 4046 tons, and at Hodbarrow 4416 tons. The joints of the cast-iron piling at Asyût were grouted.

COFFERDAMS.

For excluding water from dock works, the piers of river bridges, and other structures exposed to the influx of water during construction, timber cofferdams are often constructed with a single or double row of sheet-piling. The sheet-piling is kept in position and strengthened by horizontal timber walings bolted to the main piles, the number of rows of walings depending upon the height of the cofferdam above the surface of the ground. Cofferdams, moreover, are generally supported

[1] *Proceedings Inst. C.E.*, vol. clviii., p. 26.
[2] *Ibid.*, vol. clxv., p. 156.

at the back against the water-pressure in front, by a mound of earthwork, and by raking timber struts at intervals, butting against low piles behind.

Single-sheeted Cofferdams.—Single-sheeted cofferdams are used where the available space is small, and the height required for the dam is moderate; but considerable care has to be taken in driving the piles or panels, as the sheeting in this case constitutes the watertight barrier which has to keep out the water. The piles must be accurately sawn at their side faces, to ensure close contact between parallel surfaces, and driven in groups, if the stratum is soft enough, between two rows of walings on each side; and leakage can be reduced by caulking the joints, or under specially favourable conditions, by grooving and tonguing the piles,[1] and depositing a narrow band of clay along the outer face of the dam, where, on being exposed to the water-pressure, the piles are liable to retreat a little from the foreshore in front.[2]

Cofferdam with Puddle Wall.—The more ordinary form of cofferdam, for large works and a considerable water-pressure, consists of two rows of whole piles driven parallel to each other, about four or five feet apart, into a watertight stratum, the interval between the two lines of sheeting being filled in with puddled clay brought up in thin layers from a little below the surface of the watertight stratum, so as to form a watertight wall of clay enclosed within the sheeting, raised above the highest water-level. Walings on each side of the dam, connected horizontally by bolts through each pair of main piles in the two rows, brace together the two rows of sheeting, and prevent their spreading out towards the top under the pressure of the clay filling (Fig. 18). Numerous examples of this type could be cited, notably that used, with some modifications in detail, at Keyham Dockyard Extension, where the combined length of double and single pile dams amounted to nearly $1\frac{1}{2}$ miles, strutting on the inside with abutment piling being used. At Gibraltar Dockyard Extension the dam was supported by an inner bank of earthwork.

COFFERDAM.

Fig. 18.

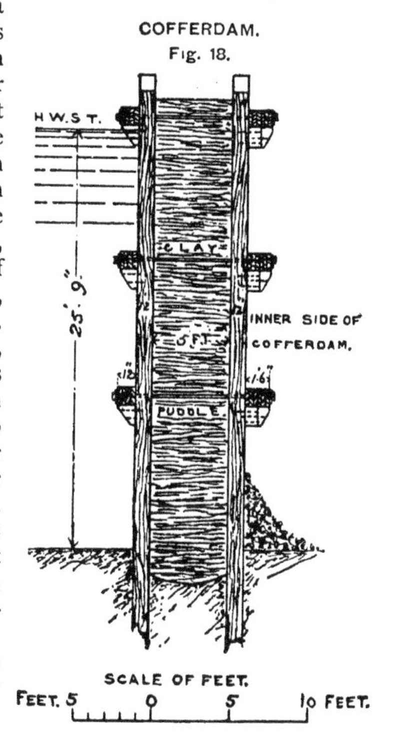

Iron or Steel Caissons.—Sometimes the site to be built upon is enclosed by one or more wrought-iron caissons, instead of by a timber

[1] *Proceedings Inst. C.E.*, vol. xxxi. p. 28, and plate 3.
[2] *Transactions of the Institution of Civil Engineers of Ireland*, vol. xiii. p. 92.

cofferdam. Thus a portion of the Thames Embankment was built within the shelter of rows of elliptical plate-iron caissons, with the view of reducing the cost of construction by using the caissons two or three times over in different parts of the work.[1] The foundations of four of the circular piers upon which the cantilevers of the Forth Bridge rest, were each formed on the rock within a plate-iron enclosure, fitted at the bottom to the sloping rock,[2] where it would have been impossible to erect an ordinary pilework cofferdam. The two large river piers, also, of the Tower Bridge were each founded within a group of twelve plate-iron caissons, strutted across with timber, which presented much less obstruction to the waterway of the Thames than an ordinary pile-work and puddle cofferdam, round the site of each pier, would have done;[3] and, moreover, the caissons could be taken to pieces, on the completion of the foundations of the pier, in a much shorter time than the removal of ordinary cofferdams would have occupied.

Earthwork Dams.—Often in dock works, when the site of the works has to be reclaimed from the foreshore of an estuary, and there is abundance of materials from the excavations, an embankment is formed round the site for shutting out the river, which provides at the same time a convenient place for the deposit of the earthwork, and is serviceable in the formation of quays. These embankments generally, unlike ordinary cofferdams, form part of the permanent work; but the earthen dam formed across St. Mary's creek in front of the Chatham Dockyard extension works, was resorted to in preference to a timber cofferdam for the sake of economy.[4]

Closure of Dams.—One of the chief difficulties in the construction of dams is the final closing of them when the range of tide is considerable, as the tidal influx and efflux produce a strong scour in proportion as the central gap is reduced in width, when the area from which the tide has to be excluded is large. With timber cofferdams, openings are provided through the dam at suitable places for the influx and efflux of the tide, till, on the completion of the dam, these openings are closed by lowering large timber panels across them.

The closing of earthen dams has to be effected by raising the bank gradually over a long length, so as to reduce the tidal scour by the large width given to the aperture. Then, instead of attempting to let out the water from the enclosure at low water of spring tides before closing the bank, it is safer to effect the raising of the bank above high tide as rapidly as practicable during the lowest neap tides, and to complete the bank to its full height before the following spring tides. By leaving water in the enclosed area, the pressure on the bank is diminished till it can be thoroughly consolidated, and the scour through the aperture is also reduced; whilst by selecting neap tides for closing the bank, the amount of material needed to raise it above high water in advance of the tide is a minimum, and the bank can be raised higher just beforehand without being exposed to undue scour.

[1] *Proceedings Inst. C.E.*, vol. liv. p. 12, and plate 2.
[2] *Ibid.*, vol. cxxi. p. 310; and "The Forth Bridge," W. Westhofen, p. 18.
[3] *Ibid.*, vol. cxiii. p. 119, and plate 4. [4] *Ibid.*, vol. xxxi. p. 26, and plate 2, fig. 1.

CHAPTER V.

FOUNDATIONS, AND PIERS OF BRIDGES.

Importance of foundations—**Foundations:** classification—Ordinary foundations, broadened base, bearing piles—Foundations in water-bearing strata, various methods of dealing with water or running sand—Congelation process through running sand—Injection of cement grout into quicksand—Well-foundations—Sinking wells, methods employed for bridge piers and dock walls—Advantages and difficulties of well-foundations—Foundations in water, various systems employed—Screw-pile foundations—Iron piles with disc at base—Iron cylinders for piers of bridges, instances—Wrought-iron caissons, instances of their employment for deep foundations of bridge piers—Compressed-air foundations, advantages, chief points of system, precautions necessary; for Antwerp quays in river Scheldt; for bridge piers; for Eiffel Tower; partial employment of system; adoption of removable caissons—**Piers of Bridges:** superstructure—Influence of pier foundations on design of bridges, instances in illustration—Provision for navigable channel, allowance required for possible future improvement in channel—Loads on foundations of bridge piers, instances.

FOUNDATIONS involve some of the most important and difficult problems with which the civil engineer has to deal, and require wide practical experience, in most cases, for their successful carrying out. The nature, indeed, of the strata to be built upon may be indicated to some extent by the geology of the district; but it should be ascertained more precisely by borings or trial pits. When the nature and condition of the strata have been adequately determined, experience, observation, and sometimes experiments are needed to choose the kind of foundation, and the method of executing it, best suited to the special conditions of each case. The piers of bridges, moreover, are so dependent, as regards their design and distance apart, upon the nature of the strata upon which they have to rest, and in the case of bridges across rivers upon the condition of the channel and bed, that they necessarily have to be considered in conjunction with foundations.

FOUNDATIONS.

Foundations may be of three descriptions, namely, ordinary foundations on land; foundations below the natural water-level, from which the water is excluded during construction; and foundations laid actually under water, or from which the surrounding water has to be kept away by special contrivances during their progress. The foundations of many structures belong to the first category, but are liable to merge into the second when the excavations have to be carried some depth below the surface to reach a sufficiently solid stratum, or when the site is close to a river or the sea; whilst the chief difficulties are generally encountered in the last class of foundations, which present the greatest variety of conditions. The stability of foundations depends upon the compactness or tenacity of the strata in relation to the load imposed by the structures erected upon them; whilst in the case of retaining or dock walls founded upon clay or slippery laminated strata, it is essential to take precautions against their slipping forward, by bestowing great care on the filling-in at the back, and by carrying down the foundations some feet below the surface in front, so as to provide a solid face for the lower part of the wall to butt against, and thus enable the wall to resist the pressure from behind.

Ordinary Foundations.—Ground sufficiently solid to support structures of moderate height is often found three or four feet below the surface of the soil, if free from the influence of water; and the building is invariably broadened out at the bottom by footings, assisted often by a batter on the face increasing the width of the work in proportion to the increase in the load to be borne at the lower part, and sometimes by stepping out the work at the back, as in the case of retaining and dock walls, thereby spreading the weight over a wider base. Where somewhat unfavourable ground is encountered, the bearing surface is readily extended by a wide layer of concrete at the bottom, which is frequently reinforced with embedded steelwork.

When the soil is soft and yielding, as in alluvial formations, bearing piles have often to be resorted to for high and heavy structures, constituting the ordinary system for the foundations of buildings in Holland; or when for any reason the driving of piles is inexpedient, the load can be distributed over a large-enough area by an extensive layer of concrete, to enable the soil to bear the weight. Bearing piles are generally connected together at their heads by a series of walings, upon which planking is sometimes laid; or the heads of the piles may be encased in a layer of concrete spread over the surface, a system which, besides effectually bonding the piles together, provides a firm, level base for the erection of the structure.

Foundations in Water-bearing Strata.—Frequently, even on land, considerable quantities of water are met with in laying foundations, owing to the excavations being carried below the line of saturation of the underground water, or the opening out of springs, or occasionally from percolation from rivers or the sea. Similar difficulties are commonly encountered, on an extensive scale, in the construction of docks, the formation of tunnels, and the laying of sewers, where most of the work has to be carried out at some depth below the surface. The amount of water to be dealt with varies greatly in different works, depending upon their position, the nature of the strata traversed, the volume of flow of the underground streams, and sometimes on the accidental tapping of a large fissure or spring. The water is ordinarily removed, during the construction of the foundations or other works exposed to it, by leading it in trenches or pipes away to a pump, by which it is raised and discharged into a suitable channel. When the water boils up under pressure at a particular spot, it can be controlled by being led into, and allowed to rise in a vertical pipe or enclosure, till the height of the column of water in the pipe counterbalances the pressure; and as soon as the adjacent foundation is completed, the pipe can be closed by cement grout, and the influx thus permanently stopped.

The danger of boils, or any other rapid influx of water into foundations which are kept dry by pumping, consists in the removal of the sand or silt, often brought along in suspension by the current of water from adjacent strata, producing underground hollows liable to cause settlement above them, which, if occurring under the works themselves, under cofferdams or embankments excluding the water, or under neighbouring buildings, may entail very serious damage. Accordingly, it is essential for the safety of foundations, or other works executed in water-bearing strata, that any continued influx of sand or silt with the water should be stopped. In carrying the London main drainage works through fine sand, a well was sunk below the foundations, beyond the line of the sewer, without resorting to pumping; and a layer of gravel, some feet in thickness, was then deposited at the bottom of the well, which enabled the water to be pumped up the well, through the gravel, from the neighbouring foundations, without disturbing the sand.[1] Another plan of dealing with sand charged with water, is to surround the site of the foundation by sheet-piling driven down to a firmer stratum, after which the water can be pumped out of the enclosure, and the sand removed to any requisite depth, or a light structure founded on the enclosed, consolidated sand.

Congelation Process through Running Sand.—Congelation of the soil by freezing liquids has been successfully employed in running quicksand, for sinking shafts, or for carrying excavations for foundations down to a firmer stratum. The freezing is effected by sinking a series of pipes vertically into the quicksand round the site to be excavated, into which a cold, unfreezable liquid is introduced and kept in circulation; and in a recent modification, liquid ammonia alone has been used, which in evaporating produces an intense cold, owing to the latent heat absorbed by it in becoming gaseous.[2] By thus congealing the ground, pumping and the resulting influx of sand and settlement of the adjacent land are avoided. Moreover, the excavations can be carried down through the hardened soil, where otherwise strong timbering of the sides or very flat slopes would be necessary. This system is very efficient; but its large cost restricts its application to those cases whose conditions are so unfavourable as to render other methods almost inapplicable.

Injection of Cement Grout into Quicksand.—By driving down a series of pipes into quicksand to the depth required for the

[1] *Proceedings Inst. C.E.*, vol. xli. p. 118.
[2] *The Colliery Guardian*, December 1, 1893, p. 960.

foundations on each side of a proposed work, and forcing water down the pipes on one side, which, following the line of least resistance, rises in the pipes on the opposite side, cavities are formed in the sand by the current between the pipes. These cavities are then filled by injecting cement grout into them down one set of pipes, after closing the outlet pipes; and thus a fairly solid floor is formed in the sand, which can be increased in thickness above or below, by raising or lowering the pipes slightly, and repeating the process.[1] Side walls also can be similarly formed by successively raising partially alternate pipes in each row, and following a similar process, thus eventually forming a solid trench, from which the quicksand can be removed, and a sewer laid in it, without encountering undue infiltration.

Well-Foundations.—Circular, brick wells, resting upon a wooden curb at the bottom tapered down to a cutting edge, and gradually built up at the top, as the sinking proceeds by the excavation of the ground below the curb through the central hollow, have been employed for centuries by the natives of India for forming foundations in sandy or silty soil. This well system has been used at sites from which the water has been excluded, of which the dock walls at Glasgow founded upon concrete wells furnish a recent instance, as well as for foundations in the water for bridge piers, weirs, and quay walls. In India, the wells are commonly built up brick by brick, thereby dispensing with plant to a great extent; but elsewhere they are generally raised by successive rings of brickwork or concrete; and iron or steel curbs have been used in place of wood in several large cylinders, to afford greater strength to the cutting edge.

Sinking Wells for Foundations.—The ground is generally removed from below the wells by some machine which can be worked from the top, and excavate under water. Sand is sometimes raised by a sand-pump, and grabs are also employed for removing sand or other loose materials; but special machines are needed for dealing with compact, adhesive strata, such as indurated silt or clay. Bull's heavy, grab bucket fitted with powerful lever arms, and other heavily-weighted grabs furnished with claws, have proved efficient in fairly stiff soil; whilst Gatmell's double "jham" grab, lowered with two strong loaded blades with their points downwards, capable of being raised to a horizontal position by projecting arms worked by chains, proved able to excavate stiff clay from the bottom of the wells of the piers of the Empress Bridge across the Sutlej, at a depth of over 100 feet (Fig. 19).[2] Occasionally divers are sent down to excavate the ground, remove obstacles encountered by the cutting edge, or to recover lost grabs; but diving work at a considerable depth in a river-bed, within the confined space of a well, has been found very dangerous to health in some cases.[3] The wells are weighted on the top to aid their descent; and the depth to which the excavations have to be carried below the curb to make the wells sink, depends on the nature of the material. In the wells of the Empress Bridge, a depth of two feet

[1] *Engineering News*, New York, vol. xxvii. p. 420.
[2] *Proceedings Inst. C.E.*, vol. lxv. p. 248, and plate 3. [3] *Ibid.*, p. 251.

sufficed in sand, whilst twelve to fifteen feet were needed in clay, the actual sinking being started by lowering the water as rapidly as practicable inside the well, thereby increasing the weight on the curb, and causing a scour of material from under the curb into the well, by the inrush of water, which also, mainly passing down from the river above along the outside of the well, facilitates its descent by reducing the skin-friction.

The quay walls of some docks have been founded, in marshy or alluvial ground, on large, square, rubble-masonry wells sunk down, by excavating inside, to a solid stratum. This system, for instance, has been resorted to in constructing the Bellot Docks at Havre, on the silty foreshore of the Seine estuary; for founding a portion of the walls of the Bordeaux Dock in the alluvial ground adjoining the Garonne, and of the Penhouët Dock at St. Nazaire,[1] alongside the outlet of the Loire estuary (Fig. 20); and for carrying down the foundations of the walls

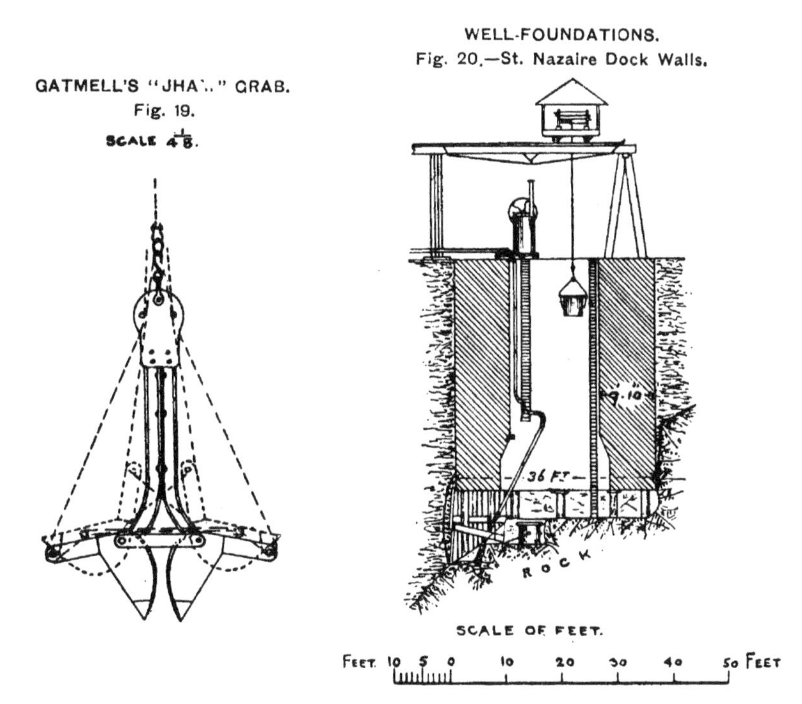

GATMELL'S "JHA'.." GRAB.
Fig. 19.
SCALE 1/48.

WELL-FOUNDATIONS.
Fig. 20.—St. Nazaire Dock Walls.

of the Third Dock at Rochefort, through fifty to ninety feet of soft alluvium, in marshy ground bordering the Charente. On completing the sinking, these wells are filled up with masonry or concrete; and the spaces left between adjacent blocks having been arched over, a continuous quay wall is erected along the top.

[1] "Ports Maritimes de la France," vol. v. p. 169.

Advantages and Difficulties of Well-Foundations.—Well-foundations possess the advantage of enabling a solid foundation to be reached, underlying a considerable thickness of mud, silt, or sand, without the necessity of enclosing the site by a cofferdam or caisson and pumping it dry; and they reduce the amount of excavation required to a minimum. They are, moreover, equally applicable in alluvial soil on land, in the silty foreshore of a river or estuary, or in the soft bed in the middle of a large river. The chief difficulties connected with them are, the occasional rapid influx of silt in passing through very soft ground fully charged with water, as happened in sinking the wells of the Rochefort dock; the hindrance presented to the sinking by embedded trunks of trees or boulders; and the liability of the well to sink irregularly, and thus to be forced out of the vertical, owing to variations in the resistance of the underlying strata, and especially in coming down on to a bed of sloping rock, a condition the wells of the St. Nazaire dock walls were exposed to, necessitating very special precautions in completing the sinking (Fig. 20). The inrush of sand, silt, and water, which is liable to occur in sinking a well through soft, water-bearing strata, occasionally filling the well and overflowing at the top, can only be prevented by keeping an excess of water in the well, which compresses the ground at the bottom, and thereby considerably augments the difficulty of excavation, or by roofing over the lower part of the well, and excavating and sinking with the help of compressed air, as found necessary in places for the dock walls at Rochefort (Fig. 32, p. 77). Large wells are less liable to diverge from the vertical than small wells of the same height. Any divergence has to be rectified by excavating on one side of the well at the bottom, by weighting the ground with stone on one side, or putting in shores to push the well back to the vertical, or by pulling the well over at the top by chains; for besides the change from the requisite position when the well leaves the vertical, the resistance to sinking is largely augmented. When a well reaches a sloping face of rock on one side, it has to be temporarily supported on the opposite side by timber piles or other form of props, till the rock can be cut away to a level bed to receive the well, or stepped down, and the well completed down to the lower part of the rock by building up underneath, termed underpinning, as effected in some parts of the St. Nazaire dock walls (Fig. 20).

Foundations in Water.—In many cases, it is possible to carry out foundations on sites covered with water by encircling the site with a cofferdam, caisson, or other watertight barrier, and pumping out the water; and then proceeding with the excavations, and the construction of the foundations of a bridge pier, quay wall, or other structure surrounded by water, in the ordinary manner. When, however, the depth of water is considerable, and especially when a very thick layer of sand or silt has to be traversed before reaching a firm, water-tight stratum, other systems have to be resorted to for carrying out the foundations, depending upon the conditions of the site, the load to be borne, and the nature and extent of the work.

In the case of breakwaters, the foundations are often spread over a wide base, and raised to a suitable height, by depositing a mound of rubble stone or concrete blocks in the water along the line of the work, upon which the superstructure is erected. If the mound is not raised to low-water level, concrete blocks are laid in rows or courses on the mound, and raised to low water, upon which the solid structure is built. If the bottom is a firm formation, such as rock, chalk, stiff clay, or boulders and shingle, the breakwater or quay wall is built up to low water, by concrete blocks on the bottom previously levelled by helmet divers

FOUNDATION OF FASCINES.
Fig. 21.—Embankment across Lake Y.

CYLINDRICAL FOUNDATIONS.
Fig. 22.—Plantation Quay, Glasgow.

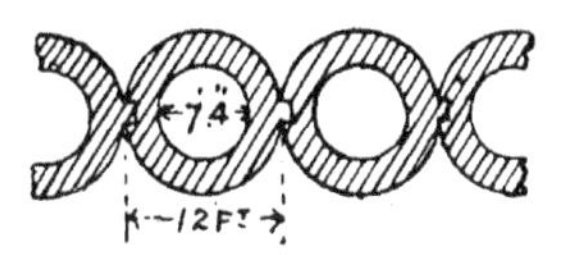

or men in a diving-bell, or by concrete-in-mass within framing, lowered in closed skips with movable bottom, or concrete in bags. Embankments have been formed in the water on very soft silt, by first depositing one or two layers of fascine mattresses, which, owing to their lightness, flexibility, and continuity, bear up the weight of the embankment in the soft ground, with only a moderate amount of settlement (Fig. 21).

BRICK WELL FOR PIER.
Fig. 23.—Empress Bridge, River Sutlej.

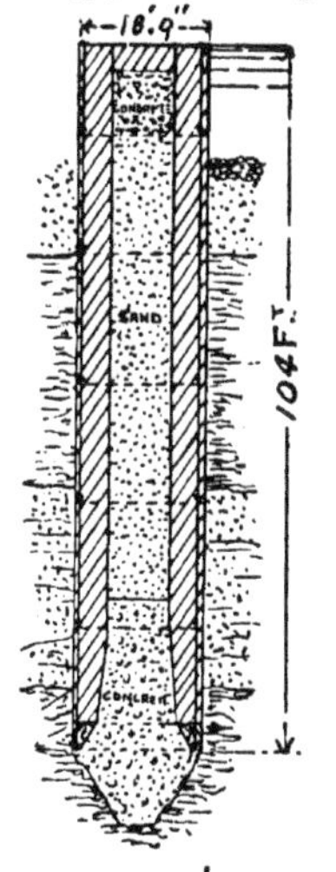

Bearing piles have been largely used for supporting the piers of river bridges on soft soil, and also for the foundations of river quays, the pilework being consolidated in very soft ground by the deposit of rubble stone round the piles. Where timber is abundant, it has also been employed in the form of cribwork for breakwaters and river piers. In the lake harbours of North America, the cribwork is floated out to its site, and sunk by being loaded with rubble stone; whilst the pockets of the cribwork for the foundations of the piers of the Poughkeepsie Bridge, across the Hudson River, were filled with concrete. Well-foundations have been used for quay walls alongside a river, as, for instance, at the Plantation Quay on the Clyde at Glasgow (Fig. 22),[1] where also triple cylinders built up in rings $2\frac{1}{2}$ feet deep, and sunk with a cast-iron shoe, were used.[2] A somewhat similar arrangement of triple cylinders has also been employed at Keyham Dockyard Extension for wharf walls where the depth from coping to rock level varied from 30 to 100 feet, through mud overlying shillet-rock.

[1] *Proceedings Inst. C.E.*, vol. xxxv. p. 188, and plate 7, fig. 1.
[2] *Proceedings Inst. Mechanical Engineers*, 1895.

Well-foundations have also been used for piers of bridges in mid-river, a good example of which is presented by the wells on which the piers of the Empress Bridge were founded (Fig. 23).[1]

Screw-pile Foundations.—Where a bridge or viaduct with moderate spans, has to be carried across a river or estuary with an alluvial bed of considerable depth, iron screw-piles have been adopted with advantage. These piles, being formed in pieces joined together, can be given any length; they present much less obstruction to the current, in proportion to their strength, than timber piles; when braced together in clusters of three, they form a rigid pier; they cause very little disturbance of the ground in being screwed down by being turned round at the top by long bars; and the projection of the screw increases considerably the bearing surface of the pile on the soft foundation (Fig. 24). This form of foundation and bridge pier combined, has been mainly used in England, India, and the United States; and the same system has been employed in erecting lighthouses on sandbanks, of which the Maplin Sands lighthouse in the Thames estuary, and the Walde lighthouse on a sandbank off Calais harbour are instances.

SCREW-PILE.
Fig. 24.
SCALE · $\frac{1}{24}$.

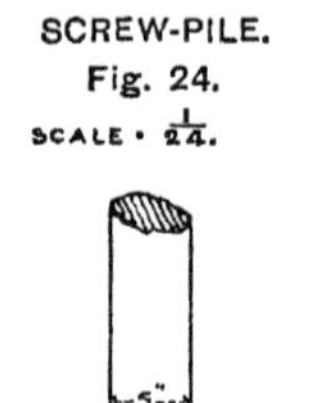

Hollow Iron Piles with a Disc at the Bottom.—Cast-iron cylindrical piles with a disc at the base, resemble screw-piles in principle, for the disc provides the additional bearing surface in this case, which in the former was furnished by the screw (Fig. 25). These piles are sunk in sand or silt by a stream of water under pressure, from a pipe lowered down inside the pile, scouring away the ground from under the disc. The piles are very readily sunk through pure sand by this process; but when the presence of silt renders the ground more compact, and the sinking consequently slow, it was found, in constructing the Leven and Kent viaducts across Morecambe Bay, where these piles were first used,[2] that by providing the bottom of the disc with sharp radial ribs, and slightly turning the pile, the ribs, by stirring up the soil, brought it more under the influence of scour, and hastened the sinking. On completing the sinking by the water-jet, a few blows on the pile from a heavy ram with a small fall, serves to bring the disc on to a solid bearing, rendering the sand compact underneath it, which had been disturbed by the jet.

Iron Cylinders for Foundations.—A plan which has often been resorted to for founding the piers of bridges in the middle of a river, consists in lowering an iron cylinder, formed of a series of superposed rings made of segments joined together, on to the bed of the river, previously levelled by dredging, and gradually sinking it to a suitable

[1] *Proceedings Inst. C.E.*, vol. lxv.
[2] *Proceedings Inst. C.E.*, vol. xvii. p. 443, and plate 10.

stratum by weighting it on the top and excavating inside;[1] and the pier is then formed inside the cylinder, by filling the enclosed space with concrete and brickwork (Figs. 26 and 27, p. 70). The cylinder in this system of foundations, resembles a cylindrical well in the way it is sunk, and a caisson in providing a water-tight skin, within the shelter of which the pier is constructed. It differs, however, from a well-foundation in forming only the outer lining of the pier, instead of an integral part of the structure; and it differs from an ordinary open caisson, in remaining in place after the completion of the pier inside. The bottom ring of

DISC PILES.

Fig. 25.—Leven and Kent Viaducts.

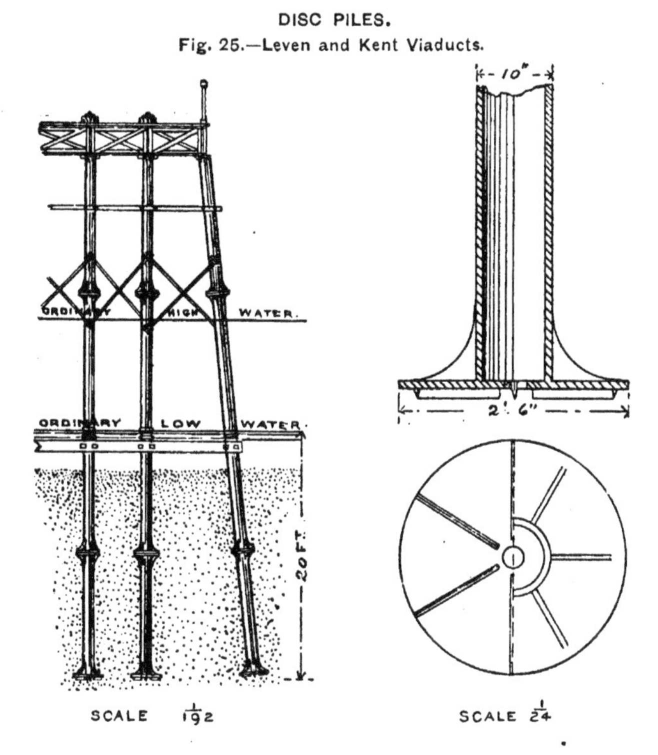

these cylinders, with a cutting edge at the base, is made thicker, and generally less high than the rest, owing to the special strain to which it is exposed. The Charing Cross and Cannon Street railway bridges, across the Thames at London, are examples of bridges with piers built within cast-iron cylinders lowered from staging and sunk by the aid of divers, and subsequently by dredging with a bag and spoon, through the surface layers of mud and gravel, into the watertight stratum of London clay, enabling the latter portion of the excavation through the clay, and

[1] *Proceedings Inst. C.E.*, vol. xxii. p. 513, and plate 17; and vol. xxvii. p. 413, and plate 15.

the depositing of concrete in the cylinders up to the ground-level, and the building with brickwork from thence to the top, to be carried out in the same manner as on land. In both these bridges, the cylinders were reduced above the surface of the ground, by a conical ring, from 14 feet to 10 feet in diameter in the Charing Cross Bridge (Fig. 27), and from 18 feet to 12 feet in diameter in the other. By this arrangement, the cylinders, which rest upon a wide-enough base at the bottom to avoid imposing an undue load on the clay foundation, present a smaller obstruction to the waterway by their reduced diameter above. The concrete, moreover, which provides a cheaper filling for the wide, lower portion of the cylinders than brickwork, can safely sustain the weight of the bridge on the enlarged area; whereas the brickwork which fills up

CYLINDRICAL BRIDGE PIERS.

RINGS OF CYLINDER.
Fig. 26.—Charing Cross Bridge Pier.
SCALE 1/100.

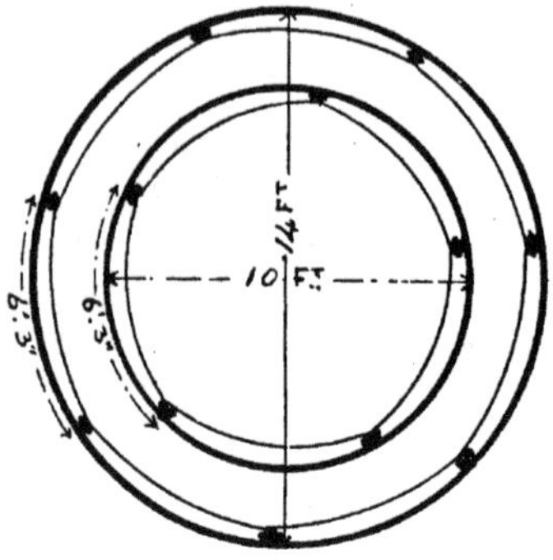

CYLINDER.
Fig 27.—Charing Cross Bridge Pier.

SCALE OF FEET

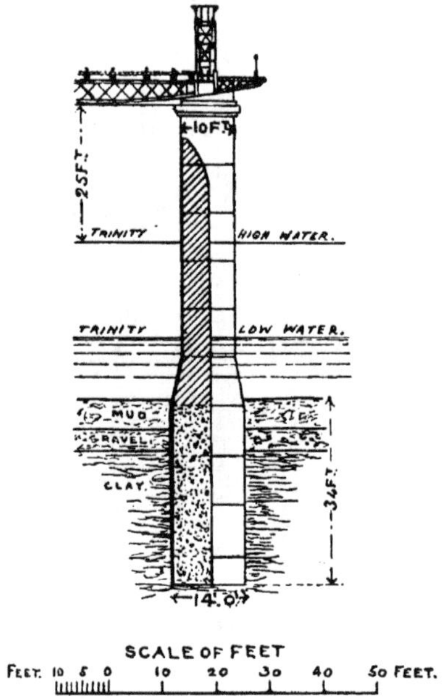

the narrower upper part of the cylinders, is capable of supporting the greater load resulting from the reduction in the section of the pier. The cylinders in each pier are connected at the top by transverse wrought-iron girders.

The piers of the original Victoria railway bridge across the Thames at Pimlico, were built within the shelter of ordinary double-sheeted cofferdams; but the subsequent extensions of the piers for widening the bridge, were founded up to low water within cast-iron cylinders sunk into the London clay, which were continued up to above high water by removable wrought-iron cylinders, to keep out the water during the sinking, and till the concrete and brickwork foundations were raised to low-water level; after which, these cylinders were removed, the spaces in the foundations between the cylinders were arched over, and the pier built in a continuous block of masonry above.[1]

The 12-feet cast-iron cylinders, with a bottom ring of wrought iron for the cutting edge, which form in pairs the foundations of the eighteen piers of the Madras Railway bridge across the River Chittravati (Fig. 28), were sunk down to the rock through a maximum depth of 78½ feet of sand, clay, sand with stones and clay, and boulder-clay.[2] The sinking through the thick top layer of sand was easily effected, the placing of the cylinders in their position being much facilitated by the bed of the river being dry for nine months in the year, though water is reached 3 feet below the bed; but the progress was slow in clay, owing to the inability of the ordinary grabs to excavate it. The boulders, however, in the stratum overlying the rock, necessitated the employment of divers for their removal, aided sometimes by dynamite for breaking up the larger boulders; and the sinking of the cylinders of four of the piers was completed by means of compressed air.

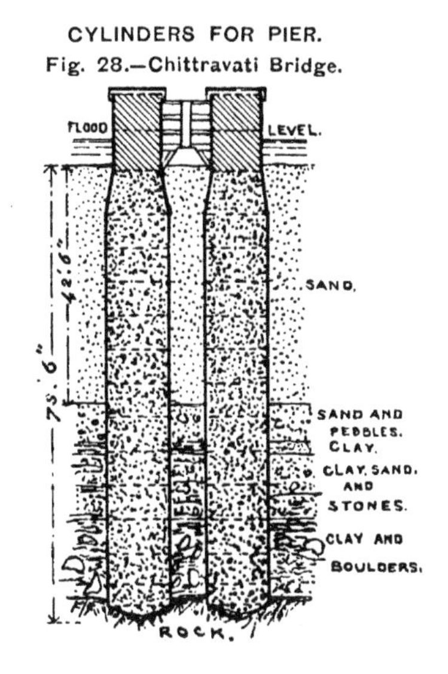

CYLINDERS FOR PIER.
Fig. 28.—Chittravati Bridge.

Wrought-iron Caissons for Foundations.—Cylindrical foundations are necessarily discontinuous; and the descent of cylinders vertically is more difficult to accomplish than that of large caissons. Accordingly, where the foundations have to be spread over a large area, owing either to the nature of the stratum to be built upon, or the considerable load to be borne, as in the case of some bridge piers, and where this has to be effected with as little obstruction as possible to the waterway, continuous wrought-iron caissons enclosing the whole site of each pier, but separated generally into sections by partitions, present advantages over cylinders; whilst these caissons are provided, like cylinders, with a cutting edge round the outside, and are sunk by similar means.

The seven piers of the railway bridge across the main channel of the Hawkesbury estuary in New South Wales, were founded by means of

[1] *Proceedings Inst. C.E.*, vol. xxvii. p. 71, and plate 3.
[2] *Ibid.*, vol. ciii. p. 138, and plate 7, fig. 1, and plate 8.

wrought-iron caissons, 48 feet long, and 24 feet wide, splayed out in the bottom 20 feet, to 52 feet and 24 feet respectively (Fig. 30), with the object of facilitating their descent, which were floated into position by help of a false bottom, and sunk through a stratum of mud, reaching a thickness of about 120 feet, to a sandy bed below, the bottom of the foundation of one of the piers being 162 feet below high water of spring tides.[1] Three vertical shafts were formed in the caisson, 8 feet in diameter and widening out at the bottom, through which the dredging was effected; and the caisson was weighted with concrete in the spaces between the shafts and the sides of the caisson (Figs. 29 and 30).

CAISSON FOR PIER.

Figs. 29 and 30.—Hawkesbury Bridge, N.S.W.

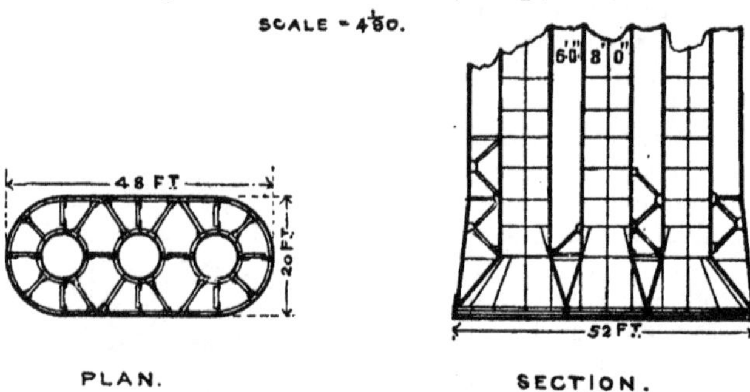

When the stratum of sand was reached, the shafts were filled with concrete; and the masonry piers were erected upon these foundations from about low-water level.

The piers of the large spans of the Dufferin Bridge crossing the Ganges at Benares, were each founded by aid of an elliptical, wrought-iron caisson, with a major axis of 65 feet and a minor axis of 28 feet, provided with a cast-steel cutting edge, and divided into three compartments by two partitions.[2] The caisson was formed with an inner shell braced to the outer skin, and the partitions also with two shells; and the intervening spaces were filled up with brickwork. The caisson was lowered on to the bed of the river between two pontoons, its height varying from 10 to 50 feet, according to the depth of water at the site; and the brickwork of the outer lining, and of the partitions, was continued above the top of the caisson, with an increased thickness by corbelling over inside, being built up as the sinking progressed, to keep the top of the structure well above the water. These foundations, indeed, only differ from ordinary well-foundations in the iron caisson lining for the bottom portion, and the elliptical form of the foundation with its three hollow compartments; for the lining of brickwork, the dredging inside

[1] *Proceedings Inst. C.E.*, vol. ci. p. 3, and plate 1.
[2] *Ibid.*, vol. ci. p. 16, and plate 3.

the hollows for sinking, and the building up of the brickwork on the top as the sinking proceeds, constitute the essential principles of well-foundations. Three of the piers of this bridge rest upon stiff clay at only a moderate depth below the river-bed, as this stratum is little subject to scour, and has been adequately protected by rubble stone deposited round the piers; but the foundations of all the other piers, in the sandy bed of the other part of the channel, and in the sandy right bank beyond the low-water channel, have been carried down considerable depths below the surface, mostly through sand, to secure them from any chance of being undermined by scour during floods, or from changes in the position of the main channel. The foundations of the two deepest piers were sunk about 80 to 90 feet in the sandy bed of the main channel, to a depth of 140 feet below the low-water level of the river, and 190 feet below the highest flood-level; whilst the two piers on the sandy right bank, nearest to the low-water channel, were sunk through 145 feet and 165 feet of sand to depths of 171 feet and 183 feet below flood-level. The increased scour resulting from the reduction in the waterway produced by the piers, has increased the maximum depth of the river at the site, from 37 to 65 feet in the dry season, and from 100 to 120 feet during the floods.

The two piers of the railway bridge across the River Húgli at Húgli, were each founded within a wrought-iron caisson, 66 feet long, with semicircular ends, 25 feet wide and 108 feet high, divided into three open compartments by two box-shaped, watertight partitions, 15 feet wide, providing the buoyancy required for floating the caisson into position and keeping it afloat till it reached the bed of the river.[1] The bottom 16-feet height of the caisson was floated out from the bank; and the remainder was built up in rings, 4 feet high, lowered from staging on two pontoons as the caisson descended. The river has a depth at the site of the piers of about 30 feet at the lowest low water; and the caissons had to traverse about 60 feet of silt before reaching the yellow clay, into which they were carried about 10 to 12 feet. The excavation at the bottom of the caissons was effected by a boring tool working in each compartment; and the material was removed from the caisson through a siphon by aid of an ejector, and discharged into the river. The sinking was assisted by the weight of the successive rings as they were added, and the 3-feet brick lining built on a series of iron ledges round the semicircular ends, by filling the buoyancy partitions with concrete below and brickwork above, and, as soon as the caisson was embedded in the clay, by pumping out some of the water. The caisson was finally filled for about the lower half of its height with concrete, and then with brickwork up to the top, which emerged some feet out of water.

These examples suffice to indicate the unfavourable conditions under which caissons have been successfully employed for foundations, and the great depths to which they, as well as cylinders, have been sunk. The caissons of the Hawkesbury Bridge afford an instance of

[1] *Proceedings Inst. C.E.*, vol. xcii. p. 75, and plate 1.

caissons being given an increased width at the bottom, with the object of reducing the skin-friction in sinking; but vertical sides appear to have the advantage of rendering the caisson less liable to cant over during its descent.

Compressed-air Foundations.—In sinking wells, caissons, or cylinders through running sand or very fine silt, trouble is apt to occur through the sudden influx of the surrounding material; and difficulties are experienced when foundations have to be laid on a shelving bed of rock at a considerable depth below the surface, or when the cutting edge comes in contact with boulders or trunks of trees. Moreover, with open foundations sunk through water-bearing strata, and not reaching a sufficiently watertight stratum for the water to be pumped out, the stratum on which the foundation rests cannot be thoroughly examined; and the bottom layers of concrete for filling up the central hollow have to be deposited under water. By the use, however, of compressed air, sand and silt can be kept out; the excavation at the bottom of the caisson, or the levelling of the rocky bed can be effected in the usual manner; boulders, trunks of trees, or other obstructions under the cutting edge can be dealt with, the foundations can be laid bare, and the whole of the concrete filling can be deposited out of water. The only objections to the use of compressed air are, that it becomes trying to the workmen at considerable depths below the water-level, and that it is liable to prove more costly than the open-air methods when the conditions are not specially unfavourable for these methods.

The essential part of the compressed-air system is the bottomless caisson, not less than 6 feet high, forming the working chamber filled with compressed air, in which the men excavate the material at the bottom of the chamber, and thus gradually cause the caisson to descend till it reaches a firm-enough stratum to bear the weight of the structure, and extends to a sufficient depth in the sandy or silty bed of a river to be secure from scour; after which it is filled up with concrete. The caisson is generally constructed of wrought iron; but occasionally wood is used instead, as for instance the caissons for the piers of the Brooklyn and Harlem bridges at New York. The roof of the working chamber is made very strong, in order to support the structure built on it within the shelter of the plate-iron sides erected round the roof at the top, and raised above the water. Access is obtained to the working chamber through one or more vertical shafts erected over apertures in the roof, each provided with an air-lock at the top or bottom, for the passage of men and materials from the open air into compressed air, and back again. The chamber is lighted with electric light to avoid the smoke, absorption of oxygen, and heat of ordinary lamps in the compressed air and confined space. To work with safety under compressed air, the men should be healthy and temperate, and the air kept pure and frequently renewed; and the insensibility which occasionally comes over a man on emerging direct from a high pressure into the open air, and is sometimes attended with dangerous or even fatal results, can usually be completely cured by placing a man thus seized in a chamber of pure compressed air at a lower pressure, which is slowly reduced to the normal atmospheric pressure.[1] (See p. 225.)

The foundations of the Antwerp quay walls in the River Scheldt are probably the most extensive continuous works which have been carried out under water by compressed air. The caissons, with the working chamber at the bottom, 6¼ feet high, were over 82 feet long and 29½ feet wide, and from 8½ to 19½ feet in height, according to the depth of water at the site where they had to be founded.[2] The caisson, having been built on shore, was floated out and placed in position between two

COMPRESSED-AIR FOUNDATIONS.

Fig. 31.—Antwerp Quays.

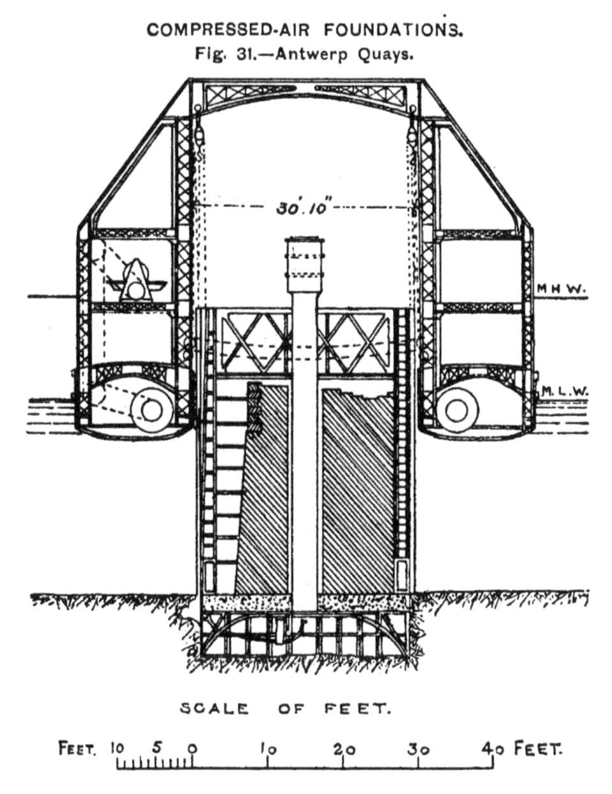

barges moored in the river and connected together above, containing the air-compressing machinery and materials; and the plate-iron coffer-dam, carried out by the barges, was then fastened to the top of the caisson (Fig. 31). The caisson was then gradually sunk by filling with concrete the portion of the caisson on the top of the roof of the working

[1] *Proceedings Inst. C.E.*, vol. cxxx, clxv.

[2] *Memoires de la Société des Ingénieurs Civils*, 1881 (1), p. 398, and plate 23; and "Travaux Publics, Ouvrages exécutés au moyen de l'Air Comprimé," H. Hersent, p. 175, and plate 4, figs. 6 and 7.

chamber, which was strengthened by girders to support the weight of the wall, and building the wall on the concrete within the shelter of the plate-iron sides till the caisson touched the bottom at high tide. The position of the caisson was then carefully adjusted, being lightly suspended by chains from the overhead staging on the barges, aided by the introduction of some compressed air below; and an additional weight of wall was built up directly the caisson had been finally lowered into position, to counteract the upward pressure which the compressed air would produce in the working chamber. Compressed air was then pumped into the working chamber; and the workmen entered the chamber through the large central air-lock and shaft, and carried down the caisson, with the superposed 81-feet length of wall, by excavating the ground under the chamber till a solid stratum was reached, below the layer of silt forming the bed of the Scheldt in front of Antwerp. Six smaller shafts, in addition to the central shaft, were carried up from the roof of the working chamber, two small ones for the discharge of the excavated materials, and four for filling the chamber with concrete on the termination of the sinking, openings being left in the wall for these shafts, which were subsequently filled with concrete after the removal of the shafts and plate-iron sides for another length of wall. The original caisson forming the sides and roof of the working chamber, was the only portion of the plant left in place; and the remainder served for several lengths of wall, since the construction of a length of foundations up to a little above low water, by this system, only occupied 25 days. The aperture necessarily left between two adjacent lengths of wall, was closed at the face and back by timber panels, and the space filled up with concrete, so that a continuous quay wall could be erected on the top.

The working chamber for the foundations of the piers of river bridges, has to be made large enough for the erection of the pier above it. Thus the hexagonal, wrought-iron caissons on which the piers of the St. Louis Bridge across the Mississippi were founded, had each a working chamber 82 feet long, 61 feet wide, and 9 feet high, with seven shafts giving access to it, having their air-locks at the bottom of the shafts;[1] and the two piers were founded on rock, at depths of 76 feet and 102 feet respectively below low water. The timber caissons on which the two piers of the Brooklyn bridge over the East River were founded, had working chambers 167 feet long, 102 feet wide, and 10 feet high, with nine shafts of access, and were sunk to depths of 78 feet and 45 feet respectively below high water. In this case, tiers of balks were raised high enough on the roof of the chamber to emerge out of water when the caisson touched the bottom, enabling the masonry pier to be built upon it without any cofferdam round it. The central pier of the Harlem River Bridge at New York, for which a level bed had to be formed in a sharply-sloping face of rock, was founded by compressed air in a working chamber 104 feet long, 54 feet wide, and 7 feet high, with timber sides 3 feet thick, and a roof 6 feet thick, upon which the masonry of the pier was built;[2] and access to the chamber was provided by a

[1] "A History of the St. Louis Bridge," C. M. Woodward, p. 201, plates 7 and 13.
[2] *Scientific American*, April, 1887, p. 244.

shaft with an air-lock at the top for the men, and another shaft with an air-lock at the bottom for materials. Six of the circular piers supporting the huge cantilevers of the Forth Bridge, rest on foundations carried down to depths of from 64 to 89 feet below high water by compressed air, two having been founded on a bed cut in a face of sloping rock, and the other four having been sunk 33 to 49 feet through mud and clay into boulder-clay.[1] The circular plate-iron caissons used for these foundations were 70 feet in diameter, with a working chamber at the bottom 7 feet high; and they were floated out, and sunk in position by putting concrete into the upper portion above the roof of the chamber. When the sinking was completed, the working chamber was filled with concrete as usual; and the concrete in the upper part of the caisson was raised to low-water level, and the masonry piers built upon it.

WELL-SINKING BY COMPRESSED AIR.

Fig. 32.—Rochefort Dock Wall.

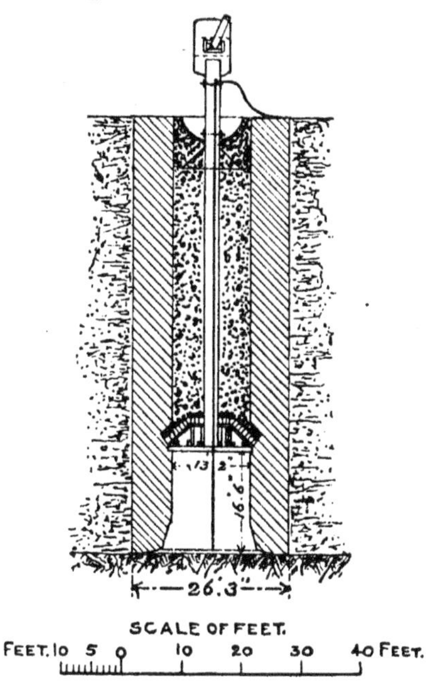

Compressed air has been sometimes used for foundations sunk through water-bearing strata on land, of which the foundations of two of the piers forming the base of the Eiffel Tower are instances, as, in order to reach a gravel bed in the case of the two piers nearest the Seine, it was necessary to carry the foundations $16\frac{1}{2}$ feet below the ordinary water-level of the river. These foundations were laid 33 feet below the ground by four wrought-iron caissons, 49 feet long and 13 feet wide, for each pier, sunk by the aid of compressed air.[2]

Occasionally compressed air is only used as a final resource, when the open-air system is greatly impeded by boulders or other obstructions, or by frequent inrushes of sand or silt and water. Thus the sinking of four out of the eighteen pairs of cylinders of the Chittravati Bridge, was completed by compressed air, owing to the difficulties in dealing with numerous boulders at a considerable depth; and the masonry wells on which the walls of a dock at Rochefort were founded, were designed so that a vaulted masonry roof, with a central circular opening, could be readily built, $16\frac{1}{2}$ feet from the bottom, forming a working chamber

[1] "The Forth Bridge," W. Westhofen, pp. 22–31.

[2] "Le Tour Eiffel de 300 mètres," Max de Nansouty, p. 18.

(Fig. 32), whenever, in sinking through 50 to 90 feet of alluvium, a greater influx of silt or water occurred than could be coped with by excavation and pumping, and the sinking completed by compressed air.[1]

In order to reduce the cost of the compressed-air system for foundations in shallow water, the caissons have been made movable in some cases, being raised by screw-jacks as the masonry is built up, a plan adopted for the foundations of two piers of the Garrit bridge over the Dordogne, for some of the foundations of the Cep bridge over the Charente, and for the foundations of the outer portions of the two breakwaters protecting the harbour of La Pallice at La Rochelle.[2] By this means, the caisson forming the working chamber can be used over again for successive lengths of foundations, instead of remaining embedded in the masonry which considerably augments the cost in the case of foundations in small depths of water; and, accordingly, this system enables compressed air to be used for works in shallow water at a reasonable cost.

Piers of Bridges.

The difficult part of the piers of river bridges has already been dealt with under the head of foundations; whilst the upper portion of these structures which is exposed to view, though on that account requiring neatness in construction, affords no cause for anxiety, and little scope for engineering skill. In some cases, as for instance where disc piles, screw-piles, or cylinder piers are adopted, the foundation and upper part of the pier form a single structure; whereas in other cases, solid masonry superstructures of reduced thickness are erected on the top of the broader cylindrical, brick well, or other foundations. The upstream end of the pier is tapered off with a pointed cut-water to a little above the water-level, to direct the current with as little eddy as practicable through the waterways between the piers; and a cut-water has to be formed at both ends of the pier in a tidal river, where the current flows alternately in both directions. An extended cut-water, or ice-breaker, has to be built on the upstream side of piers in rivers exposed to severe frosts, to break up the masses of floating ice coming down in the spring, so that they may pass through the openings of the bridge, and not heap up against the piers and block the river; and piers are often protected from the injurious blows of floating *débris* by timber pilework and fenders.

Influence of Pier Foundations on the Design of Bridges.—The design of a bridge depends primarily on the spans selected for its openings; and the spans again are determined by the number and positions of the piers. Since, however, the cost of a bridge varies approximately in proportion to the square of its span, it is evident that where the piers are of small height and their foundations easy, small spans and numerous piers provide the most economical form of viaduct.

[1] *Annales des Ponts et Chaussées*, 1884 (1), p. 150, and plate 13, figs. 3 and 4.
[2] *Ibid.*, 1889 (2), p. 461, and plates 50–54.

When, on the contrary, the piers have to be made a considerable height, and more especially at sites where subaqueous foundations have to be carried down to a considerable depth below the bed of the river, it becomes expedient to adopt large spans in order to reduce the number of piers. The low brick viaducts carrying railways above ground through towns, or across lands subject to floods, and also the higher viaducts across valleys with larger spans, afford instances of numerous piers; and the same system is resorted to where iron viaducts have to be supported by disc piles or screw-piles in a stratum of sand or silt of indefinite depth, for under these conditions, the weight which can be borne by the piers in the soft foundation without settlement, can only be kept within the restricted limits by adopting a large number of piers supporting small spans, of which the Morecambe Bay viaducts furnish a notable example, with spans in general of only 30 feet between the disc-pile piers. Large spans, on the other hand, of between 300 and 500 feet, have been adopted for many bridges across rivers, in order to reduce the number of river piers, of which the Saltash, St. Lawrence, St. Louis, Moerdyk, Húgli, and Hawkesbury bridges are instances; whilst in a few cases, it has been possible to avoid altogether building river piers at unfavourable sites, by adopting spans sufficient to bridge over the river or gorge, as in the bridges over the Niagara and Zambesi gorges, the Douro bridge at Oporto, the central arch of the Garabit viaduct in France, and the Sukkur Bridge across the Indus. Moreover, owing to the exceptional obstacles in portions of the East River and the Firth of Forth to the construction of piers, spans of unprecedented magnitude were resorted to for the Brooklyn and Forth bridges. Accordingly, the nature of the foundations for the piers, and the general conditions of the site affecting their height, have a very important influence on the spans that should be adopted for a bridge crossing a river, estuary, or arm of the sea.

Provision for Navigable Channel.—In crossing the navigable channel of a river or estuary, the piers of the bridge have to be arranged so as not to encroach unduly on the channel, and to leave an ample width for the passage of vessels. Accordingly, in a bridge designed to have several openings of small span, one or two openings of larger span, affording an adequate waterway between the piers at the navigable channel, have to be provided. Moreover, where no special local conditions of a deep gorge or valley, or high adjacent land, of themselves necessitate a high bridge, the piers must, nevertheless, be raised sufficiently high to afford the requisite headway for vessels under the bridge at the highest navigable water-level of the river. In making these provisions, full allowance must be made for any reasonable prospective improvement in the navigable channel, and increase in the size of the vessels navigating the river; and, therefore, besides allowing an ample waterway and headway through the openings of any bridge erected across a navigable channel, it is necessary that the foundations of the piers of these openings should be carried down far enough below the bed of the river to secure them from being undermined by any requisite deepening of the channel.

Loads on Foundations of Piers of Bridges.—The weight that has to be borne by bridge piers necessarily augments in an increasing ratio with an increase in the span; whilst the actual weight imposed upon the foundations increases also with the depth to which the piers have to be carried, owing to the increased height of the pier. Provision, accordingly, has to be made for the weight due to large spans and a considerable depth of foundation, by giving the pier a larger base, depending on the limit of weight which can be safely borne by the stratum on which the pier rests. The maximum load imposed on the deep sandy stratum of Morecambe Bay by the disc piles, reached 4 tons per square foot, 5 tons having been found by experiment to be the limit of the safe load; whilst the load borne by the compact London Clay under the large Tower Bridge piers is the same. Greater loads, however, than 4 tons per square foot have been safely placed upon the London Clay by piers of other bridges across the Thames at London, namely, $4\frac{1}{2}$ tons at Charing Cross Bridge, $4\frac{3}{4}$ tons at Blackfriars Railway Bridge, and nearly 6 tons per square foot at Cannon Street Bridge. Considerably greater loads, moreover, have been placed on much less compact strata by bridge piers founded at great depths below the surface, without producing settlement, as, for instance, $8\frac{1}{2}$ tons per square foot on hard sand at the Gorai Bridge in the Ganges delta, 9 tons on sand at the Hawkesbury Bridge, and on yellow clay at the Húgli Bridge, and $11\frac{1}{5}$ tons per square foot on sand at the Dufferin Bridge across the Ganges at Benares. These latter weights, however, are probably in excess of the actual loads, as no allowance has been made for skin-friction and buoyancy, which would have a material influence in reducing the weight of the piers resting on the foundation, at the great depths below the surface to which they have been sunk.

CHAPTER VI.

ROADS, AND STREET-PAVING.

Importance of roads—**Roads:** laying out ordinary, in mountainous districts; formation, metalling; rolling, with horses, and steam; maintenance, wear, influence of materials and strata, effects of traffic, repairs—**Street-paving:** necessity for, in towns; forms—**Stone paving:** cobble-stones; stone setts, materials for, sizes, instances; foundations; joints; cost; merits and defects—**Wood paving:** dimensions of blocks, laying; joints; duration; cost; advantages of hard woods; merits and defects—**Brick paving:** countries making use of, foundations; qualities needed; cost, merits and defects—**Asphalt paving:** composition of asphalt, natural, mastic, Trinidad; laying; cost; merits and defects—**Conclusions about street-paving:** relative advantages of the different pavings; kinds used in some European and American cities—**Footpaths:** country, materials employed; town, forms of paving used and their relative merits, curbs.

ROADS have been deprived by railways of their commanding position as the chief means of communication, and the pioneers of civilization; but though railways have absorbed most of the long-distance traffic, and furnish the most rapid method of opening up new countries, roads have, nevertheless, acquired an enhanced importance from the great growth of traffic produced by the railways, combined with the rapid increase of population, and the demand for improved facilities for transit. The great growth, also, of cities and towns, the progress of sanitary science, and the requirements of a higher civilization, have necessitated improvements in the paving of streets subjected to large traffic.

ROADS.

Laying out Roads.—The great roadmakers of ancient times, the Romans, used to lay out roads in a straight line from point to point, which sometimes resulted in a road going over a hill which a moderate *détour* would have avoided, as, for instance, the old Dover road in continuation of Watling Street, which passes over the summit of Shooter's Hill between Blackheath and Welling. Possibly this course was adopted as the easiest in a forest-clad country, or for strategic reasons; but modern practice has tended more and more to seek easy gradients, in preference to directness of route, for country roads which are generally laid along the natural surface of the ground, as economical considerations preclude the levelling of these roads by cuttings and embankments, except to a very limited extent at specially steep places. An example of this change of practice is afforded near Grasmere, in Westmoreland, where three coach-roads may be noticed; the oldest, long ago abandoned, being the most direct, and traversing the spurs of the hills protruding into the valley; the next, at a lower elevation, contouring to some extent the spurs, and consequently more circuitous; and, lastly, the present coach-road in the valley, which, though considerably longer, has comparatively easy gradients. Roads are well adapted for following the contours of a hilly country, as sharp turns are admissible; and zigzags are sometimes resorted to for reducing the gradient of a steep ascent.

In mountainous districts, roads should be laid out so as to obtain the minimum amount of variation of level practicable; and long steep gradients should be provided with fairly flat short pieces at intervals, to afford the horses resting-places in the ascent, and to reduce the injury to the road caused by the descent of the drainage in the event of a stoppage of the cross drains. The road has to be carried gradually up the valley leading to the dividing ridge, following along one of the side slopes, so as to make the gradient as uniform as possible in ascending to the pass through the mountain range.

The gradients of a road must largely depend on the nature of the country traversed; but the natural gradients of the land may often be materially lightened by adopting a circuitous course, and occasionally by a moderate amount of cutting and embankment. Telford, in the main roads he laid out, endeavoured usually to restrict the maximum or ruling gradient to 1 in 30, by making *détours* where necessary;[1] but sometimes gradients of 1 in 18, or even 1 in 15, have to be allowed. The sacrifices which should be made in increased length of road, or earthwork in cuttings and embankments, must clearly be determined by the difficulties of the district, and the importance of the vehicular traffic.

Formation of Road.—On firm ground, it is only necessary to dig a shallow trench or drain along each side of the road, to level small inequalities and fill up hollows, and to raise the road about 4 to 6 inches towards the centre, according to its width, for the sake of drainage, before laying on the materials forming the lining of the road, which have to stand the wear and tear of the traffic. The width adopted for a road varies, according to its importance, from about 20 feet up to 35 feet for the chief main roads. The surface of a road in cross section should be a very flat ellipse, or preferably two straight lines sloping down to the water-table or gutter on each side, joined by a short curve in the centre, so that rain falling on the road may run off to the side drains; and the removal of this water is facilitated by a longitudinal gradient,

[1] "Report of the Commissioners for repairing the Roads between London and Holyhead," 1816; and "Reports of Mr. Telford upon the Holyhead Road," 1819 and 1820, plates.

causing it to run down the side drains to the cross drains formed at intervals, leading it into the ditch alongside each fence bordering the road.

Mountain roads, cut into the side of a valley, are generally given a cross slope from the outside down towards the inside, the drainage from the road, and from the side of the hill, coming into a channel along the inner edge of the road, and being discharged at intervals through culverts passing under the road to the outer slope.[1] By this means the road is relieved from the drainage water of the slope above it, which has to flow across the road where the cross slope dips toward the outside, and from the gradual disintegration which may result from the flow of water over the outer edge of the road; but the cross drains must be given an ample capacity, to prevent the possibility of the road becoming a sort of waterway during exceptional rainfall. Slips down the hilly slope on the inner side of a mountain road, may to some extent be prevented by retaining walls; and the road has to be protected at certain places from avalanches, or the descent of *débris*, by a shelter of timber or masonry, or by carrying the road inside a tunnel across exposed slopes.

Where the ground is soft, a layer of large stones or pitching on edge is required to consolidate the foundation for the road; and where the road has to traverse marshy land or a bog, the land should be thoroughly drained before the road is formed over it, or layers of fascines may be deposited on the marsh, which serve to support the road.

Metalling Roads.—Broken stone is the material commonly employed for the coating of main country roads, and for streets with moderate traffic in towns, gravel being only used for roads having little traffic. McAdam, in the early part of the nineteenth century, pointed out that roads are best made and repaired with broken stone, when the stones are hard, angular, and of fairly uniform size;[2] and they should be broken sufficiently small to pass through a ring not exceeding 2½ inches in diameter. This system of forming roads with broken stones of definite size, termed road-metal, deposited in a layer 6 to 10 inches thick over the whole surface, has been called macadamizing after its inventor, and has been universally adopted. The interstices between the broken stones are best filled up by spreading small gravel on the top, with a small proportion of the gritty scrapings off macadamized roads in wet weather, which serve to consolidate the road. In byroads, only about the central 12 to 15 feet are metalled.

Rolling of Macadamized Roads.—Formerly, even in many of the London streets, the consolidation of the rough layer of broken stones was left to be gradually effected by the wheels of the vehicles passing over it; and during dry weather the rough stones remained loose for a long time, and were very prejudicial to horses and vehicles, as well as very onerous to traction. At the present day, in London and other

[1] *Proceedings Inst. C.E.*, vol. xxxviii. p. 81.

[2] "Remarks upon the Present System of Road-making, with Observations, with a View to the Introduction of Improvements in the method of Making, Repairing, and Preserving Roads," J. L. McAdam. Bristol, 1816.

cities and towns, and even in many country districts, the consolidation of the layer of road-metal, or macadam, is accomplished by a steam-roller, with the aid of a suitable quantity of clean gravel and sand, and ample watering. Rollers drawn by six or eight horses were often used formerly for rolling macadamized roads, as the rollers are comparatively cheap, and being about a third of the weight of steam-rollers, cost less to transport from place to place. Horse-rollers, however, are awkward to turn; and they are less effective than the wider and heavier steam-rollers, which travel in either direction; whilst the metalling is somewhat disturbed by the horses' feet. Moreover, the cost of traction is considerably less with steam-rollers, in proportion to the work done, than with horse-rollers; and on steep gradients, steam-rollers are far more advantageous.[1] Accordingly, steam-rollers are decidedly preferable to horse-rollers, provided there is sufficient work to keep them fairly occupied; and their relative economy in working soon compensates for their large initial cost.

A newly-formed macadamized road, levelled by a steam-roller, provides a smooth, solid carriage-way, not only far more easy for traction, but also more durable than unrolled layers of stones, which are worn and disturbed by the wheels and horses' feet before becoming incorporated into the roadway. Soon after, however, the opening of such a road, some of the sand works up on to the surface under the compression of the traffic, producing mud in wet weather, which should be promptly removed.

Maintenance of Macadamized Roads.—The period of duration of a road in good order depends primarily upon the amount of traffic passing over it, and the hardness and toughness of the material used for the road-metal; and it is also affected by the nature of the foundation, and its being kept free from mud in wet weather by frequently sweeping with a large revolving broom drawn by a horse, enabling it to dry quickly, and preventing the formation of puddles. On country roads, the loss by wear of a road metalled with good material varies in thickness, according to the traffic, between about $\frac{1}{8}$ inch and 1 inch in a year; but in streets of towns with large traffic, the wear may reach about 4 inches in a year. The best materials for a road are basalt, whinstone, granite, and other hard, tough, primitive rocks unaffected by the weather; broken flints are less suitable on account of their brittleness; and limestones, which are often used, though hard, are liable to disintegration by wet and frost. Gravel is only suited for roads with small traffic; and round, unbroken shingle should never be used, as the roundness of the stones prevents their binding into a compact coating for the road. On metalled streets and roads with large traffic, it is true economy to employ the best materials, as their large cost is more than compensated for by the reduction in labour on repairs, and the advantages to the public of not having the traffic frequently impeded or stopped for remaking the road. In country districts, the materials are generally obtained from the nearest available

[1] *Annales des Ponts et Chaussées*, 1882 (1), pp. 437 and 659.

quarry or gravel-pit; and very frequently the quality of the materials, and the nature of the substratum are indicated by the condition of the roads, for the maintenance of roads is necessarily far more onerous over clayey or chalky strata, with only indifferent material readily available for metalling, than on sandy, gravelly, or rocky strata, with good stone for metalling close at hand.

In forming roads through low-lying valleys, the surface should be raised above the ordinary flood-level of the adjacent river or stream; and trees should not be allowed to overhang the road, as the drip from them, and the exclusion of sun and air from a road, keeps it damp and soft, and therefore more subject to wear and less easy for traffic.

Where the traffic is large, the layer of road-metal is gradually worn down till it becomes too thin to support the traffic; and hollows are liable to be formed at the weakest places, producing puddles of water in wet weather, which further injure the road. Moreover, the greatest traffic occurs along the centre of the road, which, accordingly, becomes flat, and is no longer properly drained. The repair of main roads and streets should be effected by a fresh layer of road-metal spread over the whole metalled surface, and by reforming the road again, with more material in the centre, to its original section. The surface of the road should be cleared of mud, and with advantage slightly roughed with a pick, or preferably in towns by a scarifier armed with projecting teeth, before the new layer is deposited, to bind the new metalling to the old; and the stones of the new layer may with advantage be rather smaller in size than the first coating, and the interstices between them should be filled with the coarse road-sweepings saved for the purpose, consisting in wet weather mainly of crushed road-metal.

In country districts, the traffic keeps to the centre of the road, which is worn into hollows in soft places in the centre by the horses' feet; and a rut is formed on each side by the wheels. The covering of the road with a continuous layer of fresh road-metal would be too costly for roads of minor importance; and the sides of the road are little worn in comparison with the centre. Filling the ruts with stones is objectionable, as the vehicles tend to form fresh ruts alongside, and are apt to scatter the stones in crossing the narrow line. Accordingly, on the approach of the winter, broad, thin patches of stones are placed at intervals across each rut alternately, as well as on the worst hollows in the centre of the track. These patches divert the traffic to some extent towards the sides, till they are gradually incorporated in the road; after which, patches are laid down in the intervals, and the metalling is thus gradually renewed at a moderate cost, and without the strain on the traffic which a continuous covering of unrolled metalling would produce. Where, however, the importance of the road admits of the cost of repairing it with a coating of broken stones and rolling it, this system should be adopted.

STREET-PAVING.

In towns where the traffic in the central districts and main thoroughfares is large, macadamized roads wear so rapidly, and need such frequent repairs, that the cost of renewals and the hindrance to traffic become so great as to necessitate a more durable lining for the principal streets. Moreover, in the poor, densely populated quarters of large cities, the dirt and refuse absorbed by metalled roads, in the midst of dirty, ill-fed inhabitants, are a source of danger to health; whilst the dust and dirt produced by these roads are objectionable in the richer quarters, where it is expected that the discomforts attending the cooped-up life in large towns will be reduced to a minimum. Accordingly, under these conditions, a less absorbent coating, less liable to be ground into dust and dirt, and more easily cleansed than macadam, is essential.

Different Forms of Paving.—A paving of stone setts, laid in rows across the road, upon a solid foundation, has been extensively used for many years in streets having a very large traffic. Wood has also been adopted for the coating of streets, in different forms. Thus, in countries where timber is abundant and stone is difficult to obtain, plank roads have been formed—a system frequently adopted in America, and still more commonly resorted to for footpaths, of which many instances are still found in such comparatively old cities as Quebec, Montreal, and Toronto; but the life of a planked road is short, owing to wear and decay. Another form of wood paving, often found in Canadian cities, consists of cylindrical blocks cut from round logs of moderate size, and bedded in sand with their fibres vertical, which makes a fairly good roadway for a moderate traffic when newly laid down; but in two or three years this kind of paving becomes very irregular. A more durable paving of cylindrical blocks has been laid down in the United States, by filling the joints with cement; but the wide interstices between these blocks necessitate the employment of a large proportion of cement. The ordinary form of wood paving in streets, consists in lining the surface of the road with rectangular, wood blocks laid on a firm foundation across the street, like stone setts. Hard bricks have been long employed in Holland for paving streets and roads, and have sometimes been used elsewhere, within recent years, instead of stone setts or wood blocks. Rock asphalt, founded on a layer of concrete, is another paving which has been widely introduced in streets having the largest traffic—as, for instance, in the city of London and in New York.

STONE PAVING.

Cobble-stone Paving.—Round cobble-stones bedded in the road formation furnish the simplest form of stone paving, and were largely employed formerly in various provincial towns in England; but such a paving is very rough, quite unsuited for heavy traffic, and very difficult to clean, though it affords a fair foothold for horses.

Quality and Sizes of Stone Setts for Paving.—The best material for stone setts is a very hard, tough stone, not liable to wear very smooth, such as granite and other primitive rocks, composed of quartz, felspar, and hornblende in various proportions.[1] On steep gradients, somewhat softer stones, such as millstone grit or other stones of similar quality, specially selected owing to their retaining a rough surface under wear, should be used, so as to ensure a good foothold for horses. Formerly setts from 6 to 9 inches wide, 8 to 14 inches long, and about 9 inches deep, were commonly employed in London for paving streets with large traffic. These wide setts were heavy, and therefore little subject to disturbance, and expedited the laying by reducing the number of rows; but they generally wear smooth and become rounded, so that horses tend to slip on them, and on inclines are liable not to be arrested by the joints. Though this class of paving is still in existence in London, the best paving is now laid with setts not exceeding 3 inches in width, so as to bring the joints closer together; and the depth has been reduced from 9 inches, with a maximum of 12 inches, down to between 6 and 7½ inches, being accompanied with a corresponding reduction in the length, which should not exceed about 9 inches. More than fifty years ago, cubes of Mount Sorrel granite, with dimensions of only 3 to 4 inches, were employed for paving the departure approach to Euston Station, and subsequently a part of Watling Street in the city, with remarkably satisfactory results. The foundation was formed of three 4-inch layers of gravel and chalk, of diminishing degrees of coarseness, thoroughly rammed and consolidated, upon which the stones, 3 inches wide, from 3 to 4 inches deep, and 4 inches long, were laid in rows on a 1-inch bed of fine sand, and brought to a level surface by the blows of a 55-lb. wooden rammer, the joints being filled in with fine gravel. This pavement afforded a good foothold for horses with its numerous joints, was easy for traction with its even surface, and caused comparatively little noise under a heavy traffic; whilst its wear, after 3½ years of the very heavy traffic along Watling Street, was hardly perceptible where it had been left undisturbed.[2] Though the small length of the stones prevents the possibility of obtaining the bond between the rows which the longer stones commonly employed afford, nevertheless this paving, owing to the number of its joints, is well adapted for inclines, where a good foothold for the horses is of primary importance.

Foundations for Stone Paving.—The success of the Euston pavement was due in a great measure to the firmness of its foundation, in which respect the earlier stone pavements were often deficient. Sometimes the stone setts have been laid upon the old macadamized surface of the road; but unless this surface happens to be very solid, the paving settles unequally, producing irregularities and hollows in the surface,

[1] *Proceedings Inst. C.E.*, vol. lviii. p. 4.
[2] *Ibid.*, vol. ix. p. 216.

which are injurious to the vehicles, dangerous for the horses, and damaging to the setts. The foundations, accordingly, of a stone pavement should always be very carefully formed with consolidated layers of suitable materials, finished off with a solid, even surface, before the paving is put on; and in streets with very heavy traffic, the foundation should consist of a layer of Portland cement concrete, not less than 6 inches thick, formed with a smooth surface to the exact profile of the road, and left to set for about ten days before the laying of the setts is commenced. A layer of $\frac{3}{4}$ to 1 inch of sand should be spread on the top of the concrete foundation, as a cushion for the setts, to prevent them, when rammed down for forming an even surface, from injuring the surface of the concrete, and to render the traffic over the setts less noisy. When a consolidated macadamized layer, or an old pitched-stone foundation reaches very nearly the level of the foundation surface of the road to be paved, bituminous concrete is used with advantage, as, unlike cement concrete, it is not liable to crack when very thin.

Joints of Stone Setts.—The setts are laid across the street in parallel rows as close together as practicable, with about three longitudinal rows on each side along the curb of the footway, forming the gutters leading the water to the gullies placed at suitable intervals; and the joints between the setts are filled up with sand, gravel, lime, cement grout with gravel, or pitch with gravel. Sand and gravel fill up the joints effectually; but they admit of the percolation of water down to the foundation, which, if consisting partially of the ordinary soil, may be thereby disintegrated; and they also absorb impurities off the road. Grouting the sand or gravel in the joints with lime or cement may provide a fairly impermeable roadway; but its imperviousness is liable to be impaired by the traffic being turned on to it before the grout has had time to set, or by any movement of the setts under the shocks of the traffic; and settlement may result from the infiltration of water, unless there is a foundation of concrete. Pitch, liquefied by boiling with creosote oil, poured into the joints, completely fills the interstices between the gravel, and forms an elastic, impervious joint, not liable to crack, like grout, from the vibration of the traffic; and, consequently, this asphaltic kind of joint serves to form the most impervious and durable paved roadway.

Cost of Stone Paving.—The cost of laying down any form of paving necessarily varies with the nature of the foundation, and the quality of the material used; but with granite setts laid upon a concrete foundation, the cost in some of the principal cities of the United States, in Toronto, and London, appears to have amounted, on the average, to between 14*s.* 4*d.* and 18*s.* 9*d.* per square yard.[1]

Merits and Defects of Stone for Pavements.—Stone paving, well laid with hard setts on a solid foundation, wears very slowly and uniformly under very heavy traffic, and is therefore an economical pavement. Moreover, when jointed with pitch on a concrete foundation,

[1] "Highway Construction," A. T. Byrne, New York, p. 72.

stone paving is impervious and clean; whilst with thin setts, it is suitable for gradients up to about 1 in 16. All pavements suffer from frequent partial disturbances, for getting at gas and water pipes, and underground tubes used for various other purposes, as well as occasional interferences for the repair or enlargement of sewers; and stone paving presents less impediments to the repairs thereby necessitated, than some other pavements. Subways for the pipes and tubes under the streets, would obviate these inconveniences; but their cost has hitherto prevented their general adoption.

The chief objections to stone paving are its noise, its resistance to traction, its hardness and roughness for quiet and light traffic, its slipperiness if formed with wide setts of hard stone, and its want of cleanliness if laid with permeable joints on a soft foundation.

WOOD PAVING.

Dimensions and Laying of Wood Blocks.—A concrete foundation, which is desirable with stone setts, becomes essential when the paving is formed with wood blocks; for the wood blocks are soon destroyed by the puddles and jolts resulting from irregular settlement. The ordinary dimensions for wood blocks, laid with their fibres vertical, are 9 inches in length, 6 inches in depth, and 3 inches in width; though occasionally they differ from these by an inch more or less in length and depth. They are laid on the concrete foundation, about 6 inches thick, finished off to the proper profile of the street, in rows at right angles to the traffic; and this arrangement is preserved for vehicles turning at the intersection of streets, by laying some of the rows, on each side of the turning, diagonally in opposite directions, or in some instances at cross streets, by laying the whole of the central portion common to the two roads, diagonally at right angles to the turn round each of the four corners. Two or three rows of wood blocks are laid longitudinally alongside the curb on each side, to form a gutter. A space has also to be left at each side, of between 1 and 2 inches in width, to allow for the expansion of the blocks, after being laid down, from their absorption of moisture, which is greatest with soft, dry woods, amounting in extreme cases to 1¼ inches in 8 feet, and small in the case of Australian hard woods and creosoted pine. The open joint thus left must be filled up with sand or clay, which is gradually raked out as the expansion progresses, till it reaches its limit in a year or two, when the remaining space should be filled up permanently like the other joints.

Joints of Wood Blocks.—In laying the rows, the requisite intervals for the intervening joints are secured by inserting thin wooden laths between the rows, or by driving small projecting studs into the side of each block, so as to keep them sufficiently apart. The joints are filled with Portland cement grout, or with pitch; and as it has been found that the wide joints of ⅜ to 1 inch formerly adopted, are not needed for providing adequate foothold, the cement-grout joints are now made about ¼ inch thick; whilst in the case of pitch, the blocks can be laid close together, with a minimum joint of $\frac{1}{16}$ inch. These thin joints, by reducing the intervals between the blocks, prevent the spreading out of the fibres of the wood under the traffic. The advantages of filling the joints with pitch are, that the blocks can be put closer together than with cement grout; that the traffic can be admitted on to the paving directly the pitch has cooled, whereas the cement grout would need some days' rest to set; that the pitch joints are elastic, adapting themselves somewhat to expansion or contraction; and that they are more impervious than cement grout in the event of a very severe frost, which, by causing the blocks to shrink, would make the cement crack. On the other hand, pitch joints are said to be subject to evaporation, and consequent decay, in hot, dry weather; but this defect, if it exists, can be remedied by pointing the joints with cement grout, down to about an inch below the surface, over the pitch joint; and the pitch joint is on the whole generally preferred. Before a street paved with wood is opened for traffic, a thin layer of gravel is spread over the surface, which, being pressed into the wood by the horses and wheels, preserves it and roughens it; and gravel or sand has to be sprinkled over wood paving when, in damp weather, the surface becomes greasy from the film of mud brought by the traffic from the neighbouring macadamized streets.

Duration of Wood Paving.—The wear and life of wood pavements are necessarily proportionate to the weight of the traffic, and vary according to the quality of the wood used for the blocks; but they are subject to modification by the nature of the traffic, peculiarities in the climate, the width of the joints, and the general soundness or otherwise of the blocks. The annual wear of several wood pavements in the City and West End of London, was found to vary between $\frac{1}{18}$ and ½ inch; whilst the traffic in these streets per day of 16 hours, ranged between 279 tons per yard in width in Sloane Street, and 1360 tons in Fleet Street.[1] The average life of a wood pavement of soft-wood blocks is about seven years; but some wood pavements, by relaying and renewals, have lasted several years longer; whilst others have had to be entirely renewed in five years, or even less, when the wood has proved defective.

Cost of Wood Paving.—The cost of laying wood paving with a concrete foundation, in several streets in London, has ranged from 10*s.* 6*d.* up to 18*s.* per square yard; whilst wood paving, varying in quality and with different foundations, in various cities in the United States, has cost from 4*s.* up to 15*s.* per square yard;[2] and in Toronto, the cost has varied from 4*s.* 7*d.* per square yard for wood blocks on a gravel foundation, up to 11*s.* for tamarack on a concrete foundation.

Advantages of Hard Woods for Paving.—Experiments with various kinds of woods in the streets of Sydney, have demonstrated the very superior durability of some of the hard woods of Australia, especially tallow wood, spotted, blue, and red gum, and black butt, some of the blocks laid down having shown a maximum wear of only $\frac{1}{16}$ inch,

[1] *Proceedings Inst. C.E.*, vol. lxxviii. p. 280.
[2] "Highway Construction," A. T. Byrne, p. 93.

after sustaining for eight years a daily traffic of about 750 tons per yard of width.[1] The life of this hard-wood paving, if properly laid on a good foundation, has been estimated at a minimum of twenty-one years. The greater slipperiness of this dense wood may necessitate making the joints not less than $\frac{1}{4}$ to $\frac{3}{8}$ inch in width; and the greater noise observed in some cases with these hard woods, could be easily remedied by a cushion of sand on the concrete foundation. These dense woods, however, are not nearly so subject to expansion by absorption of moisture as soft woods; and their density is also valuable in preventing their absorbing impurities from the surface of the streets. The cost of the hard woods for a long time prevented their adoption for pavements in Europe, but they have now come into very general use; and their freedom from absorption, and their long life, should lead to their extensive employment in large towns.

Merits and Defects of Wood for Pavements.—The great merit of wood is that it is the least noisy of all pavements; whilst it also provides a fairly good foothold for the horses, is very suitable and pleasant for quick and light traffic, and has a neat appearance. Wood cannot, however, be used for gradients of over 1 in 25 to 1 in 20; and the softer woods, when uncreosoted, absorb impurities and are not easily cleansed. Moreover, soft-wood paving is liable to wear unevenly, forming hollows in which water collects, hastening the decay and leading to the rapid wear of the adjacent blocks by the jolts of the passing vehicles, necessitating costly repairs; whilst its comparatively short life involves complete renewal about every seven years, with the accompanying stoppage of the traffic. Most of these objections, however, can be greatly mitigated by the use of hard woods, at an increased initial cost. A wood pavement is not well adapted for repairs necessitated by disturbances for getting at pipes, or by unequal wear, for it is difficult to make the surface of the new work uniform with the old.

Brick Paving.

Employment of Bricks for Pavements.—Bricks have been used to a very limited extent in Great Britain for paving streets; but brick pavements have been long and extensively used in Holland, where stone of any kind has to be brought from a great distance; and they have also been laid down within the last quarter of a century in several cities in the United States. Bricks have been laid on foundations of sand or gravel; but to ensure their durability under heavy traffic, it is most important, as in other forms of paving, that they should be laid on a solid bed of concrete, and their joints filled with cement grout or pitch.

Qualities of Bricks for Paving.—Bricks for paving, generally made of about the size of ordinary building bricks, should be hard, tough, and non-porous. Bricks which absorb water are softer than close bricks, and are disintegrated by frost; whilst vitrified bricks, which have been

[1] *Proceedings Inst. C.E*, vol. xciii. p. 368; and vol. cxvi. p. 265.

frequently used in the United States, though hard, are liable to be slippery and to crack. The bricks are laid on edge in rows at right angles to the line of traffic, as in the case of granite setts and wood blocks. The quality of the bricks necessarily depends on the nature of the clay obtainable in the district; and the adoption of a brick pavement must be in great measure regulated by this consideration. Good brick pavements have lasted for long periods in Holland, where the absence of stone and the dampness of the country render them specially suitable; whilst vitrified and other kinds of bricks have proved durable in America, where in the prairie regions, the scarcity of stone and timber, and the cost of asphalt render brick the only available paving.

Cost, Merits, and Defects of Brick Paving.—The cost of brick paving in the United States appears, on the average, to approximate to that of wood, ranging between 4*s.* 2*d.* and 11*s.* 8*d.* per square yard with various foundations;[1] whilst the life of good brick paving is much longer than that of wood paving. Brick paving is liable to wear unevenly, unless the bricks are very carefully selected, owing to differences in their hardness; and it is less durable than granite paving, and is less easily repaired; but, on the other hand, it is cheaper, less noisy, more easily cleaned, offers less resistance to traction, affords a better foothold for horses, and is smoother for vehicles than stone paving. Transverse strength and endurance in the "rattler" are tests commonly applied to paving bricks.

ASPHALT PAVING.

Composition of Asphalt.—Various compositions mixed with pitch, and sometimes the pitch poured between the joints of paving blocks, have been called asphalt; but the true asphalt, employed for paving carriage-ways in Europe, is a natural bituminous limestone composed of carbonate of lime impregnated with bitumen, the proportion of which should lie between 7 and 12 per cent. for the asphalt to be suitable for paving, since with too little bitumen the asphalt is liable to crack, and with too much it becomes soft under the sun's heat. The quarries from which the best asphalt for paving is obtained, are situated in the Val de Travers in Switzerland, Seyssel in France, and Limmer in Hanover. Mastic asphalt is made by heating natural asphalt with 7 to 8 per cent. of pure bitumen, to a temperature of between 390° and 480°, causing it to melt. It is then formed into round, flat blocks, about 1 foot in diameter and 4 inches thick, in which form it is conveyed to its destination; and it has to be remelted with bitumen in order to use it.[2]

An artificial asphalt is used in America, owing to the cost of carriage of the natural asphalt, which consists of bitumen obtained from deposits at Lake Brea in Trinidad, mixed with sand and some pulverized limestone, in the proportions of 12 to 15 per cent. of bitumen, 83 to 70 per cent. of sand, and 5 to 15 per cent. of pulverized carbonate of lime.

[1] "Highway Construction," A. T. Byrne, p. 153.
[2] *Proceedings Inst. C.E.*, vol. xliii. p. 290.

The bitumen, after the removal of the earthy matter contained in the crude deposit, is heated with a heavy paraffin oil, which gives it great tenacity, and converts it into a strongly cementing substance, which binds the artificial mixture into a solid mass. This Trinidad asphalt is less slippery than natural compressed asphalt, owing to the large proportion of sand in its composition; but this relative roughness must also render it more subject to wear.

Laying Asphalt Paving.—As the layer of asphalt forming the paving surface is not more than 2 to 2½ inches thick, it is of the utmost importance for its maintenance that it should rest upon a thoroughly solid foundation, not subject to settlement under the traffic; and, consequently, a sound concrete foundation, from 6 to 9 inches thick, is essential for an asphalt paving. Compressed asphalt, formed with the powder of the natural rock pulverized in a disintegrator, is now always used in Europe for the paving of streets, as more easily laid down, and more uniform in quality than mastic asphalt, which is reserved for footpaths. The asphalt powder is heated to about 260°, and is then spread on the thoroughly set and dry concrete foundation, in a layer about 3 inches thick, which is compressed into a compact mass, 2 inches thick, by ramming it with heated iron punners and rolling.[1] The surface is lightly sprinkled over with sand before the admission of the traffic, only a few hours after the completion of the work. It is very important that there should be no moisture on the surface of the concrete when the asphalt is laid on, for the heated powder converts the water into steam, which in escaping forms cracks in the asphalt.

The asphalt made with the aid of Trinidad bitumen is laid on in two coats: the first, or cushion coat, being mixed with rather more bitumen than the surface coat, and laid on hot, and consolidated by a roller to a thickness of half an inch; and then the surface coat, having the proportions given above, is spread on the first coat to a sufficient thickness to form with the first coat, when consolidated, a layer from 2 to 2½ inches thick.

Cost of Asphalt Paving.—As the sources of asphalt suitable for street-paving are very limited, the cost of paving with it varies with the distance of the locality from the nearest source of supply; and the nature, thickness, and facility of supply of the materials of the concrete foundation, also materially affect the cost. The cost per square yard of asphalt paving in the principal cities in the United States, varies between 10*s*. 5*d*. and 18*s*. 9*d*., though the average is less than 12*s*. 6*d*., which is the average cost at Toronto;[2] whilst at Montreal, it ranges between 14*s*. 3*d*. and 16*s*. 6*d*., and has been estimated at from 15*s*. 7*d*. to 18*s*. 9*d*. in London, 16*s*. 8*d*. to 17*s*. 11*d*. in Paris, and as averaging 14*s*. 7*d*. in Berlin.

Merits and Defects of Asphalt Paving.—A paving of compressed asphalt, laid upon a good foundation, furnishes a very smooth,

[1] "The Construction of Carriage-ways and Footways," H. P. Boulnois, p. 69.
[2] "Highway Construction," A. T. Byrne, p. 114.

easy roadway, quite impervious, very readily cleaned by watering, and repaired without any difficulty. Next to granite, it is the most durable pavement, its wear being very slight under the heaviest traffic, and also uniform. Moreover, except for the clatter of the horses' feet on its hard surface, it would be almost noiseless. These qualities, in any comparison of the relative merits of the chief paving materials, give it an apparently very marked superiority over the other pavements. It labours, however, under two very serious defects, namely, great slipperiness in damp weather, when coated with a greasy film of mud brought on it by the traffic from macadamized streets, and an absence of proper foothold, which, though comparatively immaterial on level streets, except for starting and pulling up abruptly, owing to the ease of traction, renders it quite unsuitable for gradients exceeding 1 in 60 to 1 in 50 as the extreme limit. The first defect is most felt in moist climates like that of Great Britain, and can be to a great extent mitigated by ample watering directly it becomes damp; but the second defect prohibits its use on steep gradients.

Conclusions about Street-paving.

For country roads and streets in the outskirts of towns, macadam provides the only adequately economical coating for supporting the wear and tear of the traffic. In the back streets of large towns, where the traffic is comparatively small, but an impervious pavement is expedient on sanitary grounds, an artificial asphalt paving may be formed by pouring pitch over a layer of road-metal, so as to fill up the interstices; and then a thin layer of smaller stones and chippings is spread on the top and rolled, which, becoming incorporated with the pitch forced up from below by the pressure of the roller, completes the filling of the spaces between the larger stones, and forms a solid, uniform surface which soon becomes hard, and, being smooth and impervious, is readily cleansed.

In the main thoroughfares and central portions of cities and large towns, the large traffic, the desire for comfort of the richer inhabitants, and the want of sun and air in narrow, crowded streets, prohibit the use of macadam; and the choice generally lies between stone paving, wood, or asphalt. None of these pavements can be regarded as perfect. Stone should be preferred where the traffic is exceptionally heavy, the gradients steep in places, and the noise of little importance—as, for instance, alongside docks, in streets lined with warehouses, and at goods stations. Asphalt may advantageously be substituted for stone in the business portions of cities and towns, where the reduction of the noise of a large traffic is an inestimable boon to the busy occupants of offices, and the cost of the paving is of comparatively little moment owing to the high ratable value of the adjacent property, provided always that the gradients are very moderate. The slipperiness, moreover, of asphalt becomes of less consequence where the congested traffic is necessarily somewhat slow, and where there is no need to economize the water for cleansing the streets. The imperviousness also, rapidity in drying, and cleanliness of asphalt render it specially suitable for the narrow, ill-ventilated streets and courts of an overcrowded poor district. Wood is best adapted for the richer residential quarters of a town, where quiet and quickness of transit are of primary importance, where the traffic is moderate, and where the width of the streets, and the airiness and good sanitary conditions of the houses, minimize any deficiency in salubrity arising from the absorptive qualities of soft woods; whilst this objection could be almost entirely obviated by using creosoted soft woods, or the hard, close-grained Australian woods. Brick blocks may be employed instead of stone setts where stone of suitable quality is difficult to procure, provided suitable clay can be obtained in the neighbourhood for manufacturing bricks of adequate and uniform hardness without being slippery. The change from one kind of pavement to another, such as sometimes occurs twice in a single street in London, is unquestionably trying to horses, and increases their tendency to fall on a slippery pavement. In the event of motor-cars superseding to any large extent traction by horses, the reduction in the number of horses traversing the streets would correspondingly diminish the objections to wood and asphalt pavements.

Stone setts still form the principal paving in the busiest thoroughfares of London; but asphalt has been introduced in some of the main centres of business in the city and elsewhere, and in some narrow streets; whilst wood has been chiefly adopted for paving in the West End, and in the principal streets of the shopping and residential portions of the metropolis. In Liverpool, the very heavy goods traffic, and the gradients of the streets rising from the river, have naturally rendered stone paving the most suitable for the main streets. Granite setts constitute the main paving of New York; but of late years asphalt paving has been largely extended; and the present policy is to adopt asphalt for streets which are likely to become main thoroughfares, and in the poorer and more densely populated portions of the city.[1] Wood, on the contrary, which is so often used for paving in the western cities of America, where timber is abundant, and which has been extensively employed for paving in Chicago, Detroit, and Toronto, for instance, finds no favour in New York. Asphalt is the ordinary paving for the best quarters of Paris; and it also is the principal form of paving in Berlin, Vienna, and other cities on the Continent, where the climatic conditions are more favourable than in England. Asphalt paving, moreover, constitutes an important portion of the paving of other American cities besides New York—as, for instance, Buffalo and Washington; whilst it has been introduced into several other cities—such, for example, as Philadelphia and Brooklyn, in which cobble-stone pavements have largely predominated. Brick paving appears to have been gradually introduced to a limited extent in several American cities within the last quarter of a century; but its widest application is found in Amsterdam, Rotterdam, the Hague, and other places in Holland, where long ago bricks were first used for paving.

[1] "Reports of Department of Public Works, City of New York, June, 1895," p. 32, and map.

FOOTPATHS.

Footpaths should always be given a little fall towards the water-table or gutter at the side of the road, to carry off the rainfall, the amount of fall depending on the nature of the surface of the path; $\frac{1}{8}$ of an inch in a foot being sufficient for concrete or asphalt, $\frac{1}{4}$ inch for flagstones or bricks, and somewhat more for gravel or sand, which also is best obtained by raising the footpath in the centre. The width of footpaths should be regulated according to the amount of traffic, so far as the width allotted to the street admits.

Country Footpaths.—In country districts, footpaths are generally made with gravel or sand on the top, and coarser material underneath; and sometimes the surface is formed with stone chips, or with cinders. Through villages and near stations, tarred paving is often adopted; and the footpath is protected by a stone curb on the outside.

Town Footpaths.—In towns, good footpaths become a necessity, and have to sustain a considerable wear in crowded streets. The paving is formed of flagstones, concrete, compressed asphalt, mastic asphalt, bituminous concrete, or brick, laid upon a solid foundation. Occasionally granite slabs, slate, or planks are employed; but they do not form satisfactory pavements, for the granite wears too smooth in time; slate also becomes slippery; and planks, though cheap where timber is abundant, soon become irregular and decay. Flagstones from 2½ to 3 inches thick, laid with mortar, form a good paving, provided they are solidly bedded, and if they do not laminate with moisture or frost, or wear unevenly, leading to the formation of puddles. Concrete made with Portland cement affords a good, durable pavement; but it must be laid with joints about 6 feet apart, or in slabs, as changes of temperature produce unsightly cracks across a long continuous layer of concrete. Compressed asphalt, not less than 1 inch thick, laid on a concrete foundation, provides an impervious, very durable footpath; and mastic asphalt, heated with bitumen and mixed with sand, furnishes a good economical footpath. Concrete also can be used for footpaths, in which bitumen serves as the cementing material; but if pitch is used in place of bitumen, the paving becomes soft and sticky in hot weather, and is not durable under a heavy traffic. The advantages of these asphaltic pavements are that they are impervious and devoid of joints, and yet, owing to their elasticity, they are not liable to become cracked under changes of temperature like continuous concrete pavements. Bricks roughened on the surface with a chequered pattern, if hard and uniform in quality, form a convenient, durable paving; but the number of their joints, and their tendency to wear unevenly, render them generally a less good paving than the other kinds described, though in some localities economical considerations may necessitate their employment.

Strong well-bedded stone curbs are always laid along the outer edge of the footpaths through towns; and they should be not less than 3 feet in length, and should be raised 3 inches at least above the highest point of the gutter, so as to effectually secure the footpaths from the incursions of vehicles, and not more than about 7 inches above the lowest part of the gutter, to avoid too deep a step for pedestrians crossing the road. The curbs, moreover, should have a minimum depth of 9 to 12 inches, so as to extend well below the gutter, and therefore not be liable to tip over at the top towards the road. The top width of the curbing, which should not be less than 4 inches, is usually increased in proportion to the width or importance of the street, the heaviness of the traffic, up to between 8 inches and 1 foot; and in the principal thoroughfares of large towns, it is made of granite very hard stone, to withstand the shocks to which it is subject.

PART II.

RAILWAY, BRIDGE, AND TUNNEL ENGINEERING.

CHAPTER VII.

LAYING OUT AND FORMATION OF RAILWAYS

Contrast between roads and railways—Preliminary considerations for railways—Surveys and trial levels; use of maps, theodolite, levelling, reconnaissance in new country, levels with barometer, advantages of tacheometer in rough country—Ruling gradient: importance, examples—Limit of curvature: object of curves, examples, methods of denoting curves—Compensation of gradients for curvature—Switchbacks: description, instances, disadvantages—Laying out a railway—Setting out a railway, setting out curves with theodolite, tangential angle—Adjustment of earthwork in cuttings and embankments, for crossing roads, limits of depth and height—Side cuttings for embankments; across low ground, in side-lying ground—Formation width, and side slopes: instances—Calculation of earthwork: by tables, example, extension of tables, earthwork not indicated on longitudinal section, cutting partly in rock—Fencing, varieties—Provisions against slips in cuttings and embankments; ordinary, causes of slips in cuttings, drainage and protection of slopes; causes of slips in embankments, methods of prevention—Remarks on laying out railways—Culverts: object, forms, construction.

Railways differ from roads in providing a track for locomotion offering less resistance to traction than the smoothest roadway, by means of elevated rails; in following a more direct course, at the expense of heavy cuttings and embankments; and in resorting to easier gradients than roads, with the object of facilitating rapid transit. Railways, accordingly, necessitate much more numerous and extensive engineering works than roads, such as cuttings, embankments, culverts, bridges, viaducts, and tunnels, which, whilst constituting comparatively ordinary works in railway construction, are, with the exception of culverts and bridges, especially across rivers, rarely resorted to in forming roads, which follow the surface of the ground as nearly as practicable, and adopt a devious course to secure easy gradients for the main routes of traffic.

Preliminary Considerations.—The population and the probable traffic in the district to be served by a railway, as well as the nature of the country to be traversed, are important factors in determining the kind of railway that should be constructed. A railway that has to be carried through a populous district, and connects important towns situated at a moderate distance apart, should be constructed in a solid, durable manner, with as easy gradients as reasonably practicable, and more with a view of providing for rapid transit and economical working than of keeping down the capital cost. In laying out the line, the accommodation of intervening towns or villages should be arranged for, provided no very material increase in the distance between the termini is thereby involved, such as might lead to the construction of a competing line. Where, on the contrary, the railway is a pioneer line, to be constructed for developing the resources of an unsettled tract of country very sparsely inhabited, like the Canadian Pacific Railway, and the Bulawayo Railway, or for rendering outlying, newly acquired territories more secure from hostile attacks, as well as for promoting trade, like the Berber, Khartoum, and Uganda railways, the primary essentials are rapid construction and a minimum capital cost; whilst the consolidation of the line, the easing of steep gradients, and the replacing of timber trestles by more permanent structures, must be deferred till the traffic on the line has sufficiently developed to justify further expenditure in improving the railway. Except in a few backward countries, such as China, most of the long railways of the future will be pioneer or strategical lines, since the main lines in most of the settled, populous countries have already been constructed.

Surveys and Trial Levels.—Before deciding on the exact course of a railway, it is necessary to obtain as accurate a knowledge as practicable of the general nature and levels of the country to be traversed. The preliminary survey, however, which is a comparatively easy task in an open, settled country, more especially where a general survey of the district has been previously made, becomes an arduous, and even hazardous, undertaking in unexplored regions, where unhealthy swamps, dense forests, and rugged mountain gorges, of unknown extent, may have to be traversed, as in determining the location of the Oroya Railway up the deep gorges and across the precipitous cañons of the Pacific slope of the Andes, and in selecting routes for the western railways of North America, through the passes of the Rocky Mountains and Pacific ranges.[1]

When a map to a fairly large scale, with contour-lines of uniform heights, has been made of a settled locality, such as some of the 6-inch Ordnance maps of the United Kingdom, it is possible to lay down approximately a suitable line on the map; and then a walk over the ground, and a few trial levels at the more important points, suffice to determine the route along which the more detailed survey should be made. In the absence of a sufficiently large and recent survey of the locality, a traverse survey should be made with a theodolite, consisting of a series of straight lines inclined at suitable angles so as to follow closely the general course of the proposed railway, of which the angles and bearings are determined by the theodolite, and the lengths measured by a 66-feet, or 100-feet chain, of one hundred links. Whilst chaining along the lines set out by the theodolite, sufficient offsets are measured at right angles, on each side of the line, to fix the positions of houses, fences, and ponds, and the courses and widths of roads, paths, streams, and ditches, in close proximity to the proposed line; so that, in finally laying down the course of the railway on the plotted survey, interference with valuable property and unnecessary severance may be avoided, and any requisite diversions of roads and streams may be arranged. Then by taking levels at every change of inclination of the ground along the line followed, and where roads, streams, and ditches are crossed, and also at right angles to the line at intervals on side-lying ground, a longitudinal section of the ground can be plotted, with an enlarged vertical scale to magnify the irregularities of the surface, which serves as a guide for the gradients suitable for the line, and also cross sections enabling the change of level of the ground resulting from a sideways deviation of the line, on transverse sloping ground, to be determined.

In unexplored districts proposed to be opened up by a railway, it is essential to make a preliminary reconnaissance in order to determine the best general course of the line; and where mountain ranges have to be traversed, regular explorations have to be made, to discover the most feasible passes over which the line can be carried. Often in rugged, forest-clad districts, observations with a compass of the bearings, determinations of relative heights with an aneroid barometer, and very rough estimates of the distances travelled, have to suffice in the first instance; but, where feasible, chain measurements and observations of the slope of the ground with a clinometer should also be resorted to. In determining heights with an aneroid, serious errors, due to sudden changes in the atmospheric pressure, may be avoided to a great extent by keeping a precisely comparable barometer at the starting-point, the changes of which during the day are noted, to serve as corrections of the indications of the barometer taken with the surveying party. It is also advisable to restrict these observations, if possible, to periods of still weather; for during stormy and gusty weather, the pressure may vary owing to local conditions, such as exposure to, or shelter from the wind, and afford misleading indications of changes in elevation at short distances apart.

When the general direction of the line has been decided by the reconnaissance, the location of the railway, and the longitudinal section along the line selected can be determined, in the open prairie, by a powerful transit theodolite and chain measurements, and by levelling along the line as under ordinary conditions; but through the ravines and over the rugged passes of mountain ranges, a tacheometer, with its additional horizontal hairs in the eye-piece and special adjustment,[1] should be used, as it enables the distances and levels of any number of points, within sight and range of the powerful telescope, to be determined without measurements or shifting of the instrument. By placing the tacheometer in a succession of positions commanding a sight

[1] Measurement of distances by a tacheometer, is based on the principle that the apparent height of an object varies inversely with its distance from the observer; and by so arranging the interval between the horizontal hairs, placed at equal distances on each side of the central line, and the measuring angle of the instrument, that the interval bears an exact ratio, usually $\frac{1}{50}$ or $\frac{1}{100}$, to the distance of the staff from the centre of the instrument; then by observing the height of staff comprised between the hairs, the actual distance of the staff from the tacheometer is known at once.

of the course to be followed, it is only necessary for the staff-bearer, with a staff marked with broad divisions which can be seen at a distance, to gain access to the requisite points, and to get obstacles intervening between him and the observer cleared away, in order that a preliminary survey of the proposed course, with the level of each point, may be obtained, on which the line to be followed by the railway can be approximately laid down. In traversing very rugged country, it is frequently desirable to survey an alternative route before deciding definitely on the course to be followed. These preliminary surveys suffice to determine the general course of the line, and enable a rough estimate to be made of its cost; but it is generally expedient to defer fixing the exact location of the line through uninhabited districts, till the extension of the railway puts greater facilities for completing the work, within the reach of the surveying party proceeding in advance of the completed railway.

Ruling Gradient.—After completing the preliminary surveys which indicate the elevations and depressions of the country to be traversed, the first points to be determined are the inclination of the worst, or *ruling*, gradient admissible, and the radius of the sharpest curve that may be safely introduced. The decision on these questions must largely depend on the object and importance of the railway, and the physical features of the district. The main object kept in view in selecting a route for a railway, is to avoid high elevations and abrupt changes in level of the land. Accordingly, where a hilly ridge has to be surmounted, the point of minimum elevation is chosen, if possible, for the crossing; and the ascent is made as uniform as practicable by gradually rising up one of the valleys, and descending by the valley on the opposite side of the ridge; and these valleys, down which a stream or river always passes discharging the rainfall of the district, become narrower and steeper on approaching the dividing ridge, and therefore less favourable for railway construction. For main lines, accordingly, in populous countries, the steeper gradients or circuitous course required in the upper part of a valley to surmount a mountain pass, are avoided by piercing the ridge by a tunnel, a solution of the difficulty which cannot be resorted to for pioneer lines, on account of the cost and the time occupied in the construction of a long tunnel. Easy gradients, moreover, through a rugged country necessitate heavy cuttings and embankments, and expensive bridges and other works, rendering the cost prohibitive in the case of railways through undeveloped districts, or even for branch lines in thinly-populated, settled countries devoid of minerals. A ruling gradient, consequently, of 1 in 300, adopted for the earlier main lines of England, or even 1 in 100 (termed a 1 per cent. grade in America), could not possibly be adopted in rugged country for pioneer or branch lines, or even for main lines.

The importance of the ruling gradient is that it determines the weight which the locomotives ordinarily used on the line can draw up it, and consequently the length of the train. In the case, however, of some mineral lines, where the trains descend loaded from the mines or quarries in the hills to the port of shipment, and are drawn up light, the ruling gradient against the returning empty train may be made much steeper than against the loaded train. Thus, for instance, with waggons weighing 30 per cent. of their total weight when loaded, the ruling gradient may be 1 in 50 in the first direction, and 1 in 170 in the other, for the same tractive force.[1] Where a railway extends over a vast tract of country, like the Canadian Pacific Railway with a continuous length between Montreal and Vancouver of 2906 miles, and where, therefore, the locomotives are changed several times on the journey,[2] the ruling gradient and the type of locomotive employed may be modified according to the nature of the country. Thus on the Canadian Pacific Railway, the ruling gradient has been limited to 1 in 132 in the flat wheat-growing districts and undulating prairie west of Winnipeg, and only increased to 1 in 105 in ascending the Missouri Coteau;[3] whilst in the mountain division, the ruling gradient has been raised to 2·2 per cent., or 1 in $45\frac{1}{2}$, with only one steeper gradient of 4·5 per cent., or 1 in 22·2 for $3\frac{1}{4}$ miles in descending the western slope of the Kicking-Horse Pass, which was intended when constructed to be merely temporary, and will give place to the easier gradient when the original location has been completed. Special heavy locomotives, with three or four pairs of driving-wheels coupled, and weighing 53 tons without their tender, are used on this division of the line; and one of these locomotives, pushing at the rear of the train in mounting the steepest gradients, assists the ascent of the train consisting of about twelve or thirteen long cars. The ruling gradient, indeed, of the mountain division of the Canadian Pacific Railway, crossing the Rockies and the Selkirks at altitudes of 5296 and 4300 feet respectively above sea-level, are, with the one exception given above, by no means unusual; for the same ruling gradient of 1 in $45\frac{1}{2}$ has been adopted on the western lines of the United States, crossing the Rockies at higher elevations and descending the steep Pacific slope; whilst several important railways in Europe, as well as others in North and South America, have steeper ruling gradients in crossing high mountain ranges, as, for example, 1 in 44 on the Brenner, 1 in 37 on the St. Gothard, 1 in $33\frac{1}{3}$ on the Mont Cenis, 1 in $32\frac{1}{2}$ on the Arlberg, and 1 in 25 on the Denver and Rio Grande, the Vera Cruz and Mexico, and the railways ascending the Peruvian Andes.[4] Though a Baldwin locomotive, with a weight of about 40 tons on three pairs of coupled driving-wheels, has been able to draw a train weighing about 40 tons up an incline of 1 in 12, at a rate of $8\frac{2}{3}$ miles an hour, on the Cantagallo Railway in Brazil, a gradient of 1 in 25 may be regarded as the practical limit for drawing up trains with locomotives by adhesion alone; and, except under specially difficult conditions, a ruling gradient of about 1 in 45 should be treated as the proper limit for ordinary railways in hilly districts.

[1] "Railways and Locomotives," J. Wolfe Barry and F. J. Bramwell, p. 19.
[2] The locomotives run over sections of about 250 miles in length along the flatter portion of the railway, and about 125 miles in the mountain division.
[3] *Proceedings Inst. C.E.*, vol. lxxvi. p. 270.
[4] "Achievements in Engineering," L. F. Vernon-Harcourt, pp. 42-49.

Limit of Curvature.—Railways, in traversing mountain passes, could not in places accomplish the ascent, even with the steep gradients mentioned above, without increasing the length of the line at the worst places by a devious course, consisting in some parts of loops or spirals (Fig. 33), as well as winding round projecting spurs to avoid large expenses in excavation in passing through narrow gorges. This winding course, and more particularly these loops and spirals necessitate sharp

LOOPS AND SPIRALS ON RAILWAYS.

Fig. 33.—St. Gothard Railway.

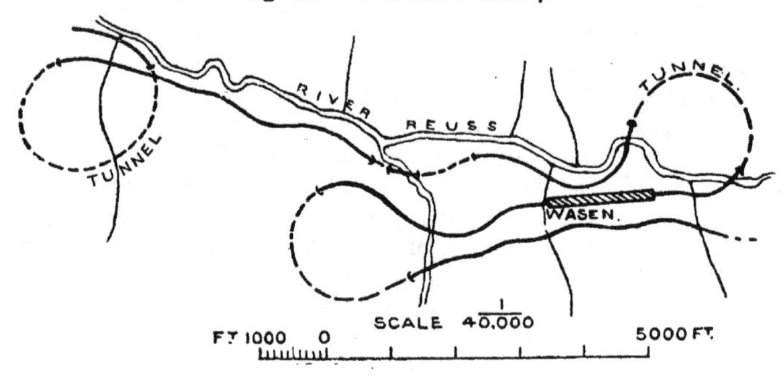

Canadian Pacific Railway.

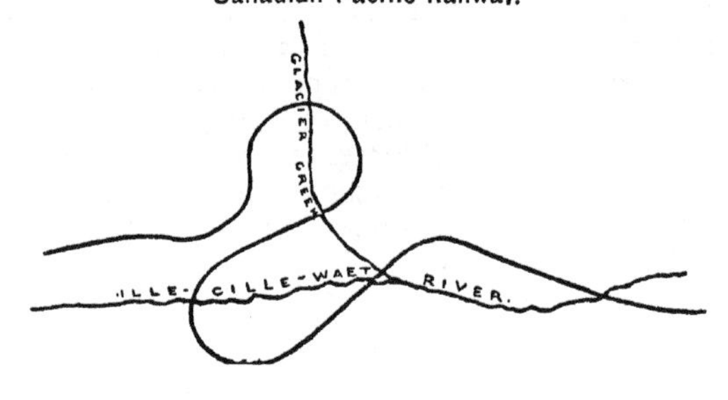

curves; and the main object aimed at, of reducing the cost of the works by keeping the line as near the surface of the ground as practicable, by contouring in rising up the slopes of the valley, is best attained by adopting the sharpest admissible curves. Curves of 40 to 30 chains radius, which are about the necessary limit for main lines on which trains have to run at a high speed, are not sharp enough for carrying a railway through rugged country at a reasonable cost. Thus in Europe, curves of 17 chains radius were resorted to on the Mont Cenis Railway, $14\frac{1}{6}$ chains on the Brenner, 14 chains on the St. Gothard, 10 chains on the Arlberg, and $9\frac{1}{2}$ chains on the Semmering;[1] whilst in America, the Pacific lines of the United States have curves

[1] *Proceedings Inst. C.E.*, vol. xcv. p. 278.

of $8\frac{2}{3}$ chains; and the mountain section of the Canadian Pacific Railway has a similar limit of curvature, with the exception of a short curve of 4 chains radius contouring a spur in the Wapta valley, on the western slopes of the Rockies, in substitution for a tunnel through the spur, which has been abandoned owing to damage from slips. On the Oroya Railway, along the most difficult part of the ascent of the Andes, the sharpest curves are 6 chains in radius; and on the railway ascending from Vera Cruz to the city of Mexico, curves of 5 chains have been resorted to, which may be regarded as the practical limit for railways of the ordinary gauge of 4 feet $8\frac{1}{2}$ inches, though still sharper curves have been adopted on narrow-gauge lines.

In America, instead of denoting a curve by its radius in feet, or, as usual in Great Britain, by the radius in chains of 66 feet, the curve is commonly indicated by the angle subtended at the centre by a chord of the curve 100 feet in length. Thus a curve of 1° is a circular arc in which half the chord, or 50 feet, divided by the radius is equal to sin $\frac{1}{2}^\circ$, giving a radius of nearly 5730 feet, or about 87 chains; and a 10° curve has a radius of $573\frac{2}{3}$ feet, or $8\frac{2}{3}$ chains, the limit of curvature on the Pacific lines of North America.

Compensation of Gradients for Curvature.—As a sharp curvature necessarily increases the resistance to traction, a steep gradient must be reduced on a sharp curve, in order that a locomotive may not experience a greater difficulty in dragging a train up an ascent round the curved portion, than along the straight parts of the line; otherwise a heavy train which could just be drawn up the gradient on the straight, would be brought to a standstill on a curve. The compensation for curvature on the mountain section of the Canadian Pacific Railway, has been effected by reducing the gradient 0·03 foot in 100 feet for each degree of curvature, or 0·3 foot per cent. for a 10° curve.[1] On the prairie section of the line, the larger compensation of 0·05 foot in 100 feet for each degree of curvature was adopted, doubtless owing to the increased resistance to traction produced by curves where the speed of the trains is greater.

Switchbacks.—On some American lines, the ascent of the steep upper portions of valleys towards the summit of a mountain pass, has been more economically effected by means of occasional switchbacks, or back shunts, forming angular zigzags, the train being drawn up into a sort of siding, from which it passes out in a reverse direction along another line, making a very acute angle to the line on which it reached the siding, thereby climbing up a steep slope without the intervention of a curved loop for reversing its direction. This expedient has been extensively used in the higher parts of the Oroya Railway, in combination with loops (Fig. 34, p. 108); and there is an instance of this arrangement on the main line between Bombay and Calcutta, near the summit of the Ghats, not far from Bombay. The reversal of direction, however, of the train at the switchback, converting the front

[1] These and other particulars about the mountain section of the C.P.R. were given me at Vancouver by Mr. H. J. Cambie, the engineer in charge of that portion of the line.

of the train into the rear, involves a stoppage for shunting the locomotive, unless the objectionable plan of pushing the train up to the following switchback is allowed, or unless the gradients are so steep, and the train so heavy, as to lead to the use of a second locomotive temporarily in the rear during the ascent, an arrangement which needs

LOOPS AND SWITCHBACKS ON RAILWAYS.

Fig. 34.—Oroya Railway, Peru.

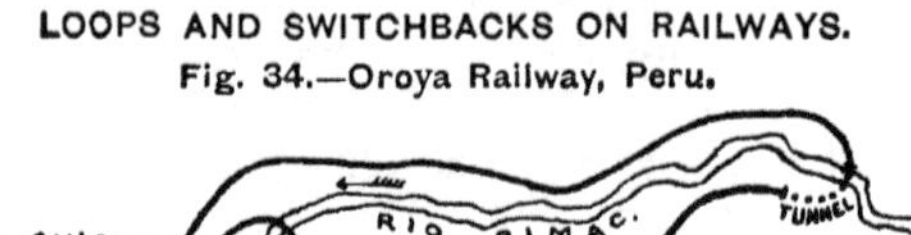

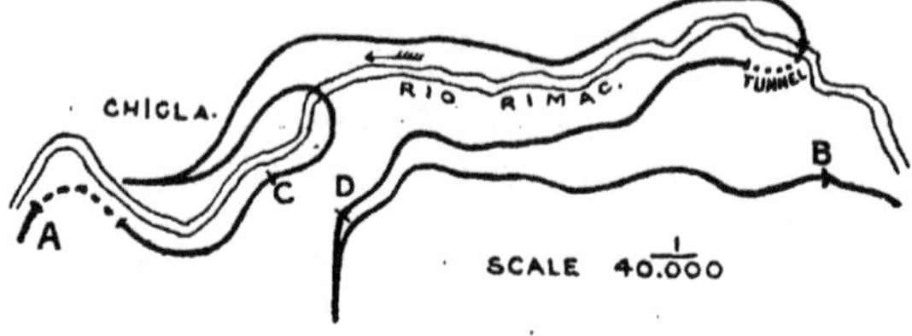

care to secure the train against the possibility of being telescoped, in the event of the locomotives at the two ends acting in opposition.

Laying out a Railway.—When the ruling gradient and the limit of curvature have been determined, the line of the railway can be laid out more precisely on the survey, in accordance with the indications furnished by the levels. In laying down curves on the plan, a piece of straight should always be inserted between curves turning in opposite directions, for a sudden reversal of the curvature would necessitate the abandonment of the superelevation of the outer rail near the point of contrary flexure, an arrangement which greatly adds to the ease of motion and safety of trains in running round sharp curves. In undeveloped countries, considerable latitude can be allowed in the final selection of the route, in order to secure the most suitable and economical line, and one which is capable of improvement as the traffic increases; but in populous districts, the choice is often confined within very restricted limits, involving costly works in traversing hilly country.

Setting out a Railway.—After laying down the course of the railway on the survey, the centre line is set out on the ground with a theodolite, and pegs are driven in at every chain; and levels are taken along the line as set out, to serve as a working longitudinal section, final modifications being introduced where the further inspection of the ground and the indications of the levels appear to render them expedient.

The setting out of the curves is best effected by placing the theodolite over the point from which the curve starts, setting the theodolite at an angle to the centre line of the straight portion of the railway produced, equal to the tangential angle of the special curve, and marking the point on the line to which the theodolite is set, which is the selected chord's length away from the tangent or starting-point of the curve (Fig. 35). This is the first point set out on the curve; and the succeeding points are obtained by turning the theodolite successively through a series of similar angles, and marking the point on each of the lines thus set out by the theodolite in succession, which is a chord's length distant from the preceding point on the curve. This is readily effected in practice by one man holding the end of a tape at the starting-point in the first instance, and at the last point

SETTING OUT A RAILWAY CURVE.
Fig. 35.

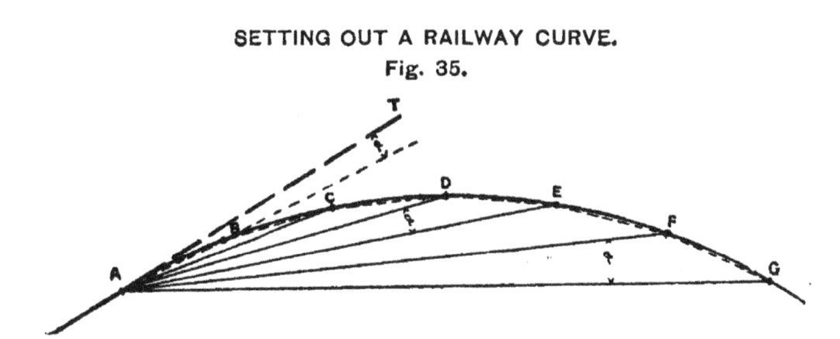

set out on the curve afterwards, and another man holding a pole upright at the required distance along the tightly stretched tape, and shifting his position, as directed by the observer at the theodolite, till he brings the pole into the exact line to which the theodolite is set. When any obstacle prevents any further setting out from the starting-point, the theodolite is transferred to the last point set out on the curve; and after setting the theodolite at twice the tangential angle from the last chord produced (one angle for the direction of the tangent line, and the second for the next point on the curve), the same operations are repeated till the setting out of the curve is completed. This simple method of setting out a circular arc is based on the fact that equal angles from any point on the circumference of a circle subtend equal arcs. A general formula for the tangential angle, a, of any circular curve is $a = \sin^{-1}\dfrac{\frac{1}{2}\text{chord}}{\text{radius}}$, which is readily obtained from the consideration that the tangential angle, a, in Fig. 36, is equal to half the angle subtended by the chord AB at the centre of the circle C, namely ACD, as both angles are complements of the angle DAC. In the case of a curve given in degrees, the tangential angle for chords of 100 feet is half the angle of the curve, so that with a 1° curve, the tangential angle is 30′, and with a 10° curve, 5°. As, however, it is advisable to use short chords with sharp curves, a 10° curve should be set out with chords not exceeding 25 feet in length, for which the tangential angle is 1° 15′.

TANGENTIAL ANGLE.
Fig. 36.

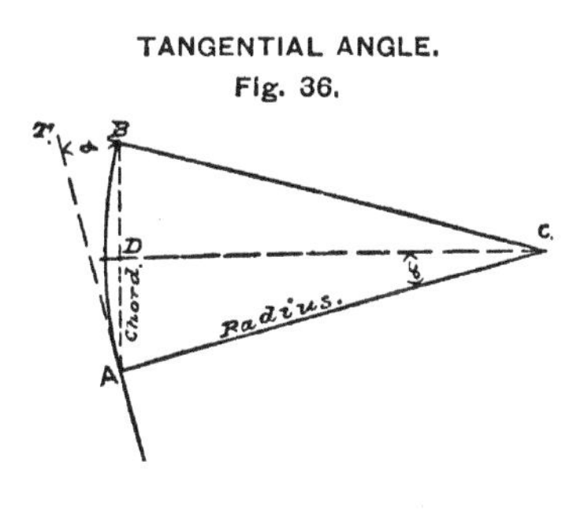

Adjustment of Earthwork in Cuttings and Embankments.—When the longitudinal section of the route, as finally settled, has been plotted, the levels and gradients of the proposed railway are drawn on it, so as to determine the amounts of the earthwork required for the cuttings and embankments, and the general nature and dimensions of the other works, such as culverts, bridges, viaducts, and tunnels. The rail-level is generally shown on the longitudinal section as the datum in accordance with which all the works have to be regulated; but except as regards headway, level crossings, and junctions, the formation level, which is reckoned as 2 feet below the top of the rails or rail-level, is the most important, as affecting the depth to which the cuttings have to be excavated, and the height to which the embankments have to be raised all along the line, and therefore should always be put on the working longitudinal section, which furnishes the principal indication of the extent of the earthwork. Under ordinary conditions, the levels of the line should be so adjusted that the excavation from the cuttings may just suffice to form the adjacent embankments (see Fig. 38, p. 113); for an excess of excavation at one part of the line, even if balanced by an excess of embankment at another part some distance off, would involve considerable delay in the disposal of the earthwork, in addition to the extra work of transport, as the intervening cuttings, embankments, and bridges would have to be carried out first, or might necessitate running the surplus excavation to spoil, and the making up of the deficient embankment with side excavation from the nearest cutting, entailing additional cost in earthwork and land occupied. Where public roads cross the line and level crossings are prohibited, the railway has to be carried sufficiently above or below the roads to provide an adequate headway for the bridges; or the roads must be raised or lowered adequately for the purpose, which involves an addition to the earthwork for forming the approaches. Pioneer lines are exempt from any such provisions; and even in long-settled parts of America, the railways pass on the level through large towns, and the only bridges required are for crossing rivers or traversing side ravines.

Where deep valleys and high ridges are encountered, the large amount of earthwork in the slopes of high embankments and deep cuttings so increases the cost, and such works become so difficult to maintain, that even with fairly good soil, it becomes cheaper to build viaducts and construct tunnels when the difference in level between the surface of the ground and formation level exceeds 60 feet; but when the excavations consist of clay or other strata liable to slips and disintegration by wet, it is economical to resort to tunnels and viaducts for smaller differences in level, in proportion to the treacherous nature of the strata. Tunnels, moreover, though very costly structures, save the cost of the land above them, of which only the underground rights have to be purchased, and also the construction of bridges for roads passing over them, which must be added to the cost of the earthwork in reckoning the relative cost of open cutting and tunnel. Viaducts, also, which are often combined with tunnels in carrying a direct railway, with easy gradients, across the lines of valleys and the intervening ridges of hills, as illustrated by parts of the London and Brighton Railway, save the cost of bridges over any roads passing along the valley, and a high bridge across the river or stream at the bottom of the valley, as well as the formation and maintenance of a high embankment.

Where a railway traverses a very open and exposed district, in latitudes subject to long and severe winters, snow-drifts are sure to collect in any confined depression; and in such localities, it is essential to avoid cuttings as far as possible, in spite of the necessity thereby involved of adopting steeper gradients for surmounting rising ground, and of forming approach embankments from side cuttings in the lower ground. This course had to be adopted on the section of the Canadian Pacific Railway crossing the comparatively flat, exposed prairies of Assiniboia and Alberta, where deep cuttings in the undulating plains would have inevitably resulted in the stoppage of the trains in winter by deep snow-drifts; and even the shallow cuttings which were formed for the line, had to be obliterated by cutting away the high ground for some distance on each side of the track.

Side Cuttings for Low Embankments.—As the cost of the earthwork for forming an embankment increases with the distance the material has to be conveyed, or, as it is termed, the *lead*, the low embankment needed for raising the railway above the flood-level in traversing an extensive, low, flat plain, can generally be more cheaply constructed by material excavated alongside the line, or side cutting as it is called, than by conveying the excavation for a long lead from the nearest cuttings. Moreover, on side-lying ground, where the line may appear to run approximately along the surface on the longitudinal section, the material excavated on one side from the upper part of the slope, serves to form the embankment on the other side on the lower part of the slope, in order to provide a level track for the railway (Fig. 37).

EARTHWORK ON SIDE-LYING GROUND.
Fig. 37.

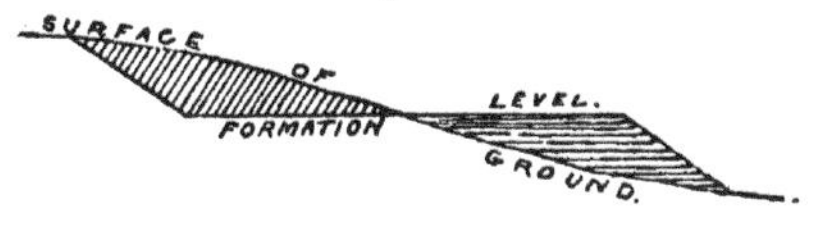

Formation Width, and Side Slopes of Cuttings and Embankments.—Before calculating the earthwork in the cuttings and embankments, it is essential to determine the bottom width of the cuttings, and the top width to which the embankments should be formed, as a base for the ballast of the permanent way. This formation width varies with the number of lines of way, and the gauge of the railway; but for the standard gauge of 4 feet 8½ inches, the average widths ordinarily assumed are 30 feet for a double line, and about 18 feet for a single line.

The side slopes, which more largely affect the amount of earthwork in deep cuttings and high embankments, than the formation width, have to be varied with the nature of the soils. These slopes are always denoted by the relation of the horizontal width to the vertical height taken as unity, the width increasing with the flatness of the slope. Experience has shown that in cuttings, rock will stand, according to its compactness and durability, at slopes of from about ½ up to ⅛ to 1, gravel 1 to 1, dry sand 1¼ to 1, and compact soil and well-drained clay at 1½ to

1; whilst slopes of 3 to 1, or even flatter, are necessary for wet clay, or clay intersected with thin layers of sand. In embankments, where the slopes are formed and consolidated by tipping, instead of the strata remaining in their natural compact state as at the sides of cuttings, the angle of repose of the materials is generally flatter, especially in the case of rock, which, in an embankment, should be given a minimum slope of 1 to 1, and preferably 1¼ to 1; whilst 1½ to 1 to 2 to 1 furnish suitable limits for the slopes of embankments formed with the better class of soft materials; but wet clay tipped from a height, especially during wet weather, will sometimes hardly stand at any slope.

Calculation of Earthwork.—The depth of the formation level in the cuttings, and its height in the embankments, from the surface, are generally figured at each chain along the longitudinal section, being the difference at each point between the level obtained by levelling on the ground along the centre line, and the formation level as calculated from the gradients. The central portion of the earthwork can be readily obtained, being merely the mean of the depths or heights between any two points in a cutting or embankment, multiplied by the length of the portion under consideration and the formation width. The contents, however, between the central rectangular block and the side slope on each side, with variable depths or heights, form a series of frustums of triangular pyramids, which must each be calculated. Accordingly, the earthwork is more readily and accurately calculated by tables like Bidder's,[1] where the contents of one foot in width of the central block for a chain in length, and the contents of the two frustums for side slopes of 1 to 1 and a chain in length, are given in cubic yards for every possible variation in the depths or heights, at the two ends of the chain length, of not less than one foot, between zero and 50 feet. After adding up the figures obtained from the table, in the column representing the contents of the several slices, 1 foot in width, of the central block for each chain in length of the cutting or embankment, and also the figures in the column containing the corresponding total contents of the side portions for a slope of 1 to 1, it is only necessary to multiply the total of the first column by the number of feet in width of the formation, and the total of the second column by the proportion of the horizontal width to the height of the slopes. The sum of these two results gives the total contents of the cutting or embankment, between the surface and formation level, in cubic yards. The table only gives the contents of a 1-foot slice of the central block, and of the side frustums, for the given depths or heights at the two ends, for a chain in length; but the calculations can, if necessary, be accelerated by taking any longer suitable lengths, and merely multiplying the figures obtained from the table by the length of each portion measured in chains, as illustrated by the simple instance of finding the contents of the cutting and embankment indicated in the longitudinal section, (Fig. 38), as given in a tabular form on page 114.

[1] "Tables showing the Contents of Excavations, Areas of Slopes, etc.," George P. Bidder.

Bidder's tables are compiled for the unit of a chain of 66 feet; but similar tables could readily be computed for lengths of 100 feet. Moreover, the tabular results are only given up to 50 feet of depth or height; but they can be extended to greater depths or heights by dividing the dimensions, when one or both of them exceed 50 feet, by such a number as will bring them within the limit of the table, and then multiplying the tabular figure representing the 1-foot slice of the central block by the said number, and the figure for the side portions by the square of the same number. Thus, for example, if the

RAILWAY CUTTING AND EMBANKMENT.

Fig. 38.—Longitudinal Section.

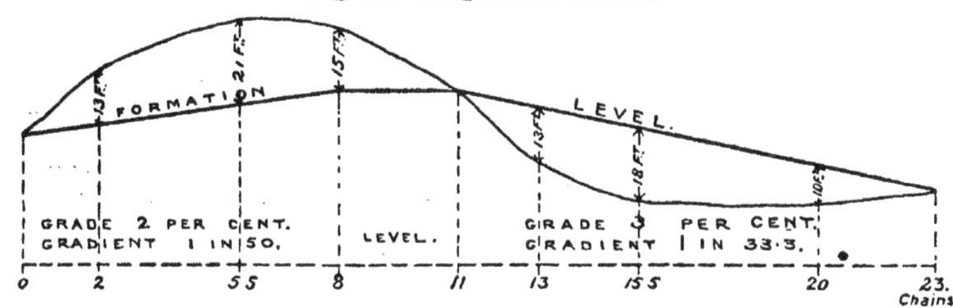

depths at the two ends of a portion of a railway cutting were 60 and 40 feet, the tabular figures for 30 and 20 feet are 61·1 and 1548, which multiplied respectively by 2 and 4 give 122·2 and 6192 as the proper tabular numbers in cubic yards for a cutting, one chain long, diminishing uniformly in depth from 60 feet to 40 feet. Moreover, if a portion of a cutting for a ship-canal increased regularly in depth from 90 feet to 120 feet, it is only necessary to multiply the tabular figures 85·6 and 3015, corresponding to depths of 30 and 40 feet, by 3 and 9 respectively, to obtain the required tabular numbers 256·8 and 27,135, which multiplied by the length of the cutting in chains, and by the bottom width of the canal in feet, in the first case, and the proportionate width of the side slopes to the height in the second case, would give the contents of this section of the canal cutting in cubic yards.

When the contents of the several cuttings and embankments have been calculated, by aid of the depths and heights to the formation level shown on the longitudinal section, they are inserted in figures on the section, over the portions of the earthwork to which they refer, so as to indicate the amount of excavation in each cutting, and the extent to which it can be disposed of in the adjacent embankments. Besides, however, the earthwork shown upon the working longitudinal section, there is the earthwork in cuttings and embankments for forming the railway on side-lying ground, as indicated in Fig. 37, p. 111, and also for lowering or raising roads for passing under or over the railway, which has to be calculated from the cross sections, and inserted on the longitudinal section at the respective points to which the several portions appertain. In order to make the excavations and embankments fairly

CUTTING.

Formation width, 30 feet. Slopes 1¼ to 1.

Length in chains.	Depths at ends in feet.	Tabular number. Cubic yards.	Contents, one foot wide, of central block. Cubic yards.	Tabular number. Cubic yards.	Contents of pyramidal portions at sides, 1 to 1 slopes. Cubic yards.	
2·0	{0, 13}	15·9	31·8	138	276	Cubic yards.
3·5	{13, 21}	41·6	145·6	720	2520	10,269 6,677
2·5	{21, 15}	44·0	110·0	799	1997	16,946
3·0	{15, 0}	18·3	54·9	183	549	
			342·3 30		5342 1¼	Total contents of cutting, say **16,950** cubic yards.
		Central block	10,269	Side portions	6677	

EMBANKMENT.

Formation width, 30 feet. Slopes 1¾ TO 1.

Length in chains.	Height at ends in feet.	Tabular number. Cubic yards.	Contents, one foot wide, of central block. Cubic yards.	Tabular number. Cubic yards.	Contents of pyramidal portions at sides, 1 to 1 slopes. Cubic yards.	
2·0	{0, 13}	15·9	31·8	138	276	Cubic yards.
2·5	{13, 18}	37·9	94·7	592	1480	9,510 7,378
4·5	{18, 10}	34·2	153·9	492	2214	16,888
3·0	{10, 0}	12·2	36·6	82	246	
			317·0 30		4216 1¾	Total contents of embankment, say 16,900 cubic yards.
			9510		7378	

balance one another in actual practice, the estimated contents of the embankments should be a little in excess of those of the cuttings, for it is impossible to pack the materials in the embankments as tight as they were naturally compacted together in the cuttings; the interstices in an embankment after consolidation, as compared with a cutting, having been found to vary from about 8 per cent. for loose materials, such as sand, gravel, or clay, up to 20 per cent. for blocks of chalk and rock.

When the earthwork has to be calculated from a parliamentary longitudinal section, which only gives the rail-level, it is necessary to add two feet to the scaled depths in the cuttings, and to deduct 2 feet from the heights of the embankments, in order to obtain the dimensions to the formation level. Moreover, when the excavation in a cutting traverses an upper layer of soft soil and an underlying stratum of rock, the side slopes have to be flat down to the rock, and steep through the hard stratum. In such a case, the line of the top of the rock above the formation level should be indicated on the longitudinal section; and the portion of the cutting below this line, being wholly in rock, can be calculated in the ordinary manner with suitable steep side slopes. The portion, however, of the cutting in soft soil above the rock, must be estimated separately, owing to its flatter side slopes; and as the width of the soft portion of the cutting at the surface of the rock varies with the height of the line of the top of the rock above the formation level, owing to the sloping sides of the rock cutting, the earthwork in the cutting through the soft soil above the rock must be calculated in short lengths, as in the case of a cutting of varying formation width, the contents of the 1-foot slice of the central block being multiplied by the average width of the bottom of the soft cutting, on the top of the rock, for each length.

Fencing.—After the requisite land has been acquired for a railway, the width of which varies with the formation width, the depth of the cuttings, and the flatness of the slopes, and before the excavations are commenced, it is necessary in settled districts to fence the strip of land along each side so far as may be required at once for the works, in order to prevent injury to persons or cattle straying on to the works. Oak post-and-rail fencing was generally used in the early days of railways; but wire fencing, with oak or iron straining-posts at intervals, is often now preferred as being cheaper. By planting a hedge of young thorns inside the rail fence, a good barrier is provided by the time the wooden fence begins to decay. A good and cheap protection for the line, used frequently in Ireland, consists in digging a ditch along the boundary on each side, making a mound along it on the inside with the excavated material, which is coated with the sods taken off the surface along the line of the ditch, and inserting strong round stakes on the top of the mound, leaning over towards the ditch, along which two or three lines of wire are stretched.

In undeveloped countries, fencing can be dispensed with; but where ranches are traversed, and also through agricultural districts where timber is abundant, rough wooden fencing serves as a barrier for the railway.

Provisions against Slips in Cuttings and Embankments.—The methods by which cuttings are excavated and embankments formed have been already described in Chapter IV.; but special precautions have to be taken to prevent slips in the slopes of cuttings and embankments, where the material consists of clay or other soils subject to be rendered soft and slippery when exposed to the action of wet. The two most important provisions for averting slips in treacherous material, in addition to an adequate flattening of the slopes, are efficient drainage to prevent wet getting through the ground to the slopes, and the protection of the surface of the slopes from the weather. This latter provision is always made in cuttings and embankments of soft soil, by removing the turf or soil from the surface over the cutting and under the embankment before commencing these works, and spreading it on the slopes immediately after their completion, sowing grass seeds, and encouraging the growth of plants, so as to protect the slopes from rain and frost, and from the formation of cracks under the heat of the sun, and to bind the ground together by means of their roots.

Slips in clay cuttings may, for the most part, be traced to two causes. The surface layer of clay alongside the cutting is liable to crack in dry, hot weather, forming sometimes a longitudinal fissure a short distance back from the top of the slope; and when water fills this fissure in wet weather, it exerts a hydrostatic pressure on the block of clay near the edge of the slope, tending to push the block towards the cutting, in which direction it is unsupported. This action, moreover, is intensified by the expansion of the water in being turned into ice during a frost; and when the pressure at the back has detached the block from the adjacent stratum, the water percolates through the fracture along the base of the block, and, rendering the surfaces of cleavage slippery, leads eventually to the fall of the block of clay into the cutting, giving the slope somewhat the form shown in Fig. 39. Sometimes a stratum of

SLIP IN RAILWAY CUTTING.
Fig. 39.

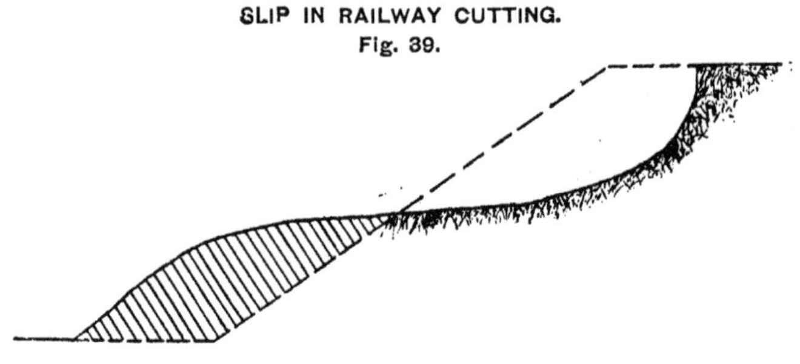

clay is intersected by thin layers of sand; and when water percolates from above into one of these seams of sand, it flows through the sand and finds an outlet at the face of the slope of the cutting. The sand is thus gradually carried away with the water; and the upper layer of clay, becoming unsupported by the removal of the seam of sand, settles down, and, becoming detached at the back, slides along the moist surface of the underlying layer of clay into the cutting. The slopes of cuttings are specially subject to this latter form of slip when the strata dip towards the cutting, as the washing out of the sand and the sliding forward of the detached upper layer are facilitated in proportion to the inclination of the dip.

Efficient drainage of the land alongside a cutting liable to slips is necessary, in order to prevent water getting near the slopes. When a cutting traverses very treacherous ground, it has occasionally been found necessary to weight and protect the slopes with dry stone pitching, laid upon a thin bed of small stones or other hard pervious layer, to serve as a rubble drain for conveying any water from the slope to the bottom of the cutting, where it is carried across the base of the pitching, by outlets built at intervals, into the drain at the side of the cutting. This is, however, a costly expedient; and a similar protection may to some extent be obtained by using turf instead of the pitching, overlying a thicker bed of the permeable layer. Often rubble drains, consisting of trenches filled with rubble stone, are formed at intervals down the face of slopes, with branch drains running into them at an angle, with the object of draining the slopes. This system of draining the slopes is very commonly resorted to when a slip has occurred in a cutting; and the hollow left by the fallen material is filled up to the line of the slope by hard, loose material, such as chalk, rubble stone, or slag, to keep up and protect the steep face near the top of the slope, without unduly weighting the lower part of the slope. Where strata of very variable and slippery nature are encountered, interspersed with thin beds of sand, in a district of heavy rainfall, it may prove advisable to resort to tunnelling in place of a cutting, even when depths of only 40 to 30 feet are reached.

Slips of embankments are mainly due to the disintegration of the bottom portion by the action of water percolating through the mass, or to the want of cohesion between the central portion and the side slopes when side tipping has been resorted to. On side-lying ground, moreover, the natural slope at the base facilitates slips down the slope under the action of water; and sometimes the mere weight of the embankment occasions a landslip of the stratum on which it rests, upon a disconnected, sloping, slippery stratum below.

To prevent water remaining at the base of an embankment, a longitudinal rubble drain has to be formed under the centre of the embankment, with cross rubble drains at intervals leading the water with a fall to the ditches at the side; and with exceptionally bad material, a layer of rubble stone has in addition been sometimes spread over the whole base of the embankment. Moreover, where the only material available is clay or other treacherous material, the tipping should not be carried on during wet weather; and the embankment should be formed throughout its width by end tipping, all side tipping being prohibited; whilst the best materials should be reserved for facing the slopes, and ashes may with advantage be incorporated with the treacherous material. Forming the embankment with a series of consolidated layers would be too costly a precaution to be ordinarily adopted; but where a very high embankment has to be constructed, it may advantageously be formed in successive stages from 15 to 20 feet in height. The liability of an embankment to slide sideways down a somewhat steep slope on side-lying ground, can be prevented by forming horizontal benches in the slope underneath the embankment, in combination with efficient drainage and deep rubble trenches down to a firm stratum; and an embankment should not be placed upon a site where there is any prospect of its weight leading to a landslip.

Remarks on Laying out a Railway.—The extent to which

sharp curves and steep gradients enable heavy works to be dispensed with, is sufficiently exemplified by the fact that it has been possible to carry a railway from Bombay over the Ghats, and across India to Calcutta, with only one or two very short tunnels, and that the Canadian Pacific Railway has been carried across the Rockies and the Selkirks, and through the Albert, Fraser, and other cañons, with a less length of tunnelling than is found on most of the main lines of England. Moreover, the construction of the western lines of North America has been much accelerated by a very extensive use of timber trestle viaducts at the outset, instead of forming embankments and erecting steel bridges with masonry piers, leaving these permanent works to be carried out towards the close of the life of the timber structures, when the facilities afforded by the railway enable the earthwork for the embankments to be readily tipped from the wooden trestles, and the materials for the bridges to be conveyed to the site, which would have involved very serious delays and difficulties during the first formation of the line, owing to the absence of any means of conveyance to most parts of the works, except along the completed line.

Culverts.—Where any ditch, small watercourse, or stream is crossed by a railway embankment, a provision has to be made for the unimpeded passage of the water conveyed by it, underneath the embankment, by some form of pipe or culvert. Glazed earthenware drain-pipes, laid in a trench in the solid ground and jointed with puddle or cement, serve for carrying the drainage water of ditches or small water-courses under low embankments; and these pipes, in common with all other passages for water, must be of sufficient capacity to convey away the maximum flow of water that can come to them, otherwise, after an unusually heavy rainfall, the water would accumulate on the upper side of the embankment, and might imperil its stability. Cast-iron pipes are used where a considerable weight has to be supported; and wooden troughs are employed where timber is abundant, and other materials costly. Where stone is at hand, a cheap form of culvert, requiring no skilled labour, consists of a bottom bed of flat slabs, with rubble-masonry side walls supporting a roof of flat stone slabs, thereby forming a rectangular masonry culvert of dimensions proportioned to the flow and the height of the embankment above it.

CULVERT UNDER EMBANKMENT.
Fig. 40.

The most ordinary form of culvert under a railway embankment is built of brickwork or masonry, arched over at the top. Cylindrical or barrel-shaped brick culverts are commonly adopted for small flows of water under embankments of some height, with two or more rings of brickwork, each 4½ inches thick, according to the load to be borne. Larger culverts are built as shown in Fig 40, with an invert at the bottom, resting on a flat foundation of brickwork, masonry, or concrete; whilst small wing-walls at the sides of the mouth of the culvert at each end, keep back the earthwork at the toe of the embankment from impeding the outlet. Culverts for large streams approximate to small bridges, and are constructed with a strong arch overhead where the embankment is high, so as to support the incumbent weight of earth in addition to the moving load of a passing train; and usually an invert is built at the base to strengthen the side walls against the pressure of the earthwork behind them, and to secure the bottom, at the same time, from scour.

The foundations of culverts under high embankments must be made very stable, as any unequal settlement of the culvert under the greater weight of the central part of the embankment, than under the diminishing slopes towards the ends, would involve the fracture of the culvert; and the escape of water through the fracture into the heart of the embankment, would soften the adjacent mass, exposing the embankment to settlement and slips.

CHAPTER VIII.

ARCHED BRIDGES.

Works of construction for railways and roads—Types of bridges: classification under five types—Dead and moving loads on bridges: definitions, estimated moving loads—Arched bridges: materials used in construction; stresses, their nature, their determination—Brickwork and masonry arches for small spans, construction, ordinary types, their dimensions—Skew arches, increased span, arrangement of courses—Large masonry bridges: instances—The Adolph Bridge at Luxemburg, and others—Large metal arched bridges: examples; description of the St. Louis Bridge, the Douro and Garabit viaducts, and other arched bridges; provisions for changes of temperature—Erection of metal arches, exemplified by St. Louis, Douro, Niagara Falls and Zambesi bridges.

IN addition to the cuttings and embankments, with their culverts and drains, which occupy such a prominent place in the formation of railways, various other very important works, involving far greater skill in their design and execution, have to be carried out for enabling railways and roads to surmount the physical obstacles which are frequently encountered in their course. Thus bridges have to be constructed for carrying railways and roads over and under roads and railways, and over rivers and canals, exhibiting considerable variety in form to suit very varied conditions; viaducts have to be erected for enabling railways to cross deep valleys at a high level, in order to secure easy gradients; and tunnels have to be formed to take railways through projecting spurs, high ridges, or across mountain chains, and occasionally for passing under rivers and beneath large cities. These works of construction, termed "works of art" by continental engineers, will be dealt with successively in the following six chapters.

Types of Bridges.—Though more variety has been exhibited in the designs of bridges than in any other branch of railway construction, owing to considerable differences in the spans and heights required, and in the conditions under which the erection of bridges has to be carried out, there are only four or five distinct types under which all bridges may be grouped. These types are arched, suspension, girder, continuous girder, and cantilever bridges. The general principles, forms, and construction of each of these types will be briefly considered in succession; and, besides some ordinary examples, notable instances of bridges of large span of each type will be given in illustration. In arched bridges all the parts are in compression, and in suspension bridges all the parts are in tension; whereas, in the other three types, some parts are subject to tension, and others to compression.

Dead and Moving Loads on Bridges.—Every bridge has to support two distinct kinds of loads, namely, the *dead load*, consisting of its own weight and the railway or roadway which it carries, and the *moving load*, consisting of the trains or vehicles and pedestrians which pass over it. The dead load is permanent, and imposes definite fixed strains on the bridge; whereas the moving load is variable, has only from time to time to be borne by the bridge, and may in rapid succession be evenly or unevenly distributed over the bridge, as in the case of trains passing quickly across it. The dead load, however, increases approximately as the square of the span, owing to the additional weight of material per lineal foot that must be put into a bridge, in order to enable it to resist the augmented strains due to an increase in the span; whereas the moving load only increases proportionately with the length of the bridge; and in the case of a railway bridge for a double line, it merely does this up to the length of the longest train, or the weight of two of the heaviest trains running on the line, placed in the most unfavourable position. Accordingly, in bridges of large span, the weight and importance of the dead load become much greater than those of the moving load as the span is augmented; and the possible extension of the span of bridges eventually reaches a limit, owing to the impracticability of designing bridges capable of bearing their own weight beyond a certain span according to the type selected, rather than on account of the strains due to the relatively small moving load which bridges are constructed to accommodate. These considerations indicate that those types of bridges should be best adapted for long spans, which concentrate their greatest stresses and greatest weight of metal near their piers, like cantilevers, rather than those which are liable to be somewhat heavier in proportion to their length near the centre of their span than towards their extremities, as in the case of ordinary girders.

The moving load for railway bridges of moderate span is generally reckoned at from 1 to 1½ tons per lineal foot per line of way, according to the weight of the locomotives employed on the line. For very short spans, the moving load with heavy engines might reach a maximum of about 3 tons per lineal foot; whereas a moving load of 1 ton per foot would suffice, under similar conditions, for spans of 300 feet. On roadway bridges, the maximum possible load is often assumed to be a densely-packed crowd of people all over, estimated to weigh 70 to 80 lbs. per square foot, which would inevitably be a slowly moving load.

Aqueducts for canals are the one form of bridge in which the ordinary moving load of the traffic is replaced by the permanent load of water contained in their troughs; for the canal barges passing along them do not impose any additional weight on the structure during their transit, as they displace a volume of water equivalent to their own weight.

Wind-pressure, which constitutes another form of moving load, acting sideways, assumes an enhanced importance in bridges of large span, owing to the large surfaces they present, the great leverage exerted by the wind at a distance from the piers, and the additional weight of metal which has to be inserted in the bridge to enable it to withstand this pressure during a strong gale.

ARCHED BRIDGES.

Materials for Construction.—Next to wooden beams and pile and plank bridges, the arch is one of the earliest forms adopted for bridges over rivers. Some of the early arches were built of wood; and timber was used for arched bridges of fairly large span in the eighteenth century, before the introduction of cast iron. Masonry, brickwork, and concrete can only be utilized for the superstructure of bridges by forming them into arches, owing to their inability to withstand tensile stresses; and cast iron was employed for the early railway bridges of larger span, in the form of arched ribs, on account of its great strength when only subjected to compressive stresses. Arches of brickwork or masonry are very commonly used for railway and road bridges of moderate span, where the available headway is adequate, on account of their simplicity, durability, and moderate cost when these materials are readily procured; while concrete, plain or reinforced, is used for spans of considerable magnitude, but bridges of very large span, in which the use of cast iron is for various reasons inadmissible, are now almost always constructed of riveted mild steel, with cast steel in subordinate details.

Stresses in Arched Bridges.—The wholly compressive stresses in an arch consist of a horizontal pressure at the crown, combined with vertical pressures due to the weight of the bridge and the moving load, increasing, in the case of the fixed load, from the centre where it is zero, to the abutments, at each of which half the total weight of the bridge and moving load, and also the horizontal pressure, have to be supported. The resultant pressures of the horizontal pressure and the increasing vertical pressures represented by the loads comprised between the crown and successive points of the half-arch, are gradually augmented, and deflected downwards by degrees along the arch, from the horizontal position at the crown to a direction approximately at right angles to the springing at the abutments, where the thrust attaining its maximum has to be borne directly by the abutments. When once the horizontal pressure has been ascertained, or the direction of the thrust at the springing has been determined, it is only necessary to calculate the weights of successive sections of the bridge, and to add to them the corresponding moving loads, in order to obtain both the intensity and direction of the successive resultant pressures in the respective sections of the bridge between the crown of the arch and the springing. Thus, if the horizontal pressure has been ascertained, and is represented by the horizontal line AC in Fig. 41, drawn to a definite scale, and if the weights of the successive sections into which the bridge has been divided for purposes of calculation, with their corresponding moving loads, are indicated by the vertical lengths **AB**, **BD**, **DE**, **EF**, and **FS**, drawn to the same scale, **AS** being the total load on the half-arch, then the lines **CB**, **CD**, **CE**, **CF**, and **CS** represent the resultant pressures at the several points **B**, **D**, **E**, **F**, and **S** of the arch, both in magnitude and direction. If, on the contrary, the direction of the thrust at the springing **S**, namely, **CS** in Fig. 41, has been determined, then, since the vertical line **AS** represents the total load on the half-arch to the given scale, the horizontal line **AC**, drawn from **A** and cut at **C** by the line **SC** drawn in the direction of the line of thrust at the springing, corresponds on the same scale to the horizontal pressure at the crown of the arch; and the resultant pressures at the several points of the arch are indicated by the lines **CB**, **CD**, etc., as before, in magnitude and direction.

PRESSURES ON ARCH.
Fig. 41.

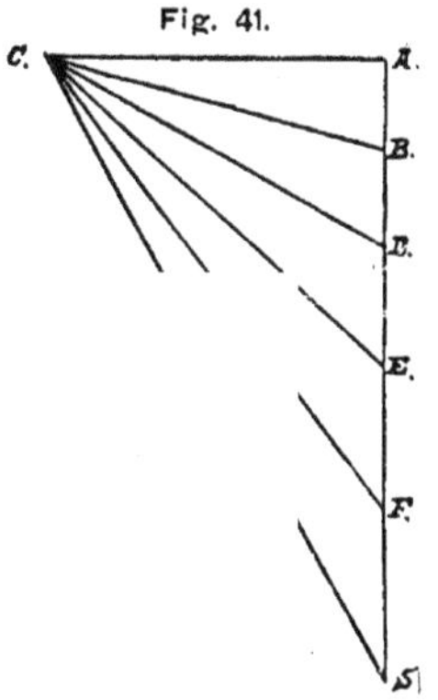

The horizontal pressure at the crown of an arch is due to the wedge-like effect of the load near the centre tending to depress the crown, and is counteracted by the resistance presented by the abutments at the springings, and transmitted along the arch to the crown; but the general value of this pressure in a simple arch, and its variation with the ratio of the rise of the arch to the span, will be best indicated by considering briefly the horizontal tension at the centre of a suspended chain, which is the precise converse to the arch. If a simple chain hung from **A** and **B** supports a load W at **C** (Fig. 42, *a*, p. 124), the tensile stress along **AC** and **BC** is $\frac{W}{2} \times \frac{1}{\sin \alpha}$, and the horizontal tension at **C** is $\frac{W}{2} \times \frac{1}{\tan \alpha}$; and since **DC**, the dip of the chain, or D, divided by the half-span, $\frac{1}{2}$S, is equal to $\tan \alpha$, the horizontal tension at **C**, given in terms of the span and the dip, becomes $\frac{WS}{4D}$. When, however, the load W is distributed evenly in a horizontal line along the chain **ACB**, the suspended chain assumes the form of a parabola (Fig. 42, *b*); and the horizontal tension at **C** becomes $\frac{WS}{8D}$, or if w is the weight per unit of horizontal length $\frac{wS^2}{8D}$, and is the same as the bending moment in the centre of a simple beam having a similar span and depth. Now, inverting the equilibrated suspension chain, as in Fig. 42, *c*, the simple arch is also in equilibrium under a uniformly distributed load, if it corresponds in its parabolic form to the curve of the bending moments; and its horizontal thrust at the crown is likewise $\frac{wS^2}{8R}$, where R is the rise of the arch, which is borne either by abutments at **A** and **B**, or by a horizontal tie **AB** in Fig. 42, *c*, whose tensile stress is equivalent throughout to the horizontal thrust, and which forms the bottom member of the bow-string girder **ACB**, having an arched top member wholly in compression. The horizontal thrust can be reduced, and consequently the weight of an arch near its crown can be diminished, by increasing the proportion of the rise of the arch to its span, just as a beam or girder of any given span is strengthened by increasing its depth.

A simple arch, unlike a suspension chain, is unable to adapt itself to a varying load; and therefore an arch has to be stiffened to maintain its equilibrium under a moving load, or even wind-pressure, causing

DIAGRAMS OF LOADED SUSPENSION CABLES AND ARCH.

Fig. 42.

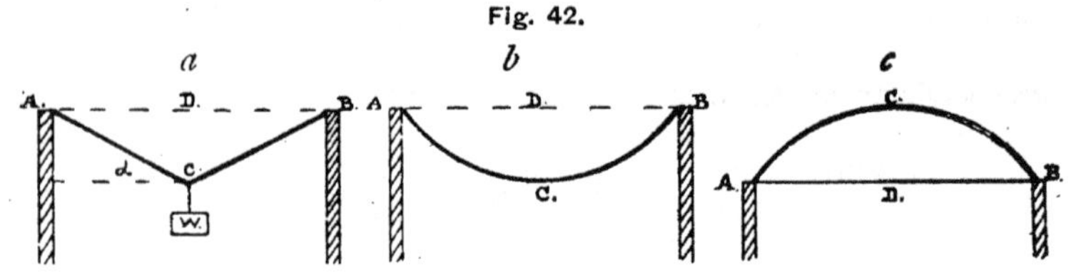

variations in the distribution, amount, and direction of the stresses. Moreover, even in suspension bridges, the fixed load is not quite uniformly distributed throughout the span; for, owing to the increasing curvature of the chains towards the piers, the greater strength which has to be provided in the same direction, and the greater length of the suspension rods or cables towards the ends of the span, the weight per unit of length increases to some extent from the centre to the ends. This is still more the case with a properly stiffened arch, in which the arch itself has to be materially strengthened in passing from the point of least pressure at the crown to the place of maximum pressure at the springings; and the spandrils which weight the haunches of the arch, and support the roadway above, have to be increased in height towards the abutments to make up for the depression of the arch. Accordingly, the value $\frac{WS}{8D}$ obtained above for the horizontal tensile stress at the centre of a suspension bridge, though commonly adopted in practice, is in reality somewhat in excess of the actual value; and the similar expression $\frac{wS^2}{8R}$ must only be regarded as an approximation to the horizontal thrust at the crown of an arch, which is reduced in proportion as the load w per unit of span is increased between the crown and the springings, the effect of which is to increase the vertical pressures on the abutments and adjacent portions of the arch, and correspondingly to relieve the crown.

The form of arch in which the horizontal pressure at the crown apparently is reduced to its minimum value, namely, where the rise is equal to half the span, or a semicircular arch, though often adopted for viaducts, does not in reality reduce this pressure to the extent which might be supposed at first sight; for a horizontal pressure at the crown could not possibly be resolved into a vertical thrust acting at right angles to a horizontal springing. Consequently, the lower portions of a semicircular arch in reality constitute part of the abutments or piers; the actual efficient rise is not as large as it seems; and a segmental arch, with a large rise, might equally well be adopted. Similarly, a suspension bridge cannot advantageously be given a considerable dip in relation to its span, for the greater the dip the greater are the alterations in the curvature of the chains under a moving load; and, moreover, a great dip involves correspondingly high towers for supporting the chains. Accordingly, in these instances, as indeed in many others, the proportions of rise and dip, or depth, to span, which appear to offer the most favourable economical conditions from a particular point of view, are not practically attainable.

Arched Bridges of Small Span.—Bridges of small span over or under a railway are generally built, in arched form, of brickwork or masonry where these materials are readily procured, when the headway is adequate, and the crossing is not very much on the skew. On completing the abutments or piers on each side, a timber centering is erected, consisting of uprights supporting a row of ribs formed at the top to the curve of the arch, across which a flooring of planks is fastened, flush on the top with the underside of the proposed arch. On this flooring, courses of brick on edge or stone are laid in mortar at

BRIDGE UNDER RAILWAY.

Fig. 43.

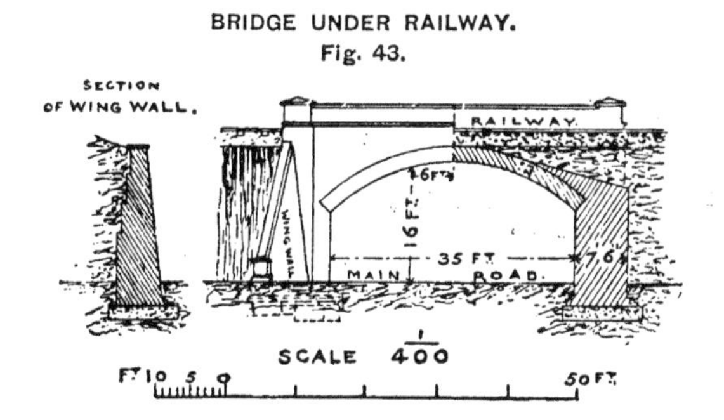

right angles to the faces of the arch, commencing on each side at the springing and meeting in the centre, forming eventually a series of rings of 4½-inch brickwork, or one or more rings of masonry according to the span of the arch. As soon as the arch is completed, the supporting centering is slightly lowered by easing some wedges or letting out sand from boxes on which the uprights rest, in order that the arch may settle down to its proper bearings; and after allowing sufficient time for the mortar to set thoroughly, the centering is removed. Additional brickwork or masonry is built on the haunches of the arch towards the abutments, as shown on the sectional elevations of the bridges in Figs. 43 and 44, in order to provide for the increasing thrust, and to direct the curve of pressures in the arch suitably towards the springing, which is further aided by the weight of the filling between the face walls of the bridge for the formation of the roadway or railway passing over it.

The two most common types of brickwork railway bridges are shown in Figs. 43 and 44, in the first of which the bridge carries a railway, running on an embankment, over a main road, and in the second a roadway is taken over a railway which is in a cutting. Wing-walls in the first case, projecting from each end of the abutments to the

BRIDGE ACROSS RAILWAY CUTTING.
Fig. 44.

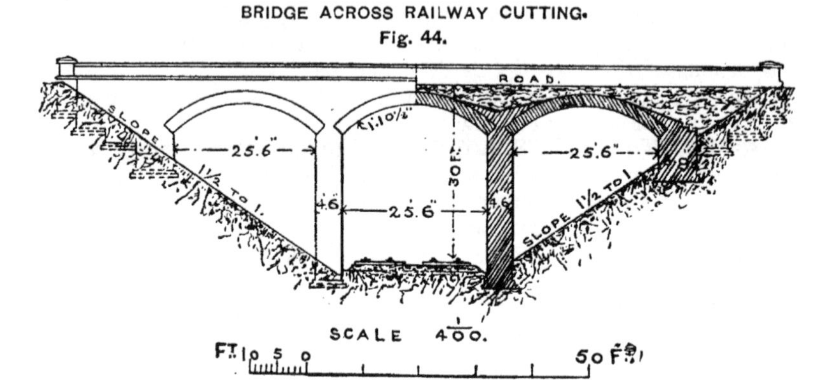

toe of the slopes of the embankment on each side, keep back the earthwork, like a retaining wall, from encroaching on the road; and side arches in the second case carry the bridge over the slopes of the cutting, the abutments of which, being below the natural surface of the ground, can be founded considerably above the bottom of the cutting. As, in the latter bridge, the horizontal thrust of the central arch on each intermediate pier is counterbalanced by the similar thrust of each side arch, these piers have only to be made adequately thick to bear the vertical load due to half the arch on each side of them. When a railway traverses a cutting of sound rock, a very cheap form of bridge can be built by throwing an arch across the cutting, abutting against the rock on each side, and thereby dispensing with abutments and piers, as shown in Fig. 45.

BRIDGE IN ROCK CUTTING.
Fig. 45.

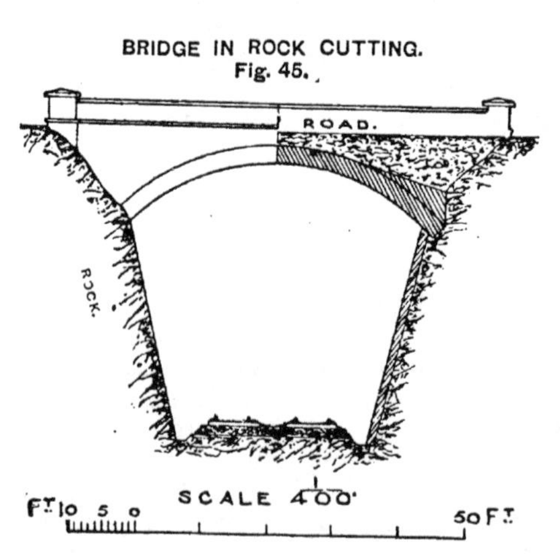

The following rules have been framed for finding the principal dimensions for arched bridges of brickwork or masonry, having spans of between 25 and 70 feet, in relation to the span, namely, rise one-fifth of the span, thickness of arch one-eighteenth, of abutments one-fifth to one-fourth, and of piers one-sixth to one-seventh of the span.[1] The regulation dimensions for railway bridges over roads in Great Britain are, 35 feet span for main (old turnpike) roads, with a headway of 16 feet for the central 12 feet (Fig. 43, p. 125); 25 feet span for ordinary public roads, with a headway of 15 feet for the central 10 feet; and 12 feet span for private roads, with a headway of 14 feet for the central 9 feet. The arch is generally made a segment of a circular arc; for an elliptical form, though affording more headway by being flatter towards the centre, necessitates a thicker arch to provide a similar strength. The minimum span for a bridge over a railway of ordinary gauge, is 24 feet for a double line and 13 feet for a single line; and a headway of 14½ feet above the rails for all structures is usually sufficient; but on the Canadian railways, a headway of 7 feet above the top of the cars is required to allow the brakesmen to pass over them in safety.

Skew Arches of Masonry or Brickwork.—When the crossing which has to be spanned by a bridge is not at right angles, the bridge has to be built on the skew; and the amount of skew increases in

SKEW ARCH, SHOWING LINES OF COURSES.
Fig. 46.

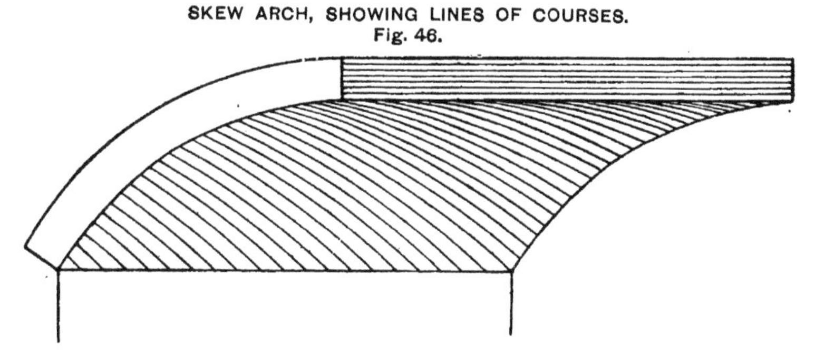

proportion to the divergence of the angle of the crossing from a right angle. The span of the bridge is increased by crossing on the skew, being equal to the span at right angles divided by the sine of the angle of skew (or the angle which the line of the roadway of the bridge makes with the abutments), so that with the large skew of 30°, whose sine is 0·5, the span would be doubled. In arched bridges, moreover, of masonry or brickwork, not only is the span increased, but the courses, which still have to be laid in lines at right angles to the faces of the arch, no longer run parallel to the springing on each side, but are laid at an angle to it in proportion to the skew, and, changing their level in traversing the centering, assume the form of a spiral, instead of a straight line as in a square arch (Fig. 46). The lines of the courses of a skew arch can be laid down on the centering, by bending a flexible

[1] "Pocket-book of Useful Formulæ and Memoranda for Engineers," 23rd Edition, Sir Guilford L. Molesworth, p. 108.

lath over the centering at right angles to the faces, and drawing a series of lines parallel to the lath, the thickness of the courses apart; or by making a flexible template exactly corresponding to the curved floor of the centering, laying it flat down, drawing on it a number of parallel lines at right angles to the sides representing the faces of the bridge, and the width of the courses apart, and then transferring these lines to the floor of the centering. Sometimes in brick arches, when the skew is slight, the courses are laid parallel to the springing, and the end bricks are left to project or are cut square to the face; but in ordinary cases, the bricks are cut to the requisite angle at the springing. In masonry arches, however, the springing stones may with advantage be so dressed with wedge-shaped projections, as to let two or more courses of stone in the arch abut square against them; and the same arrangement can be applied to brick arches by adopting springings of masonry.

Large Masonry Bridges.—Though masonry and brickwork cease to be suitable where bridges of really large span, exceeding 300 feet, have to be built, owing to the great weight of these materials in comparison with their strength, and the large, heavy centerings needed for their erection, some fair-sized, arched masonry bridges were erected before wrought iron had been employed in compression, or steel was available. For instance, London Bridge, built of granite in 1824–31, has a central arch with a span of 152 feet, a rise of $29\frac{1}{2}$ feet, and a thickness at the

LARGE MASONRY ARCHED BRIDGE.

Fig. 47.—Antoinette Bridge, France.

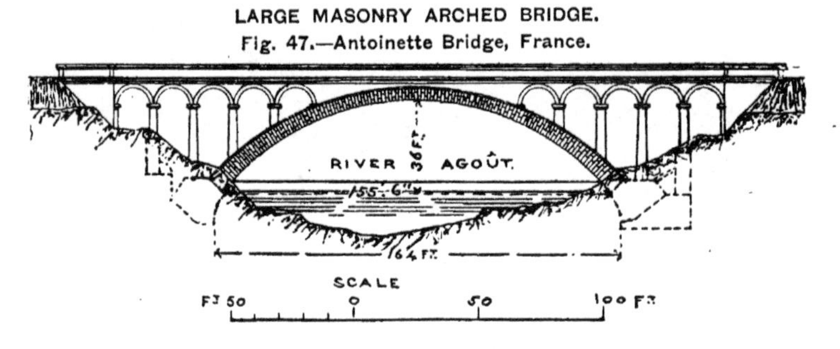

crown of $4\frac{3}{4}$ feet; and these dimensions in the concrete Alma Bridge, over the Seine at Paris, are $141\frac{1}{2}$ feet, 28 feet, and 4 feet 11 inches, and in the single masonry arch of the Grosvenor Bridge, erected over the Dee at Chester in 1827–32, 200 feet, 42 feet, and 4 feet respectively, with a thickness at the springings of 6 feet. Several large masonry bridges have been erected somewhat recently in France, carrying railways across moderately deep, narrow river valleys with a single arch, where stone was readily procured, and the banks consist of firm rock on each side, but where intermediate piers would have been difficult to found, and would have been obstructions in rivers subject to rapid and high floods.[1] One of these bridges, erected in 1882–4, crossing the Agoût near Vielmur, is shown in Fig. 47, abutting against a rocky stratum on each bank. Its

[1] *Annales des Ponts et Chaussées*, 1886 (2), p. 409, and plates 37 and 38.

arch has a radius of $101\frac{3}{4}$ feet, a span of $155\frac{1}{2}$ feet, a rise of 36 feet, and a thickness of 5 feet at the crown, and $7\frac{1}{2}$ feet at the springings; and the bridge, which has a total length of 293 feet, cost £8900. The centering used for the erection of another of these bridges in 1882–4, across the Ariège near the village of Castelet, is given in Fig. 48.[1] This granite

CENTERING FOR ERECTION OF LARGE MASONRY ARCH.

Fig. 48.—Castelet Bridge, France.

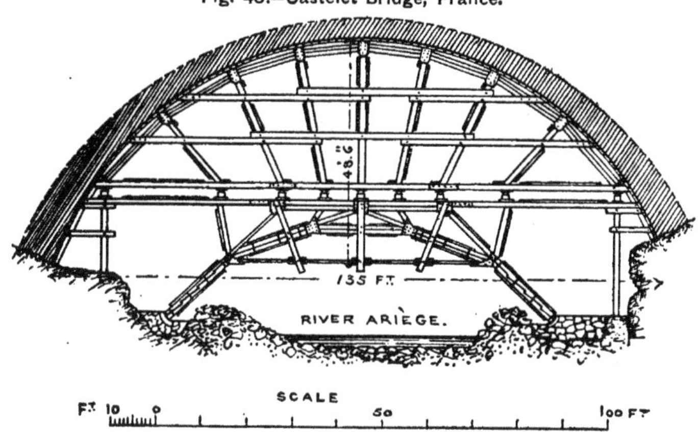

bridge, springing from a solid stratum of mica-schist, has a length of 218 feet; its arch has a radius of 73 feet, a span of 135 feet, a rise of 46 feet, and a thickness of 4 feet at the crown, and $8\frac{1}{4}$ feet, on the average, at the springings; and the cost of the bridge was £8200. Two of the largest of these bridges, the Gour-Noir across the Vezère[2] and the Lavaur across the Agoût, have each a single arch with spans of 197 and 202 feet, and their costs amounting to £9236 and £19,230 respectively. The Adolph Bridge at Luxemburg, span 277·7 feet, rise 53·15 feet, 138 feet above river level, $4\frac{3}{4}$ feet thick at crown, and 7 feet at springing, formed by two arches of hard local stone side by side, the intervening space being spanned by armoured concrete, 55 feet wide between parapets, the largest masonry arch existing, was completed in 1903.

To ensure the stability of a brickwork or masonry arch, the bridge must be so designed that the line of the centre of pressures shall keep within the middle third of the arch under any possible distribution of the moving load; for if any portion of this line passes outside the middle third, the arch is subjected to tension at the side furthest removed from the line of resultant pressures at that part, which the mortar joints are not adapted to sustain. The distribution of the pressure at any part of an arch, in relation to the position of the centre of pressures, is indicated approximately by the diagrams (Fig. 49) of a portion of an arch **ABCD**,

[1] *Annales des Ponts et Chaussées*, 1886 (2), plate 36.

[2] "Exposition Universelle à Paris en 1889: Notices sur les Modèles et Dessins relatifs aux Travaux des Ponts et Chaussées," p. 650.

divided longitudinally into three bands of equal width by the two dotted lines **EF, GH.** When the centre of pressures reaches the edge of the central band, coinciding for example with the upper dotted line **EF** at **P** in the first diagram, the distribution of pressure on the section **IK** of the arch is represented by the shaded portion **KLM**, the pressure reaching a maximum at the upper surface of the arch at **I**, and diminishing from thence to zero at the under surface at **K.** If, however, the centre of pressures passes outside the middle band **EF, GH**, in any part of the arch, as, for instance, at **P** between **AB** and **EF** in the second diagram,

DISTRIBUTION OF PRESSURES IN ARCH.
Fig. 49.

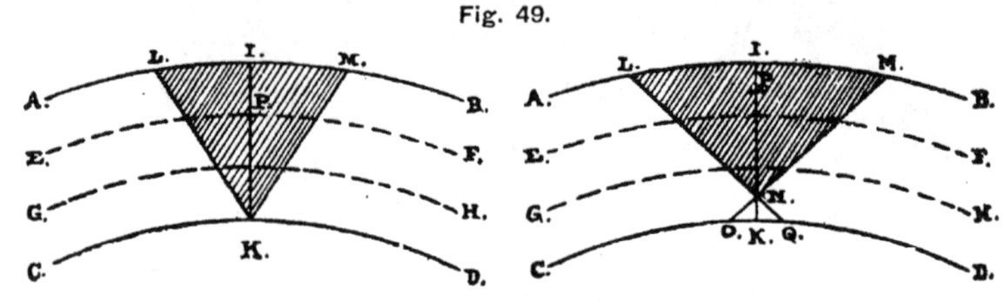

the pressure is intensified at **I**, and becomes zero within the arch at **N**; and tension occurs below **N**, amounting to **OQ** at the under surface at **K.** The occurrence of the conditions indicated in the second diagram of Fig. 49, must be prevented by so adjusting the permanent loads of the bridge in the design, and increasing, if necessary, the thickness of the arch, that the line of resultant pressures due to the dead load shall fall sufficiently within the central band, throughout the arch, as not to be forced out of it by the utmost practicable unequal distribution of the moving load.

Metal Arched Bridges of Large Span.—The capacity of iron and steel to withstand tension as well as compression, thereby dispensing with any need of keeping the line of the centre of pressures within the middle third of the arch, and their lightness in relation to their strength as compared with masonry, render them the most suitable materials for arched bridges of large span.

Sunderland Bridge across the Wear, having an arch of 236 feet span and 34 feet rise, completed in 1796, and Southwark Bridge across the Thames, with a central arch of 240 feet span and 24 feet rise, completed in 1819, are instances of large cast-iron arched bridges erected long before the adoption of wrought iron and steel for such structures.

The Victoria Bridge, carrying two railways across the Thames at Pimlico, marks a transitional period; for the four arches of the original bridge, built in 1859–60, having each a span of 175 feet and a rise of 17½ feet, were made of wrought iron; though subsequently a cast-iron arched bridge was erected from the designs of the same engineer, with a span of 200 feet and a rise of 20 feet, for carrying the Coalbrookdale Railway across the Severn; whilst the widening of the Victoria Bridge in 1865–66 was again effected by wrought-iron arches.

The St. Louis Bridge, built in 1867–74, conveying a railway, and a roadway overhead, across the Mississippi, with a central arch of 520 feet span and a rise of 47 feet, and two side arches of 502 feet span with a rise of $43\frac{2}{3}$ feet (Fig. 50, p. 132), was the first bridge of large span constructed of steel; and the safe stresses on the metal in this bridge were estimated at $12\frac{3}{4}$ tons on the square inch in compression, and 10 tons in tension.[1] Each span of the bridge is borne by four arched ribs composed of two steel tubes, one above the other, 12 feet apart centre to centre and connected by lattice work, each tube having an external diameter of $1\frac{1}{2}$ feet, and increasing in thickness from $1\frac{1}{6}$ inches at the crown to $2\frac{1}{2}$ inches at the springings. The cost of this bridge, including the approaches, was £1,361,800.

The Washington Bridge, over the Harlem River at New York, built in 1886–89, is another instance of a large steel arched bridge, having two spans of 510 feet with a rise of $91\frac{2}{3}$ feet. Each arch is composed of six ribs, with an effective depth of 12 feet; and each rib is formed by thirty-four segments.[2]

A crescent-shaped form was adopted for the wrought-iron arches of large span in the Oporto and Garabit viaducts, across the rivers Douro and Truyère, where the railway in each case passes at a considerable height above the river, and the arch springs from a low masonry abutment on each bank, necessitating a considerable rise in the arch, and consequently rendering it specially suitable for a large span (Figs. 51 and 58, pages 132 and 136); whilst the Grünenthal Bridge, carrying a railway and a road across the Baltic Canal, has a similar form. The railway viaduct across the Douro at Oporto, built in 1876–77, known as the Maria Pia Bridge, has a length of 1158 feet; and its central large arch over the river has a span of 525 feet, a rise to its underside of 123 feet, and a clear headway in the centre of $168\frac{1}{2}$ feet above low-tide level, a height at the crown of $32\frac{4}{5}$ feet, tapering off to a pivot at each springing, and a width at the top between the ribs of 13 feet, fanning out to 49 feet at the springings.[3] The still larger and higher arch of the Garabit Viaduct across the lower part of the valley of the River Truyère, erected in 1880–84, has a span of $541\frac{1}{3}$ feet, a rise of 168 feet, a height at the crown of $32\frac{4}{5}$ feet, like the Douro arch, and tapering similarly off to the springings, and a width between the ribs of $20\frac{3}{5}$ feet at the top, fanning out to $65\frac{3}{5}$ feet at the springings;[4] whilst the height of the rail-level above the lowest part of the valley is 403 feet. The total length of this viaduct is 1715 feet, and its cost was £126,900 (Fig. 51, p. 132).

The Grünenthal Bridge, which carries a railway and a road side by side over the Baltic Canal, erected in 1892, with a span of $513\frac{1}{2}$ feet and a rise of $70\frac{1}{2}$ feet, though having a crescent-shaped arch like the Douro and Garabit viaducts, with a height of $12\frac{1}{2}$ feet at the crown,

[1] "History of the St. Louis Bridge," C. M. Woodward. St. Louis, 1881.

[2] "The Washington Bridge over the Harlem River," W. R. Hutton. New York, 1889.

[3] "Notice sur le pont du Douro à Porto (Pont Maria Pia)," G. Eiffel; and *Mémoires de la Société des Ingénieurs Civils*, 1878, p. 741, and plates 128–131.

[4] *Mémoires de la Société des Ingénieurs Civils*, 1888 (2), p. 112, and plates 172–175.

tapering off to a pivot at the springings, differs from them in the position of the roadway, which, being carried across at a lower level than the crown of the arch, though affording a headway of 138 feet above the

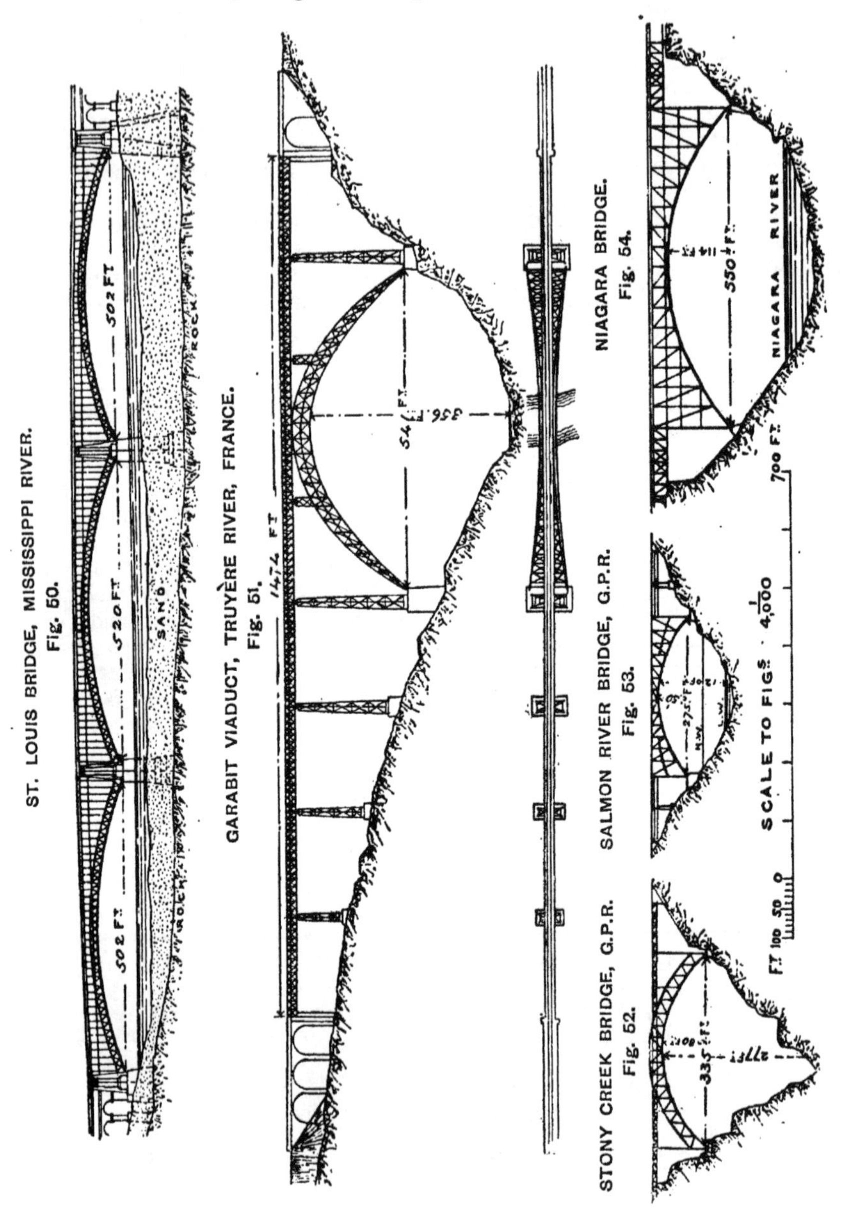

ST. LOUIS BRIDGE, MISSISSIPPI RIVER. Fig. 50.

GARABIT VIADUCT, TRUYÈRE RIVER, FRANCE. Fig. 51.

STONY CREEK BRIDGE, G.P.R. Fig. 52.

SALMON RIVER BRIDGE, G.P.R. Fig. 53.

NIAGARA BRIDGE. Fig. 54.

water-level of the canal, is suspended from the arch in the central part of the span, and supported over it at the sides.[1] This arch, moreover, was erected by the aid of scaffolding built up in clusters from the bottom of the canal excavations; whereas the Douro and Garabit arches were built out from the springings without scaffolding, till the two portions could be joined in the centre of the span (Fig. 58, p. 136). The great depth of the Douro and Garabit arches at the crown, was designed to enable the arch to resist the changes in the strains resulting from variations in the distribution of the moving load, which, in the case of the Douro viaduct, is greater per unit of length than the weight of the arch; whilst the large increase in width of the arch at the springings, distributes the weight on the abutments, and gives the arch stability against the wind-pressure, which has a considerable effect on such high structures in a narrow gorge.

Some other arched metal bridges of large span have been built more in accordance with the ordinary forms of arched bridges, where the depth of the arch increases from the crown towards the springings, without, in most cases, abandoning the freedom of movement at the springings. The Stony Creek steel arch, indeed, erected on the mountain section of the Canadian Pacific Railway, at a cost of £20,200, is of the same depth throughout (Fig. 52, p. 132), like the ribs of the St. Louis Bridge; but the strength of metal arches is readily augmented from the crown towards the springings, without necessarily increasing the depth of the arch, by merely increasing the thickness of the metal in successive panels. In some cases, however, the depth of the arch is increased, in proportion to the deviation of the curve from the horizontal, by spandril uprights and bracing, of which examples are furnished by the steel arch across the Salmon River on the Pacific section of the Canadian Pacific Railway, which cost £12,100 (Fig. 53, p. 132), and by the large steel arch over the Niagara River, which replaced the Niagara Suspension Bridge in 1897, carrying a double line of railway on the higher level, 258 feet above the river, and a roadway underneath, and which has a span of 550 feet and a rise of 114 feet, and cost about £104,200 (Fig. 54, p. 132). In other cases, the depth of the arched ribs is gradually increased to a moderate extent from the crown to the springings. An example of this type of metal arch is afforded by the Levensau Bridge, built on staging over the Baltic Canal in 1894, carrying a railway and a roadway side by side, and having a span of $536\frac{2}{5}$ feet, a rise of 64 feet, a depth of arched ribs of $10\frac{1}{2}$ feet at the crown, increasing to $18\frac{1}{3}$ feet at the springings, and giving a clear headway in the centre of 138 feet above the water-level of the canal;[2] and the Müngsten Bridge, completed in 1897, is a still larger example of the same type, with a span of 558 feet, and attaining a height of 351 feet above the bottom of the valley.[3] The second arched bridge of large span over the Douro at Oporto, erected in

[1] "Geschichte des Nord-Ostsee-Kanals," Carl Loewe, Berlin, 1895, p 30, and plate 11.

[2] "Geschichte des Nord-Ostsee-Kanals," Carl Loewe, plates 12 and 13.

[3] *Zeitschrift für Architeckten und Ingenieur wesen*, Hanover, 1898, vol. xliv. cols. 53 and 54.

1881–85, designated the Luiz I. Bridge, is similar to the Levensau and Müngsten bridges in the increase of the depth of its arched ribs towards the springings; but it differs from them in having to provide two roadways across the river at very different levels,[1] so that, whilst one roadway is carried over the arch, the lower one has to be suspended from it (Fig. 55). This bridge, which cost nearly £100,000, has a total length of 1278 feet between the abutments; its arch over the river has a span of 566 feet, a rise of 146 feet, a depth of 26¼ feet at the crown, increasing to 55 feet measured vertically at the springings; and the width between the ribs is widened out from 19⅔ feet at the crown to

METAL ARCHED BRIDGE WITH HIGH AND LOW LEVEL ROADWAYS.

Fig. 55.—Luiz I. Bridge, Oporto.

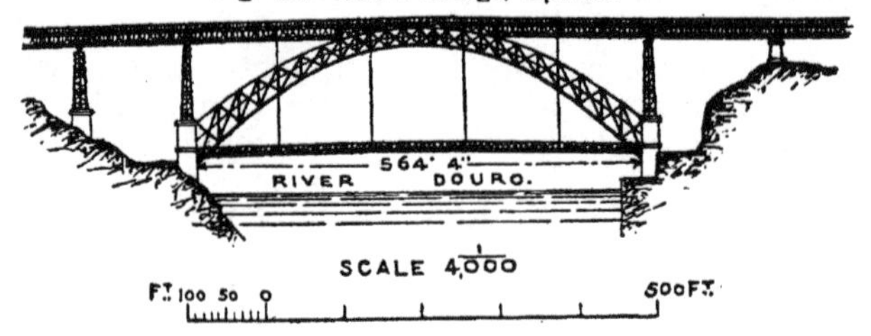

52½ feet at the springings, as in the first bridge. The upper roadway is 204 feet above the river, for the accommodation of the upper parts of the town; and the lower suspended roadway is 164 feet below the upper one. An arched steel bridge of the exceptional span of 840 feet was erected across the Niagara River a short distance below the Falls

STEEL ARCHED BRIDGE OF VERY LARGE SPAN.

Fig. 56.—Niagara Falls and Clifton Bridge.

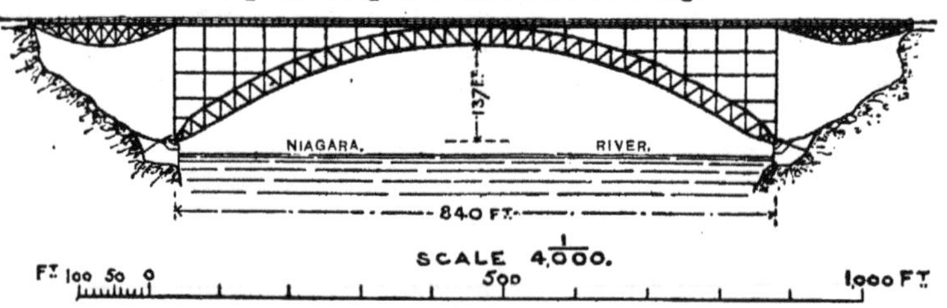

in 1898, assisted to a slight extent by the light suspension bridge which previously spanned the gorge, springing from the rock some distance down the slopes at the sides, whereby the main span was reduced from the span of 1268 feet of the suspension bridge to 840 feet, with girders over the side slopes of 190 feet and 210 feet span, the rise of the arch being 137 feet, and the depth of the arch at the crown 26 feet[2] (Fig. 56).

[1] *Mémoires de la Société des Ingénieurs Civils*, 1886 (1), p. 58, and plates 112 and 113; and *Engineering*, vol. xlii. pp. 7 and 12.

[2] *Engineering*, vol. lxvii. pp. 540, 578, 633, and 700.

The bridges over the Rhine at Bonn and Dusseldorf are fine examples of braced arches with suspended roadways. The former has a central span of 614¼ feet, with one side span on each bank of 307 feet; the latter has two central spans of 594½ feet, with one side span of 198 feet over the right bank, and three spans of 208, 187, and 164 feet over the foreshore on the left bank.

The Victoria Falls Bridge over the gorge of the Zambesi, about 700 yards below the cataract, is a two-hinged spandrel-braced arch of open hearth acid steel 500 feet span, centres of bearings, with a rise of 90 feet, with two side spans, on the left bank of 62½ feet, and on the right of 87½ feet span. The depth of the braced arch at the crown is 15 feet, and at the abutment 105 feet. The bridge carries a double line of railway of 3½ feet gauge, the width between parapets being 30 feet, and was opened to traffic in September, 1905.[1]

Provisions in Metal Arches for Changes in Temperature.—A metal arch necessarily rises or falls slightly with an increase or decrease of temperature; and when the arch is fixed at its springings, of which the widened Victoria Bridge, Pimlico, and the St. Louis Bridge are instances, the stresses due to the alteration in form have to be borne by the elasticity of the metal. The alteration in the form, however, of a metal arch with changes in temperature is facilitated, and the resulting stresses reduced, by leaving each rib free to rotate slightly on its springings; whilst this arrangement has the further advantage of enabling the arch, on completion, to settle properly on its bearings Sometimes the provision for rotation made by causing the rounded ends of the ribs to rest in segmental cast-iron shoes, as adopted for the original Victoria Bridge and the Severn Valley Bridge, has proved practically only serviceable for the settlement of the arch on completion, owing to the friction of the surfaces in contact being too great to be overcome by the stresses due to changes in temperature. In more recent arched metal bridges of large span, the requisite hinging at the springings has been more efficiently arranged, by causing the ribs to rest upon cylindrical steel pins, as in the arched Blackfriars Railway Bridge across the Thames, and the Washington Bridge over the Harlem River, or by tapering down the ribs to pivots at the springings, on which they are free to revolve, of which the Garabit, Douro, and Grünenthal arches are examples. Sometimes, also, when the arch is not tapered to a pivot, it is, nevertheless, made to bear on a curved surface at the springings, on which it is left free to move, an arrangement adopted for the arches of the Niagara, second Douro, and Levensau bridges.

Erection of Large Metal Arches.—The feasibility of building out large metal arches from the springings without the aid of scaffolding, till they meet in the centre of the span, constitutes one of the principal advantages of this type of bridge for carrying a railway or roadway at a high elevation over a deep, narrow valley, or across a broad, rapidly flowing river. The steel tubes forming the upper and lower members of each rib of the arches of the St. Louis Bridge, were built out in 12-feet lengths from the springings, with their intermediate bracing, for

[1] *Proceedings Inst. C.E.*, vol. clxx.

a quarter of the span on each side, being supported like a bracket by their attachments to the springing; and the remaining quarter portions to the centre of the span were gradually extended, by supporting the projecting portions by a number of iron bars passing over a timber tower erected on the pier, and fastened to the corresponding projecting length of rib on the opposite side of the pier, or to the shore viaduct at the abutments,[1] the whole forming a sort of temporary, balanced cantilever extending out on each side of the pier, or from the abutments (Fig. 57, The tension of the long, coupled bars supporting the projecting ribs, was adjusted by compensating for changes in length, due to a rise or fall of temperature, by raising or lowering the supports of the bars on the top of the towers, which was effected by regulating the pressure on a hydraulic plunger.

The Garabit arch, and the two arches, over the Douro were similarly erected, panel by panel, from the springing on each side, being supported by steel wire cables passing over a temporary trestle

BUILDING OUT ARCHES FROM PIERS.

Fig. 57.—St. Louis Bridge.

Scale $\frac{1}{4000}$.

erected on the top of the pier at each end of the arch, and fastened to the girders of the side spans, which were anchored to the abutments[2]

ERECTION OF HIGH ARCH FROM SPRINGINGS.

Fig. 58.—Maria Pia Bridge, Oporto.

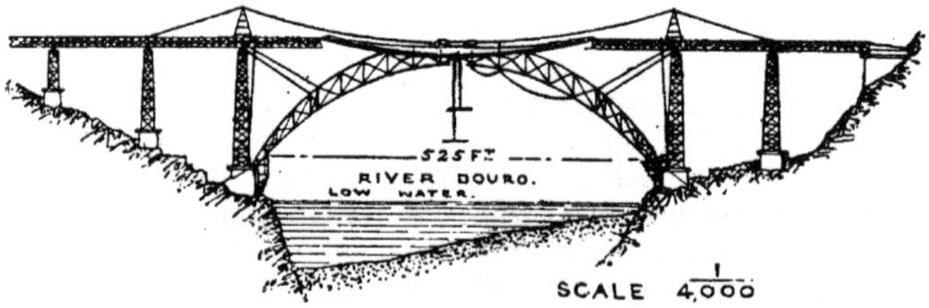

SCALE $\frac{1}{4,000}$

(Fig. 58). The ribs of the arch of the second bridge over the Douro were given a greater depth at the springings than at the crown, instead

[1] "A History of the St. Louis Bridge," C. M. Woodward, p. 160, and plates 37 and 42–45.

[2] "Notice sur le pont du Douro à Porto (Pont Maria Pia)," G. Eiffel, plates 4 and 5.

of being tapered, in order to facilitate the erection of the first panels, which could be thus effectually secured to the abutments (Fig. 55, p. 134); for the pivoted ends of the ribs of the first Douro bridge had rendered the earlier part of its erection somewhat difficult, owing to the very limited space available for temporary supports on the edge of the river-bank. Other arched bridges spanning rapid rivers or deep ravines, where scaffolding would have been very costly, or impracticable as in the case of the Niagara Bridge (Fig. 54, p. 132) over the Niagara Rapids, have been erected in the same manner.

The method by which each half-span of the Niagara Falls and Clifton arch was supported during erection, is shown in Fig. 59, where each of the two arched ribs was kept in position by a series of diverging ties connected to, and held back by, a main horizontal tie anchored landwards to the solid rock, and adjusted by a toggle which, by the aid of a screw, could relax or tighten up the main tie.

BUILDING OUT LARGE ARCH FROM ABUTMENTS.

Fig. 59.—Niagara Falls and Clifton Bridge.

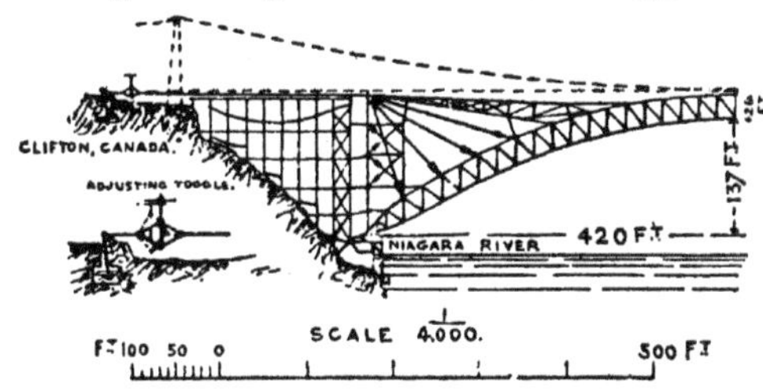

The erection of the Victoria Falls Bridge was preceded by the construction of the cable-way required for the transport of materials across the gorge, of 870 feet span, carrying 10 tons and worked electrically. The cable was 8½ inches in circumference, with an ultimate breaking stress of 270 tons.

The erection of the bridge was carried out by the cantilevering out of each half span from the opposite sides of the gorge, the structure being held back by steel wire cables with screw adjustments to anchorages in wells sunk in the solid rock.

CHAPTER IX.

SUSPENSION BRIDGES.

Single span, and with short side spans, usual forms; economy of several spans—Suspension and arched bridges contrasted—Advantages and disadvantages of suspension bridges—Original form with chains of long, flat links, or wire cables; illustrated by Menai and Friburg suspension bridges—Wire cables and flat-link chains compared—Forms of wire cables—Auxiliary cables: for large spans, instances, their advantages—Examples of principal types ot large span: Brooklyn, Williamsburg, Manhattan bridges—Stiffened suspension bridges: Lambeth, Francis-Joseph, and Albert, Chelsea, bridges; comparison of systems—Stiffened suspension chains: methods adopted; examples at Vienna and Pittsburg—Provisions for movement of chains, and against wind: arrangements on towers for motion of chains or cables; instances of roadway of bridges injured by storms, remedies adopted—Remarks on suspension bridges: great value formerly for large spans; recent examples; modification of type and increase in cost by stiffening; utility of system under certain conditions.

SUSPENSION bridges have been frequently constructed with a single large span, especially where a deep gorge with precipitous sides has to be crossed; but when traversing a river with sloping banks, piers in the river near each bank enable the main span of the bridge to be reduced, leaving much shorter side spans to stretch from the piers to the shore. These are the two ordinary types of suspension bridges; but occasionally, when the suspension system has been resorted to for crossing a wide river, two or more river spans have been adopted. A suspension bridge having several spans is, indeed, more economical per unit of length than a bridge with a single similar span; for the anchorage cables with their anchorages, required on the land sides of the piers to support the stresses of the cables carrying a bridge having only a single span, suffice equally for a suspension bridge having several similar spans.

Suspension and Arched Bridges contrasted.—A suspension bridge is the exact converse of an arched bridge, for all its members are in tension; the towers supporting the cables take the place of the springings of an arch; and the anchorage cables at each extremity, bearing the whole tension due to the weight of the bridge and its load, are the counterparts of the abutments which sustain the whole thrust of the arch. Moreover, the tensional stresses in the cables of a suspension bridge are horizontal, and at a minimum at the lowest point of the dip of the cables, which in a fully loaded bridge is at the centre of the span, and increase by the sum of the weights of the successive sections of the bridge with their respective loads, to a maximum over the supports, just as in the case of the compressive stresses in an arch between the crown and the abutments. Accordingly, it is only necessary to obtain the horizontal tensile stress in the centre of the span of a suspension bridge, as previously indicated, on p. 123, and to form a graphic diagram similar to Fig. 41, p. 123, inverted, in which this horizontal tensile stress, and the vertical stresses due to the weights of the several portions of the bridge starting from the centre, with their respective loads, are drawn to a given scale, in order to obtain the resultant tensile stresses at the several points of the suspension cables both in magnitude and direction.

The towers of a suspension bridge of several spans, like the piers of a viaduct with several arches, have only to bear the vertical loads due to half the span on each side; for the horizontal stresses due to the span on one side, are counterpoised by those on the other over the towers and piers, so that those stresses have to be provided against solely at the extremities, by the anchorages in the one case, and the abutments in the other.

Advantages and Disadvantages of Suspension Bridges.—As all the parts of a suspension bridge are in tension, each portion is able to support the maximum stress to which the metal can be safely subjected; the several parts naturally take the positions imposed by the stresses, and suited to sustain them; and no additional pieces or thicknesses, involving an increased weight, have to be provided to enable the several parts of the structure to withstand flexure. Accordingly, the metal is used to the best advantage in suspension bridges; and by employing the metal in the form of wire, which possesses a much higher degree of tenacity than the metal in its other forms, a remarkably light, strong type of bridge can be constructed, at a relatively moderate cost, which is specially adapted for large spans. Moreover, by the help of a suspended platform stretched across a deep gorge or a wide river, a suspension bridge can be erected without scaffolding; and on this account, as well as owing to the lightness and economy of such bridges, several of the earlier bridges of large span, in unfavourable situations, were built on this principle.

Suspension bridges, however, of the ordinary types, in consequence of their lightness and flexibility, are not well adapted for the rapid passage over them of a heavy moving load, such as a railway train; for in the case of bridges of moderate span, the moving load is large in proportion to the weight of the bridge; and the cables, accordingly, readily alter their curvature with the varying distribution of the moving load, producing injurious oscillations in the roadway. Consequently, though suspension bridges have been often adopted for carrying roads over broad rivers or deep ravines, as the traffic on roads is less concentrated and less quickly shifting than trains, these bridges have been generally regarded as unsuitable for railways. Nevertheless, in the case of very considerable spans, the weight of a suspension bridge becomes so large in proportion to the moving load, that the oscillation due to a passing train is much reduced; and it can be rendered immaterial if the precaution is taken of making the train traverse the bridge at a slow speed. Thus the Niagara Suspension Bridge, built in 1852–55 with a span of 821 feet, at a cost of £80,000, was able to convey the railway and road traffic safely over the Niagara River for forty years, with merely the restoration of its original wooden platform in steel in 1880, and without injury to its four suspension cables, each 10 inches in diameter and composed of 3640 iron wires, and having a dip of 60 feet. The Brooklyn Bridge also, at New York, carries two lines of rope railways, as well as two roadways on each side and a wide central footway; and the Point Bridge at Pittsburg has two lines of tramways on it. The employment, however, of suspension bridges for railways is exceptional; and the other examples which will be described serve exclusively for road traffic. Moreover, light suspension bridges in exposed sites are liable to be injured in severe gales by the action of the wind on the roadway, mainly owing to the vertical oscillations produced by gusts; and there are two or three recorded instances of suspension bridges having broken down under the stresses produced by a troop of soldiers marching in step along the roadway.

Original Form of Suspension Bridges.—The earliest type of suspension bridge of large span consisted of chains of flat iron links connected by bolts (Fig. 60), or of wire cables, anchored at the two extremities, passing over high towers, and hanging across the space to be spanned between the towers, from which the horizontal platform of the bridge was suspended, on each side, by a series of vertical bars or cables at equal distances apart. Notable instances of this type are afforded by the Menai Suspension Bridge, erected across the Menai Straits in 1819–25, and the Friburg Suspension Bridge, constructed over the deep valley of the River Sarine in Switzerland in 1832–34. The Menai Bridge is carried by sixteen chains, in sets of four, one above the other, each composed of five parallel rows of flat iron bars (about 10 feet long, 3½ inches wide, and 1 inch thick), widened out at each end for the holes through which the connecting bolts pass, from which the suspension rods are hung. These chains have a span of 570 feet between their supports, and a dip of 43 feet; and the bridge, which gives a clear headway for vessels of 102 feet above high water, cost £120,000 with its approaches.[1] The Friburg Bridge was originally borne by two wire cables on each side of the roadway, each of the four cables consisting of 1056 wires in twenty strands, giving it a diameter of 5½ to 6 inches.[2] These cables have a span between their supports of 880 feet, and a dip of 63 feet; and they were overhauled in 1880, when two more cables were added. The bridge is at a height of 167 feet above the river; and, owing to the lightness and strength of the wire cables, it only cost £24,000 in the first instance.

[1] "Life of T. Telford, C.E.," written by himself, p. 570, and plate 70.
[2] *Annales des Ponts et Chaussées*, 1835 (1), p. 3, and plate 90.

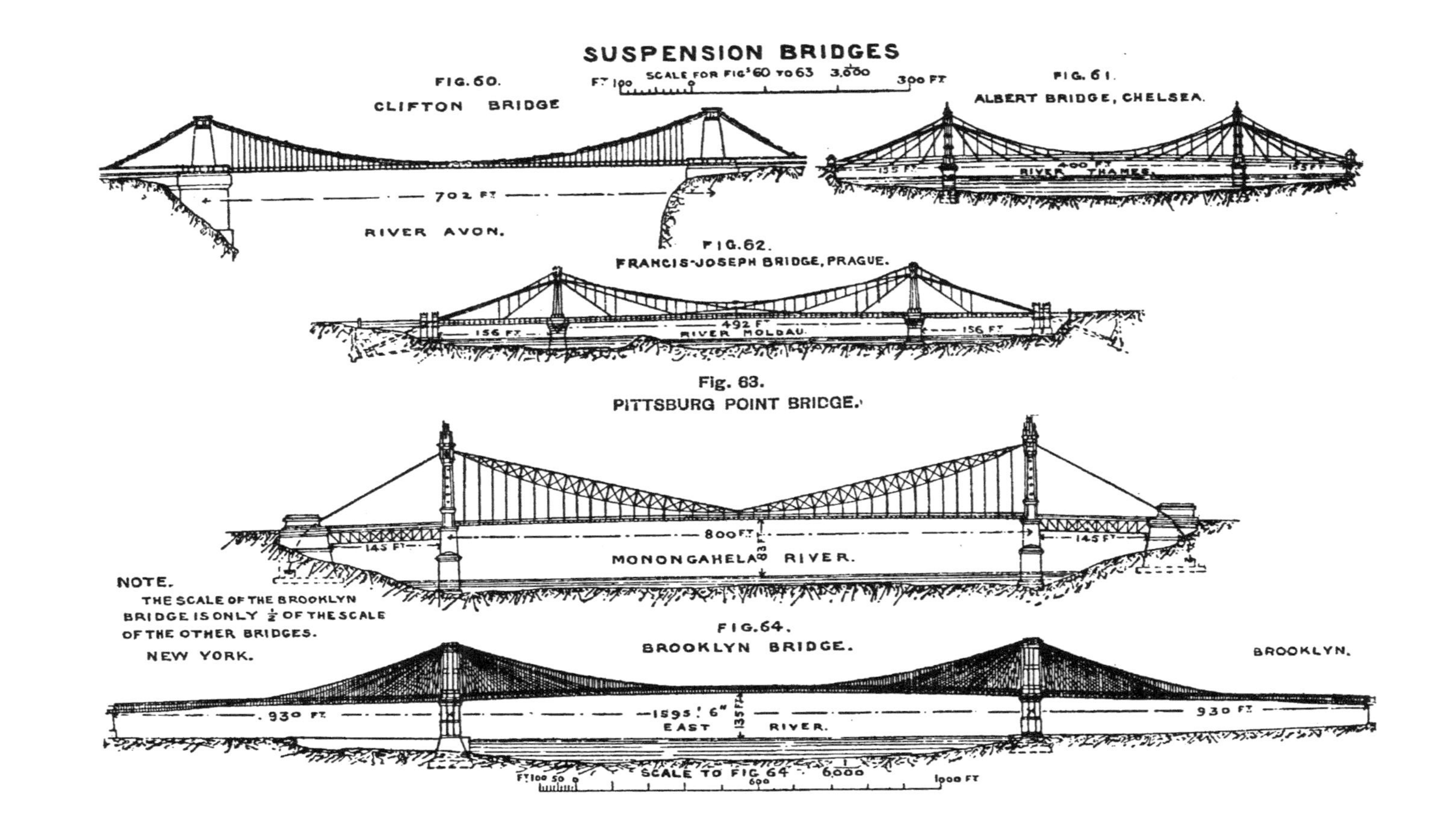
SUSPENSION BRIDGES
SCALE FOR FIG'S 60 TO 63
FT 100
0
300 FT
FIG. 60.
CLIFTON BRIDGE
702 FT
RIVER AVON.
FIG. 61.
ALBERT BRIDGE, CHELSEA.
135 FT
400 FT
RIVER THAMES.
135 FT
FIG. 62.
FRANCIS-JOSEPH BRIDGE, PRAGUE.
156 FT
492 FT
RIVER MOLDAU.
156 FT
Fig. 63.
PITTSBURG POINT BRIDGE.
800 FT
83 FT
145 FT
145 FT
MONONGAHELA RIVER.
NOTE.
THE SCALE OF THE BROOKLYN
BRIDGE IS ONLY ½ OF THE SCALE
OF THE OTHER BRIDGES.
NEW YORK.
FIG. 64.
BROOKLYN BRIDGE.
BROOKLYN.
930 FT
1595' 6"
135 FT
EAST RIVER.
930 FT
SCALE TO FIG 64
FT 100 50 0
600
1000 FT

Wire Cables and Flat-Link Chains compared. — The economy and lightness secured by the adoption of wire cables, led to the erection of numerous roadway bridges across rivers in country districts, which could not have been otherwise provided, and enabled large spans to be resorted to for crossing rivers and ravines at a comparatively early period; but in other respects the flat-link chains possess some decided advantages. The chains, made up of a series of flat bars, can be readily painted, any damage quickly observed, and defective links replaced by new ones; whereas wire cables are liable to gradual and invisible deterioration by rust, especially at their anchorages, from moisture penetrating through the interstices between the wires, unless most efficiently protected from damp on erection; and when any portion becomes damaged, the whole cable has to be renewed. Moreover, a chain of links can easily have its strength increased, in proportion to the increase in the stress between the centre of the span and the supporting towers, by augmenting the number of links or their thickness; whilst a wire cable has to be given the maximum strength required at the towers, throughout its whole length.

Forms of Wire Cables.—Wire suspension cables are formed, either by binding together a cylindrical bundle of parallel wires at intervals, and fastening several bundles together for a large cable; or by twisting the wire into cables as in ordinary ropes, using sometimes several strands of wire ropes to form the cable. The object of the first system is to maintain the tenacity of the wire unimpaired, by avoiding the bending of the wire; but the second system provides a much more flexible cable, with greatly reduced interstices, rendering the erection easier, and offering less opportunity for rusting, without apparently materially impairing the tenacity of the wire by the twisting, if the manufacture of the cable is carefully conducted, and making the tension of the wires more uniform. Experiments, indeed, indicate that a cable of twisted wire is decidedly stronger than one with parallel wires, as well as more durable.

Suspension Bridges with Auxiliary Cables.—When later on suspension bridges were designed for carrying heavier weights, as in the case of the Niagara Railway and Road Bridge, and for still larger spans, as, for instance, the Covington and Cincinnati Bridge over the Ohio, completed in 1867, with a span of 1057 feet between its supports, and the Niagara Falls Bridge, erected in 1867–69, with a span of 1268 feet, auxiliary cables or oblique stays were added, branching out from the top of the towers like a fan, and supporting the roadway at a series of points for a considerable distance out from the piers (Fig. 64, p. 141). The main cables are thereby relieved from having to support the side portions of the span; and their strength is thus in a great measure concentrated on the support of the central portion of the bridge. These auxiliary cables also stiffen the structure considerably, and reduce the undulations of the main cables, and consequently of the roadway, resulting from unequal loading of the bridge, and rapid variations in the position of the moving load. When also, in crossing over a river, piers are placed in the river to reduce the central span, and moderate side spans are introduced between the piers and the banks, these side spans can be supported by the auxiliary cables and thus balance the side portions of the main span, and at the same time assist in reducing the strength required for the anchorages of the main cables. The Niagara Railway and Road Bridge was provided with sixteen auxiliary cables at each end, which were connected with the anchorage cables at the top of the towers; but the thirty-eight auxiliary cables over each tower, supporting the side portions of the central span of the Covington and Cincinnati Bridge, support also each side span of 281 feet; and the same advantageous system has been adopted in many other cases, and notably on a large scale at the Brocklyn Suspension Bridge (Fig. 64, p. 141). In the above instances, the vertical suspending cables have been continued right up to the towers, though the auxiliary cables in the Niagara Bridge stretched out 250 feet from the towers on each side, and in the Covington and Cincinnati and Brooklyn bridges, extend out 280 feet and about 400 feet respectively; but in the more recent suspension bridges erected in France, the auxiliary cables, introduced there for the first time in 1879, have been wholly relied on for supporting the side portions of the main span, and also the side spans; and the vertical suspending cables have been dispensed with along these parts of the bridge. A good example of this arrangement is furnished by the St. Ilpize Bridge, erected over the River Allier in 1879 (Fig. 65), which, with a clear span of 224 feet between the cast-iron

SUSPENSION BRIDGE WITH AUXILIARY CABLES.

Fig. 65.—St. Ilpize Bridge, France.

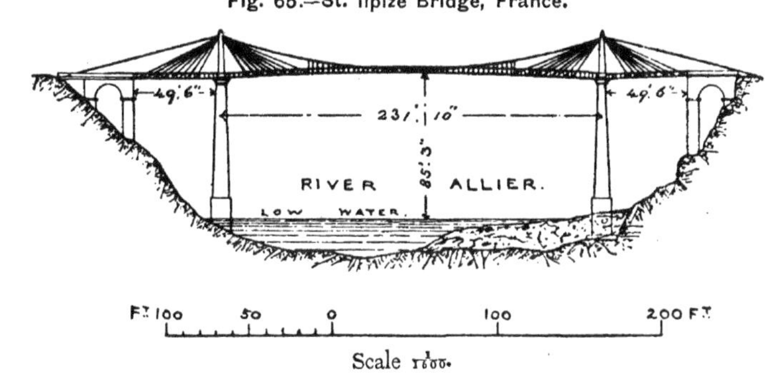

Scale $\frac{1}{1600}$.

towers, has its roadway supported over the central span for about 50 feet out from the towers on each side, as well as the side spans of similar length, by sixteen auxiliary wire cables passing over each of the towers.[1] The central 124 feet of the roadway is borne by the two main wire cables, being hung from vertical suspension rods encircling the cables; and this roadway bridge, which has a width between the parapets of 13 feet, cost only £2800. In the Lamothe Bridge, erected over the River Allier in 1883–84, with a single clear span of 377 feet, where auxiliary cables were also used for supporting the ends of the

[1] *Annales des Ponts et Chaussées*, 1885 (2), p. 662, and plate 51.

bridge near the towers, the main portion of the bridge was carried by five wire cables on each side, to enable renewals or repairs to be executed by the removal of a cable without stopping the traffic; and this bridge, with a width of 18 feet, cost £7280.

Examples of Principal Types of Large Suspension Bridges. —The two leading types of suspension bridges, namely, the bridges borne solely by the suspension chains or cables from which they are hung by vertical suspenders, and the bridges which are supported and stiffened by auxiliary cables as well, are suitably illustrated by the Clifton Suspension Bridge and the Brooklyn Bridge (Figs. 60 and 64, p. 141). In both types, bridges of large span are strengthened and stiffened by introducing longitudinal girders in the roadway as parapets, or as divisions between the roadways and the footpaths, which distribute any unequal load between several suspension rods or cables.

The Clifton Suspension Bridge stretches across the gorge, bordered by steep cliffs, through which the River Avon flows below Bristol, with a span of 702 feet between the supports of its chains, and at a height of 248 feet above high water in the river.[1] The bridge, 31 feet in width, is carried by three chains on each side, having a dip of 70 feet, the two sets of superposed chains being 20 feet apart; and the chains, composed of rows of links, 24 feet long in the centre of the span where they are horizontal, and increasing in length in proportion to the increasing inclination of the chains towards the towers, have a total sectional area of 440 square inches in the centre, increasing to 481 square inches at the towers, the horizontal strain at the centre of the span having been estimated at 2094 tons, with the assumed maximum moving load of 70 lbs. per square foot, or 600 tons, on the bridge. The joints of the three chains in each tier are so arranged in relation to each other, that the suspension rods, about 2 square inches in section, attached in succession to the $4\frac{5}{8}$-inch bolts connecting the links in each of the three chains, are evenly spaced about 8 feet apart right along the bridge, over the girder parapets of the roadway, to which they are fastened. The chains are supported over the towers by wrought-iron saddles resting on cast-steel rollers, to allow for the movements due to changes of load and temperature; and they are thence carried down to massive underground anchorages on each side, each chain being fastened to a large cast-iron anchorage plate bedded upon a brick arch bearing against solid rock. The erection of the bridge was carried out in 1862–64, the towers having been built many years before; and the chains were put together across the gorge on a suspended platform borne by eight iron-wire ropes, six underneath the planking, and two above at the sides, serving also as handrails. On completion, the bridge was tested by placing on it a distributed load of 500 tons of stone, which produced a temporary deflection of 7 inches in the centre. The maximum moving load on the bridge is three-eighths of the total suspended load with the bridge fully loaded. The bridge is much exposed to strong gales blowing along the gorge, which slightly deflect

[1] *Proceedings Inst. C.E.*, vol. xxvi. p. 243, and plate 10.

the roadway laterally, and also the land chains, and produce a vertical undulation of the roadway which has been estimated to have a maximum range of about one foot.

The Pest Suspension Bridge, erected across the Danube in 1842–49, has a central span of 685 feet, and two side spans of 297 feet; and it is carried by four chains, each formed of rows of ten and eleven bars alternately, the bars being 12 feet long, and increasing in thickness from $1\frac{1}{12}$ inches at the centre to $1\frac{1}{4}$ inches at the towers. The Kieff Suspension Bridge, erected in 1848–53 across the Dnieper, has four spans of 440 feet, and two side spans of 225 feet; and its roadway is suspended from four chains composed of eight rows of bars, 12 feet long, 11 inches wide, and 1 inch thick.[1]

The Brooklyn Suspension Bridge, erected over the East River in 1870-83 for connecting Brooklyn with New York, has a central span of $1595\frac{1}{2}$ feet between the supports of the cables on the towers, affording a headway in the centre of 135 feet above high water, and two side spans of 930 feet between the centres of the piers and the face of the abutments (Fig. 64, p. 141).[2] The roadway, 85 feet in width, comprising a central, elevated footway, two wire-rope railways, and two carriage-ways on the outer sides (Fig. 66), is borne by four galvanized steel-wire

ROADWAYS ON BROOKLYN BRIDGE.

Fig. 66.—Cross Section near Centre of Bridge.

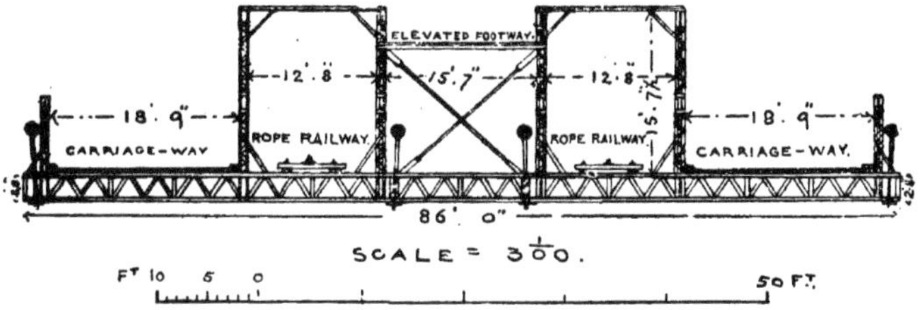

cables, $15\frac{3}{4}$ inches in diameter, having a dip of 128 feet, and by one hundred and eight auxiliary cables passing over each of the towers, and connected to the roadway of the side spans and the adjoining side portions of the roadway of the central span.[3] On the completion, in 1876, of the river piers and the towers built upon them, rising 277 feet above high water, a continuous, travelling wire rope was passed over the towers, and stretched across the river, by means of which a suspended platform was constructed, on which the strands of the main cables were laid together in succession, and the suspenders eventually fastened from which the roadway was hung. The auxiliary and suspending cables are made of wires twisted together; but the four main cables are

[1] "Life of C. B. Vignoles," O. J. Vignoles, pp. 331 and 361.

[2] It has been necessary to draw the elevation of the Brooklyn Bridge to half the scale of the other bridges given on p. 141, in order to bring it within reasonable limits of size.

[3] *Engineering News*, New York, vol. x. pp. 20, 241, and 252.

each composed of 5296 wires laid side by side without any twist, to ensure the maintenance of their full strength, being grouped into nineteen strands bound together to form one cable. The cables are supported on the towers by saddles borne on rollers, to allow for the movements of the cables due to changes in loading and temperature; and each cable is fastened at its two ends to anchor chains formed of sets of bars turning gradually at their joints into a vertical position, and held down at their extremities by a 23-ton anchor-plate embedded horizontally in a mass of masonry, weighing 60,000 tons, constituting the abutment of the bridge on each bank. The roadway is stiffened by six longitudinal trusses running the whole length of the bridge, two of them forming the outer parapets of the carriage-ways, and the four main ones going along each side of the two rope railways (Fig. 66, p. 145); and the cross girders, $7\frac{1}{2}$ feet apart, carrying the roadways, are fastened to the undersides of the longitudinal trusses. The total weight of the bridge is 14,680 tons, and the moving load only 3100 tons, little more than one-sixth of the total load; whilst the dead and moving loads of the central span are 6740 tons and 1380 tons respectively, the latter being rather more than one-sixth of the total load. The total length of the bridge, including $2533\frac{1}{2}$ feet of approaches, is 5989 feet, or about 1 mile 236 yards; and the cost of the bridge was £2,437,500, exclusive of land.

The Williamsburg Suspension Bridge across the East River, New York, has a central span of 1600 feet, and a width of 118 feet, provision being made for two railway and four trolley car tracks, two roadways, two side walks and two cycle tracks. Four steel-wire cables are used, $18\frac{3}{4}$ inches diameter, supported on steel towers, 310 feet high, from masonry to top of cables, erected on masonry piers. The total length of the bridge with approaches is nearly 6000 feet.

The Manhattan Suspension Bridge, also across the East River, New York, about 3000 feet long between anchorages, and nearly 10,000 feet long with approaches, has a centre span of 1470 feet, and two side spans of 725 feet each, at a clear height of 135 feet above mean high water. This bridge carries four lines of railway, and four trolley-car tracks, a roadway, and two footways. The towers, 291 feet in height, and containing upwards of 5000 tons of steel-work, stand upon solid masonry pedestals 134 feet by 68 feet.

Stiffened Suspension Bridges.—The want of rigidity which the old forms of suspension bridges necessarily exhibit, has been to a great extent remedied by stiffening the roadway with longitudinal girders, as in the Clifton Bridge, and with the assistance as well of auxiliary cables or sloping stays in the Niagara, Cincinnati, Brooklyn, and other bridges. Whilst, however, longitudinal trusses of some form have generally been adopted in the more modern suspension bridges of any importance, for stiffening the roadway, other systems of increasing the rigidity have occasionally been resorted to in place of auxiliary cables. Thus at Lambeth Bridge, erected across the Thames in 1862–63, with three clear spans of 268 feet, at a cost of £48,000, the latticed frames constituting the suspenders from the two wire cables on each side of the roadway, are braced together by diagonal bars in pairs, which are connected at the bottom with the longitudinal girder bordering the roadway on either side.[1]

In the Francis-Joseph Bridge, erected over the River Moldau at Prague, with a central span of 492 feet between the supports on the towers, and two side openings of 156 feet, completed in 1868, the system of sloping stays has been extended so as to constitute the main supports of the bridge, with the object of augmenting the rigidity of the structure.[2] These stays, six in number on each half of the central span, are formed of flat links, 14 feet long, 4 inches broad, and 1 inch thick, being supported at their joints by suspenders from an upper curved chain on each side, whose only function is to keep the long stays in a straight line, as the roadway is supported by the straight stays at equal intervals of about 82 feet (Fig. 62, p. 141). In the spaces between the stays, the roadway, 31 feet in width, is borne by a continuous longitudinal girder on each side, to which the stays are attached. Owing to the distance apart of the points of support in this bridge, as compared with the intervals between the suspension rods in ordinary suspension bridges, the girders had to be made specially strong, which, however, adds to the stiffness of the bridge. As only two stays are required for supporting each of the side spans in the centre, the other four stays passing over each tower, together with the upper supporting chains, are carried down direct to shore anchorages on each bank. The total cost of the bridge, including land, was £57,000.

The Albert Bridge, built over the Thames at Chelsea in 1869–72, with a central span of 400 feet between the supports on the towers, and two side spans of 155 feet between the centre of the piers and the face of the abutments, though similar in type to the Francis-Joseph Bridge in being borne by oblique stays of flat bars, with supporting upper wire cables 6 inches in diameter, differs from this bridge in some important particulars.[3] Though the main span of the Albert Bridge is 92 feet less than in the previous bridge, the stays on each side are in groups of four instead of three, and the centre of the continuous girder on each side of the bridge is connected with the curved upper cable; whilst the suspension rods from this cable, 20 feet apart, for supporting the stays, are carried down to the longitudinal girder (Fig. 61, p. 141). The points of support, accordingly, of this bridge afforded by the stays, are only 40 feet apart, instead of the 82 feet in the Prague bridge; and as these stays only extend out 120 feet from the centre of the piers along the main span, they are all counterbalanced, and serve to support the side spans by being similarly attached to the longitudinal girders on the land side of the towers. Anchorage chains and anchorages for the stays are, consequently, dispensed with; and the two curved supporting cables are

[1] "A Record of the Progress of Modern Engineering, 1863," W. Humber, p. 42, and plates 26–29.

[2] *The Engineer*, vol. xxvi. pp. 343 and 380.

[3] *Ibid.*, vol. xxxvi. pp. 281, 298, and 322; and for illustrations, pp. 288, 301, and 316.

also fastened to the ends of the longitudinal girders, so that vertical anchorages only have been provided for keeping down the ends of the girders when, with the central span loaded and the side spans unloaded, they tend to rise. The cost of this bridge, with its approaches, appears to have amounted to the very large sum of £143,000.

Both the Albert and Francis-Joseph bridges are relieved from the undulations resulting from the deflections of the chains or cables in the ordinary suspension bridges, in so far as the stays can be kept straight; and they approximate in principle to the upper portion of a cantilever bridge. The Albert Bridge possesses the advantage of supporting the longitudinal girders at much closer intervals than the other, and is more efficiently braced, with its more numerous stays and prolonged suspension rods; whilst it presents more similarity to suspension bridges with auxiliary cables, of the types of the St. Ilpize and Brooklyn bridges, than the Francis-Joseph Bridge does.

Stiffened Suspension Chains.—Another plan of stiffening suspension bridges consists in stiffening the chains themselves, so as to render them less subject to deflection under moving loads. This has been effected by putting one chain a little distance above another chain, and bracing the two chains together vertically, like the upper and lower members of the rib of an arch inverted, such as the arch of Stony Creek Bridge (Fig. 52, p. 132); and it has also been accomplished by constructing two suspended trusses on the chain, sloping down from the towers and hinged together in the centre of the span, with each half of the chain as the bottom member of the truss (Fig. 63, p. 141).

The first system was adopted for two railway bridges crossing the Danube Canal at Vienna, with a clear span of 255 feet, built in 1860 and 1864 respectively; the two chains on each side being placed about 4 feet apart, and connected vertically by Warren bracing, as in the Stony Creek Bridge. Some error was made in testing the strength of the metal used for the first bridge, which, consequently, was made too light, and was replaced by an arched bridge in 1884; but the Aspern Bridge, erected four years later, was duly proportioned to the real strength of the metal, and proved satisfactory.

In the bridge erected over the Monongahela River at Pittsburg Point in 1875–77, with a span of 800 feet between the supports on the towers, and two side openings of 145 feet span, the link suspension chains have been stiffened by building trusses, 22 feet high in the middle, on each half of the chains after their erection.[1] The chains are formed of sets of 11 to 14 flat bars, 20½ feet long, 8 inches wide, and from 1 to 2 inches thick, with a dip of 83 feet, anchored direct from the saddles on the towers, 180 feet above low water, down to the abutments, by chains of 12 and 13 bars alternately, as the side openings are spanned by independent lattice girders; and the chains were made of sufficient strength to support the unloaded roadway alone, the trussing being added to provide for the moving load. The straight upper member of the truss, forming the chord to the curve of each half chain, is of

[1] *Engineering News*, New York, vol. iii. p. 220; and vol. iv. p. 89.

inverted channel shape, and is connected to the chain at the points from which the suspenders are hung, by rectangular and diagonal bracing; whilst every fifth suspender is stiffened. (Fig. 63, p. 141). The trusses are also braced together transversely, to stiffen them against wind-pressure. When the central span of the bridge was tested with a full complement of loaded carts and people, weighing altogether 474 tons, there was a deflection of only 4 inches in the centre; and with the test load on only half the span, the maximum deflection of the loaded half was $2\frac{3}{4}$ inches, and the maximum rise of the other half was $1\frac{1}{8}$ inches. The bridge carries a central roadway 21 feet in width, and a footway on each side $6\frac{1}{2}$ feet wide; and accommodation is provided on the roadway for two lines of tramway, and a track for a narrow-gauge railway. The total length of the bridge is 1245 feet; and its cost was £109,375.

Stiffened suspension chains have also been employed for supporting the side spans of the Tower Bridge, 270 feet in width (Fig. 101, p. 185).[1]

Provision for Movement of Chains, and against Wind.— The motion of the chains or cables over the towers, owing to alterations in curvature due to the moving load or changes in temperature, are generally provided for by placing the supporting saddles, on the top of the towers, on cast-iron frames borne on rollers, so that they move with the chains or cables. In the St. Ilpize Bridge, the movement was allowed for by placing the saddles upon rocking frames with a curved base. With a suspension bridge having several spans, the motion is increased when one span happens to be fully loaded and the other spans unloaded. At Lambeth Bridge, this disadvantage was obviated by fixing the cables on the top of the towers, thereby introducing the more objectionable result of making the cables exert a longitudinal pull upon the towers when the bridge is unequally loaded.

Owing to the lightness and flexibility of suspension bridges, their roadways have not unfrequently been injured, in very exposed situations, by exceptional gales. Thus the roadway of the Menai Suspension Bridge was rendered impassable, and several of the suspension rods were broken, by a storm in January, 1839; the roadway of the Roche-Bernard Bridge over the river Vilaine in France, supported by four wire cables, with a span of 651 feet between their supports and a dip of 50 feet, was wrecked during a gale in October, 1852; and the roadway of the Niagara Falls Bridge, 190 feet above the river, was carried away by a tornado in January, 1889, leaving the steel-wire cables and their anchorages uninjured, the bridge having been rebuilt, with new cables and two steel trusses, 12 feet deep, for carrying the roadway, in place of wood, in the previous year. The wind, rising from below in a gorge, relieves the chains to some extent of the weight of the roadway, and produces undulations in a light platform, which attain their maximum about halfway between the piers and the centre of the span; and also, when blowing straight along a narrow gorge, it causes a slight sideways deflection of the bridge. In some cases these movements have been

[1] *Proceedings Inst. C.E.*, vol. cxxvii., plate 1, fig. 1.

provided against by attaching a series of oblique cables to the underside of the roadway at intervals, fastened at the other ends to the piers of the bridge, or to the rocky sides of the gorge when the headway is ample, as for instance at the Niagara suspension bridges, or by counter cables, as added in the restoration of the Roche-Bernard Bridge, and also by lateral cables on each side in a horizontal plane. Sometimes the suspending cables have been placed further apart over the towers than in the centre of the span, as for instance at the Covington and Cincinnati Bridge, where the two cables, 1 foot in diameter, are 59 feet apart over the towers and converge towards the centre of the middle span, where they are only 36 feet apart. The stiffening, however, of the roadway by longitudinal trusses together with strong decking, and wind-ties of wire cables inserted under the roadway, as at the Brooklyn Bridge, furnish more satisfactory protection against injuries by gales.

Remarks on Suspension Bridges.—In several instances, suspension bridges have rendered the important service to engineering of enabling larger spans to be surmounted than previously achieved, and in places where other forms of bridges would at the time have been deemed impracticable, of which the Menai, Friburg, Covington and Cincinnati, Niagara Falls, and Brooklyn have furnished examples in succession; whilst for many years the Niagara Railway and Road Bridge was the bridge of much the largest span traversed by trains. Though, however, the great progress effected in the design of bridges, under the stimulus of the increasing requirements of railway construction, and with the assistance of the greatly cheapened production of steel, together with the unsuitability of ordinary suspension bridges for railway traffic, and the important repairs of roadways and renewals of cables found necessary in many of these bridges, have deprived suspension bridges of their former commanding position, so that the second railway bridge erected over the Niagara rapids was a cantilever, the old Niagara Bridge has now been replaced by an arched bridge, and the great span of the Brooklyn Bridge has been exceeded by the cantilevers of the Forth Bridge, it would be a mistake to suppose that the construction of suspension bridges is a thing of the past. Undoubtedly, the serious weakening of wire cables by internal rust, leading to the fall of some road suspension bridges in France and the insecurity of others many years ago, caused the abandonment of these bridges for a time there; but in 1879 the construction of cable suspension bridges in France was again commenced on improved methods, securing better protection for the wires, supporting the bridge from several detachable cables of improved manufacture, separating the anchorage cables from the suspending cables at the towers, so that one of either could be removed and renewed independently, and making the anchorages themselves open to inspection, and putting them out of reach of damp. In the United States also, though in several cases suspension bridges have been replaced by more rigid and durable structures, nevertheless this type of bridge is still being constructed, as for example the bridges erected across the Ohio in 1896 at East Liverpool and Rochester, and notably the great Williamsburg and Manhattan

Bridges across the East River, New York, which have been already referred to. The East Liverpool Bridge has a central span of 705 feet, and side spans of 420 and 360 feet; and the girders, 20 feet in depth, are connected with the cables at intervals of 15 feet, which rest as usual upon saddles on the towers borne on rollers, the expansion of the girders being provided for at the towers.[1] The Rochester Bridge, the cost of which was limited to about £36,500, precluding the erection of a rigid bridge, has a central span of 800 feet, and side spans of 416 and 400 feet; but in this case the wire cables were fastened on the top of the towers; and the expansion of the stiffening girders, 28 feet in depth at the towers and 18 feet at the centre of the large span and at the ends, is provided for in the middle and at the ends, these unusual arrangements having been resorted to with the view of giving greater stiffness to the bridge.[2]

Bridges with stiffened chains, such as the Pittsburg Point Bridge, cease to be simple suspension bridges with all the parts in tension; for the upper member of the truss, though in tension when only the other half span is loaded, is in compression when the half span it supports is loaded (Fig. 63, p. 141). Bridges also of the type of the Francis-Joseph and Albert bridges (Fig. 61 and 62, p. 141) approximate almost as much to cantilevers as to suspension bridges; and in so far as rigidity is secured in a suspension bridge, there is a corresponding approach to the cost of a rigid bridge. In some cases, moreover, an arched bridge springing from near the base of a cliff, can cross a ravine with such a much smaller span than a suspension bridge at the top, that the arched bridge may prove the most economical, as for instance the Niagara Arched Bridge (Fig. 54, p. 132), which provides double the width of railway and roadway afforded by the old Suspension Bridge, at an increased expense of only between one-third and one-fourth of the first cost of the original bridge.

There are, however, many sites where, owing to the unfavourable conditions and the smallness of the traffic, the cheapness obtained by the suspension system is almost indispensable, of which the tourist roadway bridge below Niagara Falls furnished for nearly twenty years a striking example, having been rebuilt exactly as before after the wrecking of the roadway by the storm of January, 1889, though at last superseded, in 1898, by an arched steel bridge of 840 feet span (Fig. 56, p. 134); and the suspension bridges recently erected across the Ohio, referred to above, are notable instances of the value of the system. Moreover, both metal arch and suspension bridges possess the advantages over ordinary girder bridges, of being capable of erection without scaffolding, by building out in the one case, and by the aid of a suspended platform in the other, and of having their heaviest portions close to their supports.

[1] *Engineering News*, New York, vol. xxxvii. p. 198.
[2] *Ibid.*, vol. xxxvii. p. 194.

CHAPTER X.

GIRDER BRIDGES.

Girders independent structures—Bowstring girders, form, examples—**Girder Bridges over Single Openings:** classification of girders ; simple beam, bending moments, breaking weights ; girders, proper arrangement of materials, variations in bending moment, in shearing stresses, functions of flanges and web or latticed bracing ; calculation of stresses, in flanges, in bracing , girder bridges of small span, forms adopted, provision for expansion and contraction ; lenticular girder bridge, form of, Saltash Bridge ; tubular girders, Conway Bridge, Britannia Bridge, objections ; lattice-girder or truss bridges of large span, forms of, provisions for unequal loading, steel for reducing weight, riveted connections, pin connections, best forms for struts, depth in relation to span, weights of various forms ; large pin-connected girders, Covington and Cincinnati Bridge, Hawkesbury Bridge ; erection on scaffolding, with the aid of temporary girders, by floating out—**Continuous Girder Bridges:** stresses ; advantages, Lachine Bridge ; erection by rolling out, Bouble and Credo viaducts.

GIRDERS, unlike arches or suspension cables, need no extraneous support in the form of abutments or anchorages, but merely rest upon their piers at each end. Single girders, however, require scaffolding or supports for their erection ; and, consequently, they are not suitable for crossing valleys at a considerable height, or for spanning deep ravines, which, as indicated in the last chapter, have been satisfactorily effected by metal arches and suspension bridges.

Bowstring Girders.—The form of girder to which the term "bowstring" is applied, is in reality a metal arch, in which the thrust is borne by a horizontal tie joining the two ends, constituting the string of the bow, which converts the arch into a self-contained structure resting upon its supports at each end, and no longer needing abutments to bear the thrust. The roadway in this case is generally suspended from the arch ; and light bracing is introduced between the verticals, to provide for the unequal distribution of the load. The Newcastle High-level Bridge, which carries a railway above and a roadway below, is a good example of a cast-iron bowstring-girder bridge, having six spans of 125 feet raised 83 feet above high water, which was opened in 1849. This type of girder is unsuitable for large spans, and is now rarely used; but the principle is sometimes resorted to for station roofs, of which the finest example is the roof of St. Pancras Station in London, which has an unimpeded span of 240 feet, and a clear height in the centre, above the platform level, of 96 feet, where the ties counteracting the thrust of the arch pass under the lines of railway and platforms, consisting of the girders supporting the floor of the station over the cellarage below.[1]

GIRDER BRIDGES OVER SINGLE OPENINGS.

Girders must be separated into two distinct divisions, namely, girders across single openings, and continuous girders spanning two or more openings. The first form of girder is constructed according to a variety of types, rising from a simple beam with a small span, up to braced girders of considerable depth, spanning openings reaching a maximum width of between 500 and 600 feet.

Simple Beam.—When a horizontal beam AB is supported at its two extremities A and B, and is loaded with a weight W at any point C, the bending moment at C, due solely to this weight, is $M = W\frac{AC \times BC}{AB}$, and at any other point D of the beam, between B and C, is $m = W\frac{AC \times BD}{AB}$. The bending moment M becomes a maximum when AC = BC, or the point C and the load W are in the centre of the span, and $M = \frac{1}{4}WL$, where L = AB; and in this case, if the distance of any other point D from the nearest support, or BD, is x, then the bending moment at D is $m = \frac{1}{2}Wx$, and diminishes uniformly from $\frac{1}{4}WL$ at the centre of the span to zero at the supports. When the load W is uniformly distributed, M becomes $\frac{1}{8}WL$, or $\frac{1}{8}wL^2$, where w is the weight per unit of length; and m is $\frac{W}{2} \cdot \frac{AD \times BD}{AB}$ at any point D, or $\frac{1}{2}wx(L - x)$, the bending moments on the beam varying, as shown by the shaded portion **AEB**, (Fig. 67), from a maximum $\frac{1}{8}WL$ in the centre of the span to zero at the supports at **A** and **B**.

DIAGRAM OF BENDING MOMENTS.

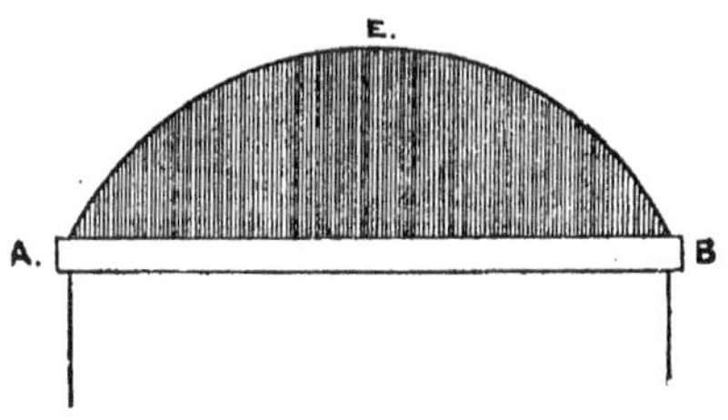

Fig. 67.—Distributed Load.

The breaking weight in cwts. of a horizontal, rectangular beam, supported at each end and carrying a distributed load, is given by the formula $\frac{8KBD^2}{L}$, where B is the breadth, D the depth, and L the length of the beam in inches, and K is a coefficient varying with the nature of the material, being 68 for wrought iron, 22 for African oak and 15 for English oak, 13 to 10 for pine, and only 7 for elm.[2] The formula shows the importance of depth in a beam for increasing its strength; for whereas two beams of similar

[1] *Proceedings Inst. C.E.*, vol. xxx., plate 8, fig. 1.

[2] "Pocket-book of Useful Formulæ and Memoranda," Sir Guilford L. Molesworth, p. 136.

breath and depth, laid side by side, have double the strength of a single beam, a beam of the same breadth, but double the depth, is four times as strong as the original beam, though the amounts of material in the two beams and in the deep beam are the same.

Girders.—The arrangement of the material in a rectangular beam is unfavourable for the strength of the beam, for its resistance to the bending moment does not merely depend on its sectional area, but on its sectional area multiplied by the square of the radius of gyration, or its moment of inertia. Now, the radius of gyration is the distance of the centre of gravity of the section, on each side of the neutral axis, from this central axis; and therefore the strength of the beam or girder is increased by putting the mass of the section, on each side of the neutral axis, as far from the centre as practicable. This is effected in wooden trusses by putting the main longitudinal timbers at the top and bottom of the truss, and connecting them together by wooden struts and iron tie-rods; but it is more thoroughly accomplished in wrought-iron or steel girders, where the mass of the metal is concentrated in the top and bottom flanges, as far as possible from the neutral axis. These flanges usually consist of a set of long, thin, flat plates riveted together and placed horizontally; and in small girders, the flanges are kept apart, and depth thus given to the girder, by a vertical web-plate connected to the flanges at the top and bottom by angle-irons, which strengthen the flanges (Figs. 69–72, p. 158); whilst in large girders, the function of the web is performed by various systems of lattice work, composed of struts and bars united to channel-shaped flanges by rivets or bolts (Figs. 74–77, p. 160).

The bending moment of a uniformly loaded girder is a maximum in the centre of the span, and diminishes towards the piers, according to the formula given on p. 153 for a beam, and as indicated in Fig. 67. Consequently, the sectional area of the flanges must be made largest in the centre of the girder, and may be gradually reduced towards the ends by diminishing the number of the plates; or, the sectional area of the flanges being maintained, the depth of the girder may be reduced in proportion to the reduction in the bending moment, as partially effected in the Kuilenberg and Covington, and Cincinnati bridges, by curving the top flange (Figs. 75 and 77, p. 160).

The shearing stress of a girder uniformly loaded, or the force tending to shear or buckle up the web-plate or lattice bars which give the girder its essential element of depth, increases regularly from zero in the centre of the span to a maximum at the edge of the piers, where it amounts to half the distributed load. The web-plates, accordingly, of small girders have to be strengthened and stiffened towards the ends of the girders by T irons or angle-irons, and at the ends, and occasionally in larger girders at other points, by gusset plates as well as angle-irons (Figs. 71 and 72, p. 158); whilst the struts and bars of large girders are made successively stouter from the centre of the span to the abutments.

The web or lattice bracing, giving depth, and therefore strength, to the girder, transmits the stresses between the centre of the span and the abutments. Considering the stresses of a straight girder, commencing at the abutments, the vertical shearing stress at these points of the girder is borne by the lattice bracing; and the horizontal component of the stresses on this bracing is transmitted to the flanges at each point of connection, till, in the centre of the span, the vertical shearing stress disappears, and the total horizontal stresses are borne by the flanges, being compressive in the top flange, and tensile in the bottom flange. Starting, on the other hand, at the centre of the span, the horizontal stresses due to the bending moment are borne by the flanges; but the vertical stresses due to the load have to be taken up by the lattice bracing at each point of connection, the total vertical stresses, equal to half the whole load, being eventually transmitted by the bracing to each abutment.

Calculation of Stresses in Girders.—In order to design a girder of any given span, the stresses on the flanges and on the lattice bracing or the web-plate have to be ascertained. The stresses on the top and bottom flanges in the centre of a uniformly loaded girder having a given span and depth, are the same whatever may be the form of the girder or the system of lattice bracing adopted; but the stresses on the lattice work vary with the form of the girder and the nature of the bracing. The calculation, therefore, of the sectional area of the flanges of a girder in the centre of the span, will apply generally to any form of girder spanning a single opening; and this sectional area should be maintained constant throughout the span, if the flange which does not carry the roadway is so curved as to render the depth of the girder proportionate to the bending moments, but should be reduced towards the piers, in a girder of uniform depth, with the decrease in the bending moments. The calculations, however, of the stresses on the lattice bracing are modified by the form of the girder, and by the arrangement and inclination of the bracing.

The sectional area of the flanges in the centre of a girder, having a length L, an effective depth D, and a uniformly distributed load W, may be obtained as follows. The safe resisting strength of the girder in the centre of the span must be equal to the bending moment, or $\frac{SI}{\frac{1}{2}D} = \frac{WL}{8}$, where S is the stress which the metal can bear without injury, and I is the moment of inertia of the flanges. Let x be the sectional area of the top flange in square inches, and S_c the safe compressive stress in tons per square inch; and let y be the sectional area of the bottom flange, and S_t the safe tensile stress. Assuming that the distance of the centre of gravity of each flange of the girder from the neutral axis may be approximately represented by $\frac{1}{2}D$, the above equation becomes for a single flange—

$$\frac{S_c \frac{1}{4} D^2 x}{\frac{1}{2}D} = \frac{1}{2}\frac{WL}{8}, \text{ or } S_c D x = \frac{WL}{8};$$

and therefore the sectional area in square inches of the top flange, $x = \frac{WL}{8DS_c}$,

and the sectional area in square inches of the bottom flange, $y = \frac{WL}{8DS_t}$, where W is the distributed load in tons. For iron, $S_c = 4$, and $S_t = 5$; whilst in the case of steel, the value allowed is $6\frac{1}{2}$ for both. When cast-iron girders were employed, the bottom flange had to be made much larger than the top flange, owing to the relatively small power of cast iron to resist tensile stresses; but in wrought-iron girders of moderate span, the top and bottom flanges have often been made of equal sectional area, for the rivets, which may be included in the sectional area of the top flange with a compressive stress, have to be deducted from the sectional area of the bottom flange subjected to tension, and approximately neutralize the advantage gained from the greater tensile strength of wrought iron.

One of the simplest forms of lattice bracing for girders is the isosceles triangle type, employed in the Warren girder; and the calculation of the stresses in this girder will suffice to indicate the distribution and transmission of the stresses in a lattice girder, and the general method of determining the stresses. A diagram of half the span, **AI**, of a Warren girder is given in Fig. 68, where the roadway is carried on

STRESSES IN WARREN GIRDER.

Fig. 68.

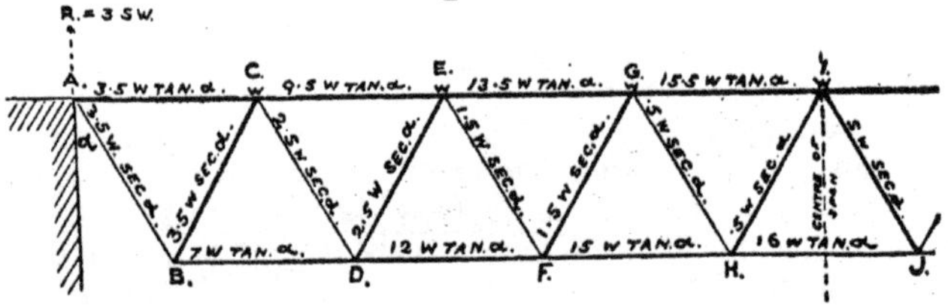

the top of the girder, and uniform loads w are assumed to be imposed at the points of junction of the bracing with the top flange. The reaction R at the abutment **A**, supporting the weight of the girder, is equal to half the total load, or $3{\cdot}5\,w$; and as the girder is in equilibrium at **A**, the tensile stress along the tie **AB** is $\frac{3{\cdot}5w}{\cos\alpha}$, or $3{\cdot}5w \sec\alpha$, where α is the angle at which the bracing is inclined to the vertical, and the compressive stress along the portion of the top flange **AC** is $3{\cdot}5w \tan\alpha$. The compressive stress along the strut **BC** is the same as the stress along **AB**, namely $3{\cdot}5\,w \sec\alpha$; whilst the tensile stress along the portion of the bottom flange **BD** is the sum of the horizontal components of the stresses along **AB** and **BC**, namely $7w \tan\alpha$. The stress along the tie **CD** is $(3{\cdot}5w - w)\sec\alpha = 2{\cdot}5w \sec\alpha$, and along **CE** is $3{\cdot}5w \tan\alpha + (3{\cdot}5w + 2{\cdot}5w)\tan\alpha = 9{\cdot}5w \tan\alpha$; and the remaining stresses on the girder, indicated in Fig. 68, are obtained in a similar manner. The stresses on the lattice bars can also be obtained by starting at the centre of the span, where, resolving the vertical weight w which is borne by the struts **IH** and **IJ**, the stress along each of these struts is $0{\cdot}5w \sec\alpha$, and the stress along the tie **GH** is the same; whilst the stress along the strut **GF** is $(0{\cdot}5w + w)\sec\alpha = 1{\cdot}5w \sec\alpha$, the stresses progressing according to the series 1, 3, 5, 7, etc., for any number of pairs of struts and ties.

A web-plate may be regarded as a continuous series of lattice bars subjected both to compression and tension, and transmitting the stresses to the flanges at every point of their length. Web-plates are serviceable for small girders where suitable lattice bars would be small, their workmanship relatively costly, and their connections with the flanges somewhat inconvenient; but these plates cannot be exactly proportioned, like lattice bracing, to the different stresses they have to resist, for their thickness in the centre of the span cannot be safely reduced below about $\frac{1}{4}$ inch, owing to the large surface exposed to corrosion; and the necessary stiffening of the web-plates towards the abutments, to bear the increasing compressive stress, is only roughly provided for by the vertical T irons or angle-irons usually introduced.

Girder Bridges of Small Span.—The form adopted for girder bridges depends largely upon the available headway and the span. The cheapest form of railway girder bridge, where the headway admits, is obtained by placing a girder directly under each rail, with a small girder at each side of the bridge for carrying the parapet (Fig. 69, p. 158). In this way, the rolling loads come exactly over their supports; and no material is expended in cross girders, rail-bearers, or other means of carrying the load at the side of the main girders, and in adding to the strength of these girders to carry this extra weight. When the available headway is very small, trough girders are adopted for small spans (Fig. 70), so as to reduce the depth between the level of the rails and the under side of the bridge to a minimum. For larger spans, the system of main girders supporting a series of cross girders on their bottom flanges is adopted (Fig. 71, p. 158), thereby enabling the supporting main girders to be made of any requisite depth without affecting the headway below, but involving the additional cost of the cross girders and rail-bearers, as well as additional metal in the main girders to carry this extra load, which are dispensed with in the system shown in Fig. 69. The depth, however, between the rail-level and the underside of the bridge can be considerably reduced if necessary, and the weight of flooring to be borne by the main girders may be materially diminished, by introducing a central main girder between the lines of way, and substituting trough flooring plates (Fig. 72) for the cross girders, rail-bearers, and curved roadway plates shown in Fig. 71.

Expansion and contraction have to be provided for in all metal girder bridges, by bolting down one end only of each girder to the bearing bed-plate on one abutment, and leaving the other end quite free to move, either on a series of iron or steel rollers, or on a perfectly smooth bed-plate, for wrought iron expands 0·000006 of its length, on the average, for each degree Fahrenheit.

Lenticular Girder Bridge.—There is a peculiar form of girder constituting a sort of combination of the arch and suspension systems, in which the thrust of the arch is counterbalanced by suspension chains hanging below, which in their turn obtain their anchorages by their connection to the extremities of the arch. Accordingly, this type of girder, which has been termed lenticular on account of its shape, combines the strength of the arch with that of the suspension bridge in a self-contained structure, avoiding any waste of material in abutments,

CROSS SECTIONS OF GIRDER BRIDGES.
Scale 8 feet to 1 inch.

Fig. 69.—Bridge with Girders under Rails.
40 Ft. Span.

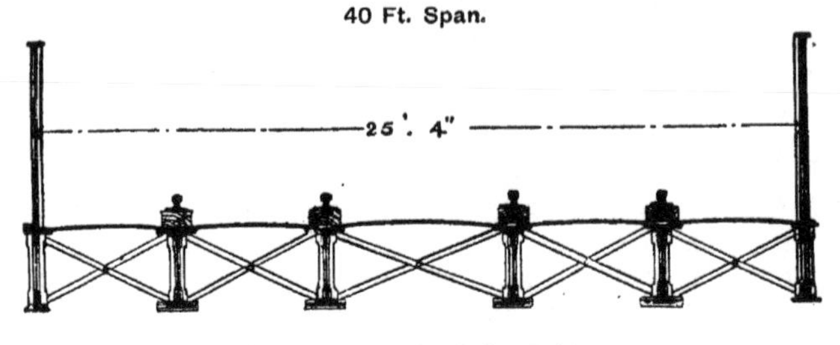

Fig. 70.—Trough Girder Bridge.

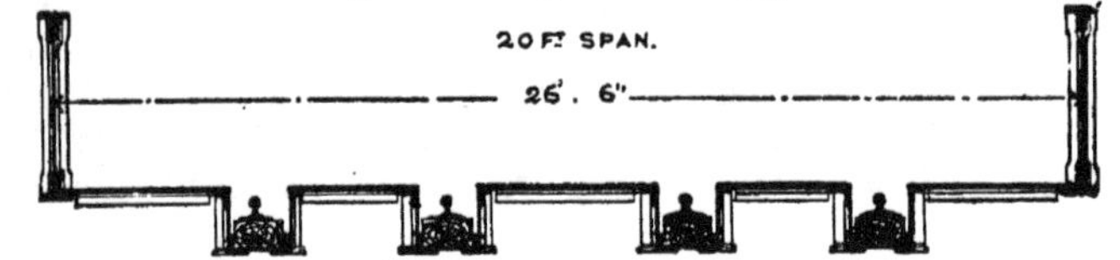

Fig. 71.—Bridge with Main and Cross Girders.

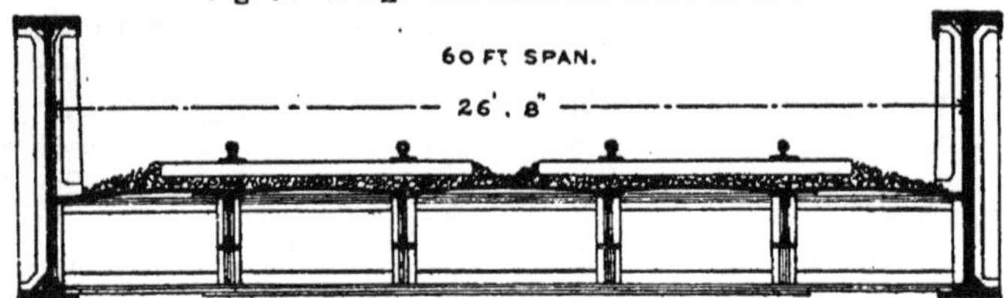

Fig. 72.—Bridge with Three Girders.

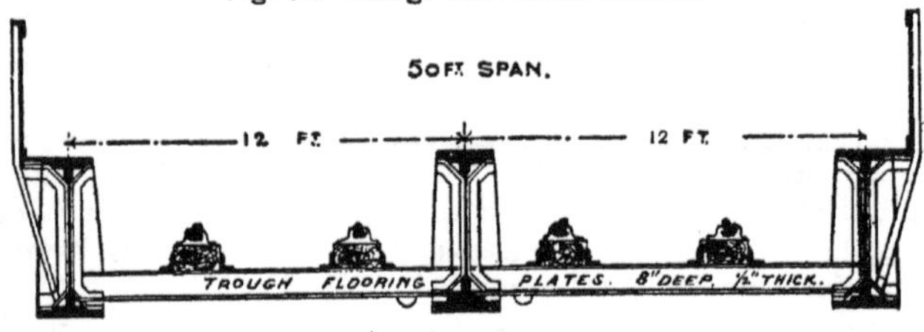

tie-bars, or anchorages, and therefore should constitute an efficient and economical system. Only a few bridges, however, have been constructed on this principle, of which the first and best-known example is the Saltash Bridge, erected across the River Tamar in 1853–59, with two spans of 455 feet between the supports of the girders, and leaving two clear openings of 436 feet between the piers, with a headway of 100 feet above high water of spring tides (Fig. 73, p. 160). The upper, arched portion of the girder consists of an elliptical tube, $16\frac{3}{4}$ feet wide and $12\frac{1}{4}$ feet deep, formed of riveted wrought-iron plates, and stiffened by internal diaphragms and bracing; whilst the suspension chain below, which is strongly braced to the tube, is composed of a row of fourteen and fifteen bars alternately, 20 feet long, 7 inches wide, and 1 inch and $\frac{15}{16}$ inch thick respectively.[1] The depth of the girder in the centre of the span, between the centre of the tube and the centre of the suspension chain, is $56\frac{1}{4}$ feet, or one-eighth of the span. The cross girders, carrying a single line of railway, are suspended by bars from the chains of the two girders spanning each opening.

Tubular Girders.—The first wrought-iron girders were made in the form of rectangular, hollow tubes, with plate-iron sides and cellular top and bottom flanges, and were erected across the River Conway, alongside Conway Castle, in 1848, for the Chester and Holyhead Railway. Conway Tubular Bridge has a clear span of 400 feet, and consists of two parallel tubes for the up and down lines, placed a clear 9 feet apart, 15 feet in width, and increasing in height from $22\frac{1}{2}$ feet at the ends to $25\frac{1}{2}$ feet in the centre.[2] The Conway end of each tube was fastened down to the abutments; but the other ends rest upon rollers and gun-metal balls, to allow for the expansion and contraction of the tubes, which at extreme temperatures might have a range of nearly $2\frac{1}{2}$ inches in a length of 424 feet. A camber, or slight rise towards the centre, was given to the tubes, amounting to $6\frac{1}{2}$ inches at the centre in the first tube, and $8\frac{1}{2}$ inches in the second, a plan generally adopted with girders to allow for subsequent deflection due to the weight of the girder and the rolling load. The permanent deflection of the tubes in the centre appears to have been about $1\frac{3}{4}$ inches in excess of the camber. The cost of the Conway Bridge was £145,200. Similar tubes were employed for the Britannia Tubular Bridge, carrying the same railway across the Menai Straits, with two central spans of 459 feet, and two side spans of 230 feet; but though the tubes were floated out and raised into position separately across each of the large spans in 1849–50, like the Conway tubes, the tubes of each of the two lines were subsequently connected across the piers, so as to form two continuous girders extending right across the straits.

The cellular flanges, though very suitable in form for resisting compressive stresses, are with difficulty protected efficiently against corrosion; whilst the two plate-iron sides of the tubes, acting as double web-plates, though giving great stiffness to the tubes, amount to one-third of the total weight of the tube. Accordingly, the cellular system of flanges was superseded by layers of plates, stiffened in the top flange by

[1] "A Treatise on Cast and Wrought Iron Bridge Construction," W. Humber, p. 231, and plates 78–80.

[2] "The Britannia and Conway Tubular Bridges," Edwin Clark, vol. ii. The effective span of the Conway Bridge has been since reduced.

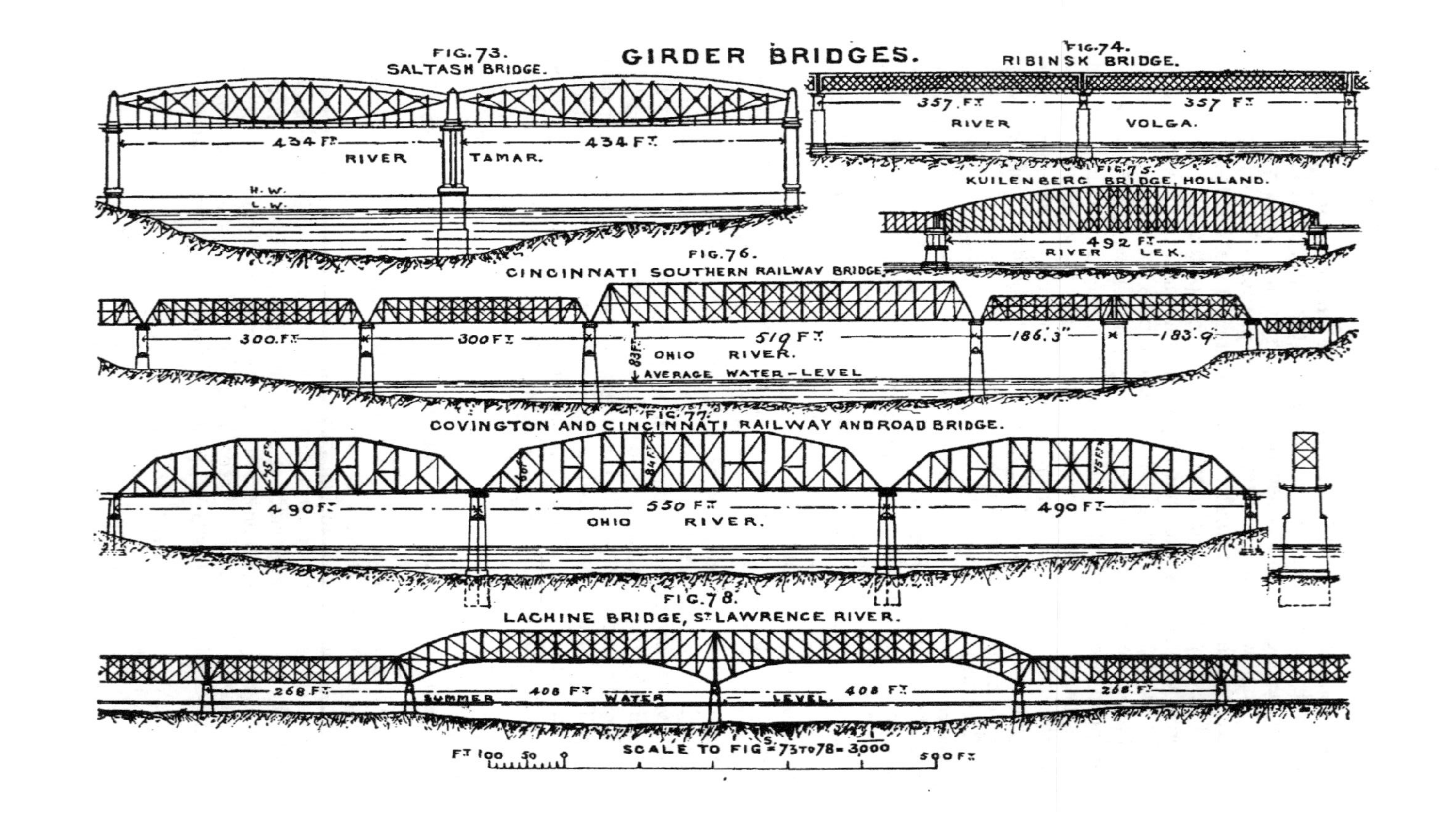
GIRDER BRIDGES.
FIG. 73.
SALTASH BRIDGE.
434 FT.
RIVER
TAMAR.
434 FT.
H.W.
L.W.
FIG. 74.
RIBINSK BRIDGE.
357 FT.
357 FT.
RIVER
VOLGA.
FIG. 75.
KUILENBERG BRIDGE, HOLLAND.
492 FT.
RIVER LEK.
FIG. 76.
CINCINNATI SOUTHERN RAILWAY BRIDGE.
300. FT.
300 FT.
519 FT.
186.3"
183.9"
83 FT.
OHIO RIVER.
AVERAGE WATER-LEVEL
FIG. 77.
COVINGTON AND CINCINNATI RAILWAY AND ROAD BRIDGE.
490 FT.
550 FT.
490 FT.
OHIO RIVER.
FIG. 78.
LACHINE BRIDGE, ST LAWRENCE RIVER.
268 FT.
408 FT.
408 FT.
268 FT.
SUMMER WATER LEVEL.
FT. 100 50 0
SCALE TO FIGS. 73 TO 78 = 3000
500 FT.

projecting ribs, in the long tubular bridge erected across the St. Lawrence at Montreal in 1854–59, having a central span of 330 feet, and twenty-four side spans of 242 feet;[1] and tubular girders, with solid plate-iron sides, have been abandoned in favour of a lattice web, with bars proportioned to the strains at the different parts of the girder, considerably lighter than solid web-plates, and offering much less surface to the wind.

Lattice-Girder or Truss Bridges of Large Span.—Formerly a common type of construction for lattice-girder bridges consisted of parallel top and bottom flanges of plate iron, closely braced with diagonal struts and ties, constituting in fact a multiplex Warren girder, wherein the points of connection of the bracing with the flanges were put fairly close together, thereby stiffening the girder and enabling the dimensions of the lattice bars to be kept very moderate, as illustrated by the Ribinsk Bridge over the Volga, having spans of 357 feet between the centres of its piers (Fig. 74, p. 160). Though bridges of moderate span are still constructed in accordance with this type, more recent practice in bridges of large span, consists in resorting to greater depths for the girders in proportion to the span, and placing the struts vertical, thereby reducing their length, whilst retaining the diagonal position for the ties, as exemplified by the Kuilenberg Bridge, and the Cincinnati Southern Railway Bridge,[2] (Figs. 75 and 76, p. 160), having large spans of 492 feet and about 509 feet, and depths of girders of $65\frac{1}{2}$ feet and $51\frac{1}{2}$ feet, or about two-fifteenths and one-tenth of the span respectively.

The reduction in depth of a large girder towards its extremities has the advantage of reducing the length of the struts and ties where the stresses on them are augmented, and has been to some extent effected in the Kuilenberg and Covington and Cincinnati bridges (Figs. 75 and 77, p. 160); but this reduction cannot generally be carried conveniently to the extent of being made proportionate to the reduction in the bending moment, since the lowness of the girders at the ends would necessitate the abandonment of the overhead bracing across the roadway towards the extremities of the girders. This overhead bracing is usually introduced for stiffening the high girders of large spans, and more particularly for increasing their resistance to wind-pressure, as indicated in the cross section of the Covington and Cincinnati Bridge (Fig. 77, p. 160).

The cross bracing which is shown in the central portion of the girders of the Kuilenberg and Cincinnati Southern Railway bridges, would be unnecessary if the moving load was always uniformly distributed; but the loading of one half only of the bridge throws a compressive stress on the diagonal ties for a certain distance beyond the centre on the unloaded portion of the bridge, which has to be provided against by the introduction of the cross bracing along the part of the girder in which the reversal of the stresses is liable to be produced by unequal loading. In the ordinary diagonal, lattice girders, the central

[1] "Construction of the Great Victoria Bridge in Canada," J. Hodges.
[2] "Report on the Progress of Work and Cost of Construction of the Cincinnati and Southern Railway," G. Bouscaren, Cincinnati, 1878.

tie-bars are constructed similarly to the struts, so as to be adapted to resist the compressive stresses which unequal loading is liable to impose on them.

In girders of large span, it is very important to reduce the dead load to be carried, and consequently the weight of the girder, by every means consistent with the preservation of the requisite strength and stability of the girder. The introduction of steel, with its considerably higher tenacity and resistance to compression, has enabled the weight of girders to be materially reduced; and therefore the sectional areas of the parts of steel girders may not merely be reduced in proportion to the greater strength of steel as compared with wrought iron, but can be still further reduced owing to the diminution in the weight to be borne. Moreover, the forms of the flanges, the design and arrangement of the bracing, and the nature of the connections very materially affect the weight of the girders. In Europe, the flanges are generally formed of horizontal flat plates, with projecting vertical plates at the sides connected by angle-irons, so as to constitute more or less of a trough-shaped section, to the sides of which the bracing is riveted; whilst in America, the bottom flange is often formed of a series of bars, the bracing being usually joined to these and to the top flange by bolts, or, as it is termed, pin-connected. The system of riveted plates, which is suitable for the top flange in compression, becomes somewhat wasteful in the bottom flange in tension, owing to the loss of area due to the rivet-holes and the additional weight of cover plates over the joints; and consequently, economy of material is effected by the use of eye-bars connected by bolts, similarly to the chains of a suspension bridge. For the same reason, the employment of pin-connected bars for the tension members of the bracing, instead of riveted plates, reduces the amount of metal. In fact, a comparison of the weight of the large span of the Kuilenberg Bridge, with girders formed of riveted plates, and of the main span of the Cincinnati Southern Railway Bridge, as increased to correspond to a double line of way, whose girders are built up with tension bars and pin-connections (Figs. 75 and 76, p. 160), shows that an economy of about 27 per cent. has been gained in the weight of the latter bridge by the modifications adopted,[1] in spite of the greater central depth of the Kuilenberg Bridge, and the reduction in the weight of the bracing towards the ends by the diminution in the depth.

The tie-bars in a girder can be made of any convenient form, provided their sectional area is adequate to sustain the tensile stresses to which they are subjected; but as the struts have to withstand flexure, like weighted columns, the metal of which they are formed should be adjusted so that its section may afford the largest practicable moment of inertia about their central axis; and therefore, though channel-irons and pairs of bars braced together are sometimes employed for struts, the most suitable forms are box-shaped sections or circular tubes.

An increased depth of girder produces a corresponding reduction in the stresses on the flanges, and consequently in their requisite weight;

[1] "A Practical Treatise on Bridge-Construction," T. Claxton Fidler, p. 349.

but as an increased depth involves an augmentation in the length of the braces, necessitating more particularly greater stiffness in the elongated struts, the ratio of the depth of a girder to its span is now commonly made between one-tenth and one-eighth of the span. These proportions, however, manifest a considerable advance on the ratio of one-sixteenth adopted for the earliest wrought-iron girders of large span; whilst the ratio of the depth to the span in the large span of the Kuilenberg Bridge is two-fifteenths, and in the peculiar girders of the Covington and Cincinnati Bridge (Fig. 77, p. 160), with the exceptional, central, clear span of 533 feet for single girders, the depth only slightly exceeds one-sixth of the span.

The reduction in the weight of the girders, so important for large spans, which has been effected by increasing the depth and by adopting open bracing for the web, is demonstrated by a comparison of the weights for similar spans of the earlier and later bridges. Thus the weight of iron in the large 330-feet span of the Victoria Tubular Bridge at Montreal, completed in 1859, was 686 tons; whereas the lenticular girders of the Mainz Bridge for a span of 345 feet, completed in 1862, weigh 359 tons; each of the fourteen 349-feet spans of the Moerdyk Bridge, completed in 1871, weighs 447 tons; and each span of the Kentucky River Bridge, carried by pin-connected girders, erected in 1877, across three openings of 375 feet, weighs 425 tons.[1] One of the tubes of the Conway Bridge, erected in 1848, across a clear opening of 400 feet, weighs 1112 tons; whilst each 397-feet span of the close lattice Dirschau Bridge over the Vistula, completed in 1856, weighs 838 tons; and the 396-feet span of the pin-connected Louisville Bridge, erected in 1870, over the Ohio, weighs 623 tons. Lastly, one tube of each of the two 460-feet spans of the Britannia Bridge, erected in 1849–50, weighs 1553 tons; whereas each of the 455-feet spans of the Saltash Bridge, completed in 1859, weighs only 945 tons (Fig. 73, p. 160); and the main span of the Cincinnati Southern Railway Bridge, exceeding the large spans of the Britannia and Saltash bridges by about fifty feet, carried by pin-connected girders 515 feet long,[2] erected in 1877, weighs 1176 tons, (Fig. 76, p. 160). The Victoria (Jubilee) Bridge across the St. Lawrence at Montreal was reconstructed in 1897–98, and deep pin-connected steel girders of the modern type on the existing and modified masonry piers, carrying a double line of railway and roadways, replaced the old single-line wrought iron-tubular plate-web girders, erected some forty years before. The total weight of ironwork in the old bridge was about 9000 tons; that in the new, about 22,000 tons.

The different types of girders commonly employed in America, exhibiting certain variations in the arrangement of the bracing, are sometimes distinguished by the names of the persons who first introduced them, so that there are the Howe, Whipple, Linville, Pratt, etc., trusses; but probably the type most frequently adopted is the Linville truss, which differs little from the Whipple truss, of which the Cincinnati Southern Railway Bridge furnishes an example (Fig. 76, p. 160).

[1] *Proceedings Inst. C.E.*, vol. liv. pp. 194 and 246, and plates 8 and 9.
[2] *Ibid.*, plate 7.

Large Pin-connected Girders of Varying Depth.—The steel girders across the large spans of the Covington and Cincinnati Bridge, exhibit a form of single bracing differing more widely than usual from the more ordinary American types (Fig. 77, p. 160). These girders are notable in spanning a larger central opening than any other unconnected girders, in being given a greater depth than usual in proportion to the span, and in having to carry two outer roadways with a clear width of 11 feet, and a footway 5 feet wide outside each roadway, in addition to two lines of railway in the centre between the two girders.[1] The upper polygonal member of these girders, under compression, is formed of a box-shaped girder with three solid webs; and the straight bottom member in tension is composed of rows of bars, increasing in number from ten at the piers to eighteen in the centre. The struts, which are indicated in the illustration by darker lines than the ties, are formed of suitable box, plate, and braced girders; whilst the ties are made of bars; and the struts, ties, and wind-bracing are connected to the top and bottom members by steel pins.

The greatest stresses to which the members of these girders are liable to be subjected in tension, are 10,000 lbs. or 4·46 tons per square inch for wrought iron, and 16,000 lbs. or 7·14 tons per square inch for steel. For the upper steel members under compression, the sectional areas have been so designed that, wherever the ratio of the length between the pins to the least radius of gyration does not exceed 50, the maximum stress is 14,000 lbs. or 6·25 tons per square inch; but where this ratio exceeds 50 in the top members, and also for the struts, the allowable maximum stress was determined by the fraction $14{,}000 \div 1 + \frac{1}{20{,}000} \cdot \frac{l^2}{r^2}$, where l is the length of the column, and r the least radius of gyration to the same unit of measure. The ratio of the length of a strut to its least diameter was not allowed to exceed 45; and an allowance of 50 per cent. on the working stresses was made for wind-pressure. Provision was made for the expansion and contraction of the girders over a range of 150° Fahrenheit; and the free end of each girder rests on steel rollers.

The girders of the three large spans were erected on scaffolding, or false work as it is termed in America, within the short period of ten months in 1888, in spite of delays and injuries from unusual floods in the river. The total weight of metal in these spans is about 4465 tons, which, assuming that the weight in each span is proportional to the square of the span, may be reckoned as 1725 tons for the central span, and 1370 tons for each side span. Considering the very unusual width of railway, roadways, and footways carried, these weights are very moderate, a result due to the use of steel, the simple character of the bracing, the depth of the girders in the centre, and the reduction of their depth towards the ends.

The Hawkesbury Bridge in New South Wales, designed and erected by an American Company about the same period, and opened in 1889,

[1] *Transactions of the American Society of Civil Engineers*, vol. xxiii. p. 47, and plates 10, 14, 15, and 16.

is carried by girders very similar in type, but only 410 feet long and 58 feet deep at the centre, spanning seven openings across the Hawkesbury estuary, and resting on piers placed 416 feet apart centre to centre[1] The cost of this bridge, with piers founded at a great depth (see p. 72), reaching a maximum of 120 feet below the river-bed, was £327,000.

Erection of Girders spanning Single Openings.—The only methods available for erecting single, unconnected girders, are building on temporary scaffolding, or floating into place; for the system of building out from the piers, so successfully resorted to for arched metal bridges, cannot be employed for erecting unconnected girders.

Erection on scaffolding is the simplest method, and is commonly

SCAFFOLDING FOR ERECTING GIRDERS.

Fig. 79.—Kuilenberg Bridge.

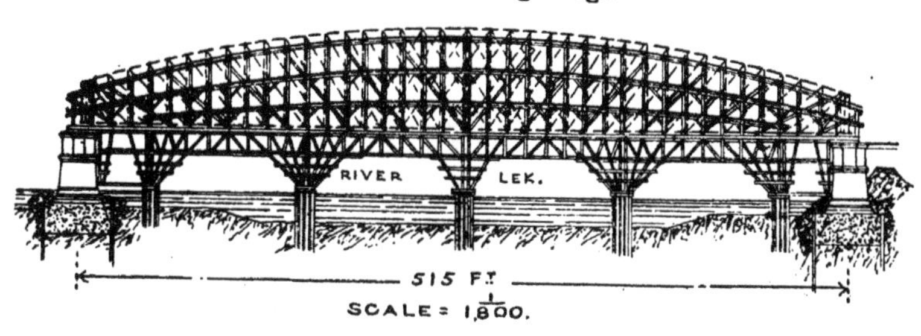

resorted to when the location of the girders is not very high, or the river not very deep, the scaffolding being supported on piles driven into the bed of the river. An unimpeded waterway, however, has often to be left for the discharge of the river, or for navigation; and this has to be provided for, either by only erecting the scaffolding across one or two of the openings at a time, in the case of a bridge of several spans across a river, or by clustering the timber supports together, and spanning the intervals by wooden trusses, as in the erection of the large span of the Kuilenberg Bridge across the River Lek[2] (Fig. 79). Occasionally, when a large opening has to be left, temporary iron girders, supported on timber stagings serving as piers, are used to provide a platform on which the permanent girders are put together, as adopted for the erection of the central span of the Königswart Bridge across the River Inn[3] (Fig. 80), where the timber structures on which the temporary girders rested were also utilized for the erection of the high central piers of the bridge.

Floating completed girders out into position on pontoons has been extensively adopted, especially where high staging would otherwise be required in an exposed position, as for instance in the erection of the girders of the large spans of the Britannia and Saltash bridges; also where

[1] *Proceedings Inst. C.E.*, vol. ci. p. 2, and plate 1, fig. 1, and plate 2, fig. 2.
[2] "Travaux Publics en Hollande," P. Croizette-Desnoyers, plate 28, fig. 1.
[3] *Zeitschrift fur Baukunde*, vol. i., 1878, p. 223, and plates 15 to 17.

the river to be crossed is deep, and the site open to strong gales, as at the Moerdyk Bridge across the Hollandsch Diep; and, lastly, where a river is liable to high floods, or staging would inconveniently obstruct naviga-

TEMPORARY GIRDERS FOR ERECTION OF BRIDGES.

Fig. 80.—Königswart Bridge.

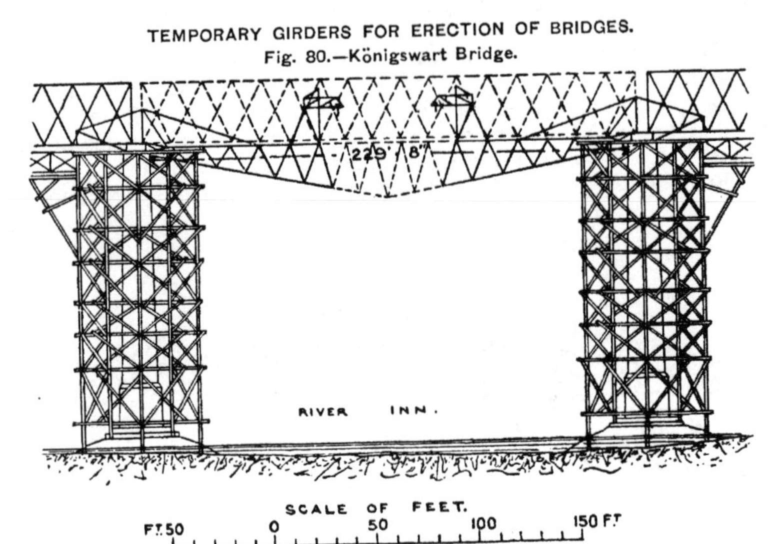

tion. The four tubes of the two large spans of the Britannia Bridge were successively floated on pontoons into their positions between the piers in the Menai Straits,[1] and the large Saltash girders were similarly

FLOATING OUT GIRDERS.

Fig. 81.—Moerdyk Bridge.

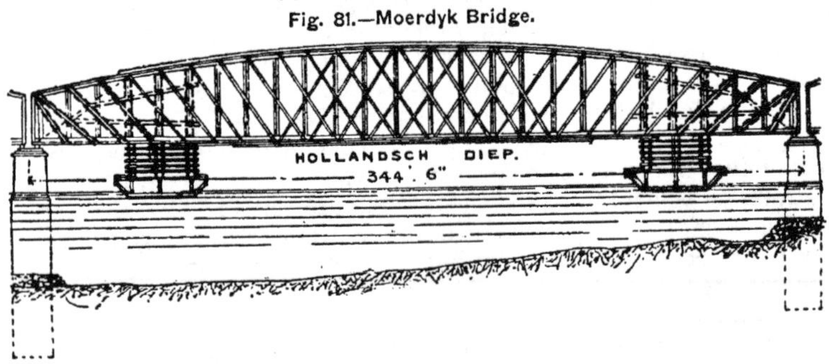

floated into place;[2] and the tubes and the girders were gradually raised by hydraulic presses to their requisite heights of $103\frac{3}{4}$, and 100 feet

[1] "The Britannia and Conway Tubular Bridges," Edwin Clark, vol. ii. pp. 675 to 696, and plates 24, 25, and 34.

[2] "A Treatise on Cast and Wrought Iron Bridge Construction," W. Humber, vol. i., frontispiece.

respectively, above high water of spring tides. The river at the site of the Moerdyk Bridge, carrying the Antwerp and Rotterdam Railway, reaches a depth of 43 feet at low water; and the twenty-eight girders spanning the fourteen clear openings of 328 feet, constructed on staging overhanging the bank of the river, were carried out by two flat-bottomed, timber pontoons, 115 feet long and 34 feet wide, and deposited near high water in their exact positions on their piers (Fig. 81). The girders spanning the seven 400-feet openings of the Hawkesbury Bridge in New South Wales, were erected successively in pairs, for a complete span, on staging raised upon a long timber pontoon, which was sunk upon a gridiron in shallow water near the shore of a sheltered bay; and when a span was finished, valves were opened and the water let out of the pontoon at low tide on the first favourable opportunity; and, the

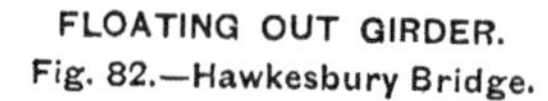
FLOATING OUT GIRDER.
Fig. 82.—Hawkesbury Bridge.

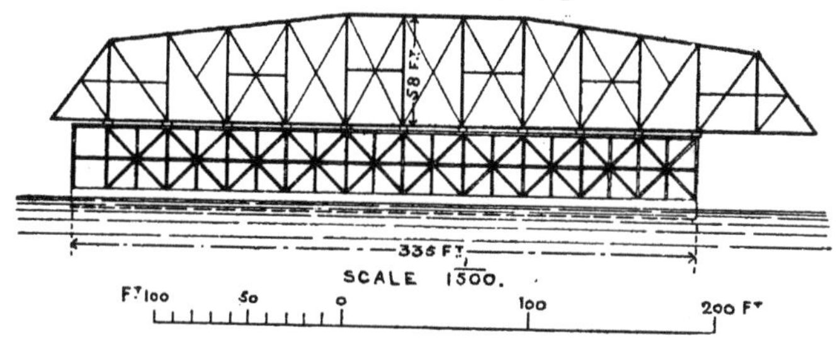

valves having been closed again, the pontoon floated with the rising tide [1] (Fig. 82), and being hauled into position between two piers, the span was gently deposited in its place on the piers as the tide fell. The staging on the pontoon enabled the girders to be placed on the piers at the requisite height to provide a clear headway of 40 feet above high water of spring tides; and the overhang given to the span over one end of the pontoon, was designed to allow for the depositing of each of the shore spans on the abutment on each bank.

The chief objections to the erection of girders by floating out are the large cost of the necessary plant, and the great care and experienced gang of men required for taking out the girder and bringing it into its exact position between the piers; but the cost of the system, in comparison with staging, is much diminished when the same plant can be utilized for several spans; whilst the selection of a calm day greatly facilitates the manœuvres.

CONTINUOUS-GIRDER BRIDGES.

When girders, instead of spanning singly a series of openings, are made continuous over two or more spans, the distribution of the stresses

[1] *Proceedings Inst. C.E.*, vol. ci. p. 9, and plate 2.

is considerably modified, as indicated by the diagram showing the bending moments in a uniformly loaded girder of uniform strength spanning two adjacent, equal openings (Fig. 83.) The girder in this case may be regarded as made up of three girders **AB**, **BD**, and **DE**, as the bending moments become zero at the points of contrary flexure of the girder, **B** and **D**, the positions of which depend on the form and strength of the girder, and the distribution of the moving load.

Stresses in Continuous Girders uniformly loaded.—In the special instance exhibited in the diagram (Fig. 83), the points of inflexion **B** and **D** are two-thirds of one span distant from the abutments, and one-third from the central pier **C**, so that the three divisions are assumed to be of equal length. Under these conditions, one-third of the total load on each span is borne by each abutment, and four-thirds by the central pier; whilst since the stresses on the portions **AB** and

CONTINUOUS GIRDER OVER TWO EQUAL SPANS.

Fig. 83.—Bending Moments, Load Distributed.

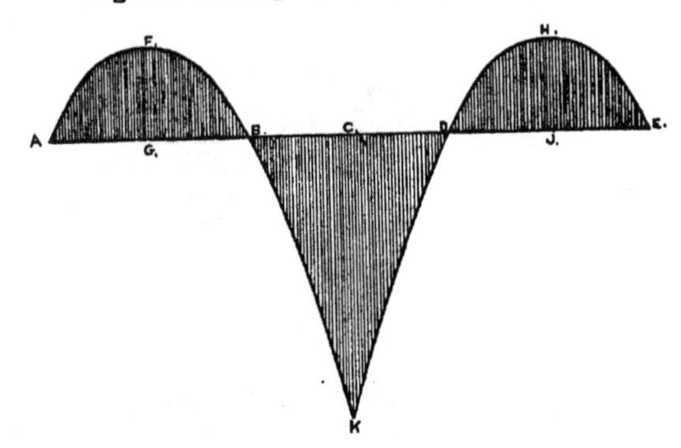

DE correspond to those borne by a girder having only two-thirds of the actual spans **AC** and **CE**, the maximum bending moments, **FG** and **HJ** in Fig. 83, halfway between **AB** and **DE**, are only four-ninths of those of detached girders spanning similar openings of width L, namely $\frac{1}{18}w\text{L}^2$; but the bending moment **CK**, over the central pier, is three times this amount, namely $\frac{1}{6}w\text{L}^2$. The shearing stress also, which is $\frac{1}{2}w\text{L}$ at each end of a detached girder, is only $\frac{1}{3}w\text{L}$ at **A** and **E** in the continuous girder, but amounts to $\frac{2}{3}w\text{L}$ over the central pier.

Advantage of Continuous Girders.—By making a girder continuous over two or more spans, it is practically subdivided, at the points of inflexion, as regards the stresses into $2n - 1$ parts, where n is the number of spans, each necessarily of less length than the width of the spans, with the result that the stresses, and consequently the weights of the parts over the openings are reduced, and concentrated upon the piers where they are supported. Accordingly, the employment of continuous girders enables a series of large spans to be bridged with less weight of metal than required for detached girders, more particularly when the depth of the girder is increased in proportion to the stresses, especially over the piers. Nevertheless, though continuous girders have often been resorted to for a series of moderate spans, they have not been extensively adopted for large spans, partly owing to the intensification of the stresses which would occur in parts of the girder if a settlement took place in one of the piers, and partly on account of the difficulties experienced in erecting a long continuous structure, when the size of the girder and the width of the spans preclude the rolling out of the girder. The experience, however, obtained in sinking cylinders and caissons to great depths for the river piers of bridges, seems to render it possible to do away with the first objection by securing the stability of the piers.

The Lachine Bridge, by which the Canadian Pacific Railway crosses the St. Lawrence a little distance above Montreal, is a fine example of a steel, continuous-girder bridge, in which the depth of the girders has been increased over the piers of the two large channel spans of 408 feet[1] (Fig. 78, p. 160). This bridge is nearly a mile in length, and cost about £200,000.

Erection of Continuous-Girder Bridges.—Continuous girders with straight, flat, bottom flanges, can be conveniently and expeditiously put in place by constructing the girder gradually on the ground near one extremity of the bridge, and rolling it out by degrees over the abutment and piers. This system is particularly advantageous where a bridge is designed to cross a river in several spans of moderate size, or where a viaduct has to be carried across a deep gorge. The only objections to this cheap and simple method of erection, are the necessity of stiffening the girder, in view of the special stresses to which it is subjected in its erection, with strong lattice bars placed somewhat close together, and the danger of injuring the girder in the process of rolling it out by the sudden and varying stresses produced, quite different from those it has been designed to bear in its final position. These temporary stresses, however, can be largely reduced, by supporting the projecting end by wire ropes passing over staging erected successively on the abutment and piers, and anchored to the inner part of the girder, and by strengthening the girder during erection by additional stiffening pieces.

The erection of the River Bouble viaduct in France furnishes an instance of this system of rolling out, adopted for a viaduct of considerable height, together with the utilization of the projecting girder as an overhanging staging, from which, by the help of a crane at the extremity, the high iron piers were built up[2] (Fig. 84, p. 170); and a similar combined process of pushing out the girders and erecting the piers in advance from them, was previously resorted to in the construction of the Friburg Bridge for a railway across the valley of the Sarine, at a height of about 280 feet above the river. Sometimes the rolling out of a girder has been facilitated by adding temporarily a strong, triangular piece at the

[1] *Engineering News*, New York, vol. xviii. p. 276.
[2] "Traité de la Construction des Ponts," R. Morandière, plate 328, fig. 5; and *Annales des Ponts et Chaussées*, 1870 (1), p. 126, and plate 214, fig. 1.

end (Fig. 85), as for example in the erection of the Credo viaduct over the Rhone,[1] which by enabling the support of the pier to be reached

ERECTION BY ROLLING OUT.

Fig. 84.—Bouble Viaduct.

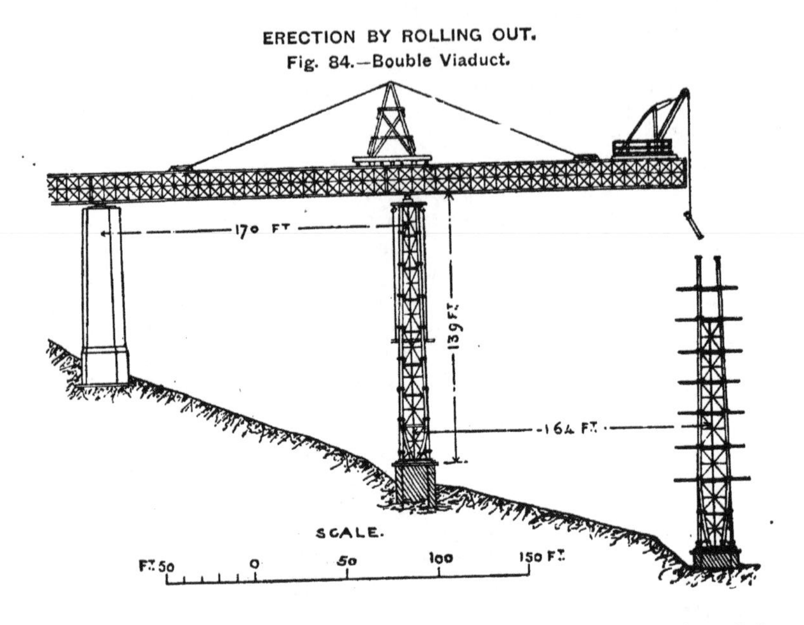

some distance in advance of the actual girder, relieves the girder sooner from the reversed stresses imposed upon it in rolling out, and

ROLLING OUT CONTINUOUS GIRDERS.

Fig. 85.—Credo Viaduct.

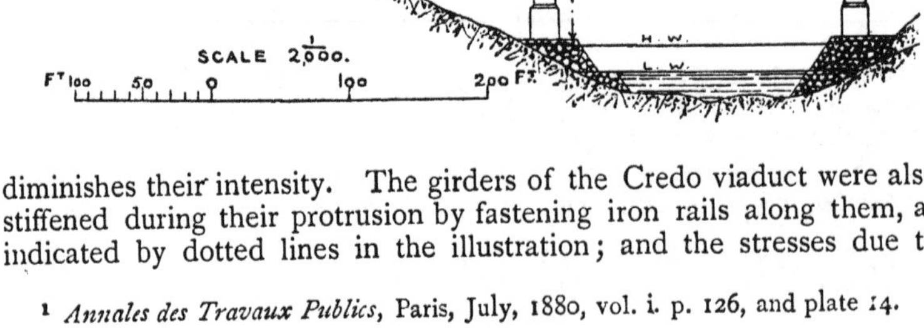

diminishes their intensity. The girders of the Credo viaduct were also stiffened during their protrusion by fastening iron rails along them, as indicated by dotted lines in the illustration; and the stresses due to

[1] *Annales des Travaux Publics*, Paris, July, 1880, vol. i. p. 126, and plate 14.

the dipping of the projecting part of the girder, and the resulting raising of the inshore portion, were mitigated by weighting the inshore part of the girder where it tended to rise.

When a continuous girder is erected in separate spans which have to be subsequently joined together, a course followed with the tubes of the Britannia Bridge over the Menai Straits, a simple junction by means of plates is not sufficient to transform the separate parts into a continuous girder; but the tops of the separate pieces have to be brought close together by tipping them up at their further ends, or by some other means, so that the top flanges may be rigidly connected so as to form a really continuous girder, and bring the points of contrary flexure into the approximate positions they were assumed to occupy in designing the girder.

CHAPTER XI.

CANTILEVER AND MOVABLE BRIDGES.

Cantilever Bridges: form and principle; stresses in cantilevers; forms with instances; compared with continuous girders, Kentucky Bridge; erection by building out, Forth Bridge, and other instances; examples of cantilever bridges, Niagara, Fraser, Poughkeepsie, Memphis, Sukkur, and Húgli Jubilee bridges, Forth Bridge, Quebec Bridge, Blackwell Island Bridge—**Movable Bridges:** types; swing bridges, different forms, Brest Bridge, Thames River Bridge, Hawarden Bridge, mode of working; traversing bridges, at Milwall and Barry Docks, telescopic Victoria Bridge over river Dee; Bascule bridges, early forms, Tower Bridge; Lift bridges, instances, Halsted Street Bridge, Chicago; rolling lift bridges, Chicago, Cleveland; floating bridges, Howrah Bridge across River Húgli.

CANTILEVER BRIDGES.

CANTILEVERS as applied to bridges, consist essentially of girders projecting out from their piers, like brackets, across a portion of the span, and supporting at their extremities independent girders whose span can in this way be considerably reduced. The projecting girder, bearing a load at its extremity, is kept in equilibrium by being strongly anchored down at its shore end, or by supporting another girder on its other end extending equally out over a second large span. The principle of the cantilever is not novel, for some centuries ago small rivers in India and Japan, too wide to be crossed by a single, long beam, were spanned by embedding timbers deeply in each bank, which supported central timbers on their projecting ends; but the application of this principle to the design of metal bridges of large span is quite recent, the first true cantilever bridge having been erected in 1883 across the Niagara Rapids below the Falls, near the old railway suspension bridge.

Stresses in Cantilevers.—Each cantilever arm of a bridge, extending out from a pier, is subjected to two distinct loads, namely, half the load of the girder supported by the cantilevers, and imposed at the extremity of each arm, and also the load borne by the cantilever arm itself. The first load, consisting of half the weight W of the central girder and its moving load, produces a bending moment on the cantilever arm of length l, which increases regularly from zero at the extremity up to a maximum of $\frac{W}{2}l$ at the pier, and a shearing stress $\frac{W}{2}$ which is uniform throughout the arm, represented graphically in Fig. 86. The second load, consisting of the weight of the cantilever arm and the moving load, if assumed to be uniformly distributed and equal to w per unit of length, causes a bending moment, increasing from zero at the extremity according to a parabolic curve represented by $y = \frac{wx^2}{2}$, up to a maximum of $\frac{wl^2}{2}$ at the pier, and a shearing stress augmenting regularly from zero at the extremity up to a maximum of wl at the pier, as indicated by the diagrams in Fig. 87. A cantilever bridge is

DIAGRAMS OF STRESSES ON CANTILEVER.

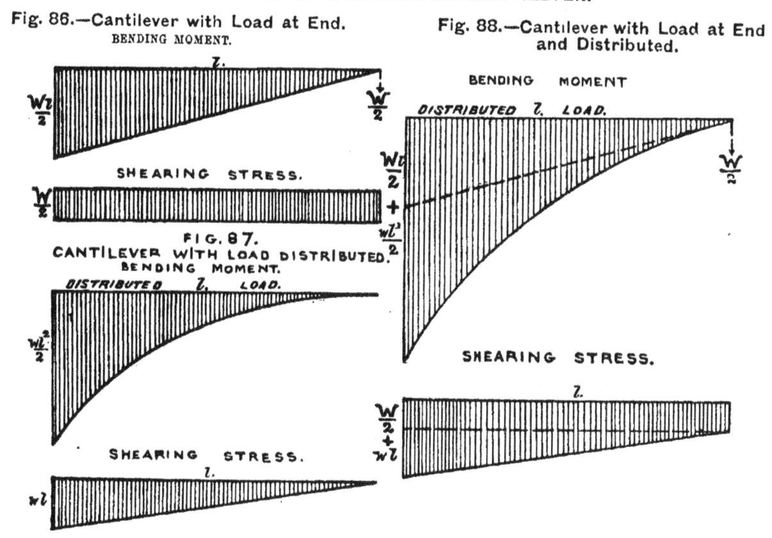

Fig. 86.—Cantilever with Load at End.

FIG. 87. CANTILEVER WITH LOAD DISTRIBUTED.

Fig. 88.—Cantilever with Load at End and Distributed.

liable to have its cantilevers subjected to both loads simultaneously, in which case the bending moments and shearing stresses are represented by a combination of the preceding diagrams, as shown in Fig. 88; and the bending moment and shearing stress attain a maximum at the piers of $\frac{W}{2}l + \frac{wl^2}{2}$, and $\frac{W}{2} + wl$, respectively.

Forms of Cantilever Bridges.—The most distinct form of cantilever bridge is where symmetrical arms extend out on each side of the piers, the outer arms over the main span supporting a central girder, and the inner arms, stretching over smaller side spans, being anchored down to the bank on each side, so as to provide a counterpoise to the weight of the central girder with its load. This is the form of the Niagara Cantilever Bridge, and of the shore cantilevers of the Forth and

Poughkeepsie bridges (Figs. 89, 90, and 92). By this arrangement, the span of the central girder is greatly reduced; and the shore cantilever arms balancing the arms of the main span, serve also to bridge over the shore spans, just as the anchoring cables of the Brooklyn Bridge support the two side spans (Fig. 64, p. 141). This advantage is lost when the cantilevers are only used for supporting the girders in the centre of a large span, as at the Sukkur Bridge (Fig. 93), where the cantilevers, as well as the central girders, have to be counterpoised by the land ties on each side.

When a cantilever bridge has more than one large span, as the central cantilever cannot be anchored down, it is necessary to provide for unequal loading, either by widening out and strengthening the central cantilever, as done at the Forth Bridge (Fig. 89); or where there are three or more river spans, by erecting a deep, central, continuous girder over two piers, between two cantilever spans, with a cantilever arm projecting out beyond the piers at each extremity, as placed across a central span in the Memphis and Húgli Jubilee bridges (Figs. 91 and 94), and across two spans between three cantilever spans in the Poughkeepsie Bridge (Fig. 90). Generally the ordinary girder, forming the adjunct to the cantilever, is supported centrally between two cantilevers; but occasionally the girder extends over a shore span of the bridge, and is supported at one end on the shore abutment, of which the Memphis and Húgli Jubilee bridges furnish instances (Figs. 91 and 94).

The increased depth over the piers necessitated by the bending moments and shearing stresses of cantilever bridges, reaching a maximum over their supports, is effected downwards in deck bridges, such as the Poughkeepsie and Niagara bridges (Figs. 90 and 92), and upwards in through bridges, where the train passes between the girders, as in the Memphis and Húgli Jubilee bridges (Figs. 91 and 94); and where the requisite depth is very great, as for the Forth and Sukkur bridges, the cantilevers are carried both above and below the roadway of the bridge (Figs. 89 and 93). Moreover, these last two very high cantilevers have been strengthened against wind-pressure by widening them out considerably towards their base, and especially over their piers, as shown on the plans of these bridges.

Cantilever Bridges compared with Continuous Girders.— The great advantages of reducing the effective span and weight of the girders in the central portion of large openings, and concentrating the chief weight and stresses on the piers, are shared by cantilever and continuous-girder bridges alike. Moreover, the bending moment which disappears at the point of inflexion in continuous girders, also becomes zero at the ends of the cantilevers; whilst the upper member of a cantilever is in tension, and the lower is in compression, precisely like the portions of a continuous girder over the piers between the points of inflexion. The chief difference between the two systems lies in the fact that, whereas the point of inflexion in continuous girders is changed in position by alterations in the distribution of the moving load, the place where the bending moment disappears is fixed in cantilever bridges by the

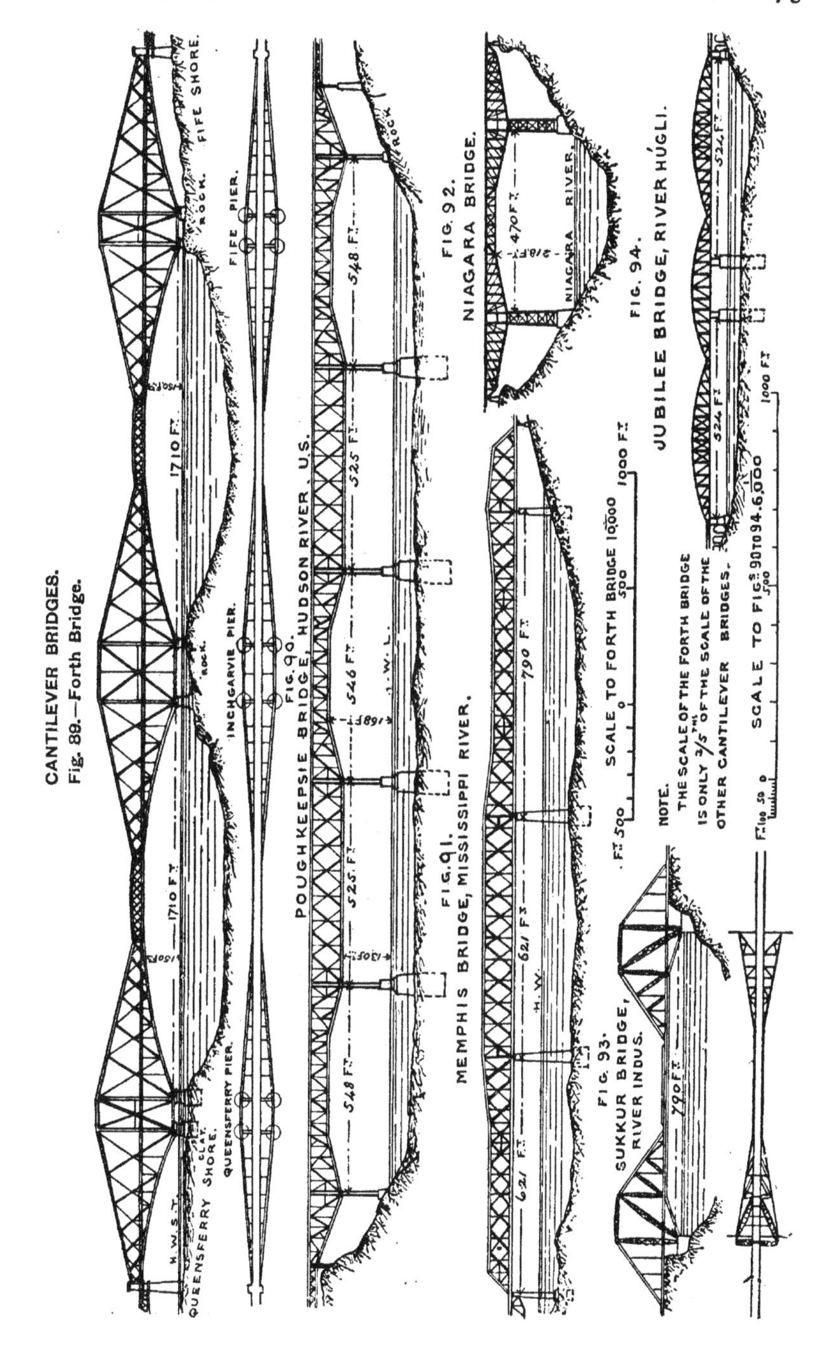
CANTILEVER BRIDGES.
Fig. 89.—Forth Bridge.
FIFE SHORE.
ROCK.
FIFE PIER.
150 FT.
1710 FT.
INCHGARVIE PIER.
ROCK.
1710 FT.
150 FT.
QUEENSFERRY PIER.
CLAY.
H.W.S.T.
QUEENSFERRY SHORE.
FIG. 90.
POUGHKEEPSIE BRIDGE, HUDSON RIVER, U.S.
ROCK.
548 FT.
525 FT.
546 FT.
168 FT.
H.W.L.
525 FT.
130 FT.
548 FT.
FIG. 91.
MEMPHIS BRIDGE, MISSISSIPPI RIVER.
790 FT.
621 FT.
H.W.
621 FT.
FIG. 92.
NIAGARA BRIDGE.
470 FT.
218 FT.
NIAGARA RIVER.
FIG. 93.
SUKKUR BRIDGE,
RIVER INDUS.
790 FT.
FIG. 94.
JUBILEE BRIDGE, RIVER HUGLI.
524 FT.
524 FT.
SCALE TO FORTH BRIDGE 1/10000
FT. 500 0 1000 FT.
NOTE.
THE SCALE OF THE FORTH BRIDGE
IS ONLY 3/5THS OF THE SCALE OF THE
OTHER CANTILEVER BRIDGES.
SCALE TO FIGS. 90 TO 94. 1/6000
FT. 100 50 0 500 1000 FT.

discontinuity or hinging of the central girder at the end of the cantilever. The insignificance, however, of this difference as regards the stresses under ordinary conditions, is illustrated by the Kentucky River Bridge, which was designed and erected, in 1877, as a continuous-girder bridge across three spans of 375 feet, but after erection had its bottom chords severed at the points of inflexion, 75 feet on the shore side of the two river piers, so that the shore 300 feet of the girders on each side became only hinged to the central portions, 525 feet in length, extending out beyond the river piers[1] (Fig. 95). This arrangement, fixing permanently the positions of the points of inflexion and the zero points of the bending moments, was adopted so as to avoid the variations in the stresses which would have been produced by the rising or falling of

CONTINUOUS GIRDER CONVERTED INTO CANTILEVER.

Fig. 95.—Kentucky River Bridge.

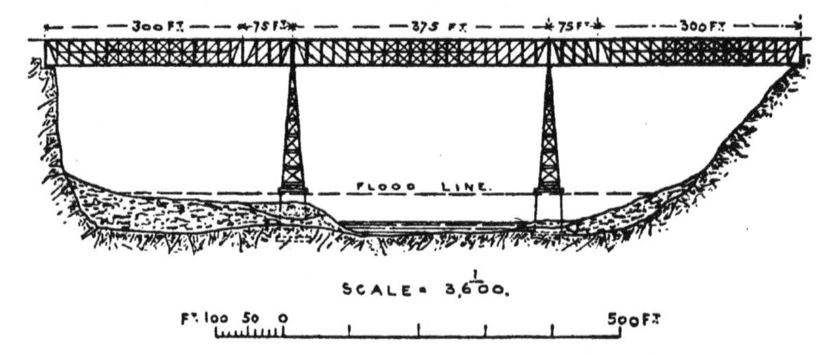

the high, iron, river piers under changes of temperature, whilst the abutments remain unaltered; though it appears that the variations in the stresses from this cause would have been quite small.[2] As this modification of the continuous girder may be regarded as converting it, in a certain sense, into a cantilever bridge of the type of the Húgli Jubilee Bridge, the Kentucky River Bridge has sometimes been called the first metal cantilever bridge erected; but the Niagara Bridge is a more perfect type of cantilever. The Kentucky River Bridge, however, shows how close is the resemblance between continuous-girder and cantilever bridges.

Continuous girders are the most suitable for small or moderate spans, where the piers are not liable to settlement, as providing a more rigid construction; whereas the cantilever system is much better for very large spans, in which the large dimensions of the parts ensure rigidity, for it is unaffected by slight settlements of the piers, and contraction and expansion is readily provided for at the discontinuous points at the ends of the cantilevers.

Erection of Cantilever Bridges.—Building out from the piers

[1] *Proceedings Inst. C.E.*, vol. liv. p. 183, and plate 8.

[2] "A Practical Treatise on Bridge-Construction," T. Claxton Fidler, p. 148.

on each side is specially applicable for the erection of symmetrical cantilever bridges, such as the Forth and Niagara bridges (Figs. 89 and 92, p. 175); for the cantilever portions are subjected in the process of erection to exactly similar strains to those which they are designed to bear when completed; and it is only the relatively short, central girder which does not share this advantage. This system of erection would also be equally suitable for the portions of a continuous girder extending out on each side of the piers to the points of contrary flexure, provided the piers were made as wide as in the case of cantilever bridges erected in this manner. The erection of the Forth Bridge by building out symmetrically and simultaneously from each of the three piers, is partially illustrated at three different stages in Fig. 96; and the Niagara

BUILDING OUT CANTILEVERS.

Fig. 96.—Forth Bridge.

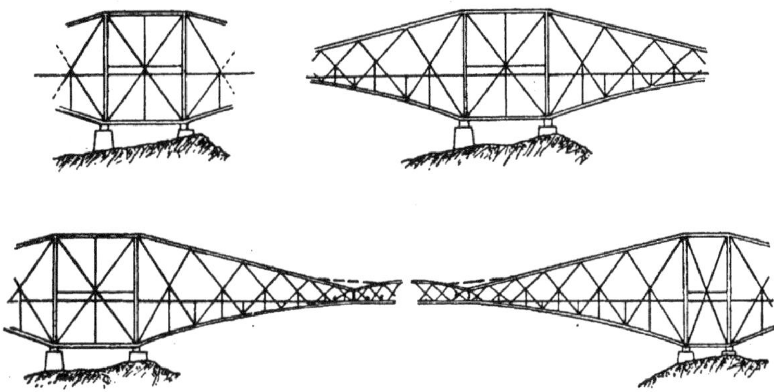

Bridge was similarly built out from its two metal piers over the river, till at last the central girder was joined in mid-air at a height of about 220 feet above the foaming Niagara rapids.

The Sukkur Bridge,[1] and the cantilever spans of the Poughkeepsie and Memphis bridges were also erected by building out (Figs. 90, 91, and 93, p. 175), being counterpoised in the first case by the land ties, and in the other bridges by the girders of the intermediate spans; but these intermediate girders, and the central girders over the outer 621-feet span of the Memphis Bridge had to be erected by aid of temporary staging. In the case of the Húgli Jubilee Bridge (Fig. 94, p. 175), the central halves, 180 feet in length, of the central girders were erected on staging extending out from the piers, and only the end 90 feet of the projecting cantilever ends on each side were built out; whilst the shore girders were rolled out along the approach viaducts, the outer end of each girder being at the same time supported and floated out by a pontoon, on which staging had been erected, till it was finally deposited on the end of its cantilever.[2]

[1] *Proceedings Inst. C.E.*, vol. ciii. p. 124, and plate 6.
[2] *Ibid.*, vol. xcii. p. 85, and plate 1, figs. 1 and 4.

Cantilever Bridges of Large Span.—The concentration of the greatest weight and stresses of a cantilever bridge over the piers and the sides of the span, and their corresponding reduction over the central portion, combined with the self-contained and balanced character of the structure with true cantilevers, and its adaptability for building out, render the cantilever system specially suitable for bridging large spans.

By building a pier at the edge of the Niagara River on each side, at the foot of the slopes of the gorge, it was possible to reduce the span across the river, from the 821 feet of the old suspension bridge, down to 470 feet, the side slopes of the gorge being spanned by the shore portions of the cantilever[1] (Fig. 92, p. 175). The bridge consists of a shore cantilever arm, 195 feet long, on each side, erected on staging and anchored down to a masonry pier on the bank, a panel 25 feet long over each tower, and two river arms, 175 feet long, supporting an intermediate girder between them, 120 feet in length, and 26 feet deep. Each masonry, anchorage pier, with the ironwork, weighs about 890 tons; whereas the lifting force at the shore end of each cantilever amounts only to about 270 tons under the most unfavourable distribution of the load. The upper members of the shore arm being mainly in tension, but being subject to some extent to compressive stresses by the loading of the arm itself, are composed chiefly of eye-bars, 8 inches wide, and from $1\frac{1}{4}$ to $1\frac{7}{8}$ inches thick, with a compression member introduced between them, formed of two web-plates, angle-irons, and lattice-work; whilst the upper chords of the pier panels and river arms consist wholly of eye-bars. The lower members of the cantilevers, being in compression, are composed of plates and angle-irons. Owing to the short time allowed for construction, only the pins and large compression members were made of steel, wrought iron being used for the remainder as more readily procurable of reliable quality. The foundations of the piers were commenced in April, 1883, and the bridge was completed in the following December.

A precisely similar design had been previously prepared for a somewhat smaller bridge to carry the Canadian Pacific Railway across the Fraser River; but this bridge, which has dimensions between the banks and the centres of the piers, of 105, 315, and 105 feet, and carries the railway at a height of 125 feet above low-water level in the river, was only erected in 1885.

The Poughkeepsie Bridge, commenced in 1886, crosses the Hudson River with five large spans, three of which are formed by cantilevers,[2] with a minimum headway under the intermediate girders of 130 feet at high water, and a maximum headway of 168 feet under the central girders, 212 feet long, supported by the cantilevers, each 160 feet in length (Fig. 90, p. 175). The shore arms of the side cantilevers, extending over the slopes of the banks, are 200 feet long. The two side cantilever openings attain a span of 548 feet between the centres

[1] *Transactions of the American Society of Civil Engineers*, vol. xiv. p. 499, and plates 49 to 63 and photograph.

[2] *Ibid.*, vol. xviii. p. 199, and plates 40 to 59.

of their piers; whilst the span of the two intermediate girders, extending over two piers with projecting cantilever arms, is only 23 feet less.

The Memphis Bridge, the steel superstructure of which was erected between March, 1891, and May, 1892, crosses the Mississippi with three large spans, the channel cantilever span being similar in arrangement to the side cantilever spans of the Poughkeepsie Bridge, but reaching a span of 790 feet between the centres of its piers, and having the double Warren girder system of bracing continued along the cantilever span, as well as in the intermediate girder[1] (Fig. 91, p. 175). As the intermediate girder is subject to a reversal of strains from the loading of its projecting cantilevers, and the central girder in the channel span was built out from its cantilevers, the bottom members of these girders, as well as the bottom members of the cantilevers, were made stiff with plates and angle-irons to resist compression; and this stiffness was extended to the bottom member of the side girder in the 621-feet span, making the bottom members stiff throughout, in order to add to the stiffness of the bridge and reduce vibrations and deflections. The top members, however, of the cantilever arms consist of a series of eye-bars, as well as the parts of the bracing in tension. Provision has been made for expansion and contraction by sliding joints at the ends of the cantilevers; and the intermediate girder is free to move on rollers on the pier adjoining the channel span, and is fixed on the other pier. The central girders have a span of 621 feet between the centres of their piers; and the side spans are bridged over by suspended girders in both cases, 452 feet long, and one cantilever arm over the 621-feet span, and two arms over the channel span, each 169 feet long; whilst a cantilever arm, 226 feet long, stretches over the slope of the river-bank on each side.

The steel cantilever span of the Sukkur Bridge, the erection of which was commenced in April, 1887, and completed in February, 1889, is remarkable, for being a cantilever bridge having only a single span, for the height of its cantilevers which amounts to about a fifth of the span, and for the large dimensions of the members of the cantilevers (Fig. 93, p. 175). This span of the bridge consists of two single cantilevers extending out 310 feet from their supports on each side, and carrying a central girder span of 200 feet at their ends, leaving a clear opening of 790 feet between the abutments on the banks of the deep Rori branch of the River Indus at Sukkur.[2] The cost of this cantilever portion of the bridge appears to have been about £185,300. The adoption of the uneconomical form of a cantilever bridge for spanning a single opening, could only be justified by the impossibility of maintaining staging in this channel of the river for a sufficient period for the erection of an ordinary girder bridge, necessitating erection by building out, as well as the much greater size of span than had at that time been bridged by disconnected girders.

The Jubilee Bridge erected across the River Húgli at Húgli in

[1] *Transactions of the American Society of Civil Engineers*, vol. xxix. p. 573, and plates 1 to 19.

[2] *Proceedings Inst. C.E*, vol. ciii. p. 123, and plates 4 to 6.

1884–86, is notable as a peculiar, symmetrical form of cantilever bridge, with a central cantilever, 360½ feet long, supported on two river piers, leaving a clear span between them of only 95½ feet, and two side spans of 520 feet in the clear, bridged by the projecting cantilevers extending 118 feet out from the piers, and by girders, 420 feet long, resting on the ends of the cantilevers and the abutments[1] (Fig. 94, p. 175). This arrangement of the spans, instead of three equal spans, enabled the pier adjoining the deep navigable channel to be built in comparatively shallow water, and the cantilever to be erected with only the aid of staging jutting out from the piers in a fan-like form; and the large spans on each side have the advantage of affording a wide navigable channel for the larger vessels alongside the right bank, and a second wide shallow channel on the opposite side, for the small craft, which prefer that side owing to the slower currents. The bridge, which carries a double line of railway, with its approach viaducts, cost £261,200.

The Forth Bridge crosses the two deep channels into which the Firth of Forth is divided by Inchgarvie Island opposite Queensferry, with two spans of 1710 feet, affording a clear headway of 150 feet at high tide for the central 500 feet of each span (Fig. 89, p. 175).[2] This steel bridge consists of three huge symmetrical cantilevers, each resting on a group of four piers built on the shore sides of the channels and on Inchgarvie Island (see p. 77), having tapering cantilever arms projecting out 680 feet on each side from the central columns erected on the piers.[3] The shore arms of the side cantilevers extend over the foreshores of the Firth, and join the approach viaducts on each bank; whilst the other arms extend over the main spans, and together with the arms of the central Inchgarvie cantilever, support at their extremities the two pairs of lattice girders which span the central 350 feet of the large openings. As the outer arms of the shore cantilevers have to bear half the weight of the central girders and of the train loads which may pass over them, the shore arms have been made heavier than the others; and additional weight has been added at their extremities sufficient to counterpoise, with an excess of 200 tons in each case, half the weight of the central girders with their maximum train load, so as to ensure the maintenance of the balance of these cantilevers. The length of the central portions of the shore cantilevers, resting on the piers, is 155 feet; but the central part of the Inchgarvie cantilever has been made 270 feet long, for though the cantilever arms are symmetrically loaded in this case with half the weight of the central girders at each end, the cantilever is liable to be loaded unequally by the centre of one of the main spans being fully loaded with two passing trains when the other is empty, which has been provided for by an additional length of support, as it cannot be counterpoised. The central portions, or towers of the

[1] *Proceedings Inst. C.E.*, vol. xcii. p. 73, and plate 1.

[2] The elevation and plan of the Forth Bridge on p. 175, have had to be drawn to the scale of three-fifths of that of the other cantilever bridges, to bring it within the limits of the page.

[3] "The Forth Bridge," W. Westhofen, reprinted from *Engineering*, February 28, 1890; and *Proceedings Inst. C.E.*, vol. cxxi. p. 309.

cantilevers, rise to a height of 330 feet above their piers; and their width across, which is 33 feet at the top, increases to 120 feet at their base; and the sides of the cantilevers widen out similarly downwards in proportion to their height, to strengthen the bridge against lateral wind-pressure, which, with the assumed maximum of 56 lbs. on the square foot, would amount to 2000 tons against one of the large spans. The bottom compression members of the cantilevers are formed of straight lengths of tubes, 12 feet in diameter, and 1¼ inches thick, connected together at a slight angle to each other, so as to present a curved outline; and they converge horizontally from 120 feet at the towers to 32 feet at their extremities (Fig. 89, p. 175). Rocking columns were interposed between the ends of the central girders and the extremities of the Inchgarvie cantilever arms, to allow for longitudinal expansion and contraction; whilst the shore arms of the side cantilevers are left free to slide on their abutments. The chief strains on a bridge having these unparalleled spans, are clearly due to the large dead weight which has to be borne; for the ordinary maximum moving load on the Forth Bridge, of 800 tons, is only one-twentieth of the weight of one of the large spans. The foundations for the piers were commenced early in 1883; and the bridge was opened for traffic in March, 1890, having cost altogether, with its approach railways and other expenses, about £3,000,000.

Great as are the dimensions of the channel spans of the Forth Bridge, the length of the main span of the cantilever bridge proposed to be constructed over the St. Lawrence near Quebec would have surpassed them by 90 feet. This bridge was originally planned to consist of one main channel span of 1800 feet, with a clear height of 150 feet above high water, and two anchor spans of 500 feet each. The trusses had a maximum depth of 315 feet on centres of main posts and 96¾ feet over anchor piers. The bridge was designed to accommodate two railway tracks, two highways, two electric car tracks, and two sidewalks. A disastrous collapse of that portion of the bridge which at that date was in process of erection took place, however, in August, 1907.

The great bridge over the East River at Blackwell Island, New York, consists of two continuous lines of pin-connected trusses, 60 feet apart, and having an aggregate length of 3724 feet 6 inches.

Commencing from the New York side the spans are as follows:—one shore span of 469 feet 6 inches, one channel cantilever span of 1182 feet, one anchor span over Blackwell Island of 630 feet, one second channel cantilever span of 984 feet, and one shore anchor span on the Long Island side of 459 feet. The bridge, which is double-decked, has been erected on the cantilever principle, the headway above low-water level being 135 feet.

Movable Bridges.

When a navigable river, canal, or dock has to be traversed at a low level by a bridge for a road or a railway, it is necessary to provide for the passage of masted vessels or steamers by making the bridge movable.

Types of Movable Bridges.—There are four distinct types of movable bridges, namely, Swing Bridges, Traversing Bridges, Bascule Bridges, and Lift Bridges; whilst when a floating bridge resting on boats or pontoons is placed across a river, part of the central portion is made readily detachable, so that a clear passage can be opened when required for vessels or barges navigating the river.

Swing Bridges.—A bridge balanced and swinging round a quadrant of a circle horizontally on a pivot, is the most common type of movable bridge for crossing a waterway, three forms of which have been constructed. Formerly the most common form of swing bridge, especially when cast iron was the material generally used, consisted of two counterbalanced halves turning round on pivots on the abutments and meeting in the centre of the span, of which the wrought-iron bridge across the River Penfeld at Brest, giving a clear width between its abutments of 350 feet, erected in 1861, is probably the largest example[1] (Fig. 97). This form of bridge has to be counterbalanced by a heavy tail-end on the shore side of each pivot; and its swing portions are really cantilevers,

SWING BRIDGE IN TWO HALVES.

Fig. 97.—Brest Harbour.

though they sometimes derive a certain amount of support when closed across the waterway by the abutting of the extremities. together at the centre.

The most economical form of swing bridge for the amount of open waterway provided, is where the river or canal is divided into two channels by a central pier, on which a symmetrical swing bridge turns stretching across two equal openings when closed. This form of bridge is a continuous girder across two openings when closed, and a cantilever when open; and therefore, whether open or closed, the greatest stresses are concentrated over the central pier where the superstructure can be advantageously given a greater depth. A fine example of this form of swing bridge was erected over the navigable channel of the Thames River at New London in 1888–89 (Fig. 98), affording two clear channels on each side of the pier carrying the bridge, each $227\frac{1}{2}$ feet in width, which can be opened under favourable conditions in $2\frac{3}{4}$ minutes.[2] Each half in this case balances the other half when swinging round, thereby avoiding the necessity of using any additional material merely as a counterpoise.

The commonest form of swing bridge in the present day, especially

[1] *Annales des Ponts et Chaussées*, 1867 (2), p. 265, and plates 150 to 153.

[2] "Construction of the Thames River Bridge at New London," A. P. Boller, New York, 1890.

across locks and entrances to docks, and at other places where only one channel has to be opened for navigation, consists of a long swing portion spanning the opening, and a shorter tail-end made specially heavy and

SWING BRIDGE ACROSS TWO EQUAL SPANS.

Fig. 98.—New London, U.S.

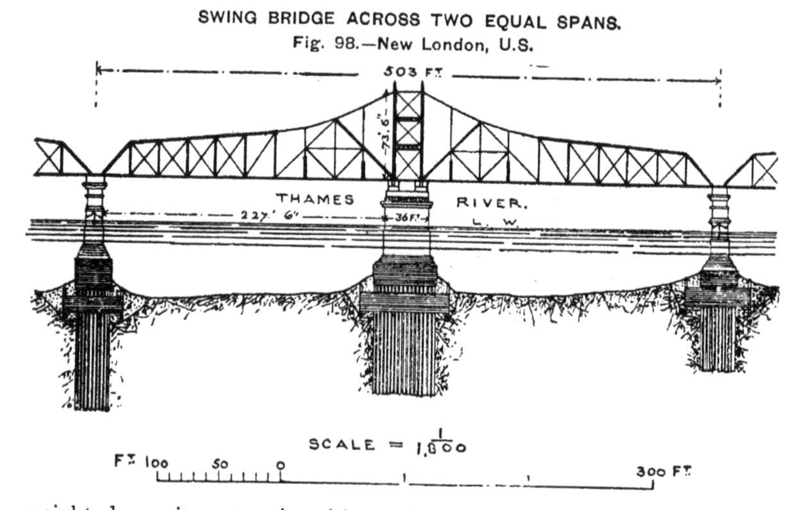

weighted near its extremity with cast-iron kentledge, so as to rather more than counterbalance the swing portion during the opening and closing of the bridge. The Hawarden Bridge, which carries a double line of railway across the Dee, some miles below Chester, and was completed

COUNTERPOISED SWING BRIDGE.

Fig. 99.—Hawarden Bridge.

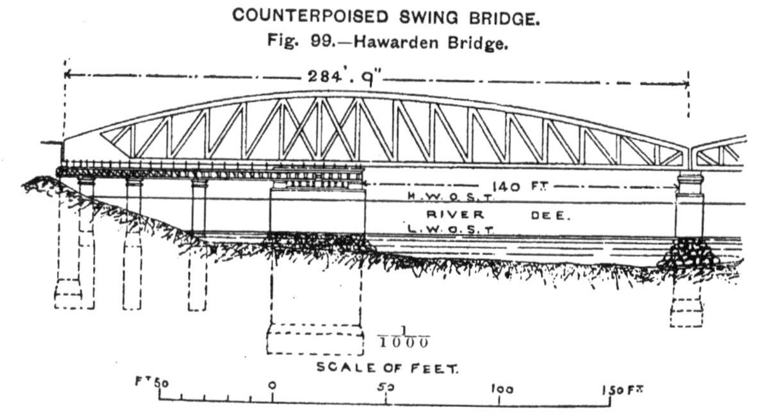

in 1889, affords a large example of this form of swing bridge,[1] giving a clear opening for navigation of 140 feet, in which, moreover, unlike swing bridges at docks, the tail-end is utilized as a portion of the bridge (Fig. 99). The two steel lattice girders which carry the movable portion,

[1] *Proceedings Inst. C.E.*, vol. cviii. p. 304, and plate 10.

have a length of $284\frac{3}{4}$ feet, divided unequally at the pivot into a length of $168\frac{1}{4}$ feet for the swing part, and $116\frac{1}{2}$ feet for the tail-end, and a maximum depth at the point of support of 32 feet, decreasing to $9\frac{1}{2}$ feet at their ends. The tail-end is counterpoised with a balance weight of $113\frac{4}{5}$ tons, giving it a preponderating weight of $8\frac{4}{5}$ tons over the swing portion. The movable portion, weighing 753 tons, can be turned into line with the river over its pier, fully opening the channel, in $2\frac{1}{2}$ minutes.

Swing bridges are generally worked by hydraulic machinery. The bridge being first slightly raised by hydraulic pressure in the press under its pivot, and resting also lightly on rollers at the extremity of its tail-end, is turned on a water-centre by chains worked by hydraulic rams, by which means a heavy bridge can be rapidly opened and closed. A large swing bridge, spanning two openings, carried by girders 500 feet long, and weighing 2000 tons, recently erected over the St. Louis River near the head of Lake Superior, is opened in 2 minutes by electric motors.[1]

Traversing Bridges.—This type of movable bridge, rolling backwards and forwards across an opening, is only resorted to when the quay on either side of the opening is not long enough to accommodate a swing bridge when opened. Under these circumstances, a traversing bridge, provided with a counterpoised tail-end like a swing bridge, can be raised from its supports, and drawn back on rollers along one of its approaches till it leaves the opening clear, without encroaching at all on the limited quay. The illustration, Fig. 100, shows a traversing bridge erected many

TRAVERSING BRIDGE.

Fig. 100.--Millwall Docks, London.

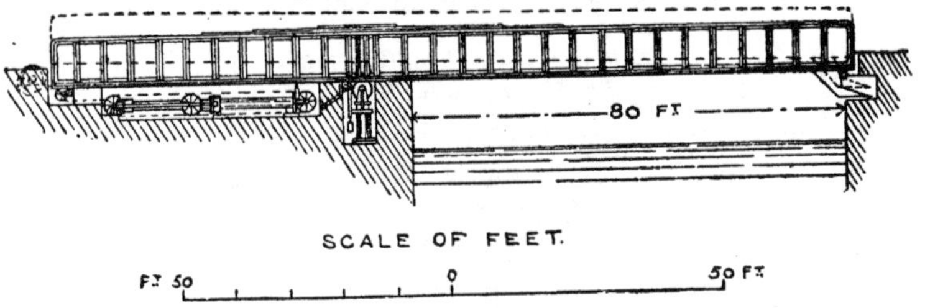

years ago to span an opening of 80 feet between two narrow projections serving to form a division between the two large basins of the Millwall Docks, London, whose only connection is through the opening.[2] The dotted lines in the figure indicate the extent to which the bridge has to be raised to enable it to be drawn back over its approach clear of the opening; and the movements are effected by hydraulic machinery placed below the quay under the tail-end of the bridge. A similar bridge, carried by steel girders $154\frac{2}{3}$ feet long, and running on steel rollers, has been erected at the Barry Docks, for spanning the passage between the dock and basin, 80 feet in width.[3]

[1] *Electrical World*, New York, 1897, vol. xxx. p. 7.
[2] *Proceedings Inst. C.E.*, vol. l. p. 86, and plate 4, fig. 22.
[3] *Ibid.*, vol. ci. p. 145.

Bascule Bridges.—The drawbridges across the moats of ancient fortresses were primitive forms of bascule bridges, turning on a horizontal hinge; and wooden bascule bridges of small span, with overhead counterpoise beams, are much used in Holland for crossing the numerous canals of that country. This type of bridge has not been much employed elsewhere; though a cast-iron bascule bridge in two halves, forming an arch when closed, was erected in 1839 for carrying the North-Eastern Railway across the Ouse at Selby, with a clear span of 45 feet,[1] which was only replaced by a wrought-iron swing bridge in 1891. The Tower Bridge, opened in 1894, spanning a central opening of 200 feet, with two balanced cantilever leaves revolving on horizontal pivots on the piers into a vertical position for opening the bridge to navigation, and meeting in the centre of the span when the bridge is closed for the passage of the road traffic,[2] is considerably the largest bascule bridge hitherto erected (Fig. 101), the next largest, at Rotterdam, spanning an opening of only

BASCULE BRIDGE IN TWO HALVES.
Fig. 101.—Tower Bridge, London.

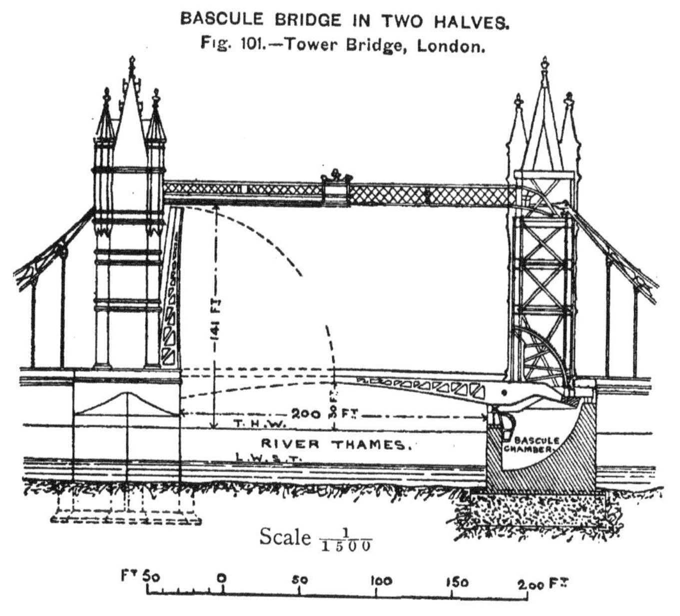

79 feet. Each leaf is carried by four steel girders, 162 feet long, placed 13½ feet apart, the bascule portion being 112½ feet long, and the tail-end, at the back of the pivot, 49½ feet in length. The tail-end is ballasted with 75 tons of cast iron, and 290 tons of lead, owing to the limited space available, so as to counterpoise the bascule portion, and bring the centre of gravity of the leaf, which weighs 1070 tons, on to the centre of the solid steel pivot, 21 inches in diameter and 48 feet long, on which the

[1] *Proceedings Inst. C.E.*, vol. lvii. p. 3, and plate 1, fig. 1.
[2] *Ibid.*, vol. cxxi. p. 320; and vol. cxxvii. p. 35, and plates 1 and 2.

leaf revolves. Each bascule, when raised, leaves a clear opening in a line with its pier, limited only above by the high-level footway, affording a headway of 141 feet above high water of spring tides; whilst the tail-end falls into a quadrant recess provided for it in the pier. The bascules are raised and lowered by pinions moved by hydraulic power, working in segmental racks fastened to the tail-ends; and the interruption to the road traffic at each raising of the bascules, is only six minutes on the average, and has sometimes been as little as three minutes.

The bascule system is very suitable for the Tower Bridge, where a swing bridge when opened would have occupied too much space in the middle of a crowded river, and where the bascules when raised are protected by the towers; but raised bascules form an obstruction on a quay, and present a large surface to the wind; whilst a deep recess has to be provided to receive the tail-end when the bascule is raised, and a heavy counterpoise has generally to be added to balance the bascule.

Lift Bridges.—Sometimes, when very little space can be obtained for a movable bridge, the bridge is only made just long enough to span

HIGH LIFT BRIDGE.
Fig. 102.—Halsted St., Chicago, U.S.

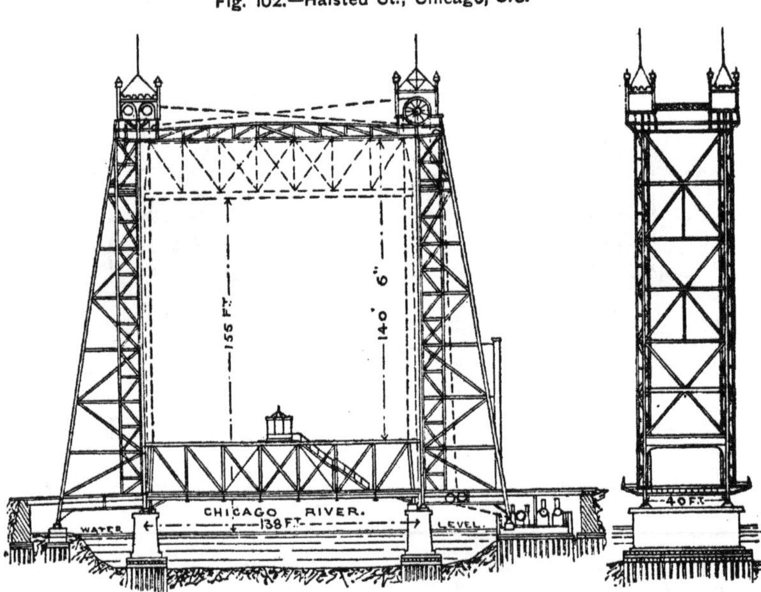

the waterway, and is lifted vertically by chains at the four corners to a sufficient height to afford the requisite headway for vessels navigating the waterway. Instances of this type of bridge are rare, but a lift bridge was erected in 1849 over the Surrey Canal, for the passage of a branch of the Brighton Railway;[1] two of these bridges cross the Ourcq Canal

[1] *Proceedings Inst. C.E.*, vol. ix. p. 303, and plate 13.

in Paris,[1] and two traverse the Oswego Canal at Syracuse; whilst the largest of these bridges was recently constructed at Chicago, for conveying the traffic of Halsted Street across the Chicago River (Fig. 102). This latter bridge, having a span of 130 feet, and carrying a roadway of 36 feet and two footways of 7 feet in width, is lifted by steam-power to a height of 155 feet above the river, by means of steel-wire cables travelling round wheels carried by a platform supported on the top of two light, steel, lattice towers, 200 feet high, erected at each end of the bridge, and guiding its movements.[2] The lifting of this bridge which, with its working appliances, weighs 600 tons, is facilitated as usual by counterbalancing weights; and masted vessels pass under it when raised to its full height.

Rolling Lift Bridges.—A type of bascule bridge in use in the United States and elsewhere known as the Scherzer Rolling Lift Bridge consists of a cantilever or bascule, having a portion of its land end shaped to the segment of a circle and rolling on a flat bearing, a familiar illustration of the mode of action being found in the rolling of the runners of an ordinary rocking chair upon the floor. These bridges may be either of the single or double leaved type, and have been constructed to span a channel of 275 feet in width at Chicago in two leaves, or as a single leaf of 230 feet in length, at Cleveland, Ohio.

Floating Bridges with Movable Portion.—Bridges of boats moored across the Rhine, Danube, and other rivers, from the nature of their construction, can be readily made to open at a suitable part of the channel for the passage of vessels. The Howrah Floating Bridge connecting Calcutta with Howrah across the Húgli, constructed in 1873–74, is a fine example of this class of bridge, 1530 feet long between its abutments, and providing a roadway 48 feet wide, and a footway on each side 7 feet in width.[3] The bridge is carried by twenty-eight rectangular iron pontoons, 160 feet long and 10 feet wide, coupled in pairs to ensure stability, and moored at both ends; and the bridge is opened at slack tide by running back drawbridges on each side of the two central sections, warping these sections upstream, disconnecting them, and drawing the adjacent section on each side clear of the opening of 200 feet thus formed. As the bridge is raised sufficiently by staging on the pontoons to afford a headway through the navigable openings of 20 feet for the numerous native craft, it has only to be opened two or three times a week for the passage of large vessels. This bridge cost £182,000; and it was preferred to a fixed bridge, owing to the difficulties and cost of founding piers in the silty bed of a river where obstructions in a rapid current lead to a great amount of scour, and the time the construction of a fixed bridge would have occupied.

[1] *Annales des Ponts et Chaussées*, 1886 (1), p. 709, and plate 20.
[2] *Engineering News*, New York, vol. xxxi. p. 320.
[3] *Proceedings Inst. C.E.*, vol. liii. p. 3, and plates 1 and 2.

CHAPTER XII.

VIADUCTS AND TUNNELS.

Viaducts: in place of embankments and bridges; materials used, and forms, examples; remarks, uses of, definition—**Tunnels**, in place of cuttings advantages, in rugged country, for special purposes, classification—**Ordinary Tunnels with Shafts**: influence of strata; shafts, advantages, construction and form, position; excavation and timbering, English system; Belgian system; central core system; Austrian method of timbering; American system of timbering; iron framing; through headings, and side drifts; difficult conditions, with examples; sections of tunnels, objects of different types with instances; construction of lining, order of, with different systems, precautions necessary; drainage in tunnels, provisions; cost of tunnels, varying with strata, various examples; remarks, uses of tunnels, enlargement or second tunnel—**Long Alpine Tunnels**: special conditions; position and elevation, internal heat; gradients; construction, headings, special provisions for Simplon Tunnel, lining; strata traversed; rate of driving headings; cost; remarks, cooling in tunnels, ventilation, advantages of Simplon route.

VIADUCTS are resorted to where, in traversing deep valleys, the embankments would become too high to be safely maintained; and tunnels have to be employed for piercing high ridges, where a somewhat abrupt rise of the ground causes the surface to be so much above the formation level consistent with suitable gradients, as to render an open cutting impracticable or too costly.

VIADUCTS.

The employment of viaducts for crossing valleys or gorges depends upon the height at which the railway has to be carried above the bottom of the valley, and the nature of the materials available for embankments. If chalk or rock can be readily obtained from adjacent cuttings, embankments may be formed up to a height of about 60 feet; but with less solid materials, and especially with clay or other treacherous soils, it would be quite unsafe to raise an embankment to that height, for fear of settlement and slips. A viaduct possesses the advantage of providing a passage for the stream or river invariably found flowing along the bottom of a valley, and also for any roads passing along the valley, thereby dispensing with the construction of bridges for the purpose, or the provision of a culvert across the base of a high embankment for the discharge of the stream draining the valley.

Materials and Forms for Viaducts.—Timber has been sometimes used for viaducts, especially for lines traversing forests in undeveloped countries, where wooden trestles and trusses have frequently been temporarily adopted in preference to embankments, as a more expeditious means of carrying the railway across hollows in the first instance; and timber trusses, carried on masonry piers, were employed originally for the viaducts on the Cornish extension of the Great Western Railway, which, however, have now been replaced by iron girders.

Viaducts are generally constructed of a series of brick or masonry arches, or of metal girders supported on high iron or masonry piers; and fine examples of these two distinct types are furnished by the arched, brick viaduct, 915 feet long, carrying the Scarborough and Whitby Railway over the Esk valley near Whitby,[1] in thirteen spans of 55 to 65 feet, at a height of 120 feet above the bed of the river (Fig. 103, p. 190), and the steel viaduct, 2180 feet long, which conveys a single line of railway across the Pecos River cañon in Texas[2] at an elevation of 321 above the river at its low stage, with a central cantilever span of 185 feet, two spans of 85 feet adjoining the central span, and twenty-two spans of 65 to 35 feet, including the minor spans across the higher ground on each side of the deep cañon (Fig. 104). The Esk Viaduct carries its railway over two railways running along the valley, as well as across the tidal River Esk in three spans at the bottom of the valley; and the Pecos Viaduct crosses the Pecos river, when at a high stage, by its three main spans, the central one being made specially large to ensure the passage of drift during floods.

A remarkable example of the use of long viaducts in railway construction is to be found in the Key West Extension of the Florida East Coast Railway from Miami to Key West, 156 miles in length, where reinforced concrete arches of 50 to 60 feet span are used to an extent aggregating 6 miles in length, in viaducts constructed in the open sea and connecting various coral islands of the Florida Keys.

Remarks on Viaducts.—The necessity for viaducts is most frequent when railways have to be carried in a direction at right angles to the general course of adjacent valleys, in traversing somewhat hilly country, so that ridges and depressions are successively encountered, rendering an alternation of tunnels and viaducts inevitable. Viaducts are also needed where railways, running along the side slopes of deep valleys, have to cross over from one side of the valley to the other side for the sake of economy in construction.

The term "viaduct" is commonly applied to structures resembling long bridges, chiefly remarkable for the number of their moderate spans and the height of their piers, and stretching for most of their length across the slopes of a valley, though serving also to cross the river at the bottom of the valley, and any roads, railways, or canal which may run along the

[1] *Proceedings Inst. C.E.*, vol. lxxxvi. p. 303, and plate 9.
[2] *Ibid.*, vol. cxx. p. 314, and plate 7.

VIADUCTS.

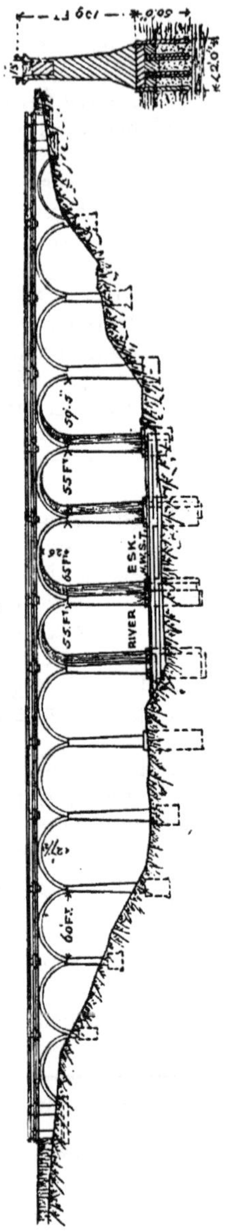

Fig. 103.—Esk Viaduct, Whitby.

Gradient 1 in 56·61.

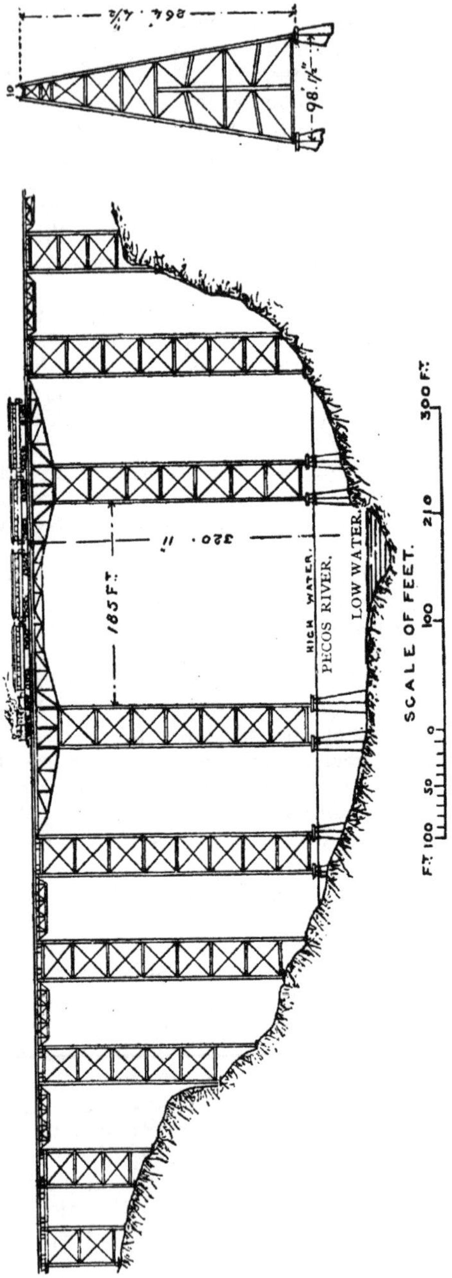

Fig. 104.—Pecos Viaduct, Texas. S. P. Railway.

SCALE OF FEET.

valley. The Garabit Viaduct has been thus denominated, on account of its crossing the valley of the Truyère with ten spans altogether; but strictly it consists of an arched bridge of large span, with an approach viaduct on each side. The new Tay Bridge which has replaced the bridge which was overthrown by a gale at the end of 1879, has been called the Tay Viaduct,[1] probably to distinguish it from the ill-fated original structure; but as it crosses the wide Firth of Tay, it is practically a river bridge throughout, though its great length of 10,527 feet, and its eighty-two spans over the firth, thirty-one of them on the Dundee side of only about 65 feet span, afford some grounds for calling it a viaduct; whilst its thirteen large spans of 227 and 245 feet over the central channel, and its twenty-three spans on the southern side of 145 and 129 feet, give it the importance of a bridge of many spans rather than a viaduct. It is, indeed, impossible to draw a very precise distinction between a viaduct and a bridge of several spans; but it appears advisable to denote a structure crossing a river with one or more large spans, together with smaller spans, as a bridge, reserving the term "viaduct" for a structure which for the most part traverses a valley with a series of spans, none of which are large.

TUNNELS.

Tunnels are usually resorted to for traversing high ground, or piercing steep ridges or mountain ranges, wherever the depth of formation for the canal or railway below the surface, makes the cost of excavating a deep cutting, with its necessary slopes, approximate to the expense of tunnelling. The depth at which a tunnel should be substituted for a cutting is about 60 feet in firm soil; but the suitable depth varies with the nature of the strata to be traversed, for where rock is reached, a deeper cutting with steep side slopes might prove more economical than a tunnel; whereas in slippery and unstable strata, through which very flat side slopes would be necessary for a cutting, and slips might be apprehended, tunnelling would prove expedient at a much less depth. When the estimates for a cutting and a tunnel are about equal, and the material from the cutting is not required for embankments, the tunnel should be preferred; for it saves the cost of the wide strip of land required for the cutting, less the price of the underground rights and the small amount of land needed for shafts and spoil-banks; it dispenses with any bridges for roads passing across the line; and it avoids severance of the property traversed.

In laying out lines for developing uninhabited districts, deep cuttings and tunnels are avoided to a remarkable extent by a circuitous course following the main valleys, and the free use of steep gradients and sharp curves, so as to reduce the cost and expedite the completion of the railway; and it has thus been possible to carry the Canadian Pacific Railway across Canada, from Montreal to Vancouver, with only a very

[1] *Proceedings Inst. C.E.*, vol. xciv. p. 99, and plate 5.

few short tunnels. For main lines, however, in populous districts, connecting important centres of trade, where a direct route, good gradients, and easy curves are important to provide for a large and quick traffic, tunnels are often unavoidable in rugged country, especially where the course of the railway is at right angles to the lines of the valleys. Thus the railway from Bombay to Calcutta, passing for a great portion of its length along the basin of the Ganges, has only one short tunnel between the Western Ghats and Calcutta, and the Great Western Railway, running along the Thames Valley, has no tunnel between London and the Box Tunnel near Bath; whilst the Great Northern passes through a long succession of tunnels soon after leaving London, and the Brighton Railway is notable for its long tunnels and viaducts.

Tunnels of great length have of necessity been resorted to for the main lines traversing the steeper Alpine ranges; and tunnels have occasionally been used for passing under rivers, where there have been objections to the erection of a bridge, and also recently for the deep underground lines improving the means of communication in London.

Tunnels may advantageously be divided into three distinct classes, namely, (1) Ordinary Tunnels, the construction of which is aided by sinking shafts in the line of the tunnel; (2) Long Alpine Tunnels, which, owing to their great depth beneath the surface, can only be bored from each extremity through the hardest rocks; and (3) Subaqueous Tunnels, constructed under rivers, estuaries, and occasionally lakes.

Ordinary Tunnels with Shafts.

When the position and course of a tunnel has been determined, its centre line is accurately set out along the surface, as in the case of a railway in open cutting, to enable the sites for sinking the shafts to be located with precision over the centre of the tunnel, and to serve as a guide for the setting out of the tunnel itself below.

Influence of Nature of Strata traversed.—The most accurate information possible should be obtained of the strata to be traversed by a tunnel, by means of geological maps of the district, and a series of borings along the line of the tunnel, aided, in very important works, by the advice and investigations of an experienced geologist; whilst very valuable additional evidence is afforded by the appearance of the actual strata traversed by the shafts. Upon the nature of the strata, indeed, depends the form that should be given to the section of the tunnel, the amount of timbering required to prevent falls or settlement, the prospect of encountering water, which is largely affected by the angle of the dip, the thickness of the lining that should be adopted, and consequently the difficulties in execution and the cost of construction.

Shafts for Tunnels.—The number of shafts which it is advisable to sink on the centre line of a given length of tunnel, has to be determined by the depth of the tunnel below the surface, which regulates the cost of each shaft, the nature of the difficulties likely to be met with in execution, and the time available for the construction of the tunnel. Each shaft adds to the cost of the tunnel in proportion to its depth; but it greatly expedites the work by adding two faces from which the tunnel can be driven; it affords an outlet for the excavated materials, and means of access for building materials; it provides a pumping shaft in case of a large influx of water; and it facilitates the correct alignment of the tunnel, by enabling the exact position of the central line to be fixed in it below, by hanging heavy plumb-bobs down it, on each side, from the central line on the surface. In long tunnels also, some of the shafts are advantageously retained, after the completion of the tunnel, as ventilating shafts to afford exits for the smoke and foul air resulting from the passage of the trains; but in such cases, they must be built up at least 8 feet above the surface, on the finishing of the works, to avoid accidents.

The brick or masonry lining of a shaft is gradually carried down, as the excavation proceeds, by underpinning, namely, excavating in sections under the finished work, and inserting props, excavating and building up between the props, and lastly completing the brickwork or masonry in place of the props; or the lining may be built upon a curb with a pointed shoe, and gradually lowered bodily by excavating under the shoe at the bottom, and building up the lining at the top as it descends, like the ordinary process of well-sinking. Care must be taken in the first system to make each length of underpinning perfectly solid, and bring it close up to the lining above it, so as to afford a perfect support and connection to the upper work; and sinking of the lining by the second system must be performed perfectly evenly, for with an unequal descent, the lining jams and sticks fast; but the descent may be sometimes facilitated by pouring water down between the lining and the excavated pit. Shafts are often lined and strutted with timber in America, and also elsewhere when required merely during construction; but when the stratum traversed is very unstable and full of water, iron tubbing is employed with advantage. Shafts, moreover, which have to be permanently maintained, are sometimes timbered as the excavation proceeds, and subsequently lined.

Permanent shafts have usually been given a circular form in England, generally ranging in diameter from 8 to 15 feet, though occasionally an elliptical shaft, as at the Hoosac Tunnel, or a square shape has been adopted; whilst an oblong, rectangular form, which can readily be divided into two or three sections by cross pieces, to serve respectively for hoisting and pumping shafts, is preferred in America.[1]

The shaft has to be stopped at the level of the top of the tunnel; and in order to support its lining, when the excavation for the tunnel is effected below it, till it can be permanently supported round the side of the opening formed in the arch of the tunnel to maintain the connection between the tunnel and the shaft, the lining has to be propped up by beams standing on the bottom of the excavation for the tunnel underneath the shaft, or occasionally, when the floor of the excavation

[1] "Tunnelling, Explosive Compounds, and Rock Drills," H. S. Drinker, New York, 1893, p. 700.

is yielding, the lining of the shaft has been temporarily suspended by rods secured to long stout timbers resting on the ground at the top of the shaft, a system followed for the shafts of the Saltwood Tunnel in sand.

The shafts have sometimes been sunk outside the line of the tunnel, more particularly in France and Belgium, and occasionally in America. Except, however, for shafts required to keep a tunnel dry by pumping after construction, or for artificial ventilation by the help of fans, the balance of advantages appears to rest with the shafts in the centre line of the tunnel; for though these expose the line in the tunnel to accidents occurring in the shaft over it during construction, and do not keep the tunnel and shaft distinct, or afford the same space as side shafts, they are far more useful for setting out the tunnel; they are less costly; they are more conveniently placed for facilitating the work; and they avoid the dislocation of the strata at the side of the tunnel produced by the sinking of a side shaft, which has occasionally, in treacherous soils, led to a slip, causing injury to the tunnel during its construction.

Excavation and Timbering for Tunnels.—The excavation for a

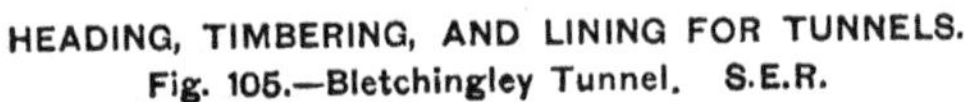
HEADING, TIMBERING, AND LINING FOR TUNNELS.
Fig. 105.—Bletchingley Tunnel. S.E.R.

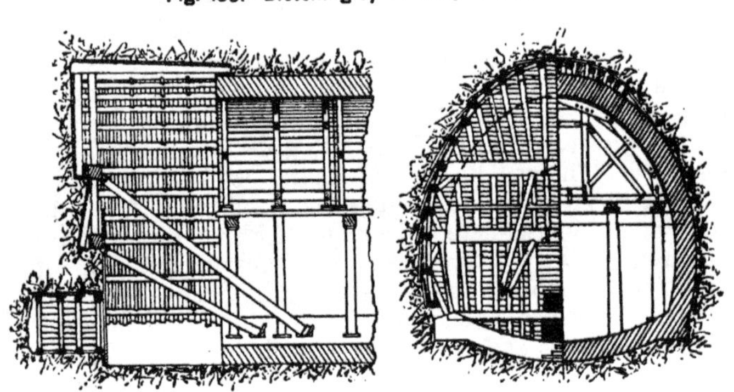

tunnel is commenced as soon as the completion of the approach cutting on each side gives access to the end faces, and from each side of the several shafts directly they have been sunk to the proper depth; and different methods have been adopted for carrying out the excavation. The old British practice was to commence by driving a bottom heading, which is strongly timbered through soft soil as it advances, and which serves to drain the superincumbent mass;[1] and the excavation of the rest of the required section for the tunnel, with its lining and timbering, is then extended upwards from the heading in short lengths, the face and sides being supported with bars, poling boards, walings, props, and raking struts, to prevent falls or settlement during the building of the lining (Fig. 105). Sometimes where the strata are dry, but are readily

[1] "Practical Tunnelling," F. W. Simms, 4th edition, 1896, p. 15, and plate vi.

disintegrated on exposure to the atmosphere, as in the case of certain clay soils, the excavation is commenced by a top heading, and the roof of the excavated section is supported at the outset[1] by bars and

HEADING, TIMBERING, AND LINING FOR TUNNELS.
Fig. 106.—Ampthill Tunnel. Midland Railway.

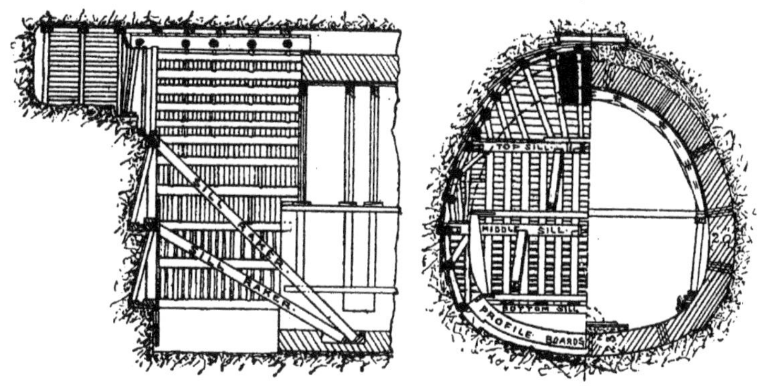

other timbering;[1] and the excavation is then extended downwards (Fig. 106). Occasionally these two methods are combined, a bottom heading being driven to provide for the drainage, and a top heading

HEADINGS, TIMBERING, AND LINING FOR TUNNELS.
Fig. 107.—Hoosac Tunnel, U.S.

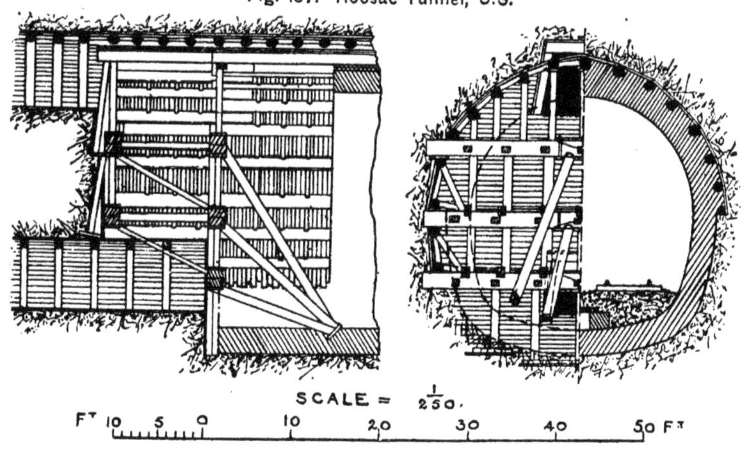

being driven at the same time to enable the roof to be at once supported by bars, a plan adopted in constructing the portions of the Hoosac Tunnel where soft soil had to be traversed[2] (Fig. 107). The amount of timbering

[1] *Proceedings Inst.* C.E., vol. cxx., plate 8, fig. 4.
[2] "Tunnelling," H. S. Drinker, p. 415.

required and its strength depend upon the nature of the soil, and its tendency to slips, falls, and settlement; and it is most important to secure the top and sides of the excavated length against any movement, till they can be supported and protected by the permanent lining of brickwork or masonry, for it is far more difficult to arrest any motion of the soil than to keep it from beginning to move. The system of excavation described above, especially with a bottom heading, and timbered as shown in Figs. 105 and 107, is commonly known as the English system, and throughout a long experience has proved perfectly satisfactory. It possesses the merits of ensuring good drainage and ventilation, of affording ample facilities for the removal of the earthwork and supply of materials, of providing a large space for the bricklayers and masons to work in, and of enabling the lining to be built up in complete lengths well bonded together. It is suitable for a variety of soft strata, ranging from soft, loose rock to treacherous strata, such as quicksand and readily disintegrated clays; but in very unstable soils, the timbering has to be considerably strengthened and put closer together, so that the available space is reduced.

Belgian Method of Excavating Tunnels.—The essential feature of the Belgian system, which has been largely used on the Continent, especially for tunnelling through loose rock, consists in opening out the excavation at the top of the tunnel first, sufficiently to construct the arch, which rests on the sides or on framing till the lower part of the excavation is removed; and the side walls are then built in sections, and by underpinning, so as eventually to take the weight of the arch. This was the system adopted for the construction of the lined portions of the Mont Cenis and St. Gothard tunnels; though in the Mont Cenis tunnel, the advanced heading was bored in the central bottom section, but was followed up by the excavation of the upper portion, and the building of the arch, which was supported by the projecting side portions below till the bottom excavation was completed and the side walls built up (Fig. 116, p. 202). The ordinary course in this system was followed in the construction of the St. Gothard Tunnel, where the advanced heading was driven at the top, as well as the first stages of the enlargement (Fig. 118, p. 202). The main advantage of the Belgian system is that the roof of the tunnel is protected at the earliest possible period, which in loose rock is the part most likely to fall, and from which most water may be expected with sharply dipping strata. The upper level, however, in the advanced portion, and the lower level behind, are inconvenient for drainage, for the removal of the excavations, and for the supply of building materials; whilst the construction of the arch has to be carried out in a confined space; and the building of the side walls by underpinning is more costly, and less favourable for producing thoroughly sound work, than when the side walls can be built up first and the arch erected upon them. The system, indeed, reduces the amount of timbering required through soft strata; and the adoption of a bottom heading, as at the Mont Cenis Tunnel, facilitates drainage, and the removal of the excavations and supply of materials; but very little, if any, timbering is generally required in dry, firm rock, and the blasting of the bottom heading tends to loosen the sides upon which the arch has to rest till the side walls are built. Several tunnels have been successfully carried out by this system, and it is specially advantageous where the roof needs immediate protection; but it is not well adapted for tunnelling through strata not sufficiently firm to support the arch securely; and the impediments to access and drainage caused by the two levels, and the underpinning of the side walls, must increase the cost in comparison with the more open system first described.

Central Core System for Construction of Tunnels.—Another system consists in the retention of a central core in the lower part of the section of the tunnel till the end, which, by bearing the props supporting the roof and side planking till the side walls and lining of the roof can be built, considerably reduces the amount of timbering required[1] (Fig. 108). Moreover, the excavations at each side of the central core for the side walls, and at the top for the roof, provide access for three narrow roads, one on each side of the core, and one on top of it, for the removal of the excavation and supply of materials. The available space, however, at each side is too narrow for the satisfactory building of the side walls and supervision of the work; and the central core is only suitable for a support to the timbering if it consists of compact, hard material. Accordingly, although several tunnels were constructed in Germany about the middle of the nineteenth century in this manner, from whence it has been termed the German system, and also in France with some modifications during the same period, it does not compare favourably with the more usual systems, or appear to merit an extended adoption.

CENTRAL CORE, WITH TIMBERING.
Fig. 108.—Wolfsberg Tunnel. Semmering Railway.

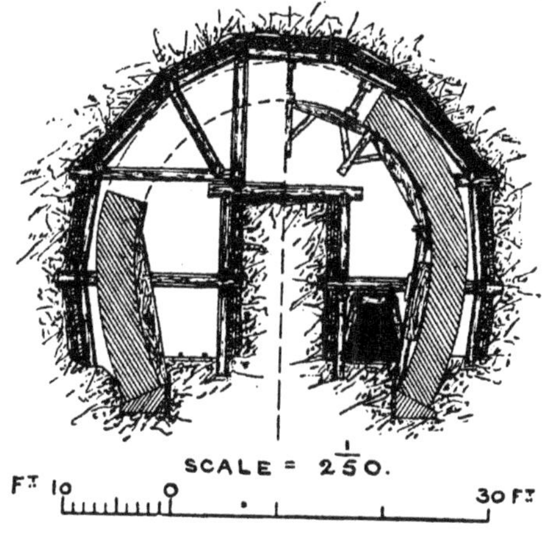

Austrian Method of Timbering for Tunnels.—The system known as the Austrian, differs mainly from the English system in the strength of its timbering, both as regards the supports and the props of the roof resembling rafters, and especially in the horizontal strutting of the sides and the staying where necessary of the framing for the invert,[2] as indicated by a comparison of Fig. 109 with Fig. 105, p. 194, so

[1] "Lehrbuch der Gesammten Tunnelbaukunst," F. Ržiha, Berlin, 1872, vol. ii. pp. 100 and 101.
[2] *Ibid.*, vol. ii. pp. 146 and 159.

that this system is specially suitable for very soft, unstable soil, where every portion of the section is exposed to considerable pressure. A bottom heading is first driven in this system, followed by a shorter top heading; and the side walls are constructed first, and secured by shorter cross struts after erection.

AUSTRIAN SYSTEM OF TIMBERING IN SOFT GROUND.

Fig. 109.

SCALE = 1/250

FT 10 5 0 30 FT

American System of Timbering Tunnels.—The special peculiarity of the American system of timbering tunnels is that the rafter truss, block arching, or stout, permanent, arched framing used for supporting the roof during construction, serves often, especially in single-line tunnels, for the permanent lining [1] (Figs. 110, 111, and 112), particularly on pioneer lines, where timber is abundant and other building materials would be very difficult to procure. Sometimes, however, the timber framing or blocking is merely employed as a support for the roof till a brick or masonry arch, or a stronger timber lining, can be put in under its shelter. Good, white oak

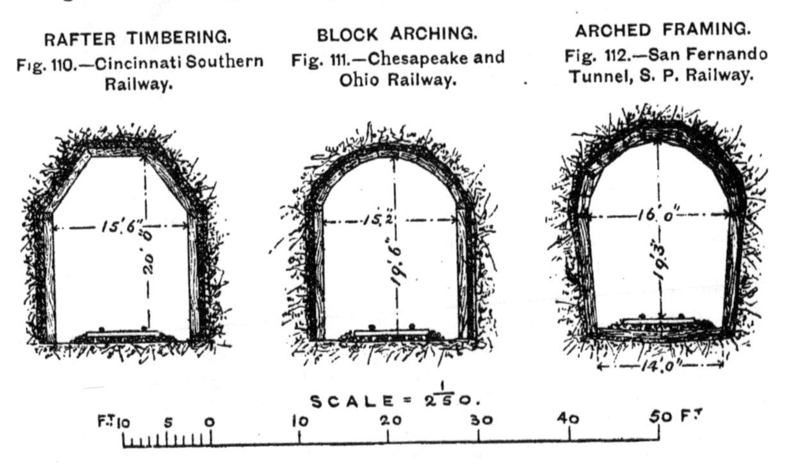

RAFTER TIMBERING. Fig. 110.—Cincinnati Southern Railway.

BLOCK ARCHING. Fig. 111.—Chesapeake and Ohio Railway.

ARCHED FRAMING. Fig. 112.—San Fernando Tunnel, S. P. Railway.

lining in tunnels where it is alternately wet and dry, or where the ventilation is bad, may last three or four years; but under good conditions, it may

[1] "Tunnelling," H. S. Drinker, p. 564.

last eight years or more. Timber arching answers well where the main pressure comes on the top of the arch, and the sides are quite stable, as in the case of loose slate or shaly rock ; but a stronger form of temporary timbering is needed in traversing treacherous clays, wet sand, or other unstable strata, as adopted in America for the Hoosac, Musconetcong, and other tunnels. Tunnelling in soft ground has, however, been rare in America compared with European practice, partly owing to differences in geological conditions, partly to much greater freedom in the choice of location in an unsettled or sparsely populated country, and partly to the steeper gradients and much sharper curves readily resorted to in rugged districts in America.

Iron Centres and Supports for Tunnels.—Iron has occasionally been employed for the centering and timbering of ordinary tunnels, on account of being stronger, more readily used several times over, and occupying less space than wood; but it is much heavier to move, and more costly under ordinary conditions. In 1824, Telford adopted an iron framing, composed of cast-iron segments bolted together, for the

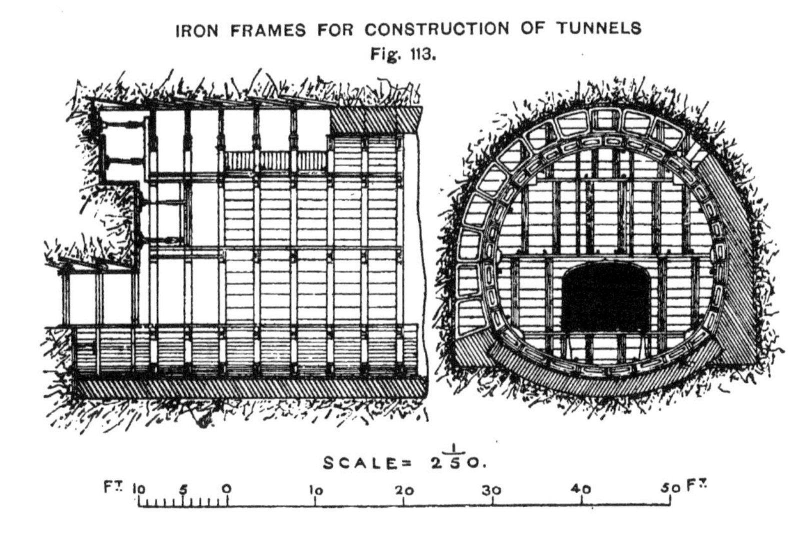

IRON FRAMES FOR CONSTRUCTION OF TUNNELS
Fig. 113.

centering of the arch, and the templates of the side walls of the second Harecastle Tunnel for the Trent and Mersey Canal, thereby dispensing with the greater part of the usual timbering ; and an extension of this system, in which the outer circumference of a second outer ring corresponds with the face of the excavation, and is removable in short segments, thereby dispensing entirely with timber supports (Fig. 113), was first adopted in executing the Naens Tunnel in 1862, and soon after in the Ippens Tunnel, through marl charged with water, and has since been employed on some other tunnels in Prussia[1].

[1] "Lehrbuch der Gesammten Tunnelbaukunst," F. Ržiha, vol. ii. pp. 334 to 337.

Through Headings, and Side Drifts.—Formerly it was a common practice to drive a heading through the whole length of the tunnel at the outset, which not merely served to drain the strata to be traversed by the tunnel, but also gave a perfect indication of the nature and condition of the strata, and fixed its exact direction. As, however, this preliminary operation, in the case of a long tunnel, would entail inconvenient delay in the completion of the tunnel, the nature of the strata is ascertained as closely as practicable by borings and geological investigations, and the heading is only pushed forward a short distance in advance of the completed work; but the thickness of the lining of the tunnel can be modified in accordance with the indications furnished by the advanced heading and the excavations for enlargement. In one of the schemes for a tunnel under the English Channel, it was proposed to drive a heading right across before enlarging, so as to make certain of the absence of large fissures in the chalk, before proceeding with the much larger expenditure of constructing the tunnel to its full size as the heading advanced.

Occasionally, when the line of a tunnel approaches in places the side of a steep hill, side drifts are driven through the side of the hill to the tunnel, in place of shafts, to obtain additional faces from which the headings for the tunnel can be driven, and to serve like shafts for the removal of the excavations and the supply of materials.

Specially Difficult Conditions for Tunnelling. — Though strata subject to disintegration or swelling on exposure to the atmosphere, and the influx of large quantities of water charged with sediment, tending to produce settlement, greatly enhance the difficulties of tunnelling, the most serious obstacles to the construction of tunnels are encountered in traversing drift, moraines, or loose *débris*, such as the rubbish tips from the slate quarries through which the Bettws-y-coed-Festiniog Railway had to be carried, and also spurs of mountain slopes composed of strata subject to slides, through one of which, composed of blue clay, gravel, and boulder drift, with intervening veins of sand, the Canadian Pacific Railway originally passed in a timber-lined tunnel, towards the bottom of the Kicking Horse Valley on the western slope of the Rocky Mountain range. In piercing loose *débris*, the whole superincumbent mass tends to move directly it is disturbed by the excavation below, and no solid support can be obtained; whilst when a slide occurs on the side of a mountain slope, the tunnel is not merely subjected to a severe sideways pressure, but is liable to be displaced bodily by the sliding downwards of the disturbed strata in which it is located, so that the above-mentioned tunnel on the Canadian Pacific Railway had to be abandoned; and the railway now runs round the foot of the projecting spur on a curve having a radius of only 262 feet, being much the sharpest curve on the whole line.

Sections of Tunnels.—The form of the cross section of a tunnel, and the extent and thickness of the lining, are regulated by the nature, condition, and dip of the strata traversed. Sometimes when tunnels traverse hard, firm rock, it is possible to dispense with any lining; but where the rock is laminated and has a great dip, promoting the influx of water, or is loose or soft, a lining of moderate thickness is necessary for protecting the line and the trains, and is also expedient when the rock is liable to suffer from exposure to the air. In long tunnels, accordingly, through rock, the lining is varied with variations in the condition of the rock, so that, as in the case of the long Alpine tunnels, some portions are left unlined, and others are arched over, the arch resting on a bench cut in the firm rock at the sides; whilst where the rock is less solid, side walls of masonry are added, increased in thickness in the worst places (Figs. 115, 117, and 119, p. 202). Where a considerable sideways pressure may be anticipated, as in soft soils, and especially through clays subject to swelling, in running sand, and with other treacherous strata, an invert has to be added at the bottom to prevent the side walls from sliding forwards at the toe, and to resist any upward pressure in very wet ground (Figs. 105, 106, 107, and 114, pp. 194, 195, and 202).

The form of the cross section of the tunnel is varied according to circumstances; for where the greatest pressure is liable to occur on the roof, and the invert serves mainly to support the side walls, as generally is the case through dry, soft soil, the roof and side walls are given a somewhat pointed elliptical form, and the invert is made fairly flat, as indicated by the section of the Bletchingley and Ampthill tunnels (Figs. 105 and 106, pp. 194 and 195). When, however, the soil is very treacherous, or very heavily charged with water, and the tunnel is at a sufficient depth below the surface not to be chiefly exposed to a settlement of the ground above it, the pressure becomes more fully distributed; and a circular, or nearly circular, section of uniform thickness, the proper form for a uniformly distributed pressure on every side, similar to a fluid pressure, is adopted with advantage. Thus in the Sydenham Tunnel through the London clay, where an ellipse with its major axis vertical was originally chosen for the arch and sides of the tunnel, suited to sustain an assumed maximum vertical pressure, with a flatter, thinner invert (Fig. 114, left-hand half-section), the sides, and the invert especially, were forced inwards considerably more than the arch by the swelling of the clay at several places, which was not effectually remedied by a thickening of the lining; and eventually a circular section, with a lining 3 feet thick, was resorted to with satisfactory results (Fig. 114, right-hand half-section).[1]

A segmental or semicircular arch, with straight side walls where necessary, provides the cheapest form through rock where the lining is only needed as a protection; a high elliptical section, with a somewhat flat invert, is required where the main pressure through soft soil consists of vertical settlement on the arch, and is advantageous for ventilation; and a circular section is advisable in very unstable, or very wet, soft strata.

Construction of Lining of Tunnels.—Where the whole section of a tunnel is excavated for a short length, according to the English system, through soft soil (Figs. 105, 106, and 107, pp. 194 and 195), the invert and side walls are built first, and finally the arch is turned, the bars

[1] *Proceedings Inst. C.E.*, vol. xlix. p. 250, and plate 6, figs. 2, 3, and 7.

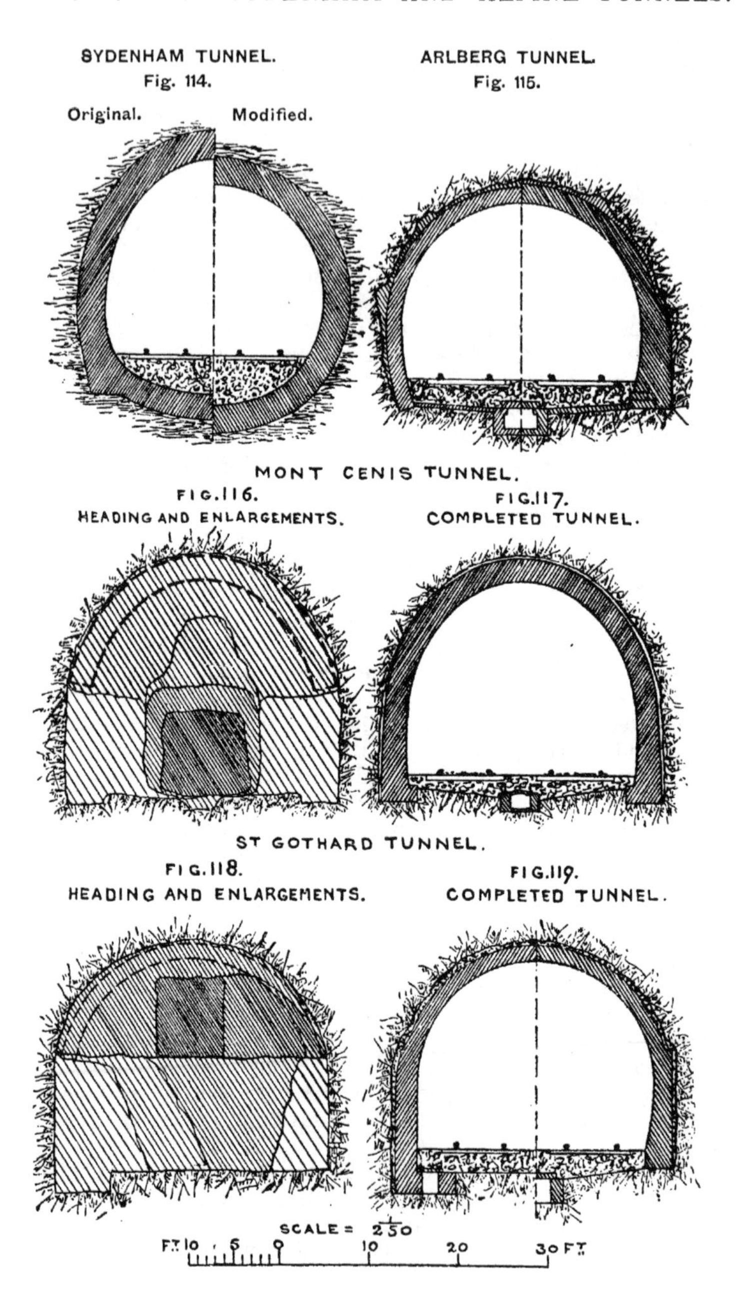
SYDENHAM TUNNEL.
Fig. 114.
Original.
Modified.
ARLBERG TUNNEL.
Fig. 115.
MONT CENIS TUNNEL.
FIG. 116.
HEADING AND ENLARGEMENTS.
FIG. 117.
COMPLETED TUNNEL.
ST GOTHARD TUNNEL.
FIG. 118.
HEADING AND ENLARGEMENTS.
FIG. 119.
COMPLETED TUNNEL.
SCALE = 1/250
FT. 10 5 0 10 20 30 FT.

supporting the roof being generally drawn out as the work proceeds, so as to be used again for another length, though occasionally, as in the Hoosac Tunnel, and where there is difficulty in drawing them, they are built in. Allowance is made in excavating soft soil for a certain amount of settlement, to ensure providing room for the arch; the arch has to be most carefully and solidly built in the confined space between the roof and the centres; and the space between the arch and the roof must be solidly packed with earth or rubble, to prevent settlement on the removal of the timbering. Where exposed to pressure, the masonry on the inner face must be specially strong and closely set in cement; and with brickwork, the two inner rings should be formed of blue bricks set in cement, bonded into the outer rings at intervals. The lining should follow as quickly on the excavation as possible, to guard against settlements, falls, and injury from exposure to the atmosphere; and the lining should be carried forward in short lengths where the ground is unstable.

With the Belgian system, the arch is built first, and the side walls subsequently by underpinning (Figs. 116 and 118); and in the French adaptation of the central core system, the same course has been pursued. This arrangement, however, is only suitable through rock, and where it is important to protect the roof at the earliest possible time.

Where iron framing is employed instead of timbering, the invert, if required, must be laid in advance, so that the frames may be erected upon it; and the side walls are next built; and, lastly, the arch is turned (Fig. 113, p. 199). In the Austrian system of heavy timbering, however, and in the German central core system, the side walls are first built, then the arch, and lastly the invert if required (Figs. 108 and 109, pp. 197 and 198). The heavy timbering of the Austrian system, for very soft ground, is kept in place till the invert is completed; but where the central core system is used, an invert should not be required, and if found necessary, should be built with great caution in very short lengths, the side walls at the same time being secured against coming forward on the removal of the core.

Drainage in Tunnels.—In determining the gradients of a tunnel, besides being governed by the difference in level of the approach lines at the two extremities, it is necessary to provide for the permanent drainage of the tunnel, as water generally finds its way into a tunnel through the drainage holes usually left at intervals in the side walls to prevent water accummulating at the back, or from percolation through the lining in places, or from drip in the unlined parts. Tunnels are, therefore, laid out with a gradient adequate for drainage, either starting from each end and rising towards the centre, or having a continuous rise or fall from one end to the other. The first system is essential for the drainage of tunnels driven solely from each end, during construction, as well as after completion; whilst the second method overcomes differences of level at the two ends with the flattest practicable gradients, but necessitates pumping from intermediate shafts during construction, as the only natural drainage is obtained for the heading driven from the lower end. When any large influx of water occurs during construction in driving the advanced heading, or from along the outside of the finished lining into the excavations, it is very important to prevent sand or silt flowing in with the water, by packing the vent with straw or other loose material, as an influx of solid matter produces hollows in the ground very liable to occasion serious settlements.

When the regular flow of water into the tunnel is small, a channel along each side wall suffices to carry it off; but generally a culvert is formed below the formation level, in the centre or on one side of a tunnel (Figs. 115, 117, and 119, p. 202), or on the lowest part of the invert (Figs. 106 and 107, p. 195), for the discharge of the water from the tunnel.

Cost of Tunnels.—The cost of constructing tunnels, which is always given in England per lineal yard, naturally varies very much according to the nature of the strata traversed, the amount of water encountered, the timbering required, the extent and thickness of the lining, and the cross section of the tunnel, together with its general depth below the surface, affecting the construction of the shafts. Tunnels have naturally been made most cheaply through firm, compact, homogeneous, moderately soft rock requiring little or no timbering, and sometimes no lining, and fairly free from fissures discharging water, such as chalk and some kinds of sandstones and limestones. In order to make the cost given of tunnels more comparable, only tunnels for a double line of way will be cited, as smaller tunnels for single lines, and especially for water conduits or sewers, would give the delusive appearance of greater economy in construction, really attributable to their smaller dimensions. The old tunnels of Guildford, Salisbury, and Petersfield through chalk, and Lydgate through the coal-measures, cost £30 per lineal yard;[1] whilst the Clifton Tunnel, constructed in 1871–74 through mountain limestone, only cost £21 4*s.* per lineal yard. Honiton Tunnel, through red marl and greensand, cost £50; the Bletchingley and Buckhorn Weston Tunnels, through clay, and the Cowburn Tunnel, through dry shale and rock, cost £72; the Blaenau-Festiniog Tunnel,[2] through bastard slate and greenstone, and for about five chains at the south end through slate *débris*, and the Totley Tunnel,[3] constructed in 1888–93, through the lower coal-measures and millstone grit, mainly black shale, slightly over 3½ miles in length, being the longest ordinary tunnel in England, cost £75; the Box Tunnel, through oolite marl, accommodating a double line of 7-feet gauge, cost £100; and the Saltwood Tunnel, through the lower greensand, cost £118 per lineal yard. Unforeseen difficulties are liable to increase greatly the cost, such as the influx of quicksand, undiscovered by the borings, in the Kilsby Tunnel, raising its cost from £40 to £125 per lineal yard, and the unexpected amount of swelling of the clay in the Sydenham Tunnel, which increased its cost from £80 to nearly £120, where in good ground the cost was only £57½ per lineal yard.

[1] *Proceedings Inst. C.E.*, vol. xxii. p. 381.
[2] *Ibid.*, vol. lxxiii. p. 164, and plate 10, fig. 1.
[3] *Ibid.*, vol. cxvi. p. 168, and plate 4, fig. 1.

The expenses of some of the most important and difficult tunnels in the United States are as follows: the Van Nest Gap Tunnel on the Warren Railway, through gneiss, £75; the Baltimore Tunnel on the Baltimore and Potomac Railway through rock, clay, and earth, £88½; the Church Hill Tunnel on the Chesapeake and Ohio Railway, through clay, £111; and the Bergen Tunnel on the Erie Railway, through dolomite, £114. The Hoosac Tunnel, the longest tunnel in the United States with a length of 4¾ miles, whose intermittent construction extended from 1854 to 1876, though traversing such hard strata as mica-schist and granitic gneiss, passed through so much decomposed rock and quicksand that its cost, enhanced by stoppages and other circumstances, amounted eventually to £248 per lineal yard.[1]

In France, the Chinon Tunnel, through grey chalk and marl, cost £35½; the Latrape Tunnel, through calcareous rock and wet, soft ground, cost £59½; and the Marseilles Tunnel from Old Port to Prado Station, through limestone with sand and thick beds of blue and red clay, and large quantities of water, cost £73 per lineal yard.

In Germany, the Singeister Tunnel, through loam, marl, and limestone, cost £48⅔; the Gotthardsberg Tunnel, through fissured dolomite, cost £61¾, and the Sommerau Tunnel in the Black Forest, through granite, cost £70½.

Several tunnels constructed on two lines in Prussia and Austria in 1872–76, by aid of machine drills, and iron framing in some cases, through old red sandstone much disturbed and intersected by beds of clay, ranging in length from 63 yards to 399 yards, cost from £43 to £77⅕ per lineal yard; whilst the Naens and Ippens tunnels, the first constructed by the help of iron frames (Fig. 113, p. 199), traversing Keuper marls and lias marls respectively, charged with water, cost £89 and £71 per lineal yard.

Occasionally the strata pierced are so unstable that great difficulty is experienced in carrying forward and permanently securing the lining in places, greatly increasing the cost, of which notable instances are furnished by the Czernitz Tunnel in Germany, passing through blue gypsum, loam, and fine veins of sand, which at first forced in the masonry in places,[2] so that its cost reached £184 per lineal yard; and the Cristina Tunnel in the Apennines, traversing shaly clay alternating with beds of limestone, and also wet, laminated clay very subject to slips, whose lining, after being dislocated in places in the latter strata, had to be made so thick, that this tunnel, though only serving for a single line, cost £145 per lineal yard.

Remarks on Tunnels.—Drifts or galleries have been extensively driven for mining operations; but the earlier tunnels of importance of modern times were constructed for navigation canals.[3] Since the introduction of railways, however, tunnels have been greatly extended in number and in length for the passage of trains. Tunnels have also been

[1] "Tunnelling," H. S. Drinker, pp. 327 and 333.
[2] "Lehrbuch der Gesammten Tunnelbaukunst," F. Ržiha, vol. ii., pp. 132 and 644.
[3] "Rivers and Canals," L. F. Vernon-Harcourt, vol. ii. pp. 362 and 363.

constructed for the conveyance of water through high ground for irrigation, for drainage, and for water-supply; whilst tunnelling is often required in the construction of intercepting or outfall sewers. The sections of such tunnels vary considerably according to the discharging capacity required and the available fall; but the principles and difficulties of their construction are the same as for railway tunnels, in which the greatest experience has been gained.

In widening a single line of railway to provide for an increasing traffic, the choice lies, with regard to the tunnels, between enlarging the existing tunnel, or constructing a second parallel tunnel. When the tunnel passes through firm rock, and especially when it is not lined, the enlargement of the existing tunnel is preferable on the score of economy; but through loose rock or soft ground, the re-construction of the heavy lining necessitated by an enlargement, and the difficulty of keeping the existing line free for traffic during the progress of the works, generally renders the construction of a parallel tunnel the most expedient course, especially as the work can be expedited by side drifts from the existing tunnel. It is necessary, however, to drive the second tunnel at a sufficient distance from the first, to secure the latter from any disturbance during the excavation for the second tunnel. Often when a railway, constructed for a single line, offers good prospects of securing a traffic necessitating its widening in a few years, it is advisable to construct the tunnels for a double line at the outset.

LONG ALPINE TUNNELS.

The special conditions which distinguish the long tunnels piercing the main dividing ridges of the Alpine chain from other tunnels carried through hard primitive rocks, are, their great length, coupled with the impossibility of expediting the work by means of shafts, owing to their great depth below the surface, so that they can only be driven from each extremity; the difficulty of ventilating such long headings during construction; and the comparatively high temperature of the rock at the considerable depths below the surface traversed by the central portions of these tunnels. The necessity of driving these long tunnels from only two faces, greatly enhanced the importance of rapidly drilling the holes in the faces for blasting the rocks, and led to the invention of percussion and rotary rock-drills worked by compressed air, which greatly expedited the progress of the headings, the air-compressors being worked by turbines actuated by the fall of water from mountain streams in the neighbourhood of each end of the tunnel.

Though it proved possible to carry a railway across the Brenner Pass, the lowest pass crossing the main Alpine range, without a summit tunnel, there is no other pass across the principal chain which could be thus surmounted in the open, by the aid of gradients and curves suitable for the speed and weight of the trains of a main continental line. In ascending the most favourable valley on each side of the mountain range, a point is reached where the general slope becomes too steep to be surmounted with the limiting gradients and curves; and it becomes necessary to pierce the dividing ridge in tunnel. The higher up the approach railway can be carried, the shorter will be the length of the tunnel; but, on the other hand, if the tunnel is commenced lower down the valley, the ruling gradient can generally be made less steep and the curves less sharp, rendering higher speeds practicable for the trains. The elevation, however, at which the tunnel is commenced, is governed to a great extent by balancing two opposing considerations. If the open railway is carried high up the mountain slope on the northern side of the Alps, in the neighbourhood of the highest peaks where the cold is greatest, it is liable to be blocked by snow in hard winters; whilst if the tunnel is commenced at a comparatively low level, it is longer, and has to be carried through rocks having a considerable internal heat owing to their depth below the surface.

Position and Heat of Long Alpine Tunnels.—Four long tunnels have been constructed through the Alps, namely, the Mont Cenis Tunnel, 8 miles long (1857–71), and the St. Gothard Tunnel, 9⅓ miles long (1872–82), across the principal range, and the Arlberg Tunnel, 6¼ miles long (1880–84), across an outlying spur in Austria;[1] whilst a fourth tunnel, the Simplon Tunnel, 12⅓ miles long, commenced in 1898, has been constructed through the main chain near the Simplon Pass. The northern end of the Mont Cenis Tunnel is 3793 feet above sea-level, and of the St. Gothard Tunnel 3638 feet, the western end of the Arlberg Tunnel 3992 feet, and the northern end of the Simplon Tunnel only 2254 feet above sea-level; whilst the other ends of the tunnels are 477 feet, 118 feet, and 278 feet higher, and 175 feet lower, respectively. The maximum height of the surface of the ground above rail-level in the tunnel is 5076 feet at the Mont Cenis, 5733 feet at the St. Gothard, 2362 feet at the Arlberg, and 7005 feet at the Simplon; and the maximum heat of the rock in the central part of the headings during construction was 85°·1 in the Mont Cenis Tunnel, 87°·4 in the St. Gothard, 62°·2 in the Arlberg, and 133° in the Simplon in proximity to the maximum height of surface above rail level.[2]

Gradients in Long Alpine Tunnels.—The tunnels have in every case been constructed with gradients ascending towards the centre, simply to secure drainage towards the outlet during construction on the higher side, and in order to overcome the difference in level between the two ends on the lower side, together with the increase in elevation due to the rising gradient from the other end. Accordingly, drainage gradients have been given, falling from the central summit to the higher end, of 1 in 2000 to 1 in 1000 in the Mont Cenis Tunnel, 1 in 2000 to 1 in 500 in the St. Gothard, 1 in 520 in the Arlberg, and 1 in 500 in the Simplon Tunnel; whilst the rising gradients from the lower end are 1 in 43·5, 1 in 172, 1 in 66, and 1 in 143 respectively.

[1] *Proceedings Inst. C.E.*, vol. xcv. pp. 257, 265, and 269, and plates 6 and 7.

[2] The increase of temperature in sinking wells or shafts into the earth near sea-level, has been found to amount to about 1 degree Fahrenheit for each 60 feet below the surface; but the fairest average of the results in the Simplon Tunnel gives 1° Fahrenheit for each 67·5 feet below the surface.

Construction of Long Alpine Tunnels.—The Mont Cenis, St. Gothard, and Arlberg tunnels were driven in a perfectly straight line from each end, so as to reduce the possibility of a divergence from the exact junction of the two long headings to a minimum; and though it was originally proposed to make the two headings of the Simplon Tunnel meet at an angle, in order to avoid passing under the highest ground of the direct route, the tunnel has now been driven in a straight line, like its predecessors, its position having been modified sufficiently to prevent a straight route reaching an extreme depth below the surface, with its increased internal heat. To secure accuracy in direction for these exceptionally long lines, the country over the route of the tunnel is carefully surveyed, the centre line is set out along the surface, and the exact length of the tunnel is ascertained; and the precise direction for each heading is given by a powerful theodolite from an observatory built on the prolonged line of the tunnel, at a convenient distance from each extremity, and in a position commanding the entrance, so that the surface line may be sighted with precision, and its direction indicated, checked, and prolonged in the heading as the driving progresses. In joining the headings in the Mont Cenis Tunnel, the direction proved quite correct, but there was an excess of 1 foot in height on the French side, owing probably to the tunnel being 45 feet longer than calculated; and in the St. Gothard Tunnel, where each heading approximated to 4½ miles in length, the errors in direction and level were 13 inches and 2 inches respectively, the total length being 25 feet less than calculated, the Arlberg Tunnel being 10 feet short of the estimated length. In the Simplon Tunnel with headings of about 6 miles in length the errors in direction and level were 8 inches and 3½ inches respectively, the total length being 31 inches less than anticipated.

The advanced heading was driven along the bottom in the Mont Cenis Tunnel, and along the top in the St. Gothard Tunnel, the successive enlargements being indicated by lighter shadings on the cross sections of the excavation (Figs. 116 and 118, p. 202); whilst the arch was built first in each case. In the Arlberg Tunnel, the advanced heading was driven at the bottom, 8⅕ feet high and 9 feet wide; and break-ups, or upward shafts, were driven in the roof of the heading, 79 feet apart on the eastern side where the rock was hard, and 216 feet apart on the western side through softer rock, from the upper part of which, top headings were driven in each direction, 7½ feet high and 6½ feet wide, till they met the headings from the adjoining shafts, whereby the progress of the enlargement was much expedited, the excavated rock being tipped down the shafts into waggons in the heading below. In the Simplon Tunnel two parallel bottom headings known as Nos. 1 and 2 were simultaneously driven for single line tunnels 55·8 feet apart centres, with cross galleries at intervals of 656 feet, the enlargement of No. 1 being effected somewhat as shewn in Fig. 116; heading No. 2 being only lined where necessary, and the completion of this second tunnel being deferred pending traffic requirements[1] (Fig. 120). The

[1] *Proceedings Inst. C.E.*, vol. clxviii.

adoption of a bottom heading in long tunnels through rock, has the advantage that by means of break-ups from the roof of the advanced heading, the enlargement and completion of the lining can be made to follow much closer on the main heading, than with a top heading as employed for the St. Gothard Tunnel. The second heading was made use of to provide for the ventilation and cooling of the first tunnel during construction, by driving 1760 cubic feet of fresh air through it per second by means of fans worked by turbines, and for the admission of a supply of cold water, which was used in the form of fine spray for cooling the current of air, and also for reducing the temperature of the rock headings and upper galleries, in addition to the cooling effected by the current of air. The second heading also

SIMPLON TUNNEL.

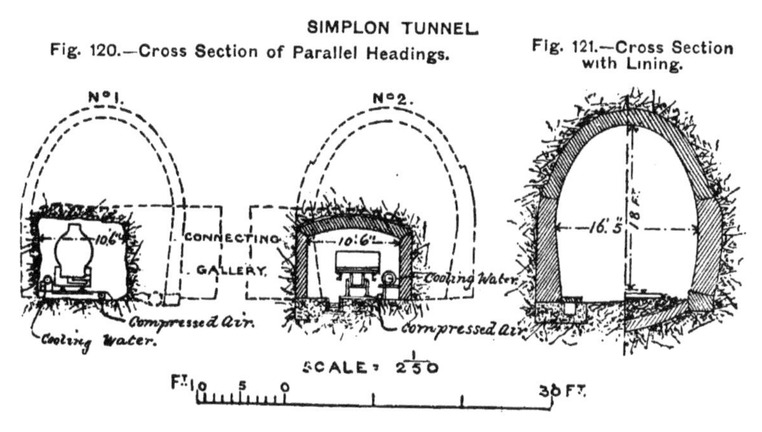

Fig. 120.—Cross Section of Parallel Headings.

Fig. 121.—Cross Section with Lining.

served as the route for the entering trains of waggons, which returned along the first tunnel when loaded with the excavations, running on lines of 2 feet 7½ inches gauge. Water under pressure was used at the moment of the explosion, for throwing the shattered rock to a distance from the face, and thereby permitting the immediate resumption of the drilling; and at the same time it served for cooling the face of the heading and the *débris*. These expedients were thus adopted in view of the great central temperature experienced, the unprecedented length of the headings, and in order to expedite the work; for the less high temperature in the shorter headings of the St. Gothard Tunnel, combined with imperfect ventilation in the moist atmosphere, in spite of the supply of air from the compressors and the use of Mékarski's compressed-air engines in place of ordinary locomotives in the tunnel, proved deleterious to both workmen and horses.

The lining of these tunnels, as in ordinary tunnels through rock, is varied according to the nature and condition of the rock, and the amount of water met with. The strata traversed by the Mont Cenis Tunnel were fairly free from water, and no timbering was required in construction; whilst a short portion of the tunnel was not lined, other portions were only arched over, and side walls were added where the rock was less firm (Fig. 117, p. 202). The rock traversed by the St. Gothard Tunnel was so disintegrated and fissured in places, that strong timbering had to be employed in its construction through these parts, especially where there was a large influx of water, and an invert introduced; whereas in some parts no lining was needed, in others arching alone was used, and in most places side walls of varying thickness and form were added[1] (Fig. 119, p. 202). The eastern portion of the Arlberg Tunnel passes through very hard rock, but along the western portion very fissured rock and large quantities of water were encountered; and, consequently, the protection of the excavated section has been varied, according to the conditions, from a mere masonry lining of the roof and sides, $1\frac{2}{3}$ feet thick, up to an arch 3 feet thick and side walls 4 feet thick (Fig. 115, p. 202), to which an invert $2\frac{1}{2}$ feet thick has been added at the worst places. The Simplon Tunnel (No. 1) is lined throughout, the section varying with the conditions of the strata, the average thickness from end to end not exceeding 2 feet 3 inches to 2 feet 6 inches, but excessive pressures met with at $2\frac{3}{4}$ miles from Iselle required a thickness of no less than $5\frac{1}{2}$ feet in granite arching. A rising of the floor occurred in several places even in rock, necessitating the construction of masonry inverts for a considerable distance.[2]

Strata traversed, and Rate of Driving Headings for Alpine Tunnels.—The long Alpine tunnels traverse the hardest primitive rocks, varying, however, considerably in composition, hardness, and compactness. The Mont Cenis Tunnel passes mainly through calcareous schist; but its northern portion traverses carbonaceous schist in the first $1\frac{1}{3}$ miles, followed by comparatively short lengths of quartz and limestone. The strata through which the St. Gothard Tunnel was carried vary greatly,[3] consisting mainly of gneiss with mica-schist and hornblende for $4\frac{1}{2}$ miles of the central portion, and of granite with gneiss, and gneiss with schist along the northern part, and of mica-schist with hornblende and quartzous schist at the southern part. The Arlberg Tunnel traverses quartzous schist, approximating to gneiss, along its eastern portion, and fissured mica-schist along its western portion. The strata through which the Simplon Tunnel has been carried consist chiefly of gneiss, mica-schist, and on the Italian side of antigorio gneiss, but in some places, particularly about $2\frac{3}{4}$ miles from Iselle, limestone was encountered highly charged with springs of cold water at from 52° to 62° F. In the vicinity of this marble a decomposed mica-schist was met with, giving rise to serious difficulties both in timbering and lining. The "Great Spring" thus encountered had a flow of about 10,500 gallons per minute, and in September, 1904, at $5\frac{2}{3}$ miles from the Italian side, hot springs occurred with a discharge of from 3000 to 4000 gallons per minute with a temperature of 114°·6F. Analysis of this water revealed an unique absence of chlorine, and appeared to

[1] "Rapport Trimestriel, No. 20, de la ligne du St. Gothard," Berne, 1878, vol. v. plates 4 and 5; and "Rapports de la Direction et du Conseil d'Administration du Chemin de Fer du St. Gothard," Lucerne, 1873–1884.

[2] *Proceedings Inst. C.E.*, vol. clxviii

[3] "Rapport Trimestriel, No. 31, de la ligne du St. Gothard," Berne, 1880, vol. viii. plate 12.

indicate a plutonic origin. The maximum flow of cold springs was found to be coincident with the melting of the Alpine snows.[1]

The progress of the headings not only increases with the softness of the rock, but also with its homogeneity; for if the rock is softer on one side of the holes being driven in the face, than on the opposite side, the boring tool diverges towards the softer side, and jams itself in the hole, so that a harder, compact rock is better for drilling in than a softer, non-homogeneous rock. At the Mont Cenis Tunnel, boring by hand was carried on at first for about five years at the northern heading, and for about three years at the southern heading, the advance at both headings being only about a mile on an average of four years; whereas the headings were joined after an average of nine years' work with the rock-drills worked by compressed air, the distance traversed being about $6\frac{2}{3}$ miles, so that the rate of advance was increased threefold by introducing the machines. The rate of progress, however, varied within a wide range according to the strata, averaging 12·9 feet per day during the month of most rapid advance through carbonaceous schist, and only 1·17 feet per day in the month of least advance when traversing quartz. The driving of the headings was commenced by hand at the St. Gothard and Arlberg tunnels, till the air-compressing machinery could be installed; but the earlier introduction of machinery and improvements in the rock-drills, and the employment of dynamite instead of gunpowder for the blasting, expedited the progress in these tunnels. Thus the driving of the straight headings at the Mont Cenis Tunnel, 7·6 miles in length, occupied 13 years 1 month, giving an average daily advance of 2·8 yards; at the St. Gothard, with a length of straight headings of 9·26 miles, the period was 7 years 5 months, making a daily progress of 6·02 yards; and at the Arlberg, with 6·33 miles of headings, the period was 3 years 4 months, or a daily advance of 9·15 yards. At the Simplon Tunnel the date of actual commencement of the excavation was the 1st August, 1898, and the meeting of the galleries took place on the 24th February, 1905.

Including Sundays, saints-days, holidays, and occasions on which work was suspended for various reasons, the average daily advance at each face was $4\frac{1}{2}$ yards, but allowing only for the actual days on which the boring machines worked, the average advance was 5·81 yards per day at each face.

This result was due to the employment of the Brandt rotary rock-drill, combined with the extent and efficiency of the means provided for thorough ventilation, the cooling of the rock, and the great facilities afforded by the parallel headings with their connecting galleries.

The permanent way was completed on the 25th January, 1906. The Tunnel was traversed on the 19th May by the King of Italy and the President of the Swiss Republic, and the formal opening took place on the 30th May, 1906. The Tunnel is equipped throughout for electric traction on the three-phase system with overhead conductors.

The tunnel through the Tauern Range between Gastein in Salzburg and Mallnitz in Carinthia on the direct route from Salzburg to Trieste

[1] *Proceedings Inst., C.E.*, vol. clxviii.

was opened in July, 1909. This tunnel, situated 3800 feet above the sea, is about 5¼ miles in length, and ranks fifth among the longest tunnels in Europe. The rock temperature was found not to exceed 75° F., but considerable difficulties were encountered during construction, owing to water and the hardness of the rock. The amount of rock excavation exceeded one million cubic metres, while 230,000 cubic metres were used in lining.

Cost of Long Alpine Tunnels.—Owing to the novel conditions under which the Mont Cenis Tunnel was constructed, its cost reached the large sum of £224 per lineal yard; but it is believed that the actual expenditure by the contractors was much less, notwithstanding the novel and experimental character of the appliances they devised for expediting the advance of the headings.

Profiting by the experience gained in the Mont Cenis Tunnel works, the St. Gothard Tunnel was executed for £142 per lineal yard; so that the second long Alpine tunnel was constructed much more cheaply and more rapidly than the first, in spite of its greater length and higher internal temperature. The Arlberg Tunnel, with the benefit of the enlarged experience derived from the two previous tunnels, and only about two-thirds the length of the St. Gothard Tunnel, was constructed for about £108 per lineal yard. The approximate cost of the Simplon Tunnel with a tunnel for a single line only, and lined throughout, but with a second heading lined where necessary, is believed to have been about £3,200,000; or at the rate of £148 per lineal yard. In comparing this rate with the cost of the St. Gothard Tunnel, consideration must be given to the considerably greater length of the headings, the higher internal temperature, the nature of the difficulties experienced, and the greater provision for ventilation and cooling.

Remarks on Long Alpine Tunnels.—Though the headings for the long Alpine tunnels have been driven in a perfectly straight line throughout, the tunnels have been connected with the approach lines on each side by short curved lengths of tunnels branching off from the main tunnel near each extremity, making the lengths of the tunnels traversed by the trains somewhat longer than the lengths of the straight headings.

The maximum heat is generally reached near the central portion of the headings for these tunnels, shortly before the junction of the headings; and as soon as there is a continuous passage for the air from end to end, the rock begins to cool, so that the highest temperature of the rock in the St. Gothard northern heading of 87°·4 in 1880, was reduced down to 74°·5 in two years and four months after the junction of the headings; and three years later, in the middle of 1885, the temperature had fallen to 73°·6. In nine years the maximum temperature of the rock in the Arlberg Tunnel fell from 62°·2 to 56°·7. The experience gained at the Simplon shewed that the temperature of the rock gradually fell as the excavation had passed any given point, and although the maximum rock temperature was 129° to 133° F., that of the air did not exceed 89°, while the large amount of fresh air introduced, together with the spraying of cold water, greatly ameliorated the conditions, the supply mains of cold water being insulated with charcoal. A ventilating current of 1410 cubic feet per second sufficed to lower a rock temperature of 122° to 86°.

The ventilation of ordinary tunnels is to some extent effected through the shafts; but even in some of these tunnels, during damp, still weather, and with a larger traffic, and especially with a train going up a steep gradient where the locomotive has to be kept fully at work, the air in the tunnel is liable to be vitiated to an injurious extent. In the long Alpine tunnels, with no outlets except at the two extremities some miles apart, similar conditions lead to still worse results, in the absence of any artificial ventilation, except a supply of compressed air which can be turned on at places by the platelayers working in the tunnel, to prevent their being overpowered by the foulness of the atmosphere. Fortunately the atmospheric conditions are usually dissimilar on the northern and southern slopes of the Alps, so that generally a current of air passes along the tunnel, which gradually carries out the noxious fumes emitted by the locomotives. Owing, however, to the large increase of traffic on the St. Gothard Railway, the natural ventilation which originally kept the air in the long tunnel sufficiently pure for the platelayers to work in without inconvenience, gradually became so vitiated at times as to necessitate recourse to artificial ventilation. This is effected by two fans, 16½ feet in diameter, which, worked by a steam-engine at the north end, force air into a chamber round the face of the tunnel, from which it is driven along the tunnel through an annular aperture, so that it drags the vitiated air in the tunnel along with it, and thereby effects a renewal of the air. In the case of the Simplon Tunnel on the occasion of the official opening, the tunnel being traversed by steam, the ventilation was found to be excellent, and the temperature ranged from 82° to 87°·5, the external temperature at Brigue being 84° and at Iselle 90°. On the return journey the external temperature at Domo d'Ossola being 90° and at Brigue 78°, the internal temperature ranged from 90° at the south end to 65° at the north On this occasion the volume of air blown into the tunnel from the north and south ends wás 3650 cubic feet per second.[1]

The comparatively low level of the Simplon Tunnel, with its extremities at an elevation of about 1500 feet less above sea-level than those of the St. Gothard Tunnel, rendering the access to the tunnel far less liable to be blocked by snow, making the ascent of the approach railways considerably shorter, and enabling the ruling gradient to be somewhat flatter, will be very advantageous to the new route across the Alps as a competing line; whilst the actual distance between Calais and Brindisi will be about forty miles shorter by the Simplon route, than by the St. Gothard, which is about 25 miles shorter than the Mont Cenis route. The Simplon Tunnel at present provides only one single line for the passage of trains, instead of the double line in the long tunnels of the two competing routes; but, nevertheless, not only will a second line be provided by the completion of the second tunnel as soon as the traffic renders it expedient, but a wider tunnel has been constructed in the central part at the summit-level, for a length of 576 yards, which will serve as a passing-place for the trains in the single-line tunnel, till the line is doubled by opening out the second tunnel.

[1] *Proceedings Inst. C.E.*, vol. clxviii.

CHAPTER XIII.

SUBAQUEOUS TUNNELS.

Peculiarities of subaqueous tunnels; system of construction; driven by headings, and lined with brickwork, illustrated by Severn and Mersey tunnels; difficulties resulting from land springs; Severn and Mersey tunnels, gradients, drainage, ventilation, remarks—Tunnelling through clay under lakes, illustrated by Chicago tunnels under Lake Michigan —Iron shield used for Thames Tunnel, progress and cost of work, intended for road, used by railway—Annular shield, used for Cleveland Tunnel under Lake Erie; with diaphragm and cast-iron tube for Tower Subway, advantages of system—Compressed air and shield work at the Hudson and East River Tunnels, New York—Tunnelling under St. Clair River with shield and tube; under River Thames at Blackwall, Greenwich and Rotherhithe Tunnels; description of work—Difficulties in using compressed air for tunnelling under rivers; precautions necessary—Compressed-air Caisson disease—Concluding remarks on subaqueous tunnelling with shield, tube, and compressed air.

THE chief differences between subaqueous tunnels and ordinary tunnels are, that instead of traversing strata situated considerably above the drainage level of a valley, subaqueous tunnels have to be carried below the general water-level of the district, and therefore are not only exposed to the possible influx of water from the river or lake under which they pass, but also have to be constructed through strata liable to be charged with water; and, secondly, that they generally have to be given descending gradients towards their central portion, instead of the ascending gradients of other tunnels, in order to dip sufficiently under the deepest part of the bed of the river or estuary. Consequently, considerable volumes of water may have to be raised by pumps, placed in shafts sunk at the nearest available sites on land to the centre of the tunnel, aided in the longer tunnels by a drainage heading leading from the lowest point of the central part of the tunnel, with a falling gradient, to the bottom of a deep shaft.

Systems of Construction.—Two methods of construction have been adopted for subàqueous tunnels, namely, (1) Tunnels constructed by driving headings from shafts on the shore in the ordinary manner, the headings and tunnel being kept dry by pumps and drainage headings, of which the Severn and Mersey tunnels are notable examples;

and (2) Tunnels driven under the shelter of an iron tube, provided with a steel shield pushed forward gradually at the face as the excavation progresses, the water being kept out at the worst places by means of compressed air, a method adopted for the recent London electric railways passing at a considerable depth below the Thames, and through clay and water-bearing gravel beds underlying the metropolis, and on a larger scale for the road or foot ways under the Thames at Blackwall, Rotherhithe, and Greenwich, and also for the Hudson River Tunnels, the Pennsylvania Railway Tunnels, and the Battery and Steinway Tunnels under the Hudson and East Rivers, New York.

Subaqueous Tunnels driven by Headings and Lined.—The railway tunnels constructed under the Severn estuary in 1873–86, and under the deep outlet channel of the Mersey estuary, between Liverpool and Birkenhead, in 1879–86, 4 miles 624 yards, and 1 mile 940 yards long, respectively, were driven by headings from the shafts, 15 feet and 17½ feet in diameter, sunk on the two banks, aided by drainage headings at a lower level, and lined with brickwork (Figs. 122, 123, 124, and 125, p. 216). The Severn estuary has a width of about 2¼ miles at high tide over the site of the tunnel, with a maximum depth of 95 feet at high water of spring tides, and a minimum thickness of ground between the top of the tunnel and the river-bed of 44¾ feet; whilst the Mersey is ¾ mile wide where the tunnel crosses, there is a maximum depth of 100 feet at high water of spring tides, and a minimum thickness of ground between the top of the tunnel lining and the river-bed of 30 feet. The strata through which the Severn Tunnel passes consist of conglomerate, limestones, carboniferous beds, sandstone, marl, gravel, and sand, and have a considerable dip; whilst the Mersey Tunnel, throughout its lower portion between the two main shore shafts, 1770 yards apart, is in the new red sandstone, except for 66 yards under the river near the Liverpool shore, where the arch of the tunnel, though covered with 70 feet of soil, emerges 3 to 6 feet out of the rock into clay with a layer of sand, where doubtless a deep channel of the river ran in ancient times.

A drainage heading, 7 feet square, was driven from the Monmouthshire shaft to the lowest point of the Severn Tunnel, rising with a gradient of 1 in 500 to drain the tunnel during construction, and also after its completion, and was continued up the gradient descending from the Gloucestershire shore[1] (Fig. 122, p. 216); whereas at the Mersey Tunnel, after a trial heading had proved the absence of fissures in the sandstone rock, drainage headings, 7 to 8 feet in diameter, were driven from the two shore shafts, with rising gradients of 1 in 500 followed by 1 in 900, till they met in the centre (Fig. 124, p. 216), the Liverpool drainage heading, driven by hand, advancing 11½ yards per week on the average; and the other, excavated by a Beaumont boring machine progressed at a rate of 14½ yards per week, which machine, with certain modifications,

[1] "The Severn Tunnel; its Construction and Difficulties, 1872-87," T. A. Walker, 1888; "Achievements in Engineering," L. F. Vernon-Harcourt, pp. 95 and 100 to 106; and *Proceedings Inst. C.E.*, vol. cxxi. p. 305.

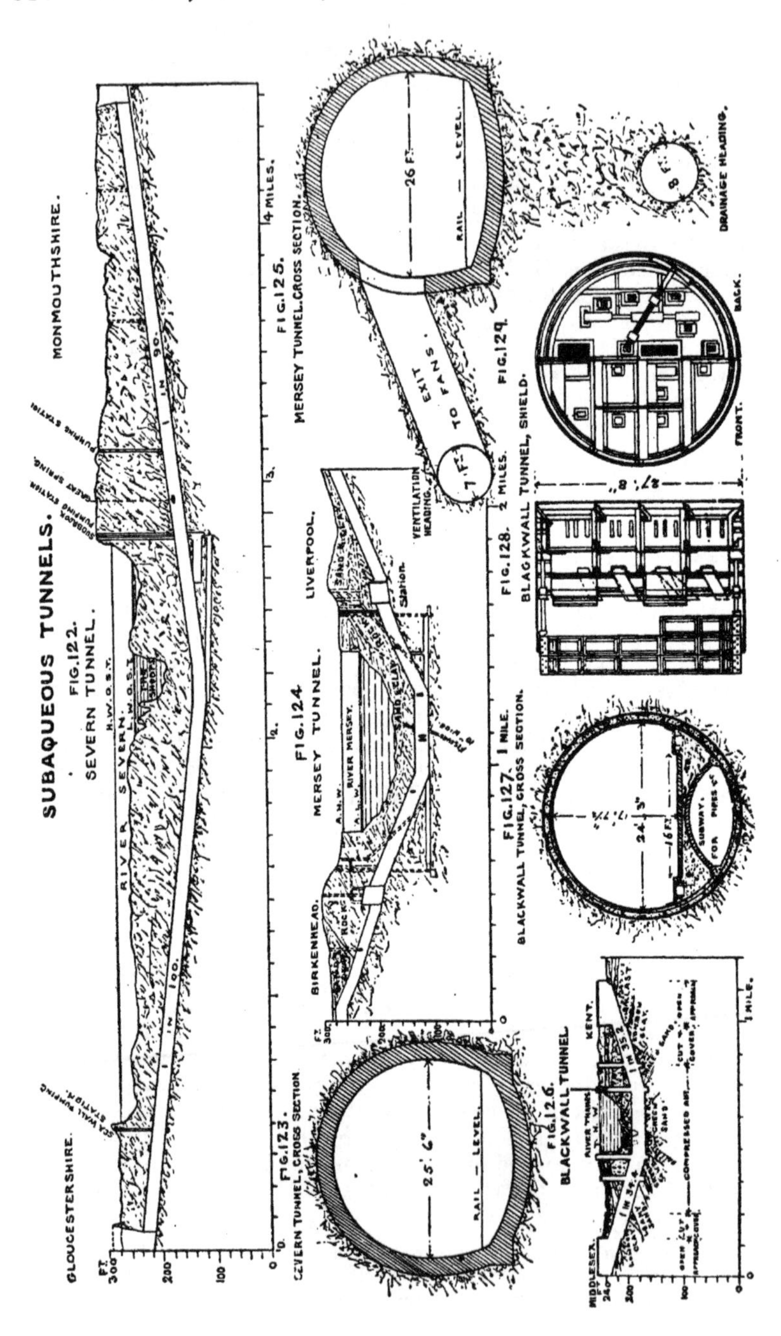

SUBAQUEOUS TUNNELS.
FIG. 122.
SEVERN TUNNEL.
GLOUCESTERSHIRE.
MONMOUTHSHIRE.
RIVER SEVERN.
H.W.O.S.T.
L.W.O.S.T.
THE SHOOTS
SEA WALL PUMPING STATION.
SUDBROOK PUMPING STATION.
GREAT SPRING.
PUMPING STATION.
1 IN 100.
1 IN 90.
FT. 300
200
100
0
1.
2.
3.
4 MILES.
FIG. 123.
SEVERN TUNNEL, CROSS SECTION.
25' 6"
RAIL — LEVEL.
FIG. 124
MERSEY TUNNEL.
BIRKENHEAD.
LIVERPOOL.
RIVER MERSEY.
ROCK
SAND & CLAY
Station.
1 MILE.
2 MILES.
FIG. 125.
MERSEY TUNNEL, CROSS SECTION.
26 FT.
RAIL — LEVEL.
EXIT TO FANS.
7 FT.
VENTILATION HEADING.
8 FT.
DRAINAGE HEADING.
FIG. 126.
BLACKWALL TUNNEL
MIDDLESEX.
KENT.
1 IN 34.
COMPRESSED AIR.
OPEN CUT
1 MILE.
FIG. 127.
BLACKWALL TUNNEL, CROSS SECTION.
24' 3"
16 FT.
SUBWAY FOR PIPES &c.
FIG. 128.
BLACKWALL TUNNEL, SHIELD.
27' 8"
FIG. 129.
FRONT.
BACK.

gave a much greater speed in subsequent headings.[1] Both tunnels were subsequently constructed by means of bottom headings with break-ups, the Mersey Tunnel headings being drained through boreholes to the drainage headings; and the shore portions of the tunnels were advanced by the aid of additional shafts. At the Severn Tunnel, the rock was bored by drills worked by compressed air; and the blasting was effected by cotton powder or tonite in both tunnels, on account of its comparative freedom from noxious fumes. The headings of the Severn Tunnel, joined in 1881, proved to be in perfect line, 3370 yards, out of 4048 yards between the main shore shafts, having been driven from the Monmouthshire side, the actual time occupied in driving having amounted to 4 years 10 months; and the divergence in the lines of the Mersey headings was only 2½ inches where they were joined, 1115 yards from the Birkenhead shaft, the difficulty in setting out the line having been increased by the pumping shafts being away from the centre line.

Difficulties resulting from Land Springs.—The greatest difficulty in constructing the Severn Tunnel resulted from the piercing of a large spring, in 1879, in driving the inland heading from the Monmouthshire shaft, about 1½ furlongs landwards of the estuary, which rapidly flooded the works and arrested their progress for nearly 14 months, till the flow of water could be controlled by pumping, and the spring shut off. The spring burst out a second time in 1883, and again flooded the works for a short time; and eventually sufficient pumping power was established in a shaft sunk by the side of the tunnel, at the site of the spring, to deal with the whole flow. A leakage into the works did, indeed, occur in 1881 from a pool in the estuary near the Gloucestershire shore; but it was soon stopped with clay deposited at low water in the bed of the pool. In constructing the Mersey Tunnel, also, more water was met with under the land than under the river, though one fissure was met with near the centre of the tunnel, filled up with clay and disintegrated sandstone. The remarkable freedom from water in both cases under the river, must be attributed to the closure of the interstices in the strata by the silt of the river. A watertight door for dividing the long heading of the Severn Tunnel in two parts in case of flooding, left open in the confusion on the bursting out of the large spring, was closed under water by a diver carrying a knapsack of compressed oxygen, which greatly facilitated the pumping out of the water; but flood doors provided in the Mersey headings, were not required.

Gradients and Drainage of Severn and Mersey Tunnels.—The Severn Tunnel has been given descending gradients of 1 in 100 and 1 in 90 from each end respectively, for rather over 2 miles on each side, with a central, level portion rather more than a furlong in length (Fig. 122); and the Mersey Tunnel has falling gradients of 1 in 30 and 1 in 27 from the two main shafts towards the centre, prolonged upwards to the stations on each shore, with lengths of about

[1] *Proceedings Inst. C.E.*, vol. lxxxvi. p. 40, and plates 3 and 4; and "Achievements in Engineering," pp. 94 to 100.

$2\frac{3}{4}$ and $3\frac{1}{2}$ furlongs, and a central portion, about $2\frac{1}{2}$ furlongs in length, rising towards the centre with gradients of 1 in 900 to provide for drainage to the headings on each side (Fig. 124, p. 216). In order to keep these tunnels dry, with gradients falling towards the centre, permanent pumping machinery has to be maintained, more particularly in the shafts in communication with the drainage headings, extending to a depth of 200 feet at the Severn Tunnel, and 170 feet at the Mersey Tunnel. Pumps have been erected in five shafts at the Severn Tunnel, capable of raising 66 million gallons of water per day, more than double the maximum hitherto required; and six sets of pumps have been placed in the main shore shafts of the Mersey Tunnel, to ensure it against being flooded.

Ventilation of Severn and Mersey Tunnels.—These subaqueous tunnels descending to such a low level underground, and necessarily exposed to a damp atmosphere, are specially in need of ventilation, which has been effected in the case of the Severn Tunnel by fans working in the two main shafts, the fan in the Monmouthshire shaft having a diameter of 40 feet and a width of 12 feet. The ventilation in the Mersey Tunnel is accomplished by two fans in two shafts on each side of the river, 40 feet and 30 feet in diameter, and 12 feet and 10 feet wide, by which, with the aid of ventilation headings opening out of the tunnel, the foul air is drawn out of the tunnel into the ventilation headings, and thence to the four fans; the tunnel has, however, since 1903 been worked by electric traction.

Remarks on Severn and Mersey Tunnels.—The Mersey Tunnel Railway, about 3 miles in length, cost about £500,000 per mile, including two stations, purchase of property, and rolling-stock, equivalent to £284 per lineal yard, which, however, embraces expenses not properly chargeable to the tunnel. Subaqueous tunnels executed under the conditions of the Severn and Mersey tunnels, are costly and somewhat hazardous works; and, moreover, they involve a yearly charge in pumping for drainage, and in ventilation. Nevertheless, tunnels of this kind, like Alpine tunnels, should not be regarded merely as isolated works, but as links in long lines of communication, the utility and traffic of which they may greatly increase and extend, which is the function the Severn Tunnel has effected for the Great Western Railway system, whose lines to the north and south were formerly severed by the Severn west of Gloucester. The Severn Tunnel is the longest railway tunnel in England, being 1434 yards longer than the Totley Tunnel; but it is exceeded in length, to the extent of 696 yards, by the Hoosac Tunnel in the United States, as well as by the long Alpine tunnels previously described.

Tunnelling through Clay under Lakes.—Instances of subaqueous tunnels constructed through soft soil in the ordinary manner, are furnished by the two tunnels driven through blue clay under Lake Michigan, for supplying Chicago with pure water from the lake, at a sufficient distance from the shore to be beyond the influence of the sewage and refuse discharged into the lake from the town. The first tunnel, constructed in 1864–67, extends from a land shaft near the shore of the lake, to a lake shaft sunk in the bed of the lake and protected by timber crib-work weighted with rubble stone. The excavation was carried out from both shafts; and the tunnel was lined with two rings of brickwork, being given an inside diameter of 5 feet, and a fall shorewards of $3\frac{2}{3}$ feet in its length of 2 miles, to provide a supply of 50 million gallons of water per day.[1] The tunnel was ventilated during its progress by drawing out the foul air through an 8-inch pipe; its construction, together with its two cast-iron 9-inch cylindrical shafts, occupied 1003 days; and it cost £27 per lineal yard. The second tunnel, constructed in 1872–74, parallel to the first at a distance of 50 feet, has an internal diameter of 7 feet, in order to afford a further supply from the lake of 100 million gallons a day. It is 2 miles 83 yards long between the shore and lake shafts; its lake shaft is protected by the same crib-work as the first; and it cost £24 per lineal yard. The influx of water, and the obstruction of old timber, somewhat impeded the sinking of the land shafts through the 14-feet layer of sand overlying the clay; and a leakage of water from the first tunnel caused difficulties in sinking the second lake shaft, and in constructing the second tunnel close to the crib; but the only other obstacle encountered in making the tunnels, was the occasional escape of inflammable gas from pockets in the clay.

Iron Shield used for the Thames Tunnel.—Whilst the ordinary system of tunnelling can be employed for subaqueous tunnels, carried at a considerable depth through rock rendered watertight by long deposit of silt, and devoid of large fissures, and also through stiff clay, without exceptional difficulties, special contrivances have to be resorted to in carrying subaqueous tunnels through silt, sand, gravel, or other permeable strata. An iron shield was used in the construction of the first subaqueous tunnel in soft soil under the Thames, about $1\frac{1}{2}$ miles below London Bridge, with the object of shutting out from the works any influx of water from the river through a variable bed of clay, at a minimum depth of only 16 feet below the bed of the river. The cast-iron shield, 38 feet wide, 7 feet long, and 22 feet high, was divided into three horizontal stages, having each twelve compartments large enough for a man to work in, and excavate the face in front of the shield, which was pushed forward, as the excavation proceeded, by a series of horizontal screws resting against the finished brickwork behind. Only so many of the thirty-six compartments were opened at a time as might appear safe; and they were closed, when necessary, by horizontal boards strongly strutted at the back. Nevertheless, the river burst twice into the tunnel works through seams in the clay; and the holes formed by the influx were closed with bags of clay. The rate of progress was about 2 feet in 24 hours; but owing to the flooding and financial difficulties, the tunnel, commenced in 1825, was only completed in 1843, though only 400 yards long between the shafts (50 feet in diameter and 80 feet deep) sunk on each bank; and the cost of the tunnel reached £1137 per lineal yard. The tunnel was originally intended to provide a double roadway connecting the roads on the

[1] "Annual Reports of the Board of Public Works of Chicago;" and "Tunnelling," H. S. Drinker, pp. 911 to 939.

two sides of the river; but the approaches having never been constructed, it was only used through the shafts by foot-passengers, till, in 1866, the East London Railway purchased it for carrying a double line of railway under the river, thereby forming a connection below London Bridge between the lines on the north and south sides of the river.

Annular Shield or Cutter.—A shield made of wrought iron and annular in form, $6\frac{1}{2}$ feet internal diameter and 6 feet long, was employed for constructing, through soft clay, 140 feet of the tunnel carried under Lake Erie from Cleveland, in 1870–74, for supplying the town with pure water from the lake. This shield was forced forward by hydraulic presses so as to cut into the clay, which was prevented from coming into the shield too fast by its friction against some horizontal shelves in the front part.[1] The hinder two feet of the shield were made smooth inside and left unobstructed, so that the brick rings of the tunnel could be constructed under its shelter; and the shield was pushed forward about 16 inches at a time, so that it still overlapped the finished brickwork for 8 inches, leaving sufficient space for a length of two brick rings. This shield, accordingly, cut into the clay in front, and supported the excavated length till the brickwork could be built; but it was not adapted for contending with water-bearing strata.

Annular Shield with Diaphragm and Cast-iron Tube.—A little earlier, a somewhat similar annular wrought-iron shield was used for constructing the Tower Subway under the Thames $\frac{1}{3}$ mile below London Bridge, through stiff London Clay, with a minimum thickness of ground of 22 feet between the river-bed and the tunnel. This shield, however, $4\frac{1}{2}$ feet long, was provided with a plate-iron diaphragm in front, with a central aperture through which men could excavate the clay, and eventually pass out to form a heading in advance, along which the shield was subsequently pushed forward by screws, aided by its cutting edge at the face. The cast-iron tube forming the lining of the subway, having an internal diameter of 7 feet, was put together in successive rings, $1\frac{1}{2}$ feet long, made up of three segments and a key-piece bolted together, under shelter of the rear $2\frac{3}{4}$ feet of the shield.[2] The 1-inch space left round the tube by the advance of the encircling shield, was filled up by forcing lias lime grout into it through small holes formed for the purpose in the rings, so as to prevent any settlement of the surrounding clay. The Tower Subway was constructed in about nine months in 1869, from two shafts, 10 feet in diameter, sunk 63 feet and 56 feet respectively, below the surface on each bank into the London Clay; no difficulties were experienced in driving it, a maximum rate of progress of 9 feet in 24 hours having been attained; and its cost of about £10,000, for a length of tunnel of 450 yards, was equivalent to £22 per lineal yard.

The adoption of a cast-iron tube removes the liability to injury, to

[1] "Annual Report of the Board of Trustees of Water-works," Cleveland, 1874; and "Tunnelling," H. S. Drinker, pp. 939 to 954.

[2] "On the Relief of London Street Traffic, with a Description of the Tower Subway," P. W. Barlow, London, 1867; and *Proceedings Inst. C.E.*, vol. cxxiii. pp. 56 and 77.

which the Cleveland tunnel was exposed, by the shield bearing unduly on the extremity of the brick lining when pushed forward, and settling somewhat in the soft clay; and the grouting with lime round the outside of the tube, where the stratum traversed has sufficient consistency not to fill up immediately the void left by the shield in its advance, furnishes a valuable protection against settlement. The addition of an air-tight partition across the completed tube at the back, in a suitable position, with an air-lock providing a passage through it, enables the tube to be advanced through water-bearing strata under the protection of compressed air, and completes the system which has been very successfully employed for some subaqueous tunnels in recent years.

Compressed Air and Shield Work at the Hudson and East River Tunnels, New York.—Compressed air was employed in constructing the double tunnel under the Hudson River for connecting Jersey City with New York, which was mainly carried through silt, with a minimum thickness of ground of 15 feet between the bed of the river and the crown of the tunnel, where the maximum depth of water attains 60 feet; and iron rings, 2½ feet in length, built up in segments, were also used, the top portions being at first pushed forward to serve as a hood till the lower excavation could be accomplished, and the rings completed. The use of compressed air, however, in a horizontal tunnel, is much more difficult to control than in vertical cylinders or caissons, which constitute a sort of diving-bell; and it was found necessary to vary the pressure according to the conditions, for with too low a pressure, water leaked in through the silt, and with too high a pressure, the silt lost its compactness by the forcing back of the water, and fell in. Leaks also occurred on two occasions through the disturbed silt, flooding the works; and in traversing gravel and sand, a shield had to be resorted to, built in pieces down the face of the excavation as it progressed, and strongly strutted to keep up the face.[1] Moreover, eventually, as the iron rings, constituting the framework of the brick tunnel built inside, could not be maintained in a perfectly straight line, a small pilot tube, 60 feet long and 6 feet in diameter, was built out about 30 feet in advance of the finished tunnel which supported its hinder half; whilst the rings of the tunnel were erected in exact line by being strutted by radial braces from the pilot tube. After various stoppages due to flooding and deficiency of funds, the works, which had been commenced in 1874, were resumed in 1889, after a stoppage of about six and a quarter years, with the aid of a shield and cast-iron tube similar in principle to the Thames Subway, but on a large scale, and with the addition of compressed air. The steel shield was 10½ feet long, divided into two portions by a plate-iron diaphragm across it, 5⅔ feet inside of the cutting edge; but whilst the inner portion was kept clear for the erection of the tube, the outer portion was divided into nine compartments by two horizontal and two vertical diaphragms.[2]

[1] "Tunnelling under the Hudson River," S. D. V. Burr, New York, 1885; and "Achievements in Engineering," pp. 89 to 94.

[2] "Practical Tunnelling," F. W. Simms, 4th edition, 1896, p. 494*g*, and plate xxxi. B.

The shield was pushed forward by sixteen hydraulic rams, making the cutting edge dig its way into the soft silt, which, filling the front of the shield, was squeezed through the doors (2½ feet by 2 feet) in the diaphragm. The cast-iron tube, having an outside diameter of 19½ feet, was built up in rings under shelter of the rear part of the shield, each ring forming a length of 1 foot 8¼ inches of tube, being composed of eleven segments and a key piece, with a skin 1¼ inches thick and flanges 9 inches deep. A hydraulic erector was adopted, to put into position and support each of the heavy segments of a ring whilst bolting it to the adjacent segment. An additional length of 1900 feet of the Hudson Tunnel had been constructed in this manner, the greatest weekly advance reaching 72 feet, when the works were again stopped in July, 1891, for want of funds, leaving 1600 feet on the southern New York side to be built to complete the tunnel, which was subsequently finished and opened in 1908, with the addition of a second tube.

Simultaneously with the completion of these tubes, the Lower Hudson Tunnels from Jersey City to Fulton Street and Cortlandt Street were commenced by means of shield and compressed air. These consist of two cast-iron tubes, 16 feet 7 inches outside diameter, about 70 to 90 feet below water level, and are a mile in length between banks. The Upper and Lower Hudson Tunnels are worked with electric cars.

The Pennsylvania Railroad Tunnels are, perhaps, the most remarkable amongst the great engineering works which constitute the group of New York subaqueous tunnels. These tunnels, under the Hudson River, consist of two tubes of cast iron lined with concrete, 23 feet outside diameter and 37 feet apart centres, while those under the East River consist of four similar tubes, 4000 feet in length between banks, in two pairs about 300 feet apart, started from caissons sunk on each shore. These, as in the case of the other examples, were carried out by means of compressed air and shield, the latter being provided with hydraulic "segment erectors" on their rear face.

The tunnels carrying the New York Rapid Transit Subway under the East River, known as the Battery Tunnels, consist of two cast-iron tubes of 15½ feet inside diameter and 16 feet 8 inches outside, about 4200 feet in length between banks, with a maximum depth of about 90 feet below water level.

The Steinway or Belmont Tunnel under the East River at Forty-Second Street, consists of two tubes 15½ feet diameter inside, about 3000 feet long between banks, and having a maximum depth of about 100 feet below water level.

The East River Gas Tunnel, one of the earlier tunnels of the group, and completed in 1894, about half a mile in length, and 100 feet below water level, is for the most part excavated in rock, but compressed air was resorted to where soft mud occurred.

It is impracticable here to adequately discuss the varied difficulties encountered during the progress of these works, arising either from the soft mud of the Hudson River or the quicksands, boulders, and rock of the East River, or to describe the bold and ingenious methods by which they were successfully overcome. It must suffice to say that the entire record constitutes a notable example of the application of compressed air and shield work to the construction of subaqueous tunnels under conditions of great difficulty.

Shield and Cast-iron Tube for Tunnel under St. Clair River.—Prior to the completion of the works above described a tunnel was successfully carried out under the St. Clair River, connecting Lakes Huron and St. Clair, for joining the Grand Trunk Railway of Canada to the United States railway system at Port Huron. This tunnel, constructed in 1889–90, is 2017 yards long, of which 767 yards are under the river; it traverses blue clay sufficiently compact for compressed air to have been dispensed with; and its tube, with an internal diameter of 20 feet, is lined inside with masonry, and affords a passage for a single line of railway, which is now worked electrically.

Tunnelling under the Thames with Compressed Air, Shield, and Tube.—The tunnel constructed under the Thames at Blackwall in 1892–97, for connecting the roads on the two sides of the river, furnishes a good illustration of the successful application of the system in tunnelling through water-bearing strata at a small depth below the bed of a tidal river. The total length of the work is 6200 feet, including 1735 feet of open approaches at the two ends flanked by retaining walls; whilst out of the 4465 feet of covered way, 1349 feet were excavated in the open and then arched over, known as the "cut and cover" method, and the remaining central 3116 feet were carried out by tunnelling with a shield under compressed air, of which 1220 feet are under the river [1] (Fig. 126, p. 216). The strata traversed consisted of clay, coarse gravel, sand, shale, and a little chalk; and at one place, where the shield had to pass at a minimum depth of five feet below the bed of the river, a layer of clay was temporarily deposited from hopper barges in the bed of the river, on the intervening stratum of coarse gravel, along 450 feet of the line of the tunnel for a width of 150 feet, its maximum thickness reaching ten feet. This layer checked the escape of air through the coarse gravel, and also prevented the pressure of air from the tunnel forcing up the thin stratum between it and the river-bed. The steel shield was $19\frac{1}{2}$ feet long and $27\frac{2}{3}$ feet outside diameter; and its forward portion was stiffened by three horizontal and three vertical plate diaphragms, dividing the working face into four floors and twelve compartments (Figs. 128 and 129, p. 216). Two cross diaphragms, one about the centre of the shield and the other 3 feet behind it, shut off the face completely from the tunnel when necessary; and four air-locks provided for access to the face; whilst openings in the cross diaphragms, closed by doors and furnished with shoots in all the compartments, enabled the material from the face to be disposed of. The shield was forced forward by twenty-eight hydraulic rams; and in driving it through wet sand and gravel under the river, six additional rams were required, giving a total pressure of 5165 tons. The rear portion of the shield, for a length of $6\frac{2}{3}$ feet, was kept clear as usual for the erection of the rings, each $2\frac{1}{2}$ feet long, so that the shield, when pushed forward

[1] *Proceedings Inst. C.E.*, vol. cxxx. p. 52, and plates 2 and 3.

sufficiently for the erection of another ring, still fully overlapped the last completed ring. The rings, having an external diameter of 27 feet, and made up of fourteen segments and a key, were erected with the help of two hydraulic erectors carried on the back of the shield (Fig. 129, p. 216), which lifted the segments successively into place. The flanges were eventually buried in a filling of concrete, lined on the inside face with glazed tiles. The roadway through the tunnel is 16 feet wide, with a footway on each side; and there is a capacious subway under the road for pipes (Fig. 127, p. 216). The cost of the work amounted to £871,000, equivalent to £421 per lineal yard of tunnel and approaches; whilst the subaqueous portion of the tunnel cost about £550 per lineal yard.

The Blackwall Tunnel was followed by the construction in 1899–1902, under compressed air and with a shield, of the Greenwich Footway, consisting of a tunnel connecting two shafts, one at Poplar, the other at Greenwich, 1217 feet apart. The shafts are 43 feet external, and 35 feet internal diameter. The tunnel is lined with cast iron similar in construction to other works of the same class, 12 feet 9 inches external and 11 feet 9 inches internal diameter, the footway being 8 feet 9 inches in width, with a headway at centre of 8 feet 8 inches. Gradients of 1 in 15 from each shaft falling towards the river are connected by a gradient of 1 in 277 falling towards Greenwich.[1]

The Rotherhithe Tunnel, opened in 1908, from Rotherhithe to Ratcliff, has a total length of 6883 feet, of which 3689 feet are cast-iron tunnel, and 3194 feet are of open approach and cut-and-cover, while the subaqueous portion is 1570½ feet in length centres of shafts. The internal diameter is 27 feet giving a roadway 16 feet wide, with two footways 4 feet 8½ inches wide each, the external diameter of the cast-iron ring being 30 feet. Four shafts were sunk by means of steel caissons 60 feet diameter, two on each side of the river. The approach gradients are about 1 in 36 down to the subaqueous portion, which is on a gradient of 1 in 800, falling towards Rotherhithe. A pilot tunnel 12 feet 6 inches outside diameter, was first driven to ascertain the nature of the strata, followed by the construction of the main tunnel by means of compressed air and shield.[2]

Difficulties attending Employment of Compressed Air for Tunnelling.—The main difficulty experienced in using compressed air in driving a horizontal tube, consists in the pressure required for keeping the bottom portion of the section clear of water exceeding the pressure needed at the top, so that with a sufficient bottom pressure, the air tends to escape at the top. The tendency also of water to come in, and therefore the pressure required to resist it, varies with changes in the nature or condition of the strata; whilst a sudden large escape of air from the top, by lowering the pressure, often leads to a sudden inrush of water; and, moreover, when it occurs under the bed of a river, by loosening the intervening strata, the influx of water into the works is further promoted. In traversing very porous water-bearing

[1] *Proceedings Inst. C.E.*, vol. cl.
[2] *Ibid.*, vol. clxxv.

strata, such as the gravel beds of the Blackwall Tunnel, the openings in the diaphragm of the shield, through which the excavated materials are drawn, have to be reduced to a minimum, the gravel having been taken from the face of the Blackwall Tunnel through holes of only seven inches by four inches for many days in succession, the rate of progress under such conditions being sometimes only one foot per day. Provision also has to be made against any sudden influx of water from an escape of air or other cause, by hanging an air-tight screen a short distance back from the shield, across the upper half of the tunnel, with an air-lock at the top, through which, on the occurrence of a rapid flooding, the men could escape to the back of the screen, where, being sure of finding an air-space, they could pass along a gangway to the emergency air-lock at the top of the brick bulkhead separating the portion of the tunnel under compressed air from the rest.

Compressed Air Caisson Disease, its effects, origin, and treatment.—The maximum pressure reached both in the Hudson and Blackwall Tunnels, was about 35 lbs. per square inch beyond the atmospheric pressure; and it was found at the Hudson Tunnel, that the men, on leaving the compressed air, were liable to be suddenly seized with pains in the limbs, and sometimes paralysis; and the prevalence of this illness proved to be greater in proportion to the pressure, and the deficiency in purity of the air. A man might work in compressed air for months without feeling any ill-effects, and then one day be suddenly struck down after leaving his work. Under a pressure of 35 lbs., the amount of carbonic acid in the air should not be allowed to exceed one in 1000. The high death-rate amongst the men employed under compressed air in the Hudson Tunnel, led to the adoption of a compressed-air hospital, in which men were placed on exhibiting any signs of illness on emerging into the open air. In this chamber, filled with pure air under a moderate pressure, the men generally felt immediate relief; and by reducing the pressure very gradually, the symptoms did not reappear on reaching the atmospheric pressure. This method of treating the first symptoms of the compressed-air illness, was adopted from the commencement at the Blackwall Tunnel with successful results.

Investigations into the nature of compressed air caisson disease appear to shew that the primary cause of the malady is the absorption by the body of nitrogen, and the formation of bubbles in the blood and tissues, producing results varying with the organs affected, while experience has so far shewn that the disease may be largely mitigated by the adoption of periods of slow decompression varying in length with the air pressure, and the length of shift, combined with a sufficiently low percentage of carbonic acid in the air at the working face.

Concluding Remarks.—The annular, steel shield, with its strong cutting edge which excavates the material in soft soil under hydraulic pressure, and with its transverse diaphragm which can be partially or wholly closed against the influx of water, the cast-iron lining which is gradually erected ring by ring under the perfect shelter of the shield, after each adequate advance of the shield, and, finally, the introduction of air under pressure into the shield and front part of the completed tunnel, together constitute a system which enables tunnelling to be accomplished through water-bearing strata, even when a river overhead is in direct contact in places with the permeable strata. These strata require in such a case to be partially closed and weighted with a deposit of clay on their surface, to check the escape of the air and the disturbance of the strata, and to stop as far as possible the rapid downward percolation of the river water. The shield performs the various functions, of excavator at the face with its cutting edge and hydraulic presses; of protector of the working chamber from flooding by its transverse diaphragm, which also regulates the ingress of excavated material by variations in its openings according to circumstances; and of timbering at its rear part, by keeping up the excavations till the lining has been constructed.

CHAPTER XIV.

METROPOLITAN RAILWAYS.

Types of Metropolitan railways—**Underground Railways:** two forms of construction, differences, advantages and disadvantages of the two systems; London Metropolitan Railway, description, methods of construction, diversion of sewers, ventilation, cost; Low-level London Electric Railways, instances, methods of construction, advantages, arrangements of tunnels, dimensions, depth, ventilation, remarks; Paris Metropolitan Railway, preliminary works, course and length, description, stations; Glasgow District Railway; New York Rapid Transit Subway: Chicago Underground Freight Railways—**Overhead Railways:** on brick viaducts, instances, Berlin Metropolitan Railway, description and cost; New York and Brooklyn Elevated Railways, lengths, different types of iron viaducts, gradients and curves, cost of New York lines, modifications at Brooklyn; Liverpool Overhead Railway, length, description, swing and bascule bridges, cost; Berlin Electric Railway, description of steel viaduct, gradients and estimated cost—Comparison of overhead and underground railways—**Widening Railways:** methods, retaining walls for cuttings and embankments, construction, essential provisions for retaining walls.

WHEN railways have to be extended into large cities for the convenience of passengers, or when the growth of a city and the increase of traffic afford scope for a railway within the city itself, to expedite transit and prevent the overcrowding of the streets, two distinct methods of construction can be resorted to, namely, overhead lines like the Charing Cross Railway from London Bridge Station, and the extension of the South-Western Railway from Nine Elms to Waterloo Bridge Station, and underground lines like the Metropolitan and City railways. Underground railways avoid all interference with the traffic, by passing underneath the streets; whilst overhead railways have to be raised sufficiently for the bridges by which they cross over the streets, to leave adequate headway under them for the vehicular traffic, and are somewhat unsightly. A special type of above-ground city railways is carried on slender pedestals and girders along the streets, at a sufficient elevation not to interfere with the road traffic, of which the New York Elevated Railway and the Liverpool Overhead Railway are instances; but such railways are only suitable for the strictly commercial part of a city, as in Liverpool, where the convenience of a railway outweighs the obstruction it presents in the streets, and the discomfort of its noise.

UNDERGROUND RAILWAYS.

Underground railways have been constructed on two different principles. In the one case, the railway is carried at only a moderate depth, for the most part, below the surface, so that it can be to a great extent constructed on the "cut and cover" system; the stations are readily accessible from the streets, and are generally in open cutting; and tunnelling is only resorted to in places where the ground rises rapidly, as exemplified by the Metropolitan Railway in London. In the other case, the railway is formed in tunnel throughout, at a considerable depth below the surface; and the stations are reached from shafts in which hydraulic lifts work, a system adopted in the recent underground lines constructed in London for connecting South London, Waterloo Station, and Central London as far west as Shepherd's Bush, with the City and in other well-known lines.

Other important differences between the two systems are, that the Metropolitan Railway forms connections at several points with other railways, and, though now electrified, was for many years served by steam locomotives which ran over connecting lines outside the metropolis; whilst the low-level City railways are quite independent lines, and are worked by electricity. Moreover, in constructing a railway under a city at a moderate depth below the surface, houses have to be pulled down, or their foundations underpinned and supported, streets are temporarily interfered with, and sewers, water-mains, and gas-pipes have to be diverted; whereas railways formed in single-line tunnels lined with circular rings, at a good depth, avoid interference with the houses and streets under which they pass, except at the shafts, and keep below the level of the sewers. The first system possesses the advantages of forming very serviceable links with other railways, and of having easily accessible, and, for the most part, open-air stations; whilst the second system has the merit of being constructed at a smaller cost, owing to the slight extent to which it interferes with the property above it.

The electrification of the Metropolitan has modified the comparison formerly instituted between the high and low level systems of underground railways as regards atmospheric vitiation, the difficulties formerly experienced on the Metropolitan having been relieved by the change from steam to electricity. The stale atmosphere of low level tunnels cannot be adequately changed without the assistance of artificial ventilation, but the stuffiness arising in such situations is counteracted to some extent by the motion of the air caused by constant traffic; the Metropolitan has however the advantage of the more open nature of its construction near the surface.

London Metropolitan Railway.—The line, which passes in a sort of irregular, flattened, elliptical course under the most central portion of London, embracing the area comprised between the Tower and Kensington east and west, and Regent's Park and Pimlico north and south, commonly known as the Underground or Inner Circle Railway, runs at as uniform a depth as practicable below the surface, following approximately the natural irregularities of the ground, except where the changes in level are too abrupt, and, consequently, deep cuttings and tunnels were adopted.[1] The depth of the rail-level below the surface ranges from 9 feet in shallow cutting, up to 63 feet in tunnel; but the depth, for the most part, is comprised between 15 and 25 feet. The railway rises to a maximum elevation of about 83 feet above Ordnance datum (mean sea-level at Liverpool) at Edgeware Road Station, and descends to its lowest point of 9 feet below the same level under Victoria Street; and the steepest gradients employed for overcoming the differences in level, are 1 in 70, 1 in 75, and 1 in 100. The limiting radius of the curves on the Inner Circle is 10 chains; but a curve of $6\frac{2}{3}$ chains radius had to be inserted at the junction of the Great Northern branch.

The railway passes for a considerable portion of its length through covered ways, constructed on the "cut and cover" principle, where there are streets or very valuable property overhead (Figs. 130 to 133), actual

COVERED WAY. — Fig. 130.—Metropolitan Railway.

HIGH COVERED WAY. — Fig. 131.—Metropolitan Railway.

LOW COVERED WAY WITH GIRDER TOP. — Fig. 132.—Metropolitan Railway.

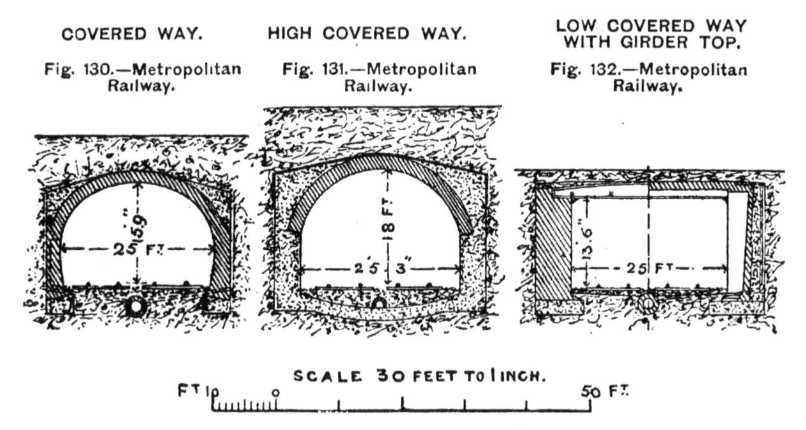

tunnelling having been resorted to only for 728 yards under the higher part of Clerkenwell, and for 421 yards under Campden Hill, where the steepest gradient occurs; and the remainder of the line is in open cutting, at the sides of which retaining walls take the place of slopes to economize land (Figs. 134 and 135, p. 230). In constructing the covered ways, after clearing the site where necessary, the timbered trench was carried out in the first portion of the line to its full width at once, and the requisite depth, the retaining walls being then built at each side, and finally the line arched over; but in the later portions, the side walls were each built in a timbered trench, 6 feet in width, and carried up to 4 feet above the springing, after which the excavation was completed to the full width down to the springing, the centering erected, the arch turned, and finally the central mass of earth was excavated to the full depth (Fig. 133). Where the depth of the railway below the surface was ample, a brick arch with a good rise was used for the covering (Fig. 131);

[1] *Proceedings Inst. C.E.*, vol. lxxxi p. 1, and plate 1.

but in some places, shallow iron girders, 6 to 8 feet apart, supporting small brick arches between them, had to be adopted to provide the

COVERED WAY UNDER STREETS.

Fig. 133.—Underpinning Houses, and Construction, London Metropolitan Railway.

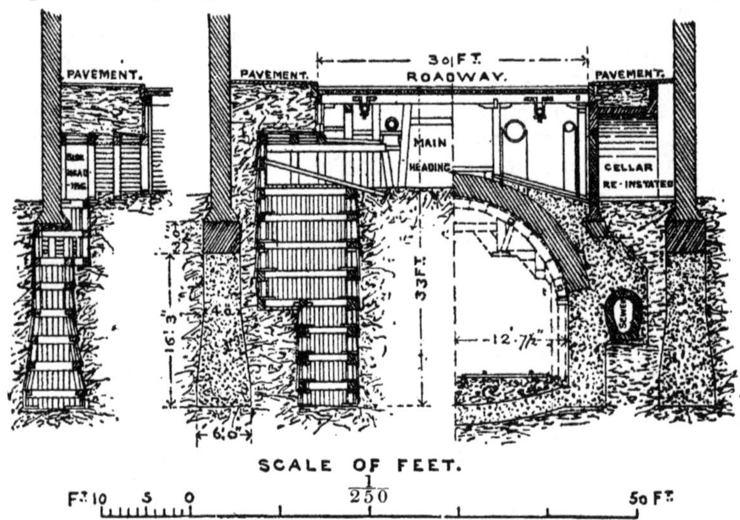

SCALE OF FEET.

RETAINING WALLS, LONDON METROPOLITAN RAILWAY.

Fig. 134.—Open Cutting with Retaining Wall.

Fig. 135.—Deep Cutting with Cast-iron Struts.

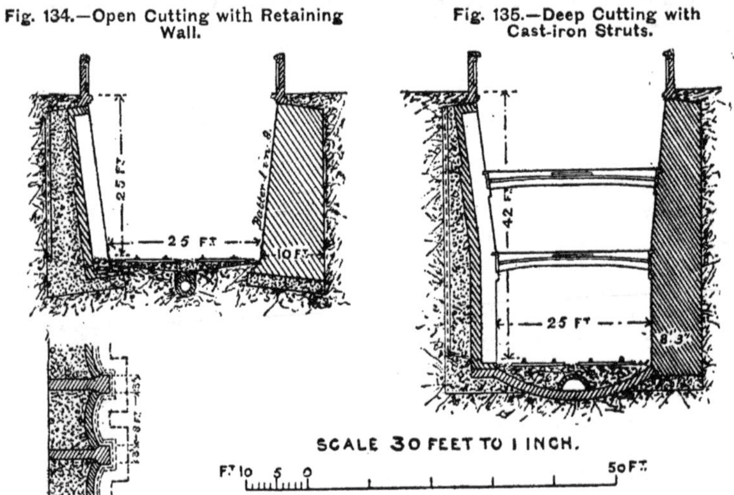

necessary headway (Fig. 132); and in many places, buildings have been erected over the covered way. The portion of the covered way along the Thames Embankment was built simultaneously with the embankment; but in other places, where the railway runs under streets, the roadway had to be maintained during the progress of the works underneath them, by means of timber balks stretching across the excavations, on which timber planking was laid (Fig. 133). Care also had to be taken to avoid disturbing adjacent buildings, by carrying out the excavations in short lengths, and following on as quickly as possible with the side walls and arch, and by underpinning the foundations in certain cases. Recesses were formed in the side walls of the covered ways at intervals of 30 feet, as shelters for workmen during the passage of a train, a system commonly resorted to in tunnels and on viaducts where the clear space at the sides is small.

At the sides of the open cuttings, provided where the railway comes near the surface away from streets, and the cost of land is not excessive, retaining walls of brickwork and concrete were built, with piers 3 feet wide and recesses of 8 feet, having a batter on the face of 1 in 8, and a thickness at the base, 5 feet below rail-level, of about two-fifths of their height (Fig. 134). Where the depth of the cutting is considerable, one or two rows of cast-iron struts, or narrow brick arches at intervals, have been introduced, enabling the thickness of the retaining walls to be reduced (Fig. 135).

The sewers interfered with by the railway, were in some cases carried over the line in cast-iron tubes or pipes supported by girders; and in other cases, the sewers were gradually lowered from some distance back, and carried under the line in a brick channel with an iron top; whilst occasionally new sewers were constructed along the covered way, to replace the intersected sewers. Gas-pipes and water-mains have been generally carried across the line in iron troughs.

Openings have been provided at the sides of the underground stations, and also in some streets, for the escape of the noxious smoke emitted by the locomotives; but objections were raised against a considerable increase in the number of openings in the streets under which the railway runs; and the working of some ventilating fans was put a stop to, on account of the vibrations produced by them. Accordingly, the adoption of electrical traction appeared to be the best means of improving the condition of the air, and the result of the substitution of steam by electricity appears to have justified this view.

The cost of the works alone of this railway appears to have amounted, on the average, to nearly £200,000 per mile of double line; but the total cost was very largely augmented by expenditure on land, financial arrangements, and various special items.[1]

Low-level London Electric Railways.—The underground railways recently constructed and projected in London, have been all designed as electric railways, carried at a sufficient depth below the surface to avoid interference with streets, buildings, and sewers, being constructed in tunnel advanced by aid of a shield with a cutting edge pushed forward by hydraulic presses, and lined with circular cast-iron

[1] *Proceedings Inst. C.E.*, vol. lxxxi. p. 10.

rings, on precisely the same principle as was adopted in the construction of the Blackwall and other subaqueous tubular tunnels described in the last chapter. The City and South London Railway, connecting Stockwell with the City at King William Street, opened in 1890, was the first railway of this type; and it has been followed by the Waterloo and City Railway, opened in 1898; the Central London Railway running from Wood Lane and Shepherd's Bush under the Uxbridge and Bayswater Roads, Oxford Street and Holborn to the Bank, opened in 1900; the Baker Street and Waterloo opened in 1906; the Great Northern, Piccadilly and Brompton Railway (Hammersmith to Finsbury Park) the Charing Cross and Hampstead; the Great Northern and City. Three of these railways, in connecting South London and Waterloo Station with the northern part of London, have to pass under the Thames, near London Bridge, Blackfriars Bridge, and Charing Cross Bridge, respectively; but as these subaqueous tunnels have been carried through the London clay underlying the bed of the river, their construction has presented no difficulty in comparison with the portions of the tunnels traversing the gravel beds, charged with water and resembling subterranean rivers, which are met with in places at some depth under London, covering depressions in the London Clay, where it was necessary to have recourse to compressed air for carrying forward the tunnel.

By constructing these low-level lines almost wholly under streets, the danger of any settlement of buildings and interference with the underground rights of private owners of property are avoided as far as possible; whilst the stations are thereby very conveniently situated alongside the main thoroughfares of traffic, which the railway accommodates and relieves.

Serious vibrations, however, have been experienced from the Central Railway, resulting to a great extent from the sleepers having been laid upon the tubes without the intervention of ballast, as well as the heavy motors employed.

Though the use of temporary shafts in the roads has been practically given up in the interests of the traffic and the owners of property along the route, the lift shafts at the side of the stations, situated on acquired land adjoining the street, which have to serve during the construction of the line for the removal of the excavations and the supply of materials, are very well placed for cartage. A temporary shaft was sunk in the river-bed, on the line of the crossing under the river, of both the City and South London Railway and the Baker Street and Waterloo Railway, to facilitate the removal of the excavations by means of river barges, instead of by carts through crowded thoroughfares.

The up and down lines of these low-level railways have been placed in separate tunnels, owing to certain advantages which this system affords.[1] Thus the ventilation of an underground line, which can have no communication with the outer air except at the stations, is promoted by the train acting like a sort of piston, and driving the air before it in a single tunnel; the space occupied by the railway can be reduced by

[1] *Proceedings Inst. C.E.*, vol. cxxiii. p. 46.

running one tunnel above the other under a narrow street, where otherwise the rights of property at the sides would have to be encroached upon; and two small tunnels can be constructed with greater safety, more cheaply, and with less excavation, than a tunnel for a double line. Moreover, with a single-line tunnel, a dip can be given to the line between the stations, with gradients adjusted to suit the load, so as to diminish the cost and increase the speed of working the trains, as effected on the City and South London Railway, where the descending gradient on leaving a station is 1 in 30, followed by an ascending gradient of 1 in 100 on approaching the next station.

The cast-iron tube forming each single-line tunnel of the City and South London Railway, has an inside diameter of $10\frac{1}{6}$ feet in one part, and $10\frac{1}{2}$ feet in the remainder, each ring having a length of 19 and 20 inches respectively (Fig. 136); whilst the rings of the tube of the

TUBULAR TUNNEL WITH CAST-IRON LINING.
Fig. 136.—City and South London Railway.

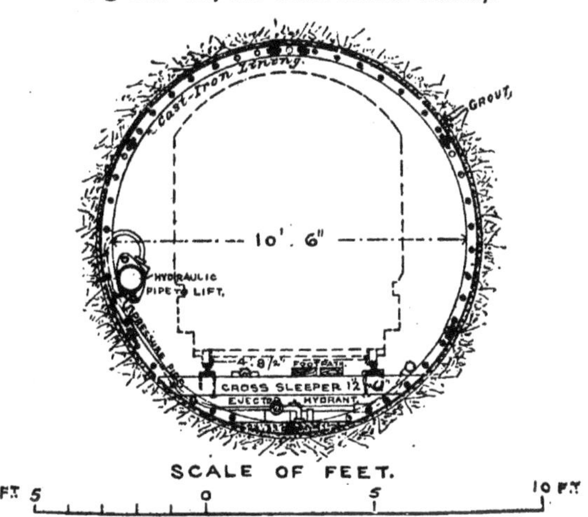

Waterloo and City Railway have inside diameters of $12\frac{1}{6}$ and $12\frac{1}{2}$ feet, and a length of 20 inches. Compressed air has been resorted to when, on approaching a stratum of gravel, the thickness of clay above the tunnel is reduced to about 5 feet; and the most troublesome water-bearing stratum is generally found in the coarse gravel lying directly on the top of the London Clay. One of the air-locks, constructed with a lining of brickwork and concrete, and iron doors, for proceeding with compressed air in tunnelling through gravel and sand for the City and South London Railway, is shown in Fig. 137; and these brick air-locks are found preferable for the men to air-locks with plate-iron sides, owing to their preserving a more equable temperature.[1] The maximum depth

[1] *Proceedings Inst. C.E.*, vol. cxxiii. plate 2.

below the water-level of the subaqueous portion of the City and South London tunnels, is 75 feet, and of the Waterloo and City tunnels, 62 feet.

The rate of progress on the City and South London Railway reached eventually a maximum of 16 feet per day at each face, and 80 feet per week for several weeks in succession. The segments of the rings of the tunnels of this railway, weighing only $4\frac{1}{2}$ cwts., could be readily

BRICK AIR-LOCK.

Fig. 137.—City and South London Railway.

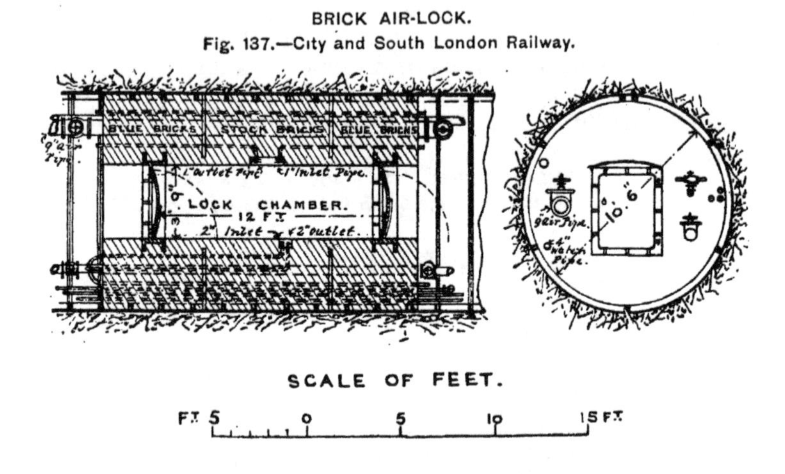

raised and put in place by six men. The stations are situated in brick tunnels of larger dimensions to give space for the platform, which were constructed by the aid of strong timbering in the ordinary manner. Hydraulic lifts are placed in the shafts, 25 feet in diameter, communicating with the stations, for the conveyance of passengers between the station and the street, the difference in level being between 51 and 65 feet. The ventilation of the railway is effected through the stations and shafts; and in order to reduce the resistance offered by the air to the passage of a train in a confined space, and the rush of air into a station produced by an approaching train under such circumstances connecting passages have been provided in places between the two tunnels. These works cost about £220,000 per mile.

The Waterloo and Baker Street Railway, opened in 1906, consists of a double line 3 miles 1 furlong in length carried in two cast-iron tunnels 12 feet internal diameter. The tubes are 23 feet apart under the Thames, diminishing to 17 feet 6 inches in Northumberland Avenue, and to zero in College Street on the Surrey side, where they are one over the other vertically.

The tunnels were driven throughout in the London Clay, except in a portion under the river, where water-bearing ballast overlying a deep depression in the London Clay, and under a head of 70 feet at high tide, was met with. The subaqueous portion was driven by the aid of compressed air and a shield in the usual way.[1]

These low-level tunnels preserve a very uniform temperature throughout the year, and appear to be little affected by the fogs above ground. The railways of this type already completed or projected within London, afford a prospect of ample means of transit underground across certain portions of the metropolis; but they cannot serve, like the Metropolitan Railway, as a means of intercommunication between the different lines of railway coming into London.

Paris Metropolitan Railway.—Though various schemes have been brought forward from time to time for the construction of underground and overhead railways in Paris, it was only early in 1898 that a circular railway following the lines of the outer boulevards, and five cross railways, also keeping to the lines of the boulevards and streets, proposed by the municipality, were authorized.[2] In the same year, preliminary works were commenced for carrying out the railway running east and west across Paris, which comprised the diversion of sewers and water-mains, and the construction of four underground galleries leading from the line of the railway to the Seine, to enable the excavations to be removed and materials brought up by river.[3] The first portion of the scheme undertaken, opened in 1900, consists of the railway crossing Paris from the Dauphine gate to the Vincennes gate, under the Champs Élysées, the rue de Rivoli, and the Boulevard Diderot, which runs underground along its whole length of nearly 7 miles (Fig. 138). The circular

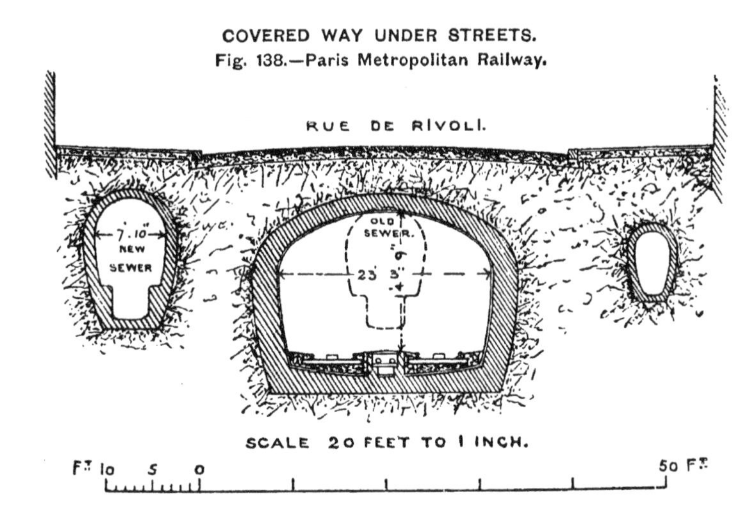

COVERED WAY UNDER STREETS.
Fig. 138.—Paris Metropolitan Railway.

railway now virtually complete, is of a mixed type, partly in tunnel or covered way, partly in open cutting, and partly on viaducts, crossing over

[1] *Proceedings Inst. C.E.*, vols. cl., clvi.
[2] *Le Génie Civil*, vol. xxx. p. 329, and vol. xxxii. p. 382.
[3] *Ibid.*, vol. xxxiii. p. 405, and plate 26.

the Seine twice; and its length is 13⅔ miles, exclusive of the portion which it shares with the first railway. Including the four other cross railways to be subsequently constructed, the total length of the proposed lines amounts to 39 miles, of which 25 miles will be underground, 7¼ miles in open cutting with retaining walls and invert, and 6¾ miles on iron viaducts. These double lines are to be laid to the standard gauge of 4 feet 8½ inches throughout, so that the carriages running over them will be able to go on to the railways they join; but the available width provided of only 20½ feet in the covered ways and open cuttings, will be too narrow to admit the rolling-stock of the ordinary railways on these lines. A gradient of 1 in 25 is proposed in rising to cross the Seine, and dipping down to go under the St. Martin Canal, but elsewhere the gradients will be flatter; and curves of 3¾ chains are to be introduced at the sharpest turns in passing from the line of one street to another.

There are to be one hundred and twenty-three stations on these railways, of which about seventy will be underground; and their average distance apart will be about a third of a mile, closer together than the stations on the London Inner Circle and Central railways, which are on the average about half a mile apart. The estimated cost of the works for these railways, undertaken by the municipality, is about £168,000 per mile; but including the cost of the plant for working them by electricity, and the rolling-stock, the estimated expenditure amounts to £204,000 per mile. The subterranean quarries which have been worked for building stone occupy extensive areas beneath Paris, and in the progress of the Metropolitan Railway a large amount of consolidation work has been found necessary to provide supports for tunnel footings at considerable depths.

Glasgow District Railway.—The underground railway which traverses Glasgow in a sort of elliptical course, of which the River Clyde, which it passes under twice, forms approximately its major axis, was opened in December, 1896; and its length is about 6½ miles. It runs under streets for a considerable portion of its length, and was formed through rock, shale, sandstone, clay, and sand, and in sand, rock, and mud at one of the river crossings, and in sand at the other.[1] Two circular tunnels, 11 feet inside diameter, and from 2½ to 6 feet apart, have been constructed for the up and down lines, laid to a gauge of 4 feet (Fig. 139), except at the stations, where the railway is covered by a single arch, or is in open cutting with retaining walls, a central platform serving both lines (Fig. 140). The railway has been so laid out that all the stations are at a depth of only 20 to 30 feet below the surface, with rising gradients generally to them on both sides, which facilitate the starting and stopping of the trains. There are gradients of 1 in 20 and 1 in 23 in dipping under the Clyde, but they are elsewhere for the most part much flatter; and the sharpest curves are 10 chains radius. The covered ways or tunnels were formed in some places, near

[1] *The Engineer*, vol. lxxxii. pp. 558 and 559; and "The Glasgow District Subway," A. H. Morton.

the surface, by cut and cover (Fig. 139); by tunnelling in the ordinary manner, and lining with brick, through rock (Fig. 140) and other softer impervious strata, where settlement was unlikely or of comparatively little consequence; with cast-iron rings by the aid of a shield through

UNDERGROUND GLASGOW DISTRICT RAILWAY.

Fig. 139.—Covered Way. Fig. 140.—Tunnel and Station.

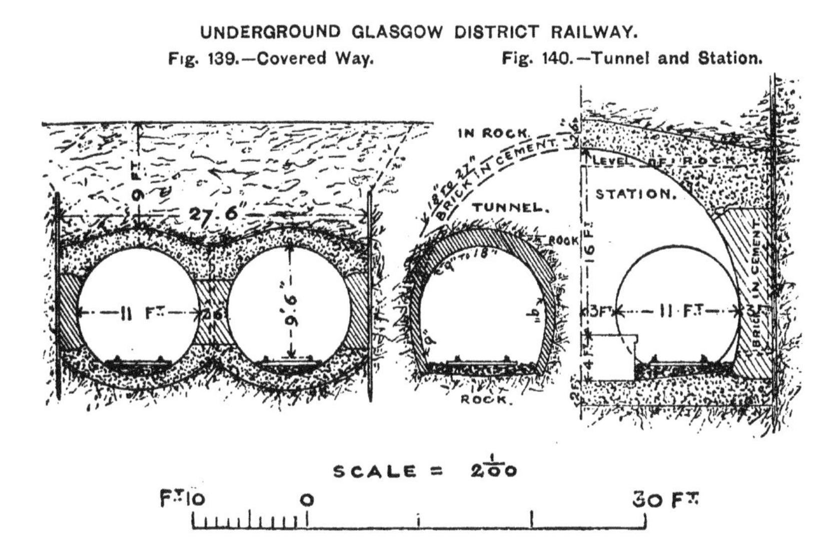

impervious material, where settlement had to be guarded against; and with tube, shield, and compressed air, through water-bearing strata at some depth below the surface, a maximum pressure of 23 to 30 lbs. on the square inch having been required under the Clyde, 55 feet below high water. The trains are drawn by a steel cable in each tunnel, at a speed of about 16 miles an hour. The stations are between half and a third of a mile apart on the average; and the two lines at these places come under a single arch, and are both served by a central platform (Fig. 140).

New York Rapid Transit Subway.—This subway, commencing in Brooklyn, runs under the East River to Manhattan Island, proceeding to 103rd Street, from thence branching in one direction over Harlem River to Bailey Avenue and in the other under the same river to Bronx Park, the combined length of the routes being 25·7 miles. The leading principles of the design of this subway were the adoption of the least depth possible below street level, and the provision of a high speed service by extra tracks in the district of maximum traffic, with a local service with short intervals between stations. Owing to the topography of the site, however, this railway includes many types of construction, such as cut-and-cover tunnelling in rock and soft ground, subaqueous tunnelling by compressed air and shield, open cutting, embankment, and steel viaduct. Steel beam and column construction with reinforced concrete are used in shallow work, while concrete arching is applied where headway permits. Cast-iron tubing for subaqueous work is similar in design to that used under the Thames. The permanent way is of Vignoles type, 100 lbs. per yard, on timber sleepers.[1]

Chicago Underground Freight Railways.—For the relief of the street traffic a system of electrical railways for the carriage of goods and mails, following the centre lines of the principal thoroughfares and constructed at a level of about 46 feet below the surface, was opened in 1904, and has been subsequently extended until the system has attained an aggregate length of 60 miles. The subways are constructed of concrete, the main tunnels being 14½ feet high, and 12¾ feet wide at base. A system of telephone cables is installed in the tunnels, which are electrically lighted. The gauge of rails is 2 feet, and the traffic is controlled by telephone.

OVERHEAD RAILWAYS.

There are two distinct types of overhead railways, namely, railways which are carried through a city on a low brick viaduct, crossing over the streets as nearly at right angles as practicable, on arched bridges or iron girders affording just the requisite headway; and secondly, elevated railways, which run along a line of streets on girders supported by iron pillars erected on the roadway. In the first case, the brick viaduct, in place of an embankment with slopes, reduces the width of the strip of land required for a railway to a minimum; and the space under the arches is serviceable for storing goods. The second system dispenses with the purchase of costly land by making use of the public roadway, interfering to some extent with the convenience of the traffic along the streets traversed; whilst it entirely disregards the interests of owners of residential property, or business premises, in the streets along which the elevated railway passes.

Railways on Brick Viaducts.—The Greenwich line, the first railway brought into London, is carried on a brick viaduct throughout, this form of construction having been probably adopted, in that early period of railway enterprise, quite as much on account of there being no cuttings along the line to furnish earthwork for a long embankment across the low-lying land between Greenwich and London, and to ensure rapidity of construction, as to reduce the expenditure on land and avoid severance of property. This system, however, possesses such obvious advantages in populous districts, that the railways penetrating into central London from the south and east, have naturally been constructed in this way, as well exemplified by the metropolitan portions of the South-Western (Fig. 141), the London and Brighton, the Chatham and Dover, the South-Eastern, and the Great Eastern railways. When, indeed, in 1864, powers were sought for the extension of the Metropolitan Railway, opposing schemes of open-air railways on viaducts

[1] *Proceedings Inst. C.E.*, vol. clxxiii.

were brought forward as the natural alternative to underground railways; but the Metropolitan District Railway was preferred by Parliament to the overhead railways proposed, mainly on account of the fairly complete connection thereby afforded between most of the terminal stations in London, and also owing to the underground line avoiding interference with buildings and streets.

The Berlin Metropolitan Railway, constructed in 1879–82 across the city for $7\frac{1}{2}$ miles east and west, with four lines of way to keep the local and fast through traffic distinct, as it connects the principal railway termini, is an overhead line for its whole length, raised about 18 feet above the ground; for the River Spree, which flows through the city, was

OVERHEAD RAILWAY ON BRICK VIADUCT.

Fig. 141.

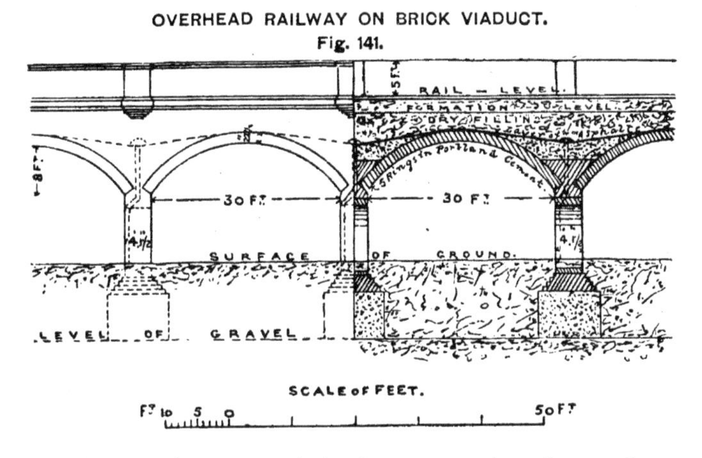

considered at the time to preclude the construction of an underground railway. The railway is carried on a brick viaduct for nearly 5 miles; on a viaduct with an iron superstructure for $1\frac{1}{8}$ miles; and outside the city, on embankments, supported in places by retaining walls, for the remainder of its length.[1] There are curves of 14 to 25 chains radius on the line; but the worst gradient does not exceed 1 in 125, for the extreme variation in level along the whole course is under 12 feet. Assuming that the four lines make the railway equivalent to a double line of twice the length, its cost amounted to about £114,000 per mile of double line.

New York and Brooklyn Elevated Railways.—An elevated single-line railway carried on an iron viaduct for three miles along a street, the girders being supported by a single row of iron columns widened out at the top, was opened in New York in 1867, and worked by a wire cable, with the object of facilitating and expediting communication between the southern, business part of the city and the northern portion of Manhattan Island, which has a length of 13 miles north and south, and an average width of only 2 miles. A few years later, this

[1] *Zeitschrift für Bauwesen*, vol. xxxiv. and pp. 1, 113, and 225.

system of elevated railways was gradually extended for a double line along some of the streets of New York; and there are now 36 miles of these railways on the island, worked by locomotives. The construction of similar railways was commenced in Brooklyn in 1879, and about 28½ miles of lines have been carried out; whilst 26 miles of elevated railways have been established in Chicago.

Three types of viaduct have been constructed in New York, according to the width and traffic of the streets traversed, and the character of the adjacent buildings, the supporting columns being placed along the street at intervals of from 37 to 44 feet.[1] Where the roadway is wide and has too large a traffic to admit of supports being placed in it, and proximity to the buildings alongside the street is of no consequence, the columns have been erected in a single row along the curb of the pavement on each side, being widened out at the top to receive the longitudinal girders, laid 5 feet apart centre to centre (Fig. 142). Where it is important to place the railway as far away from the buildings

ELEVATED RAILWAYS ALONG STREETS.

Fig. 142.—Columns in Single Row. Fig. 143.—Columns in Roadway Braced together.

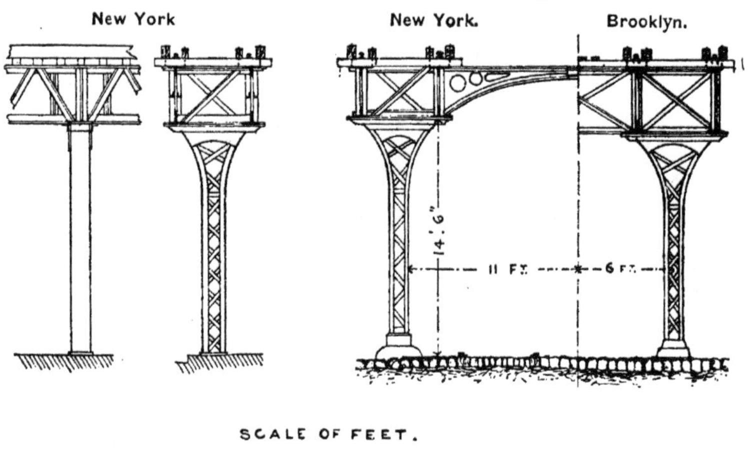

SCALE OF FEET.

FT 10 5 0 10 20 FT.

as possible, and yet the traffic is too great to put the supports in the roadway, the columns have been erected along each curb, as in the first case, but are connected by transverse girders, 28 to 45 feet long, which carry the ends of the four main girders supporting the two lines, as nearly in the centre of their span as practicable (Fig. 144, p. 241). Lastly, when the roadway is wide enough, and the traffic moderate, the columns, widened out at the top, have been placed in a double row in the roadway, and connected in pairs by arched or lattice bracing, a clear width

[1] *Proceedings Inst. C.E.*, vol. lxviii. p. 229, and plate 5.

of 22 feet being left between the rows to allow of two tramway lines running between them (Fig. 143).

The gradients of the railway conform as nearly as practicable to the varying inclination of the street, except where frequent changes of level, or steep inclinations render a modification expedient, a minimum headway of $14\frac{1}{2}$ feet being always provided under the girders for the vehicles passing below. The steepest gradient on the line is 1 in 50; and the sharpest curve has a radius of 90 feet. The height of the columns ranges generally only between 18 and 21 feet when they are in single row; where higher columns are required, the columns of the two rows are connected across the street; and where in one place the great height of 65 feet is reached, the columns are braced together in groups of four. Guard timbers are fixed on both sides of every rail, only

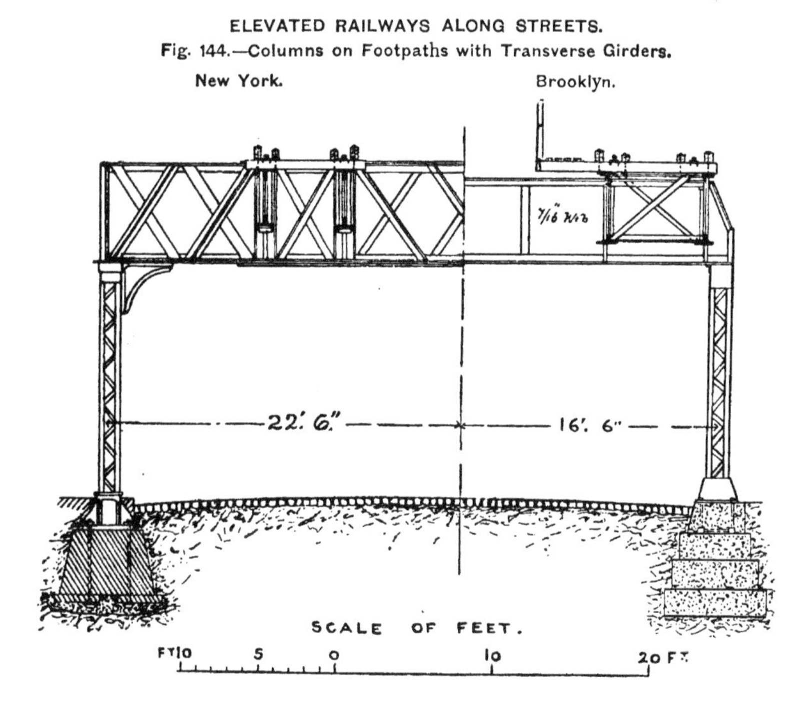

ELEVATED RAILWAYS ALONG STREETS.
Fig. 144.—Columns on Footpaths with Transverse Girders.

4 inches from each edge of the rail, to secure the train from running oft the viaduct in the event of its leaving the rails.

The cost of the New York Elevated Railways was £71,500 per mile for the works, and £83,500 per mile including the rolling-stock, nothing having been paid for the space occupied on the roadway by the columns, and no compensation having been given to the owners of houses along the streets traversed, for the deterioration in value resulting from the loss of privacy and the noise. The line is worked by heavy bogie locomotives; and the stations, placed generally at the intersection of the streets, are about $\frac{1}{3}$ mile apart.

The Brooklyn Elevated Railways resemble the New York lines in being carried along the streets on iron columns and girders, and in following approximately the inclinations of the ground, leaving a clear headway for the traffic below of 14 feet; but the chief type of construction consists of columns placed along the edge of the footpaths, supporting transverse girders, usually from 35 to 45 feet long, which carry the longitudinal girders on which the railway runs (Fig. 144), only about one-sixth of the lines being carried on columns in the roadway, braced together in pairs [1] (Fig. 143, p. 240). Plate girders also have been introduced in the most recent portions, on account of their being as economical as lattice girders for short spans, and more convenient. The steepest gradient is 1 in 50, as in New York; and the sharpest curve, in turning round a right angle to pass from one street into another, has a radius of 100 feet. The line is worked by locomotives; and the stations are about $\frac{1}{3}$ mile apart, as in New York.

Liverpool Overhead Railway.—The double line of railway running along the dock road passing at the back of the Liverpool Docks throughout their whole length, is about $6\frac{1}{3}$ miles long, exclusive of the southern extensions mostly in tunnel; and it is carried on an iron viaduct, consisting of two main girders, 22 feet apart, with a flooring of arched plates and **T** irons between them (Fig. 147, p. 243), supported by pairs of steel columns underneath them, at intervals of about 50 feet, composed of two channel-irons joined by a plate on each side, forming a box-shaped pillar which rests upon a block of concrete in made ground, so that the maximum load at the base does not exceed one ton per square foot.[2] The pairs of columns leave a clear width between them of 21 feet for the passage of two lines of dock railway; and a clear headway of 14 feet is afforded under the bracing between the pairs of columns, and 16 feet elsewhere for cross traffic under the railway (Fig. 145 and 147). The spans of the main girders, which are ordinarily 50 feet, had to be modified in several places, to avoid interference with rights of way and property, to spans varying from 30 feet up to 98 feet; and plate girders were adopted up to 75 feet span, and bowstring girders for the larger spans.

Except where the railway has to dip under a coal siding with a short gradient of 1 in 40, the gradients are easy; whilst the sharpest curve has a radius of 7 chains. A swing bridge with a double floor has been provided across the entrance to the Stanley Dock, for the passage of the dock railway underneath, and the overhead railway above (Fig. 146). Three bascule bridges have been erected on the railway, having a bascule portion $33\frac{3}{4}$ feet long, and a counterpoised tail-end of $14\frac{1}{2}$ feet, which are turned on a horizontal pivot into an almost vertical position when large boilers or other high loads have to be taken across the line (Fig. 147). The railway is worked by electricity passing along

[1] *Proceedings Inst. C.E.*, vol. cxxvii. p. 333, and plate 8.
[2] *Ibid.*, vol. cxvii. p. 54, and plate 4.

a steel bar conductor laid between the rails, a system of traction which has proved economical for the moderate and fairly uniform loads carried

OVERHEAD RAILWAY, LIVERPOOL.

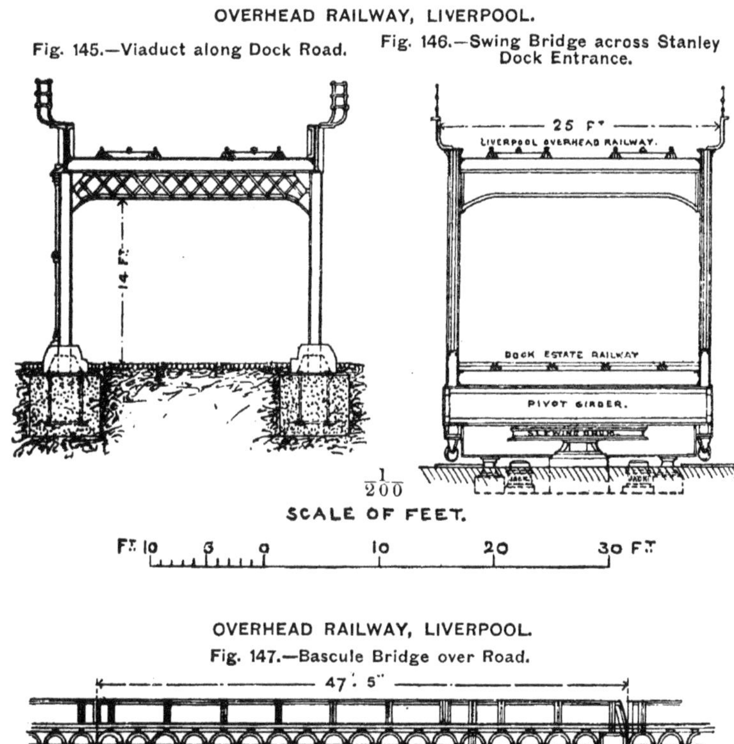

Fig. 145.—Viaduct along Dock Road.

Fig. 146.—Swing Bridge across Stanley Dock Entrance.

OVERHEAD RAILWAY, LIVERPOOL.

Fig. 147.—Bascule Bridge over Road.

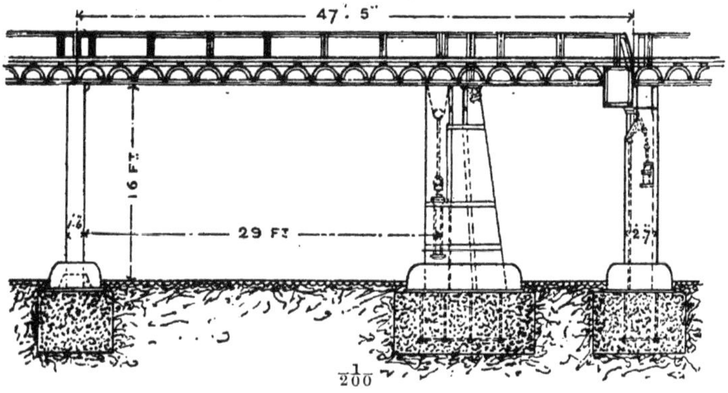

47'. 5"
16 FT.
29 FT.
1/200
SCALE OF FEET.
FT. 10 5 0 10 20 30 FT.

by the trains running at short intervals apart; and the signals also are operated by electricity, being raised or lowered automatically by the passage of the trains. The average distance between the stations is two-fifths of a mile; and the railway was opened in 1893 for a length of 5 miles, which length has since been increased to 7 miles by extensions at both ends.

The actual cost of construction, including the stations, was £63,600 per mile; but adding to this the expenditure on electric plant, rolling-stock, alterations of the dock railway and roads underneath the railway, parliamentary proceedings, and other items, the total cost must be reckoned at £94,000 per mile.

Berlin Electric Railway.—The elevated railway in Berlin, as shown in Fig. 148, is somewhat similar to the Liverpool overhead line, with the exception that the columns of each pair are only 11½ feet apart centre to centre, instead of 22 feet, so that the superstructure has been made to overhang its supports, on each side of the line of columns, sufficiently to afford a width of 23 feet between the side railings.[1] This railway consists of a double line, 6¼ miles long, carried along the streets on a steel viaduct having a watertight floor, at an elevation of about 18 feet, and leaving a headway of 15 feet underneath (Fig. 148). The supporting columns, braced together in pairs, are

OVERHEAD ELECTRIC RAILWAY, BERLIN.
Fig. 148.—Viaduct along Streets.

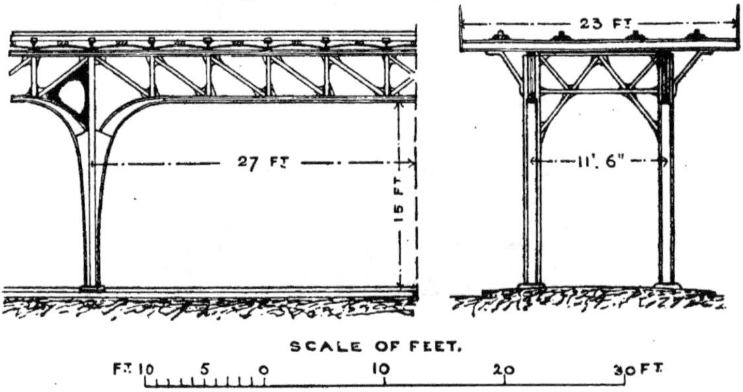

erected along the roadway at intervals of 54 feet; and, as in Liverpool, the lines are laid to the ordinary gauge, and the iron electric conductor is laid centrally between the rails of each line. The steepest gradient on the line is 1 in 38; and the maximum speed of the trains has been fixed at 31 miles an hour.

The cost of this elevated railway has been estimated at £108,800 per mile.

Comparison of Overhead and Underground Railways.—In two respects elevated railways possess decided advantages over underground lines, namely, the provision of an open-air journey for the

[1] "Die Electriskhe Stadtbahn in Berlin von Siemens und Halske," F. Baltzer. Berlin; *Archiv fur Post und Telegraphie*, 1897, p. 722; and *Engineering*, vol. lxxii. p. 75.

passengers, and their much smaller cost. On the other hand, an elevated railway in a street, though relieving the traffic, is somewhat of a nuisance to the public using the street, and very objectionable to occupiers of business or residential houses alongside, even where electric traction and a watertight floor are adopted, as in Liverpool and Berlin, instead of locomotives and open flooring, causing smoke and noise, and the dropping of water and cinders on the roadway below, as in New York and Brooklyn.

Electric traction should evidently be introduced on all elevated lines running over streets, and underground lines; whilst tubular, low-level, underground lines, such as those constructed in London and Glasgow, avoid as far as possible interference with streets, sewers, and house property, and appear more consistent with the preservation of private rights and the interests and convenience of the public, than overhead railways running along streets, except under such special conditions as the road alongside the Liverpool docks. The much smaller cost of elevated railways, by avoiding to a great extent the purchase of very valuable land, is, indeed, an important consideration in favour of their extension in preference to underground lines; but due regard for the general public advantage seems to dictate keeping the railways out of the streets, and putting them underground, except where exceptional circumstances preclude the construction of underground works.

WIDENING RAILWAYS.

The increase of traffic, especially of a local character, on railways in the neighbourhood of large cities with rapidly increasing suburbs, has necessitated in several cases the doubling of the lines, as for instance most of the main lines running into London and some of the Paris railways.

When the railway is on a brick viaduct, the widening can only be effected by purchasing a strip of land alongside of the requisite width, and building an additional width of viaduct on to the old work, care being taken to go down to a good foundation, so as to secure the new work from settling down, away from the old, when the weight and vibration of the trains come upon it.

Where a railway just outside the metropolis, or in passing neighbouring towns, was originally constructed with cuttings and embankments in the ordinary way, the adjoining land has often, owing to the extension of buildings, increased so much in value as to render it expedient to avoid the purchase of additional land for the widening of the line, as far as practicable. Moreover, slips in the slopes of newly-formed clay cuttings and embankments, which, in entailing the extension of the boundaries of the railway, are of comparatively little importance through agricultural districts, become a serious matter when leading to encroachments on valuable building land. Under such conditions, if a mere widening of the original cuttings and embankments is resorted to, it is specially important to execute this enlargement on one side only if possible, so as to disturb only one of the consolidated slopes; and in the immediate vicinity of the metropolis or large towns, it is often advisable to effect the widening by building a retaining wall on each side, to keep up the upper part of the slope, enabling the toe of each slope to be removed in the case of a cutting, and allowing an embankment to be widened at the top without entailing an extension of the toe of the slope on each side, which is kept back by a retaining wall erected along the outer edge of the slopes of the original embankment.

In order to widen a cutting by substituting a retaining wall for the lower part of the slope on each side, a trench is excavated along the slope, with its inner face flush with the upper part of the portion of the slope to be removed; and the trench is timbered strongly as it is carried down to below the formation level in the cutting, the width of the trench being increased with an increase in the depth of the cutting or in the instability of the soil. The trench is excavated in short lengths; and as soon as it has reached its full depth, it is filled in with concrete, the timbers being removed as the concrete is carried up; and as soon as the concrete has set, the lower, inner part of the slope can be removed. To provide against the accumulation of drainage water at the back of the wall, tending to disintegrate the material behind it and increase the pressure, drainage holes are formed through the wall in building it; and the sliding forward of the wall is prevented by carrying it down into the solid ground below formation level. Retaining walls for embankments are built along the toe of the slopes; and after the best available material has been raised in layers against the back of the wall up to its crest, the formation width of the embankment can be increased with slopes at the sides down to the crest of the walls.

Good drainage at the back of retaining walls; due provision against the commencement of slips from above in cuttings; filling up with a succession of thin layers, carefully punned, behind the retaining walls for embankments; and the carrying down of the foundations to a sufficient depth into the solid ground to afford an adequately firm toe for the base of the wall to press against, in the event of its tending to move forward, are essential precautions for ensuring the stability of all classes of retaining walls. The horizontally arched form of retaining wall commonly used for lining the sides of metropolitan railways in open cutting through treacherous strata (Fig. 134, p. 230), has the advantage of providing a greater width of wall and of base, and therefore a stronger wall, in proportion to the amount of materials employed, than the ordinary form of solid wall. Where retaining walls at the sides of a cutting are exposed to a considerable pressure at the back in unreliable strata, the tendency of the walls to slide forward at their toe can be prevented by the insertion of an invert, whereby the pressure against the base of one wall is counteracted by the pressure on the opposite side. Moreover, the liability of retaining walls, especially when high, to be pushed forward at the top by the pressure of the ground behind, can be prevented by struts stretching across from wall to wall (Fig. 135, p. 230), a system long ago adopted in constructing the approach to Euston Square Terminus, where cast-iron struts stretching across the four lines of way, support the upper part of the retaining wall on each side in a clay stratum.

CHAPTER XV.

PERMANENT WAY; JUNCTIONS; AND STATIONS.

Final works on railways—**Permanent Way:** definition, primary considerations; Gauge, varieties of, in different countries, standard, conditions affecting gauge, break of gauge; Narrow Gauge, saving effected by, advantages of, for curves in mountainous country, economy; Ballast, materials employed; Wood Sleepers, advantages of, timber used, longitudinal and cross sleepers, dimensions and merits, bed-plates on sleepers for flange rails; Cast-iron Sleepers, object, form; Wrought-iron and Steel Cross Sleepers, forms, disadvantages of, under certain circumstances, extensively tried, conditions affecting use; Rails, forms described, adoption of steel, weight; fastenings for rails, chairs, advantages of flange rails, object of cant; connection of rails, fish-plates; space for expansion; Provisions for Safety on Curves, super-elevation of outer rail, check rail, other precautions, bogies—**Junctions:** arrangements; switches and crossings, description of switches, their locking, rising rail, description of crossings; forms of junctions, ordinary, diversion of outer line—**Stations:** conditions relating to them; forms, arrangements of intermediate stations, arrangements of terminal stations.

AFTER the cuttings and embankments of a railway have been completed to formation level, the viaducts, bridges, and culverts built, and the tunnels constructed, the line has to be completed by the laying of the permanent way, the forming of the connections at the junctions, and the erection of the stations.

PERMANENT WAY.

Permanent way consists of the ballast, sleepers, and rails, laid on the formation level, raising the line about 2 feet above this level with rails supported on chairs, and about 1 foot 8 inches with flat-bottomed rails; and the term "permanent" has been adopted to distinguish these lines, solidly laid to bear the passage of heavy trains at a high speed, from the temporary lines laid down by the contractor for carrying out the works.

After the course of a railway has been decided upon, two points have to be determined at the outset, before the drawings for its construction can be commenced, namely, whether the railway is to have a single or a double line, and the gauge to which the rails are to be laid, as the formation width of the earthwork, and the minimum width that must be provided for the railway between the side walls of the over-bridges, tunnels, and other structures, depend mainly on these considerations. A single line is usually laid down for pioneer or branch railways in the first instance, with passing-places at the stations, a second line being subsequently added as soon as the traffic has increased sufficiently to necessitate more accommodation; but where there appears to be a good prospect of a rapid development of traffic on the line, works which are not well adapted for being widened, such as over-bridges and tunnels, are often constructed at the outset to accommodate two lines of way.

Gauge.—The gauge of a railway, or the width between the inside edges of the two rails of a line of way, has been varied somewhat in different localities, and sometimes in accordance with variations in the conditions. Thus the standard gauge of Great Britain, of 4 feet 8½ inches, to which all the railways of the country have to be laid unless another gauge has been specially authorized, is also the gauge of continental Europe, with the exception of Russia, Spain, and Portugal, and of the United States, Canada, Mexico, Brazil, Uruguay, Peru, Egypt, and New South Wales. A gauge of 5 feet was purposely adopted by Russia to prevent the possibility of the neighbouring European nations, in time of war, pouring armies by rail into the country. A guage of 5 feet 3 inches was chosen in Ireland, Victoria, South Australia, New Zealand, and for some lines in Brazil; whereas 5 feet 6 inches are the gauge of the railways in Spain and Portugal, Ceylon, Argentina, Chili, of the main lines of British India, and some lines in the United States and Brazil. Narrower gauges have been adopted as the standard gauge in a few countries, as for instance 3 feet 6 inches for the railways of the Cape Colonies, Queensland, West Australia, Tasmania, and Japan, and for secondary or light railways, with widths ranging between 3 feet 6 inches and 2 feet, examples of which are furnished by the secondary railways of Norway and Egypt, and the light railways of South Australia and New Zealand, with a gauge of 3 feet 6 inches, of India and Uganda with a gauge of 3′ 3⅜″ (metre), of Mexico and some light railways in England and Ireland with a 3-feet gauge, and several light railways on the Continent, in India, and in England, with gauges ranging from 2 feet 6 inches down to 2 feet, the latter being the gauge of the well-known Festiniog (actually 1 foot 11½ inches) and Darjeeling railways in mountainous districts.

The standard English gauge of 4 feet 8½ inches has been proved by experience to satisfy all the requirements of large, heavy, and rapid traffic; for not only is it the gauge which has been most largely adopted, but it is also the gauge of the railways which carry the most numerous, quickest, and heaviest trains. The safety of trains appears to depend more upon a well-laid permanent way than upon the width of the gauge; and though a wide gauge must give a train greater stability in running along a straight line, it is unfavourable for the passage of trains round curves, owing to the greater difference in length and curvature between the outer and inner rail with an increase in the width of the gauge. Whatever standard gauge, however, may be adopted in a country as best suited to its conditions, it is very important that the network of the main lines and branches should be laid as far as practicable to the same gauge, so that the rolling-stock may traverse any part of the system. A change in the gauge, or, as it is termed, a break of gauge, necessitates a change of vehicles by the passengers, and a transhipment of goods, involving inconvenience and delay.

Narrow Gauge.—In some countries, certain parts differ so essentially from the rest, in population, resources, and physical conditions, that the choice lies between a cheaper form of railway with a narrow gauge, or no railway at all. Sometimes also in outlying districts, the traffic requirements are so different from those served by the main system of railways, and the interchange of goods and products is so moderate, that a break of gauge is of comparatively little importance, and is dictated by economical considerations. The saving effected by the adoption of a narrow gauge, consists in the reduction of the formation width, and consequently a diminution in the width of the strip of land required for the line; in the span of the over-bridges, and the width of the under-bridges; and a reduction in the earthwork ot the central portion of the cuttings and embankments (though leaving the contents of the slopes unaltered), the amount of ballast, and the length of the cross sleepers, in proportion to the narrowing of the gauge. It is evident, therefore, that in order to compensate amply for the inconvenience and cost of transhipment involved in a break of gauge, it is important for the modified gauge to be made notably narrower than the standard gauge, so as to gain all the economy practicable by the change; and, consequently, a gauge of 2½ feet appears preferable to a 3½-feet or a metre gauge. Moreover, a narrow gauge, though necessitating the purchase of special rolling-stock, has the advantages of enabling the rolling-stock to be fully adapted to the requirements of the district, and the load proportioned to a lighter form of permanent way. In mountainous districts, a narrow gauge possesses the great advantage of enabling much sharper curves to be introduced than would be possible with the standard gauge, whereby the cost of a railway passing through rugged country can be largely reduced. Thus, for instance, the sharpest standard curves on the mountain section of the Canadian Pacific Railway, with a gauge of 4 feet 8½ inches, have a radius of 573 feet, which is also the minimum radius admitted for curves in rugged country in India, under exceptional conditions, with the 5½-feet gauge; whereas curves of 358 feet radius are allowed on the metre gauge in India, and curves of 83 feet radius have been resorted to in some places on the Festiniog Railway in North Wales with a gauge of 1 foot 11½ inches, and of 70 feet radius on the Darjeeling Railway in the Himalayas with a gauge of 2 feet. The influence of gauge on cost, is indicated by a comparison of the cost per mile of some of the more recent railways in India, constructed to three different gauges, which averages £7200 for the 5 feet 6 inches gauge, £4500 for the metre gauge, and £1800 for the 2 feet 6 inches gauge.[1]

[1] "Light Railways at Home and Abroad," W. H. Cole, p. 176.

Ballast.—The objects of ballast are, to serve as a bed for the sleepers, so as to distribute the shocks to which they are exposed by passing trains and prevent their cutting into the soft parts of the formation; to enable the water in wet weather to drain away rapidly from under the sleepers, and thereby preserve them; to keep the sleepers firmly in position; and also when the ballast is heaped up over them, as is now sometimes done, to protect them from the sun. Ballast should consist of clean, broken stone, shingle, gravel, broken bricks, slag, or other hard material not subject to disintegration by shocks and moisture; and the coarser material should be laid at the bottom, with the finer above, forming a layer not less than 18 inches thick, and extending out about 18 inches beyond the ends of the cross sleepers. Where ballast cannot be procured within a reasonable distance, as on the prairies of North America and the African deserts, and in laying pioneer lines in undeveloped districts, the sleepers are laid directly on the ground, with packing of sand or of the best soil available; and by putting the sleepers closer together in districts abounding in timber, and where the climate is dry, a sufficiently solid track is thereby obtained.

Wood Sleepers.—The materials used for sleepers are timber, cast iron, wrought iron, and steel. Timber sleepers are most commonly employed, as they are cheap, more particularly in carrying a railway through undeveloped forests; they form an elastic, moderately quiet track; and they are fairly durable if made of sound, well-seasoned timber, generally protected against decay by creosoting under pressure or other preservative method. Hard woods are preferable for sleepers where they are readily obtained, as they are less subject to injury under a heavy traffic; but Baltic pine, sawn on the spot to requisite dimensions, is commonly used on European railways; whilst larch, if grown to a sufficient size, forms durable, cheap, semicircular sleepers; and white oak has been largely adopted for sleepers in America with satisfactory results.

Two forms of wooden sleepers have been employed, namely, longitudinal sleepers and cross sleepers. Longitudinal sleepers, in conjunction with bridge rails, form a very even track, the rails being supported by the sleepers along their whole length, so that a lighter section of rail can be adopted (Fig. 149); and in the event of a vehicle leaving the rails, it may run along the sleepers for some distance without injury. Moreover, with the original broad gauge of 7 feet of the Great Western Railway, the longitudinal sleepers laid down, 12 inches wide and 6 inches thick, were cheaper than cross sleepers. To maintain the gauge, however, with longitudinal sleepers, they have to be connected at intervals by cross pieces of timber, which were made $4\frac{1}{2}$ inches wide and 6 inches deep on the broad gauge. The connection also of bridge rails has to be effected by a plate under the rail, which forms a less satisfactory joint than the fish-plates connecting

LONGITUDINAL SLEEPERS AND BRIDGE RAILS.

Fig. 149.

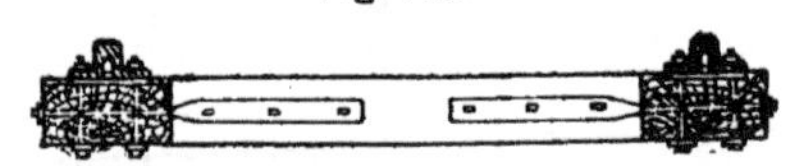

other forms of rails; and the repairs and renewals of longitudinal sleepers are more costly and troublesome than of cross sleepers. The cost, moreover, of longitudinal sleepers is greater for the standard gauge of 4 feet $8\frac{1}{2}$ inches, now so extensively adopted, than that of cross sleepers. Accordingly, cross sleepers are at the present time almost exclusively resorted to where timber sleepers are used.

Cross sleepers, or ties as they are called in America, from their connecting the two rails of a line of way together at their proper distance apart, vary somewhat in their dimensions, and in the distance they are placed apart, according to the traffic on the line. Thus on the main lines in England, the usual dimensions of timber sleepers are 8 feet 11 inches long, by 10 inches wide, by 5 inches thick; though on lines with very frequent, heavy traffic, like the Metropolitan Railway, the cross dimensions are increased to 12 inches by 6 inches, and in lines of light traffic, reduced to 9 inches by $4\frac{1}{2}$ inches. On the Canadian Pacific Railway, the wooden cross sleepers are 8 feet long, 6 to 8 inches wide, and 6 inches thick. The standard sizes of the wooden cross sleepers on several of the main lines of railway in the United States, are 8 to $8\frac{1}{2}$ feet in length, 6 to 9 inches in width, and 6 to 7 inches in thickness.[1] The sleepers are generally placed closer together on the American railways than in Europe, more especially where the ballast is of poor quality or has to be dispensed with, so that instead of being laid 3 feet apart, and 2 feet at the joints, centre to centre, as usual in England, numbering about 1800 sleepers per mile, they are put only

METAL BED-PLATE UNDER FLANGE RAIL.

Fig. 150.

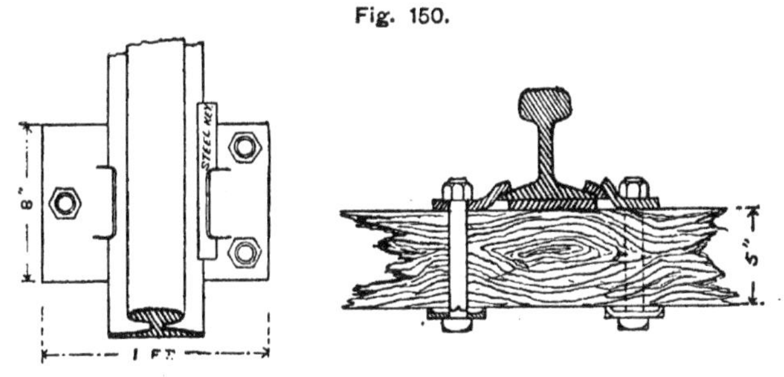

2 feet to 1 foot 8 inches apart in America, giving about 2240 to 3200 sleepers per mile. The sleepers, however, are being placed closer together now in England in parts where the traffic is exceptionally large.

The life of a wood sleeper varies between about five and eighteen years, according to the quality of the timber or its protection, the climate, the traffic, and its exposure to damp. Where the traffic is very heavy, and especially on sharp curves and steep gradients, the indentation of the flat-bottomed rails into the sleepers, or the working

[1] "Railway Track and Track Work," E. E. Russell Tratman, New York, p. 29.

loose of the fastenings, determines the life of the sleepers; but by placing metal bed-plates under the flat-bottomed rails at the sleepers, thereby affording a considerably larger bearing surface, and a much firmer attachment between the rail and the sleeper, the indentation of the sleepers and the damage at the fastenings are prevented (Fig. 150).

Cast-iron Sleepers.—In some countries, the attacks of insects, such as the white ant in India, and an unfavourable climate, render wooden sleepers unsuitable from the rapid injury they undergo. Accordingly, many years ago cast-iron, elliptical, pot or bowl sleepers were adopted under these circumstances instead of wood, connected in pairs by an iron tie across the line, to maintain the gauge. Chairs are cast on these pot sleepers to receive a double-headed rail (Fig. 151), or clips for fastening on a flat-bottomed rail (Fig. 152); and holes are provided in the upper part of the bowl, through which the packing of

CAST-IRON POT SLEEPERS.

Fig. 151.—Indian State Railways.

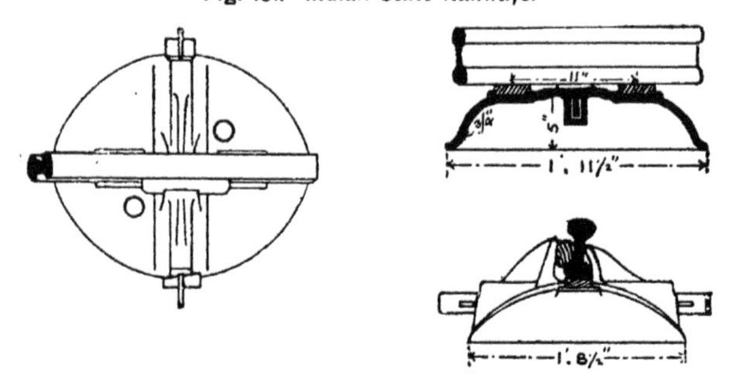

Fig. 152.—Indian Midland Railway.

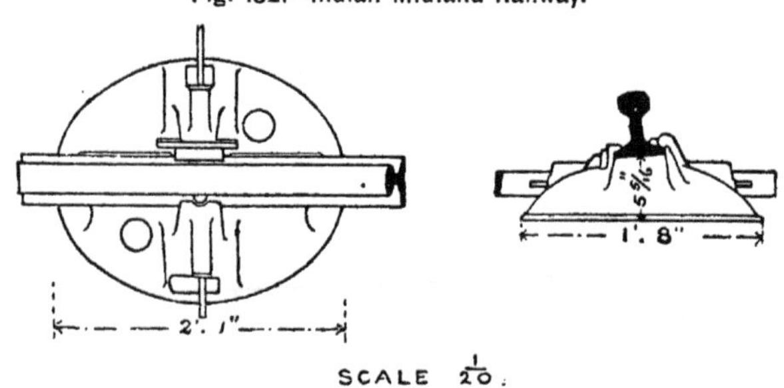

SCALE $\frac{1}{20}$.

the ballast, so as to fill the hollow underneath the bowl, can be completed, in order to both support the bowl and keep it firmly in place. These pot sleepers, therefore, serve the double purpose of sleeper and chair, and they do not readily rust; but being of cast iron, they are liable to breakage from shocks.

Wrought-iron and Steel Cross Sleepers.—The tendency of

cast-iron sleepers to fracture, and the gradual increase in some parts in the price of timber, and its liability to decay, led to the employment of wrought iron for sleepers, in the form of an inverted trough. Several attempts were made to design satisfactory forms of wrought-iron longitudinal sleepers several years ago;[1] but, as in the case of timber sleepers, metal cross sleepers have proved preferable, formerly made of wrought iron, and more recently of steel. Though all these metal cross sleepers have been made of the inverted trough shape, closed at the ends, so as to be embedded and kept immovable in the ballast, their precise form, and the connections provided for the rails resting upon them, have been considerably varied by different designers. The accompanying illustrations (Figs. 153, 154, and 155) show three different forms of steel cross sleepers. In the first case (Fig. 153), ordinary chairs holding double-headed rails are fastened on to the steel cross sleepers by bolts, a felt pad being inserted between the chair and the sleeper to prevent the clatter of metal against metal in the vibrations produced by the trains; but in some types of steel sleepers, the chairs form part of the sleeper. The two other drawings (Figs. 154 and 155) show steel cross sleepers differing in shape, and also having different arrangements for fastening the flat-bottomed rails to them.

Steel sleepers last, in most cases, considerably longer than wooden sleepers; but their first cost is also much larger. In localities, however, where the soil is impregnated with saline deposits, steel sleepers rust rapidly, which, owing to the thin metal used, soon causes a serious deterioration in their strength, prohibiting their use under such conditions; whereas cast iron is far less readily attacked. Steel sleepers are less easily packed and replaced than wooden sleepers; and where steel sleepers have been injured by a train running off the line, it has been found very troublesome to remove them and restore the track. As such accidents are more liable to occur at points and crossings than elsewhere, the use of wooden sleepers at these places has been advocated, as much easier to replace when damaged, on railways where metal sleepers are otherwise used.[2]

Wrought-iron and steel cross sleepers have been tried upon a great number of railways in different parts of the world, and extensively adopted in countries where the conditions are unfavourable for wooden sleepers, such as Africa, India, Mexico, and Brazil; and it is probable that their use will extend with improvements in their form, the cheapening of their manufacture, and the gradual reduction in the available supplies of timber, coupled with an increasing demand for sleepers with the extension of railways. Similar reasons have called attention to the use of concrete in sleepers, combined with steel in various ways and in an experimental stage. The advisability of introducing steel sleepers on a railway must depend upon the climatic conditions, the cost of timber suitable for sleepers, the expenses of maintenance, and the period of renewals. Unless steel sleepers can be obtained of such a

[1] *Proceedings Inst. C.E.*, vol. lxvii. p. 3.
[2] *Ibid.*, vol. cxvii. p. 313.

quality as to make the cost of their maintenance notably less than that of wood sleepers, and their life considerably longer, then, so long as wood sleepers remain at their present prices, the greater first cost of

STEEL SLEEPERS.

Fig. 153.—Midland Railway.

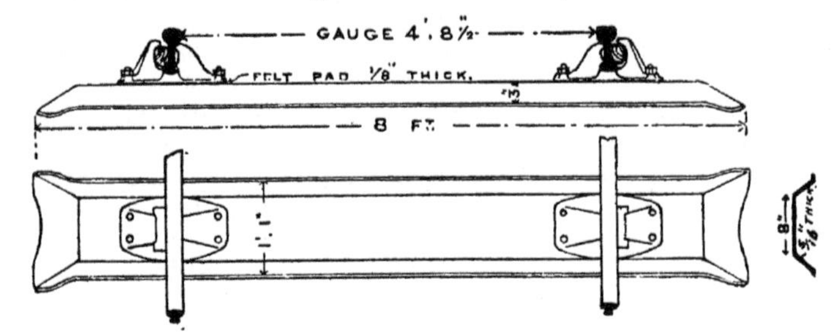

Fig. 154.—Simplon Railway.

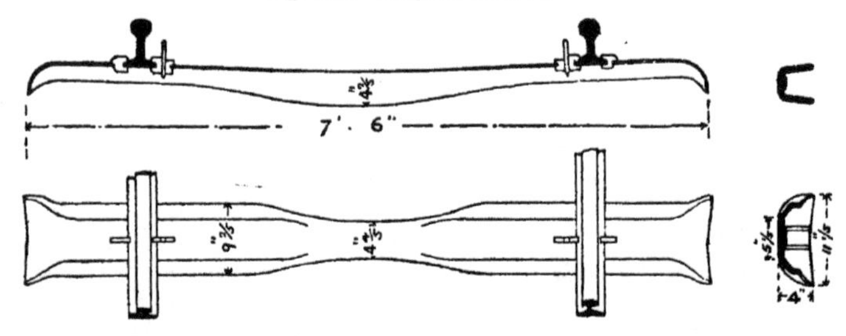

Fig. 155.—New York Central and Hudson River Railway.

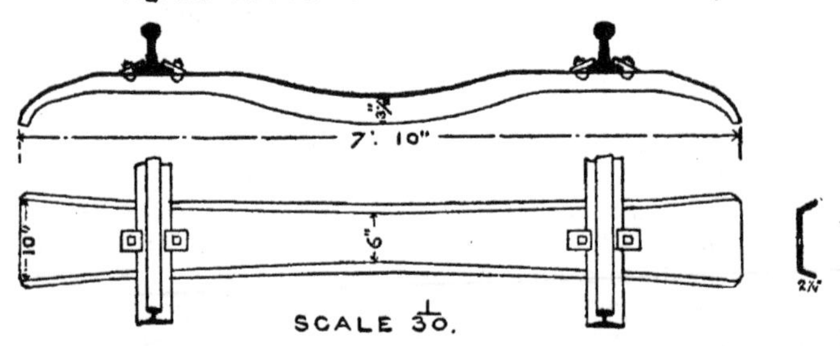

SCALE 1/30.

steel sleepers will prevent their superseding wood sleepers where the conditions are not unfavourable for timber.

Rails.—Four forms of rails have been extensively used, namely, the double-headed rail, the bridge rail, the flat-bottomed or flange rail, and the bull-headed rail, though only the last two are now in very general use for railways.

The double-headed rail (Fig. 151, p. 252, and Fig. 153) was originally adopted with the idea that when the surface of the one head had been worn down by the traffic, by turning the rail upside down in the cast-iron chairs which supported it, a new surface would be provided by the other head, thereby doubling the life of the rail. In spite, however, of various precautions, the surface of the lower head always became indented at its supports, being hammered by the passing trains against the chairs, so that the turned rail presented an uneven surface, jarring to the trains and subject to rapid wear; in addition to which, the reversal of the strains in the two heads causes a rapid deterioration; and, accordingly, the turning of the rails has had to be abandoned.

The bridge rail (Fig. 149, p. 250), so called from its resemblance in section to a bridge, was very convenient with the longitudinal sleepers suited to the broad gauge of the Great Western Railway; but with the disappearance of the broad gauge, and the discontinuance of longitudinal sleepers, the advantage of the bridge rail has ceased, except in such special cases as railways with a single line laid in a cast-iron tubular tunnel, like the Central London Railway, where longitudinal sleepers are more conveniently supported on the tube than cross sleepers.

Flat-bottomed, or flange rails (Fig. 156) possess the advantages of being fastened directly to the sleepers without the intervention of chairs, and of possessing great lateral stiffness, as well as having considerable vertical strength, though somewhat less than the deeper double-headed rails. This form of rail, being more economical than a rail supported on chairs, has been very extensively adopted, being the ordinary form in America and the colonies, and on most foreign railways, with the exception of the older lines in Europe.

FLANGE RAIL WITH FASTENINGS.
Fig. 156.

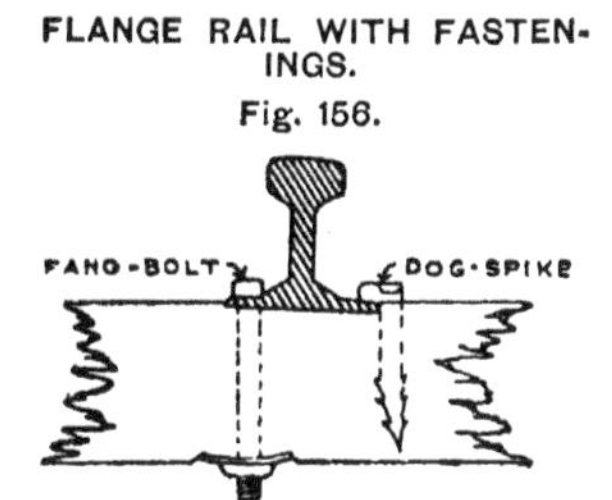

The bull-headed rail, with its lower portion made smaller than the working head of the rail (Fig. 157), has superseded the double-headed

BULL-HEADED RAIL WITH CHAIR.
Fig. 157.—90 lb. Rail.

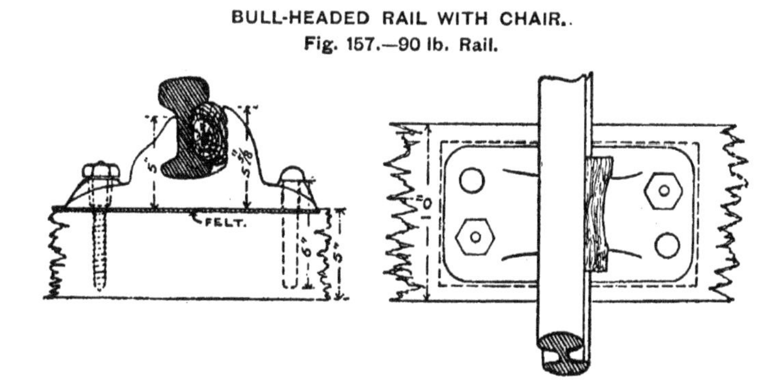

rail on English railways; for the object of making the two heads alike ceased as soon as it became recognized that the rail could not be advantageously turned; and the additional section required for the heavier rails which have been brought into use, has been concentrated on the upper head, which has to bear the wear and tear of the traffic.

Rails which were formerly made of wrought iron, have been made of steel since the cheapening of its manufacture by the Bessemer process, which has increased their life, notwithstanding the increase in the weight of the engines and carriages, and in the number of the trains on most lines, and the great friction imposed on the rails by the introduction of continuous brakes. The upper surface of the head of rails is now generally made somewhat flat, to afford a larger bearing surface for the wheels than on the rounder surface formerly provided; but the actual shape of the rails is varied somewhat according to the special predilections of the different railway companies, who have their rails rolled to a particular standard template, in lengths generally of between 24 and 30 feet, though occasionally reaching 60 feet. The weight of rails is commonly stated in pounds per lineal yard; and whereas in 1860–70, the standard weight of rails for railways in England was 75 lbs. a yard, the weight has been gradually increased, with the increased weight on the driving-wheels of locomotives and the augmentation in the traffic, up to 90 lbs.; and rails of 100 lbs. per yard and over, have been used for very heavy traffic.

Fastenings for Rails.—Double-headed and bull-headed rails have to be supported in cast-iron chairs, into which they are wedged by hard-wood keys (Fig. 157). The large chairs now used on the principal English main lines, weighing from 44 to 50 lbs., afford a bearing on the sleepers of from 107 to 112 square inches, as compared with the 50 to 60 square inches of bearing of flange rails; and these chairs with their wide base transversely to the rail, give it considerable rigidity sideways. The keys or wedges which fix the rail in the chairs, are generally driven in on the outer side of the rail, so that the rail, when pushed outwards by the flanges of the wheels of a passing train, press against the wooden keys and keep them in place; whilst the wood forms a more elastic cushion than the cast-iron jaw of the chair. Occasionally, however, the keys are placed on the inner side of the rail, an arrangement which, though more convenient for the platelayers, who can inspect the wedges of both rails when walking along inside the track, forms a less easy road, and leads to the loosening of the keys by the passage of the trains.

The spikes, screws, and bolts by which the chairs and flange rails are attached to the sleepers, are more liable to work loose with a flange rail, and the sleepers to be injured by the pressure of the flanges upon them, than when wide chairs, with a large bearing surface at their base, are employed. Accordingly, bull-headed rails with chairs furnish the strongest and most durable track for a very large and heavy traffic; but the flange rail, fastened direct to the sleepers, merely by dog-spikes, and by fang-bolts near each end in many cases, but preferably by a fang-bolt and dog-spike to each sleeper on opposite sides alternately (Fig. 156), has been preferred under all other conditions, on account of its smaller cost and the rapidity with which it can be laid; whilst sleepers put closer together where timber is cheap, and the addition of bed-plates under the flange rails at the sleepers, render a track laid with flange rails almost as strong and durable as a chair road.

The rails of a line of way are given a cant inwards of about 1 in 20 (Fig. 151 to 154, pp. 252 and 254), so that the tyres of the wheels, made slightly conical with the object of facilitating the travel round curves, may have as wide a bearing on the rails as possible. With double-headed and bull-headed rails, this cant is effected by the shape given to the chairs (Fig. 157, p. 255); and with flange rails, it is accomplished by adzing the top of the sleeper at the rail-bed to the proper inward slant (Fig. 156, p. 255), or by inclining the surface of the bed-plate or steel sleeper on which the rail rests (Fig. 154, p. 254).

Connection of Rails.—Adjacent rails are always connected together by means of a pair of fish-plates, on the two sides, at the ends of the rails, bolted together by four fish-bolts passing in pairs through holes

FISH-PLATES WITH DOUBLE-HEADED RAIL.

Fig. 158.

FISH-PLATES WITH 90-LB. FLANGE RAIL.

Fig. 159.

in each rail. Formerly the fish-plates were always inserted in the hollow between the upper and lower part of the rails on each side (Fig. 158); but on lines with heavy traffic, where the weight of the rails has been increased, the fish-plates have also been strengthened by being made deeper, as shown on Figs. 159 and 160, making the joint much stronger

DEEP FISH-PLATES WITH BULL-HEADED RAIL.

Fig. 160.

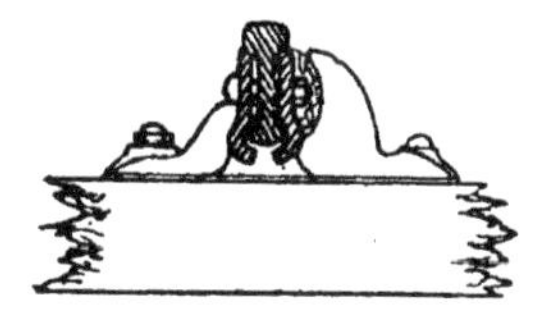

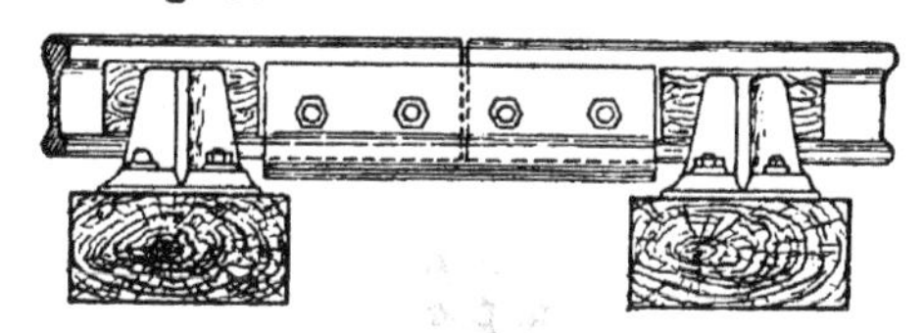

than formerly. A small space has to be left between adjacent rails to allow for expansion by heat, the proper width of the interval depending upon the length of the rails, the temperature at the time of laying, and the extreme range of temperature in the locality. Rails 30 feet long, laid in winter at a temperature of 100° F. below the possible maximum, would require a space of slightly over $\frac{1}{4}$ inch to be left between them. The holes in the rails for the fish-bolts are made slightly elliptical to allow of the movements of expansion and contraction of the rails.

Provisions for Safety on Curves.—The tendency of the wheels of a train, on entering a curve at high speed, to retain their straight course, and therefore to mount the outer rail of the curve, and to leave the inner rail, has to be provided against round sharp curves. The tendency to mount the outer rail is guarded against by elevating this rail above the other, in proportion to the sharpness of the curve, the width of the gauge, and the maximum speed at which trains are intended to travel round it, so that the vehicles are somewhat tilted inwards and thus run more easily round the curve, just as a skater leans over, when going on the outside edge, in proportion to the sharpness of the curve he is cutting on the ice. The formula commonly adopted in practice for finding the suitable super-elevation of the outer rail is—

$$\left.\begin{array}{l}\text{Super-elevation of outer}\\ \text{rail in inches}\end{array}\right\} = \text{gauge in feet}\ \frac{(\text{velocity in miles per hour})^2}{1{\cdot}25\ \text{radius of curve in feet}}.$$

On very sharp curves, a guard or check rail is laid close alongside the inner edge of the inner rail, leaving only just sufficient space between for the flanges of the wheels to pass freely round. This check rail prevents the wheels of the vehicles from leaving the inner rail, and thereby assists the super-elevation of the outer rail in keeping the wheels from mounting the outer rail. With flat-bottomed or bull-headed rails, the check rail is laid in special double chairs carrying the inner rail; and with flange rails, an angle-iron is commonly bolted on to the sleepers, with its vertical portion close alongside the inner rail to serve as the check rail.

In laying out very sharp curves, it is advantageous to increase the curvature gradually from the straight line, so that the elevation of the outer rail, which must be effected gradually, may be proportional to the curvature, and be completed by the time the full curvature is attained. Moreover, to ensure the provision of the proper super-elevation, it is essential to introduce a short piece of straight line between two sharp reverse curves. The long, rigid wheel-base of the older class of rolling-stock is liable to widen out the gauge on sharp curves; and iron tie-bars are, accordingly, sometimes introduced to maintain the gauge on very sharp curves. The use of pivoted bogies, so universal on the long American cars, by substituting a short wheel-base near each end of the car, in place of the long, rigid wheel-base of the ordinary European railway carriages, greatly aids in making a train run easily round sharp curves, so that in most cases the use of check rails has been dispensed with on the curves contouring the spurs of the Selkirk range on the Canadian Pacific Railway.[1]

JUNCTIONS.

At junctions, where a branch line has to diverge from the main line, a curve has to be formed at the commencement of the branch line to produce the required divergence; special rails have to be introduced with pointed extremities, termed switches or points, to turn the train on

[1] For an important discussion on the whole subject, see *Proceedings Inst. C.E.*, vol. clxxvi., 1909.

to the branch line; and contrivances, known as crossings, have to be laid down to enable the flanges of the wheels of the branch train to pass across the rails of the main line, through narrow gaps left in them, without leaving the line of way on which the train is intended to travel. The main-line trains also have to pass the rails of the branch line which go across the main line, so that special precautions have to be taken to prevent the trains at a junction leaving the rails, or following the wrong line, or splitting so that portions of the same train get on to different lines.

Switches and Crossings.—The simplest form of junction is shown on Fig. 161, where a single line of railway is represented branching off from another single line. The switches are two inner rails

SINGLE-LINE RAILWAY JUNCTION.

Fig. 161.

tapering in width to a knife-edge, connected together by a tie-bar, and so arranged and pivoted, that when one switch is moved close against its adjacent rail at its narrow end, the other switch is moved out from its rail about four inches, so as to leave an ample space for the flanges of the wheels travelling on the outer rail to clear its point. By this means, a train pursues its course along the main line, or is diverted to the branch line, according to the position of the switches. To a train travelling from right to left in Fig. 161, the switches are facing points; but to a train going in the opposite direction, they are trailing points. In the latter case, the train merely forces open the switch to its left if the switches are wrongly placed, unless they are locked; but a train getting on the wrong side of facing points not placed properly, or not opened sufficiently, proceeds on the wrong line. The splitting of a train in two has been caused by the switches being moved during the passage of the train. Facing points, accordingly, are a source of danger; but ingenious contrivances have been devised to prevent accidents. The exact position of the switches has been secured by making holes in the connecting bar, through which a bolt can only be passed when the switches are fully opened or fully closed; and till this bolt has locked the switches in their precise position, the signals cannot be lowered for the passage of a train. The splitting of a train by the inadvertence of a signalman, has been prevented by causing the same operation as the unbolting of the switches, to raise a rail on rockers on the inside of one of the rails, this rail being made longer than the greatest interval between two pairs of wheels of a train; and when a train is passing, the rail in being raised comes in contact with the flanges of the wheels, so that the operation cannot be completed, nor the switches set free to move, till the train has passed.

The manner in which the necessary gap in the rails is arranged for the passage of the flanges of the wheels across them at the crossing **C**, is shown on Fig. 161, p. 259, where the two converging rails from the switches spread out at their extremities on each side of the point, from which the diverging rails of the two lines start beyond the gap; and the passage of the flanges of the wheels across the gap is controlled by check rails alongside the outer rails on the two sides, which by keeping the outer flanges close to the outer rail, prevent the inner flanges from straying from their proper rail in passing the gap.

Forms of Junctions.—The ordinary form of junction with a main line is shown on Fig. 162, where both the main and branch lines are double. This involves a greater number of crossings than the single-line junction, but the principle is precisely the same. Such an arrangement, however, necessitates the blocking of the main line when

DOUBLE-LINE RAILWAY JUNCTION.

Fig. 162.

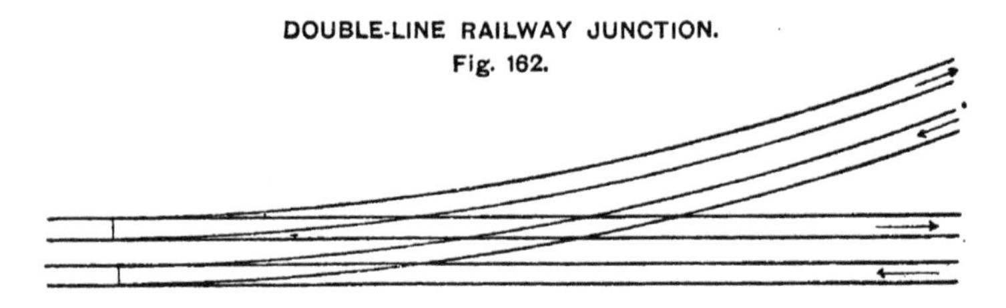

a branch train is entering or leaving the junction. Moreover, the trains along the branch line must travel slowly round the curve till they are clear of the main line, as no elevation of the outer rail of the curve is possible in crossing the main line.

Where the traffic on the main line is very large, the passage of the

BRANCH LINE CARRIED UNDER MAIN LINES.

Fig. 163.

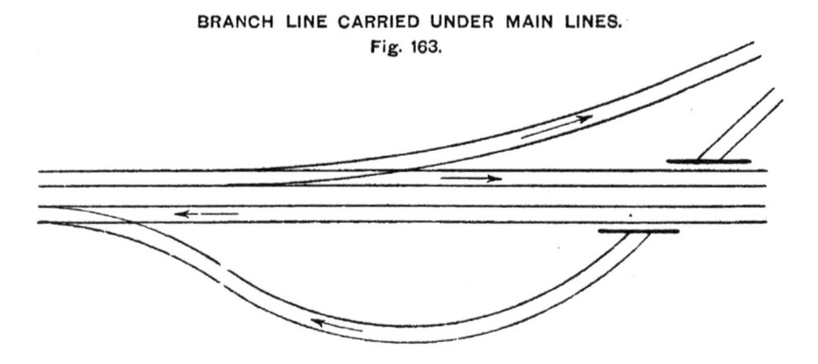

outer branch line right across the main lines is sometimes avoided by the arrangement shown in Fig. 163, where the far branch line is made to diverge on its own side, and then by curving round and passing by means of a bridge under or over the main line, comes eventually alongside the other branch line beyond the junction. This course, though involving the cost of a bridge and some additional land, reduces the crossings of the rails, in the case of double lines, from six to two, and considerably expedites the traffic; and the system is especially valuable where, with widened main lines, the two additional central lines for the fast through trains would otherwise have to be crossed by the outer branch line..

STATIONS.

Stations should be provided at convenient intervals apart for the district the railway is intended to serve, and especially at the most suitable places for giving the resources of the locality access to a market, and attracting the population. The best positions also for stations should be given some weight in deciding upon the course of a railway and its level at these places.

Conditions relating to Stations.—A station should be placed as near a main road as possible, and also as close as practicable to the town or village it is designed to accommodate. It is also expedient that the line should be level at a station; whilst the most advantageous arrangement for the approaches is where easy gradients fall away on each side, assisting the starting train and pulling up the incoming one; and a position with gradients descending to the station on each side, should be avoided. A station should, moreover, be situated where the railway is close to the natural surface, for stations in cuttings or on embankments are inconvenient, costly, and difficult of access; and curves should be avoided in the neighbourhood of stations where feasible, as an uninterrupted view for some distance on both sides is of value in view of the security and regulation of the traffic.

Forms of Stations.—A station has to be adapted in size to the traffic it has to accommodate. A small, intermediate, roadside station on a railway of single line, is only provided with a booking office, which serves also as a waiting-room, and the necessary adjuncts, a platform on one side, and a short siding, and also with a second line in suitable places for trains going in opposite directions to pass. On a double line, there is a platform on each side; but only a small shelter is provided on the second platform. At the larger intermediate stations, two additional lines are often provided at the sides, so as to separate the slow and quick traffic, the slower, stopping trains being diverted on to the side lines alongside the platforms, whilst the quick trains pass unimpeded along the two central lines. Frequently also in such cases, island platforms are placed in the centre between the two main lines, on which the passengers by the quick trains can occasionally alight, or from which they can embark, access to the central platform being obtained by an overhead footbridge or a subway. These larger stations, also, are provided with improved accommodation, and more ample sidings, in proportion to their importance; and the platforms are sheltered by an overhanging roof along the central portion of their length. A cross-over road is, moreover, introduced between each pair of rails, to transfer carriages or waggons from one line to another (Fig. 164).

Stations on the continent are generally devoid of platforms, which is very inconvenient for getting into or out of the ordinary European form of high railway carriages, with their primitive means of access; and under such conditions, the system of steps furnished at each end of the American cars, carried even lower down, should be resorted to.

CROSS-OVER ROAD.

Fig. 164.

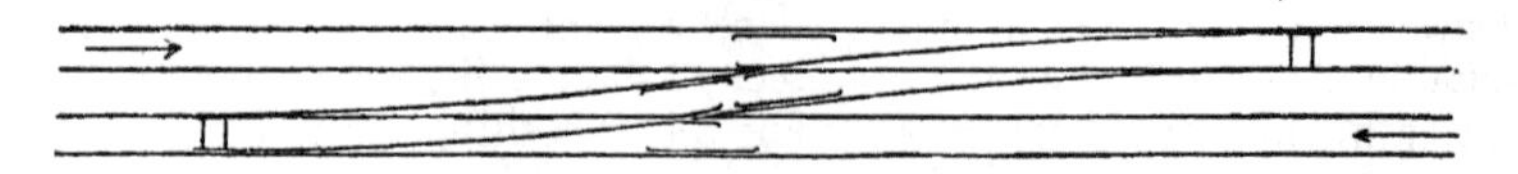

Terminal stations should be placed close to, but not immediately abutting on, great central thoroughfares, and also where land for extensions can be obtained without difficulty; and owing to the enhancement of the value of land by its propinquity to a station, an ample area for future enlargements should be secured alongside all important stations, if possible, at the outset.

An old form of large terminal station consisted of a long setting-down roadway on one side, separated from a long departure platform, parallel to it, by booking-offices, luggage, waiting, and refreshment rooms, and the other necessary conveniences; and on the other side, a long arrival platform, with a roadway for cabs and carriages adjoining. This arrangement was very convenient for long-distance trains, and so long as the traffic was not too great for all the trains to start from the one platform and arrive at the other. With the development, however, of suburban traffic, an increase in the number of lines for trains to start from and arrive at, with platforms alongside, became imperative; and, accordingly, central lines and platforms have been added, reached from the end of the station, or by a subway, bridge, or movable platform; or lines have been introduced at the side, towards the outer end of the station, for the local traffic, of which Victoria, Paddington, and Liverpool Street stations in London are instances. The other common arrangement of terminal stations consists in placing the booking and other offices at the entrance end of the station, with an approach yard in front; and after passing through the booking offices to an open space beyond, access is gained to the several platforms from which the express and local trains start. This system is exemplified by Charing Cross, Cannon Street, Fenchurch Street, and the North and South Waterloo stations in London, and by numerous stations on the Continent, as for instance Frankfort on the Main, where the luggage is kept to a great extent out of the way of the passengers by subways and lifts. Sidings on which main-line trains can be made up and wait till their platform is clear, should be provided as near the station as practicable; but local trains often start again from the same platform soon after their arrival, necessitating merely a siding for the locomotive.

At a large terminus, the goods station is distinct from the passenger station, being usually placed where an ample space can be provided for sidings for sorting the trucks and marshalling the trains, and for the various appliances and conveniences for the unloading, storage, and despatch of goods.

CHAPTER XVI.

LIGHT RAILWAYS.

Definition of light railways—Conditions necessary for light railways—Gauge, considerations affecting it; cost of railways having different gauges, varieties adopted; advantages of narrow gauge—Laying out light railways, considerations affecting route; object of Light Railway Act of 1896—Cost of ordinary and light railways compared; cost of construction in various countries; economy of narrow-gauge lines in India, Belgium, and Sweden; cost of light railways in the United Kingdom, instances of cheap lines—Light railways in the United Kingdom, effect of Act of 1896; gauges and lengths of standard and narrow-gauge lines—Mono-rail railways—Lartigue system, Listowel and Ballybunion; suspended cars on elevated rail—Light railways in Belgium, arrangements for construction, gauges, limiting curves and speeds, advantages—Light railways in India, lengths of standard and narrow-gauge lines open and sanctioned; value of narrow gauge, unexpected development—Extent and progress of the railways of the world, in the different quarters of the globe; fields for future extension.

LIGHT railways are branch, or local lines passing through agricultural, sparsely populated, or undeveloped lands, which, owing to the very moderate prospects of traffic in the districts they traverse, must be constructed in a light, cheap manner, in order to afford them a chance of yielding a fair return on their capital cost. The expenditure on the land required for the line must necessarily be moderate, and can occasionally be entirely dispensed with, by laying the line along an unused strip at the side of the public highway. Light railways, moreover, should be relieved from the costly and onerous conditions of road bridges and safety appliances, imposed, in the interests of the public, on ordinary railways over which frequent trains run at a high speed.

Conditions necessary for Light Railways.—A light railway occupies a sort of intermediate position between an ordinary railway and transport by vehicles along the high-roads; and its object is to facilitate and expedite locomotion, and to economize the cost of carriage in outlying rural districts not served by ordinary railways.

The primary necessity for a light railway is cheapness of construction, for in the absence of State aid, such as is furnished by some of the European Governments, money can only be attracted with certainty and regularity to undertakings which offer good prospects of providing a reasonable rate of interest on the capital expended; and the moderate earnings of a local line must be counterbalanced by a compensating economy in the cost of the works. This economy can only be effected by using lighter rails, lighter structures where bridges or viaducts are indispensable, and level crossings in place of bridges wherever possible, by dispensing with fencing where practicable, by utilizing the public roads where convenient, and by avoiding all unnecessary expenditure on stations and safety appliances. To obtain these concessions, it is necessary to reduce the limiting load upon a pair of wheels, and consequently to employ lighter locomotives, and also to limit the speed of the trains.

Gauge of Light Railways.—A light railway is not necessarily a narrow-gauge railway, though so far as reduction in gauge diminishes cost of construction, it is an advantage in this respect to adopt a narrow gauge. The question of gauge really depends primarily on the position of the light railway in regard to standard lines, the character of the country, and the length of the line. If a light railway can be readily connected with a standard line, if the country which it traverses is flat, and the length of the line is short, it is expedient to lay it to the standard gauge so as to avoid transhipment, which, under the conditions of a short line and flat country, would compensate to a large extent for the larger cost of the wide gauge. When, on the contrary, the country is rugged and the line long, a narrow gauge should be adopted, as the reduction in first cost far more than compensates for the expense of transhipment. Also for lines passing through poor districts, a small first cost is much more important than saving of transhipment.

The following table, published by the International Railway Congress, indicates how the gauge affects the cost of railways, of similar character, under different conditions of country traversed :—

COST PER MILE.

Nature of country.	Standard gauge, 4 feet 8½ inches.	Metre gauge, 3 feet 3⅜ inches.	Narrow gauge, 2 feet 6 inches
	£ £	£ £	£ £
Level	2,392 to 3,987	1,595 to 2,551	1,260 to 2,057
Slightly undulating .	3,652 to 5,582	2,392 to 3,987	1,595 to 2,392
Very undulating . .	4,785 to 7,179	3,652 to 4,785	2,057 to 3,190
Slightly hilly . . .	6,380 to 9,571	3,987 to 5,582	2,392 to 3,987
Very hilly . . .	8,742 to 11,163	4,785 to 7,178	3,652 to 5,582
Slightly mountainous	10,365 to 12,760	6,380 to 8,774	4,785 to 6,380
Very mountainous .	11,963 to 15,950	7,977 to 11,164	5,183 to 7,976

According to this table, the cost of a railway with a narrow gauge of 2½ feet, which is rather more than half the cost of a railway of standard gauge across level country, is less than half the cost under the other conditions, and appears to be specially low through hilly country; but the figures given in the table do not indicate nearly the same remarkable economy of the 2½-feet gauge, as compared with broader gauges, estimated to have been realized in the construction of Indian railways, quoted on page 249.

There has been a considerable variety in the gauges adopted for light railways, as was the case in early days with ordinary railways, at least twelve different narrow gauges having been used for light lines, in addition to the standard gauge, ranging from 3 feet 6 inches down to 1 foot 8 inches. The narrow gauges most commonly employed have been 3 feet 6 inches, the metre gauge 3 feet $3\frac{3}{8}$ inches, 3 feet, 2 feet 6 inches, and 2 feet. As, however, light railways are generally situated in quite distinct districts, without any prospect of intercommunication, there is not the same objection to differences of gauge for light railways, as for ordinary branch railways which are in connection with the main lines of the country. Nevertheless, there is no advantage in the multiplication of gauges with only minute differences in width; whilst economy results from adherence to one or two standard types of narrow gauge, and the adoption of a gauge admitting of an interchange of rolling-stock with a neighbouring light line to cope with a temporary emergency. The metre gauge so advantageously used for several of the secondary railways of India, is a notable reduction in width from the standard gauge of 5 feet 6 inches; but gauges of 2 feet 6 inches and 2 feet have also been resorted to in India for the distinctly light lines of the country. Moreover, a gauge of 2 feet 6 inches appears a more suitable reduction for narrow-gauge lines in a country where the standard gauge is 4 feet $8\frac{1}{2}$ inches; whilst a gauge of 2 feet might be adopted, under exceptional conditions, in mountainous districts.

The selection of a narrow gauge for a light railway possesses the important advantages, that with a narrow gauge, combined with a limited speed, sharp curves can be introduced, greatly reducing the cost of the earthworks in hilly country, and enabling the line to be adapted to the sharp turns of an ordinary road; that the railway takes up a smaller width, and therefore can be more easily laid along the high-road; that a light rolling-stock, suited in size to the particular requirements of the district and a small, varied traffic, can be provided, thereby reducing considerably the loads to be hauled, and the proportion of the dead load to the paying load; and that portable branch sidings can be laid across the fields right up to farms in the neighbourhood of the line, along which the farmers can readily draw their crops on to the railway, whereby the cost of carriage of agricultural produce is materially reduced. Moreover, a narrow gauge secures a light railway from the liability of having its rails and bridges overweighted, by the heavy locomotives and rolling-stock of the main line being run over it.

Laying out Light Railways.—The objects of a light railway are to serve the district through which it passes as efficiently as possible, and to effect this purpose with the smallest expenditure practicable. The selection of a fairly direct route between the two extremities of the line, required in connecting large centres of trade, must be abandoned; and the route must be chosen with the view of collecting as much traffic as possible along the line, and with strict regard to economy in construction. The line should be made as nearly a surface line as possible, by the aid of sharp curves, so as to avoid a large expense in earthworks, and to cross roads on the level. By following the line of the main roads, the ordinary route of the traffic of the locality is secured, and the purchase of land is avoided. Moreover, where it is necessary to diverge from the high-road, since the railway is designed to benefit the district, the necessary land should be sold at its ordinary market value. A light railway in fact, instead of being treated as an interloper, and a fair object to get money out of, should be regarded as a friend to be assisted in every possible way. If light railways are to be largely extended in the United Kingdom, they must be promoted and aided by the inhabitants and landowners of the district they are designed to accommodate, thereby interesting the localities they pass through in their financial success, and avoiding all unnecessary expenditure. The Light Railways Act of 1896 was passed in order to simplify and cheapen the procedure for authorizing the construction of light railways, and, under certain conditions, to dispense with onerous restrictions with regard to their construction and working, as the light railways constructed in Great Britain and Ireland previously to that time had, for the most part, cost too much to offer a prospect of yielding a reasonable return on the capital expended; and therefore there was no inducement, under the existing circumstances, to invest capital in the extension of such lines. The greatest measure, however, of economy can be attained by a judicious laying out of the line, so as to avoid heavy works, whilst selecting the most convenient route for the district.

Cost of Ordinary and Light Railways compared.—The capital expended in the construction of more than half of the mileage of the railways of the world has averaged £11,832 per mile; but the average in different countries has ranged between somewhat wide limits. Thus in Europe, the United Kingdom heads the list, with a capital expenditure on the construction of railways of £48,393 per mile; and of the large European countries, France comes next with £25,232 per mile; Germany, £20,275; Spain, £18,338; some of the railways of Italy, £18,443; Austro-Hungary, £17,554; and Russia, £14,741. Norway and Sweden possess the cheapest railway system in Europe, with an expenditure of only £7111 and £5903 respectively per mile;[1] whilst in the smaller European countries, the State railways of Holland and Belgium, and the railways of Switzerland, have been the most costly, reaching £39,768, £26,245, and £21,059 per mile respectively. In India, which contains the greater part of the railway mileage of Asia, the expenditure on the whole of the railways opened, up to the end of 1899, amounted to £8068 per mile (taking the rupee at its settled standard value of 1*s.* 4*d.*), the cost being greatly reduced by the economy effected by the metre and other narrow-gauge lines.[2] In North America, the railways of the United States have cost £12,221 per mile, and in Canada £11,595 per mile; whilst in South America,

[1] *International Railway Congress, Bulletin*, vol. xiii., 1899, p. 1290.

[2] "Administration Report on the Railways of India for 1899–1900," Simla, 1900, part 1, p. 9.

the expenditure per mile has been £36,860 in Brazil, £10,555 in Argentina, and only £8285 in Chili. In Africa, the railways of the Cape Colony have cost £10,369 per mile; and in Australasia, the cost of the railways per mile indicates considerable differences in the various colonies, being £14,443 in New South Wales, £12,496 in Victoria, £8580 in Tasmania, £7895 in New Zealand, £7314 in South Australia, £7210 in Queensland, and only £3937 in West Australia.[1] In comparing, however, these expenditures on railways, it must be borne in mind, that most of the railways in England have a double line, and on most of the main lines for some distance out of London, four lines of way; that, unlike several railway systems on the Continent, almost the whole of the railways of the United Kingdom have been made heavy lines of the standard gauge; that the main lines have had to be made suitable for a rapid and heavy traffic; and that in a thickly populated country, the price of land has been high, and parliamentary expenses exceptionally heavy. The cost of railways in Norway, New Zealand, and South Australia has been reduced by the adoption of light lines for extensions, with a gauge of 3 feet 6 inches; whilst the use of this gauge for the railways of Cape Colony, Queensland, West Australia, and Tasmania, has made the cost of these railway systems lower than the average.

The narrow-gauge railways of West Australia alone, have been constructed within the limits of cost suitable for light railways with a very moderate traffic; but a comparison of the total expenditure on construction in India per mile of line open, on the standard lines, the metre-gauge lines, and the narrow-gauge lines, amounting to £10,388, £4848, and £2158 respectively, or per mile of single track, £8014, £4283, and £2041, shows what a great economy can be effected in construction by reducing the gauge and modifying the character of the railway. The standard railways of Belgium have cost about £26,245 per mile, about the same as in France; but the metre-gauge light railways, laid with 43-lb. rails, have been constructed for about £2700 per mile. The State Railways of Sweden, with a length of 2031 miles, cost £8215 per mile; whilst the 3866 miles of light railways constructed by private companies, portions of which are laid to narrow gauges ranging between 4 feet and 1 foot 11½ inches, have only cost £4705 per mile; and some of the lines laid with 35-lb. rails to a narrow gauge, have only cost about £2000 per mile.

Individual lines necessarily vary considerably in their cost of construction, according to the character of the country, the kind of traffic that has to be provided for, and the terms upon which the capital is raised, so that occasionally a nominally light, or narrow-gauge railway costs more than a standard railway constructed under more favourable conditions. Thus the Festiniog narrow-gauge railway, 14 miles long, has involved a capital expenditure of about £10,800 per mile; but it carries a very heavy traffic from the Welsh slate quarries, and passing through a mountainous district, it has been solidly built and laid with 50-lb.

[1] *International Railway Congress, Bulletin*, vol. xiii., 1899, p. 1291.

rails, and it is very well equipped. The cost, however, of the Southwold Railway, 9 miles long, with a 3-feet gauge, through flat country, and devoid of mineral traffic except coals for local consumption, which amounted to £8500 per mile, must be attributed to financial operations and the onerous regulations of the Board of Trade. Very few, indeed, of the light and narrow-gauge railways carried out in Great Britain before 1896, have been constructed for a capital expenditure of less than £5000 a mile, and only four in Ireland.[1] The cheapest of these are the Corris Railway in North Wales, 11 miles long, with a gauge of 2 feet 3 inches, which cost only £1814 per mile; the Talyllyn Railway, 6½ miles long, with the same gauge, which cost £2143 per mile; the Torrington and Marland Railway in Devonshire, 8 miles long, with a 3-feet gauge, which cost £2500 per mile; the Clogher Valley Railway, 37 miles long, with a 3-feet gauge, costing £3286 per mile; and the mono-rail Listowel and Ballybunion Railway, 9 miles long, costing £3666 per mile. The light railway also from Barnstaple to Lynton, with a gauge of 1 foot 11½ inches, opened in 1898, cost about £2614 per mile.

Though the secondary and light railways of France, carried out by State aid, have for the most part proved costly undertakings, some narrow-gauge lines have been constructed cheaply, as for instance the Caen, Dives, and Luc-sur-mer Railway, 23½ miles long, having a 2-feet gauge laid with 30-lb. rails riveted to steel sleepers, which cost £2143 per mile, and the Pithiviers-Torcy Railway, 19 miles long, and laid along the side of the high-road to a 2-feet gauge, but with only 19-lb. rails, which cost £1551 per mile.[2]

The Anglo-Chilian Nitrate Railway from Tocopilla to Toco, a distance of 55 miles, through very rough country, laid with 40-lb. rails to a gauge of 3 feet 6 inches, with a ruling gradient of 1 in 25 and curves of 181 feet radius, and rising with the aid of loops and switch-backs, 4847 feet in a distance of 33¾ miles, cost approximately £3807 per mile, a moderate expenditure through such difficult country.[3]

Light Railways in the United Kingdom.—The extension of light railways in the United Kingdom has been very greatly restricted by the large cost of most of the light lines constructed under the regulations in force previously to 1896; but the passing of the Light Railways Act in that year has greatly encouraged schemes for their construction, several of which have been authorized. The estimated cost, however, of many of these lines appears to be hardly low enough, except under specially favourable conditions of traffic, to ensure the financial success needed to encourage the further extension of these railways, so as to develop fully the resources of the country, and provide as cheap carriage for goods and produce as is secured for some of the continental countries of Europe by the fostering care of their Governments.

[1] "Light Railways for the United Kingdom, India, and the Colonies," J. C. Mackay, pp. 96 and 106.

[2] *Le Génie Civil*, 1894, vol. xxv. p. 169.

[3] *Proceedings Inst. C.E.*, vol. cxv. pp. 327 and 331.

Though a departure from the standard gauge has generally been discouraged in Great Britain, especially by the principal railway companies, and the broad-gauge lines of the Great Western Railway were converted to the standard gauge in 1892, whilst some of the light railways in England and Ireland have been laid to the ordinary gauge, nevertheless various gauges have been adopted for some of the short light lines in England and Wales, as shown by the accompanying table,[1] giving the gauges of the railways, with their lengths, in the United Kingdom in 1900:—

	ft. in.	ft. in.	ft. in.	ft. in.	ft. in.	ft. in.	ft. in.	ft. in.	ft. in.	ft. in.	ft. in.
Gauge	1 11½	2 3	2 4	2 4½	2 6	2 9	3 0	3 6	4 0	4 8½	5 3
Miles	50	18	3	9	4	7	419	7	21	18,541	2764

Seven miles, however, of a 4-feet gauge in Scotland, for the Glasgow District Railway, and 410 miles of 3-feet gauge in Ireland, are the only deviations from the standard gauges in those countries. The 3-feet gauge is by far the commonest narrow gauge in the United Kingdom, owing to its extensive adoption for the narrow-gauge ordinary (191 miles) and narrow-gauge light railways (219 miles) of Ireland; whilst amongst the short narrow-gauge railways in England and Wales, the gauge which has been most used is 1 feet 11½ inches, laid for a length of 50 miles.

Mono-rail Railway on the Lartigue System.—The single double-headed rail of the Lartigue System (Fig. 165) is supported on

MONO-RAIL RAILWAY.

Fig. 165.—Lartigue System.

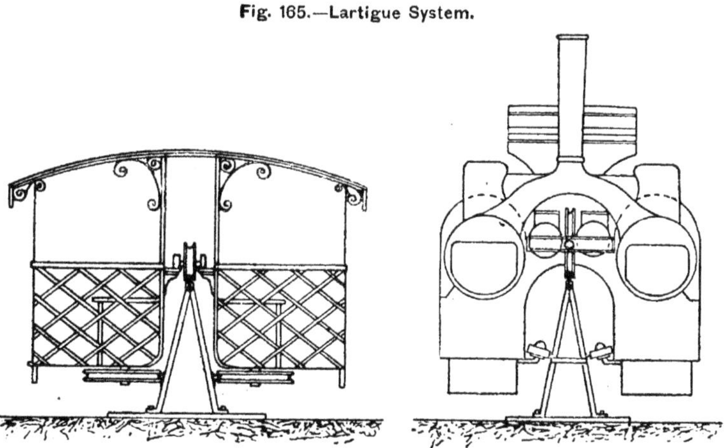

angle-iron trestles, about 3 feet above the ground; and these trestles, placed about 3 feet apart, carry a guide rail on each side, about 18

[1] "Railway Returns for England and Wales, Scotland, and Ireland, for the year 1900," London, 1901, pp. v, 78, 80, 82, and 84.

inches below the main rail, to maintain the vehicles in position. The carriages of the train extend downwards, like panniers, on each side of the trestles, being balanced on the central rail along which they travel.[1]

On the Listowel and Ballybunion Railway in Ireland, 9 miles long, opened in 1888, the trestles are fastened to steel sleepers resting on wooden sleepers; and a movable length of trestlework has to be shifted to enable the train to change its line, and has to be turned out of the way to open a passage for vehicles at a level crossing. This line, which has curves of only a chain radius, gradients up to 1 in 45, and cost £3666 per mile with rolling-stock, is worked by locomotives, having a boiler and funnel on each side of the central rail to maintain the balance; whilst the driver and stoker have to stand on opposite sides of the rail.[2] The system was first used as an elevated horse tramroad, for the transport of produce in Algeria, where surface rails are liable to be buried in drifts of sand; and since the successful application of locomotives for working the system in Ireland, short mono-rail lines have been constructed in France, Russia, and Peru.

The maximum ordinary speed attained on the Ballybunion line is only from 24 to 30 miles an hour; but in some trials with electric traction on an experimental line near Brussels in 1897, a maximum speed of about 90 miles an hour was reached, the limit that could be obtained with the power available. As a result of these experiments, the construction of a mono-rail line between Liverpool and Manchester has been authorized, to be worked by electricity, for providing an express passenger service with an estimated speed of about 110 miles an hour. A special advantage of the mono-rail system for running safely round curves at a very high speed is that, besides doing away with the difference in curvature of the two rails of an ordinary line, the centre of gravity of the train can be placed well below the rail on which the train runs, and thus, with the aid of the inner guide rail, secure the train from any danger of leaving the rail, though a considerable pressure is imposed on the inner guide rail by a train in rounding a sharp curve.

Suspended Cars on elevated Mono-rail.—A novel form of railway, completed in 1900, passing through Barmen, Elberfeld, and their suburbs, in Rhenish Prussia, 8¼ miles long, is carried on girders resting on elevated frames, placed about 95 feet apart, erected over the River Wupper for 6⅕ miles, whose course is followed by the railway for this distance, and over streets or roads for the remainder of its length, under which the cars travel, hung from trollies running on a single rail laid on the top of the girder, as shown by the illustrations[3] (Fig. 166). This railway is, accordingly, both an elevated line suitable for affording rapid means of communication through crowded towns, and also a new form of mono-rail railway, adapted for electric traction at a fairly quick speed round sharp curves, and in which the centre of gravity of the suspended cars is considerably below the elevated rail on which they

[1] *The Engineer*, vol lxii. p. 223.

[2] "Light Railways for the United Kingdom, India, and the Colonies," J. C. Mackay, p. 104, and plate xxii.

[3] *Engineering*, vol. lxix. pp. 412, 438, and 501.

run. The chief peculiarities of the line are the adoption of the course of the river for the line of the railway for a considerable length, and the suspension of the cars from a mono-rail. The maximum gradient is 1 in 22; the sharpest curves on the regular line have a radius of $4\frac{1}{2}$ chains; and the speed is limited to 31 miles an hour. The trains consist of two to four cars, each car being $37\frac{1}{2}$ feet long, $6\frac{1}{2}$ feet wide, and weighing 14 tons with its load of 50 passengers, suspended from two bogies 26 feet apart, each running on two tandem wheels, 35 inches in diameter, on the rail. Each car carries its own motor overhead; and

SUSPENDED MONO-RAIL RAILWAY.

Fig. 166.—Elberfeld and Barmen Railway, Prussia.

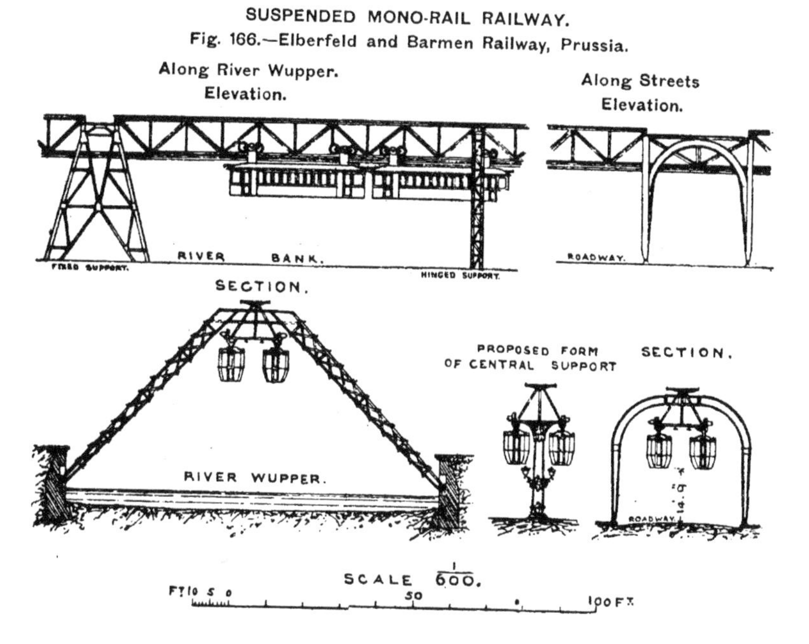

it leaves a clear headway of $14\frac{3}{4}$ feet above the roadway. The stations on this line are only about two-fifths of a mile apart; but the average speed, including stoppages, is about $20\frac{1}{2}$ miles an hour.

Light Railways in Belgium.—Narrow-gauge light railways have been more systematically developed in Belgium since 1885, considering its small area, than in any other country. These railways are constructed and worked by a special company, the necessary funds being subscribed by the State, the Provinces, and the Communes, aided to a small extent by private individuals. Of these railways, amounting altogether to about a thousand miles in length, more than three-fourths are laid to the metre gauge; and of the remainder, most have been laid to a gauge of 3 feet 6 inches, the gauge of the Dutch light railways, owing to their proximity to Holland, and the consequent prospect of their forming connections with its light lines; whilst small lengths of the

Belgian light railways have been laid to the standard gauge, where it was important to avoid transhipment.[1] They have been laid for the most part along the roads, though where the roads are paved, the expense is considerably increased; but in some cases land has been purchased for the purpose. The elevation of the outer rail for sharp curves on roads, is effected by lowering the inner rail as well as raising the outer one, so as to keep the centre of the track level with the road; but as the speed is limited to 18½ miles an hour in the country, where the minimum radius is 246 feet, and to 6⅕ miles an hour through towns, where sharp curves are often necessary, the elevation required is not excessive. These railways in Belgium, which are being continually extended, have proved on the whole financially successful; and they have been very valuable in conveying crops and market-garden produce to market, and thereby stimulating their production, as well as developing the other resources of the districts through which they pass, and furnishing a cheap means of transit for the population.

Light Railways in India.—In spite of the objections raised by so many engineers in 1873 to the break of gauge involved in the adoption of the metre gauge for the secondary railways of India,[2] these lines have been so extensively constructed since that period, that they seem likely before long to equal the standard-gauge railways in mileage. On the 31st December, 1907, there were 15,821 miles of railway open of the standard gauge, 12,613 miles of the metre gauge, and 1576 miles of special gauges; the additional railways constructed since 1900 being 2151 miles standard, 3117 miles metre, and 979 miles special gauge, respectively.[3] Considering the remarkable economy in construction effected by the adoption of the metre and other narrow gauges in India, as recorded on page 265, it is evident that if the Government of India had deferred to the arguments pressed upon them against a break of gauge, the railway system of India would have been of necessity greatly restricted, and some of the railways, such as the Darjeeling Railway, could not have been carried out. The financial results of railway construction in India, exhibit very clearly the value of supplementing the main standard lines of a country by light narrow-gauge lines, penetrating into the more sparsely populated and less accessible districts, and serving as very useful feeders of traffic to the main lines.

The statement made by Mr. Thornton of the India Office, in his Paper on "The State Railways of India" in 1873, that the construction of 10,000 miles of railways in India was contemplated by the Government, in addition to the 5000 miles at that time open or in progress, to complete the railway system of India, was so strongly contested in the discussion on the paper, that he eventually admitted that it was an overestimate, and that only 3000 miles of the 10,000 miles referred to had been actually marked out.[4] Nevertheless, in 1900, twenty-seven years

[1] *The Engineer*, vol. lxxix. pp. 259, 279, and 325; and vol. lxxxvi. p. 204.

[2] *Proceedings Inst. C.E.*, vol. xxxv. pp. 229 to 438.

[3] Administrative Report by the Railway Board, 1907.

[4] *Proceedings Inst. C.E.*, vol. xxxv. pp. 214, 217, and 452.

later, not only had an additional length of about 8670 miles of lines of standard gauge been constructed, and 814 miles more were in progress or sanctioned, but in place of the 10,000 miles of narrow-gauge railways which Mr. Thornton advocated, 10,093 miles had already been opened for traffic, and 2212 miles more were in progress or sanctioned.

Extent and Progress of the Railways of the World.—According to the bulletin of the International Railway Congress, the following are the miles of railways in the five quarters of the world, which were open for traffic at the end of 1897, the increase in these lengths during the four preceding years, and the percentage of the whole lengths this increase represents in each case[1]:—

Quarter of the globe.	Railways open in 1897.	Increase between 1893 and 1897.	Percentage of increase.
	Miles.	Miles.	Per cent.
Europe	163,514	15,328	10·3
America	236,364	12,143	5·4
Asia	30,922	6,692	27·6
Africa	9,910	2,218	28·7
Australasia . . .	14,300	1,128	8·5
Totals . . .	455,010	37,509	8·9

These figures show that, though America possesses considerably the largest railway mileage, the increase during recent years has been greater in Europe. Moreover, whilst the recent additions to the railways of Asia and Africa are large in proportion to their previously existing mileage, Asia with only about 10,600 miles of railways outside India, and Africa with barely 10,000 miles of railways, offer vast fields for future railway enterprise, considering the very extensive areas comprised in these two quarters of the globe, and the large natural resources in both these regions awaiting development.

A brief reference to the Capetown-Cairo railway must here, and for the present, suffice. This great railway, in progress of construction on the 3 feet 6 inches or South African gauge, reached Bulawayo, 1362 miles from Capetown, in 1897, and has since been carried forward in its northerly progress towards Lake Tanganyika as far as Broken Hill, 2017 miles from Capetown, and 375 miles north of the Zambesi Bridge, which was opened to traffic in 1905. (See p. 135.)

[1] *International Railway Congress, Bulletin*, vol. xiii., 1899, p. 1289.

CHAPTER XVII.

MOUNTAIN RAILWAYS.

Definition of mountain railways—Limit of gradient for adhesion alone—System adopted for steep inclines, central rail, rack, cable—Central-rail system; as designed for crossing the Mont Cenis; on the Cantagallo Railway; on the Rimutaka Incline, New Zealand—Ladder-rack railways: first rack incline; Mount Washington Railway; Vitznau-Rigi Railway; gauge, gradients, curves, and cost of various lines laid with ladder-rack—Double side-rack, on Pilatus Railway—Flat bar rack, thin single bar; Telfener thick single rack, on St. Ellero-Saltino Railway, advantages—Strub rack, on Jungfrau Railway—Abt rack, description, advantages; particulars of railway laid with Abt rack; instances of these railways worked by electricity—Racks on steep inclines for extension of railways, importance of system in conjunction with ordinary railways; steep incline with rack at Trincheras in Venezuela, and Usui Railway in Japan; rack on steep gradients of ordinary railways, instances of lines worked by adhesion and by rack; ratio of weight of locomotive to load drawn—Remarks on rack railways, superiority of Abt rack; central-rail and rack systems contrasted; value of rack for extending railways in mountainous districts, reference to Chamonix Railway.

Definition proposed for Mountain Railways.—Though railways passing through mountainous districts, with heavy works and steep gradients, such as the Brenner, the Mont Cenis, and the St. Gothard railways, are sometimes called mountain railways, the term "mountain railway" might with advantage be restricted to those lines on which other modes of traction have to be resorted to, beyond the mere adhesion of heavy locomotives to the rails, on account of the very steep gradients necessitated by an exceptionally rugged country, or the considerable heights which have to be surmounted in a limited distance.

A ruling gradient of 1 in 25 was adopted for the railway ascending from Vera Cruz to Mexico, rising nearly 8000 feet in the first hundred miles to reach the high central plateau, and for a branch line of the Denver and Rio Grande Railway, ascending 3675 feet in 25 miles to attain the summit of the Marshall Pass, having an altitude of 10,850 feet above sea-level; and the same ruling gradient was maintained for the Peruvian railways surmounting the steep Pacific slopes of the Andes, more especially the Oroya Railway which mounts from the seacoast at

Callao, the port of Lima, to a summit-level of 15,645 feet in a length of 104½ miles. This gradient of 1 in 25 may, indeed, be regarded as the steepest gradient up which regular trains are commonly drawn by the most powerful locomotives by adhesion alone. Accordingly, when steeper gradients have been deemed necessary, owing to the peculiar character of the country, the nature of the elevation to be surmounted, or special economical considerations, other systems have been resorted to for facilitating the ascent.

Systems adopted for surmounting Steep Inclines.—The most important cases are where steep inclines are introduced in places on an ordinary railway, for surmounting which, two systems have been employed, namely, the Fell system, with a central, elevated, double-headed rail laid sideways, which is gripped by horizontal wheels on each side, which greatly augment the adherence, and the Riggenbach, Abt, and other systems, with central racks, in which cog-wheels work, whereby the adhesion of the ordinary driving-wheels is greatly assisted in drawing a train up the incline, and the descent of the train is kept under control. In tourist lines ascending the steep sides of mountains for the sake of the views, a cog-wheel working in a central rack is generally used as the sole means of propulsion up the inclines. Lastly, where the ascent is steep, straight, and fairly short, a cable is employed for hauling up the vehicles, resembling in principle the inclines worked by ropes in mines, a system which has also occasionally been adopted for steep inclines on ordinary railways.

Central-rail, Fell System.—The central-rail system was first adopted for crossing the Mont Cenis pass, by a railway laid mainly along the road between St. Michael and Susa, a distance of 48 miles, having a gauge of 3 feet 7⅝ inches, and surmounting a difference of level of 5300 feet between Susa and the summit, with a total variation in level between its termini of about 9900 feet.[1] The ruling gradient was 1 in 12, the average gradient about 1 in 17, and the central rail, raised 7½ inches above the ordinary rail-level, was laid along all gradients exceeding 1 in 25; whilst the minimum radius for the curves was 2 chains. The greatest train load carried over the Mont Cenis Fell Railway, was 36 tons; and the heaviest locomotives employed on it weighed 26 tons. In this system, the grip of the horizontal wheels on the central rail, not merely secures sufficient adherence to mount steep inclines, but also serves as a very effective brake in the descent, and keeps the locomotive firmly on the line in going round sharp curves.

On the Cantagallo Railway in Brazil, to which the plant of the Mont Cenis line was sent on the closing of that line when the tunnel was opened in 1871, an incline of 1 in 12, having a length of 6¼ miles, was worked satisfactorily on the same gauge round curves of 2 to 5 chains radius by the same system for several years;[2] but some years ago, Baldwin locomotives, weighing 45 tons, were substituted, which draw

[1] *Proceedings Inst. C.E.*, vol. xcv. p. 252.

[2] *Report of the British Association*, 1870, Liverpool Meeting, Transactions of Sections, p. 216.

up a load of 40 tons, at a speed of about 8 miles an hour; though the central rail continued to be used for applying the brake on the descent.[1]

The Rimutaka incline, on the Wellington and Featherston Railway in New Zealand, with a gradient of 1 in 15 for 2½ miles, and a total rise of 869 feet, opened about 1879, having a gauge, like the rest of the railway, of 3 feet 6 inches, and curves of 5 chains radius, was laid with a central rail (Fig. 167); and the traffic on the incline has been worked

CENTRAL RAIL. FELL SYSTEM.

Fig. 167.—Rimutaka Incline, New Zealand.

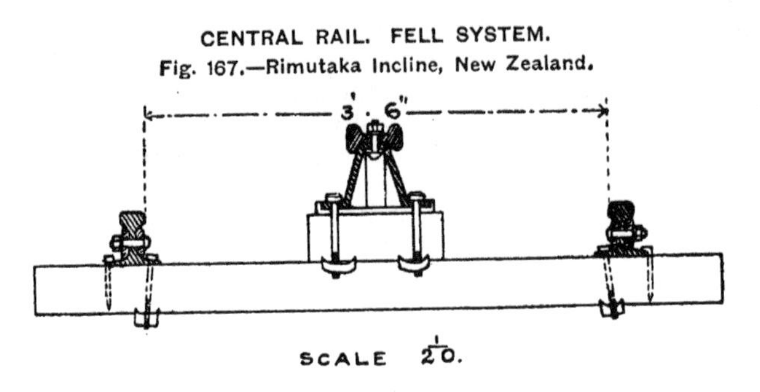

continuously by locomotives with horizontal wheels gripping the central rail.[2] Each engine, weighing about 36 tons, can draw a maximum train-load of 70 tons up the incline; and in order to avoid an undue strain on the draw-bars, the three engines employed for taking up a heavy train, are so distributed between the carriages as to enable each to draw its own load. The moderate speed of about 5 miles an hour, has been adopted for drawing the trains up the incline, in order to prevent undue wear and tear, to economize fuel, and to increase the security of the traffic. The system has proved very safe and satisfactory, and remarkably well adapted for running round sharp curves; whilst the saving in cost of construction by adopting the incline on this particular railway, instead of a more circuitous course to obtain flatter gradients readily surmounted by ordinary locomotives, was estimated at £100,000, or £5000 a year at the rate of interest charged on loans in the colony.

Ladder-rack for Mountain Railways.—Since the opening of the Mount Washington Railway, with a central rack, in 1869, numerous rack railways have been constructed in Switzerland and other mountainous districts, in which the most important differences have been the modifications adopted in the rack, combined also with variations in the steepness of the greatest inclination surmounted, depending on the conditions of the site, in the gauge selected, and in the limiting curves resorted to.

A solid, central rack was introduced for the first time in 1847, on an incline of the Madison and Indianapolis Railway of standard gauge, in the United States near Madison, which is 1⅓ miles long with gradients of 1 in 16½ to 1 in 17; for the most powerful Baldwin locomotives of

[1] *Proceedings Inst. C.E.*, vol. xcvi. p. 175.
[2] *Ibid.*, vol. lxiii. p. 52, and plate 4; and vol. xcvi. p. 137.

that period could only draw two cars up this incline.[1] Trains were drawn up the incline by special locomotives, carrying a cog-wheel working in the rack, from 1847 up to 1868, in which latter year the rack system was abandoned in favour of Baldwin locomotives, each weighing 50 tons, having five pairs of wheels coupled, and capable of drawing a load of cars and coal, weighing 137 tons, up the incline at the rate of 6 miles an hour.

The rack railway, however, which was the precursor of the numerous Swiss mountain railways for tourists, was the line of ordinary gauge, 3 miles in length, constructed up to the top of Mount Washington in New Hampshire in 1866–69, rising altogether a height of 3600 feet, with a ruling gradient of 1 in 3, and curves having a minimum radius of $7\frac{1}{2}$ chains.[2] The rack in this case was formed in lengths of 10 feet, with two parallel angle-irons, 4 inches apart, connected by a series of round, wrought-iron bars constituting the teeth of the rack, which resembles a ladder laid on the ground. The locomotives, provided with a central cog-wheel working in the ladder-rack, draw the vehicles up the mountain at a rate of about 3 miles an hour.

The first rack railway carried out in Europe up a mountain slope, was the Vitznau-Rigi Railway, constructed from Vitznau, on the Lake of Lucerne, to the summit of the Rigi in 1869–73, rising 4472 feet in its course of $4\frac{1}{3}$ miles, with a ruling gradient of 1 in 4 for about a third of its length, and never less than 1 in 6 except at the stations, together with curves of $8\frac{1}{2}$ chains radius.[3] The wrought-iron rack is of the open ladder form, but with trapezoidal bars in place of round ones, affording a wider bearing to the cogs of the driving-wheel (Fig. 168, p. 278); but the wear of these bars has been greater than with the round ones; and when one or two bars are broken, a 10-feet section of rack has to be renewed. The line is laid to the standard gauge; the rack and rails are kept rigidly in place by cross sleepers fastened to longitudinal timbers; and the whole framing is prevented from sliding gradually downhill, by being secured at intervals to masonry foundations built firmly into the rock or solid ground. By means of spur gearing on the locomotive, actuating the axle of the central cog-wheel on each side, the power of the driving cog-wheel is increased, and a low speed maintained. The locomotive on these mountain lines is always placed below the carriages, so as to push them up the inclines and control their descent, the speed of the trains on the Rigi line being limited to between 3 and 4 miles an hour. Owing to the steep gradient, a vertical boiler was in the first instance provided for the locomotives; but subsequently on this railway, as well as on the more recent mountain lines, boilers have been adopted which are horizontal when the locomotives are on the average incline. The driving cog-wheel, and the other cog-wheels fitted to the locomotive and carriages, are furnished with powerful brakes which, when applied, keep the cogs firmly engaged in the rack, so as to arrest the descent of the

[1] *Transactions of the American Society of Civil Engineers*, vol. vii. pp. 68, 71, and 72.
[2] *Proceedings Inst. C.E.*, vol. xcvi. p. 260.
[3] *Ibid.*, vol. xxxvi. p. 106, and plate 18.

train; and an air-brake acting on the piston of the locomotive, serves to regulate the downward speed. Strong hooks attached under the locomotive and carriages, encircle the top flange of each side-piece of the rack, and thus secure the train from leaving the rails or being blown over by the wind (Fig. 168).

Several other mountain railways have since been constructed with a similar rack, known as the Riggenbach rack, to surmount very similar gradients, but for the most part laid to narrower gauges and with sharper curves, thereby cheapening the cost of construction. The Arth-Rigi

LADDER-RACK.
Fig. 168.—Rigi Railway.

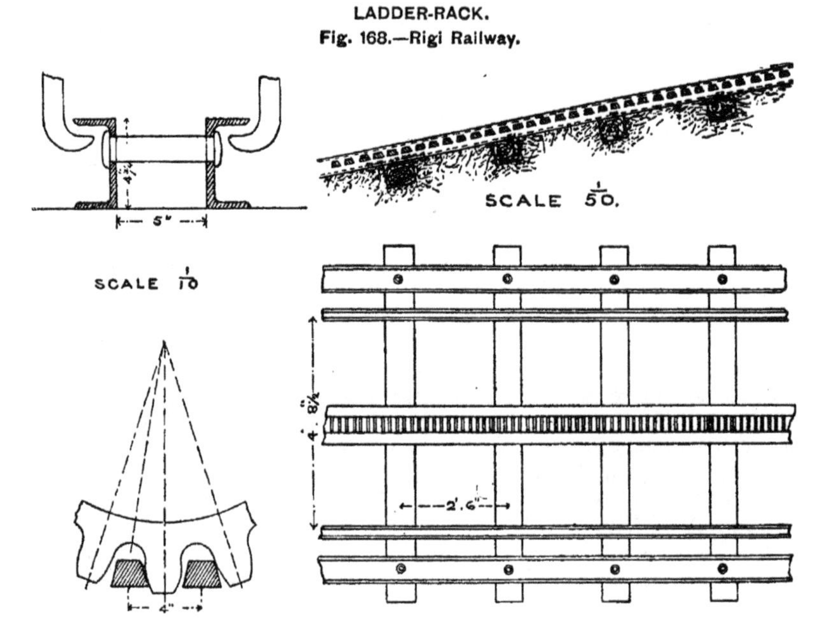

Railway, $5\frac{3}{5}$ miles long, with a rise of 4360 feet, and a maximum gradient of 1 in 5, opened in 1875, is laid to the standard gauge; but the Drachenfels and Rüdesheim-Niederwald railways in Germany, opened in 1884, and the Salzburg-Gaisberg Railway in Austria, opened in 1887, having ruling gradients of 1 in 5, 1 in 5, and 1 in 4, and curves of 10 chains, 15 chains, and $7\frac{1}{2}$ chains radius respectively, are laid to the metre gauge with iron cross sleepers; whilst the Schynige-Platte and Wengern-Alp railways in Switzerland, opened in 1893, $4\frac{3}{5}$ miles and $11\frac{1}{5}$ miles long, and rising 4593 feet and 4150 feet respectively, with a maximum gradient of 1 in 4, and sharpest curves in the latter case of 3 chains radius, are laid to a gauge of 80 centimetres, or 2 feet $7\frac{1}{2}$ inches, with steel cross sleepers.[1] The Vitznau-Rigi Railway cost £26,208 per

[1] *Proceedings Inst. C.E.*, vol. cxx. pp. 21, 41, and 42.

mile, the Arth-Rigi Railway £27,385, the Drachenfels Railway £10,600, the Salzburg-Gaisberg Railway £19,840, the Schynige-Platte Railway £25,110, and the Wengern-Alp Railway £15,738 per mile.[1] These are tourist lines; and the train-load carried up, exclusive of the locomotive, ranges between $10\frac{1}{2}$ tons, the load on the Vitznau-Rigi Railway, and $14\frac{1}{5}$ tons on the Rüdesheim-Niederwald Railway.

Double Side-rack on Locker System.—A steel, rack rail with teeth on each side, in which horizontal cog-wheels work, was adopted for surmounting the exceptionally steep inclines of the Pilatus railway, averaging 1 in 2·8, and attaining 1 in 2·08 in some places (Fig. 169),

DOUBLE SIDE-RACK AND TRAIN.

Fig. 169.—Pilatus Railway.

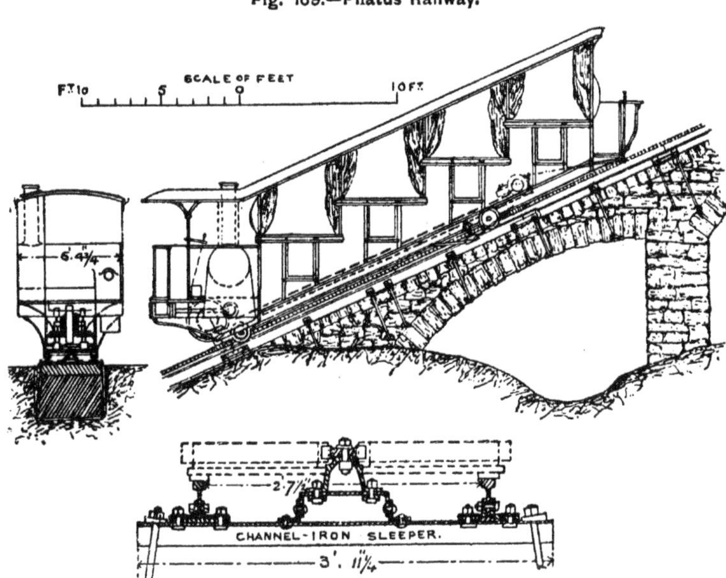

preliminary trials having proved that the ladder-rack was unsuitable for such gradients. This Locker system presents some resemblance to the Fell central-rail system (Fig. 167, p. 276), with the advantage of the greater tractive and controlling power secured by the rack and cog, in comparison with the adhesion on the central rail, enabling much steeper gradients to be surmounted, and to be descended in safety.

The Pilatus Railway, opened in 1889, starts from Alpnach, on the Lake of Lucerne, and rises 5363 feet in its length of $2\frac{3}{4}$ miles, half of the line being straight, and half being on curves of 5 to 4 chains radius; and it is laid to a gauge of 2 feet $7\frac{1}{2}$ inches, with iron sleepers bolted to

[1] *Organ für die Fortschritte des Eisenbahnwesens*, 1898, vol. xxxv. plate xxv.

longitudinal, parallel masonry walls on each side.[1] The driving cog-wheels are actuated by spur gearing; and the two pairs of cog-wheels are controlled by hand brakes, which suffice to regulate the descent of the train, or to stop it if necessary. An air-brake acting on the pistons of the locomotive, as adopted previously on the Rigi Railway, furnishes additional control of the train on its descending journey; and if at any time the speed in descending becomes more than 3 miles an hour, a reserve automatic brake comes into action. To prevent the train being blown off the rails during the violent gales to which the mountain is exposed, clips fixed to the carriage are turned under the outside of the head of each rail. The compartments of the carriage are arranged in steps to accommodate them to the incline, as on the steep inclines of rope railways; and the seats are, moreover, pivoted so as to adjust themselves to the changes in inclination (Fig. 169). The cost of the Pilatus Railway amounted to £32,260 per mile, owing to the heavy works necessitated by the steepness of the incline and the exposed situation. Though the line is worked in perfect safety, there is a considerable vibration, indicating that, where practicable, such steep inclines are better adapted for cable traction.

Flat Bar Toothed Rack.—Another form of rack consists in cutting the edge of a flat, steel bar, so as to provide a uniform row of teeth on its upper side; and the strength of the rack can be increased for steeper gradients, by increasing the thickness or the number of the bars. The rack is thus formed by a series of solid bars, with teeth shaped to the most convenient form for the working of the cog-wheel in them. This simple form of rack, consisting of successive lengths of single bars joined at their ends, and laid in the centre of the track, have been employed on the flatter gradients of several rack railways, where the Abt system of two or more such bars, laid so that their teeth are not in line across the track, is resorted to on the steeper parts of the lines. Though the Abt system of teeth breaking joint has been usually adopted for the steep gradients of a rack railway laid with solid bars, the expedients of merely increasing the width of the teeth by two or more bars placed close together, or by actually forming the teeth of the bars of increased width, have been occasionally resorted to.

The St. Ellero-Saltino Railway, the first purely rack railway built in Italy, was constructed in slightly over four months in 1892. This railway rises 2765 feet in a length of 5 miles; and it is laid to the metre gauge, with a ruling gradient of 1 in 4·55, and curves having a minimum radius of 3 chains. The rack on gradients not exceeding 1 in $8\frac{1}{3}$, consists of two angle-steel bars riveted together, 4 to 6 feet long, with teeth formed in them[2] (Fig. 170); but for steeper gradients, up to the maximum of 1 in 4·55, two flat, steel bars are introduced between the angle bars, increasing the thickness of the teeth and the rigidity of the rack, which latter can be still further augmented by introducing a distance piece between the angle bars, so as to form two

[1] *Engineering*, vol. xliii. p. 444; and *Zeitschrift des Vereines deutscher Ingenieure*, 1887, p. 1117.

[2] *Proceedings Inst. C.E.*, vol. cxvii. p. 278, and plate 9.

or three parallel racks with a small interval between them, in which the cog-wheel works with a widened bearing. This Telfener rack is simpler in construction, and cheaper than the Riggenbach and Abt racks; but it does not possess the special advantage of the Abt rack, of thoroughly engaging two or three successive teeth of the cog-wheel at the same time. The favourable conditions of the site, enabling tunnels to be

TELFENER RACK.

Fig. 170.—St. Ellero-Saltino Railway.

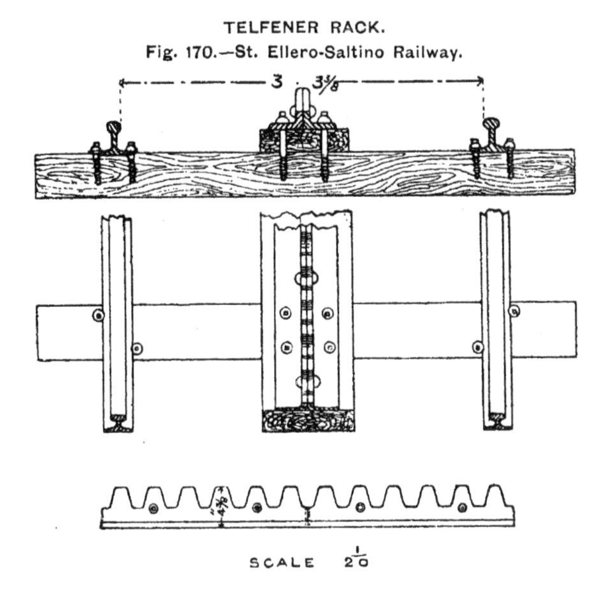

dispensed with, combined with the introduction of sharp curves, and the adoption of a cheap rack, brought the cost of construction down to the remarkably small amount of £3622 per mile. The speed of the trains ranges from $5\frac{1}{2}$ to $4\frac{1}{3}$ miles an hour, according to the gradients, and averages 5 miles an hour.

A more complicated form of single rack, resembling a flat-bottomed rail in its low portion, and widened out considerably for the teeth at the top (Fig. 171, p. 282), called the Strub system, after its designer,[1] has been recently introduced on the Jungfrau Railway, which is laid to the metre gauge, and was opened in 1899, the motive power being electricity generated by waterfalls on the mountain. This line rises 6657 feet in a length of $7\frac{3}{5}$ miles, with gradients ranging from 1 in $14\frac{1}{3}$ up to 1 in 5; and the upper $6\frac{1}{5}$ miles are in tunnel; whilst the final ascent to the summit is effected by a vertical lift of 241 feet. The central, rack rails, $11\frac{1}{2}$ feet long, are joined together at their ends by fish-plates, like ordinary flat-bottomed rails. A brake is provided, which encircles and grips the widened-out head of the rack. The wide base of the Strub rack

[1] *Organ für die Fortschritte des Eisenbahnwesens*, Wiesbaden, 1897, vol. xxxiv. p. 151, and plate xx.

gives the rack considerable lateral rigidity, and enables chairs to be dispensed with for fastening the rack to the metal sleepers (Fig. 171); but this rack lacks the simplicity and economy of the Telfener rack

STRUB RACK.

Fig. 171.—Jungfrau Railway.

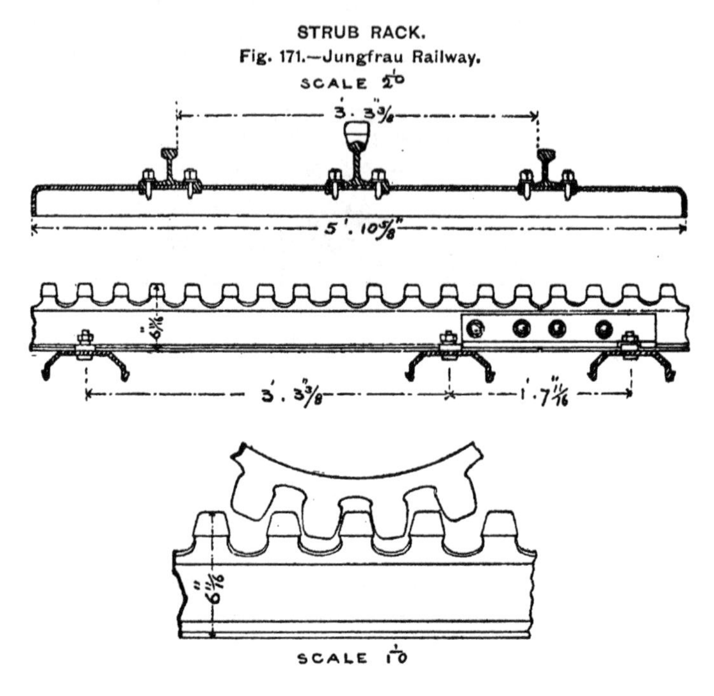

described above (Fig. 170, p. 281), and the special merit of the multiple Abt rack (Fig. 172).

Racks on Abt System.—The Abt system consists essentially of two or three steel, rack bars, from $\frac{11}{16}$ inch to $1\frac{13}{16}$ inches thick, and 2 to $4\frac{1}{3}$ inches deep, placed nearly two inches apart, and so arranged that the teeth are not opposite each other, but as it were break joint (Fig. 172), causing the cog-wheels to engage in a tooth in front on one rack before leaving the tooth behind on the adjacent rack, which renders the motion smoother, and increases the security of the trains in descending, besides proportioning the strength of the rack to the steepness of the gradient by the addition of one or two bars. Though this system has less lateral stiffness than the ladder-rack, its plates can be more readily bent to sharp curves; and its superior merits in other respects has led to its adoption on numerous mountain railways since its introduction in 1882.

The Generoso and Rothorn railways in Switzerland, $5\frac{2}{3}$ miles and $4\frac{4}{5}$ miles long, rising 4326 feet and 5515 feet, with ruling gradients of 1 in 4·55 and 1 in 5, and constructed in 1889–90 and 1891 respectively, are laid to a gauge of 2 feet $7\frac{1}{2}$ inches with cast-steel sleepers, and provided with a double Abt rack, in which cog-wheels on the driving axles work; and the sharpest curves on these lines have a radius of 3 chains. The cost of the Generoso Railway was £13,597 per mile, and of the Rothorn Railway £16,561 per mile.[1] The train-loads carried up are 9 tons and $8\frac{1}{2}$ tons respectively; and the speed up and down on the steepest gradients of the Generoso Railway is $3\frac{3}{4}$ miles an hour, and 5 to $6\frac{1}{4}$ miles on the easier gradients. The Glion-Naye Railway is a very similar line, $4\frac{4}{5}$ miles long, with five tunnels, rising 4209 feet, with gradients reaching a maximum of 1 in 4·55, and curves

ABT TRIPLE RACK.

Fig. 172.—Hartz Mountain Railway.

Scale 20 feet to 1 inch.

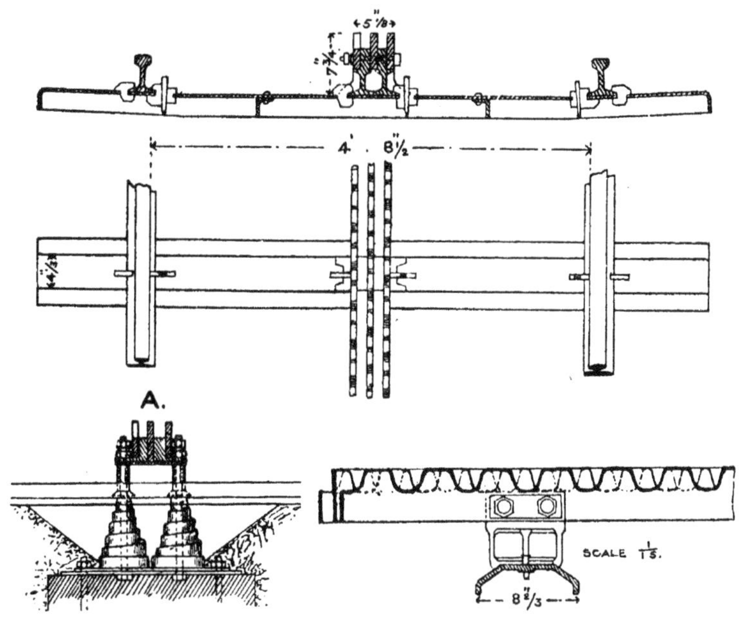

of 4 chains radius, laid to a gauge of 2 feet $7\frac{1}{2}$ inches, and with a single or double rack according to the gradient. It was opened in 1892, and cost £18,033 per mile; and it possesses the special interest of forming a continuation, on another system, of the cable railway between Territet and Glion, and leading to the health resort of Caux, 3580 feet above sea-level, beyond which it is a mere tourist line like the preceding ones, with trains running only in the summer, accomplishing the journey between Glion and Naye in about $1\frac{1}{2}$ hours.

The system has also been extended to mountain lines in several other countries, as, for instance, the Manitou and Pike's Peak Railway in Colorado, of standard gauge, rising 7552 feet in a length of $8\frac{3}{4}$ miles,

[1] *Organ für die Fortschritte des Eisenbahnwesens*, Wiesbaden, 1898, vol. xxxv. p. 140, and plate xxv.

with a maximum gradient of 1 in 4, and curves of $5\frac{3}{4}$ chains, provided with two lines of steel, rack bars, from $\frac{4}{5}$ inch to $1\frac{1}{4}$ inches thick, in lengths of $6\frac{3}{4}$ feet, and constructed in 1890 at a cost of £11,409 per mile; the Aix-les-Bains and Revard Railway in France, the Schafberg Railway in Austria, and the Montserrat Railway in Spain, $5\frac{3}{4}$ miles, $3\frac{3}{4}$ miles, and 5 miles long, laid to the metre gauge, the steepest gradients being 1 in $4\frac{3}{4}$, 1 in $3\frac{9}{10}$, and 1 in $6\frac{2}{3}$, and the sharpest curves $3\frac{3}{4}$ chains, 4 chains, and $3\frac{1}{4}$ chains radius respectively;[1] and the Snowdon Railway in Wales, rising 3140 feet in a length of $4\frac{5}{8}$ miles, with a maximum gradient of 1 in $5\frac{1}{2}$, and curves of 4 chains radius, laid to the fairly common Swiss gauge of 2 feet $7\frac{1}{2}$ inches, with two lines of steel, rack bars, from $\frac{3}{4}$ to 1 inch thick, in lengths of nearly 6 feet, and constructed in 1895–96 at a cost of £13,617 per mile.[2]

Instances of the application of electricity as the motive power on mountain railways laid with the Abt rack, where water-power is readily available for generating the electrical current, are furnished by the Mont Salève Railway near Geneva, and the Gornergrat Railway ascending from Zermatt. These railways, constructed in 1891 and 1896–98 respectively, have lengths of $5\frac{3}{8}$ miles and $5\frac{4}{5}$ miles, with rises of 2363 feet and 4600 feet, and are laid to the metre gauge, with gradients of 1 in 4 and 1 in 5, and a double line of rack; and they cost £12,714 and £20,804 per mile respectively.

In all these rack railways, special care is always taken to anchor the track firmly down into the solid ground, so as to prevent its creeping gradually downhill under the pressure of the cog-wheels on the rack.

Racks on Steep Inclines for the Extension of Railways.—The tourist railways referred to above, though interesting as showing how light loads can be conveyed up steep mountain slopes, have no commercial importance, are open for the most part only in the summer, and cannot be regarded as forming part of the regular railway system of a country. The introduction, however, of a rack for surmounting the steep inclines occasionally found necessary in carrying a railway through mountainous country, or for extending railway communication along the upper parts of gorges in rugged districts, and across the water-parting of mountain torrents, where gradients surmountable by mere adhesion become impracticable at any reasonable cost, invests the rack system with an importance in the problem of railway extension, which it could not otherwise claim. In some instances, a single steep incline provided with a rack enables a sudden rise of the land to be surmounted, forming a connection with the ordinary railway below and above, of which the Trincheras incline on the Puerto Cabello and Valencia Railway in Venezuela, and the Usui Railway in Japan, furnish typical examples; whereas in mountain valleys, especially in the neighbourhood of the dividing ridge between two river basins, the rough and variable character of the gorge may necessitate the adoption of steep inclines requiring a rack in several places, with intervening

[1] *Schweizerische Bauzeitung*, Zurich, 1893, vol. xxii. p. 140.
[2] *Engineering*, vol. lxi. pp. 427, 442, and 479.

gradients surmountable by adhesion alone, as illustrated by the Hartz, Eisenerz-Vordernberg, Brünig, Bernese Oberland, Visp-Zermatt, and Transandine railways.[1]

The ground rises from the Caribbean Sea at Puerto Cabello to an elevation of 1950 feet at Entrada, in a distance of 25¼ miles; and it was found that a railway for connecting these places, to be worked by adhesion alone, having to commence its rise near the seacoast with a ruling gradient of 1 in 40, and involving very heavy works in spite of very frequent and sharp curves, could not have been constructed at a reasonable cost. Accordingly, a route was selected where the railway following along the bottom of a flat portion of a river valley, rises only 420 feet in the first 18 miles, whence it ascends about 780 feet in the next 5 miles to Trincheras, with a ruling gradient of 1 in 28½ and curves of 4½ chains radius, and then accomplishes the final rise of 750 feet in 2⅕ miles, by a steep incline having a gradient of 1 in 28½ to 1 in 21 for the first 4000 feet, and 1 in 12½ for the remaining 7700 feet to Entrada;[2] and the incline is provided with a triple Abt rack (Fig. 172, p. 283), formed of three parallel bars, 1$\frac{9}{16}$ inches apart, bolted firmly together, in which two cog-wheels work, with a special ring of teeth for each bar. The sharpest curves on the incline have a radius of 7½ chains, the gauge of the railway being 3½ feet; and the locomotive, weighing 40 tons, which effects the ascent by means of its cog-wheels alone, can push up four waggons weighing altogether about 68 tons when loaded, which is also the train-load which an ordinary locomotive can bring up the steep gradients of 1 in 28½ below the incline.

A rise of 1830 feet in a direct length of 5 miles, at the Usui pass, separated the railways which by 1888 had been completed from the east and west coast of the main island of Japan, to the central mountain range which stretches down the whole length of the island, and as far as the foot and summit of the pass respectively, on each side. A survey in 1889 of the rugged intervening district, proved the possibility of laying out a line 15¼ miles long, 4½ miles of which would have been in tunnel, with a ruling gradient, like the approach line, of 1 in 40 and curves of 10 chains radius, at a cost of about £20,000 a mile. It was, however, determined in 1890 to adopt a direct incline, laid with the Abt rack, to test the system in view of the construction of other mountain railways in Japan, and to diminish the cost of the works by reducing the length of the line from 15¼ miles to 7 miles, with nearly 5 miles of steep incline. The incline of 1 in 15 has been constructed in two portions, 2·45 miles and 2·41 miles long respectively, with a short length of level between them to provide a passing-place for the trains.[3] This connecting link, constructed in 1891–92, which, owing to the very rugged nature of the district traversed, especially by the upper incline, necessitating 2¾ miles of tunnel and several viaducts for spanning the ravines, cost £42,667 per mile, including four locomotives

[1] *Proceedings Inst. C.E.*, vol. cxx. plate 2; and *Schweizerische Bauzeitung*, vol. xxii. p. 140.

[2] *Proceedings Inst. C.E.*, vol. xcvi. p. 120, and plate 4.

[3] *Ibid.*, vol. cxx. p. 43, and fig. 1.

with two cog-wheels each for working in the triple Abt rack, so that little economy was effected by the inclines over the estimated cost of the circuitous adhesion line; but the working and maintenance of 8¼ miles of line with heavy gradients were saved. The line is laid to the gauge of 3 feet 6 inches, the standard gauge of Japan; and the sharpest curves on the incline have a radius of 13 chains. The first locomotives, weighing 33½ tons, pushed a train-load of 60 to 70 tons up the incline at a speed of 4·7 miles an hour; but more powerful locomotives were subsequently constructed, which, weighing 46 tons, can push up a load of 100 tons at a similar speed.

Where several steep gradients laid with a rack are interspersed between easier gradients readily surmountable by adhesion, combined locomotives have been adopted, which can draw a train by adhesion in the ordinary manner, and are provided in addition with cog-wheels which are only driven over the heavier gradients laid with the rack. This mixed system possesses the great advantages of enabling the gradients to be accommodated to the varying slopes of the ground in mountainous districts, and at the same time allowing the traffic to be worked with locomotives by adhesion over all the suitable gradients, reserving the cog-wheels for the steep gradients where a greater tractive force becomes necessary. By thus introducing a rack where requisite, steep gradients can be readily resorted to, and the works thereby considerably lightened, so that railways can be extended into rugged districts, and along mountain valleys, where the provision of gradients suited to adhesion alone would not be financially practicable. The Hartz Railway in Germany, between Blankenburg and Tanne, and the Eisenerz-Vordernberg Railway in Austria, are instances of the mixed system of adhesion and rack applied to lines of the standard gauge; whilst the Brünig, Bernese Oberland, and Visp-Zermatt railways in Switzerland, and the Transandine Railway between Mendoza and Santa Rosa, on the line connecting Buenos Ayres and Valparaiso, furnish examples of its employment with the metre gauge.

On the Hartz Mountain Railway, nearly 19 miles long, eleven sections, with an aggregate length of 4⅓ miles, are laid with a triple Abt rack of steel bars, 4¼ inches deep, ¾ inch thick, in lengths of 8⅔ feet, 1⅜ inches apart, and raised 2¾ inches above the rails (Fig. 172, p. 283).[1] The rack is laid on gradients of 1 in 22 to 1 in 16⅔, with curves of 10 chains radius; whilst on the ordinary portions of the line, the ruling gradient is 1 in 40, and the sharpest curves 9 chains radius. The railway was constructed in 1884–86, and its cost was £10,458 per mile. A locomotive weighing 56 tons can draw a train of 120 tons up the steep inclines at an average speed of 6⅔ miles an hour. The Eisenerz-Vordernberg Railway, opened in 1891, has 9 miles laid with a rack in a total length of 12⅖ miles, on gradients of 1 in 14 along the straight portions, reduced to 1 in 15⅖ on curves of 9 chains radius. The double Abt rack is formed of steel bars 1 inch thick; and the locomotives,

[1] *Zeitschrift für Bauwesen*, Berlin, vol. xxxvi. p. 71, and plates 17 and 18; and *Proceedings Inst. C.E.*, vol. xcvi. p. 131.

weighing 55 to 59 tons, can draw a train of 100 to 120 tons at an average speed of 6⅓ miles an hour throughout the line.[1] This railway, with a much greater proportionate length of gradients laid with a rack than the Hartz Railway, appears to have cost £45,160 per mile. The liability of the cog-wheels to cause a jolt on entering or leaving the rack, has been minimized by placing the last lengths of the racks upon springs (Fig. 172, **A**, p. 283), which contrivance also reduces considerably the wear of the teeth of the cog-wheel and rack.

The Brünig and Bernese Oberland railways, 40 miles and 14½ miles long, have lengths of 5⅗ miles and 3 miles laid with a ladder-rack, on gradients considerably steeper than the ruling gradient of 1 in 40 on the ordinary portions, and reaching a maximum of 1 in 8½ and 1 in 8¼, with sharpest curves of 6 chains and 5 chains radius respectively; and they cost £9668 and £7500 per mile.[2] The Visp-Zermatt Railway, 21¾ miles long, has six sections, amounting altogether to 4¾ miles, laid with a double Abt rack, from ¾ to 1 inch thick according to the gradient, which attains a maximum of 1 in 8, and curves of 5 chains radius; whilst the ruling gradient on the ordinary portions of the line is 1 in 35¾, and the sharpest curves 4 chains radius. The railway cost £9693 per mile; and locomotives of 29 tons can draw a train weighing 45 tons up the steepest gradients on the line. The steepest portion of the mountain section of the Transandine Railway has 17⅖ miles laid with the Abt rack in a length of 31 miles, whereby gradients of 1 in 12½ are surmounted, with curves of 10 chains radius; whereas the ruling gradient on the adhesion parts of the line is 1 in 40, in which, however, curves of 5¾ chains radius are introduced. Locomotives of 42 tons draw trains weighing 60 tons over this railway.

The ratio of the average load drawn, to the weight of the locomotive on several of these combined adhesion and rack lines, ranges between 2½ and 1, depending necessarily upon the maximum gradient adopted; whereas in the case of the purely rack mountain lines, where much steeper gradients are resorted to, the weight of the locomotive often exceeds that of the train it pushes up, and in a few instances nearly reaches twice the weight of its train.

Remarks on Rack Railways.—Though several mountain lines, and some combined adhesion and rack railways, have been laid with the ladder or Riggenbach rack, the Abt rack appears to be the better system, owing to the facility with which its strength is adjusted to an increased gradient by augmenting the thickness or number of the bars; the greater accuracy of the form of the teeth cut out of a solid bar, than in a ladder-rack; the greater adaptability of bars to curves, and the greater uniformity of the strains on sharp curves; and the freedom from accumulations of snow in the teeth of narrow bars, which are liable to clog the ladder-rack.

Traction by means of a rack and cog-wheels possesses the advantage of not being affected by the weather, which has an important influence

[1] *Verhandlungen des Vereins für Eisenbahnkunde*, Berlin, 1892, p. 14.
[2] *Schweizerische Bauzeitung*, vol. xxv. pp. 60, 70, 76, and 125.

on traction by adhesion; and though the Fell central-rail system is better adapted for the safe and easy passage of trains round very sharp curves, its dependence on adhesion for its tractive force exposes it to variations in efficiency from atmospheric changes.

The most valuable feature in the rack system of traction is its application, in combination with adhesion, to occasional steep inclines on ordinary railways, whereby it is possible to extend railway communication into districts which appeared formerly to be debarred by nature from such facilities for traffic, on account of the impossibility of providing gradients throughout, capable of being surmounted by the adhesion of the locomotives alone, at any reasonable cost. The gradients of railways constructed for working on the combined system, and intended for regular traffic, should be kept as far as practicable within reasonable limits, not exceeding if possible about 1 in 15, so that the weight or speed of the train drawn may not be unduly reduced, reserving the very steep inclines for lines ascending mountain slopes with light loads, and for tourist railways like the Brünig and Visp-Zermatt lines.

In addition to the examples given above of railways which have been constructed on the mixed system, an instance may be given of an Alpine railway which illustrates the value of the system in penetrating regions which, without its aid, might have well been considered inaccessible to railways, namely, the line which has been constructed through the narrow rocky gorges of the upper valley of the Arve, and under the shadow of Mont Blanc, to Chamonix. The railway from Geneva has been gradually extended without difficulty along the flat, though narrowing valley of the River Arve, as far as Le Fayet St. Gervais; but a little beyond this village, the valley changes abruptly in character to a succession of narrow, rapidly rising gorges, along which the ascending high-road has had to be cut out of the side of the steep rocky slope, and through which the line 10·8 miles long, the steepest gradients being 1 in 11 and 1 in 12½, is carried by means of side cuttings in the rock, tunnels, retaining walls, and bridges. This railway, which serves to connect the existing railway with the comparatively flat, wide valley of Chamonix, is electrically worked by turbine-driven dynamos at two generating stations, deriving their power from affluents of the River Arve; and it is proposed eventually to extend this railway to Martigny across the dividing ridge of the Arve and Rhone valleys, thereby providing railway communication through some of the wildest portions of the Alpine region.

CHAPTER XVIII.

CABLE RAILWAYS.

Conditions favourable for cable traction—Uses and limits of cable traction—Cable traction on steep gradients, formerly, superseded by locomotives—Cable incline on San Paulo Railway; description, arrangement for trains to pass, method of working, brakes—Cable system for mountain and cliff railways; suitable sites, gradients, arrangements, motive power—Mountain cable railways; instances; descriptions of Stanzerhorn, Look-out Mountain, Vesuvius, and St. Salvatore cable railways; and their modes of working—Steep-incline and cliff railways with water counterpoise; general arrangements; descriptions of Giessbach, Territet-Glion, Marseilles, and Clifton Rocks Railways; their methods of working, and cost—Suburban cable railways, worked by stationary engines; at Lyons, Havre, and Lausanne, descriptions, and methods of working—Remarks on cable and mountain railways: respective advantages of cable and rack systems.

TRACTION by cable is advantageously adopted where, owing to the peculiar configuration of the locality, a very steep incline is required for a railway, in order to surmount a considerable elevation in a comparatively short distance; and the system is specially suitable where the incline by itself suffices to accomplish the desired communication, as, for instance, cliff and other steep railways which provide a direct and short connection between places fairly close to one another, but differing considerably in elevation.

Uses and Limits of Cable Traction.—Recourse has been occasionally had to cable traction where it has been possible to effect a very considerable economy in the cost of construction of an ordinary railway, by introducing in its course a steep incline for surmounting an abrupt rise of the land, in place of following a much longer, winding route, in order to obtain gradients suitable for traction by adhesion alone. Under such conditions, it merely furnishes an alternative system of traction to the central-rail system, as employed on the Rimutaka incline (p. 276), and to the rack system as adopted on the Trincheras incline and the Usui Railway (p. 284), and on other railways, as described in the previous chapter. The cable system is used to the best advantage when the incline is very steep, and fairly straight; whereas the rack system is the most suitable for railways partly worked by adhesion, but passing along portions of their length through rugged country, in a winding course, with gradients not exceeding about 1 in 12, and also for tortuous, mountain railways with gradients up to about 1 in 4.

Cable Traction for Steep Gradients on Railways.—In the early days of railway working, before the great tractive power of heavy locomotives had been fully developed, cable traction was employed for several years on some of the steeper gradients. Thus, for instance, a cable was at first used for drawing the trains from Camden Town to Euston up a gradient of 1 in 66; and owing to the reduction in adhesion of the driving-wheels to the rails, by the condensation of the steam from the locomotives in the tunnels between Edgehill and Liverpool, cables were employed for a long time for drawing the trains up three gradients of 1 in 90, 1 in 56, and 1 in 48, each about 1½ miles in length.[1] Cable traction was also very naturally resorted to at first for drawing trains up an incline of 1 in 27 in approaching Oldham near Manchester, 1½ miles long, and an incline of 1 in 21 and 1 in 18 at Aberdare Junction, half a mile in length; but subsequently locomotives worked the traffic up these inclines, weighing 27 tons, and drawing a train of 50 tons at the rate of 15 miles an hour up the incline at Oldham, and weighing 36 tons and drawing a train of 50 tons up the Aberdare incline, which was eventually altered to 1 in 40.[2]

Cable Incline on the San Paulo Railway in Brazil.—An interesting example of a steep incline worked by a cable, introduced for surmounting an abrupt rise between two sections of an ordinary railway, is furnished by the Serra do Mar incline of 1 in 9¾, rising 2557 feet in about 5 miles, on the San Paulo Railway in Brazil, having a gauge of 5 feet 3 inches.[3]

The railway starts from Santos, situated on a large inlet from the Bay of Santos, and after traversing a flat, swampy plain for 13½ miles, it reaches the foot of the Serra, which, with its rapid rise, constitutes a sort of inland cliff separating the lower part of the line from the upper portion, which passes over a high undulating plateau to the interior of the district, with a ruling gradient of 1 in 40 and curves of 17 chains radius. The incline was constructed in four separate lengths, between 1$\frac{1}{10}$ and 1⅓ miles long, with a flat portion, 250 feet long, at the top of each length, on which the stationary steam-engines for working the cables are situated; and curves of from 80 to 30 chains radius were introduced when the line was constructed in 1860–66, though after a landslip in 1872, curves of 15 chains radius were resorted to in repairing the injured portion, without impairing the efficiency and safety of working. By dividing the incline into sections, the length of the cables is reduced and the capacity for traffic is increased, a passing-place being provided at the centre of each section for the ascending and descending trains, which, being each attached to one end of the cable, to some extent counterpoise one another.

[1] *Proceedings Inst. C.E.*, vol. xv. pp. 367 and 371.
[2] *Ibid.*, vol. lxiii. p. 127. [3] *Ibid.*, vol. xxx. and vol. clxiv.

The upper portion of each section of the incline is laid with three rails, the centre one being run over by the inner wheels of each train, but with their flanges on opposite sides; and on reaching the central passing-place, the central rail branches out into two to form the two inner rails of the diverging lines, thereby dispensing with switches, and forming a very convenient arrangement for ensuring the passing of the counterpoised trains of cable railways along their respective lines without any mechanism, and without entailing the provision of a separate line throughout for each train. Along the upper half of each section of the incline, a double row of pulleys, placed in the centre of each track on each side of the central rail, guide the ascending and descending portions of the cable, the pulleys being put 10 yards apart on the straight lines, and 5 to 7 yards apart on the curves; whilst a single row of pulleys suffices for guiding the single portion of the cable which travels up and down the lower half of each incline; and consequently the two lines of way at the central passing-place are brought into a single line below, the passage of the trains at the point of change being regulated by self-acting switches. The cable is composed of six strands, each formed by seven steel wires; and in hauling up a train of three loaded waggons and a brake van, weighing 40 tons, whilst a similar train is descending, the cable is subjected to a maximum strain of 4½ tons, only about one-eighth of its breaking strain.

On reaching the top of one section of the incline, the train readily passes down the gradient of 1 in 75, 250 feet long, separating the top of one incline from the foot of the next. As the whole ascent is ordinarily accomplished in an hour, four trains could be passed up, and four down in an hour; and as the ascent could be readily accomplished in 45 minutes, the capacity of the incline for traffic could at any time be increased by one-fourth. To ensure complete control of the trains on the incline, a brake can be applied to the flywheel of each of the stationary engines at the top of each section, whereby the motion of the cable is arrested; and, moreover, in addition to the ordinary brakes for all the wheels of a train, the brake van is supplied with a pair of clip brakes encircling the head of each rail, each brake being worked by a pair of long levers, and, when brought together at their lower extremities by the action of screws turned by a wheel, they grip the rails so firmly as to stop the train within a distance of a few yards. These clips are also useful for stopping a descending train at the proper place on the top of each incline, when the rails are greasy in damp weather.

The traffic has been conducted on this incline with regularity and safety; and to meet increasing requirements the duplication of these inclines was commenced in 1896 and completed in 1901. An easier gradient of 1 in 12½ was obtained, the length of the new inclines being 6½ miles, and the rise 2606 feet. The incline consists of five equal lengths of 1¼ miles, with bank-heads between each. The track has three rails, one being common to the up and down lines, except at passing places. The ropes are endless, traversing horizontal return wheels 14 feet diameter. The rope $1\frac{11}{16}$-inch diameter has a breaking strength of 96 tons and a working load of 12 tons. A steam locomotive brake van, containing the picking-up gear, rope nipper, hand brakes, automatic vacuum brake, and emergency clip-rail brakes, is placed at the lower end of the train.

Cable System for Mountain and Cliff Railways.—Cable traction is applied to the best advantage on those short, steep lines which connect places at very different elevations, and are also used at seaside resorts for communication between the sea-shore and the top of the cliffs. Some of these railways surmounting the steep lower slope of a mountain to reach a level plateau, may be regarded as mountain

CABLE RAILWAY, TERRITET-GLION, SWITZERLAND.

Fig. 173.—Longitudinal Section.

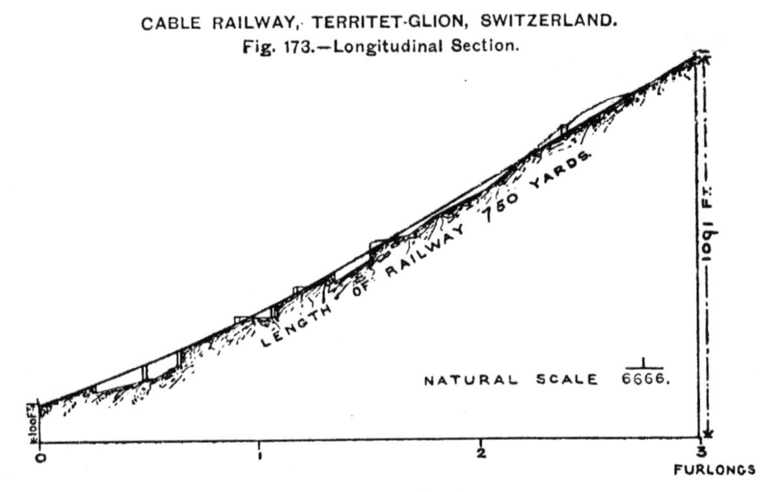

Fig. 174.—Cross Sections.

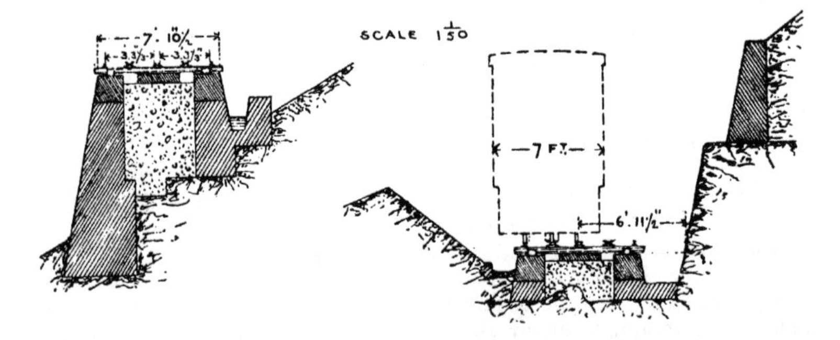

railways, owing to the special method of traction, though only to a limited extent, or as the first stage of a regular mountain railway in the case of such a line as the Territet-Glion Cable Railway (Figs. 173 and 174), where the ascent is continued above Glion by a rack railway up to the Rochers de Naye, reaching an elevation of 6485 feet above sea-level, of which also the Lauterbrunnen-Grütsch Electric Cable Railway, followed by a rack line to Mürren, is another example.

Generally, however, the system is not suited for ascending mountains to considerable heights, not so much on account of the length of cable required, for this can be obviated by dividing the line into sections of suitable length, but mainly owing to a line with much lighter works being usually attainable by a circuitous course. Nevertheless, the Stanzerhorn Railway furnishes a notable instance of a cable railway ascending a mountain, divided into three sections to reduce the length of the cables and increase the capacity of the line for traffic, which also enabled an easier route to be adopted, and allows of a higher speed along the flatter lowest section of the line.

Numerous short cable railways and cliff railways have been constructed, with gradients ranging for the most part between 1 in $8\frac{2}{3}$ and 1 in $1\frac{2}{3}$, consisting of an ascending and a descending carriage attached to the ends of the cable, counterpoising one another approximately according to their relative loads, and passing at the centre of the incline (Fig. 175). The motive power is often conveniently provided, where there is a good supply of water, by filling a tank under the floor of the carriage at the top of the incline with water, so as to overbalance sufficiently the

CABLE RAILWAY, TERRITET-GLION, SWITZERLAND.

Fig. 175.—Passing-Place.

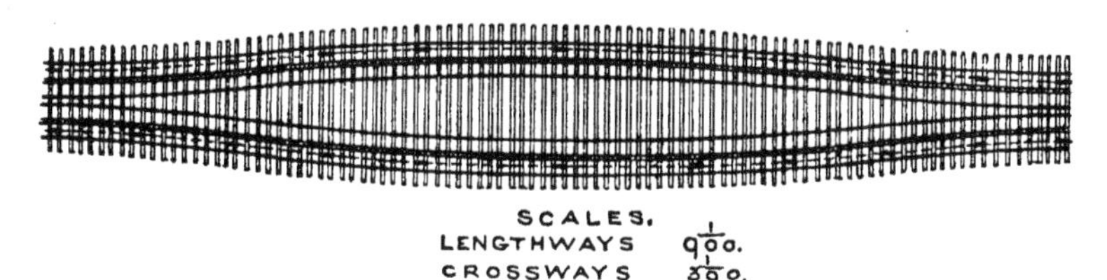

SCALES.
LENGTHWAYS $\frac{1}{900}$.
CROSSWAYS $\frac{1}{300}$.

carriage with its passengers at the bottom of the incline, from whose tank the water has been discharged; and the travel of the carriages is frequently controlled by laying a Riggenbach or Abt rack in the centre of the track, so that the revolving cog-wheels under the carriages fitting in the rack, when stopped by a brake, become firmly fixed in the rack and arrest the motion, thereby effectually supplementing the other brakes (Figs. 174 and 175). Occasionally, where water-power is readily available, it is used to generate electricity; and the cable is set in motion by means of a dynamo. Where there is neither a fall of water, nor a sufficient supply of water to overweight the top carriage, steam-power has to be used for working the cable.

Mountain Cable Railways.—In addition to the Stanzerhorn Railway, to which allusion has already been made, which rises 4594 feet in a length of about $2\frac{2}{5}$ miles, attaining an elevation of 6070 feet above sea-level, only 160 feet below the summit of the mountain, there are other shorter cable lines which must be classed as mountain railways, from their ascending mountain slopes and reaching nearly the summit of their respective mountains, as, for instance, the Look-out Mountain

Railway, near Chattanooga, in the United States, rising 1170 feet in a length of 4360 feet; the Mount Vesuvius Railway ascending the great cone with a rise of 1410 feet in a length of 2730 feet; the Mont St. Salvatore Railway, near Lugano, rising 1975 feet in a length of slightly over a mile; and the Mendel Railway, in South Tyrol, rising 2700 feet, the total length on the incline being 2600 yards.

The three divisions of the Stanzerhorn Railway rise 866, 1664, and 2064 feet respectively; and in order that the traffic may be worked economically on the inclines at a uniform speed, the gradient has been steepened in ascending, so that the first division starts with a gradient of 1 in $12\frac{1}{2}$ and ends with 1 in 3·7, and the second and third divisions begin with 1 in $2\frac{1}{2}$ and end with 1 in 1·6.[1] The line is laid with a single track to the metre gauge, without a central rack, and is provided with a passing-place for the trains midway on each division; and its course is fairly straight, having only two curves on the first division with a minimum radius of $7\frac{1}{2}$ chains, two on the middle division with a minimum radius of $10\frac{1}{2}$ chains, and only one curve, of 10 chains radius, on the upper division. As the line is single, the grooved rollers or pulleys for guiding the cable, to each end of which a carriage is attached, are laid in pairs, $6\frac{2}{3}$ inches on each side from the centre of the track, on which the two portions of the cable travel; and the pulleys are $11\frac{4}{5}$ inches in diameter, and placed at intervals of 38 feet on the straight portions of the line, and $23\frac{3}{5}$ inches in diameter, and only 33 feet apart on the curves, being inclined in the latter case so as to adjust the cable to the curve. To ensure the carriages keeping to their proper line at the passing-place, the wheels of each carriage, which should be on the outside rail of the respective lines at the crossing, are provided with double flanges, whilst the wheels on the other side are devoid of flanges and very broad; and, accordingly, the flanges make each carriage keep to its through rail, forming the outer rail on each side at the passing-place, whilst the broad wheels pass without hindrance over the crossings. The cables of the three divisions, commencing with the lowest, have diameters of $\frac{9}{10}$ inch, $1\frac{3}{10}$ inches, and $1\frac{1}{3}$ inches, and are capable of supporting tensile strains up to $24\frac{1}{2}$ tons, $51\frac{1}{5}$ tons, and $54\frac{1}{10}$ tons; whilst the greatest strains to which they are subjected are $1\frac{3}{4}$ tons, $4\frac{2}{5}$ tons, and $5\frac{1}{2}$ tons respectively. The rate of transit on the lowest division is about $6\frac{1}{2}$ feet per second, and on the other two divisions about $4\frac{1}{2}$ feet per second; and the whole journey is accomplished in 45 minutes. Each carriage is provided with three brakes, two of which act automatically if the cable breaks. On each division, the cable is worked by a dynamo-motor at the top of the incline; and the electricity is transmitted to the three dynamo-motors from a dynamo at a generating station, $2\frac{1}{2}$ miles distant, where the electricity is obtained by means of water-power. The cost of this cable railway amounted to about £24,000 per mile.

The Look-out Mountain Railway reaches its elevation of 1170 feet by a gradient of 1 in $3\frac{3}{4}$, having a length of four-fifths of a mile; and it is laid to a gauge of $3\frac{1}{4}$ feet, with a central passing-place 200 feet

[1] *Proceedings Inst. C.E.*, vol. cxx. p. 11, and plate 1.

long, and a central rail common to the two lines on each side of it, above and below the passing-place, so as to avoid the necessity of any movable mechanism for guiding the carriages at the junction of the passing-place with the lines above and below.[1] The railway is fairly straight, having only three curves on it, situated on the upper portion, with a minimum radius of $15\frac{1}{3}$ chains. A carriage is attached near each end of the cable which passes round a grooved wheel or pulley, 8 feet in diameter, at the top of the incline, and is made up of six strands, each composed of nineteen steel wires; and whereas the cable, $1\frac{1}{4}$ inches in diameter, has an ultimate strength of 50 tons, the maximum strain to which it is subjected in working is only 5 tons. As the railway is worked in this instance by a steam-engine placed at the bottom of the incline, a second cable is fastened underneath the carriages to the ends of the upper cable, by means of which the working of the line is effected. Tightening pulleys on a weighted sliding frame keep the lower cable duly stretched, in spite of changes in temperature and strain.

The railway rising 1410 feet up the great cone of Mount Vesuvius, with gradients ranging from 1 in $2\frac{1}{2}$ up to 1 in $1\frac{3}{5}$, has two lines, each laid with a single central rail, on which the central, double-flanged wheels of the carriages run, the carriages being kept vertical by a pair of flat wheels on each side, inclined at an angle of 30° to the horizontal, and bearing on guide bars running along the bottom of each side of the longitudinal sleeper supporting the central rail.[2] The railway is worked by two endless cables driven by a steam-engine at the bottom of the incline, the cables being so arranged that the two ascending portions of the two cables pass close along each side of the carriage on one line, and draw it up when attached, and the descending portions similarly guide the descent of the carriage on the other line; and the direction of motion of the cables is reversed by reversing the engine. The cables have a strength amounting to ten times their working strain; the journey up or down is accomplished in 10 minutes; and powerful hand and automatic brakes ensure the safety of the traffic.

The Mont St. Salvatore Railway is divided into two sections, rising 700 feet and 1275 feet respectively, with gradients increasing from 1 in $5\frac{4}{5}$ up to 1 in $2\frac{1}{3}$ between the bottom and top of the lower section, and attaining at last 1 in $1\frac{2}{3}$ on the upper section; and there are two curves of nearly 15 chains radius on the lower end of the upper section.[3] The railway is laid to the metre gauge with a single line on each section, and a central, double Abt rack to control the speed of the train. A dynamo-motor at the middle station between the two sections, drives a drum, 13 feet in diameter, at this station, thereby winding up or unwinding a wire cable which passes from the driving drum round another drum, $11\frac{2}{5}$ feet in diameter, at the top of the second section, so that when the carriage on the lower section is being drawn up, the

[1] *Transactions of the American Institute of Mining Engineers*, vol. xvi. p. 203.
[2] *Annales Industrielles*, 1879, col. 520; and *Proceedings Inst. C.E.*, vol. lxiv. p. 399.
[3] *Schweitzerische Bauzeitung*, vol. xix. p. 35.

carriage on the upper section is descending; and the two meet side by side at the middle station for the passengers to change carriages. The power for driving the drum at the middle station, is supplied from a dynamo at a power station about 4½ miles distant, where the electricity is generated by turbines driven by a fall of water, and actuating dynamos. The steel-wire cable, 1¼ inches in diameter, has an ultimate strength equivalent to 9½ times its working strain of 5⅓ tons. The carriages travel at the rate of 3 feet per second; and the whole journey of 1 mile 76 feet, with an ascent of nearly 2000 feet, including the change of carriages in the middle, is accomplished in about 30 minutes. The cost of this railway was £24,000, equivalent to about £23,660 per mile, closely approximating to the cost per mile of the Stanzerhorn Railway.

Steep-incline and Cliff Railways with Water Counterpoise —Several cable railways surmounting a steep incline or ascending a cliff, having a carriage attached to each end of the cable, so that the ascending and descending carriages counterbalance one another when equally loaded, are worked by admitting sufficient water into a tank under the floor of the carriage at the top of the incline, to give this carriage an excess of weight over the carriage at the bottom with its tank empty, to overcome the inertia and friction, and make the carriages traverse the incline in opposite directions. The water counterpoise is varied according to the relative number of passengers in the two carriages; and the tank has to be made large enough to hold a weight of water sufficient to overbalance the upper carriage when this one is empty and the other is filled with passengers. On the short lines, a double line is laid throughout, and sometimes a counterpoised truck is used instead of a second carriage; whilst usually on the longer lines, a single line, or two lines laid close together, are laid above and below, with a central passing-place; and increased control is obtained by the addition of a rack, which necessitates placing the cable a little to one side of the centre of the track and of the carriage.

The Giessbach Cable Railway, 378 yards long, rising from the Lake of Brienz to a hotel on the top of a cliff 303 feet high, with an average gradient of 1 in 3⅖, and a maximum gradient of 1 in 3⅛, is laid with a single line to the metre gauge, and a Riggenbach rack along the centre of the track; whilst a passing-place for the carriages is provided midway along the line. For a length of 204 yards, the railway is carried on a wrought-iron bridge with five spans of 125 feet.[1] A carriage, about 36 feet long and weighing 5³⁄₁₀ tons, attached at one end of the steel-wire cable, counterbalances approximately a similar carriage at the other end; and the cable composed of seventy wires, and having a tensile strength of 30 tons, passes round a pulley at the top; and in the event of its breaking, it would simultaneously release a toothed rod, which, by falling into the rack, would stop the carriage. The actual load to be hauled up the incline varies from 6 up to 9½ tons, so that to set it in motion, the descending carriage must weigh from 7³⁄₁₀ to 10⅘ tons, necessitating the loading of its tank with from 1⅗ to 5⅘ tons of water.

[1] *Die Eisenbahn*, Zurich, vol. xi. p. 97.

The rate of motion of the carriages on the incline is limited to one metre, or about $3\frac{3}{10}$ feet, per second; and the journey is accomplished in 6 minutes. In order that the carriages may invariably run on to their proper lines, right and left, at the passing-place, the wheels of one carriage are made with their flanges on the inside of the rails, and of the other carriage on the outside, thereby rendering any movable mechanism at the points unnecessary. The cost of this railway was £5824, corresponding to a cost per mile of about £27,100.

The Territet-Glion Railway affords a typical example of a cable railway surmounting a steep, hilly slope, in order to provide direct, rapid, and cheap communication between a town and a village situated at a much higher elevation. The railway starts from Territet, adjoining Montreux, on the Lake of Geneva, and ascends 991 feet in a straight course of about 750 yards to Glion, on the Rigi Vaudois hill, with a gradient which, commencing at 1 in $2\frac{1}{2}$, passes gradually by a parabolic curve, in a length of about 287 yards, to its maximum gradient of 1 in $1\frac{3}{4}$, which is continued to the summit (Fig. 173, p. 292).[1] The railway consists of two lines of way laid to the metre gauge, with their inner rails placed close together, and a Riggenbach rack laid along the centre of each track to regulate and control the motion of the carriages (Fig. 174); and the two tracks diverge midway along the line for a length of 325 feet, giving a clear width at the centre of the divergence of about $5\frac{1}{2}$ feet between the inner rails, to allow ample space for the ascending and descending carriages to pass one another (Fig 175, p. 293). The steel-wire cable attached to the carriages runs on pulleys placed at intervals along each track, midway between the rack and the outer rail, and passes round a large pulley on the top of the incline, revolving in a plane parallel to the gradient. The cable has a diameter of $1\frac{3}{8}$ inches; and the maximum strain to which it is liable to be subjected in working, is $6\frac{1}{5}$ tons. The two carriages attached to the ends of the cable, and travelling alternately up and down in opposite directions, are constructed with the floors of their compartments rising in steps to adjust them to the steep incline, a plan resorted to on many of these cable railways, and resembling the type of carriage adopted on the steep Pilatus Railway (Fig. 169, p. 279), with the addition of a tank underneath the floor for holding the water counterpoise. Each carriage can seat forty passengers, and has room on a platform at one end for ten more to stand, or for luggage; and the carriage weighs $8\frac{2}{3}$ tons when both it and its water-tank are empty; $11\frac{4}{5}$ tons when fully laden with passengers, but with no water in its tank; and $13\frac{2}{3}$ tons with its tank containing the full load of 5 tons of water, but with no passengers. Accordingly, under the most unfavourable condition of a carriage fully laden with passengers at the bottom, and an empty carriage at the top of the incline, the counterpoise in a tank full of water gives the descending carriage a preponderance of weight of nearly 2 tons, which suffices to put the carriages in motion; and the amount of water introduced is varied

[1] "Die Drahtseilbahn Territet-Montreux-Glion," Emil Strub; and *Bulletin de la Société vaudoise des Ingénieurs et des Architectes*, Lausanne, 1893, p. 93, and plate 20.

according to the relative loads of passengers. The speed is controlled, and the carriages can be stopped on the incline, by a centrifugal speed regulator, by a hand-brake, by a brake which is brought into action by a weight descending when the cable slackens at the bottom of the incline, or in the event of its breaking at any point, and by a clutch which is automatically applied on the occurrence of a fracture of the cable. On these steep inclines, the works of construction are liable to be very heavy, owing to the very small scope for any material alteration of the gradient, or the introduction of curves (Figs. 173 and 174, p. 292); and the permanent way and rack have to be very solidly laid and anchored down to prevent any downward movement, so that the cost of these railways is necessarily large; whilst the expenses for stations and equipment on these short lines is considerable in proportion to their length. The cost of the Territet-Glion Railway amounted to about £18,600, which is at the rate of £43,600 per mile; but the railway, having a large traffic, appears to yield a good return on the capital expended.

The Marseilles Cliff Railway which ascends a steep hill, close to the town, on which the church of Notre Dame de la Garde is situated, and from which an extensive view is obtained, is remarkable for the steepness of its straight incline, which is at an angle of 59° 48′ to the horizontal, making the gradient 1 in 0·58.[1] It rises 237 feet; and the length of the incline is 275 feet, on which a double line of way has been laid throughout on solid masonry foundations, with a central rack in each track for controlling the motion of the carriage. Four flat, steel cables, $3\frac{3}{4}$ inches wide and $\frac{1}{2}$ inch thick, are attached to each of the two carriages travelling on the incline, each cable coiling round a separate drum at the top of the incline; and as the eight drums have a common axis, the carriages counterbalance each other. In spite of the steepness of the incline, the carriages are mounted on special framing, so as to make the carriage floor level throughout; and each carriage can hold seventy passengers, and has a tank underneath its floor for receiving the counterpoise water, the weight of which would amount to 8 tons if the tank was completely filled. The carriage, when empty, weighs 10 tons; and the tank of the carriage at the top of the incline is always loaded with 6 tons of water, which provides a sufficient counterpoise for setting the carriages in motion, even when the top carriage is empty and the bottom carriage has its complete load of passengers, weighing 4 tons on the average. Only a sufficient quantity of the normal counterpoise load of 6 tons of water is let out, when the carriage reaches the bottom of the incline, to set the carriages in motion, which varies according to the relative weights of the passengers ascending and descending, thereby saving the surplus water resulting from the counterpoise provided by passengers in the descending carriage, and returning it to the summit, which effects an economy, since the reservoir at the summit supplying the water has to be replenished by pumping. The descending carriage, with 6 tons of water in its tank, and a full complement of passengers, weighing altogether 20 tons, imposes a maximum strain of 17 tons on

[1] *Le Génie Civil,* vol. xxv. p. 73, and plate 5.

the four cables to which it is attached, which is only one-twelfth of their combined breaking strain. The cables travel along the incline on hard-wood pulleys placed at intervals of 36 feet; and in order to maintain a uniform tension on the four cables of each set in working the incline, the cables are connected with the pistons of four hydraulic cylinders, which are all in communication.

The Cliff Railway ascending from the roadway alongside the River Avon, below the Bristol docks, up to Clifton, a populous suburb of the city on high ground, possesses the peculiarities of being wholly in tunnel, and laid with four lines of way, having a gauge of $3\frac{2}{3}$ feet, to accommodate two pairs of ascending and descending carriages.[1] This railway rises 200 feet in a length of 450 feet, with a gradient of 1 in 2; and its tunnel through the limestone cliff bordering the Avon, is 28 feet wide and 17 feet high. The ascending and descending carriages, connected in pairs by a wire cable encircling a pulley at the top, revolving in a plane parallel to the incline, are placed upon framing so as to provide a level floor (Fig. 176); and the tank for holding the counterbalancing

CLIFF CABLE RAILWAY, CLIFTON.

Fig. 176.—Carriage on Track.

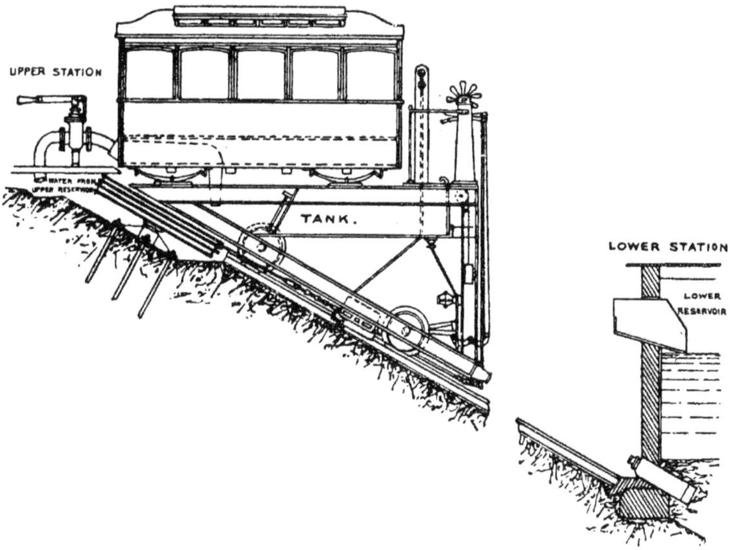

water by which the railway is worked, is contained within the framing, the water being supplied from a reservoir at the top, in quantities varying with the load to be dealt with, and discharged into a reservoir at the bottom, from which it is pumped up again. Hydraulic brakes are only prevented from gripping the rails and stopping the carriage, by

[1] *Proceedings Inst. C.E.*, vol. cxvi p. 320.

the direct intervention of the conductor; and directly his control of the brake wheel is removed, the brakes come into action.

Suburban Cable Railways worked by Stationary Engines.—Sometimes cable railways are employed for connecting a town with its suburbs, where, for instance, a town situated alongside a river is eventually extended to the neighbouring heights, as exemplified by Lyons and Havre, or where a town situated on high ground requires convenient connection with a village in the plain below, of which Lausanne and its port Ouchy, on the Lake of Geneva, furnish an example. These cable railways resemble somewhat in principle the wire-rope tramways traversing streets with steep gradients, of which some tramways at San Francisco are notable instances; but they differ from those tramways in two important respects, namely, in running over a specially constructed track instead of along a street, and, consequently also, in having the cable above the surface. The conditions of the site, and the generally flatter or varying gradients of such lines are not usually favourable for the water-counterpoise system, which has led to the adoption of stationary motors for actuating the cable.

Lyons, situated on the low-lying plain above the confluence of the Rhone and the Saone, is connected with its populous suburbs on high ground, to the north and west, by cable railways. The line to St. Fourvière and St. Just rises 320 feet in a length of 900 yards, with a gradient of 1 in 5 along its lower half, and 1 in 16½ along its upper half; and it is laid with a double line of way to a gauge of 1½ metres (4 feet 11 inches), and is carried in a tunnel for four-fifths of its length.[1] The trains, weighing 17 tons, are drawn up the incline, at a speed of 13 feet per second, by an endless steel-wire cable, 1¾ inches in diameter, and composed of eight strands of 19 wires each, carried on pulleys, and passing round a drum at the top of the incline, which is driven by a stationary steam-engine. Owing to the difference of the gradient on the two halves of the line, the tractive force required to draw up a train varies considerably on the two portions, quite irrespectively of variations in the relative loading of the ascending and descending trains. To adjust this variation, a weighted waggon is provided for each line, which only travels on the lower steep incline, being attached to its train by a cable, 450 yards long, when this train is ascending and descending the upper, flatter incline; so that this weighted waggon travels up and down the steep incline when the train on the other line is going down and up the same incline, and thereby compensates for the difference in the effective weights of the trains on the two different gradients. The cost of this line, with its equipment, was £13,360, or at the rate of £23,500 per mile.

The cable railway connecting Havre with its suburbs on a hill to the north, rises 243 feet in a length of 390 yards, with a gradient of 1 in 6⅔, on a metal viaduct 158 yards long, followed by a gradient of 1 in 2⅖ along the upper portion.[2] The line is single, except at the central passing-

[1] *Revue général des Chemins de Fer*, 1882 (2), pp. 77 and 163, and plates 6 to 10.
[2] *Le Génie Civil*, vol. xix. p. 233

place; and in order that the carriages in crossing may run on their respective lines, the wheels on one side, following the outer rail in each case, have double flanges; the wheels on the other side, following the inner rail of each line at the crossing, being flat for passing readily across the rails and cable of the other line, like the arrangement for passing on the Stanzerhorn Railway (p. 294). The cable by which this line is worked consists of six strands of 19 steel wires each, and is subjected to a maximum strain of $3\frac{2}{3}$ tons, only about one-tenth of its breaking strain; and it is guided by pulleys along the incline, and passes round a drum at the summit, 13 feet in diameter, which is driven by a steam-engine. The ascending and descending carriages, each accommodating 48 passengers, accomplish the journey in 5 minutes, including an intermediate stop of 2 minutes, making the time of transit 3 minutes; and brakes can very rapidly arrest this motion, being aided by an Abt rack laid along the track.

The Ouchy-Lausanne Railway, 1650 yards long and perfectly straight, surmounts a difference in level of 335 feet, with a maximum gradient of 1 in 9; and it is laid with a double line to the standard gauge, and is in tunnel along a portion of its length.[1] The trains are connected and worked by a wire cable passing round a drum, $19\frac{2}{3}$ feet in diameter, at the top of the incline; and the drum is driven by two turbines, $7\frac{1}{3}$ feet in diameter, turned by a head of water of 393 feet. The transit up or down is effected in about 9 minutes.

Remarks on Cable and Mountain Railways.—Cable railways possess the advantages of surmounting rapidly the abrupt differences of elevation which present in places serious impediments to intercommunication, and also of effecting the transit in an economical manner by means of counterbalancing carriages and the addition of a counterpoise of water. The actual speed of the carriages on these lines is, indeed, not rapid; but, owing to the steepness of the incline and the directness of the course, a cable railway conveys passengers from point to point far more rapidly than the journey could possibly be effected by any conveyance over a necessarily long, winding road. Though cable traction has occasionally been employed for ascending hills of moderate height, or a suitable portion of a mountain slope, or for surmounting a sudden rise of the general level of a country, its chief utility consists in affording cheap and rapid communication between neighbouring places situated at very different elevations. Moreover, the system is not adapted for the conveyance of heavy trains; and a large traffic on a through line can only be passed over a steep cable section, by taking the trains up the incline in two or more divisions. The value of these cable railways in uniting towns and their suburbs, or adjacent districts, more or less separated by physical obstacles, is forcibly demonstrated by the very large passenger traffic they attract. The direct course, however, which must be followed by these cable lines, and the importance of avoiding abrupt changes of gradient, necessitate heavy works for their construction.

[1] *Organ für die Fortschritte des Eisenbahnwesens in technischer Beziehung*, Wiesbaden, vol. xiv. p. 41.

Rack railways ascending mountains resemble light railways in keeping down the cost of works, by following along the surface as nearly as practicable by means of steep gradients and sharp curves; and they enable numbers of persons to enjoy the benefits of mountain air, and the pleasures of mountain scenery, who would be quite incapable of attempting steep ascents on foot; but such lines are not adapted for conveying heavy loads.

Rack sections on steep inclines, interspersed between ordinary gradients, and traversed by combined rack and adhesion locomotives, enable railways to be extended into mountainous districts, where gradients suitable for adhesion throughout would be either unattainable, or could not be provided at a cost at all consistent with the prospects of traffic. This combined system of rack inclines and adhesion gradients, accordingly, offers important facilities for the extension of railways into districts which might otherwise have been reasonably regarded as beyond the range of railway accommodation.

CHAPTER XIX.

TRAMWAYS.

Differences between tramways and railways—Foundations for track of tramways, materials employed—Tramway sleepers: forms, materials used—Tramway rails: kinds in use at present time; step rail, its merits and defects, used in United States; grooved rail, its advantage and disadvantage, instances of its adoption; flat-bottomed rail, value, objections to its use; flat-bottomed grooved rail, its weight, varieties of form, resistance to traction—Formation of road: forms of paving used; macadam, conditions of its employment—General design: arrangement of track, in Great Britain, in United States, arrangement of rails at passing-places—Gauges: varieties, in United Kingdom, on Continent, in North America—Methods of traction: horse, cable, steam, fireless motors, electricity—Cable traction: conditions of use, description, arrangements, merits, and defects—Steam traction: relative merits of locomotives and steam cars, instances of use—Various forms of motors: fireless locomotives; compressed-air motors; gas and oil engines; their relative merits and defects—Electrical traction: four general methods; overhead wire with trolley; underground conductors; accumulators; the surface contact or stud system; their respective merits and defects; central surface conductor on a Paris tramway, description of arrangements—Cost of construction of tramways: capital cost per mile in some countries; with different modes of traction—Cost of working: with principal systems of traction—Lengths of tramways: in North America, worked by different systems of traction, changes in traction indicated, remarkable increase in electrical traction; predominance of horse and steam traction in Europe, in United Kingdom; prospects of electrical traction—**Peculiar methods of transit on rails over water:** two methods recently adopted—Elevated car travelling along seashore: description of line from Brighton to Rottingdean, method of working, speed, cost—Suspended travelling car or transporter bridge: object, merit of system for crossing navigable waterways; examples at the River Nervion, Rouen, Marseilles, Widnes and Runcorn, Newport, Martron, Nantes, and Duluth.

TRAMWAYS possess the advantage of reducing the resistance to traction, like railways, by running with flanged wheels on smooth rails laid to an exact gauge. They differ, however, from railways in general by running along the streets and public roads, and therefore saving the purchase of land and the formation of a special route, and even from light railways using the side of the road, by running on rails flush with the surface of the road, with a groove for the flanges, instead of on elevated rails, so as not to interfere with the ordinary vehicular traffic, thereby sharing the road, along a particular line, with the other vehicles, instead of monopolizing a special track. Tramways also differ from railways in being used exclusively for the conveyance of passengers, like omnibuses, and not for the carriage of goods; and they, moreover, usually take up and set down passengers at any point on their route, and not like railways only at fixed stations, which is made easy by their very moderate speed, and rendered expedient by the comparatively short distances and populous districts they ordinarily traverse.

As tramways have their route ready-made for them in the streets and roads they pass along, the engineering interest in them is confined to the forms of rails adopted, with the character of their foundations and accessories, constituting the track, and the various methods of traction resorted to.

Foundations for Tramway Track.—Though tramways in towns are laid along streets which have been consolidated by traffic and renewals of surface, it is important to lay a solid bed of concrete under the longitudinal sleepers, or encasing the cross sleepers, so as to ensure the maintenance of the level of the rails, especially where there is a prospect of heavy traffic, and some form of mechanical traction is to be employed (Figs. 178 to 180, p. 305). Outside towns, however, in the United States, gravel and sand are commonly used for the foundation layers, to diminish the cost of construction in undeveloped districts.

Tramway Sleepers.—Timber longitudinal sleepers, resting upon cross sleepers, were used for affording the necessary continuous support for the original flat forms of stepped or grooved rails (Fig. 177, *a*); and longitudinal steel sleepers embedded in concrete are sometimes used now for carrying the grooved, flat-bottomed rails of the present day (Fig. 177, *f*). Cast-iron chairs of various patterns, laid at intervals on the foundation, for supporting shallow types of rails, connected together in pairs across the track by ties to preserve the gauge, were extensively used at one period;[1] whilst in one system, the chairs have been converted into an almost continuous, cast-iron longitudinal sleeper laid in 3-feet lengths (Fig. 177, *e*). Cross sleepers of timber and steel are very frequently used for supporting the ordinary flat-bottomed rails which are preferred in the United States for suburban tramways (Fig. 178), and the modern flat-bottomed, grooved rails so generally employed in towns[2] (Fig. 179); whilst the high, heavy rails of the flat-bottomed grooved type (Fig. 177, *g*, and Fig. 180), used with deep paving, are usually laid direct on the concrete foundation, being maintained in gauge by cross ties (Fig. 182, p. 311).

Tramway Rails.—Various forms of rails have been introduced from time to time for tramways, several of which have been abandoned as unsatisfactory; but the chief types in common use at the present

[1] *Proceedings Inst. C.E.*, vol. ciii. pp. 204 to 206.

[2] "Electric Railways and Tramways, their Construction and Operation," Philip Dawson, pp. 33 and 34.

time are given in Fig. 177. They may be classed under two groups, namely, the step rail,[1] and the grooved rail, the first three sections of rails in Fig. 177 belonging to the first group, and the four others to the second group. In each case, flat, shallow rails were first adopted, supported on longitudinal sleepers, exhibiting considerable variety in form and connections, especially in the case of the grooved rail; whilst greater

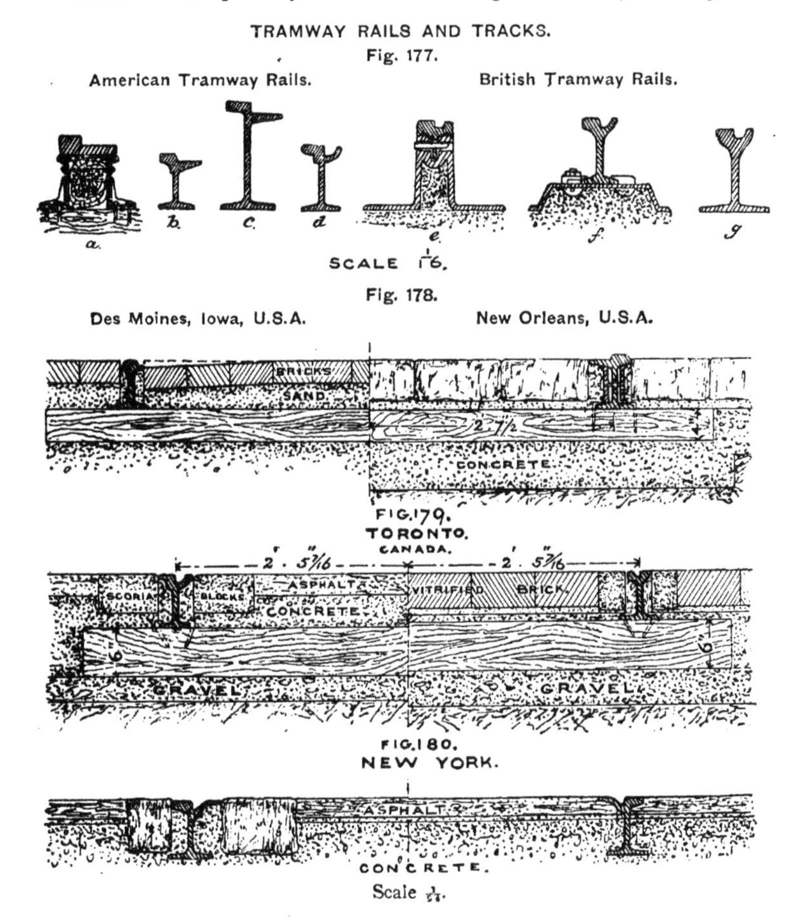

TRAMWAY RAILS AND TRACKS.

Fig. 177.

Fig. 178.

FIG. 179. TORONTO. CANADA.

FIG. 180. NEW YORK.

Scale 1/24.

simplicity and strength have been subsequently secured by the employment of flat-bottomed rails. The ordinary flat-bottomed rail used on some American tramways is merely the adaptation of a railway rail, as far as practicable, to the special requirements of a tramway, which are only imperfectly fulfilled (Fig. 178, left half).

[1] *Transactions of the American Society of Civil Engineers*, vol. xxxvii. pp. 73 and 74, figs. 1, 2, 4, 9, and 10.

The step rail with its upper surface, on which the wheels run, raised sufficiently above the surface of the road (Fig. 177, *a*, *b*, and *c*, and Fig. 178), to allow the flanges guided by the step to clear the flat surface at the base of the step, possesses the advantage of providing a clear passage for the flanges of the wheels of the tramcars, unimpeded by the dirt which is liable to accumulate in grooves, more particularly on uncared-for roads. By the projection, however, of the step above the road, this form of rail fails to fulfil the primary requirement for a tramway rail laid along a road, of not impeding the vehicular traffic on the road. Nevertheless, in the United States, as the carriage traffic is small, and carts run easily along the flat surface at the base of the step, the raised step is not generally objected to, as it would be in Europe; and it is, accordingly, used in a large number of towns, especially in the Eastern States. In the Western States, though the step rail is also used, the ordinary flat-bottomed rail is preferred (Fig. 178, left half, p. 305).

The grooved rail is commonly used in Europe, on account of its not interfering with the surface of the streets and roads; though the resistance to traction is considerably increased when the groove becomes partially filled with dirt, and consequently impedes the passage of the flanges, and especially when the dirt becomes frozen in the groove. The flat-bottomed, grooved rail has also been adopted in Canadian cities, as, for instance, in Montreal and Toronto (Fig. 179, p. 305); but in the United States, it is only employed where other forms of rails are prohibited, as in New York, where flat-bottomed, grooved rails, as shown in Figs. 177, *d*, and 180, p. 303, have been laid of late years.[1]

The railway flat-bottomed rail possesses a still greater advantage as regards ease of traction than the step rail, and it is strong and readily laid; but the space for the flanges between the rail and the central paving, is either imperfectly secured by leaving a gap between the rail and the paving-stones, which is liable to be partially closed by any displacement of the latter, or is obtained by lowering the surface of the paving on approaching the rail (Fig. 178, p. 305), which renders the rail an obstruction in the roadway.

Flat-bottomed tramway rails are connected by fish-plates, like the rails of a railway. Their weight varies within wide limits, owing to the considerable differences in their height. Thus the shorter flat-bottomed, grooved rails, resting upon sleepers, weigh generally from about 40 to 70 lbs. per lineal yard, and the high rails, bedded on the concrete foundation, from 70 to 100 lbs.; whilst the flat-bottomed rails, 7 inches high, of the South London Tramways weigh 102 lbs. per yard.[2]

A central groove has occasionally been adopted, but without any apparent advantage; whilst the ordinary side groove enables the wheels to press more directly over the central web of the flat-bottomed rail. The best form of groove is one with a vertical face adjoining the track of the wheels, and a sloping face on the inner side, so as to enable the flanges to more readily push the impeding dirt out of the groove, which is considered to be best attained by the most recent form of grooved

[1] *Transactions of the American Society of Civil Engineers*, vol. xxxvii. p. 73, figs. 2, 3, 5, and 6. [2] *Proceedings Inst. C.E.*, vol. clvi.

rail laid down in New York (Fig. 180, p. 305). The resistance to traction on a level tramway with grooved rails, has been estimated at 20 lbs. per ton on the well-kept tramways of Europe; whilst the less careful maintenance of the tramways in the United States, has been reckoned to more than neutralize the advantage to traction of the step rail over the grooved rail, making the resistance to traction 22 to 24 lbs. per ton.

Formation of Road for Tramways.—The space between the tramway rails, and for about 2 feet or more beyond on each side, is usually paved in order to maintain the track, and prevent the displacement and injury to the rails, which would occur under a heavy traffic if they protruded above a worn-down road. The best paving for withstanding a very heavy vehicular traffic, often met with in the principal streets of a large town, consists of hard stone setts, laid in rows 3 or 4 inches in width, and firmly bedded on the top of a concrete foundation. The stone must be of a quality not liable to wear smooth and therefore become slippery, but preserving a rough surface, and consequently affording a good foothold for horses, such, for instance, as Aberdeen granite with a heavy traffic, and millstone grit for less frequented roads. Cobble-stones were formerly largely used for suburban and country tramway tracks in the United States, as affording a cheap paving, and a fairly good foothold for horse traction.

Hard or vitrified bricks with a rough surface are sometimes used for paving the track of a tramway, where suitable stone is scarce or dear, and a fairly economical paved roadway is required (Figs. 178 and 179, p. 305).

Wood blocks furnish a cheap and noiseless, but less durable paving; but the blocks must be creosoted, or made of a close-grained wood, so as not to be liable to swell from absorption of moisture and dislocate the track. When the roadway is paved with hard-wood blocks on a concrete foundation, the same paving suffices for the track of the tramway.

Asphalt has often been regarded as unsuitable for tramway paving where the traffic is heavy, on account of its liability to break away at the edges alongside the rails; but this objection may be obviated by laying a row of hard blocks of stone, scoria, or brick by the side of the rails as an edging to the asphalt, strong enough to bear the shocks of the wheels running along or across the rails (Figs. 179 and 180, p. 305). Where the vehicular traffic is not very heavy, as in several of the large western towns of the United States, asphalt has been laid down in the streets alongside tramway lines, without any hard edging, with satisfactory results, even when ordinary flat-bottomed rails are employed; and a sufficient width of groove inside the rail is secured, by running a heavy car with wide flanges over the track whilst the asphalt is still soft;[1] but such a roadway could only prove durable near the rails, with the best asphalt and a moderately light traffic. In broad streets also in New York, where the traffic is light and does not follow the tramway

[1] "Electric Railways and Tramways, their Construction and Operation," Philip Dawson, pp. 28 and 35.

track, the asphalt has been carried right up to the grooved rails (Fig. 180, right half, p. 305).

A macadamized track is commonly considered objectionable for a tramway, on account of the difficulty of maintaining the level and keeping in good repair the narrow strip of road between the rails; and in a populous country district, it is certainly expedient to adopt, if possible, some cheap form of paving for tramway tracks, even outside towns. In recently settled and undeveloped localities, however, where tramways are carried beyond the limits of the new, straggling towns, with a view to their extension and the development of the district, and furnish the best means of conveyance along the roughly-formed roads, it is often necessary to rest content with macadam, as economy in construction is of primary importance for tramways under such conditions; and paving the track would render the cost of laying them down prohibitive.

General Design of Tramways.—Tramways are usually laid with only a single line in the centre of the street or road, with passing-places at suitable intervals. Along routes, however, where the tramway traffic is considerable, a double line is adopted, preventing delays in waiting at the passing-places for the car going in the opposite direction, and allowing much greater freedom in running the cars, but involving a much larger cost in construction.

In towns in Great Britain, the Tramway Act of 1870 requires that the tramway lines shall be laid along the centre of the street, and not nearer the curbstones of the footpaths than 9½ feet for a length of over 30 feet, to afford adequate facilities for the passage of vehicles. When the street is too narrow to admit of full compliance with the above conditions, one rail only is sometimes placed at the prescribed distance from the curb, and the other rail allowed to be put as much nearer as the insufficient width necessitates. In very narrow streets, however, the passage of the tramcars in both directions on one side of the street, would cause the cars running in the wrong direction as regards their side, to interfere seriously with the passage of the general traffic; and therefore, where the width is not sufficient for vehicles to pass on either side of the tramcars, the tramway has to be made with a double line, so that the cars may run on each side in the same direction as the vehicles.

In the United States, the tramways generally run along the centre of the streets in towns, and also in the centre of suburban roads less than 50 feet wide; but where the width of these roads is 50 feet or over, the tramways are often with advantage laid with a line on each side of a central carriage-way of 20 feet, with side roadways along each margin for the local traffic.[1]

The entrances to the lines of the passing-places on a single line of tramway, should be so arranged that the car may run without fail on to its proper line. This is best effected, with grooved rails, by the arrangement of rails and crossings shown on Fig. 181, in which the entering car is made to follow a straight course in leaving the single

[1] *Transactions of the American Society of Civil Engineers*, vol. xxxvii. p. 69.

line, till the flanges of its front wheels have entered the grooves in the rails of its proper line.[1]

TRAMWAY PASSING-PLACE.
Fig. 181.

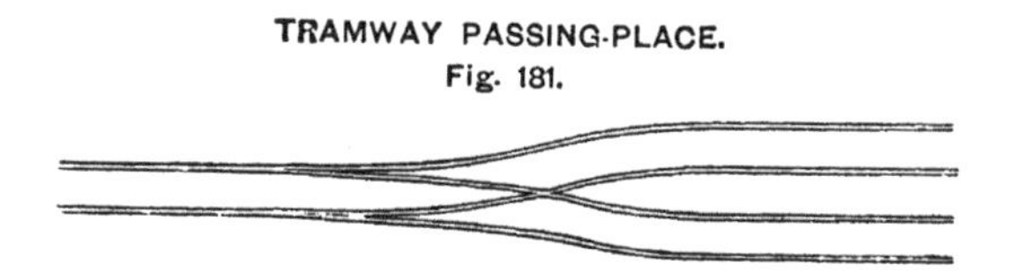

Gauges of Tramways.—The standard railway gauge of 4 feet 8½ inches has naturally been very extensively adopted for tramways; but various other gauges have also been laid down. In Great Britain, the standard gauge is the most common; but several tramways have been given a gauge of 4 feet, and also of 3 feet 6 inches, which is the gauge of the Clay Street and Sutton Street cable tramways in San Francisco, and a few a gauge of 3 feet; whilst tramways in Ireland have been laid to the Irish railway gauge of 5 feet 3 inches, to the ordinary standard gauge of 4 feet 8½ inches, to a 3-feet gauge, and one tramway to a gauge of 3 feet 6 inches.[2] The metre gauge has been resorted to on some foreign tramways, as, for instance, at Lausanne and at Belleville in Paris, and gauges of less than 3 feet in Piedmont; whilst in the United States and Canada, though the standard gauge is by far the most common, many tramways have gauges of 3 feet 6 inches, several of 3 feet, and a few of 4 feet; and some have gauges ranging between 4 feet 10½ inches and 5 feet 3 inches.[3]

The use of the standard gauge makes it possible, in special cases, to connect a tramway with a railway, provided the groove is given an adequate width; and this gauge possesses the more general advantage of coinciding with the width between most cart-wheels, so that heavy carts going along the tramway track run mainly on the rails, and therefore do not wear down the paving. With a narrow gauge, on the contrary, as, for instance, the fairly common 3½-feet gauge, the wheels on the one side of the carts often keep on one of the rails; whilst the other wheels wear down the paving at its weakest part, on the outside of the track. On the other hand, a narrow gauge reduces the cost of construction, enables the requisite width to be provided at the sides in narrower streets, and facilitates the passage of cars round the very sharp curves, of from about 75 feet down to 30 feet radius, necessitated for turning round the corners of streets, where no elevation of the outer rail is permissible.

Methods of Traction on Tramways.—Horse traction was naturally the first method adopted for tramcars, being the common system in use on streets and roads; and the tramway by greatly reducing the

[1] *Proceedings Inst. C.E.*, vol. ciii. p. 220.

[2] "Return relating to Street and Road Tramways, for the year ending the 31st of December, 1907," London. See also page 321 of this book.

[3] "Poor's Manual of the Railroads of the United States, 1900," New York, "Department of Street Surface Railroads in the United States and Canada," pp. 847–1051.

resistance to traction, enables one or two horses to do readily the work of two or more, or draw a heavier load than on roads, and also provides a much smoother motion than omnibuses. This system, however, though convenient enough along fairly level streets and for short distances, is unsuited for streets with steep gradients; and it is too slow for surburban and country lines, where long distances have to be traversed, and higher speeds are both allowable and expedient.

Cable traction for tramways was first introduced in San Francisco in 1873, where the steepness of some of the streets, reaching a maximum inclination of about 1 in 6, in the busiest quarter of the city, and their consequent inaccessibility by ordinary tramways, threatened to draw away their trade when the level portions of the city had been well supplied with horse tramways.

Steam traction was also naturally resorted to in the early days of tramways, in imitation of railways for increasing the speed and power, and reducing the cost of traction; but steam motors on streets and roads have had to be materially modified, so as to prevent the noise and visible escape of steam and smoke inherent in the working of railway locomotives, in order to avoid frightening horses and incommoding the public.

Various motors have been tried as substitutes for tramway locomotives, so as to get rid of the inconveniences entailed by burning fuel in a furnace when passing along public thoroughfares. Thus fireless locomotives have been employed, charged at first with ammonia, but subsequently with water superheated under pressure, and also compressed-air motors; whilst the gas engines more recently introduced, effect the same object in a different way.

Electrical traction has of late years become a very popular method for tramways, and has been rapidly developed, more especially in the form of the overhead wire and trolley system; but where overhead wires are considered objectionable, it has been effected by means of underground conductors, and to a small extent in special cases by means of accumulators, as the raised conductor in the centre of the track, adopted for the Liverpool Overhead Railway and other electric railways, is not available for electric tramways traversing streets.

Cable Traction on Tramways.—Cable traction can only be employed with advantage for tramcars when steep gradients occur on portions of the route, or where the traffic is considerable, so that the power imparted to the endless-wire cable running under each line of way, can be continually utilized in hauling along a succession of cars in both directions. Edinburgh, Streatham Hill, Highgate Hill, and Belleville in Paris, have furnished examples of cable tramway working on steep gradients, but in some cases are, however, in course of transformation into systems controlled by electric power; and in San Francisco, where the cars run at intervals of 5 minutes for sixteen hours a day, the cable system has been extended to the level streets.

The system comprises a longitudinal slit in the centre of each track, $\frac{3}{4}$ to $1\frac{1}{4}$ inches wide, rigidly formed by angle- or channel-irons at the top of the conduit containing the travelling cable, through which the gripper passes for catching hold of the cable, so as to communicate its motion to the car (Figs. 182 and 183).[1] The conduit under the road is lined with concrete, or iron encased generally in concrete, strengthened at short intervals by iron cross frames, and contains the pulleys supporting the

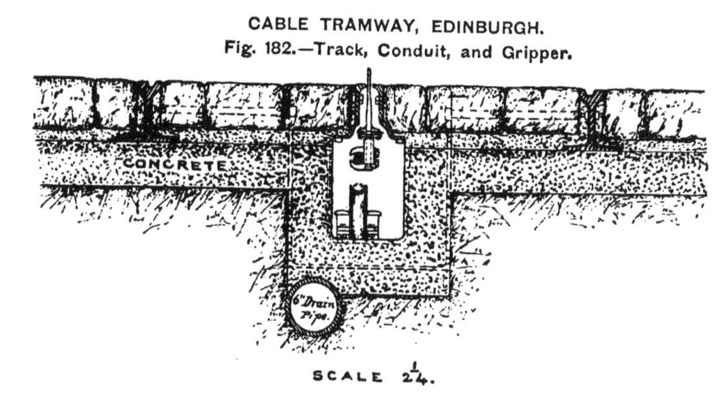

CABLE TRAMWAY, EDINBURGH.

Fig. 182.—Track, Conduit, and Gripper.

cable, being made deep enough to allow the dirt from the road, unavoidably falling down through the slit, to accumulate to some extent at the bottom without interfering with the pulleys, till it can be removed; and

CABLE TRAMWAY, BELLEVILLE, PARIS.

Fig. 183.—Track, Conduit, and Gripper.

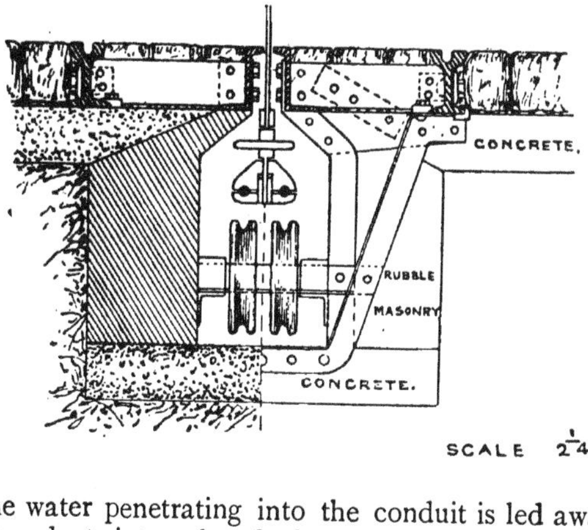

Fig. 184.—Pulleys in Conduit on Curve.

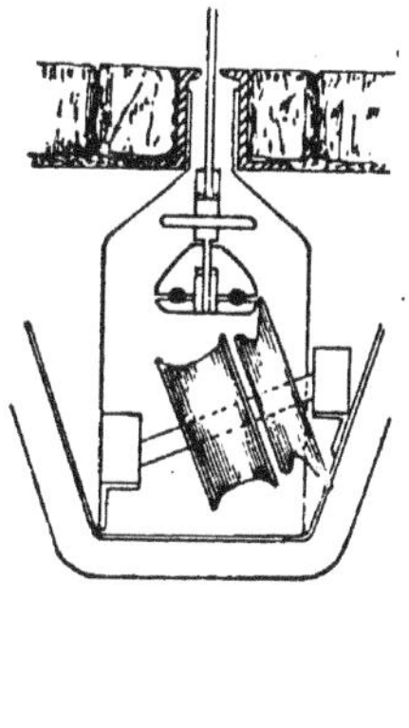

CONCRETE.

RUBBLE MASONRY

CONCRETE.

SCALE 1/24

the water penetrating into the conduit is led away by a drain. Pulleys placed at intervals of about 30 to 40 feet along the bottom of the conduit, guide and support the cable, turning on horizontal axes along the straight portions of the line, and inclined to the horizontal along

[1] *Annales des Ponts et Chaussées*, 1893 (1), p. 523, and p. 624 *bis*

curves[1] (Fig. 184); whilst at changes to a rising gradient, pulleys have to be placed at the top of the conduit above the cable, to keep it down in position whilst changing its inclination. The cable composed of steel wires wound in strands round a central core of hemp to give it flexibility, has usually a diameter of between 1 and 1½ inches; and the pulleys along the straight portions have a diameter of from 8 to 21 inches, and on curves of 16 inches to 4 feet. The cable passes at one end round two driving pulleys, or drums, of large diameter, travelling either in a parallel direction between the pulleys, or crosswise so as to wind round the pulleys in the form of the figure 8; and it passes round a large return pulley at the other end. The cable is kept taut by a pulley placed on a small trolley running on wheels, round which it turns; for the cable in shortening draws the pulley forward, which thereby raises a weight; and when it becomes slack, the weight draws the pulley back again and tightens up the cable.[2]

The gripper has to be so designed as readily to catch hold, and to leave go of the cable;[3] and it is either placed upon a special dummy car which, where the cable traction is only used for surmounting inclines on a portion of a tramway line, draws one or more cars up the steep gradients, or is located on the cars themselves, especially on tramways worked solely by cable, which arrangement has the advantage of making the cars self-contained.

Cable traction possesses the advantage of enabling the power required for working a long line of tramway to be produced at a single station, generally by means of a steam-engine, and thereby generating the necessary force economically, and proportionating it to the actual work done along the whole circuit. As, however, the cable has to be kept constantly in motion, which absorbs a considerable proportion of the whole power expended in working, the system is best adapted for a large traffic. A breakdown of the engines supplying the motive power, or a fracture of the cable, stops all the traffic till the repairs are completed, unless horses can be procured; but careful inspection during the regular periods of rest can, in great measure, prevent these accidents. The rate of motion of the cable determines the speed of the cars, and is generally comprised between 5 and 9 miles an hour, which is, on the average, notably quicker than ordinary horse traction, reckoned at 4½ miles an hour. Where different lines of cables have to be worked at different speeds from the same power station, the diameters of the driving pulleys are made different, so as to produce the respective speeds; but when a slower speed is required for a section of the line at a distance from the power station, the modification in the speed of the auxiliary cable has to be effected by pulleys inserted under the line, actuated by the main cable, which turn pulleys of smaller diameter imparting a slower speed to the auxiliary cable wound on them.

[1] *Annales des Ponts et Chaussées*, 1893 (1), p. 526.
[2] *Ibid.*, 1896 (1), p. 231, and plate 17, figs. 6–9.
[3] *Ibid.*. p. 223, and plate 16, figs. 1–11; and *Proceedings Inst. C.E.*, vol. lxxii. p. 211; and vol. cvii. p. 327, and plate 7, figs. 4–7.

Steam Traction for Tramways.—Traction by steam on tramways is effected, either by a locomotive cased round to conceal the machinery, and free from noise in working, drawing along two or more cars, or by a steam-engine placed within the car itself at one end.

A tramway locomotive is employed to the best advantage in drawing a train of three or more cars at a good speed for a long distance, and, in fact, when the conditions approximate to those of a light railway, with trains running at considerable intervals of time apart. These, unfortunately, are not the conditions applying to ordinary tramways in towns, where a very frequent service of single cars, going at a moderate speed, is the most suitable for the requirements of the inhabitants. The weight of the locomotive, moreover, necessitates the use of a heavy rail, and a solidly-laid track. Steam tramway locomotives, however, besides being able to draw heavy loads on ordinary gradients, are capable of ascending steep gradients with moderate loads, up to about 1 in 15; and though they cannot surmount the steep inclines worked by cables, they have the advantage over cables of being able to vary their speed to suit the conditions, and of proportioning their work to their load.

Tramway cars carrying their steam-engines, have the merits of being self-contained, and economical in working; but they bring the boiler, and any fumes and vibration that may accompany their working, close to the passengers. They are employed on the South Staffordshire tramways, which are also worked with locomotives and electric cars, in Paris and its environs, Berlin, Moscow, and some other continental cities.

Steam traction has been most largely developed in Italy, for working numerous lines of tramways branching out into the country for long distances from Milan, Turin, Mantua, Cremona, Bologna, and other large towns,[1] with a total length approximating to two thousand miles, on which the limit of speed has gradually been raised from 8½ to 12, or even, in some cases, 15 miles an hour; so that these tramways serve the purposes of light railways, whilst preserving the characteristic features of tramways, in following public roads, and keeping their rails level with the surface, so as not to interfere with the road traffic. Belgian steam tramways come next in length, notwithstanding the small area of the country, with less than half the above mileage, followed by France and the United States, each having under a third of the mileage of Italy. In the United Kingdom, the length of steam tramways was about 280 miles in 1900; but in 1908 the mileage had fallen to 64, and the number of locomotives from 589 in 1898 to 64 in 1908.

Various Forms of Motors for Tramways.—Fireless locomotives, in which the boiler is supplied with water heated to a high temperature under pressure, and thus stores up a large volume of steam, which is gradually let out for working the locomotive, were first employed on some tramways in New Orleans in 1873, and have since been adopted on some suburban tramways going out of Paris, Lyons, Lille, and Marseilles.[2] Whilst, however, fireless locomotives are free from the

[1] *Proceedings Inst. C.E.*, vol. cxix. plate 9.

[2] *Annales des Ponts et Chaussées*, 1878 (2), p. 261; and "La Traction Mécanique des Tramways," R. Godfernaux, Paris, p. 61.

inconveniences of a furnace and smoke, they possess the serious defect of a gradual decrease in power in transit, and are accordingly unfitted for ascending gradients towards the end of their course; and they cannot cope with considerable variations in the load to be drawn, owing to their inability to afford an increase of power when required.

Compressed-air motors, drawing their charge from a central station of supply, like fireless locomotives, have been employed upon tramway cars for traction. The air under pressure is stored up in a series of cylindrical reservoirs carried on the car, so that the requisite pressure can be more fully maintained than with a fireless locomotive, and is regulated in the admission of the air to the motor according to circumstances; two or three of the reservoirs are kept in reserve to assist, in case of the failure of the others, or the occurrence of a steep gradient; and the pressure supplied is adapted to the conditions of the line.[1] This system of traction is in operation on some tramways in Paris and its suburbs, in some other parts of France, and in Switzerland. The reservoirs of compressed air, however, with the hot-water reservoir for raising the temperature of the air before its admission to the motor, with their accessories, in addition to the motor itself, render the weight to be carried by the car even greater than when accumulators are used for electrical traction, and the air-compressing machinery is costly; so that the conditions are not favourable to a large extension of the system, in spite of its cleanliness and convenience.

Improvements in gas and oil engines have rendered them convenient motors, and economical for small powers; they are compact and self-contained, requiring only a reservoir of gas, which, when compressed, occupies a small space; and they are free from the disadvantages of the furnace, heat, and smoke of locomotives. They have therefore been very naturally tried for the working of tramways; but they labour under certain defects. In order to keep down their weight, the best forms of gas and oil engines are worked at a high speed, which should be kept fairly regular; and the intermittent character of the explosions necessitates the addition of a somewhat heavy and cumbrous flywheel, to obtain a uniform rotary motion; and whilst the regular motion required for the engine has to be adjusted to the variable speed of tramway cars by means of toothed wheels, the flywheel increases the weight to be carried and reduces the available space for passengers. Moreover, to avoid the difficulty experienced in starting these engines at each stoppage, the engine is merely put out of gear when the car is stopped, and continues working, causing a useless consumption of gas, and continuing the vibrations due to the running of the engine, which are an objectionable feature in the system. The cylinder also of the engine has to be kept cool by a current of water, involving the addition of a reservoir of water at the top of the car; and the engine is not free from smell. These inconveniences have hindered the extension of this form of motor for tramways; but, nevertheless, some tramways are worked by gas engines, as, for instance, between Blackpool and Lytham,[2] in Dresden, and at Dessau near Berlin.

[1] "La Traction Mécanique des Tramways," R. Godfernaux, p. 93.
[2] *The Engineer*, vol. lxxxii. p. 66.

Electrical Traction on Tramways.—Traction by means of electricity is effected in four different ways on tramways, according to circumstances, namely: (1) by an overhead wire along which a trolley or travelling pulley runs, attached by a long pole to the top of a tramway car, whereby the electrical current is conveyed from the wire to a dynamo-motor on the car; (2) by underground conductors, where overhead wires are objected to, communication between which and the car is provided by a suitable connector hanging down from the car, and passing through a longitudinal slit formed at the top of the underground conduit containing the conductors; (3) by electric accumulators carried on the car; and (4) by the surface contact or stud system.

The overhead-wire system is the simplest, cheapest, and most convenient, and consequently is by far the most common. The copper wires for the up and down lines of tramways are supported, either by wires stretched across the street at intervals from the walls of the buildings, or from poles on each side (Fig. 185); or by bracket arms projecting over the street, carried by a row of standards in the centre of the street or

ELECTRIC OVERHEAD-WIRE TRAMWAYS.

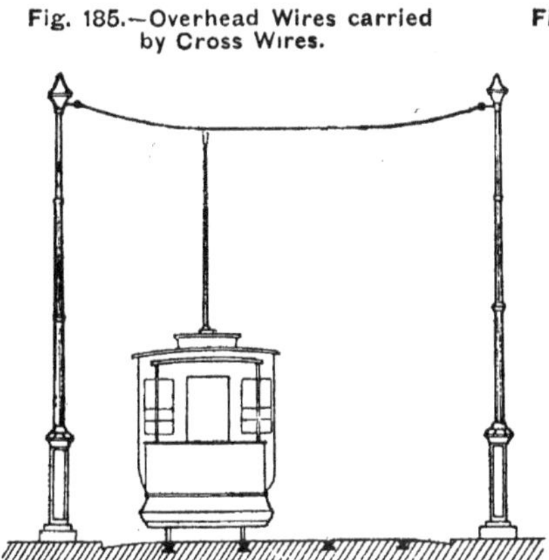

Fig. 185.—Overhead Wires carried by Cross Wires.

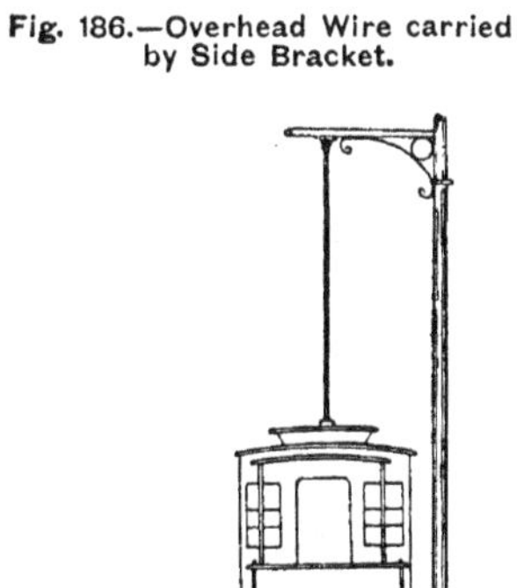

Fig. 186.—Overhead Wire carried by Side Bracket.

road, where there is ample space; or from standards on each side (Figs. 186 and 187). Sometimes the trolley-pole is connected with the car in such a flexible manner as to allow of considerable variations in the position and height of the trolley, so that much more latitude can be permitted in the course of the wire round curves, of which the South Staffordshire electric tramways furnish an example.[1] As the rails of the tramway serve as the conductor for the return current, they must be given an adequate section and be efficiently connected by copper wires, so that the current, which should be given a low potential, may not be liable to find a line of less resistance in the gas, water, or other iron pipes under the road, resulting in their damage. This system possesses the economical

[1] *Proceedings Inst. C.E.*, vol. cxvii. pp. 290–292.

advantages belonging to the generation of the power at a central station, with the consequent occasional inconvenience of a breakdown stopping the traffic: the cars carrying only the motor are light, their speed can be readily varied, and steep gradients can be surmounted; and the passengers are exempt from heat, smoke, smell, and vibrations in every form of electrical traction.

Though overhead wires furnish the best system in every respect for electrical traction in suburban districts, they sometimes become difficult to erect satisfactorily along tortuous, narrow streets, except by the aid of several additional standards and guy ropes, which are obstructive and unsightly; whilst, besides the æsthetic objections under such conditions, interference with telegraph and telephone circuits might possibly occur, and the liability to accident from the breaking of a wire is much increased in a narrow, crowded thoroughfare. Accordingly, to obviate these defects, the electrical conductors are occasionally placed in conduits underneath

ELECTRIC OVERHEAD-WIRE TRAMWAYS.

Fig. 187.—Overhead Wires carried by Central Bracket.

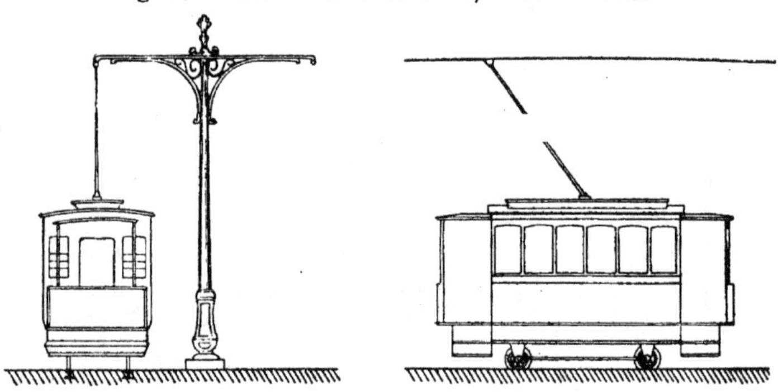

the track, similar to those used for cable tramways (Figs. 188 and 189). In these cases, the electricity is generally supplied by an insulated iron conductor of relatively large section, affording an ample surface of contact for the connector taking the place of the trolley; and the return current is provided for by a second conductor in contact with the opposite side of the connector. Underground conduits are necessarily much more costly than the supports for overhead wires, the leakage of current is liable to be greater, and maintenance and repairs are more difficult; and, consequently, underground conductors are only used when overhead wires are prohibited by the authorities. In some instances, underground conduits are adopted in the centre of a city, and overhead wires in the outskirts and suburbs; and if the cars are provided with both a connector and a pole and trolley, they pass readily from one system to the other. The conduit system has been extensively adopted in the South London Tramways, and is also in operation in New York, Washington, Berlin, and, in combination with the trolley system, at Bournemouth.

Accumulators charged with electricity, which are carried on the cars to supply power to the motors, rendering the cars self-contained, have been used in Paris and Hamburg in preference to underground conduits. The great weight, however, of the accumulators, notably increasing the load to be drawn, the cost of changing them and of re-charging, the rapid wear of their plates, and their low efficiency for ascending steep gradients, in spite of the mechanism enabling the supply of current to the motor to

ELECTRIC UNDERGROUND CABLE AND CONNECTOR.
Fig. 188.

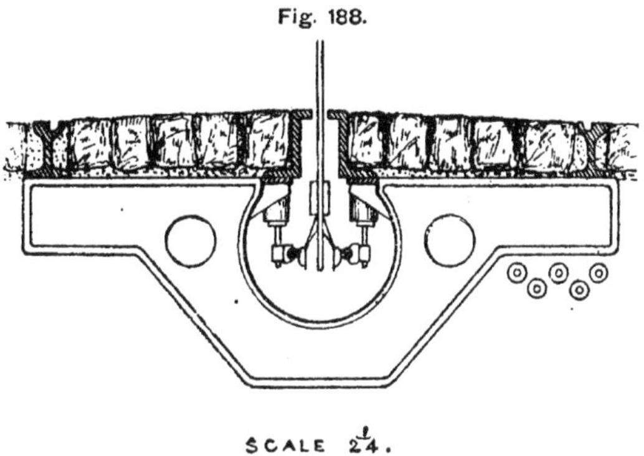

ELECTRIC UNDERGROUND CONDUCTOR AND CONNECTOR.
Fig. 189.

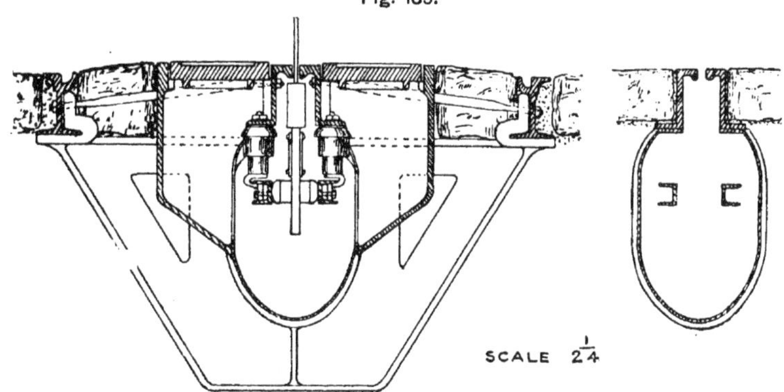

be regulated, are impediments to their extended use. Nevertheless, the system has been advantageously applied to tramways which, passing mainly through suburbs where overhead wires are unobjectionable, traverse for a certain distance the heart of a crowded town, in which the erection of these wires is prohibited; for by using accumulators on this section of the line, it is possible to avoid the cost of construction and other inconveniences of underground conductors.

The sole conditions under which an exposed, central conductor is admissible for a tramway, are, firstly, an insulated conductor with its upper surface flush with the surface of the road; and, secondly, this conductor so laid in short sections as to be free from current, with the exception of the section which the car is actually traversing, so as to avoid the possibility of pedestrians or animals receiving electric shocks in crossing the track. This system was applied in 1894 to a tramway in Lyons, nearly two miles long, and in 1896 to a tramway in Paris, running from the Place de la Republique to Romainville, about $4\frac{1}{3}$ miles in length. In the latter line, insulated steel blocks, placed $8\frac{1}{5}$ feet apart, united in pairs by a metal wire, are connected with distributors placed under the pavement at intervals of 312 feet, which derive their current from the main conductor coming from the generating station, and serve, with the aid of an angle-iron collector under the car, $10\frac{5}{6}$ feet long, and therefore always in contact with one of the blocks, to supply the current to the car through the block it is over, and to cut off the current from the block directly the car has left it.[1]

Cost of Construction of Tramways.—Considering that no land has to be purchased for the track of tramways, and no works for levelling the track, or for crossing roads, streams, or rivers, have in general to be carried out, it might naturally be supposed that the cost of construction of tramways would be small. This, however, is not the case, except where the tramways are laid in country districts along the metalled roads without any special provision for the paving of the track, as for instance in Italy, where the steam tramways referred to above, including land and buildings for stations, and rolling-stock, cost only about £2600 per mile. Assuming the cost of construction, equipment, and other charges to be represented by the capital liabilities, the average expenditure on tramways per mile, up to 1898, amounted to £19,540 in the United States, £8670 in Canada, and £14,600 in the United Kingdom; whilst in France, in 1892, it was £12,440 per mile, with a tendency to a decrease in the cost per mile, owing to the extensions consisting mainly of country lines.

The cost of tramways necessarily varies considerably according to the method of traction for which they are constructed, the condition and situation of the streets or roads traversed, in regard to cities or country districts, and the regulations imposed by the local authorities. The actual cost of construction, per mile of single line, of several of the earlier tramways in the United Kingdom, ranged between £3000 at Leicester, and £5500 in Liverpool and London;[2] whereas the capital cost of the tramways of the United Kingdom up to 1900, give an average total outlay of £11,195 per mile of single line actually open, and an expenditure of £9006 per mile of single line on the works, including land and buildings;[3] and the estimated cost of supplying electrical traction to 200

[1] *L'Éclairage Électrique*, vol. vii. p. 222.

[2] "Tramways: their Construction and Working," D. K. Clark, 2nd edition, pp. 115, 158, and 234.

[3] "Return relating to Street and Road Tramways for the year ending the 30th of June, 1900," p. 31.

miles of single line of tramways, proposed by the London County Council, in London and its suburbs, is £11,500 per mile for lines with overhead wires, and £14,500 per mile for lines with underground conductors.

In the United States, the cost of construction of tramways for horse traction in the States of Pennsylvania and Massachusetts, has averaged £3830 per mile of single line, ranging between about £1300 for unpaved tracks, and £6400 for paved streets. The cost, however, of the tramways constructed for electrical traction in these States has amounted to about double those used for horse traction;[1] whilst cable tramways in the United States, laid with a double line, have cost from £15,600 to £41,700 per mile, and might be constructed under ordinary conditions for about £21,000 per mile of double line, including the cable.[2] The cost of some of the tramways in New York has greatly exceeded these figures; but a mile of double line of city tramway, with a well-paved track, can be constructed in the United States for about £9700.[3]

In France, the expenditure on the tramway systems of the principal cities and suburban districts has varied greatly, having amounted, for a double line in most cases, to £27,050 per mile in Marseilles, £24,250 in Paris, £13,400 in Toulon, £12,760 in Havre, £6700 in Lyons, and only £1150 per mile of single line at Versailles.[4] The conversion of a horse tramway in Lyons into an overhead-wire electric tramway, cost only £1020 per mile for consolidation and equipment.

The cost of construction of the track of a mile of single line of tramway, with an underground conduit and the electrical conductors, has ranged at Blackpool, Washington, Berlin, and Budapest, between £4500 and £6400, and about twice these amounts for a double line.

Cost of working Tramways with Various Systems of Traction.—The cost of traction varies with the local conditions, as well as with the system adopted; and therefore the most reliable comparisons of the various systems are made with tramways in neighbouring localities, and somewhat similar in their conditions. The expenses of different modes of traction on the Chicago tramways, in 1894, was $10\frac{1}{4}d.$ with horses, $8\frac{3}{4}d.$ with electrical traction, and $6d.$ with cable traction, per car mile; whilst in the United States generally, the prices were $9\frac{1}{5}d.$, $7\frac{1}{3}d.$, and $5\frac{4}{5}d.$ per car mile respectively. Horse traction is always found to be more costly than mechanical traction; and from a comparison of the expenses of working and repairs with different systems on the North Staffordshire and Birmingham tramways in 1893, traction by electric accumulators was the most costly, amounting to $11\frac{3}{5}d.$ per car mile, followed by an average of $7\frac{1}{2}d.$ by locomotives, $4\frac{1}{5}d.$ by cable, and slightly over $4d.$ per car mile for electric traction with overhead wires.[5] The working expenses of mechanical traction in a city such as Paris, has been reckoned at $6\frac{3}{5}d.$ per car mile with gas, $6\frac{2}{5}d.$ with compressed air,

[1] *Annales des Ponts et Chaussées*, 1896 (1), pp. 91 and 112.

[2] "Street Railways: their Construction, Operation, and Maintenance," C. B. Fairchild, New York, p. 132.

[3] *Ibid.*, pp. 317 and 318; and *Transactions of the American Society of Civil Engineers*, vol. xxxvii. p. 120.

[4] *Annales des Ponts et Chaussées*, 1896 (1), p. 114.

[5] *Proceedings Inst. C.E.*, vol. cxvii. p. 293.

$5\frac{1}{5}$*d.* with accumulators, 5*d.* by locomotives, $4\frac{3}{4}$*d.* with overhead wires, and $4\frac{1}{7}$*d.* by cable.[1]

Lengths of Tramways worked by Respective Systems.—The most notable feature in the working of tramways of recent years, is the great increase of electrical traction with overhead wires in the United States, the birthplace of tramways in their modern form, accompanied by a considerable decrease in horse traction, as shown by the following statistics.[2]

UNITED STATES TRAMWAYS.

Methods of traction.	Horse.	Electric.	Steam.	Cable.	Total.
Year.	Miles.	Miles.	Miles.	Miles.	**Miles.**
1893	3,497	7,466	566	657	**12,186**
1897	947	13,765	467	539	**15,718**

UNITED STATES AND CANADIAN TRAMWAYS.

Methods of traction.	Horse.	Electric.	Steam.	Cable.	Total.
Year.	Miles.	Miles.	Miles.	Miles.	**Miles.**
1891	5,442	3,009	1,918	660	**11,029**
1898	663	16,306	535	460	**17,964**

The tramways of Canada in 1898 comprised 9 miles of horse tramways, 634 miles of electric tramways, and 30 miles worked with steam and other modes of traction, the heading under which steam is comprised in the returns for 1897 and 1898 being denoted as various. The above tables manifest the remarkable development of electrical traction, which increased in extent more than fivefold in seven years; whilst horse traction shows signs of becoming quite insignificant, if it does not wholly disappear. Cable traction has decreased somewhat, owing to the great preference exhibited for electrical traction; but it is not likely to be superseded on steep inclines. Steam and other motors were greatly affected at first by the introduction of electricity for tramways; but the length of tramways worked by them has remained fairly stationary of late years.

Though horse traction and steam traction have hitherto maintained a very decided predominance on European tramways, extending over more than 4000 miles in each case, electrical traction is being gradually extended on the tramways in France, and has been more

[1] "La Traction Mécanique des Tramways," R. Godfernaux, p. 361.

[2] *The Street Railway Journal*, New York, vol. vii. p. 538; vol. x. pp. 33 and 147; vol. xiii. p. 684, table; and vol. xv. p. 101, table.

largely developed in Germany; but the total length of tramways in Europe worked by electricity, which has been given as 840 miles in 1896,[1] was quite insignificant in comparison with the extent of these tramways in North America, and considerably less than their extension in a single year in the United States.

In Great Britain and Ireland electricity is rapidly superseding other forms of traction on tramways.

The total length of route open for public traffic in 1898 was 1064 miles; in 1908 this total had risen to 2464¼ miles, of which 2286 miles were worked by electric traction, 52½ by steam, 27½ by cable, 94¼ by horses, and 4 by gas motors.

The total number of horses employed was, in 1898, 38,777; but in 1908 the total number was only 5288, showing a reduction of 33,489 horses in ten years.

Steam locomotives had also, as previously stated, fallen from 589 in 1898 to 64 in 1908; while the total number of cars in 1908 was 12,049, only 1141 of which were non-electric.

The total number of passengers carried for the year ending 31 March, 1908, was approximately 2625 millions by all systems, while the quantity of electrical energy employed was nearly 432 millions of Board of Trade units.

The tramways throughout the United Kingdom, as stated on p. 309, are laid to various gauges, and of the total of 2464¼ miles open to public traffic in 1908, 55½ per cent. were of 4 feet 8½ inches gauge, 24½ per cent. of 3 feet 6 inches gauge, 7¼ per cent. of 4 feet 7¾ inches gauge, 7¾ per cent. of 4 feet gauge, and 5 per cent. of gauges of 5 feet 3 inches, 4 feet 9 inches, and 3 feet.[2]

PECULIAR METHODS OF TRANSIT ON RAILS OVER WATER.

Two methods have been recently adopted for conveying passengers over water, by means of cars transported along rails, which cannot properly be classed either under railways or tramways. The first system consists of a platform and covered saloon, raised upon four high, tubular legs resting upon wheels running on rails laid on the bed of the sea foreshore, below high-water mark, between Brighton and Rottingdean; and in the second system, a car is suspended from a truck running along rails laid on a high-level suspension-bridge platform spanning a river.

Elevated Car travelling along the Seashore.—Four lines of rails have been laid down, fastened to blocks of concrete bedded in the chalky or rocky foreshore, about 3 feet apart, between Brighton and Rottingdean, a distance of 3 miles, on which an elevated car ran with its platform raised 24 feet above the bed of the foreshore, which is covered by the tide to a maximum depth of 14 feet at high water.[3] The two outer rails are placed 18 feet apart, to give stability to the high car in its exposed position: and each inner rail is laid 2 feet 8½ inches inside the corresponding outer rail, thereby forming two lines of way of this gauge,

[1] "The Dictionary of Statistics, 1899," M. G. Mulhall, p. 814.
[2] Return of Street and Road Tramways and Light Railways, 1908.
[3] *Engineering*, vol. lxii. p. 711.

upon which the four four-wheeled trucks carrying the four supports ran in pairs (Fig. 190). The supports consist of steel columns, 11 inches in diameter, braced together and carrying brackets at the top. The line has a maximum gradient of 1 in 300; and the sharpest curve has a radius of 40 chains. The raised platform, 45 feet long, and 22 feet wide, carried a saloon which had seats inside, and on the roof, for 150 passengers.

SEASHORE RAILWAY.

Fig. 190.—Brighton to Rottingdean.

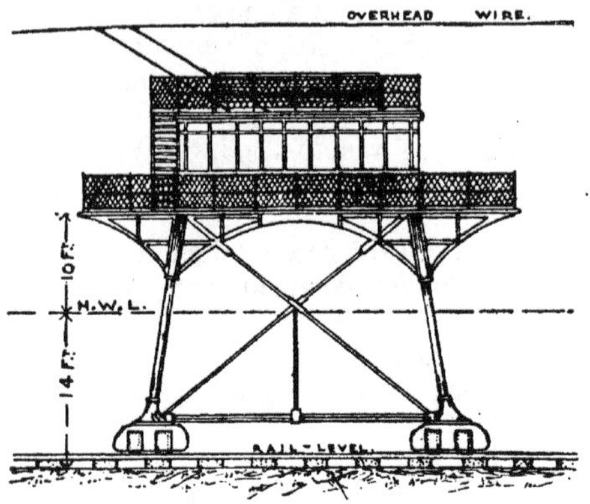

The car was propelled by two shafts passing down two of the legs and actuating the wheels, driven by electric motors placed on the platform, which were supplied with current by two trolley wires from a generating station at Rottingdean. Rods for working brakes applied to the wheels of two of the trucks, pass down the other two legs, in order to regulate the motion. The speed of transit was about 6 miles an hour; and the wetting of the rails when the tide covers them, by greatly reducing the adhesion of the wheels, and therefore favouring slipping, if the wheels are driven hard, placed a limit on the speed attainable. The line was opened in November, 1896; but considerable damage was done to the line, the cars, and the landing-piers by the sea during a severe storm in the following month, which was duly repaired. The line, with its equipment and landing-piers at each end, cost about £10,000 per mile to construct.

Suspended Travelling Car.—In order to enable passengers and goods to be transported across a navigable waterway used by masted seagoing ships, without the erection of a movable bridge at places where its river piers might impede navigation, and at the same time provide ample headway for the vessels, without the long, inclined approaches necessary for a high-level bridge, a plan has been devised of hanging a car from a truck travelling along rails on a suspended, high-level platform (Fig 191). The car, by being suspended at a low level, dispenses with any need for high approaches, and by travelling rapidly across the waterway, causes a very temporary obstruction, which can be readily kept out of the way when vessels are passing the place of crossing. The suspended platform is placed at a sufficient height to afford the requisite headway; a heavy roadway is dispensed with, as only the passage of the truck on a line of rails has to be provided for; and the car with its contents is the only movable load to be supported in addition to the truck and the hanging gear, so that the skeleton bridge can be of light construction. The car, in fact, fulfils the objects of a commodious ferry, without the inconveniences of the currents, waves, and changes of level experienced when crossing on a tidal river in an exposed situation.

The first construction of this type was erected across the River Nervion, near its mouth facing the Bay of Biscay, and opened in 1893, affording a clear headway of 147 feet, where the width of the river between the quays on each bank is 525 feet,[1] which is traversed in one minute by the suspended car, carrying 150 passengers, and weighing nearly 40 tons with its load, accessories, and truck (Fig. 191).

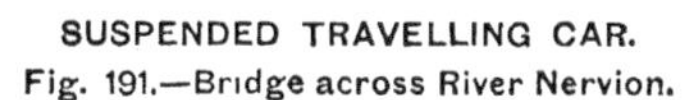

SUSPENDED TRAVELLING CAR.
Fig. 191.—Bridge across River Nervion.

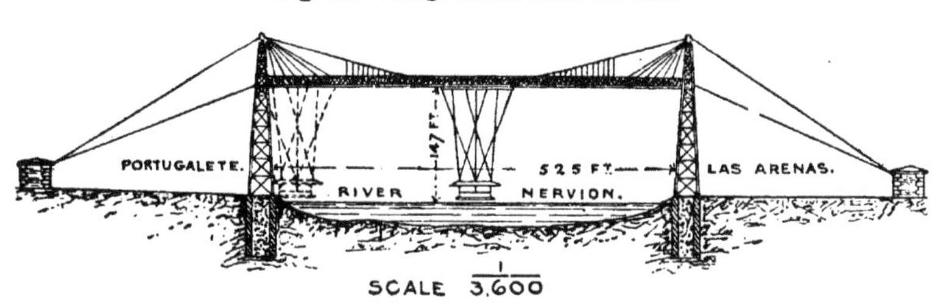

A similar structure was erected across the Seine, at Rouen, in 1898–99, for providing an easy means of communication between the quays on the two banks at the port. In this case, the approaches to a high-level bridge, in addition to the cost of construction, would have involved the purchase of very valuable property on the Rouen side of the river, and have interfered with the traffic along the quays on both banks; whilst a movable bridge would to some extent have impeded the free waterway of the port. The river at the site of the bridge is 469 feet wide; and steel, lattice towers erected on the banks, 217 feet high, support a skeleton suspension bridge, affording a clear headway of 167 feet, along which the sixty rollers of the trolley carrying the thirty steel cables bearing the travelling car, run on four rails.[2] The suspended car, 33 feet long and $42\frac{2}{3}$ feet wide, has a central carriage-way $26\frac{1}{4}$ feet in width, with footways on each side. The car, when empty, weighs 20 tons; and the trolley running on the bridge, together with the electric motor, weighs about 18 tons; and as the maximum load liable to be borne by the car is 36 tons, the moving load carried by the bridge is 74 tons, without the suspending cables. The car is so hung that its floor is level with the quays on each side of the river; and it crosses the river in half a minute.

The Transporter Bridge across the old harbour at Marseilles has a central span of 541 feet with a clear headway of 164 feet above Low Water. The supporting towers are 284 feet in height above the same datum.

[1] *Le Génie Civil*, vol. xxiii. p. 229.

[2] "Atlas des Voies Navigables de la France, 2e Série, 5e Fascicule, Navigation entre Paris et la Mer," Paris, 1899, p. 41, and plate 23, fig. 52, and plate 37.

The transporter car is 33 feet in length by 39 feet in width, and is suspended at the level of the adjacent quays.

The Widnes and Runcorn Transporter Bridge opened in 1905, which crosses the River Mersey and the Manchester Ship-Canal, has a central span of 1000 feet, centres of towers, with four approach spans of 55½ feet on the Widnes side, and one span on the Runcorn side. The bridge is of the ordinary cable suspension type with stiffening girder, and the traffic is carried between low level approaches on each side by a car 55 feet long and 24½ feet wide suspended from a trolley 77 feet long and articulated, running on the lower flanges of the stiffening girders, and propelled electrically. The headway under the stiffening girder is 82 feet above High Water, and the height of the towers above the same datum to the centre of the cable is about 170 feet. The headway in this case was mainly determined by that available under the London and North Western Railway Bridge at Runcorn which is only 75 feet at High Water.

The Transporter Bridge over the River Usk at Newport, Monmouthshire, is of the suspension cable and stiffening girder type, with a central span of 645 feet centres of towers. The clear headway above High Water level to underside of girders is 177 feet, and the height of the towers above level of approach road is 242 feet.

The platform car is 33 feet long and 40 feet wide, suspended from a travelling frame 104 feet long propelled electrically.

Similar transporter bridges are found at Martron, span 461 feet, height above water level 180 feet; at Nantes, span 465 feet, height above water level 165 feet; while a transporter bridge at Duluth, Minnesota, with a span of 393¾ feet, and height above water level of 135 feet, has a supporting girder of the rigid trussed type in lieu of suspension cables.

PART III.

RIVER AND CANAL ENGINEERING, AND IRRIGATION WORKS.

CHAPTER XX.

CHARACTERISTICS, FLOODS, CONTROL, AND REGULATION OF RIVERS.

River basins—Variations in rainfall, and in discharge of rivers—**Characteristics of Rivers:** influence of strata and forests on floods, torrential and gently-flowing rivers; fall of rivers, and descent of detritus, change in fall, origin and course of detritus; influence of lakes; river channel, suited for average discharge—**Floods of Rivers:** periods and effects, rise of Nile, winter and summer floods; warnings of floods, arrangements for predictions—**Control of Rivers:** methods; afforesting mountain slopes; arrest of detritus, by dams across torrents; removal of obstructions from river-bed; enlargement of channel, increasing capacity for discharge; straight cuts, increasing fall, aided by catch-water drains; embanking rivers, advantages, objections to high embankments of raising flood-level and increasing damage of breaches; raising of river-bed due to embankments, results, instances of disastrous consequences; pumping for preventing inundations—**Regulation of Rivers:** natural defects of rivers; removal of hard shoals; improvement of channel at bends, cause of shoal at crossing, and means of lowering it; regulation of channel by dykes, only applicable to large rivers, forms of dykes and objects, instances of regulated rivers; remarks on regulation works.

RIVERS owe their existence to the rain which, falling on various distinct drainage-areas, is carried to the sea along the channels which were originally formed, and are maintained by its discharge, and wind through the lowest part of the valleys which they drain. The size of rivers depends upon the extent of their drainage-area or basin, and the amount of rain which, falling on this area in a given period, finds its way into the river. The areas of river basins vary with the extent and configuration of the country, ranging from the insignificant drainage-areas of small streams flowing straight into the sea, up to large portions of vast continents, such as the Volga basin in Europe extending over 563,000 square miles, the Obi basin in Asia of 1,300,000 square miles, the Nile basin in Africa of 1,500,000 square miles, and the Amazon basin in America of 2,250,000 square miles. The rainfall also varies considerably, not only in different parts of the world, ranging from zero in some deserts up to a yearly average of 474 inches at Cherrapunji in Assam, but sometimes even within the limits of a single river basin; for the average annual rainfall in such a relatively small and uniform basin as the Seine, is under 24 inches in some parts and over 48 inches in others. Moreover, in tropical countries, there is very commonly a rainy season and a dry season, the rivers being in high flood during the first period, and very low, or in some cases running nearly dry, in the second period. Even in temperate regions with a very variable climate, the year may be divided, as regards river floods, into two fairly distinct periods, namely, in the northern hemisphere, the warm season from May to October, when floods are rare and never very high, owing to the great reduction in the rainfall which actually reaches the river, caused by the active evaporation which prevails during this period; and the cold season from November till April, when, in the absence of evaporation, most of the rain which falls finds its way into the river. Rivers, however, which derive most of their flow from glaciers, have their floods in the summer, when the glaciers and snow of high mountains are being melted by the heat.

CHARACTERISTICS OF RIVERS.

Influence of Strata and Forests on the Floods of Rivers.— The permeability or impermeability of the strata forming the surface of the basin of a river, and the extent to which vegetation covers it, profoundly affect the nature of the floods of the river. Where impermeable strata constitute the basin, such as clays, lias, and primitive rocks, most of the rain which falls over the drainage-area when evaporation is inactive, flows off rapidly from sloping ground into the river. The rain sinks, on the contrary, into permeable strata, such as sand, gravel, and chalk, and is both removed from the influence of evaporation, and delayed in its passage to the river. Accordingly, rivers draining impermeable strata have floods following quickly on the rainfall, which rise rapidly to a considerable height, and also subside quickly on the cessation of the rain; whilst the floods of rivers draining permeable strata rise more slowly, and are not so high, but continue for a longer period. Torrential rivers, moreover, differ from gently-flowing rivers, not only in the sudden rise and height of their floods, but also in their dry-weather flow; for rain falling on flat, or only slightly sloping impermeable strata, is subject to rapid evaporation in warm weather; and therefore the flow of torrential rivers is liable to fall very low in warm, dry weather, unless they derive their supply from glaciers and mountain snows. The summer rain, however, falling on the permeable strata of the basin of a gently-flowing river, is less exposed to evaporation, and therefore is able to some extent to replenish the river; whilst the springs supplied by the percolation of winter rains through the permeable strata, assist in maintaining the summer discharge of a gently-flowing river.

Rain falling on bare mountain slopes flows rapidly down to the valley, and notably augments the floods of the torrential river draining the mountain valley. Forests and vegetation covering the hillsides retard the descent of the rain, and thereby regulate the flow of the river. Extensive clearings of forests in hilly regions have, indeed, proved very prejudicial, in increasing the torrential character of a river and the height of its floods.

Fall of Rivers, and Descent of Detritus.—The fall of a river corresponds to the general slope of the valley along which it flows; and as rivers usually rise in mountains, and traverse plains before reaching the sea, the fall is generally rapid near the source of the river, and becomes quite gentle towards its mouth. Since the rate of flow depends mainly upon the fall, rivers often commence their career in mountainous regions, as rushing torrents with a small but rapid flow, and end their course as gently-flowing rivers meandering through flat, alluvial plains, with a large discharge collected from their various tributaries, and a slow current.

In addition to water, many rivers bring down a considerable quantity of detritus, mainly collected from the disintegration of the surface of mountain slopes by the weather, frost, and glaciers, the smaller particles being carried in suspension by the current, and the larger *débris* being rolled along the river-bed. Rushing torrents in high flood roll boulders down the steep slopes; but the rivers in emerging on to the plains, have their powers of transportation greatly reduced, owing to their diminished fall; and the boulders which succeed in reaching the base of the mountains, in spite of constant abrasion in their descent, are either left on the plain at the change of fall, or are gradually triturated against one another by the river floods, till the pieces become sufficiently reduced in size to be carried along by the enfeebled current. Eventually, the *débris* brought down is so reduced by attrition, or sifted in its course, that only sand, clay, and silt are discharged at the outlet.

Influence of Lakes on Rivers.—Lakes exercise a twofold influence on the rivers which traverse them, for they receive all the *débris* brought down by a river from above, which settles in the still water to the bottom of the lake; and by spreading the flood-waters of the river falling into them over their wide expanse, their water-level is little raised thereby. A river, therefore, issuing from a lake is free from silt, and has a regular flow. Thus the River Neva deposits its burden of alluvium in Lake Ladoga, thirty-six miles above its mouth; and the River St. Lawrence has a very regular flow, on account of its being fed by the large lakes of North America.

River Channel.—The channel or bed of a river has been naturally formed in bygone times by the scour of the rain-water which it carries down; and sometimes a rocky barrier checks this scour, forming rapids; whilst changes in the fall of the bed produce frequent alterations in the section of the channel, and in the rate of flow. High floods do not occur frequently enough, and are not sufficiently confined to the channel, to scour out an adequately large channel to carry off the whole of their discharge; and though river floods temporarily deepen an alluvial bed, deposit occurs on the slackening of the sediment-bearing current as the floods abate, and neutralizes the previous erosion. Accordingly, a river channel is adapted by nature for conveying an average discharge from its drainage-area; and when floods come down, the rivers overflow their banks, and inundate the adjoining low-lying lands. Rivers, moreover, in traversing flat, alluvial plains, usually adopt a winding course which, besides checking the flow, increases the length of the channel, thereby reducing its fall, and consequently its capacity for discharge; and the progressive erosion of the concave bank in the bends gradually augments the sinuosities of the river.

FLOODS OF RIVERS.

Periods and Effects of River Floods.—Rivers in tropical countries with well-defined rainy and dry seasons like India, and rivers which derive their supply from tropical rainfall such as the Nile, have distinct periods of flood and of the dry stage. Thus the Nile at Cairo begins to rise at the end of June or the beginning of July, and rising rapidly throughout July and August, attains its maximum height at the commencement of October, soon after which it falls at a gradually decreasing rate, till it reaches its lowest level in June. The actual rise and fall, however, exhibit fluctuations from year to year.

In temperate regions, where the climatic changes are much less regular, the floods of rivers are irregular as regards their period, duration, and height, with the exception that though the rainfall may be quite as large in the summer as in the winter, evaporation prevents floods being as frequent or as high in the summer as in the winter.

Egypt owes its fertility to the alluvium which the Nile floods spread over the land, and the flood-water with which it is irrigated; and a low Nile inflicts serious loss on the cultivators of the soil. Winter floods in temperate zones are also beneficial to some extent in certain cases, when flowing gently on to the land. Winter floods, however, may be prejudicial on low-lying lands, and are liable to be disastrous when towns are situated at a low enough level to be exposed to their inundations. Summer floods are most injurious to growing crops and grass.

Warnings of the Approach and Height of Floods.—A flood in the lower part of a river is the result of a combination of the floods of its upper portion and of its tributaries, in so far as they may coincide in their arrival at the place under consideration. In order, accordingly, to predict the time of arrival, and the height of a flood at a given place, it is only necessary to ascertain the rise and height of the upper river and the tributaries by means of fixed gauges at suitable points, and to observe the length of time floods of definite heights take in passing down from each of these points to the station where the warning is to be given. Then on the arrival of a flood in the upper valley, the height of the river and tributaries at certain times at the several points are telegraphed to the main station, and in combination with the observed periods of the transit of the floods, enable the greatest flood-rise at the station, and the time of its arrival, to be calculated beforehand, and the necessary warnings issued. This system of predicting floods was first adopted on the Seine in 1854; and it has been since established on other continental rivers, and also on the Ohio.

CONTROL OF RIVERS.

The rain which falls over the basin of a river must find its way to the sea along the channels provided for it by nature; whilst, in large basins, the flood discharge is liable occasionally to be so large, owing to an excessive rainfall and a rapid succession of floods, heavy rain falling on melting snow, or an exceptional coincidence in the arrival of the floods of various tributaries, that it could not be completely controlled. Nevertheless, precautionary measures can be adopted with great advantage to mitigate floods, by regulating their flow, by preventing a rapid descent of detritus from obstructing the channel on reaching the plains, by improving the facilities for discharge, and by protecting important sites, by embankments, from inundation.

Afforesting Mountain Slopes.—Bare mountain slopes are subject to denudation by the effects of the weather, and by the washing away of the soil by the rapid descent of the rainfall. Under such conditions, the rain finds its way with increasing rapidity into the river, augmenting the torrential character of its discharge; whilst a greater amount of detritus is carried down by the torrent into the main river, thereby promoting the obstruction of the channel as it emerges from the hills, or devastating the neighbouring plains with the *débris*. These progressive deteriorating effects may be prevented by promoting the growth of trees and vegetation in mountainous regions, and thus regulating the flow of the river by retarding the influx of the rain, and protecting the surface of the slopes from degradation. The injurious results of the reckless clearing of forests in the upper part of river valleys, carried on formerly without check in parts of the Continent and America, have been recognized; and steps have been taken by the Governments for the reafforesting of those regions.

Arrest of Detritus carried down Torrential Channels.—Torrents descending mountain valleys with a rapid fall, erode their channels, occasionally cause landslips, and carry down large quantities of detritus into the plains below. By constructing solid dams at intervals across their channel, the velocity of their flow and their erosive action are checked, and the amount of detritus carried down by them is materially reduced. In some mountain valleys in California, such large quantities of mining *débris* were sent down in the streams, that dams have been erected in places across the narrow valleys, forming hollows in which the *débris* can accumulate, and is thus no longer carried down to the plains to the detriment of the river channel.

Removal of Obstructions from River-bed.—The mitigation of floods is greatly assisted by the removal of obstructions from the channel, so as to render it as efficient as practicable for the discharge of flood-waters. Rocky shoals unaffected by scour, a contracted channel, boulders, trunks of trees, masses of vegetation, such as the "sudd" which obstructs the Nile above Khartoum, the wide piers of old bridges, solid weirs, mill-dams, and fish-traps present serious impediments to the passage of floods, and raise the flood-level in proportion to the obstruction they offer. By the systematic removal of all such obstacles, the height of floods on a river may be notably reduced; but the improvement of one section alone, by facilitating the discharge through this section, would increase the height of the floods in the portion immediately below it.

Enlargement of Channel and Straight Cuts.—An obvious method of increasing the discharging capacity of a river, is the enlargement of its channel. This expedient, however, is too costly for general adoption, and is only advisable where very valuable interests are at stake; and, moreover, it is open to the objections, that the channel will be liable to silt up soon to its original section if much sediment is brought down by the river; that if the river is deepened, the water-level will be proportionately lowered in dry weather to the prejudice of the adjacent land; and if it is merely widened, the shallowness of the river at its low stage will be augmented.

Straight cuts, by removing the worst bends of a river, reduce its length, and therefore improve the fall of the channel, and consequently its discharging capacity; but unless the natural fall is very slight, and the banks are protected, irregularities are produced in the flow promoting the formation of shoals below the cuts, and the sinuosities are liable to be formed again by the renewed erosion of the banks. Where, however, a river traverses flat, low-lying lands with a very slight fall, the straightening of the river throughout to gain all the available fall, and the provision of straight subsidiary channels with independent outlets, for supplementing the deficiency in the discharging capacity of the main river, are required for preventing inundations. This system has been long adopted with success for the drainage of the Fen districts of England, where the land, having to a large extent been originally reclaimed from the sea, is at such a uniformly low level that very little fall is available; the straightened channel of the River Witham, for instance, having a fall of only four inches in a mile between Lincoln and the sea.

Where rising ground borders very flat plains, the rain falling on the high ground flows down quickly to the plain below, and aggravates the floods of the river traversing the low-lying land. The floods thus occasioned have been mitigated by forming catch-water drains along the slopes of the higher land, so as to collect the water before its descent to the plains, and utilizing the greater fall from the high ground, convey it to the river lower down, and thus relieve the river above, or the inundated plains, of this water.

Embanking Rivers.—The formation of a continuous line of embankment along each side of a river, protects the adjacent land from inundations, by retaining the flood-waters within the embanked channel; and it is sometimes combined with the rectification of the river, as exemplified by the embanking of the straightened River Witham for the thirty miles between Lincoln and the sea. Low embankments suffice to retain the comparatively moderate summer floods in temperate regions;

and these embankments, being overtopped by the winter floods, do not interfere with the spreading over the land of the fertilizing alluvium brought down by these floods. Generally, however, the embankments are designed to be high enough to protect the land completely from inundation; and this provision is essential where towns have been built considerably below the flood-level. The system of embanking possesses the advantage of leaving the ordinary water-level of the river quite unaltered, whilst providing an enlarged channel for the discharge of floods in proportion as the embankments are set back from the river banks (Fig. 192). In theory, high embankments, well set back, afford a perfect protection against floods; but in practice, some serious objections to their employment have become manifest.

Flood embankments should be raised above the highest flood-level, be perfectly continuous along the parts to be protected, and thoroughly watertight. The confinement, however, of the flood-waters within embankments raises the flood-level, especially as an adequate setting back to obviate this would involve a considerable loss of land; and on many rivers an exceptional flood, due to an unfavourable combination of causes, occurs about once in ten years, and overtopping the embank-

EMBANKING RIVER FOR PREVENTING INUNDATIONS.

Fig. 192.

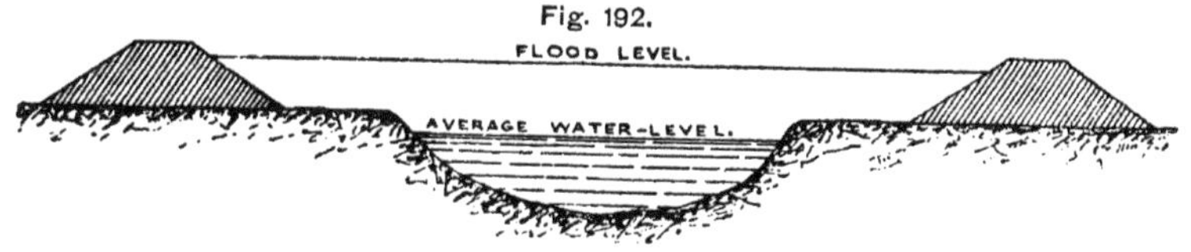

ment soon creates a wide breach, and, rushing out with great velocity, devastates the surrounding country. The clearing also of forests, and the extension of land-drainage, have, in some cases, gradually raised the height of the floods; the flood-level of the River Po, for instance, having risen about seven feet in the last two centuries from these causes, combined with the restriction of its flood-waters by the extension of its embankments. The fertile, alluvial plains bordering the Mississippi below the confluence of the Missouri, have been protected by a very extensive system of embankments, termed levees, which have been gradually prolonged and consolidated; but these levees have not yet been made quite continuous, breaches occur yearly at weak places when the river is in flood, and during exceptional floods, occurring at intervals, the levees are sometimes overtopped at several points and wide crevasses formed, liberating very large volumes of water. The last-named injury might, indeed, be prevented by an adequate raising of the embankments; but the cost involved in a general raising might possibly exceed the loss from occasional damage. The greater, however, the height of an embankment which is overtopped, the greater is the injury inflicted by the increased rush and volume of water liberated.

The high, rapid floods which come down the torrential River Loire at intervals of about ten years, usually burst through weak places in the embankments, which have been constructed for protecting the Loire valley below the confluence of the River Allier, and inundate the adjacent plains. To avoid the injuries caused by the sudden outburst of these uncontrollable floods, it has been proposed to form openings in the embankments at suitable places, amply protected from scour by pitching, and at a sufficiently high level to retain the floods in ordinary years, so as to provide a gentle means of escape for the exceptional floods, and thus render comparatively harmless the unavoidable, occasional inundations caused by them.

Raising of River-Bed resulting from Embankments.—The most serious objection to the prevention of inundations by embanking a river, is the raising of the river-bed by the deposit of alluvium, which occurs on some rivers in consequence of the confinement of the flood-waters. A river coming down from the mountains with a rapid flow, and carrying along a large quantity of detritus, can no longer transport this material as soon as its velocity is checked on emerging on to flat plains; and a portion of the sediment is carried by the inundations on to the land. When, however, such a river is embanked in its course through the plains, the sediment, as well as the flood-waters, is wholly confined within the embanked channel, and, consequently, is sooner or later deposited according to its density, thereby raising the river-bed, and also the flood-level. In some instances, the increased height of the floods, resulting from the raising of the river-bed, has led to the raising of the embankments to prevent inundations, which has naturally been followed by further deposit in the channel, necessitating corresponding additions to the height of the embankments. Some rivers descending from the hills in Japan, have had their beds and embankments so raised above the land by repetitions of these processes in traversing flat plains on their way to the sea, that, in certain cases, in constructing railways across the plains, it has proved easier to carry the railway in a tunnel under the river, than to erect the high bridge, with long approaches, which would have been required for crossing over the raised river. This system, however, cannot be continued indefinitely; and sooner or later a high flood overtops, or forces a passage through a weak place in the embankments; and the river pours out from its elevated position with devastating effect over the surrounding plain. The Yellow River, in China, is a remarkable example of a sediment-bearing river, whose floods have been controlled by embankments periodically raised, and which occasionally bursts through its banks, and suddenly inundates a large tract of country, with disastrous consequences to life and property.

Pumping for preventing Inundations.—Occasionally lands lie so low in relation to the sea-level that it is impossible to drain them by gravitation, as for instance the Haarlem Meer reclamation and numerous other reclaimed polders in Holland, and also some portions of the Fen country. Under these conditions, the requisite fall has to be created by raising the drainage water by pumping, and discharging it over the protecting embankments enclosing the low-lying land, into a watercourse or drain at a suitable level for conveying it to the sea, or from which it can eventually be again raised if necessary to ensure its outflow. By this means, lands can be completely secured from submergence, if properly enclosed by watertight banks, and ample pumping power is provided; but the method is only applicable to lands which are valuable enough to yield a fair profit after defraying the cost of pumping.

Regulation of Rivers for Navigation.

Rivers in their natural condition generally exhibit considerable variations in flow, notable irregularities in depth, and a somewhat winding course, particularly through alluvial plains; and, consequently, rivers are often inconvenient for navigation at their low stage, more especially in the upper portion of their course, where they are frequently torrential in character, with a rapid fall and a small dry-weather flow. A river, on the contrary, is better suited for navigation in its lower part, on account of its channel and discharge being larger, owing to the greater extent of basin drained, and its flow more uniform with a smaller fall, and the greater variety of meteorological conditions met with over an extensive drainage-area. Rivers at their low stage are liable to consist of a succession of shoals and deep places, or rapids and pools; an enlargement in width, or a subdivision into two channels, is accompanied by a reduction in depth; and the channel close alongside the concave bank in a bend, which is kept deep by the current in its direct course impinging against the concave bank, becomes shallow in passing across the river from one concave bank to the next, on the opposite side in the bend below (Fig. 193).

WINDING RIVER CHANNEL.
Fig. 193.—Differences in Depths.

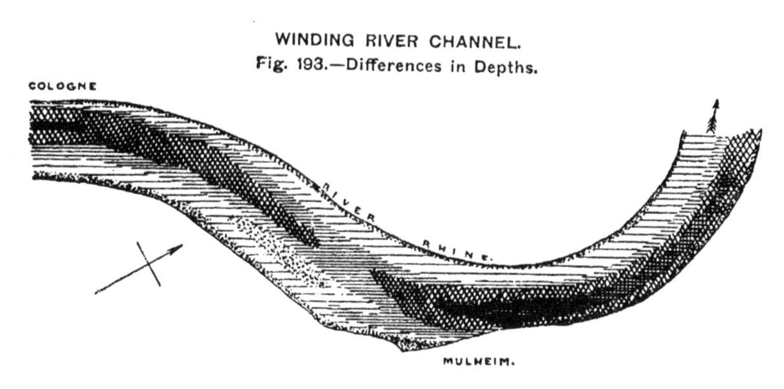

Removal of Hard Shoals.—Where deficiencies in depth and rapids are due to the existence of hard shoals across the bed of a river, which the current is incapable of scouring away, such as coarse gravel, stones, boulder-clay, or rocky reefs, the depth may be increased and the flow regulated by removing these obstructions by dredging, aided by blasting in the case of rocky ridges. The removal, however, of these hard shoals, though effecting decided local improvements in the channel, do not generally of themselves suffice to provide an adequate navigable depth throughout, more especially as the deepening of the channel across the shoals, by facilitating the discharge of the waters previously dammed up to some extent by the shoals, lowers the water-level at the low stage above the site of the shoals, which is liable to create a deficiency in depth over the shoals higher up in the neighbourhood of the improvements.

Soft shoals, being the consequence of the configuration of the channel, cannot be permanently removed by dredging alone; for if the causes which produced them remain unaltered, they are sure to reappear again, unless the dredging is constantly renewed.

Improving River Channel at Bends.—The winding course commonly adopted by rivers flowing through alluvial plains, originated by differences in the hardness of the materials constituting the banks, or by accidental obstructions deflecting the current, possesses the defect in respect to navigation, that the rapid current hugging the concave bank in the bends, forces vessels to keep close to this curving bank; that the navigable channel becomes increasingly tortuous, owing to the progressive erosion of the concave bank by the current; and that a shoal is invariably encountered at the crossing between the deep channels in two successive bends (Fig. 193, p. 335).

The increasing tortuosity of the river, augmenting the awkwardness and length of navigation, can be arrested by protecting the concave banks from erosion with fascines, rubble stone, piles, or dipping cross dykes; whilst the bends can be actually eased by constructing a longitudinal training dyke in front of each concave bank, extending somewhat into the river in the centre of the bend so as to form a flatter curve (Figs. 194 and 195).

The shoal at the crossing between two bends, which is due to the spreading out of the current in the absence of any bank directing and concentrating its flow, such as the concave bank provides in the bends, and to oscillations in its course from variations in its velocity, can be lowered by contracting the width of the channel between the bends, and thereby concentrating the current and increasing its scour over the shoal (Fig. 193, p. 335).

Regulation of Channel by Dykes.—Those rivers only are suitable for improvement for navigation by regulation alone, which possess a good flow, and, consequently, a fair depth of water at their lowest stage; and therefore this system is only applicable to the lower portions of large rivers draining extensive basins, such as the Rhine, the Elbe, and the Mississippi, or in a smaller measure to such a river as the Rhone, which, though having a moderate-sized basin and rather a large fall, is fed by rivers having their floods at very different periods, so that its discharge never falls very low. The objects of regulation works are, to render the width of the low-water channel, and the slope of the water surface more uniform; to concentrate the flow in dry weather in a single channel, and to direct this flow into a central course; and to give the low-water channel a less tortuous direction, and a better navigable depth, by easing the bends and reducing the width of the channel at the low stage.

The regulation is best effected by a judicious combination of longitudinal dykes, cross dykes, dipping cross dykes, and submerged dykes, these works being constructed, according to the local conditions, of fascines, timberwork, rubble stone, pitching, masonry, or concrete.[1] Longitudinal dykes are employed for training the river in easier curves round sharp bends, and for shutting off indents, being connected with the bank behind by cross dykes to prevent the current in flood-time scouring out a channel at the back of the longitudinal dykes (Figs. 194

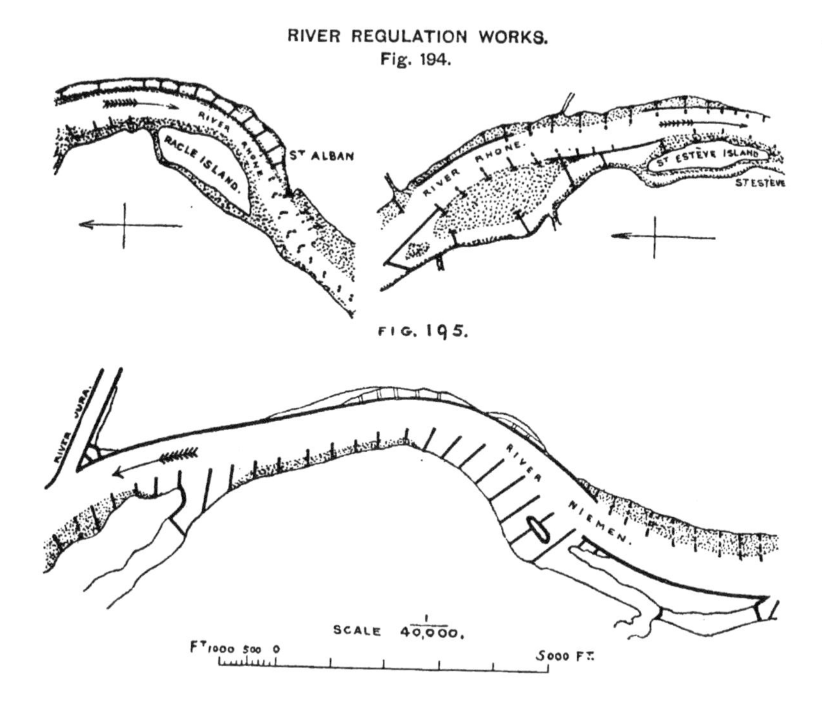

RIVER REGULATION WORKS.
Fig. 194.

FIG. 195.

and 195). Low cross dykes, besides protecting the back of longitudinal dykes, are also used for closing secondary channels during the low stage of the river, and for preventing these channels being reopened by the rapid current in flood-time; and they are further employed for restricting the low-water channel; and being made to slope down towards the central channel, and directed so as to point somewhat up-stream, they concentrate the current at the higher stages of the river into the low-water channel. Dipping cross dykes, projecting into the low-water

[1] *Proceedings Inst. C.E.*, vol. cxviii. p. 6, and plate 1, figs 1 to 5; and vol. cxxix. p. 263, and plate 6; and "Rivers and Canals," 2nd edition, 1896, L. F. Vernon-Harcourt, vol. i. p. 60, and plate 3, figs. 2 to 6.

channel, and dipping under water towards their extremities, serve to contract the navigable channel at the low stage; and, being constructed so as to point somewhat up-stream, they direct the low-water flow and the principal scour into the low-water channel. Submerged dykes are placed across the pools in the river-bed, with their tops well below the navigable depth, with the object of checking the ready flow through these deep hollows, and thereby raising the water-level of the river during the low stage over the shallows above the pools; so that the navigable depth is improved over the shoals by rendering the fall of the water surface more uniform by regulating the river-bed, just as the depth of a channel is regulated by regulating its width (Figs. 194 and 196). These submerged dykes are usually constructed of two

MODIFICATION OF WATER-LEVEL OF RIVER BY SUBMERGED DYKES.

Fig. 196.

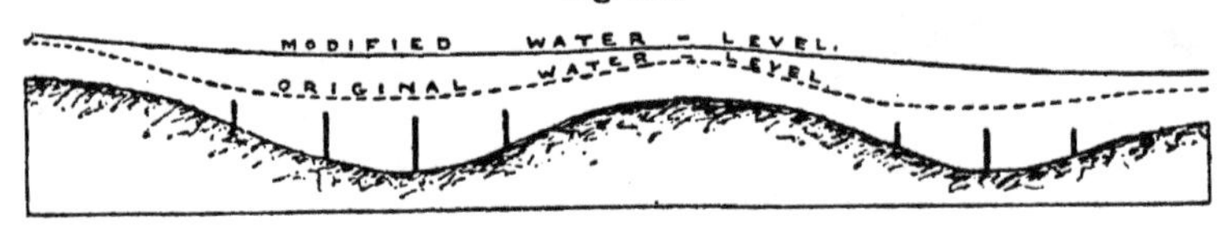

straight, dipping portions meeting at an angle in the centre of the channel, with the point on the up-stream side, so that they also, like the cross dykes, may guide the main current into a central course. Sometimes dipping cross dykes are put out into the channel in front of a longitudinal dyke which eases the curvature at a bend; and they, in such a case, protect the longitudinal dyke from scour along its concave side, still further reduce the bend of the navigable channel, and by diverting the current into a more central direction, make vessels keep a straighter course further off from the longitudinal dyke. These regulation works permanently fix and improve the low-water channel; whilst owing to the low level to which the dykes are kept down, and the deepening of the main channel, the flood channel is practically not interfered with, though floods are prevented from altering the course of the main channel.

The Rhine below Strassburg, the Elbe from its entrance into Germany, the Niemen through Prussia (Fig. 195), and the Rhone below Lyons (Fig. 194), have been systematically regulated by dykes; and the Mississippi and other large rivers of North America have been trained and deepened by similar methods.

Remarks on Regulation Works.—As the improvement of isolated shoals is liable to modify the condition of the river in their neighbourhood, it is generally advisable to deal with the portion requiring amelioration as a whole, and to regulate it in a systematic manner, with the view of securing a definite minimum navigable depth throughout. It is rarely practicable to obtain any great increase in depth by regulation works, except possibly at some solitary, specially defective spot; but an increase in the low-water depth of even three or four feet, furnishes an important improvement in the navigable condition of a large river. The rapid current of the Rhone might have seemed to preclude its improvement by regulation works; but the large quantities of stones and gravel carried down by the river between Lyons and Arles, prevented its improvement by canalization—a method which would have secured the important advantage of greatly facilitating the up-stream navigation, which is seriously impeded by the opposing current. Large gently-flowing rivers are best adapted for improvement by regulation works; and the regulation of rivers, though not facilitating the passage of the up-stream traffic, possesses the important merit of leaving the navigation of the river perfectly open, and its water-level practically unaltered except at its lowest stage. Moreover, the works, though they should all form part of a definite, comprehensive scheme of regulation, may be carried out gradually.

CHAPTER XXI.

CANALIZATION OF RIVERS.

Improvement of depth of rivers by canalization—**River Locks:** object, general form, and working; position of locks, suitable sites, varieties, distances apart; sizes and arrangements of locks, conditions, standard dimensions in France, on River Main, different arrangements—**Weirs:** classification—**Solid Weirs:** materials used, objections; oblique and angular weirs, object, description, examples—**Draw-door Weirs:** object, construction, example; draw-doors sliding on free rollers, advantage, instances in Ireland, description of Richmond weir, overhead bridge necessary, facility of raising—**Movable Weirs:** object, three classes, navigable passes—1. **Frame Weirs:** general form, four types, design of first three types; needle weir, description, contrivances for working, large example on Great Sandy River; sliding panel weir, advantageous mode of closing weir, instances; rolling-up curtain weir, description of curtain, on Lower Seine, advantages and defects; suspended frame weir, object of modified form, description and method of working—2. **Shutter Weirs:** general form; bear-trap, description, improved form, across drift pass on Ohio, inconveniences; shutter with trestle and prop, description, methods of lowering and raising, defects in regulating discharge, remedies, examples, working, remarks—3. **Drum Weirs:** description; drum weirs in France, on River Marne; drum weirs in Germany, on River Main, at Charlottenburg; drum weirs in United States, novel form on Osage River, low cost—Remarks on movable weirs.

RIVERS which possess a very variable flow, or whose water-level falls very low during the dry period, and are therefore not suitable for improvement by regulation works, can have their navigable depth permanently increased by canalization. By placing dams, known as weirs, across the river at suitable intervals, the water is retained above them at a higher level, and the river is thereby converted into a series of nearly level reaches, forming a set of long steps in place of the usual slope of the surface of a river, resembling a canal in this respect; whilst the difference of level between two successive reaches is surmounted by a lock, as in a canal (Fig. 197). Owing to the descending slope of the river-bed seawards, the weir has to hold up the water to a higher level at the lower end of the reach than is necessary for the required depth at that part, to provide for the gradual shoaling of the channel up-stream, due to the rise of the bed up-stream being greater than the very slight inclination of the water surface of the held-up river in dry weather; whilst the requisite depth at the upper end of the reach may be conveniently obtained by dredging. Canalized rivers have been sometimes called still-water navigations, for they largely reduce the descending current by concentrating the fall of the water at the weirs; and by this means they materially facilitate the up-stream traffic.

River Locks.

Locks on rivers do not of themselves actually aid in increasing the navigable depth of the channel, which is effected by the weirs, whose

CANALIZATION OF RIVERS.

Fig. 197.—Canalized River Main. Longitudinal Section.

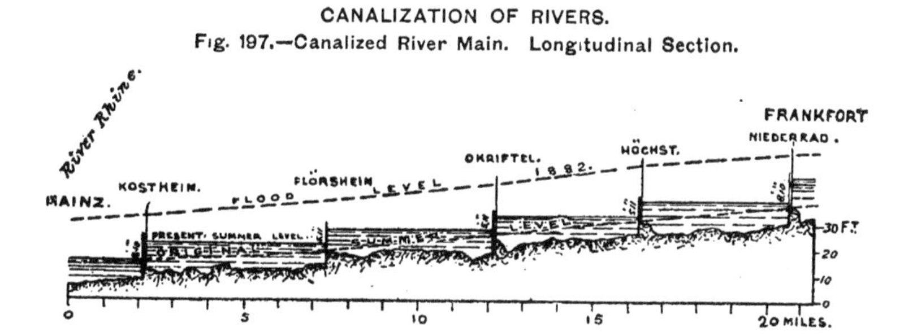

Fig. 198.—Lock and Weir on Canalized River.

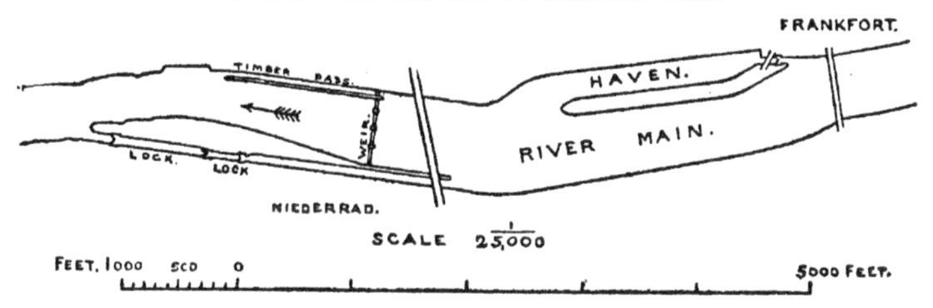

function is to hold up the water in dry weather, and to regulate the flow; but the locks constitute essential adjuncts to the improvement works, in providing for the passage of vessels from one reach to the next when navigation along the main channel of the river is barred by weirs. River locks are closed by a pair of gates, meeting at an angle pointing up-stream and abutting against a sill, at each end of the lock-chamber, into which vessels are admitted from the upper or lower reach, and in which, with both pairs of gates shut, they are lowered or raised to the extent of the difference in level between the reaches, by letting out water from the chamber into the reach below, or by introducing water from the upper reach into the chamber for vessels proceeding up the river.

Positions of Locks on Rivers.—The best position for a lock is in an artificial channel formed across the land at a projecting bend, as exemplified by the Niederrad Lock on the canalized River Main just below Frankfort (Fig. 198), for by this arrangement, the lock can be constructed in the solid ground away from the river, the connecting channel being only excavated on the completion of the lock; the lock in no way interferes with the river channel; and the approach to the lock is placed well out of the influence of the currents caused by the fall of water at the weir. Locks, however, have been often placed in the smaller of two channels formed by an island in the river, which, though involving more trouble in the foundations, is equally convenient for navigation in separating the approach to the lock from the weir (Fig. 200, p. 345); but in order that the closing of the minor channel by the lock may not impede the discharge of the river, the main channel, across which the weir is placed, should be correspondingly enlarged. Sometimes, where the river has a straight course and a gentle flow, the lock is put alongside one of the banks; and the weir extends across the rest of the channel (Fig. 202, p. 345), a system which has been adopted for the locks on the Upper Seine above Paris, with the precaution that the weir is erected across the river at the down-stream end of the lock, so that the up-stream entrance to the lock is placed as far as practicable from the weir;[1] and the approach to it from above can be further guarded, if necessary, by a jetty.

The distance apart of the locks has to be regulated by the average fall of the river-bed; and, consequently, in ascending a river, the locks have to be placed closer together. Thus an available navigable depth of 10½ feet has been obtained on the Lower Seine, from the tidal limit up to Paris, by locks and weirs about 13½ miles apart on the average; whereas the navigable depth on the Upper Seine, between Paris and Montereau, of 6½ feet, has only been secured by locks and weirs at average intervals of 4⅔ miles; and the average distance apart of the locks and weirs on the Main, between its confluence with the Rhine and Frankfort, affording a navigable depth of 8¼ feet, is also 4¾ miles.[2] Eventually, in going up a river, the fall becomes too great, the channel too small, and the detritus brought down too large, to enable a river to be improved by canalization at a reasonable cost.

Sizes and Arrangements of Locks on Rivers.—The width at the entrances to a lock, and the available length between the lock-gates at the two ends of the lock-chamber, fix the size of the vessels which can navigate a canalized river; and therefore these dimensions must be determined by the largest-sized vessel which might be expected to use the improved waterway. The depth over the sills of the locks must be at least equal to the available depth in the river; and it may with advantage be made two or three feet deeper, if there is a prospect of the river being eventually deepened to this extent by dredging, to accommodate a growing traffic and vessels of larger draught. Sometimes, however, an increase in depth is obtained by raising the weirs;

[1] "Rivers and Canals," 2nd edition, 1896, vol. i plate 4, fig. 3.
[2] *Ibid.*, plate 4, figs. 1, 2, and 6.

and this raising of the water-level improves correspondingly the depth over the sills of the locks.

The standard dimensions of the locks on the main lines of inland waterways (both rivers and canals) in France, are 126⅓ feet in length, 17 feet in width, and 6½ feet depth of water over the sills, so as to accommodate barges of 300 tons; but the locks on the Lower Seine, with a depth of 10½ feet on their sills, and serving for vessels of 1000 tons, have been made 39⅓ feet wide at their entrances, widened out inside to 55¾ feet, and from 463 to 722 feet long according to the traffic (Fig. 199). The locks on the Main, when the canalization was completed in 1886, were 279 feet long and 34½ feet wide, admitting the Rhine steamers of

DOUBLE LOCK ON RIVER.

Fig. 199.—Suresnes Lock, River Seine.

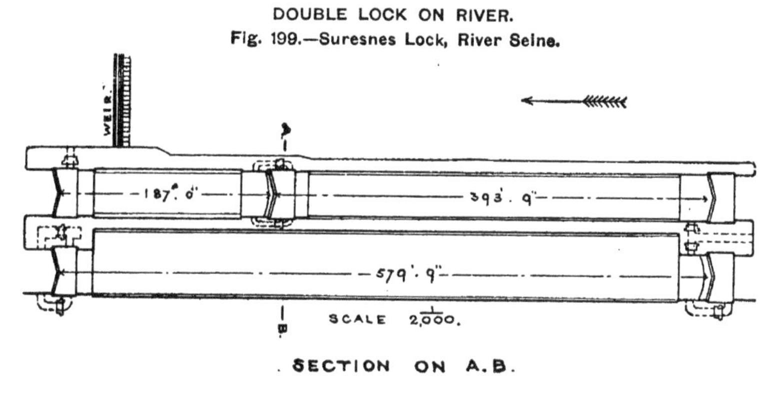

SECTION ON A.B.

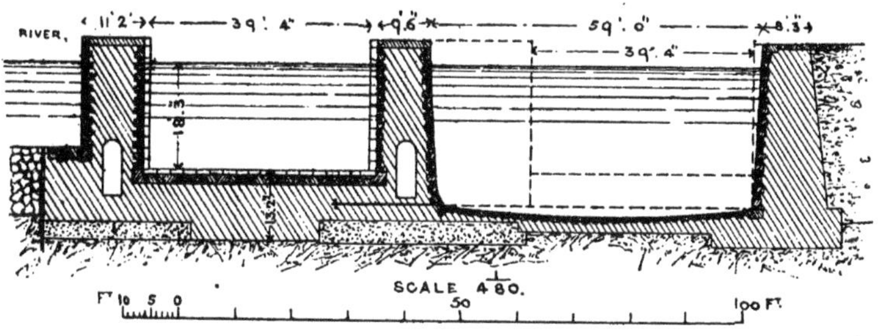

1300 tons; but within the last few years a third pair of gates has been put in the side channel, affording an additional length of 820 feet, into which six of the largest Rhine barges can be admitted two abreast, with their tug in the original lock (Fig. 198, p. 341).[1] Moreover, as the sills of the locks on the Main were placed 8¼ feet below the water-level retained by the weirs, or 1¾ feet lower than the available depth in the river, it has been possible to increase the navigable depth recently from 6½ to 8¼ feet, merely by dredging in the upper part of the reaches.

[1] "Rivers and Canals," 2nd edition, 1896, vol. i. pp. 69 and 103, and plate 4, fig. 5.

The widening out of the lock-chamber within the entrances, enables more vessels to be locked at the same time. Sometimes two locks, or occasionally three locks of different sizes are provided, as for instance at Poses on the Lower Seine, to suit a variety of traffic and expedite the passage of vessels; whilst an intermediate pair of gates is often introduced in a long lock, to save time and water in locking small vessels. Sluices in the gates suffice for filling and emptying small locks; but large sluiceways in the side walls are usually employed in large locks, or where the difference in the upper and lower water-levels is considerable, to avoid undue loss of time in locking. The locks at Suresnes on the Lower Seine (Fig. 199, p. 343) illustrate the use of two locks of different sizes, the widening out of the lock-chamber of the large lock, the introduction of an intermediate pair of gates in the smaller lock, and the employment of sluiceways in the side walls.[1] The difference in level between two adjacent reaches, which constitutes the rise or fall at a lock, is varied according to the local conditions; but it is generally comprised between 5 and 14 feet. The discharge of a river at its lowest stage usually supplies ample water for locking, unless there is an excessive leakage through the weir.

WEIRS.

Three distinct classes of weirs have been employed for the canalization of rivers, namely, solid weirs, draw-door weirs, and movable weirs. A weir is essentially a barrier placed across the channel of a river to raise its water-level in dry weather; but the three classes of weirs differ with regard to their capabilities of regulating the flow of a river, and permitting the discharge of floods. Solid weirs form very efficient, permanently-fixed dams for maintaining the river above them at a higher level, requiring no looking after, unless they should happen to be damaged by a high flood; but the channel being closed up to the top of the solid weir, the river has to rise above the crest of the weir before it can commence to discharge its drainage waters; and, consequently, the river has to rise higher in flood-time than before its channel was closed, for a considerable distance above the weir. Draw-door weirs, consisting of panels or lifting gates sliding in grooves, and closing openings between piers, can be partially opened for the passage of floods by raising the gates. Movable weirs consist of temporary barriers placed across a river in dry weather, which can be entirely removed from the channel on the approach of floods.

SOLID WEIRS.

Solid weirs formed of rubble stone with sheet-piling, masonry, crib-work weighted with stones, or fascine-work (Figs. 201 and 203), possess the advantages of simplicity, strength, and durability; but

[1] *Annales des Ponts et Chaussées*, 1889 (2), p. 50, and plates 34 and 35.

whilst improving the depth of a river for navigation, they render the channel less free for the primary object to which it owes its existence, namely, the discharge of the rainfall, for they bar the lower part of the channel as much in flood-time as at a low stage, and consequently increase the height of floods.

Oblique and Angular Weirs.—The obstruction presented by solid weirs to the discharge of floods is so obvious, that efforts have been made to minimize this defect by placing the weir in such a position as to increase the length of its crest. One device consisted in building

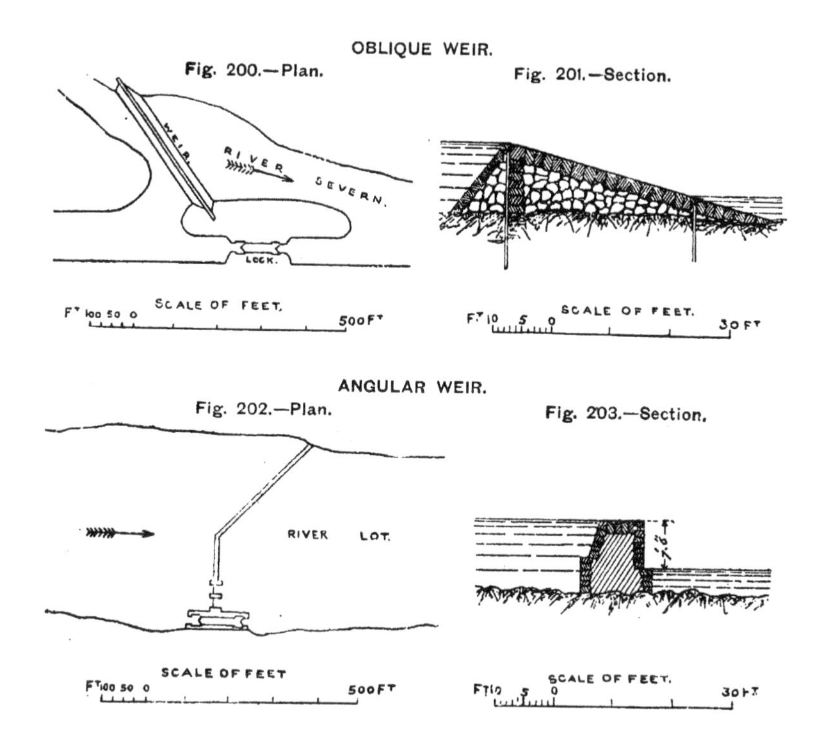

OBLIQUE WEIR.

Fig. 200.—Plan. Fig. 201.—Section.

ANGULAR WEIR.

Fig. 202.—Plan. Fig. 203.—Section.

the solid weir obliquely to the channel, so as to increase its length, as exemplified by weirs on the River Severn (Fig. 200); but as the direction of the current follows the line of the channel, and does not approach such a weir at right angles to its crest, the discharge over an oblique weir differs little from that over a weir built straight across a river at right angles to the stream. The only effectual way of increasing the discharge over a solid weir is by placing it across a wide part of the river, or by keeping its crest below the required water-level, and raising its height temporarily, during the dry season, by placing planks edgewise along the top.

Angular or segmental weirs, with the angle or the convex face placed up-stream, have been sometimes used for directing the water falling over the weir into a central course, and also probably with the idea of improving the discharge. In canalizing the torrential River Lot in France many years ago, for 185 miles above its junction with the Garonne, the first weirs erected were made in two oblique halves meeting at an angle; [1] but this arrangement, by causing eddies, produced inconvenient back currents towards the down-stream end of the lock, so that the design was altered to a single oblique weir on one side of the channel, with a straight length of weir intervening between the oblique weir and the lock on the opposite side (Fig. 202, p. 345). This latter arrangement diverts the main current away from one bank, and directs it towards the channel leading to the lock.

DRAW-DOOR WEIRS.

The hindrance to the discharge of floods presented by solid weirs, has been often mitigated by introducing draw-door weirs in place of portions of the solid weirs. These draw-door weirs, with their sills placed some feet below the crest of the solid weir, are closed by wooden or iron gates sliding in grooves formed in the piers on each side; and by raising these gates on the approach of a flood, openings are provided between the piers for the discharge of the flood-waters which, accordingly, do not rise as high as when the channel is barred by a solid weir. Usually the dry-weather flow of the river passes over a length of solid weir; whilst the draw-doors are only raised when the discharge becomes large. It is evident that if a weir was wholly closed by draw-doors with their sills at the level of the river-bed, the flow could be completely regulated; and the piers between which the gates slide would form the sole impediment to the flood discharge, when the gates are fully raised. Generally, however, with the object of saving trouble in working and cost in construction, the draw-doors only form part of the weir; and their sills are placed on a solid mound several feet above the river-bed. For instance, Teddington weir at the tidal limit of the River Thames, with a total length of 480 feet, has solid weirs extending out from the banks for half its length; whilst the central half of the river is closed by plate-iron doors, 6 feet high and $7\frac{1}{3}$ feet wide, sliding in grooves at the sides of iron frames erected at intervals across the central part of the river, the sill of the weir being laid on the top of a rubble mound raised about 5 feet above the bed of the river.[2]

Draw-doors of Weirs sliding on Free Rollers.—When a draw-door or lifting gate has to be raised against a head of water on one side of it, the frictional resistance to motion is considerable, especially if the gate is a big one and the head large. To reduce this friction to a minimum, instead of making the gate move along an iron or masonry face at each side, it is allowed to slide, in a recess in each pier, upon a row of free rollers suspended vertically in the loop of a

[1] *Annales des Ponts et Chaussées*, 1865 (1), p. 151, and plates 103 and 104.
[2] "Rivers and Canals," 1st edition, 1882, vol. ii. plate 4, figs. 9 and 10.

chain, one end of the chain being attached to the gate, and the other to the side of the pier.

Weirs on this principle were erected in Ireland across the River Erne at Belleek in 1883, and across the River Suck at Ballinasloe in 1885, for controlling the drainage of the districts above them, the gates being 31 feet and 25 feet wide respectively; whilst the most notable

DRAW-DOOR WEIR AT RICHMOND.

Fig. 204.—Elevation of Weir and Footbridge.

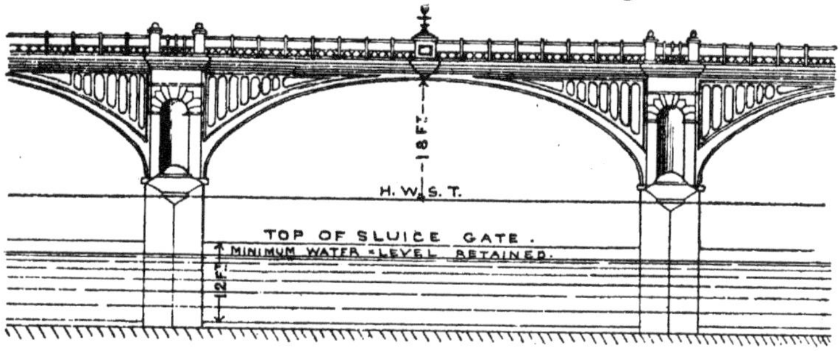

18 FT.
H. W. S. T.
TOP OF SLUICE GATE.
MINIMUM WATER-LEVEL RETAINED.
12 FT.

FIG. 205.

PLAN OF DRAW-DOORS AND PIERS

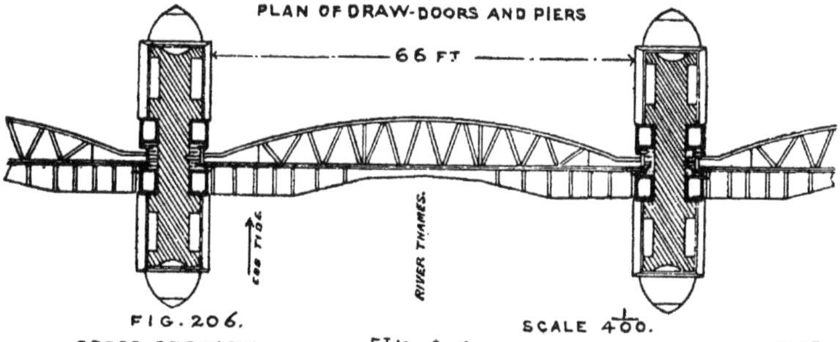

66 FT
EBB TIDE
RIVER THAMES.

SCALE 1/400.

FT 10 5 0 50 FT.

FIG. 206.

CROSS SECTION.

H.W.S.T.
MINIMUM WATER-LEVEL
AVERAGE LOW WATER L

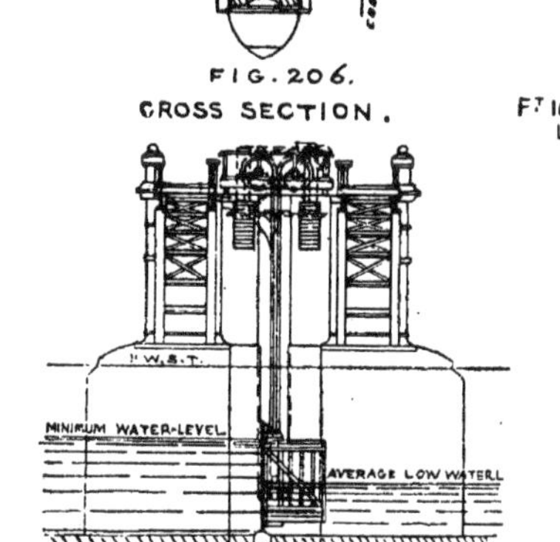

weir of this type was erected across the Thames at Richmond, in 1892–94, for maintaining the river above at half-tide level, so as to cover the mud-banks, by lowering draw-doors, 12 feet high, across the three central arches of an iron footbridge traversing the river channel with clear spans of 66 feet each,[1] and leaving the tide free to ascend the river up to Teddington, as before, when it rises above half tide, by raising the draw-doors (Figs. 204, 205, and 206). The draw-doors at

[1] *Engineering*, vol. lxi. p. 47.

Richmond, when raised by cables from the footbridge, are rotated automatically by curved grooves in the piers into a horizontal position, in the space provided for the purpose between the double footbridge, so that they may not obstruct the view of the river through the arches (Fig. 206); and free rollers are in this instance placed both on the up- and down-stream sides for the draw-doors to slide against, as the head of water with a rising tide may be on the down-stream side of the draw-doors. The draw-doors in all these instances, being large and heavy, are counterbalanced by weights, so that they are easily raised, a point of considerable importance for the Richmond weir, where the draw-doors have to be raised and lowered again each tide, except in flood-time or during low water occurring at night, when they are kept up. Each draw-door at Richmond, weighing 32 tons, can be raised in 7 minutes; and the discharge of the river is provided for, when the weir is closed, by a requisite raising of the draw-doors from their sills. The head is usually insignificant at the Richmond weir, except when the draw-doors have to be raised at low tide to regulate the discharge; but in consequence of the adoption of free rollers and counterbalance weights, the draw-doors at the Belleek weir can be easily raised, by one man, against a head of $14\frac{1}{2}$ feet.

An overhead footbridge constitutes an essential portion of draw-door weirs; and for such a weir to be used in flood-time as a navigable pass, it would be necessary to have a high enough bridge, and to raise the gate sufficiently, to afford the required headway for vessels. By the adoption of free rollers, aided by counterbalance weights, it has been possible to give draw-door weirs such wide openings as to render them practically nearly equivalent to movable weirs, in providing an almost unimpeded channel for the discharge of floods. The remarkable facility, indeed, attained in moving the draw-doors of these weirs, has made it much easier and quicker to open or close these weirs than most systems of movable weirs; and the only hindrances in the way of their more extended use, are their cost, and the high overhead bridge they require.

Movable Weirs.

Movable weirs are essentially temporary barriers placed across a river, on the approach of the dry season, to maintain the water-level for navigation throughout the low stage of the river, and removed entirely from the channel when floods come down; though these weirs are usually divided into wide bays, by masonry piers, for convenience in working; and occasionally, for the sake of economy, the sills of portions of the weirs are somewhat raised above the river-bed. These weirs may be divided into three classes, namely, (1) Frame weirs, (2) Shutter weirs, and (3) Drum weirs, the first and second of which comprise some distinct types. The weirs of the larger continental rivers are generally composed of one or more shallow bays near one bank, sometimes used for regulating the flow, and one or two navigable passes, with deep sills, in the centre or deepest channel of the river, which are traversed by vessels in flood-time when the weir has been removed, and when the lock at the side is often submerged, and consequently inaccessible, and navigation along the open river is both more convenient and quicker for the down-stream traffic.

1. *Frame Weirs.*

A frame weir consists of a series of hinged, iron frames placed across the river, supporting a wooden barrier, formed of a number of parts, which constitutes the actual weir retaining the water above it, and which is removed and replaced by men standing on a footbridge resting on the frames. There are, however, four types of frame weirs, three of which are similar in regard to their frames, but differ in the form of barrier used for closing the weir; whilst the fourth, and newest type, has a different kind of frame, and a different method for removing the frames from the channel. These four types may be denominated, needle weir, sliding panel weir, rolling-up curtain weir, and suspended frame weir. The different forms of barrier in the first three types, all rest against braced, wrought-iron frames, built up of angle-irons, channel-irons, T irons, and bar irons,[1] placed in a row across the river, at intervals generally of between three and four feet, upon a masonry apron laid on the river-bed, and hinged at the bottom so that they can be lowered sideways by chains, and lie almost flat at the bottom of the river during the flood season; whilst when the frames are raised and connected by bars, a footbridge of planks or of hinged, iron plates is readily formed along the top across the river, from which the barrier forming the weir is put in place or removed in sections (Figs. 207, 208, and 209, p. 350).

Needle Weir.—The earliest type of movable frame weir derived its name of *barrage à aiguilles*, or needle weir, from the square spars, termed needles in France, which are employed for closing the weir. These spars, about 2½ to 3, or even 4 inches square, rounded off into the form of a handle at the top, are ranged close together in a continuous row across the river, resting against a projecting sill at the bottom, and against a bar connecting the up-stream face of the frames at the top, giving the needles a little inclination up-stream towards the bottom (Fig. 207, p. 350). The discharge of the river can be regulated by removing some of the needles, or even by pushing a sufficient number forward at the top, and keeping them in position by blocks of wood.

In the earlier needle weirs, the needles being of a fairly manageable size, were readily put in place and removed by the weir-keeper; but when the height of the weirs, and, consequently, the length and cross section of the needles were increased, hinged bars were used in some cases for the needles to rest against, which could readily be released and displace a series of needles; and in other instances, iron hooks have been attached to the needles, encircling the bar against which they rest, guiding and supporting the needles in their descent for closing the weir, and enabling the needles, on being raised by a winch on the footbridge clear of the sill, to revolve down-stream on the bar into a

[1] *Proceedings Inst. C.E.*, vol. lx. plate 2, fig. 3, and plate 4.

floating position, so as to open the weir, from whence they can be removed at leisure.

Some of the largest needle weirs on the Continent have been erected across the Belgic Meuse and the Main, in depths of about 10 feet; but this type of weir has been extended to a depth of 13 feet, with needles which are in reality beams, in the weir completed at Louisa, across the

MOVABLE FRAME WEIRS.

Fig. 207.—Needle Weir. Fig. 208.--Panel Weir. Fig. 209.—Rolling-up Curtain Weir.

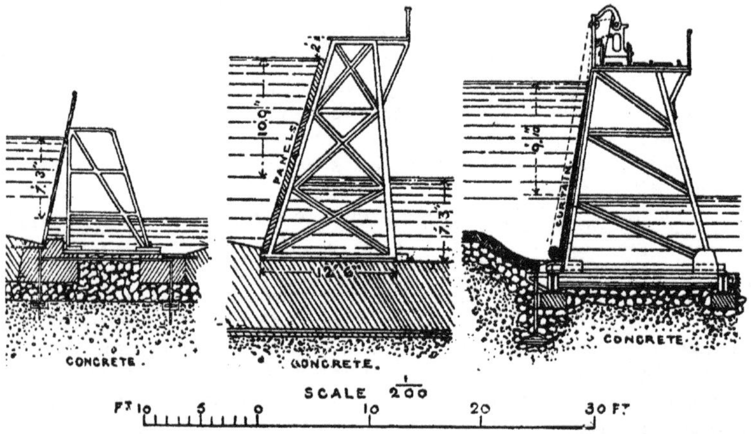

Great Sandy River in the United States, in 1897. The frames, placed 4 feet apart, are light, steel structures, so designed that when lowered on to the river-bed, they lie flat, one inside the next in front, and not partly on one another as in the ordinary frames, reducing the depth of the recess required to receive and protect them. The white pine needles are all 12 inches wide, and are $8\frac{1}{4}$ feet long, $3\frac{1}{2}$ inches thick at the bottom and $2\frac{1}{2}$ inches at the top in the shallower portion of the weir, and $14\frac{1}{4}$ feet long, $8\frac{1}{2}$ inches thick at the bottom and $4\frac{1}{2}$ inches at the top across the navigable pass.[1] The needles are put in place and removed by means of chains worked by a derrick erected on a special boat, and attached to handles provided for the purpose at the top of the needles. A shallow groove cut in each side of the needles of the pass, enables leakage between the needles to be prevented, if necessary, by the insertion of strips of rubber in these grooves.

Sliding Panel Weir.—The defects of the ordinary forms of needles, namely, the thickness not being proportioned to the pressure, the needles becoming too heavy for deep weirs, and undue leakage occurring through the numerous joints, though got over at the Great Sandy River weir, on the condition of having to use special machinery

[1] *Transactions of the American Society of Civil Engineers*, vol. xxxix. pp. 460-487, and plates 15, 16, and 17.

for closing and opening the weir, have been obviated in a more simple and manageable manner, by substituting wooden panels sliding in grooves between each pair of frames, for the needles (Fig. 208, p. 350). These panels can be varied in thickness according to the depth at which they are to be placed; there are only two or three horizontal joints in each row of panels, weighted by the panels above them; and by increasing the number of panels, or reducing the width between the frames, they can be readily adapted to any height of weir, without becoming too heavy for the weir-keeper to put in place, from the footbridge, with a boat-hook encircling the iron loop at the top of each panel, and to raise by a hook and chain worked by a crab travelling on the footbridge.

This form of weir was first used for some weirs on the River Moskowa; it was also adopted for the regulating portion of the Mulatière weir across the Saône at Lyons; and it is in operation, on a larger scale, at the reconstructed Suresnes weir across the Seine below Paris,[1] where four panels, each $4\frac{1}{4}$ feet high and 4 feet wide, slide down grooves between each pair of frames, 20 feet in height, and retain a head of $10\frac{3}{4}$ feet above the weir, in a depth of water of about 17 feet (Fig. 208).

Rolling-up Curtain Weir.—Another fairly watertight, easily worked barrier for closing frame weirs, consists of a curtain composed of a series of horizontal, wooden laths increasing in thickness downwards, and connected by watertight hinges, which is rolled up from the bottom by a pair of endless chains worked by a crab travelling along the footbridge. The curtain, about $7\frac{1}{2}$ feet wide, rests, when let down, against two frames, and extends beyond them on each side for the width of half the space between two frames, leaving only a slit, $1\frac{1}{2}$ inches wide, between it and the adjacent curtain, to afford clearance for working, through which the only leakage across the weir takes place. These curtains were first employed in 1880, for closing the new weir across the Lower Seine at Port Villez (Fig. 209, p. 350); and they have since been adopted for several of the more recent weirs on the same section of the river. They are readily raised from the bottom to any desired extent for regulating the discharge; and they can each be entirely rolled up, or lowered, in about ten minutes by aid of the crab. These rolling-up curtains provide a superior barrier for closing frame weirs than sliding panels, in respect of facility of manipulation and rapidity in opening and closing the weir; but, on the other hand, they are more costly and less durable than panels, they create a considerable scour at the bottom when partially rolled up to regulate the discharge, and they afford more scope for leakage in the interval between them, than panels properly put down.

Suspended Frame Weir.—In the ordinary forms of frame weirs just described, though the movable barriers closing these weirs can be stored away when the weir is open during the flood season, the frames have to lie flat on the bed of the river, exposed to the chance of injury

[1] *Annales des Ponts et Chaussees,* 1889 (2), p. 63, and plates 36 42.

from any heavy drift the river may bring down in flood-time, and are liable to be strained in being lowered on one another under water, and raised again. The comparative freedom, however, of these frames from serious injury, is adequately proved by their extensive employment on the Continent of Europe, and in the United States. Nevertheless, rivers bringing down large quantities of detritus might bury the frames with deposit towards the close of the flood season; and it was with the view of canalizing the Rhone several years ago, that the idea of suspended frames was conceived. Though the suggested scheme was never attempted on the Rhone, the system of frames suspended from a wide, overhead bridge, has been adopted for two of the more recent weirs on the Lower Seine, namely, at Poses and Port-Mort[1] (Fig. 210).

The frames, braced in pairs, are hinged and suspended from the

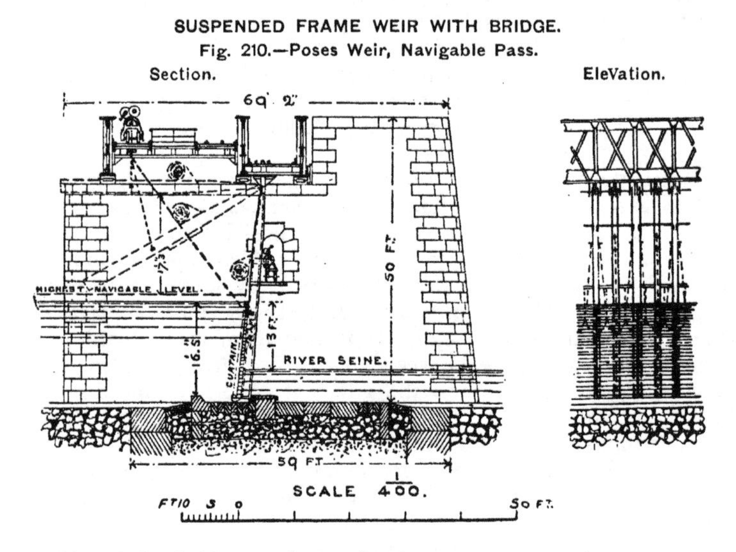

SUSPENDED FRAME WEIR WITH BRIDGE.

Fig. 210.—Poses Weir, Navigable Pass.

underside of the bridge, and can be hung almost vertically, so as to abut against a sill at the bottom, for closing the weir by the aid of rolling-up curtains, or can be lifted by a chain, worked by a movable winch from the bridge, into a horizontal position on the approach of floods, thereby removing all the movable parts of the weir out of the river. The rolling-up curtains are worked from a small footbridge formed by a series of hinged brackets let down at the back of each frame when in place. The piers carrying the overhead bridge, with spans of 99 to $106\frac{1}{2}$ feet, divide the weir into separate bays, across the shallower of which, the girders are only raised sufficiently to be out of the reach of the highest floods; but across the navigable passes, the bridge has to be placed high enough to afford the requisite clear headway of $16\frac{1}{2}$ feet above the highest navigable level, with the frames

[1] *Proceedings Inst. C.E.*, vol. lxxxiv. p. 234, and plate 3.

raised underneath the girders (Fig. 210). The curtains, which can, if necessary, each be rolled up or lowered in the deepest passes in 13 and 7 minutes respectively, rest and remain on the frames when these are raised; the raising of all the frames of the seven passes of Poses weir, with a total opening of 693 feet, can be accomplished by five men, with one electric winch, in 12 hours, and they can be lowered in 7 hours; and by using a steam winch as well, the whole of the weir can be fully opened by twelve men in 7 hours, and the frames lowered again in $3\frac{1}{2}$ hours.[1] This period of opening is, indeed, very long in comparison with the rapid lifting of the draw-doors, or gates, at Richmond weir; but the operation has only to be effected at the Seine weirs on the approach of a high flood; there is, moreover, ample time available for opening these weirs, the curtains being raised gradually in the ordinary course as the discharge increases, and the weir-keepers are on the spot; whilst the passes on the Lower Seine are considerably wider than at Richmond, and the headway which must be provided in flood-time is greater. In this form of weir, however, as in draw-door weirs, an overhead bridge is a necessity, from which other forms of weirs are exempt. The lock and weir at Poses furnish an interesting recent application of electricity, generated by a turbine actuated by the fall of water at the weir, to the working of the lock-gates and sluices, the opening and closing of the weir, and the lighting of the lock and weir at night.

2. *Shutter Weirs.*

Shutter weirs consist of a row of large panels, or shutters, hinged to the apron of the weir or to iron trestles, which when the weir is closed are kept raised against the river, somewhat inclined down-stream towards the top, by resting either against a row of reversed shutters at their upper ends, or on a series of central, iron props hinged to the centre of the back of each shutter, and fitting at their lower extremities into cast-iron shoes fastened on the apron and supporting the props (Figs. 211 and 212, p. 354).

Bear-trap Weir.—The earliest form of shutter weir, erected first in the United States, and known there as the bear-trap, and subsequently introduced into France for a small weir across the Marne at La Neuville-au-Pont,[2] consists of two shutters, or gates, abutting against one another at an angle at the top when raised for closing the weir, the up-stream one forming the weir and the down-stream one acting as a support, each shutter turning on a horizontal axis laid across the apron of the weir, so that they can be either raised to close the weir, or lowered so as to lie flat one on the other on the bed of the river. This form of weir fell into disuse for some time, owing to defects in working; but it has recently been revived again in America with improvements in the design, one of the best types being shown in Fig.

[1] "Navigation de la Seine de Paris à Rouen," A. Caméré, Paris, 1900, pp. 1-16; and "Notice sur les Nouveaux Types de Barrage appliqués sur le Basse Seine," A. Caméré, Paris, 1900, p. 7.

[2] "Rivers and Canals," 2nd edition, 1896, vol. i. p. 132, and plate 4, fig. 15.

211, which, placed across the opening of a weir 80 feet in width, has been readily raised, lowered, or put in any intermediate position, under a maximum head of 16 feet, by means of chains worked by one man turning a winch.[1] Moreover, a bear-trap weir of the older type, in

BEAR-TRAP WEIR.

Fig. 211.

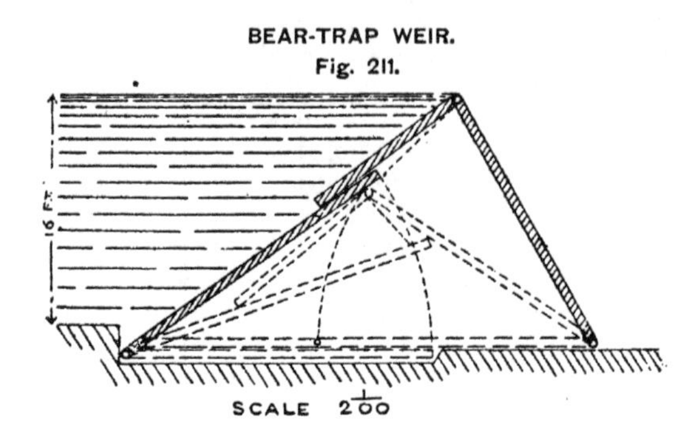

SCALE $\frac{1}{200}$

which the shutters were designed to be raised by the water-pressure admitted underneath them by culverts bringing water from the upper pool, has been erected on the Ohio across a drift pass, 52 feet wide, in the Davis Island weir, owing to its being more easily raised again at a

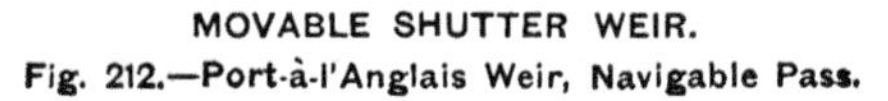

MOVABLE SHUTTER WEIR.

Fig. 212.—Port-à-l'Anglais Weir, Navigable Pass.

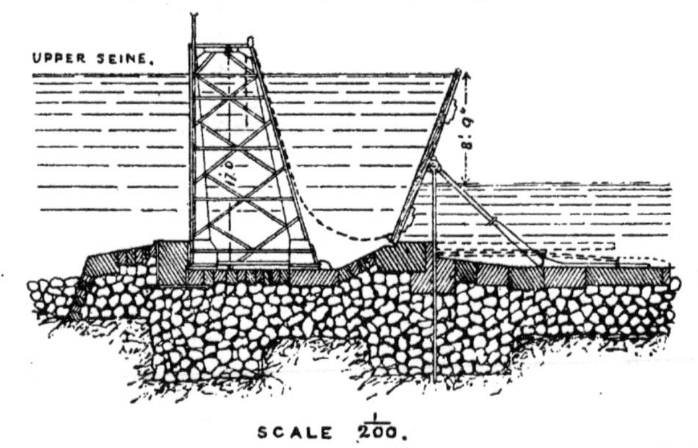

SCALE $\frac{1}{200}$.

low stage, after the drift has been let through, than the French type of shutters (Fig. 212), which close the rest of the weir. Hitherto it has not proved practicable to adopt this form of shutter weir to navigable passes, owing to the insufficiency of the head of water for raising the shutters in a deep pass; and, moreover, the considerable inclination of

[1] *Proceedings Inst. C.E.*, vol. cxxix. p. 258, and plate 6, fig. 2.

the shutters when raised, and the wide apron required for the two shutters, render the shutters very large, and the form of weir costly for deep passes.

Shutter Weir with Trestle and Prop.—The ordinary form of shutter weir used in France, Belgium, and the United States, consists of a row of wooden or iron shutters turning on a horizontal axis at the back, or down-stream side of the shutter, placed a little above the centre of pressure of the shutter, and supported on an iron trestle which is hinged to the apron of the weir, and assumes an upright position when the weir is closed, with the shutters inclining over somewhat down-stream at the top and abutting against a sill at the bottom (Fig. 212). The hinged trestle and shutter is supported in position, when raised, by an iron prop resting against a cast-iron shoe at its lower end; and when the prop is released from its shoe, either by a tripping bar laid on the apron, provided with projecting teeth at intervals, which when worked from the bank draw the props sideways out of their shoes, or by a slight pull of the props up-stream by means of chains fastened to the bottom of the shutters, the unsupported trestles and shutters fall flat on the apron over the props, as shown by the dotted lines in Fig. 212. The weir is reinstated by pulling up each shutter successively, in a horizontal position, from a special boat, or from a footbridge on movable frames, and thereby raising the trestle and replacing the prop in its shoe. The shutters, which are usually made between 3 and 4 feet wide, have necessarily small intervals left between them for clearance, through which leakage occurs, and which have occasionally been closed by wooden laths laid over the apertures on the up-stream side during the dry weather.

Originally it was supposed that the smaller shutter weirs across the shallow passes, would regulate the water-level of the river above them automatically, by the tipping down of the balanced shutters as the river rose; but the tipping occurred too suddenly, and the shutters did not right themselves till the river had been considerably lowered. Accordingly, the regulation of the discharge at a shutter weir, when not effected by a special, shallow, regulating pass, like the needle weirs alongside the shutter weirs on the Upper Seine, is accomplished, either by partially tipping some of the shutters by means of chains from a footbridge on frames, which has often been added for facilitating the working of shutter weirs, or by the opening of butterfly valves revolving on a horizontal axis, resembling small shutters, in the upper panels of the shutters, either automatically with the rise of the water-level, or by aid of a push with a boat-hook.

Some of the largest shutters of this type have been placed in the navigable pass, 340 feet wide, of the weir across the Saône at Lyons, being $14\frac{1}{3}$ feet high and $4\frac{1}{2}$ feet wide, made of iron, and having butterfly valves 5 feet high by 3 feet wide; and these shutters are readily raised and lowered from a footbridge on frames. A footbridge was provided for working the shutter weir across the Ohio at Davis Island, as the drift and gravel brought down by the river, and the width of the pass of 559 feet, precluded the adoption of a tripping bar; but the injuries to the frames of the bridge caused by the accumulations of drift and floating ice, have led to the employment of a pusher, worked from a boat, for releasing the props of this weir. The shutters of the eight movable weirs on the Great Kanawha River, are supported by rectangular iron frames hinged to the apron, in which the lower half of the shutters revolve, in place of trestles, and are worked in some cases by tripping bars, and in other instances from a footbridge; and these weirs, which, like the weir on the Ohio, have to be frequently opened at times for the passage of sudden floods, are lowered in between one and two hours, except when delayed by drift or ice, and raised again in $4\frac{1}{2}$ to 7 hours. In 1895–6, one of these weirs was raised and lowered six times within six months.

The advantage of this form of weir, like that of the bear-trap, is that it can be quickly opened, by a tripping bar or pusher, on the approach of a sudden flood, leaving the channel quite open for the discharge of the flood-waters, the descent of drift and floating ice, and the passage of vessels. The addition of a footbridge on frames to the shutter weir, for facilitating its regulation and working, renders the shutter weir more complicated and costly than the ordinary frame weir, closed by needles, panels, or rolling-up curtains (compare Figs. 207, 208, and 209, p. 350, with Fig. 212, p. 354); prolongs the period required for the complete opening of the channel; and makes the shutter weir as unsuitable for the sudden passage of masses of drift down the river, as the ordinary frame weir.

Remarks on Shutter Weirs.—Other forms of shutter weirs have been used,[1] especially the double shutters, closing some Indian irrigation weirs, in which the main shutters, hinged to the apron, rise with the stream, and have to be controlled in their ascent by hydraulic brakes; but the two forms described above are the only ones at all in common use, at the present time, in connection with the canalization of rivers.

3. *Drum Weirs.*

The drum weir is composed of an upper and an under wrought-iron paddle making a quarter of a revolution round a central, horizontal axis laid along the sill of the weir (Fig. 213). The straight, upper paddle forms the weir when raised; and the slightly larger under paddle, which is made crooked to allow of the admission of water above it when horizontal, revolves in the quadrant of a cylinder constructed below the sill of the weir, constituting the drum, and closes or opens the weir according to the adjustment of the pressure from the upper pool on its two faces, by means of see-saw sluice-gates regulating the connection, through sluices in the pier, between the upper and lower pools and the upper and lower portions of the drum. The control of the drum weir is, indeed, so perfect, that the upper paddle can be lowered and maintained at any angle between the vertical and the horizontal, for regulating the discharge of the river, and raised again against the full current through the open weir, into any desired position.

1 "Rivers and Canals," 2nd edition, 1896, vol. i. pp. 133–135 and 141, and plate 4, figs. 12, 13, and 14; and *Proceedings Inst. C.E.*, vol. lx. pp. 33 and 44, and plate 5.

Drum Weirs in France.—The shallow, regulating passes of the twelve weirs on the River Marne, constructed in 1867–69, are controlled by a row of the upper paddles of drum weirs, $3\frac{1}{2}$ feet high and $4\frac{2}{3}$ feet wide, which can be completely raised or lowered in 3 or 4 minutes, or placed at any angle, by a man on the bank working the sluice-gates. These weirs have worked perfectly, without needing repairs; but the system has not been adopted elsewhere in France, probably on account of the large cost involved in the construction of the drum, which has to be carried deeper below the sill of the weir than the actual weir rises above it.

DRUM WEIR.

Fig. 213.—Charlottenburg: Navigable Pass.

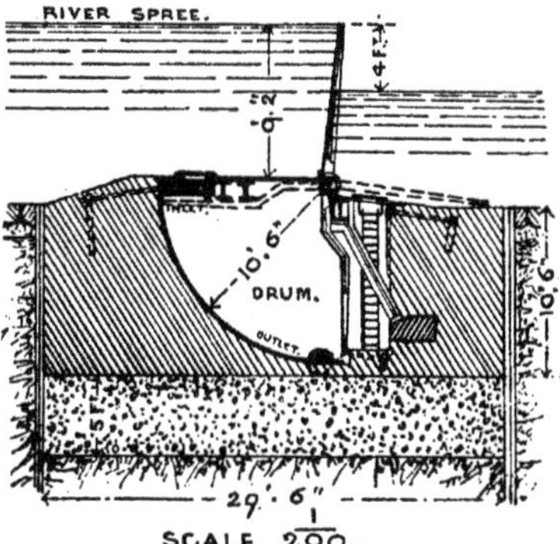

Drum Weirs in Germany.—In spite of the cost of the drum, the merits of the drum weir for regulating the discharge of a river and rapidly opening and closing a pass, have led to its adoption across the timber passes at the River Main weirs, and for a navigable pass on the River Spree at Charlottenburg.

A special channel has been provided at each of the weirs on the canalized Main for the passage of rafts of timber, situated alongside the bank on the opposite side of the river to the lock (Fig. 198, p. 341), across the up-stream end of which a drum weir has been placed, closing the pass by a single upper paddle, $39\frac{1}{3}$ feet wide, and reaching a height, when raised, of 5 feet 7 inches above the sill.[1] These weirs can be readily lowered to any extent for regulating the discharge of the river, and thereby obviate the necessity for the displacement of any of the needles of the adjacent needle weirs; and the passes can be rapidly opened for the passage of timber down-stream, and closed again without difficulty against the full rush of water through the pass.

The largest drum weir hitherto constructed closes the navigable pass of the Charlottenburg weir across the River Spree near Berlin, by means of a single upper paddle, $32\frac{3}{4}$ feet wide, with its top $9\frac{1}{6}$ feet above the sill of the weir, when raised;[2] whilst the drum has been carried down about 11 feet below the sill, and is enclosed in massive masonry resting upon a concrete foundation (Fig. 213).

Drum Weir in the United States.—A peculiar form of drum weir has been constructed on the Osage River near its confluence with the Missouri, where, instead of the paddles, a hollow, wooden, cylindrical sector, stiffened inside by iron stays, forms the weir. This cylindrical

[1] *Zeitschrift für Bauwesen*, 1888, p. 19, and plate 17; and *Proceedings Inst. C.E.*, vol. xcvi. p. 191.

[2] *Zeitschrift für Bauwesen*, 1886, p. 338, and plates 31 and 32; and *Proceedings Inst. C.E.*, vol. xcvi. p. 192.

sector, turning on a horizontal axis, is raised out of the drum by the water-pressure from the upper pool, on its under-side, to close the weir, and falls again into the drum on the removal of the pressure for the opening of the weir; but a triangular space is left between the under-side of the lowered sector and the vertical side of the quadrantal drum, through which the pressure from the water of the upper pool can be admitted for closing the weir again.[1] Provision has also been made for raising the sector to close the weir when the difference of head is insufficient, by forcing air into the lowered sector through a pipe, so as to raise it by rendering it buoyant. The crest of the weir when fully raised, will be 7 feet above the sill; and the radius of the drum, as well as its maximum depth below the sill, is 9 feet. The estimated cost of this weir is only about £30 per lineal foot, owing to its being mainly constructed of timber; whereas the cost of the Charlottenburg drum weir reached nearly five times this rate, which can only be partially attributed to about $2\frac{1}{4}$ feet greater height of weir, and the costly accessories of the overhead bridge and its supports, quite independent of the weir.

Remarks on Movable Weirs.—The value of movable weirs on rivers consists, not merely in leaving the channel, and consequently the discharge, quite unimpeded in flood-time, but also in enabling vessels to navigate the river freely during the high stages of the river, and especially during floods when the submergence of the locks, which is liable to occur on Continental and American rivers during the flood season, would otherwise put a stop to navigation.

The needle weir combines the advantages of cheapness and simplicity for weirs of moderate height; whilst in the United States, the employment of a floating derrick has enabled the system to be extended to greater depths of water. The adoption, moreover, of sliding panels or rolling-up curtains, makes the ordinary frame weir available up to the limit of height and weight at which the frames cease to be readily raised; whilst the use of suspended frames enlarges the scope of the frame weir, and increases its security and ease of working, under the condition of the additional cost of an overhead bridge with high piers.

The simple shutter weir can be rapidly opened across passes of moderate width by means of a tripping bar, lowering the shutters in successive groups, or somewhat more slowly but surely by aid of a pusher from a boat; and as it offers no obstructions to drift and floating ice, it is a convenient form of weir for rivers subject to frequent, sudden floods, and bringing down large masses of drift. The addition of a footbridge, though facilitating the regulation of the discharge and the raising and lowering of the shutters, increases considerably the cost, prolongs the time required for fully opening or closing the weir, and offers a liability to injury from the accumulation of drift against the frames of the footbridge and the shocks of floating ice.

The drum weir provides a perfect movable barrier across passes which have to be frequently and rapidly opened and closed, or for regulating

[1] "Report of the Chief of Engineers, U.S. Army, for the year 1898," part vi. p. 3546, and plate; and *Transactions of the American Society of Civil Engineers*, vol. xxxix. p. 555.

the discharge of a river, on account of the ease with which it is worked, and the precision with which its movements are controlled. The sole objection to its more extended use is its cost, owing to the depth to which the drum has to be carried below the sill of the weir; though under the special conditions of the Osage River, and the extensive employment of timber, this objection appears to have been overcome in the drum weir across that river. Moreover, where perfect control of the passage through a special, narrow pass of a weir is of primary importance, as at the timber passes on the Main, and the navigable pass at Charlottenburg, the remarkable efficiency of the drum weir in this respect may render its cost quite a secondary consideration.

CHAPTER XXII.

IMPROVEMENT OF RIVER OUTLETS OBSTRUCTED BY DRIFT, AND TIDELESS DELTAIC OUTLETS.

Changed condition of rivers towards their outlet—Tideless and tidal rivers, differences in conditions—**Improvement of River Outlets obstructed by Drift:** jetties at outlets of rivers, diversion of outlet by drift, instances, fixed and deepened by jetties, examples—**Improvement of Deltaic Outlets of Tideless Rivers:** formation of deltas, origin and position of bars; conditions affecting advance of delta, illustrated by examples; alluvium in proportion to discharge, and its nature in deltaic rivers; operations tried at deltaic outlets of tideless rivers, failure of harrowing and dredging bars, adoption of jetties; embankment of east outlet of Rhone, description, effect, reason of failure; jetties at Sulina mouth of Danube, description, effect on bar, dredging; jetties at South Pass outlet of Mississippi, description, lowering of bar, decreasing depth in front; remarks on jetty system, results at Sulina and South Pass compared.

RIVERS on approaching the sea become more uniform in their flow, have a larger discharge on account of the influx of their various tributaries and the increase in their drainage-area, and combine a smaller fall with a larger, and generally a deeper channel, than in the upper portions of their course; and they are, consequently, better adapted for navigation than higher up. They are, however, very liable to flow through a shallow, shifting channel on expanding into a wide sandy estuary before reaching the seacoast, and to be seriously obstructed by a bar at their outlet. The increase in the size of rivers in the lower part of their basin, necessitates a corresponding increase in the scale of the works required for their improvement; but, at the same time, their possible capabilities for serving as a highway for large vessels are considerably greater if properly developed, and if natural obstacles are removed by suitable works.

Tideless and Tidal Rivers.—In the neighbourhood of their outlets, rivers must be divided into two quite distinct classes, according as the sea into which they flow is tideless or tidal. In the first case, the size of the outlet channel of a river is proportionate to its discharge; and the river emerges into an inert mass of water which gradually arrests the outflowing current. Where, however, on the contrary, the sea is tidal at the mouth of a river, the tide flows up the river and ebbs out again, for a distance depending upon the rise of tide, the flatness of the river-bed, and the facility of influx, and generally maintains a channel, especially near the outlet, far larger than could possibly be formed by the unaided fresh-water discharge of the river. Accordingly, rivers flowing into tideless seas are only suited, under favourable conditions or after improvements, for affording access to large vessels, if they drain a large basin and have a fairly good discharge at their lowest stage, like the Danube, the Mississippi, and the Amazon; whereas rivers draining quite small basins, and insignificant in size above their tidal limit, may, in consequence of a good tidal rise in the sea at their mouth, and the resulting large tidal flow and ebb in their channel, be accessible in the lower part of their tidal portion to vessels of the largest draught, like the Thames, the Severn, the Mersey, and the Humber. These two classes of rivers are, accordingly, subject to entirely different conditions, necessitating quite distinct methods of improvement, with the sole exception that, in certain cases, their outlets are equally exposed to obstruction by the littoral drift, brought along the coast under the action of the prevalent winds by means of the waves, tending to form a continuous beach across the mouth of the river.

IMPROVEMENT OF RIVER OUTLETS OBSTRUCTED BY DRIFT.

Jetties at the Outlets of Rivers.—On an exposed seacoast, where there is a considerable drift of sand or shingle along the shore, in accordance with the direction of the prevailing winds, the outlet channel of the river formed by its discharge, or, on a tidal coast, by the combined fresh-water discharge and tidal ebb, is more or less obstructed by the travel of the drift, which is only prevented from forming a continuous line of sloping beach by the action of the outflowing current forcing a passage for itself across the foreshore. Where the issuing current is feeble and the drift considerable, the river is liable to be deflected from its direct outlet, and to be forced to flow parallel to the shore in the direction of the drift, till the accumulation of its waters in flood-time, or its reaching a weak spot in the barrier between it and the sea, enables it at last to force an outlet for itself across the beach. Thus the River Yare was formerly driven 4 miles southwards from its old direct outlet at Yarmouth, by the drift of sand and shingle brought along from the north by north-easterly gales (Fig. 214, p. 362); the River Adur was forced to find a new outlet, in the eighteenth century, about 3 or 4 miles to the east of its former mouth at Old Shoreham, by the travel of shingle resulting from south-westerly gales; whilst the mouth of the River Adour, falling into the stormy Bay of Biscay, was in old times shifted as much as 18 miles by the action of violent storms on the sand and shingle forming the beach. In all these instances, the shifting outlets were finally fixed by parallel timber or masonry jetties, leading the rivers direct into the sea at the most convenient available sites; whilst the jetties, by being carried across the beach, direct the scour of the issuing current so as to maintain an outlet channel leading into deep water.

The outlet also of the River Maas has been greatly improved for

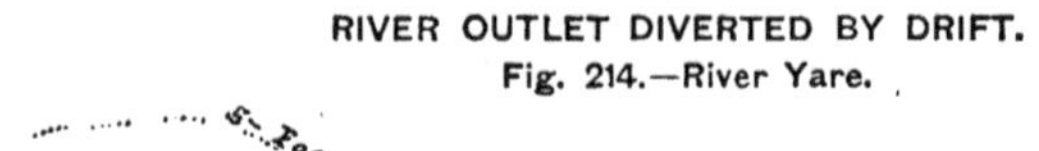

RIVER OUTLET DIVERTED BY DRIFT.

Fig. 214.—River Yare.

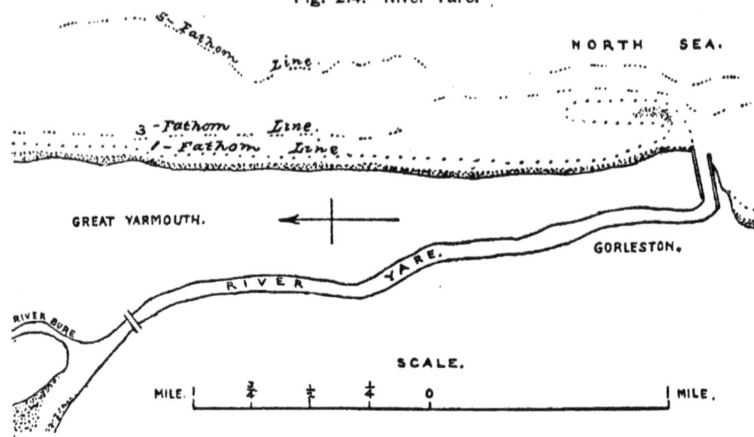

STRAIGHT CUT WITH JETTY OUTLET.

Fig. 215.—River Maas.

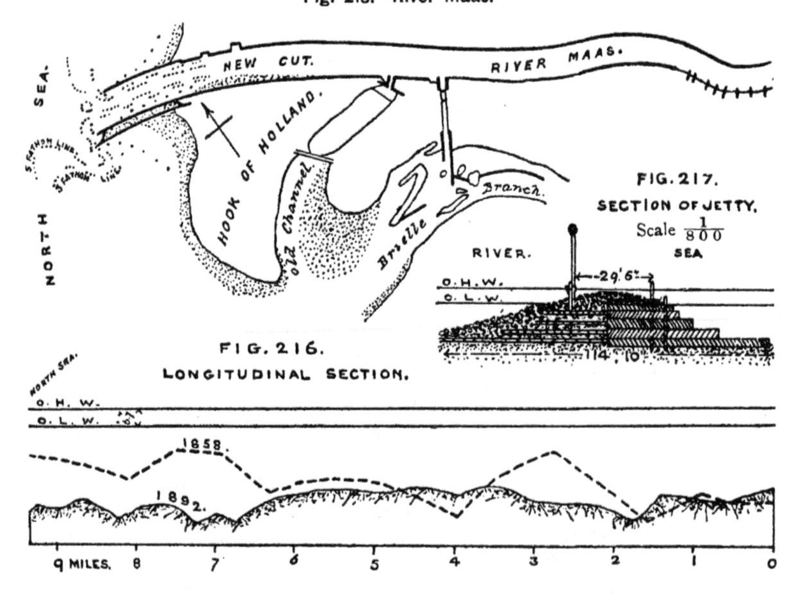

navigation by abandoning the natural, winding, shallow outlet channel, and substituting a straight cut across the Hook of Holland; and this new direct channel is led across the beach into deep water between parallel jetties constructed of fascine mattresses weighted with stone (Figs. 215, 216, and 217). In all the above instances, the fresh-water discharge of the river is reinforced by the tidal ebb and flow.

Rivers also flowing into tideless seas, which do not carry down much sediment, but are exposed at their outlets to drift, or to the natural shoaling resulting from the loss of scour when the discharge, being no longer guided by banks, is dispersed in flowing across the beach, can be materially deepened in front of their mouth, out to deep water, by concentrating the scour of the discharge between jetties across the beach. Thus the River Oder, which deposits its alluvium in some lakes before reaching the sea, has had its navigable outlet at Swinemünde deepened by jetties, aided by dredging; and the outlets of some Russian rivers flowing into the Baltic, and of North American rivers flowing into the Great Lakes, have been improved by similar means. The scour, and consequently the depth of the outlet channel, can be increased in these cases by making the width between the jetties somewhat less than the width of the river between its banks above; but such a contraction is inadmissible in the case of tidal rivers, as it would check the tidal influx, and therefore reduce the volume of tidal water entering and leaving the river.

IMPROVEMENT OF DELTAIC OUTLETS OF TIDELESS RIVERS.

Rivers which bring down a considerable amount of sediment, and flow into tideless seas, form deltas at their outlets by the deposit of this sediment when their current is arrested on entering the sea; and this deposit, by obstructing the discharge, and reducing the fall by its gradual progression seawards, causes the river to split up into shallower branch channels, spreading out like a fan in traversing the delta, through which the water finds its way by separate outlets to the sea (Fig. 218, p. 364). The water flowing through each of these diverging channels, carries down an amount of alluvium proportionate to its volume, which by its deposit after reaching the sea, causes a corresponding advance of the delta in front of the outlet. A bar, also, or ridge, stretches across the channel seawards of the outlet, where the chief deposit of the heavy alluvium occurs, over which the depth is less than higher up the channel or further out at sea; and the depth of water on the bar, accordingly, determines the accessibility of the channel for navigation. Moreover, not only is the advance of the delta greater seawards of the channel which conveys the largest discharge, but the bar also is formed further out beyond the outlet of this channel, owing to the arrest of the more powerful current, resulting from a larger discharge, being less quickly effected by the inert mass of sea-water into which it flows, than that of the smaller discharges through the minor channels; and both these circumstances have an important bearing on the selection of a channel to have the depth over its bar increased in the interests of navigation. The bars in front of the outlets travel seawards correspondingly to the delta, owing to the gradual prolongation of the banks of the outlet channels by the deposit.

Conditions affecting the Advance of a Delta.—The rate of advance of a delta depends upon the proportion of sediment brought

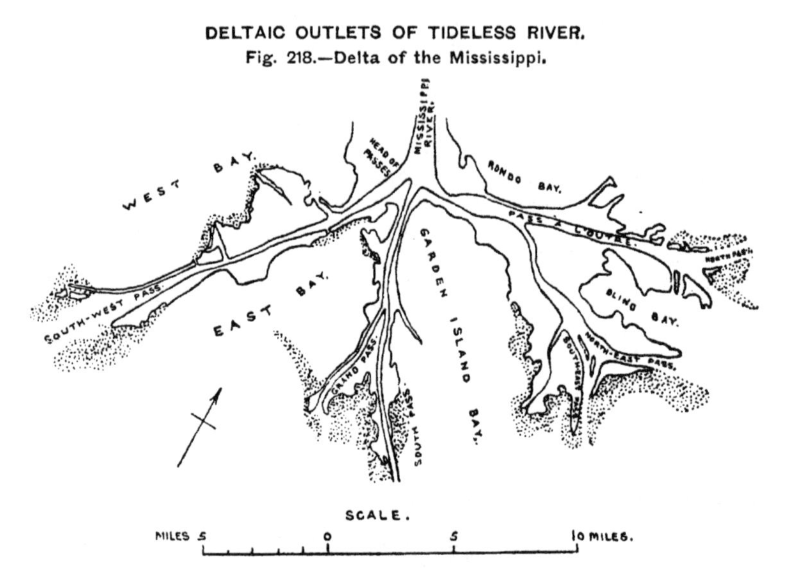

DELTAIC OUTLETS OF TIDELESS RIVER.

Fig. 218.—Delta of the Mississippi.

down by a river, and its density; the extent of coastline over which the deposit is distributed; the depth of the sea in front of the outlets; and the existence of any disturbing cause impeding the accumulation of deposit, such as a littoral current, or the action of waves along the shore, under the influence of the prevalent winds. A large volume of sediment in relation to its zone of distribution, and a high density favouring a rapid deposit, and preventing the dispersion of the material which occurs, under the action of winds and waves, when it floats for some time in the fresh water flowing along on the surface of the sea-water, necessarily promote the rapid advance of a delta. A shallow sea in front of a delta, like the Caspian Sea where the Volga flows in, leaves so little space available for the accumulation of deposit, that the progression of the delta of a river bringing down a considerable volume of alluvium is naturally rapid; whilst a deep sea at the outlet of a deltaic river, such as the Gulf of Mexico into which the Mississippi flows, retards the advance of the delta, which is further checked in the case of the Mississippi by a littoral current, and still more so at the central part of the Danube delta, where, in addition to the influence of a southerly littoral current on the light matter in suspension, erosion from wave-action takes place along a portion of the foreshore, thereby compensating to a considerable extent for the shallowness of the Black Sea in front of the delta.

Alluvium in proportion to Discharge, and its Nature in Deltaic Rivers.—The proportion the alluvium carried down bears to the discharge, has been estimated at $\frac{1}{2166}$ for the Rhone, $\frac{1}{2420}$ for the Mississippi, $\frac{1}{2444}$ for the tidal Húgli, and $\frac{1}{6700}$ for the alluvium carried in suspension in the Danube, without any allowance for sediment rolled along its bed. The alluvium consists generally of sand, clay, and silt, the heavier sand rolled along the bottom coming first to rest, and the lighter, finer particles of silt being carried out to sea, attaining a maximum distance of about 300 miles from the shore in the case of the Amazon; and whereas silt is found in deep places in the Húgli estuary, and is visible in large quantities in a finely-divided state in suspension in the river, and clay is met with in parts of the channel, the bars of the Húgli below Calcutta are almost wholly composed of pure sand.[1]

Operations tried at the Deltaic Outlets of Tideless Rivers.—Harrowing and dredging the bars in front of the central Sulina branch of the Danube, and the South-west Pass of the Mississippi, were the first measures tried for improving the navigable depth over them. The slackening current, however, which, on entering the sea, was already dropping the materials it had brought down, could not possibly carry away the additional volume of material stirred up from the bar; and dredging, though producing a slight improvement at the Sulina outlet, and increasing the depth over the bar of the South-west Pass of the Mississippi from 13 feet up to 18 feet, did not produce any permanent amelioration, as the deepened channel soon shoaled up again with fresh deposit when the dredging was suspended. The depth also at the outlet of one of the branches of the Volga delta, which has been increased from 4 feet up to 8 feet by dredging, can only be maintained by regular yearly dredging operations.

The system subsequently resorted to for lowering the bar at the Sulina mouth of the Danube, and at the South Pass of the Mississippi, and which had been previously attempted on a somewhat different plan at the mouth of the Rhone, consists of an artificial prolongation of the banks of the outlet channel by means of solid jetties on each side, extended out from the shore across the foreshore to the bar, so as to direct and concentrate the issuing current against the bar, and thus scour a deep channel through it.

Embankment of East Outlet of Rhone.—The most direct, eastern channel of the Rhone was selected for improvement; and in 1852–57 an embankment was formed along each side of this channel, closing the three southern channels branching off from the East Channel, and also two minor northern ones; and these embankments were extended a little distance seawards of its outlet to within about half a mile of the bar, thereby concentrating the whole of the discharge of the Rhone, and consequently the whole also of the alluvium brought down by the river, into a single outlet. The increased discharge thus directed

[1] "Report on the River Húgli," L. F. Vernon-Harcourt, Calcutta, 1897, pp. 9–11.

into the East Channel, and the improved scour produced by the moderate extension of the banks of this channel towards its bar, augmented temporarily the minimum depth of the outlet channel by about 5 feet, and drove the bar further out. Owing, however, to the deposit being confined to the neighbourhood of a single outlet, coupled with the unusual density of a large proportion of the sediment, the absence of any littoral current in the sheltered bay into which the East Channel discharges with its outlet facing the worst wind, and the shallowness of the sea in front of the outlet, the deepening by scour was only very temporary; and the bar soon formed again to about the same height as before, further seawards, in a more exposed situation still less favourable for navigation; whilst the advance of the delta in front of the outlet has been considerably accelerated by the concentration of the discharge of all the alluvium at a single mouth, and the bar has progressed about a mile seawards since the closure of the other outlets.

The depth over the bar only averages about $6\frac{1}{3}$ feet; and the St. Louis Canal was, accordingly, constructed in 1863–73, branching off from the Rhone about 4 miles above its outlet, and shut off from the turbid waters of the river by a lock at its entrance, with the object of providing the traffic on the river with deep-water access to the Mediterranean and Marseilles, which the ill-designed embankments along the East Channel had failed to secure. The adoption of the East Channel as the sole outlet for the Rhone was evidently a mistake, for, with the exception of its following more nearly the direction of the river above than the other channels, its natural conditions were decidedly inferior to those of the southern outlets; and if the central, southern Roustan Channel had been improved by jetties at its mouth, with its superior channel, its outlet swept by the Mediterranean littoral current flowing westwards, which, aided by wave-action, is at the present time eroding the southern face of the Rhone delta, and the good depth of the sea in front of its outlet, and if, moreover, the other outlets had been left open so that the sediment brought down by the river might be distributed over as wide an area of the sea-bottom as possible, it is fairly certain, judging from the later experience of the results of similar works at the Danube and Mississippi outlets, that a notably deeper outlet would have been obtained for the Rhone across the Roustan bar.

Jetties at Sulina Mouth of Danube.—The central, Sulina mouth of the Danube was selected in 1858 for improvement by jetties, carried out from the shore on each side of the outlet to the bar, on account of the distance of its bar from the shore being only half that of the bar of the St. George's mouths to the south, and the advance of the delta at Sulina about one-fourth the rate of its advance at the Kilia mouths to the north, owing to the discharge of the St. George's branch being four times that of the Sulina branch, and of the Kilia branch about eight times, with shallow water in front of its mouths. The works, consequently, required at the Sulina mouth were much less than those which would have been needed at the St. George's mouths, and had much better prospects of success and permanency than would have been possible at the Kilia mouths; whilst at the time the works were decided upon, the Sulina mouth was the only navigable outlet, with 9 feet over its bar.[1] The jetties, constructed in 1858–61, of rubble stone and pilework, were carried out beyond the crest of the bar into a depth of 18 feet, and were consolidated with concrete blocks in 1866–71.

These jetties, 600 feet apart at their extremities, and extending out nearly seven-eighths of a mile from the shore, by concentrating the scour of the issuing current, gradually increased the depth over the bar up to 20½ feet in 1872, which depth was subsequently maintained, as the comparatively light alluvium brought down by the river has been carried into deeper water, and driven somewhat to the south by the littoral current from the north. Deposit, accordingly, has occurred seawards of the former site of the bar, and to the south, causing an advance of the 4- and 5-fathom lines of soundings in front of, and to the south of the channel; and a progression of the southern shoreline has taken place under the shelter of the jetties. The deep channel was, consequently, being deflected towards the north by the southerly deposits, and a bar showed signs of forming again seawards of its old site, when dredging was commenced in 1894 to straighten the channel beyond the jetties and improve its depth; and by this means, a more direct channel has been obtained between the outlet and deep water, with an available depth of 24 feet, which, however, is liable to be reduced, in spite of the dredging, during years when high floods of the river bring down large quantities of alluvium. The dredging, therefore, is deferring the period when the general advance of the delta, by deposit in front of the Sulina mouth, will eventually necessitate a prolongation of the jetties to scour away again a fresh bar forming further out.

Jetties at South Pass Outlet of Mississippi.—The outlet of the South Pass of the Mississippi was selected for deepening by parallel jetties carried out to the bar, in spite of the South-west Pass possessing a deeper outlet and a better navigable channel through the delta, on account of the bar of the South Pass being less than half the distance out from the shoreline that the bar of the South-west Pass was, by reason of the discharge through the South Pass being only one-tenth of the whole, as compared with one-third through the South-west Pass, and owing to the advance of the delta being 100 feet annually in front of the South Pass, in place of the 300 feet yearly progression in front of the South-west Pass (Fig. 218, p. 364). The two jetties constructed of willow mattresses weighted with limestone, and protected with large concrete blocks at their exposed outer ends (Fig. 219, p. 368), were carried to the same distance out, 2¼ and 1½ miles in length, on the east and west sides of the channel respectively, right across the bar into a depth of 30 feet, in 1876–1879 (Fig. 220, p. 368). These jetties, placed about 1000 feet apart, were curved a little southwards towards their extremities, so that the alluvium discharged might be brought directly under the influence of the littoral current flowing from east to west across the outlet; and the concentrated current directed by the jetties rapidly scoured away

[1] *Proceedings Inst. C.E.*, vol. xxi. p. 285, and plates 5-9.

the bar, so that the depth at the outlet was increased from 8 feet up to a minimum central depth of 31 feet by 1880 (Fig. 221).[1] To maintain the required central depth of 30 feet, and 26 feet for a minimum width of 200 feet, in the jetty channel, it has proved necessary to narrow the channel to about 660 feet by an inner jetty on each side, and to resort occasionally to dredging. Moreover, though a channel with a minimum depth of 30 feet has hitherto been maintained outside the jetties by the increased scour, the channel has been deflected eastwards by the encroachment of the deposit taking place on the western side, towards which the alluvium is driven by the littoral current; and dredging has been undertaken at times for straightening the outlet channel.

The works, indeed, have been eminently successful in scouring away the bar; but as within a fan-shaped area of 1½ square miles in front of

JETTIES AT DELTAIC OUTLET.

Fig. 220.—Mississippi South Pass Outlet.

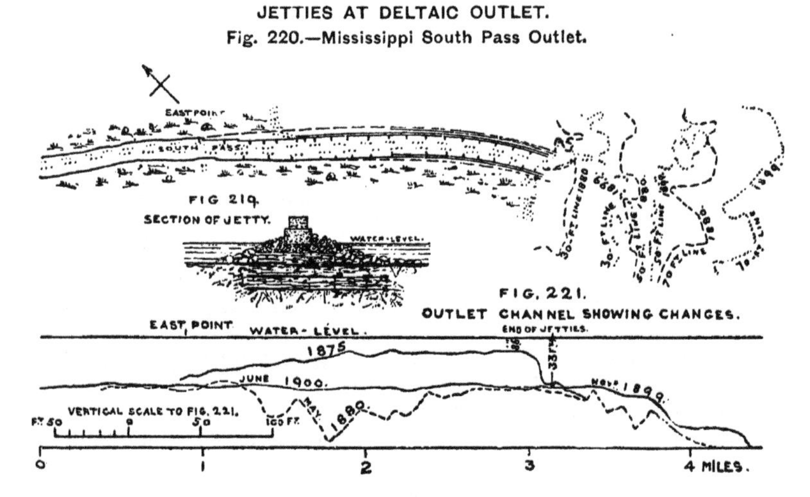

the outlet, surveyed each year, the average decrease in depth between 1876 and 1899 has amounted to 17½ feet, a maximum annual advance of 106 feet is occurring at the 70-feet line, and is even 97 feet at the 100-feet line beyond the fan-shaped area,[2] and a bar is evidently in course of formation beyond the site of the old bar (Fig. 221), an extension of the jetties will clearly be needed before long to enable the scour to cope with the advancing delta, and maintain the required depth in the outlet channel.

Remarks on the Jetty System at Tideless Deltaic Outlets.—The value of the system of scouring away the bar at one of the deltaic outlets of a tideless river, by prolonging the channel, and,

[1] "A History of the Jetties at the Mouth of the Mississippi River," E. L. Corthell, p. 94, and plate 11, and p. 202, and plate 17; and "Report of the Chief of Engineers, U.S. Army, for the year 1880," part ii. p. 1123, and charts 3-5.

[2] "Report of the Chief of Engineers, U.S. Army, for the year 1900," part iii. p. 2237, and charts 1 and 2.

consequently, the current out to the bar, has been proved by the remarkable success which has attended these works at the Sulina mouth and the South Pass. This success, however, must be partly attributed to the existence of a strong littoral current sweeping across the outlet in both these cases, and preventing, in a great measure, the alluvium from being deposited in front of the outlet; and it is also due to the lightness of the alluvium brought down by the Danube facilitating its conveyance to deep water by the concentrated current, and its dispersal by littoral drift, and to the depth of the sea in front of the South Pass providing a large space in which the heavier sediment of the Mississippi can accumulate for a considerable period, before heaping up sufficiently high on the sea-bottom to interfere with the navigable depth of the outlet channel. The lightness, indeed, of the alluvium appears to be as important a factor, when subjected to the influences of the concentrated and littoral currents, as a good depth in front of the outlet; for the smaller depth in the Sulina outlet channel has been maintained for a longer period unimpaired, and exhibits a prospect of greater permanence in the future, than the deeper channel at the South Pass outlet. The importance also of selecting a suitable outlet exposed to a littoral current, and with a good depth in front where the sediment is heavy, of carrying the jetties right across the bar as quickly as practicable, and of leaving the other outlets open, so as not to neutralize the effect of the works by accelerating the advance of the delta at the navigable outlet, has been manifested by the failure of the works at the mouth of the Rhone, which resulted from the neglect of these provisions.

Owing to the inevitable advance of the delta at the improved outlet, the jetty system cannot be regarded as a permanent method of improvement; for after the lapse of a certain period, depending on the conditions of the site, the jetties must need to be extended in proportion to the progression of the delta, in order to concentrate the issuing current again over the bar in process of formation further out.

CHAPTER XXIII.

THE IMPROVEMENT OF TIDAL RIVERS AND ESTUARIES.

Considerable differences in conditions—Tidal flow in a river, value, tidal lines, indications of *bore*, cause of *bore*, and examples—Removal of obstacles to the tidal influx, beneficial effects—Dredging for improving tidal rivers, value, results in Clyde and Tyne, on Mersey bar—Regulation and training of tidal rivers, influence of sharp bends on tidal currents and channel, examples in River Húgli, method of improvement of channel; regulation works on Maas, Nervion, and Weser, general results, increase in width seawards, rate of enlargement—Protection of outlet by breakwaters, examples, advantages—Training tidal rivers through sandy estuaries, instances of estuaries, conditions for improving channel by dredging alone; nature and object of training walls, important effects of training walls in the estuaries of the Seine and the Tees, training works in Weser estuary, construction of these training walls, value of system, improvement limited to trained channel, importance of due enlargement of channel seawards, instances of neglect of this provision—Accretion in estuaries due to training works, causes, in Seine estuary, increase of sandbanks at outlet, instances, cases of stoppage of training works on account of the resulting accretion—Remarks on training works through sandy estuaries, advantages, conditions, and limitations.

Tidal rivers exhibit a greater variety in their conditions, and involve more complicated problems for their improvement, than tideless rivers, owing to differences in the tidal rise at their mouth, the different distances to which the tidal influence extends up rivers, the differences in the proportions of the tidal flow to the freshwater discharge, and in their consequent relative influences, the twofold origin of the alluvium found in tidal rivers, and the considerable variety exhibited in the forms of the estuaries and outlets of tidal rivers. In these rivers, the tidal ebb and flow combine with the freshwater discharge in maintaining the outlet channel, and also in a gradually diminishing amount up the river as far as the tidal limit, above which the channel depends solely on the freshwater discharge; and therefore a tidal river, to be in a good condition, should gradually increase in width, and also in depth at high tide, from its tidal limit down to its outlet in the sea. Whilst, however, the freshwater discharge, though varying according to the seasons of the year, is dependent upon definite physical conditions, and therefore cannot be materially increased or diminished, but must find its way to the sea along the river, the tidal influx is wholly dependent upon the facilities offered to its progress up a river, and the space available for receiving it, and can be greatly modified by the removal of obstructions to its progress, or by placing artificial barriers in its path, or by the spaces at the sides of an estuary into which it flows, on passing up a river, being gradually filled up by accretion.

Tidal Flow in a River.—The flood tide sometimes comes into a river charged with material collected from the erosion of the neighbouring coast, or from outlying sandbanks, and put into suspension by the inflowing current aided by the waves during storms; and, consequently, proposals have occasionally been made to exclude this sediment-bearing tidal water, in order to prevent accretion in the river. The loss or reduction, however, of the tidal ebb and flow in a river, by the erection of a barrier across the channel, provided with sluice-gates for the discharge of the inland waters, has invariably led to the deterioration of the outlet channel below by accretion; whilst in a tidal river with a well-shaped channel gradually enlarging in cross section from its tidal limit to its outlet, the alluvium brought in by the flood tide is carried out again by the ebb, whose somewhat feebler flow is compensated for by being reinforced by the fresh-water discharge, especially during floods. Accordingly, the benefits gained by a river from the tidal flow and ebb providing a much greater volume of water, particularly during the dry season, and maintaining a considerably larger channel, especially towards the outlet, than the fresh-water discharge could secure, far outweigh the disadvantages which might possibly result, under unfavourable conditions, from the deposit of some of the alluvium brought in by the flood tide. Moreover, the alluvium brought down by a tidal river from inland does not mainly deposit in front of its mouth, but is carried up and down by the tidal currents and dispersed over a wide area in the sea; so that rivers with a very large discharge in the flood season, heavily charged with silt, flowing into tidal seas, though gradually forming deltas, of which the Ganges is an instance, with a tidal rise at its outlets in the Bay of Bengal of 9½ to 11 feet at springs, are less obstructed by bars at the outlet of their navigable channel, even at low water, than tideless rivers, and have the advantage of the additional depth furnished by the tidal rise.

The actual progression of the tide up and down a river is indicated by taking a series of simultaneous observations of the heights of the river at definite times during a single tide, at selected stations along the tidal portion, from which a series of simultaneous tidal lines can be drawn, to a very distorted scale, representing approximately the level of the river at the several times of observation during the flood and ebb throughout the tidal portion, as illustrated in the case of a high spring tide in the River Húgli, during the dry season, in Fig. 222. These lines, with a chart of the river, enable the volume of the tidal influx to be calculated; and the form assumed by these lines in their progress up a river, indicates the localities where the unfavourable condition of the channel impedes the upward flow of the tide. Thus the steep form of the head of the tidal influx into the River Húgli, exhibited by the diagram (Fig. 222), commencing near the outlet, becoming clearly marked at Buj-Buj, and reaching a maximum between Konnagar and Chinsurah, proves that the channel presents obstructions to the advance of the flood tide, and the existence of a *bore*, a well-known phenomenon which appears on

SIMULTANEOUS TIDAL LINES.

Fig. 222.—River Húgli, Spring tides, Dry Season.

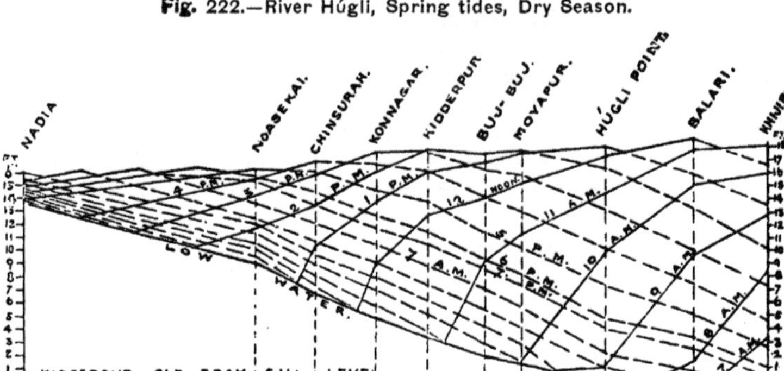

certain rivers having obstructed channels, with the first flood of high tides, intensified by wind, as a sort of crested wall of water or breaking wave travelling rapidly up the river, reversing instantaneously the ebbing current, and producing a sudden rise in the level of the river. The Severn, the Seine, the Tsien-Tang Kiang, and the Amazon, as well as the Húgli, are instances of rivers on which a bore has been observed, reaching heights of from 6 to 15 feet; and the bore is due to the retardation of the flood tide, till it has gained a sufficient head to overcome every obstacle to its upward progress with a rapid rush.

Removal of Obstacles to the Tidal Influx up a River.—Since tidal flow in a river is almost wholly beneficial, one method of improvement of a tidal river consists in removing obstacles, as far as possible, to the progress of the flood tide, thereby increasing the volume of water entering and leaving the river every tide, prolonging the tidal portion inland, and reducing the duration of slack tide to a minimum, during which period deposit mainly occurs. Thus hard shoals obstructing the channel, and unaffected by scour, should be removed, old bridges with wide piers and narrow openings should be rebuilt with enlarged waterways, and all other obstacles to the progress of the flood tide should be taken away; whilst any deepening of the channel, by facilitating the tidal influx, increases the tidal scour, and, consequently, the depth over any soft shoals below, and at the outlet. No tidal river, having a fair length, is completely filled by the tidal water to the full extent of its tidal capacity, for the tide begins to fall at the mouth before high water has been reached at the upper part of the tidal portion, as clearly indicated by the tidal diagram of the Húgli (Fig. 222); but every improvement of the channel augments the proportion of the filling, and also increases the tidal capacity by faciliating the efflux, as manifested by the lowering of the low-water line.

Dredging for improving Tidal Rivers.—Dredging furnishes a simple method of increasing the navigable depth of a tidal river, which has been increasingly resorted to, and the scope of which has been enlarged by the great modern improvements effected in dredging machinery. The extent, indeed, to which dredging can be carried is only limited by its cost, and the increasing difficulty of maintaining a channel which has been thus artificially deepened far beyond the limit of the scouring power of the currents to preserve. Dredging is much more effective in small tidal rivers, such as the Clyde and the Tyne, with a comparatively small, well-defined channel, than in large tidal rivers and estuaries, where the large quantities of alluvium brought down in flood-time are very liable to deposit, as the floods abate, in the deepened channel with its proportionately enfeebled scouring currents; and the shifting of the channel in an unstable bed, owing to seasonal changes in the flow, or the fretting action of the currents in a sandy estuary, imperils the maintenance of the depth obtained.

Dredging operations on a large scale have been most successfully applied to the improvement of the navigable condition of the small channels of the Clyde and the Tyne, where the channels leading from the Firth of Clyde to Glasgow, and from the sea to Newcastle, 19 miles and 10½ miles long respectively, which were formerly dry in places at low water, have been given an available depth throughout of about 20 feet below a lowered low water, thereby converting Glasgow and Newcastle into flourishing seaports, accessible to vessels of the largest draught.[1] Moreover, this deepening has materially improved the tidal condition of these rivers, as evidenced by the lowering of the low-water line 8½ feet at Glasgow and 3 feet at Newcastle, and the acceleration of the times of high water at these places after high tide at the mouths of the rivers, being only one hour instead of two hours later at Glasgow, and 12 minutes in place of 63 minutes at Newcastle. Accordingly, the dredging of these rivers has not merely provided a deep channel, but has also increased their tidal capacity and the volume of water entering and leaving the rivers at each tide, and, consequently, the tidal scour available for maintaining their channels. The dredging, however, in both instances has been carried far beyond the depth which the improved scour could maintain, so that dredging will always be required for removing the alluvium brought down by the rivers, which must deposit in the enlarged channel, and amounts to a considerable quantity in the case of the Clyde, which receives the sewage of Glasgow. The enlargement of the channels of these rivers has had the further incidental advantage of stopping the inundations of the low-lying lands

[1] "Rivers and Canals," 2nd edition, 1896, L. F. Vernon-Harcourt, vol. i. pp. 253 and 260, and plate 8, figs. 2 and 8.

bordering their tidal portions, owing to the lowering of the flood-level of the rivers by the ample waterway provided for the discharge of the fresh waters.

Though dredging, unaided by training works, could not be expected to produce the same improvement in depth, as achieved in the Clyde and the Tyne, in wide sandy estuaries with very shifting channels, such as the Seine and the Ribble, or in large rivers bringing down vast quantities of alluvium, and changing their channels in places with the dry and rainy seasons, as in the River Húgli above its estuary, yet it has been extended to deepening the outlet channels of river ports, of which a notable example is furnished by the deepening of the Mersey bar in Liverpool Bay. This sandy bar, about 11 miles beyond the actual outlet of the river at New Brighton, which is due to the heaping-up action of the sea tending to form a continuous beach in front of the mouth of the Mersey, a result which is only prevented by the influx and efflux of the tide into and out of the Mersey estuaries below and above Liverpool, reduced the navigable channel over the lowest part of its crest to about 11 feet below the lowest low water, thereby barring access to Liverpool, for some time before and after low water of the lowest tides, for vessels of large draught which had crossed the Atlantic at a high speed. Since 1890, however, the navigable channel across the bar has been gradually deepened by suction dredgers, of the type shown in Fig. 10; and up to 1909 upwards of 133 million tons of sand had been removed from the approaches to the Mersey, while in the year ending July 1st, 1909, 12,237,980 tons had been removed by five dredgers, a minimum depth of 30 feet having, generally speaking, been obtained. The unimproved channel across the Mersey bar was fairly stable, only shifting its position very slowly; and the effect of its deepening, in drawing a larger tidal influx and efflux through it as the channel of least resistance, should increase its stability.

Regulation and Training of Tidal Rivers.—A considerable improvement may often be effected in the depth and navigability of a tidal river by regulating its channel, so as to ease sharp bends, and do away with abrupt variations in width, and consequent irregularities in depth. Sharp bends in tidal rivers, as in non-tidal rivers, are not only inconvenient for navigation, especially in the case of large vessels, but also tend to become worse by the continued erosion of the concave bank, and check the tidal influx; and the navigable channel in a wide reach, besides being shallow, is liable to shift its position according to the alternate predominance of the fresh-water discharge in flood-time, and of the flood tide in the dry season. Changes also in the velocities of the currents, due to irregularities in the channel, favour deposit and the formation of bars in the wide parts of a river when the water is charged with alluvium.

Sharp bends possess the special disadvantage in a tidal river of producing a marked conflict in the action of the descending current of the ebb tide and freshets, and the ascending current of the flood tide, which, flowing in opposite directions, tend to form channels along opposite banks between the bends. The descending current follows the ordinary course of flow of the current in a non-tidal river, keeping close to the concave bank at a bend, and eventually crossing over to the next bend after the change of curvature of the bank it has followed, and generally constitutes the navigable channel; whereas the flood tide assumes a more direct course, and after passing round along the concave bank in a bend, only crosses over to the opposite concave bank, in the next bend above, after having cut a blind channel into the shoal projecting from the convex bank of the same bend. The channels thus formed along opposite banks, with an intervening shoal, are clearly indicated in the charts of the two worst places in the River Húgli between Calcutta and its estuary (Fig. 223 and 224), where the main ebb-tide

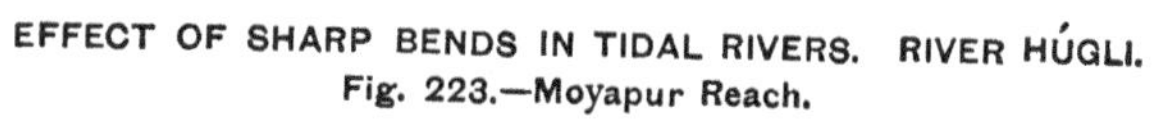

EFFECT OF SHARP BENDS IN TIDAL RIVERS. RIVER HÚGLI.

Fig. 223.—Moyapur Reach.

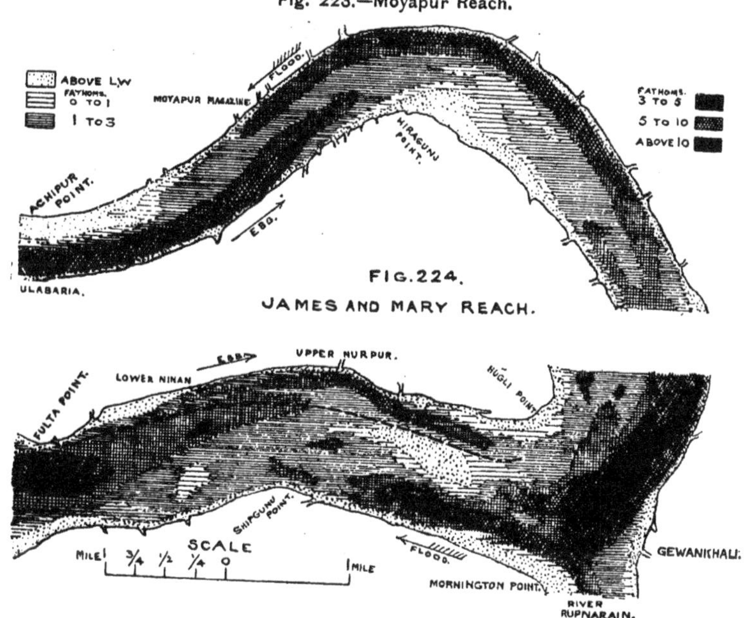

current at the Moyapur crossing (Fig. 223), keeps alongside the right bank (looking downstream) from the concave bank at Ulabaria nearly down to Hiragunj Point, passing over opposite this Point to the left bank; whilst the main flood-tide current, coming up along the right bank, and cutting into the shoal projecting from the convex shore, passes over to the left bank a little below the Point, and hugging this bank up to above the Moyapur Magazine, eventually, after cutting into the Achipur shoal, crosses the river to the concave right bank at the upper bend.[1] In the James and Mary Reach (Fig. 224), the main ebb-

[1] "Report on the River Húgli," L. F. Vernon-Harcourt, Calcutta, 1897, pp. 41, 48, and 55, and plates 6 and 7.

tide current follows the left bank from Fulta Point to Nurpur Point, where, being deflected, it passes in a widening and shoaling channel across the river at right angles to the deep flood-tide channel alongside the right bank below Gewankhali; whilst the main flood-tide current running up close along the concave right bank below the confluence of the River Rupnarain, splits on approaching Mornington Point, a portion flowing up the Rupnarain; and the greater portion, after curving round approximately at right angles to its former course, keeps still alongside the right bank of the Húgli till it crosses over near Shipgunj Point, in a shallow, ill-defined, and shifting channel, towards the left bank, somewhat in the direction of Lower Ninan. The bars across the navigable channel at the crossings in those reaches vary in position and height with the seasons, according as the freshets or the flood tides prevail; and the transposition of material with the changes in the channels are so large, and the liability of deposit of alluvium to take place towards the close of the heavily-charged freshets is so great in these reaches, which are being widened gradually along both banks by the erosion of the freshets and the flood tide respectively, that dredging, though certain to effect an improvement, would be too onerous for making and constantly maintaining an adequate navigable depth, to be a satisfactory and sufficiently certain method of improvement. A preferable course, under such conditions, would be to narrow the low-water channel of the river, which has been unduly widened by the unrestricted currents; and as the River Húgli is dependent for about two-thirds of the year upon the tidal flow for its maintenance, it would be essential to avoid all interference with the influx of the flood tide, by training the ebb-tide current into the flood-tide channel with training walls, having their tops at or below the lowest low water, and thus leaving the tidal capacity of the river unaffected, as indicated by the dotted lines on Figs. 223 and 224, p. 375.[1] By such an arrangement, the flood and ebb tides can be made to combine in deepening and maintaining the navigable channel, instead of acting in opposition by expending their force in forming different channels, on opposite sides of the river, in the widened reaches between the banks of a winding tidal river.

A systematic regulation, which is unnecessary in the Húgli, has been carried out on other tidal rivers of smaller dimensions, and with less good general depths, as for instance on the River Maas between Rotterdam and the sea,[2] on the River Nervion below Bilbao, to its mouth[3] (Fig. 225), and notably on the River Weser from Bremen to its estuary (Fig. 228, p. 378). These works consist of cuts to do away with sharp bends; occasional cross dykes to make the navigable channel follow a central course, and to close secondary branch channels; and longitudinal training walls, generally constructed of rubble stone or

[1] "Report on the River Húgli," Calcutta, 1897, pp. 88 and 92, and plate 6, figs. 3, 7, and 8, and plate 7, fig. 4, and plate 8, figs. 8 and 9.

[2] *Proceedings Inst. C.E.*, vol. cxviii. p. 25, and plate 2, figs. 12 and 13.

[3] *Ibid.*, p. 30, and plate 2, figs. 17 and 18.

fascine mattresses, to straighten the channel where necessary, and render it uniform in width, and consequently in depth (Figs. 226 and 229). In the Weser, the low-water channel has been regulated by training walls, not raised above low water (Fig. 230, p. 378), so that whilst regulating

REGULATION AND PROTECTION OF OUTLET. RIVER NERVION.

Fig. 225.—Plan. Fig. 227. -Western Breakwater, Section.

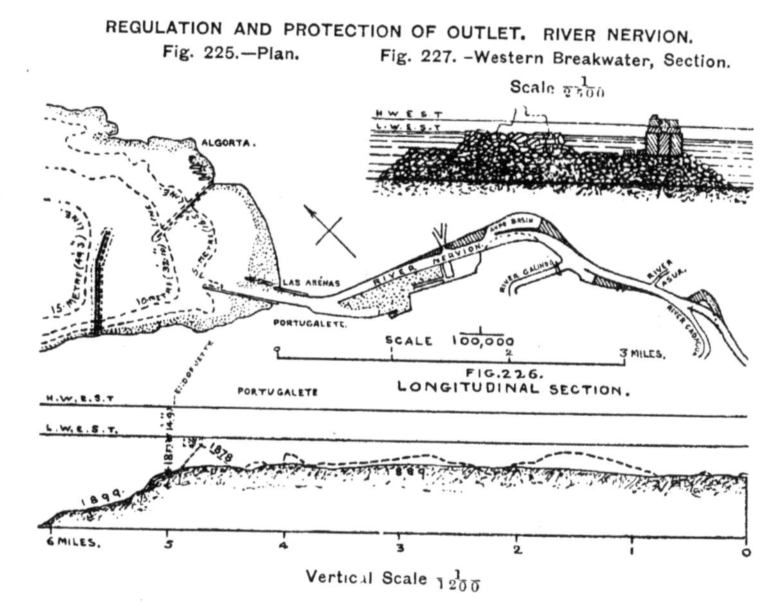

Scale 1/2500

FIG. 226. LONGITUDINAL SECTION.

Vertical Scale 1/1200

and deepening the navigable channel, the high-water channel of the river may remain unrestricted for the reception of the tidal influx as formerly.[1] These works, by regulating and deepening the channel, facilitate the influx and efflux of the tide, and by lowering the low-water line, cause a greater volume of tidal water to enter and leave the river, and extend the tidal limit (Fig. 229, p. 378). Moreover, the works are so designed as to increase the width of the river gradually towards its mouth, so that the regulated River Maas has a width of 985 feet a little below Rotterdam, increasing to 2300 feet at the outlet between the jetties in the North Sea (Fig. 215, p. 362); the River Nervion is 210 feet wide at Bilbao, and 525 feet between the jetties at its mouth (Fig. 225); and the width of the trained low-water channel of the River Weser has been gradually enlarged from about 500 feet at Bremen, up to about 3600 feet where it emerges into its estuary at Bremerhaven (Fig. 228, p. 378). The velocity of the currents is thus rendered fairly uniform, so that the abrupt reduction in velocity which occurs on entering a wide reach, with its consequent deposit of alluvium, is prevented. Dredging is also generally carried out in the shallow portions of the regulated channel, to

[1] "Die Korrektion der Unterweser," L. Franzius, Bremen, 1888, with 2 plates; and *Proceedings Inst. C.E.*, vol. cxxxv. p. 230, and plate 5, figs. 4 and 5.

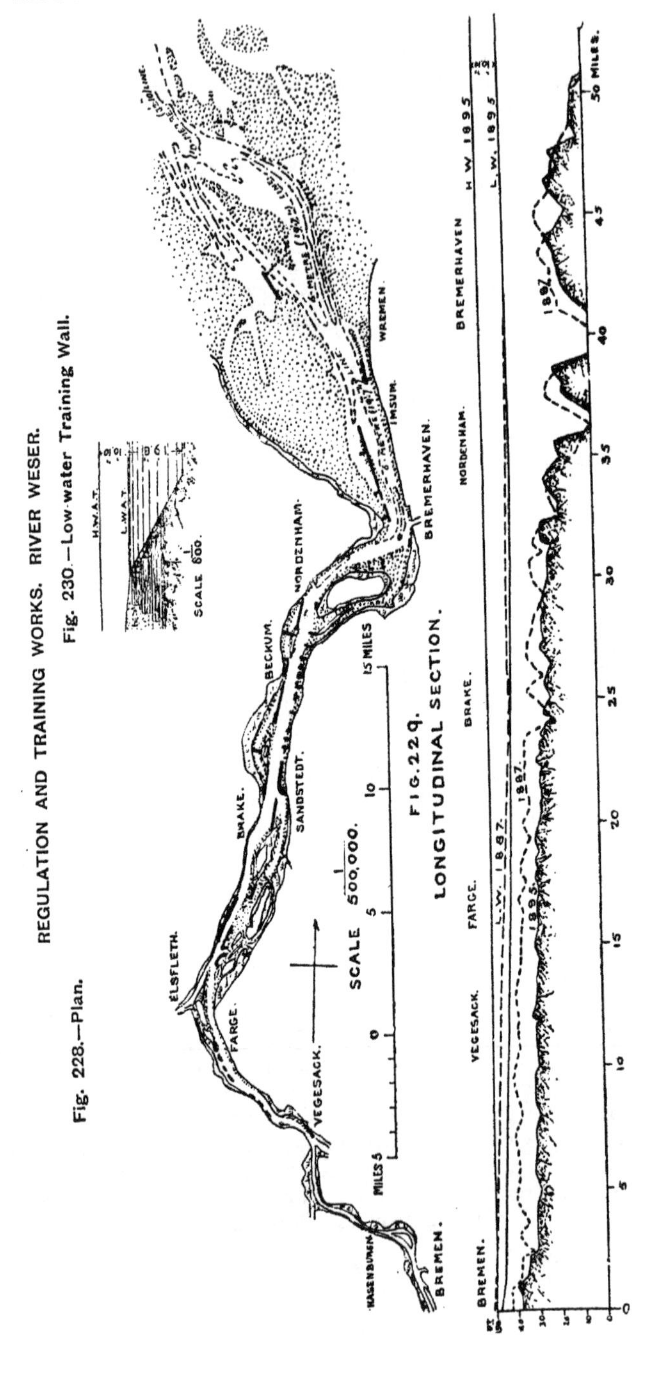

REGULATION AND TRAINING WORKS. RIVER WESER.

Fig. 228.—Plan.

Fig. 229. LONGITUDINAL SECTION.

Fig. 230.—Low-water Training Wall.

effect a greater deepening than unaided scour could produce, and also to prevent the material scoured out of the trained channel from forming a shoal in front of the mouth of the river, a result which interfered at first with the due improvement of the access to the River Maas by the cut across the Hook of Holland, whose enlargement to the proper width was originally left to the scour of the currents, which did not effect the requisite widening, and deposited a considerable part of the material it removed from the narrow cut in the wide jetty channel, so that eventually the cut had to be widened and the deposit removed by dredging, and the jetty channel narrowed by an inner southern jetty (Fig. 215, p. 362).

The rate of enlargement of the trained channels referred to above, in relation to their lengths, amounts to 1 in 80 for the Maas between Rotterdam and the sea, 1 in 75 for the Nervion from Bilbao to its mouth, and 1 in 71 for the Weser from Bremen to Bremerhaven, which ratios correspond very closely to the rate of enlargement of the Clyde of 1 in 83 between Glasgow and Dumbarton, and of the Tyne of 1 in 75 between Newcastle and its mouth; and in these cases, with the exception of the Weser, the fall of the river-bed seawards is inappreciable, and only slight in the Weser below Bremen. Accordingly, a suitable rate of enlargement for the lower portion of tidal rivers with fairly level beds, appears to be 1 in 70 to 1 in 80, with a larger ratio for rivers with beds rising inland, in proportion to the inclination, and a more rapid enlargement towards the outlet of rivers trained through sandy estuaries.

Protection of Outlet of Tidal Rivers by Breakwaters.— Tidal rivers flowing straight into the sea, without expanding into an estuary, are generally guided across the beach by projecting jetties, in order to deepen and maintain the outlet channel, and protect it from the inroad of drift along the coast, of which examples are furnished by the River Maas (Fig. 215, p. 362), the River Nervion (Fig. 225, p. 377), and the River Wear. At the mouth of the Maas, the jetties have been extended out so far beyond low-water mark that they effectually protect the outlet from the inroad of drift, and shelter the dredging operations for connecting the outlet channel with deep water; and the wide entrance between the jetties appropriate for this large river, situated well out from the shore, renders the access easy in favourable weather. Where, however, the jetties are not carried beyond low-water mark, as at the mouth of the Wear; where the drift along the coast is considerable, and tends to form a bar in front of the outlet, as at the mouth of the Tyne; where the exposure is great, as at the mouth of the Nervion; and where the small size of the river precludes the possibility of providing a wide, as well as a deep outlet channel across the beach, a harbour has been formed in some cases in front of the mouth by two converging breakwaters, within the shelter of which the outlet channel is readily deepened and the bar in front removed by dredging. Such a harbour, moreover, with an entrance of adequate width, in deep water away from the coast, can be entered by vessels during storms with much greater safety than the unprotected mouth of a river close to the line of breakers along the shore; whilst the windward breakwater arrests the drift, or causes it to travel into deep water. The mouths of the Tyne and the Wear have been protected in this manner, and their outlet channels deepened considerably by dredging within the sheltered area;[1] and converging breakwaters are constructed, projecting out from each side of the bay into which the Nervion flows, for facilitating access to the river on that stormy coast, and securing its outlet from being impeded by drift (Figs. 225 and 227, p. 377).

Training Tidal Rivers through Sandy Estuaries.—Many tidal rivers, before reaching the sea, expand into large estuaries more or less encumbered by sandbanks, of which the Thames, the Severn, the Mersey, the Humber, the Seine, and the Loire are well-known examples. In some instances, the depth is ample in the navigable channel even at low water, as in the Bristol Channel up to the mouth of the Avon, the Firth of Clyde up to Greenock, and the St. Lawrence up to Quebec; and in other cases, dredging suffices to render the navigable channel in an estuary accessible for large vessels at any state of the tide, as in the Thames up to Tilbury, the Mersey up to Rock Ferry, by dredging on the bar, and the St. Lawrence between Quebec and Montreal. Many estuaries, however, in their natural condition can only be navigated near high water; and their navigable channel has a winding, and frequently shifting course through extensive sandbanks, as exemplified by the Seine, the Loire, the Dee, the Ribble, and the Upper Mersey estuary.

The possibility of improving the navigable channel through estuaries by dredging alone, depends upon the extent of its stability, and also upon the suitability of the material forming its bed for being raised by suction dredgers. Thus the channel across the Mersey bar in Liverpool Bay, has proved capable of being largely deepened and widened by sand-pumps, owing to its fair stability of position, and the purity of the sand constituting the bar; and a like success might be anticipated from the employment of suction dredgers for lowering the bars in the Húgli estuary, where the navigable channel is to a great extent subject to similar favourable conditions.[2] Where, however, an estuary channel is very unstable, its improvement by dredging alone is impracticable; and an admixture of silt or clay with the sand largely reduces the efficiency of ordinary sand-pump dredgers, and necessitates the addition of water-jets or cutters to stir up and disintegrate the material.

Training walls consisting of a continuous mound of rubble stone, slag, or fascines, extended along each side of the navigable channel down an estuary, have been commonly resorted to for permanently fixing the channel, and also deepening it by the concentrated scour of the flood and ebb tides, together with the fresh-water discharge, in the trained channel; and the depth thus gained is usually further increased by dredging in the fixed channel. The value of training works in deepening a wandering channel through a sandy estuary, is well illustrated by the great improvement in depth effected by the training

[1] "Rivers and Canals," vol. i. plate 6, figs. 6 and 7, and plate 8, figs. 1 and 2.
[2] "Report on the River Húgli," Calcutta, 1897, pp. 76 and 94.

walls, unaided by dredging, carried out through the upper estuary of the River Seine in 1848–69, as shown in Figs. 231 and 232, p. 382, and by the deepening of the navigable channel through the Tees estuary by training works since 1852, assisted in this case by dredging in the trained channel (Figs. 234 and 235, p. 383). In the Weser estuary below Bremerhaven, the navigable channel was trained in 1891–1892 by a single training wall, 4½ miles in length, slightly diverging from the nearly parallel opposite shore, which has enabled a second training wall to be dispensed with (Fig. 228, p. 378); and an improvement and regulation of the channel further out in the estuary has been sought by closing branch channels, and thereby concentrating the scour in the main channel; while the navigable depth has been gradually increased by suction dredging. The Seine training walls are composed of mounds of rubble chalk, faced on the side of the channel by pitching or a layer of concrete, and further protected in some places against the bore by sheet-piling along the toe (Fig. 233, p. 382); the Tees training walls are mainly formed of refuse slag from the neighbouring iron blast-furnaces (Fig. 236, p. 383); and the training wall in the Weser estuary consists of layers of fascine mattresses, not raised above low-water level.

In addition to deepening the navigable channel through an estuary, and rendering it capable of being further deepened by dredging, training works, by permanently fixing and straightening the channel, make it much safer for navigation than the original shifting channel, which sometimes, as was formerly the case in the upper part of the Seine estuary, has its dangers aggravated by the appearance of a bore at certain periods, a phenomenon whose violence is mitigated by the regulation of the channel. The improvement, however, of the channel is confined to the trained portion; and where the training works have not been carried out to deep water, as in the estuaries of the Seine, the Loire, the Ribble, and the Dee,[1] the navigable channel beyond the training walls remains unstable and encumbered with shoals, as the influence of the training works extends very little beyond their termination. Accordingly, though the navigable condition of a tidal river is undoubtedly improved by training works through the upper portion of its estuary, in proportion as the interval of shallow, shifting channel, between deep water at sea and the fixed and deepened trained channel, is reduced, and therefore can be more readily traversed by vessels near high water, the improvement of the river cannot be regarded as complete till the trained channel has been extended out to deep water. Moreover, the due enlargement of the trained channel seawards, so essential, as already pointed out, for the free admission of the flood tide, and for promoting uniformity in the tidal flow, and consequently uniformity in depth, has been often neglected in training rivers through wide, shallow estuaries, as exemplified by the training works in the Ribble and Dee estuaries, and formerly in the Seine estuary, where the increase in width between the training walls towards their outlets was made very inadequate (Fig. 231); and these narrow trained channels open abruptly

[1] "Rivers and Canals," vol. i. plate 9, figs. 1, 4, 11, and 13.

TRAINING WORKS IN A SANDY ESTUARY. RIVER SEINE.

Fig. 231.—Plan.

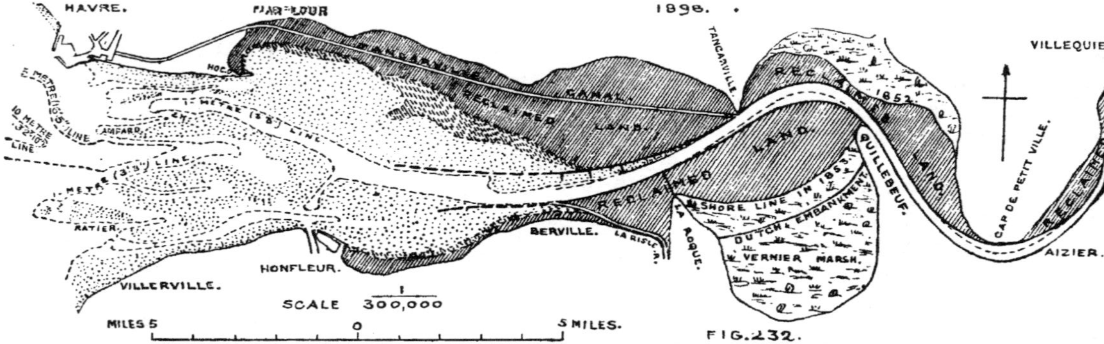

1898.

HAVRE. HARFLEUR. TANCARVILLE. CAUDEBEC. VILLEQUIER. LA MAILLERAYE. LA PIETTE. LA VAQUERIE. AIZIER. CAP DE PETIT VILLE. QUILLEBEUF. LA ROQUE. LA RISLE R. BERVILLE. HONFLEUR. VILLERVILLE. RATIER. AMFARD. HOC.

TANCARVILLE CANAL. RECLAIMED LAND. RECLAIMED LAND 1852. SHORE LINE IN 1853. DUTCH EMBANKMENT. VERNIER MARSH.

5 METRES (16'5") LINE. 10 METRE (32'10") LINE. 1 METRE (3'3") LINE.

SCALE $\frac{1}{300,000}$

MILES 5 0 5 MILES.

FIG. 233.

TRAINING WALL.

H.W.S.T. ACCRETION. SLOPE 1½ - 1. L.W.S.T.

SCALE $\frac{1}{800}$.

FIG. 232.

LONGITUDINAL SECTION.

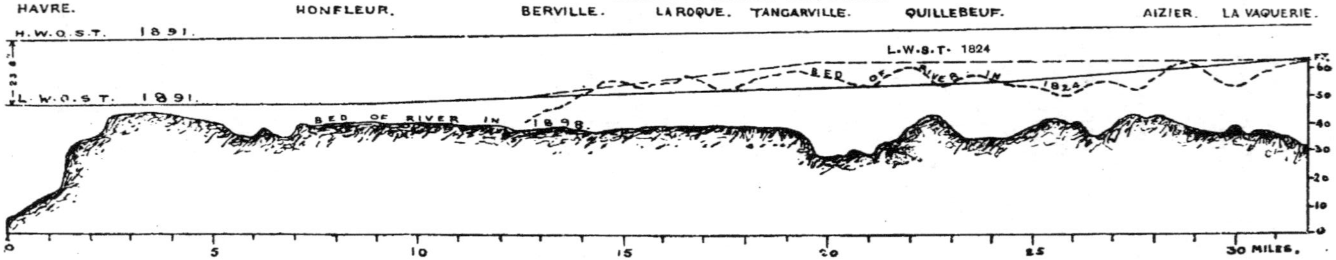

HAVRE. HONFLEUR. BERVILLE. LA ROQUE. TANCARVILLE. QUILLEBEUF. AIZIER. LA VAQUERIE.

H.W.O.S.T. 1891.

L.W.O.S.T. 1891.

L.W.S.T. 1824

BED OF RIVER IN 1824

BED OF RIVER IN 1898.

23.6

FT. 60 50 40 30 20 10 0

0 5 10 15 20 25 30 MILES.

into the middle of wide estuaries encumbered with sandbanks, instead of expanding with a trumpet-shaped outlet channel into deep water, which nature indicates as the best form for securing a good navigable depth towards the mouth of a tidal river. The trained channel of the Tees, on the other hand, has been given a very good rate of enlargement below Middlesbrough, approximating to 1 in 45 (Fig. 234); whilst the regulated and trained channel of the Weser has been given an enlarge-

TRAINING WORKS THROUGH A SANDY ESTUARY. RIVER TEES.
Fig. 234.—Plan.

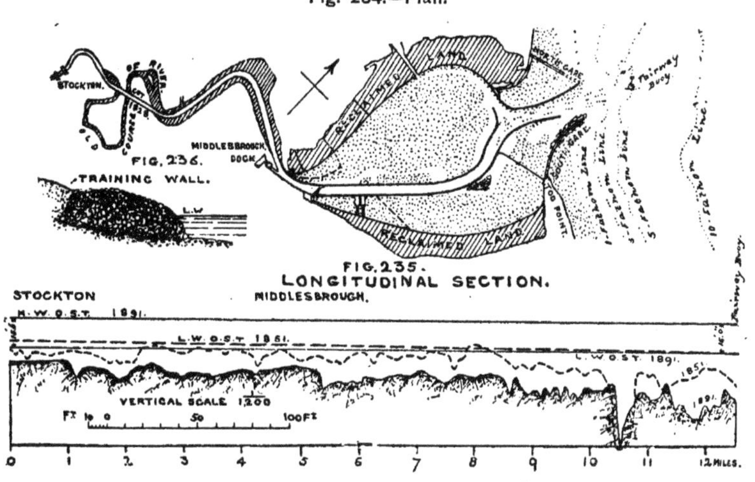

ment of 1 in 50 between Bremen and the termination of the training wall in the estuary, a marked contrast to the rate of 1 in 200 adopted for the Seine trained channel many years ago, and now admitted to be inadequate.

Accretion in Estuaries due to Training Works.—The concentration of the tidal flow and fresh-water discharge in a trained channel, whilst wholly beneficial to navigation in the channel itself, withdraws these scouring influences from the estuary behind the training walls, and, consequently, favours the deposit of materials carried down from inland by the river in flood-time, or brought in from the seacoast by the flood tide. A river with its wandering channel in a sandy estuary, disturbs in succession every portion of the estuary through which it shifts its course, and aided by the predominance of the ebbing current reinforced by the fresh water, carries seawards the materials brought down from inland, or in from the sea. As soon, however, as the channel is permanently fixed between training walls, the disturbing and scouring influences are concentrated in this channel; and the sedimentary matter brought into the estuary settles gradually in the slack water at the back of the training walls, and even along the sheltered sides of the estuary in advance of the works, as clearly exhibited in the chart of the Seine estuary where, owing to the large amount of material eroded from the coast of Normandy and brought in by the flood tide, especially during westerly gales, accretion has taken place on a very large scale since the introduction of the training walls, leading to reclamations, not merely in the upper estuary, but even along the sides of the estuary down to Harfleur on the north and Honfleur on the south (Fig. 231, p. 382). The reclamations were undoubtedly hastened in this instance by carrying the training works diagonally across the upper estuary between Quillebeuf and Tancarville, by the construction of a cross dyke from La Roque to the south training wall, enclosing the southern part of the estuary to the north of the Vernier Marsh, and by raising the training walls considerably above low water. Accretion at the sides, however, is the inevitable result of concentrating the scour in a fixed channel through an estuary exposed to the introduction of sediment; and it is necessarily accompanied by a reduction in the tidal capacity of the estuary. Nevertheless, the rate and extent of the accretion may be diminished considerably by keeping the training walls as low as possible consistently with the fixing of the channel, and duly enlarging the width between them seawards; whilst any resulting loss of scour in the channel, can be compensated for by dredging to obtain the requisite depth.

A large reduction of tidal capacity, by diminishing correspondingly the volume of tidal water entering and leaving the estuary, leads to the extension of sandbanks at the outlet, as indicated by the progression of the shoals at the mouth of the Seine,[1] the enlargement and advance of the sandbanks along the sea face of the Ribble estuary, and the remarkable deterioration in depth of the lower estuary of the Dee, which has resulted from the extensive reclamations effected in the upper estuary.[2] The depth in front of the outlet of the trained channel may, indeed, be increased by the concentrated scour, in spite of the general accretion; but the access to seaports at the sides of the estuary, or at some distance beyond the termination of the training works, may be endangered by the reduction in tidal capacity. Thus the Seine training works were stopped at Berville in 1869, for fear that the extended accretion resulting from their prolongation would shoal the approach to the port of Havre at the mouth of the estuary; and the training works in the upper Loire estuary were not carried beyond La Martinière, about 23 miles above the mouth of the estuary at St. Nazaire, lest the loss of the tidal capacity due to accretion, which had already been occasioned to some extent by the existing training works, should be so largely increased by their further extension, as to render the reduced scour at the outlet incapable of maintaining the approach channel to the port. An extension of the Seine training walls was, indeed, authorized in 1895 (see dotted lines Fig. 231, p. 382), but only subject to the condition that an entirely new approach from the sea must be provided for Havre, out of reach of the Seine accretions (Fig. 286, p. 449); and the rate of enlargement of the trained, high-water

[1] "Rivers and Canals," vol. i. plate 9, fig. 2.
[2] *Ibid.*, plate 9, figs. 13 and 14.

channel, from the commencement of the new, high, northern training wall a short distance below Tancarville, to the proposed extremity of the prolonged southern training wall, amounts to about 1 in 10.

Reclamation is progressing in the Ribble and Tees estuaries, owing to the promotion of accretion by the training works in these estuaries; and though the converging breakwaters at the Tees outlet protect its estuary from the inroad of drift from the coast, they favour the deposit of any sediment which may be introduced into the sheltered area (Fig. 234, p. 383); but in this case, there are no ports at the sides of the estuary to be considered, and any loss of tidal capacity from accretion can be compensated for by additional dredging in the trained channel.

Remarks on Training Works through Sandy Estuaries.—Training works provide a sure method of fixing and deepening a wandering channel in a sandy estuary in the interests of navigation; but the training walls should, if possible, not be raised above low water, so as to keep down the resulting accretion at the sides, and consequent loss of tidal capacity, to a minimum; and they should form a channel enlarging seawards at a rate of not less than about 1 in 80, and more rapidly on approaching the outlet, carried out to deep water. Where ports exist at the side of an estuary, it may prove necessary to limit the extension of the training works, in order to avoid impeding the access to the port by accretion, unless a new outlet on the seacoast can be provided, as is now the case at Havre. Where a port is situated on the narrowed outlet of a large estuary subject to the introduction of sediment, like Liverpool in the narrows leading out of the inner Mersey estuary, and St. Nazaire at the contracted outlet of the extensive Loire estuary, where the estuary serves as a large natural sluicing basin maintaining deep-water access to the port by tidal scour, it is necessary to relinquish the improvement of the navigable channel through the estuary by training works, lest these works, by leading to the accretion of the estuary outside the trained channel, should deprive the port at its outlet of the scouring current by which its approach channel from the sea is preserved.

CHAPTER XXIV.

INLAND NAVIGATION CANALS.

Comparison of canals and railways—Instances of large, well-situated canals —Objects of inland canals, canals connecting rivers, lateral canals in place of rivers—Construction of canals, various works, aqueducts—Sizes of canals, in Great Britain, France, Germany, Belgium, Holland, Russia, North America, requirements as regards width and depth—Supply of water, sources, methods of economizing—Methods of transferring vessels from one reach to the next, locks, flight of locks, inclines, lifts—**Canal Locks**: compared with river locks, similarities, differences; sizes, in England, standard in France, on Merwede Canal, in Canada, on St. Mary's Falls Canal; lifts of canal locks, average in England and France, large lifts on French locks, on Welland and Soulanges Canals; flights of locks, instances; reduction in time of passage through locks, by various contrivances; methods of saving water in locking, by duplicating locks, by side ponds—**Canal Inclines**: advantages, method of working; barges conveyed in cradles, instances, description of inclines; barges in caissons with water, object, descriptions; objections; Chignecto Ship-railway, object, description, unfinished—**Canal Lifts**: principle of working; hydraulic canal lift, description of, at Anderton, at Fontinettes, and at La Louvière; floating canal lift, description of, at Henrichenburg, advantages; lifts compared with locks and inclines, relative merits and disadvantages.

NAVIGATION canals provide waterways for the transit of barges through inland districts, just as railways serve for the passage of trains; and canals also, like railways, traverse high ridges in cuttings or tunnels, and valleys on embankments or aqueducts. They differ, however, from railways in having to be constructed in a series of level reaches, instead of with varying gradients, so that they have to follow the winding contours of the country as closely as possible, with a circuitous route, in order to maintain a uniform level; and the differences in level between the successive level reaches have to be surmounted by locks, inclines, or lifts. Canals, accordingly, are at a disadvantage for competition with railways, in having to follow a more circuitous course to preserve the level of their reaches at a reasonable cost, and in the delays involved in the passage of vessels from one reach to the next, as well as on account of the rate of transit being slower in water than on rails; but, on the other hand, the resistance to traction is much less on a waterway than on a railway, and the working and maintenance of a canal is less onerous. Canals offer the best advantages where the country is fairly flat, and the traffic in bulky goods considerable, or where a short length of canal serves to connect long lengths of river or lake navigation. Before the introduction of railways, canals were extended into many localities which could have been more economically served by railways; but numerous old canals, though often inadequate in size, where suitably situated, have been able to maintain a large traffic in bulky goods in spite of the keen competition of railways; and in several districts, enlarged inland canals have been constructed within recent years with good prospects of traffic, and with benefit to the country they traverse, as for instance the St. Mary's Falls Canals, the St. Clair Flats Canal, and the Soulanges Canal, in North America, the St. Denis and other enlarged canals in France, the Merwede Canal in Holland, the Dortmund-Ems Canal in Germany, and the Marie Canal in Russia.

Objects of Inland Canals.—Though the primary object of inland canals is to provide communication by water for the districts they pass through, they also fulfil two still more important purposes. Thus a canal, by connecting two river navigations across the water-parting of their basins, may open up a very long line of water communication. This is almost the sole object of the canals which have been constructed or projected in Russia; the principal rivers of France have been connected together by this means, and also with German rivers; and remarkable instances of this use of canals are furnished by the junction of the St. Lawrence, through the Great Lakes, with the Mississippi, and the uniting of the Obi and Yenesei rivers in Siberia by a canal, only 5 miles long, joining two of their tributaries, and opening out a waterway about 3300 miles in length. When, however, the ridge separating two river basins attains a high altitude, the cost of construction of the numerous locks required, and still more the delays the passage through them involves, render a canal an unsuitable means of communication.

Canals are also sometimes constructed alongside rivers as substitutes for river navigation, where a river, for a certain length, owing to falls, rapids, or shoals, is unsuited for improvement. A notable example of such a lateral canal is the Welland Canal, which furnishes a navigable connection between Lakes Erie and Ontario, since the Niagara River joining the lakes is barred by its Falls and the rapids above and below; whilst the Gloucester and Berkeley Canal, avoiding an awkward part of the River Severn, affords an instance in England of a lateral canal.

Construction of Canals.—The works for canals are similar to those for railways, with the exceptions, that in cuttings through porous strata, and on embankments, the trough for the canal has to be made watertight by a lining of clay, concrete, or other impermeable materials, that high embankments should be avoided, and that the greatest care must be taken in forming the embankments to guard against settlement or slips, which would be attended by serious damage in the case of a canal. The cheapest formation for a canal is where the level is such that the trough of the canal has only to be partially excavated, and the upper part is enclosed by a bank on each side, formed of the excavated materials, with a central puddle trench going down into the solid ground to render the banks watertight. A towing-path has to be formed along one side of the canal where traction by horses is employed. The side slopes in soft soil, near the water-level of the canal, have to be made flat, and protected by pitching, timberwork, fascines, or plants, to prevent their erosion by the wash raised by the passage of the barges, especially where the waterway is small in relation to the cross section of the barge, or the speed of navigation is fairly quick. Through towns, vertical side walls save land and provide quays.

Bridges carrying canals over roads, rivers, and railways, or aqueducts as they are termed, differ from railway bridges in having to provide a watertight trough, with a towing-path at the side if required, and in being subjected merely to the dead load of the bridge and the water carried by it, and not to any moving load. Canal aqueducts and tunnels are often restricted to the width necessary for the passage of a single barge, in order to reduce the cost; and the towing-path is sometimes dispensed with in tunnels, the propulsion being effected by the bargemen or by mechanical means.

Sizes of Canals.—The differences in dimensions of canals and their locks, constructed in former times by different companies, resembling differences in gauge on railways, have seriously impeded the development of through routes and the extension of inland navigation. The inland canals of Great Britain can only, for the most part, afford a passage for barges 72 feet long, 7 feet to 15 feet wide, and drawing 3 to 5 feet of water; and little has been done on a comprehensive scale to meet modern requirements, except on the Aire and Calder Navigation and the Weaver Navigation, which have been successively enlarged, and afford depths of 9 feet and 15 feet respectively, so that the latter can accommodate vessels of 1000 tons. In France, on the contrary, the main lines of waterways have been gradually remodelled, and necessary links constructed since 1879, so as all to conform to the minimum standard depth of 6½ feet, to be increased eventually throughout to 7¼ feet, and give access to vessels of 300 tons, 126 feet long, 16½ feet wide, and from 6 to 6⅔ feet draught (Fig. 237). In Germany, where the rivers furnish the chief routes for inland navigation, the depths of the canals connecting them range mostly between 5 and 7 feet; but the new Dortmund-Ems Canal has been given a depth of 8⅙ feet in cuttings, and 11½ feet on embankments to reduce the earthwork, with a bottom width in cuttings of 59 feet, and side slopes of 2 to 1, flattened to 3 to 1 near the water-level, and allows of the passage of vessels of 600 tons (Fig. 238). The main inland canals of Belgium have depths of from 6½ to 13½ feet; whilst the new Canal du Centre, with a depth of 8 feet, has been made large enough to admit vessels of 400 tons. The Merwede Canal in Holland, completed in 1893, has a depth of 12⅛ feet, a bottom width of 65½ feet, and side slopes of 2 to 1 (Fig. 239); whilst the lately re-constructed Marie Canal in Russia, crossing the water-parting of the Volga and Neva basins, and thus forming a link in the waterway connecting the Caspian Sea and the Baltic, affords access to vessels of 655 tons, 210 feet long, 31½ feet wide, and 6 feet draught. The lateral Soulanges Canal in Canada, for avoiding the rapids of the River St. Lawrence, and forming a new link, opened in 1900, 14 miles long, in the improved waterway between this river and the Great Lakes, for vessels of 14 feet draught, has been given a depth of 17½ feet, a bottom width of 96 feet, and side slopes of 2 to 1, protected by rubble pitching near the water-level, and allows the passage of vessels of over 2000 tons[1] (Fig. 240); whilst the new Sault-Sainte-Marie Canal, connecting Lakes Huron and

INLAND NAVIGATION CANALS.

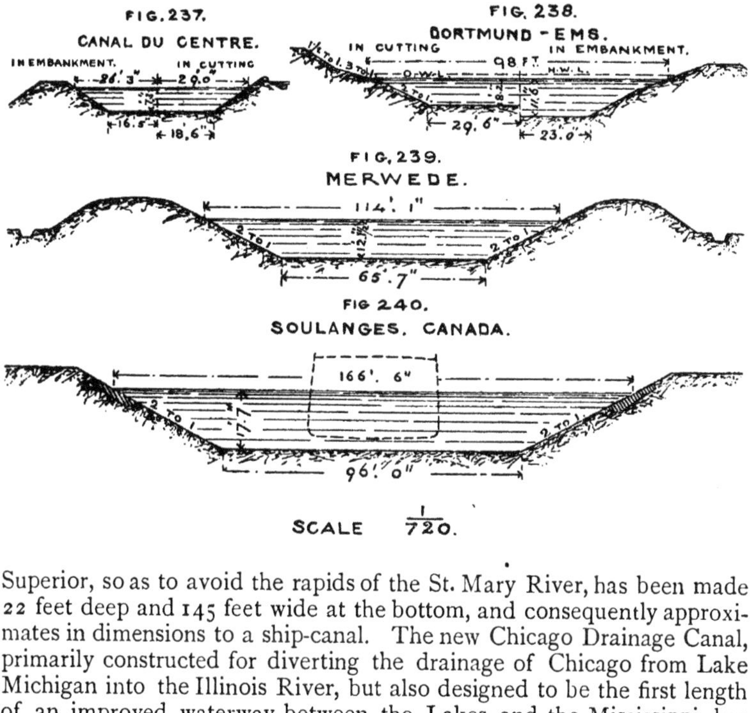

FIG. 237. CANAL DU CENTRE.

FIG. 238. DORTMUND - EMS.

FIG. 239. MERWEDE.

FIG. 240. SOULANGES, CANADA.

SCALE $\frac{1}{720}$.

Superior, so as to avoid the rapids of the St. Mary River, has been made 22 feet deep and 145 feet wide at the bottom, and consequently approximates in dimensions to a ship-canal. The new Chicago Drainage Canal, primarily constructed for diverting the drainage of Chicago from Lake Michigan into the Illinois River, but also designed to be the first length of an improved waterway between the Lakes and the Mississippi, has been given very similar dimensions, with bottom widths of 160 feet in rock and 200 feet in clay, side slopes of 2 to 1, and a depth of 22 feet.

The bottom width of a canal must be regulated by the beam of the largest vessels navigating the canal, and the speed at which they travel. Inland canals are generally made wide enough for barges going slowly to pass one another, which necessitates a bottom width somewhat in excess of twice the beam of the largest barges; but with an increased speed, a larger margin of clear space between two vessels passing one

[1] *The Engineering Record*, New York, April 8, 1899, p. 419. A cross section of the Soulanges Canal was sent me by Mr. Thomas Monro, the engineer-in-chief.

another is required. A minimum depth of 8 inches to 1 foot of water under the keel of barges of maximum draught, is essential even for low speeds; whilst 1½ to 3 feet should be provided under the keel where the vessels are intended to attain a fair speed. The width of canals must be increased along curves in proportion to the curvature, to provide for the increased difficulty of navigation; and sharp curves should be avoided.

Supply of Water to Canals.—A regular supply of water has to be provided for canals, to make good the losses due to evaporation and leakage, and the expenditure of water necessitated in transferring vessels from one reach to the next, which is proportionate to the amount of traffic passing through. Lateral canals are readily supplied, at their upper end, by the river for whose navigation they serve as a substitute; and canals traversing low-lying districts can generally obtain their supply from adjacent watercourses. On higher ground, lakes, springs, and wells sometimes provide the requisite supply; whilst the upper reaches of a canal, especially when traversing the high ridge separating two river valleys, have generally to be supplied from artificial reservoirs formed in hilly valleys by a dam across the valley, barring the outlet of the mountain stream flowing down the valley, and thus storing up its waters in flood-time for feeding the canal during the dry season.

Where the supply of water is difficult to obtain, special arrangements are made for economizing the water in locking; and occasionally the water expended is pumped back again into the upper reach, so as to be used over again.

Methods of transferring Vessels from One Canal Reach to the Next.—The well-known system of locks, as previously described with reference to canalized rivers, is the ordinary method employed for raising or lowering vessels from one reach to the next. Locks, however, are more frequently required on canals than on rivers; they have to be placed much closer together where a canal is rising to surmount the water-parting between two river basins; and occasionally several locks are placed close together end to end, in what is termed a *flight*, to overcome an abrupt and considerable change of level. Whereas the water required for locking is usually amply supplied on canalized rivers by the natural discharge of the river, its expenditure becomes an important consideration on canals, where the necessary supply has to be furnished artificially. Moreover, the delays incurred in passing through the locks become more serious on canals, where the locks are more numerous, and the differences of level to be surmounted are considerably greater than on rivers. Consequently, expedients for saving water and time in passing through canal locks have been devised.

Occasionally, where the differences in level between canal reaches are considerable, in passing through rugged country, both time and water have been saved by abandoning flights of locks, and constructing inclines between the reaches, up and down which the barges are carried on trucks with wheels running on rails on the incline.

A third system consists of a balanced hydraulic lift, by which vessels floating in a tank of water are rapidly raised or lowered vertically about forty to fifty feet, from one reach of a canal to the next, with quite a small expenditure of water and power. The cost, however, of construction of these lifts is large.

Canal Locks.

Large canal locks resemble large river locks in having a pair of gates meeting at an angle against a sill at each end of the lock-chamber, and in having large sluiceways running through the side walls from end to end, with several side openings into the chamber to expedite and tranquillize the filling and emptying of the chamber (Fig. 241).[1] Moreover, cylindrical sluice-gates fitting into a circular aperture at the bottom of the sluiceways alongside the gate recesses, leading into the main inlets at the upper end, and the main outlets at the lower

CANAL LOCK, WITH LARGE LIFT AND CYLINDRICAL SLUICE-GATES.
Fig. 241.—Canal du Centre, France.

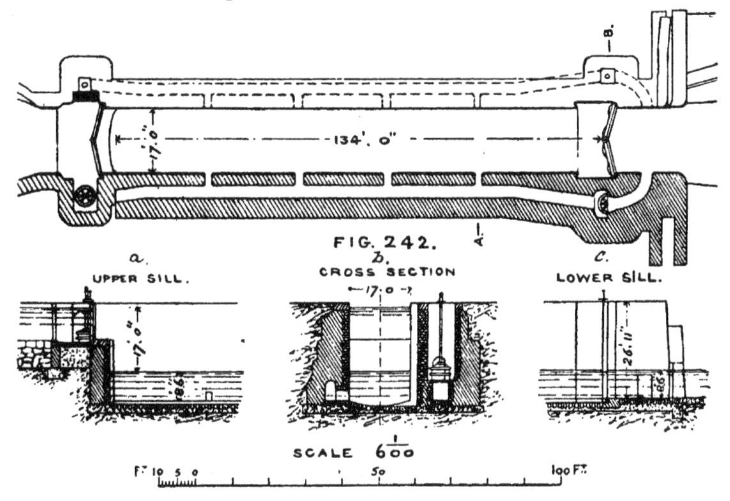

end of the lock, being balanced by the uniform water-pressure all round the cylinder, are readily raised for opening communication, or lowered for closing it; and owing to their form, these sluice-gates open a large aperture, with a comparatively small lift, for the inrush or outrush of the water into or out of the lock-chamber, and therefore greatly aid the rapid filling and emptying of the lock (Figs. 241 and 242, *b*). A canal lock, however, differs from a river lock in having its upper sill raised upon a lift-wall with its top level with the bottom of the upper reach,

[1] "Voies navigables de la France, Ministère des Travaux Publics, 1894, Canal du Centre," Atlas, plate 6.

and, consequently, having a height above the bottom of the lock-chamber equal to the lift of the lock, or the difference in level between the upper and lower pools; and the upper lock-gates are, accordingly, shorter in height than the lower gates by the amount of the lift of the lock (Fig. 242, *a*).

Sizes of Canal Locks.—The dimensions of canal locks have to be made adequate for receiving the largest vessels capable of navigating the waterways; and, consequently, the locks differ considerably in size according to the nature of the traffic and the size of the waterway which they serve. Sometimes, also, as in river locks, canal locks have their chamber widened out to allow of barges being ranged alongside one another in the lock, or are provided with an intermediate pair of gates to adjust the length of the lock to the vessel passing through, and thereby save both time and water in locking.

Many of the old canal locks in England are only about 74 feet long, and 7 feet or 14 to 16 feet wide, with 4 to 5 feet depth of water over their sills, giving passage to barges of 50 to 100 tons. On the Aire and Calder Navigation, however, with a large coal traffic, navigated by vessels 120 feet long, 18 feet beam, and 9 feet draught, the locks are 201 to 339 feet long, and 18 to $22\frac{1}{2}$ feet wide, with $8\frac{1}{2}$ to 10 feet of water over their sills; whilst the Weaver Navigation, serving the Northwich salt-mines, has double locks of different widths, with intermediate gates, longitudinal sluiceways, and cylindrical sluice-gates, the largest locks being 120 feet long and $42\frac{2}{3}$ feet wide, with 15 feet of water over their sills, and giving passage to vessels of 1000 tons. The smallest standard dimensions admitted for the locks on the main lines of waterways in France, are $126\frac{1}{3}$ feet length, 17 feet width, and $6\frac{1}{2}$ feet depth; but these dimensions are often exceeded in the most important canals, as, for instance, at the reconstructed locks on the Canal du Centre, connecting the Saône and the Loire, having an available length of chamber of 134 feet, and a depth on the sill of $8\frac{1}{2}$ feet (Figs. 241 and 242, p. 391); whilst the new Canal du Centre in Belgium has locks of similar dimensions. The Merwede Canal, connecting the River Merwede with Amsterdam, has two locks side by side at the Amsterdam end, having entrances 46 feet wide at each extremity, and a chamber 370 feet long, widened out to about 95 feet in the central 255 feet; whilst elsewhere the two locks are arranged end to end, with a pair of gates between them, being given an entrance width of $39\frac{1}{3}$ feet, and chambers about 340 feet long, widened out to about 80 feet in the central 270 feet.

The enlarged system of canals in Canada between the St. Lawrence and the Great Lakes, provides a continuous waterway for steamers, drawing 14 feet of water, bringing wheat from the western end of Lake Superior to the St. Lawrence at Montreal, built of sufficient size to be employed for trading along the New England coast in the winter when the inland waterway is frozen over (Fig. 240, p. 389); and their locks have chambers 270 feet long and 45 feet wide, which are constructed of concrete faced with dressed masonry on the new Soulanges Canal, having side walls 20 to 25 feet wide at the base, in which longitudinal sluiceways have been formed, communicating with the chamber by ten cast-iron pipes on each side, $2\frac{1}{2}$ feet in diameter, enabling the lock, having a lift of $23\frac{1}{3}$ feet, to be filled or emptied in about five minutes. Still larger locks have been recently constructed on the short canals connecting Lake Huron with Lake Superior, for accommodating the larger vessels navigating the Great Lakes, the lock on the Sault-Sainte-Marie Canal, on the Canadian side of the St. Mary River, being 900 feet long and 60 feet wide, with $20\frac{1}{4}$ feet of water over the sills, large enough to receive a lake vessel 320 feet in length, and two vessels suited for passing through the Welland and St. Lawrence Canals, 255 feet long, at the same time; whilst the lock on the St. Mary's Falls Canal, in the United States, is 800 feet long and 100 feet wide, with 21 feet of water over the sills.

Lifts of Canal Locks.—The lift, or the difference between the upper and lower water-level surmounted by the lock, varies according to circumstances. It ranges usually between 4 and 9 feet in England, and between $6\frac{1}{2}$ and 9 feet in France. The time occupied in passing through a lock is necessarily proportionate in some measure to the height of the lift; but as part of the delay is independent of the lift, and short reaches are disadvantageous, in reconstructing a portion of the Canal du Centre in France, where the locks were somewhat close together, the new locks were given a lift of 17 feet in place of $8\frac{1}{2}$ feet, reducing the number of locks by one half, and elongating the reaches correspondingly. In deepening the St. Denis Canal from $6\frac{1}{2}$ feet to $10\frac{1}{2}$ feet, to correspond with the improved depth of the Seine below Paris, two of the new locks with lifts of about 14 feet replaced two flights of two locks each; and a single lock, at the Paris end of the canal, with a lift of $32\frac{1}{2}$ feet, was substituted for two flights of two locks. The great height of the lower, single lock-gate, of about 50 feet, which this unprecedented lift would have entailed in ordinary construction, has been avoided by building an archway over the lock, affording the standard headway of $17\frac{1}{4}$ feet above the water-level of the lower pool, against which the top of the lower gate rests, and thereby enabling the height of the gate to be reduced to $32\frac{2}{3}$ feet.[1]

The Welland Canal surmounts a difference of level of $326\frac{3}{4}$ feet by twenty-six locks, giving an average lift of about $12\frac{1}{2}$ feet; whilst the lift of the single locks on the two canals joining Lakes Huron and Superior, is 18 feet, in place of the lifts of two locks originally constructed; and three out of the four locks on the Soulanges Canal have a lift of $23\frac{1}{3}$ feet, and are situated within the first mile from the lower end of the canal.

Flights of Locks.—Where a considerable change of level of the land takes place abruptly, flights or chains of locks have been frequently adopted on canals, consisting of a succession of locks in line, with the upper gates of the lower lock forming the lower gates of the next lock. Thus a rise of 203 feet at Tardebigge on the Worcester

[1] *Annales des Ponts et Chaussées*, 1893 (2), p. 45, and plate 17; and "Rivers and Canals," vol. ii. p. 380, and plate 11, figs. 1 to 6.

and Birmingham Canal, is surmounted by a flight of twenty-nine locks; there is a flight of twenty-two locks on the Somersetshire Canal, and a flight of twenty-one locks on the Leeds and Liverpool, and the Warwick and Birmingham canals; and numerous smaller flights exist on several English canals. Though flights of locks are less common and shorter on French canals, there is a flight of seven locks on the Canal du Midi, and of five locks at Fontinettes on the crowded Neuffossé Canal. Whilst, however, flights of locks occasion inevitable delays to navigation, they possess the advantage over single locks placed at intervals, of almost halving the number of gates required, and of concentrating the delay at a single spot. An increase in the lift, so as to diminish the number of locks, is also similarly advantageous.

Reduction in Time of Passage through Locks.—Large longitudinal sluiceways with numerous side openings into the lock-chamber, do not merely fulfil the important purpose, with the aid of counterbalanced cylindrical sluice-gates of between 4 and 6 feet diameter, of rapidly filling or emptying the chamber, but also serve for facilitating the entrance and exit of large vessels by providing side passages for the escape and admission of water, to balance the displacement of water caused by the movement of a vessel occupying nearly the whole cross section of the chamber. This object is further aided by making the bottom of the lock two or three feet deeper than the canal. Hydraulic or electric power expedites the opening and closing of large lock-gates; and it is also advantageously applied to the working of capstans, for hauling barges into and out of the lock. The saving in time effected by these means is of considerable importance when several locks have to be passed through, when the traffic is large enough for the capacity of the waterway to depend upon the rate of passage through the locks, and when competition exists with other means of communication.

The capacity of the Soulanges Canal for the large lake traffic which is expected to make use of it, has been greatly increased by lighting the canal, along its whole length, with electric arc lamps of 2000 candle-power at intervals of 480 feet, so as to enable navigation to be safely conducted during the night. The locks also and swing bridges on this canal, besides being efficiently lighted, are operated by electricity generated by water-power from Lake St. Francis, which also supplies the canal with water.

Methods of saving Water in Locking.—The consumption of water on a canal with a large traffic, in passing vessels through locks with a large lift or flights of locks, becomes a matter of some concern near the summit-level, where an adequate supply of water in dry weather is often difficult and costly to obtain. Vessels in being locked in succession down, consume a lockful of water each, or the volume of water measured by the area of the lock-chamber multiplied by the height of lift, less the water displaced by the vessel; and therefore water is saved by having locks of different sizes, or an intermediate pair of gates for the smaller vessels. In passing through a single lock, barges locked up and down alternately, reduce the consumption to one half; but at a flight of locks, the expenditure of water is less when several barges pass up or down in succession. Water is, accordingly, saved by having two flights of locks side by side, one for ascending, and the other for descending barges; and the consumption is also reduced by placing two single locks side by side, communication between which can be opened or closed at pleasure, so that one lock can be partially filled by the water let out of the adjacent lock, and a barge can enter the lock which has the most suitable water-level. The duplication of the locks, moreover, obviates delays, and doubles the capacity of the canal for traffic; but it involves a large increase in the cost of the works. A more economical method of saving water consists in the formation of one or two shallow side ponds close to the lock, into which the upper portion of the water in the lock is discharged on emptying the lock, and is used again in refilling the lower part of the lock.

CANAL INCLINES.

A considerable saving in time, water, and expenditure, is effected, in traversing districts where considerable differences in level between the reaches of a canal have to be surmounted, by resorting to inclines on land, instead of numerous locks, up which barges are drawn from one reach to the next. Barges are conveyed on these inclines, either on a cradle running on wheels on rails laid along the incline, or in water in a tank supported horizontally on a specially constructed truck running on the incline. Power has to be provided at the top of the incline for hauling up the barges with their cradle or truck; but the power required can be considerably reduced when the ascending barge can be counterbalanced by another barge descending at the same time. Inland navigation in this way abandons canals for short distances, and resorts to the railway system, a principle which has been further developed on a large scale in the schemes for ship-railways.

Barges conveyed on Cradles on Inclines.—Very small barges are conveyed on cars on two very steep inclines on the Shropshire and Shrewsbury canals, rising 213 feet and 73½ feet respectively. In constructing the Morris Canal in 1825–31, crossing a spur of the Alleghanies with a summit-level of 914 feet above its starting-point on the Hudson River opposite New York, twenty-three inclines were introduced, with gradients of 1 in 10 to 1 in 12, and rises of between 44 and 100 feet, the combined rises of the inclines on the two slopes being 1448 feet. The barges, 79 feet long, are carried on cradles, with eight wheels running on rails laid to a gauge of 12⅓ feet, and hauled up the inclines by wire ropes actuated by a water-wheel. A loaded barge and its cradle, weighing 110 tons, are drawn up a height of 51 feet in 3½ minutes, so that practically no time is lost in the ascent; whilst the water used in drawing up the barge is reckoned at one twenty-third of that expended in locking through a flight of locks having the same lift. Five very similar inclines were subsequently introduced on the Oberland Canal in Prussia, for the conveyance of barges of about the same size in a similar manner, the last one having been substituted for the five lowest locks on the canal.

An incline has been somewhat recently constructed for connecting the Ourcq Canal with the River Marne at Beauval, where the canal is only $\frac{2}{3}$ mile from the river, but 40 feet above it, and could not supply sufficient water for working a flight of locks. The incline, 1476 feet long, with a gradient of 1 in 25, has a line of rails laid on it to a gauge of $6\frac{1}{2}$ feet, on which laden barges of 75 tons are conveyed on wrought-iron cradles weighing 35 tons, which are drawn up the incline in 35 minutes by a wire rope worked by a turbine in the Marne (Fig. 243).

BARGE ON CRADLE.
Fig. 243.—Ourcq Canal Incline.

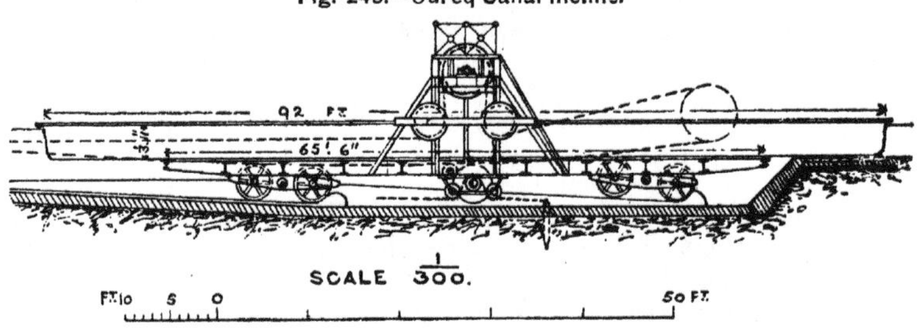

Barges conveyed on Inclines in Caissons with Water.—In order to obviate the possibility of heavily-laden barges being strained by being carried out of water on a cradle, the scheme was devised of conveying barges on an incline floating in water in a caisson, a plan first adopted in 1840 at some inclines on the Chard Canal, where two caissons filled with water, counterbalancing one another, went up and down the incline by putting an excess of water into the caisson at the top. In 1850, an incline similarly worked on a larger scale, was constructed to supplement a double flight of eight locks at Blackhill on the Monkland Canal near Glasgow, where sometimes the supply of water was insufficient for the traffic. This incline rising 96 feet with a gradient of 1 in 10, has two lines of way laid on it to a 7-feet gauge, up and down which two iron caissons filled with water travel, counterbalancing one another, each supported on a 20-wheeled truck in a horizontal position. The caisson, 70 feet long, $13\frac{3}{4}$ feet wide, and $2\frac{3}{4}$ feet deep, weighs with its truck, barge, and water, about 80 tons; and a barge can be passed up the incline every eight minutes, making a saving of from 20 to 30 minutes on the passage through the adjacent flights of locks. To avoid the oscillations which the movement of the caisson on the incline imparts to the barge when floating freely, the barge is made just to rest on the bottom of the caisson. A larger caisson was constructed in 1876 for conveying barges of 115 tons on an incline rising 39 feet with a gradient of 1 in 12, for connecting the Potomac River at Georgetown with the Chesapeake and Ohio Canal, in place of two locks. This caisson, 112 feet long, $16\frac{3}{4}$ feet wide, and $7\frac{1}{2}$ feet deep, supported on three trucks running on four rails (Fig. 244), is counterbalanced by four waggons loaded with stone running on adjoining rails, and is drawn up the incline in about three minutes by wire ropes worked by a turbine actuated by water from the canal. The weight, however, of the caisson with its trucks and load, amounting to 390 tons, damaged the line on which it runs, so that to reduce the weight, the water is let out of the caisson as soon as a loaded barge has entered it for the descent, thereby losing the advantage for which the caisson with its supply of water was designed,

BARGE IN CAISSON.

Fig. 244.—Georgetown Incline, U.S.

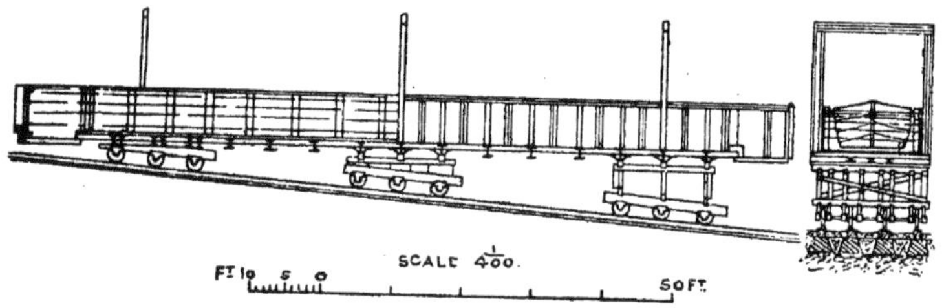

and increasing the time required for transferring a barge from the river to the canal. Four barges, however, can be readily passed up the incline every hour. On a new incline on the Grand Junction Canal, the counterbalancing caissons with their barges travel sideways.

Chignecto Ship-Railway.—An illustration of the extension of the principle of canal inclines to the conveyance of coasting vessels across necks of land without the aid of canals, marking a further incursion of the railway system into the domain of inland navigation, is furnished by the ship-railway constructed across the neck of land, only 15 miles wide, separating Chignecto Bay, an inlet from the Bay of Fundy, from Baie Verte leading through Northumberland Strait into the Gulf of St. Lawrence, to enable coasting vessels of 1000 tons register and 2000 tons displacement to avoid a stormy *détour* of 500 miles round the coast of Nova Scotia. The line is 17 miles in length, and nearly straight throughout; it is level for half its length, and on the other half the gradients do not exceed 1 in 530; and the vessels, 235 feet long, 56 feet beam, and 15 feet draught, after being raised out of water by hydraulic rams, are to be conveyed in steel cradles, in sections 75 feet long each, running on sixty-four solid 3-feet wheels on two lines of way of standard gauge, placed 11 feet apart and laid with steel 110-lb. rails, at a speed not exceeding 10 miles an hour.[1] The work was commenced in 1885; but in 1891, after about three-quarters of the work had been completed, the undertaking was stopped for want of funds.

[1] "The Chignecto Ship-Railway," H. G. C. Ketchum, *Transactions of the Canadian Society of Civil Engineers*, vol. v. p. 309, and plates 11 and 12.

Canal Lifts.

A lift furnishes another method by which barges are enabled to surmount a considerable difference of level between two reaches of a canal, both expeditiously and with only a small expenditure of water. The barge passes into an iron trough filled with water, which is closed by a lifting gate at each end, and is raised or lowered vertically; and the motive power is mainly obtained by altering the balance of the trough by letting out or introducing a certain amount of water. Two types of these lifts have been constructed, namely, a hydraulic canal lift in which two troughs, each supported by a central hydraulic piston, counterbalance one another, and go up and down alternately, owing to the preponderance of the weight of water given to the upper trough; and a floating canal lift with a single trough, whose weight is exactly supported by floats immersed in wells full of water, so that any alteration in the weight of the trough, by removing or adding water, causes the trough to rise or descend.

Hydraulic Canal Lift.—Though small suspended caissons, counterbalancing one another, were used for canal lifts in England in the earlier half of the nineteenth century, the first hydraulic canal lift was erected at Anderton in 1875, to enable barges of 100 tons to pass between the River Weaver and the Trent and Mersey Canal, where the difference of level is 50½ feet, and space was not available for a flight of locks.[1] Each of the two troughs is 75 feet long, 15½ feet wide, contains 5 feet of water, and is supported in the centre by a ram, 3 feet in diameter. To work the lift, 6 inches depth of water is removed by siphons from the bottom trough, which gives a preponderating weight of 13 tons to the top trough, causing it to descend; and when the descending trough reaches the water in the bottom of the lift pit, the final lift of 4 feet of the ascending trough is effected by closing the communication between the two presses in which the rams work, and admitting water under pressure into the press of its ram. The lift of 50½ feet is accomplished in 2½ minutes; and the transference of a barge from the river to the canal, and another barge in the opposite direction, can be effected in 8 minutes, with the expenditure of only a 6-inch layer of water over the area of the trough. As the passage of a barge through a flight of locks with the same lift occupies at least 1¼ hours, a lift evidently effects a great saving of time and water.

The hydraulic lift which commenced working in 1888 at Fontinettes on the Neuffossé Canal of France, to relieve a flight of five locks, with a lift of 43 feet, where the coal traffic to Paris by water is very large, furnishes an example of the extension of the same system to barges of 300 tons. The two troughs at Fontinettes, each 129½ feet long, 18⅜ feet wide, and containing 6½ feet depth of water, supported by a central cast-iron ram of 6½ feet diameter, working in a steel press, descend into a dry lift pit, so

[1] *Proceedings Inst. C.E.*, vol. xlv. p. 110, and plate 2.

that the preponderating weight of 1 foot depth of water, amounting to $63\frac{1}{2}$ tons, introduced into the trough at the top, continues to act in this case to the end of the descent.[1] The troughs are kept in position by guides on the central towers, and on the abutment of an aqueduct at the upper end; and the lift of 43 feet is accomplished in 4 minutes, one trough going up as the other descends (Fig. 245). A turbine actuated by a fall of water from the upper reach, stores up water under pressure in an accumulator, for raising the counterbalanced lifting gates closing

HYDRAULIC CANAL LIFT.

Fig. 245.—Fontinettes Lift, Neuffossé Canal.

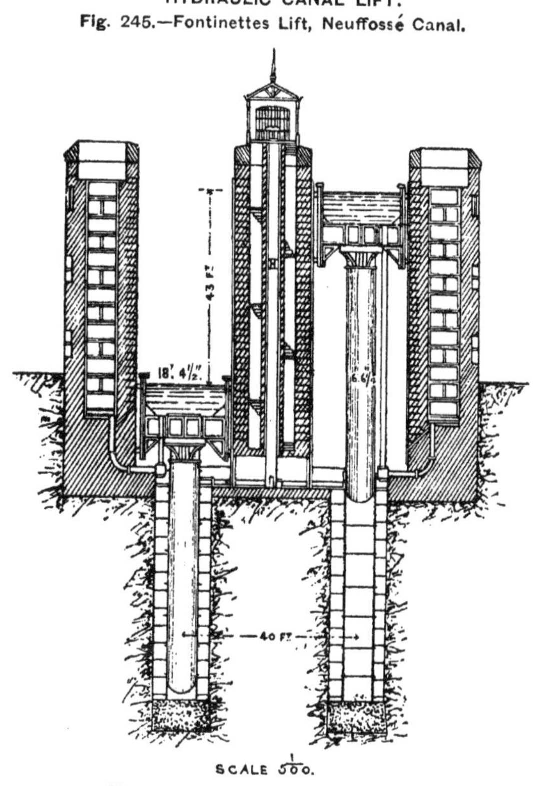

the ends of the reaches and the troughs, turning the capstans foı hauling the barges in and out of the troughs, and for assisting the raising of a trough on the commencement of work, or when it has dropped somewhat through leakage. The cost of this lift was about £68,000. The transference of the largest barges from one reach of the canal to the other occupies 19 minutes.

[1] *Proceedings Inst. C.E.*, vol. xcvi. p. 182.

Another very similar lift, for accommodating vessels of 400 tons, was completed about the same time at La Louvière in Belgium, on the works of the Canal du Centre. The troughs are 141 feet long and 19 feet wide, and are filled with water to a depth of nearly 8 feet for receiving barges drawing $7\frac{1}{5}$ feet; and guides on light lattice-iron towers at the centre and each corner, keep the troughs in position.[1] The counterbalanced weight which has to be raised is 1037 tons; and the lift of $50\frac{1}{2}$ feet is accomplished in $2\frac{1}{2}$ minutes, by admitting a layer of 10 inches of water into the trough at the top, giving it a surcharge of 62 tons; whilst a barge can be transferred from one reach of the canal to the other in 15 minutes. This lift cost £55,750; and three other similar lifts, with rises of $55\frac{1}{2}$ feet, were intended to be constructed on the canal, for surmounting a rise of 217 feet in a distance of only $4\frac{1}{3}$ miles where water is scarce; but owing to the cost, the canal works have not hitherto been completed.

Floating Canal Lift.—A large canal lift for surmounting a sudden rise of 46 feet, and accommodating barges of 950 tons, has been constructed at Henrichenburg on the Dortmund-Ems Canal, opened in 1899. Unlike the smaller, hydraulic canal lifts, this lift has only a single large trough, 230 feet long, 28 feet wide, and filled with water to a depth of $8\frac{1}{5}$ feet, which is supported by five cylindrical floats, $27\frac{1}{4}$ feet in diameter, and $33\frac{3}{4}$ feet high, formed of mild steel plates, floating vertically in deep wells filled with water, and placed in communication at the bottom by horizontal conduits, so as to maintain a uniform water-level in all the five wells [2] (Fig. 246). As the floats weigh 600 tons, the trough, with its lifting gates and supports erected on the floats, 800 tons, and the water in the trough 1600 tons, the total load moved is 3000 tons; and it is exactly in equilibrium in the middle of its course. The motion is obtained by admitting water to the trough when it is at the top, and letting out water when it is at the bottom, which operations are effected by stopping the rise of the trough a little before its water-level attains the level of the water in the upper reach of the canal, and arresting its descent a little before its water-level has come down to the water-level in the lower reach. This regulation of the position of the trough, and the maintenance of its level, as well as its support in the event of any sudden failure in buoyancy of the floats, are accomplished by two screw spindles on each side, all four being turned at the same rate by an electric motor. The trough is raised or lowered the full distance in $2\frac{1}{2}$ minutes; and all the operations required for transferring a barge from one reach to the other, and returning the trough to its original position, only occupy 15 minutes. The motion is guided and kept vertical by braced uprights on each side, near either end, supporting the overhead bridge carrying the electric motor (Fig. 246). The two adjacent lifting gates closing the trough and the canal reach at each

[1] *Proceedings Inst. C.E.*, vol. xcvi. p. 184, and plate 6, fig. 1; and "Rivers and Canals," vol. ii. p. 408, and plate 11, fig. 12.

[2] *Le Génie Civil*, vol. xxix. p. 161, and plate 11; and "Festschrift zur Eröffnung des Dortmund-Ems-Kanals," 1899, pp. 4, 8, and 12.

end, are raised together, as usual, by an electric motor; and two electric capstans haul the barges into and out of the trough. The cost of this lift was £130,000, constituting about twice the expenditure on the largest double hydraulic lifts; but the lift can accommodate much larger barges.

FLOATING CANAL LIFT.

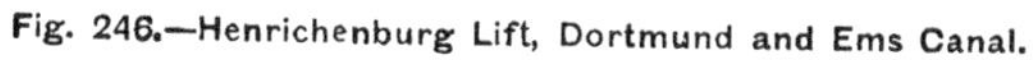

Fig. 246.—Henrichenburg Lift, Dortmund and Ems Canal.

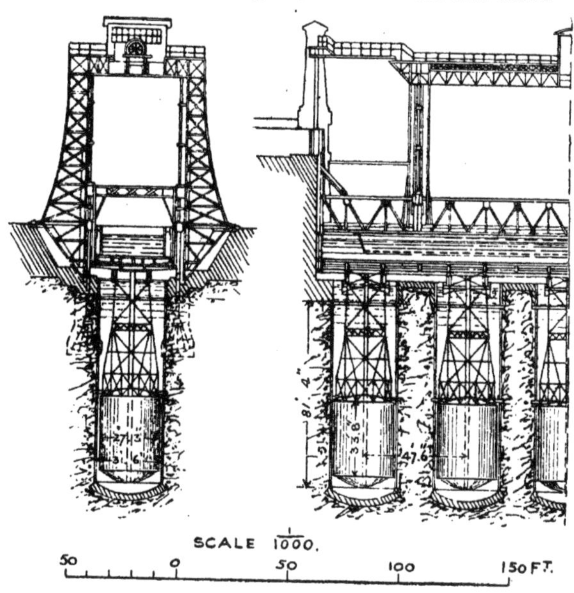

Canal Lifts compared with Canal Locks and Inclines.—Canal lifts, like locks, possess the advantage of keeping the barges floating in water, which, though effected by horizontal caissons on inclines for small barges, has not hitherto been successfully realized for large barges in this method of transit. For ordinary differences of level, and where there is an ample supply of water, locks are preferable to lifts on account of their simplicity; but where a considerable difference of level has to be surmounted and water is scarce, lifts possess a very decided advantage over locks, both in effecting a considerable acceleration in the passage from one reach to the next, as compared with a flight of locks, and also in greatly reducing the consumption of water. Canal inclines possess similar advantages over locks; but they involve raising large barges out of water, they necessitate the employment of much more power for drawing up the barges than lifts, and occupy somewhat more time in passing barges from one reach to the next. Inclines, however, are much simpler and far less costly than lifts; and they appear capable of extension to considerably greater differences of level than lifts could possibly attain, and to much larger vessels in the form of ship-railways. Lifts have not hitherto been applied to greater differences of level than 50½ feet, though greater heights might be surmounted by combining two or more lifts; but a very notable simplification in their working, and extension of their capacity, have resulted from the adoption of the system of floats supporting a single trough, instead of two troughs counterbalancing each other.

CHAPTER XXV.

SHIP-CANALS.

Special features of ship-canals : limitations imposed on these works—Sizes of ship-canals, width and depth of principal ship-canals, side slopes—Construction of ship-canals, methods of excavation, dredging and modes of deposit—Protection of slopes, provisions adopted—Supply of water, sources—Bridges across ship-canals, high-level, and movable, instances—Differences in ship-canals, principal ship-canals compared and contrasted—Locks on ship-canals, instances, dimensions, and arrangements—Special features of Suez Canal, Panama Canal locks and the Gatun Dam, Baltic Canal, Amsterdam Canal, Bruges Canal, Manchester Canal—Methods of increasing capacity for traffic, multiplying passing-places, lighting, widening—Remarks on ship-canals, instances of lateral ship-canals, necessity of favourable conditions for construction of ship-canals.

SHIP-CANALS differ mainly from inland canals in their size, on account of the sea-going vessels which they have to accommodate, and in the obstacles which the large works involved impose against their penetrating far inland, or traversing districts in which the differences of level are considerable. Moreover, the size of channel to be provided, and the volume of water contained in it, preclude the use of embankments, and therefore necessitate the construction of the canal at a low level in respect to the land throughout its length; and locks are only sparingly available, on account of the delays they involve and their cost. Lifts also for ship-canals would reach prohibitive dimensions; and if inclines were introduced, it would be as well to convert the ship-canal into a ship-railway. Accordingly, ship-canals are much more limited in their scope, and in the opportunities for their application, than inland canals; but at the same time, under favourable conditions, they constitute far more important works in most instances, not only converting large inland towns, at a short distance from the sea, into seaports, but also modifying, by cutting through isthmuses, the commercial sea routes of the world.

Ship-canals by being in cutting throughout, except where they pass through lakes, involve in their construction the disposal of enormous quantities of earthwork and dredgings; whilst their exposed entrances on the seacoast necessitate costly harbour works for sheltering them.

Moreover, ship-canals, unlike railways, only commence to be useful when they have been completed from end to end; and their prospects of traffic are often less assured, on account of the uncertainties which attend the selection of routes and ports by shipping. Accordingly, the carrying out of these large works, except as Government undertakings, is very liable to be hindered by the difficulties experienced by a private company in raising the whole of the large capital required on reasonable terms.

Sizes of Ship-Canals.—The minimum size of a ship-canal has been practically fixed by the dimensions originally given to the Suez Canal, namely, a bottom width of 72 feet and a depth of 26 feet, this width being a reduction to less than half the width at first proposed, in

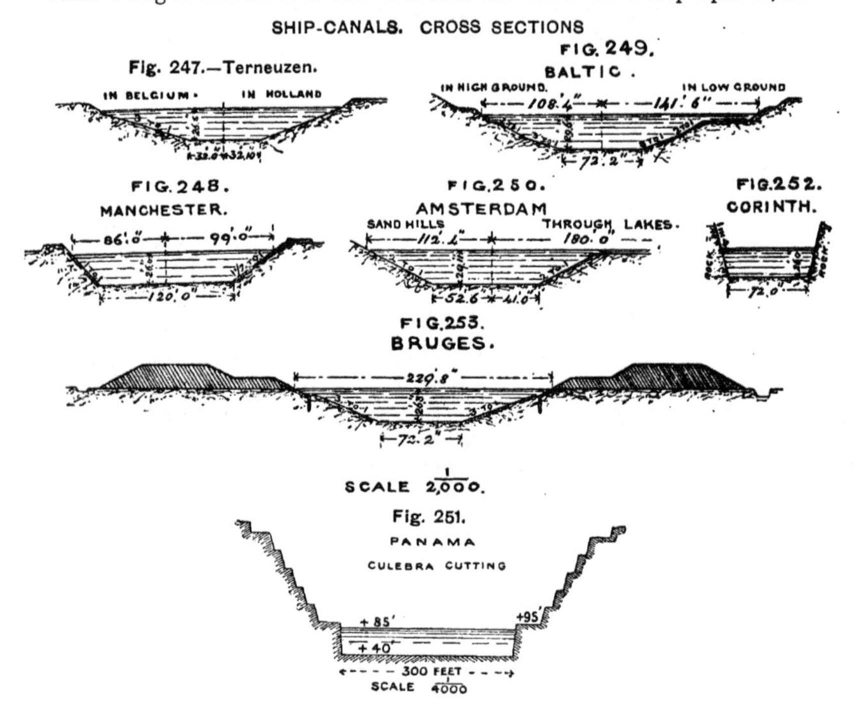

order to meet the financial necessities during construction. This bottom width of 72 feet has, indeed, proved inadequate for the large and increasing traffic passing through the Suez Canal, in spite of the provision of enlarged passing-places for vessels, and has been considerably increased; but it was also adopted for the enlarged Ghent-Terneuzen Canal (Fig. 247), the Baltic and North Sea Canal (Fig. 249), the portion of the Amsterdam Canal traversing the lakes when first constructed, the Corinth Canal (Fig. 252), and the Bruges Canal (Fig. 253). The depth of the Manchester Canal was as first constructed, Fig. 248, 26 feet, with a bottom width of 120 feet, but on completion of deepening operations, vessels drawing 27 to 27½ feet will be accommodated. The

Amsterdam Canal originally constructed with 26 feet depth through the lakes, Fig 250, has been enlarged to accommodate vessels drawing 29½ feet. The Terneuzen and Bruges Canals have 26¼ to 26½ feet depth, Figs. 247, 253, and it has been decided to reconstruct the Baltic Canal, Fig. 249, with a depth of 36½ feet in place of 29½, and a bottom width of 145 feet in place of 72. The Suez Canal, Figs. 257, 258,[1] upon the completion of works in progress is intended to have a depth of 34⅛ feet, and a bottom width of 328 feet, while the depth of the Panama Canal, now in progress as a lock canal, will be not less than 41 feet.

The side slopes of these canals are varied according to the nature of the strata they traverse. Thus through rock, the side slopes of the Corinth Canal have been made 1 in 4 below the water-level, and 1 in 5 to 1 in 10 above through firm rock, and 1 to 1, or steeper, through the sandstone rock on the Manchester Canal; whilst pitched 1½ to 1 slopes have been adopted generally on the Manchester Canal, 2 to 1 slopes on the enlarged Amsterdam Canal, 2 to 1 to 3 to 1 on the Baltic Canal, 2½ to 1 slopes at present on the Suez Canal, except through the rock cuttings, which are eventually to be flattened in very soft low-lying ground to 4 to 1, and 3 to 1 slopes on the Terneuzen and Bruges canals.

Construction of Ship-Canals.—The wide, long, and occasionally deep cuttings required to be excavated for ship-canals, necessitate the formation of a series of tracks for the passage of excavators and trains of waggons for the removal of the earthwork to the deposit tips, along terraces rising one above another in a succession of steps in very deep cuttings, like the Culebra cutting at the Panama Canal (Fig. 251, p. 404), so that the central trench and side slopes may be simultaneously excavated at various levels. Even where the cuttings are not particularly deep, their width and length involve the employment of special appliances for the vast masses of excavation to be dealt with, so that at the Manchester Canal, about one hundred excavators of different types were at work at one time in its construction, together with 223 miles of temporary railways, 173 locomotives, and 6300 waggons and trucks; whilst at the Chicago Drainage Canal works, where the cuttings approached in magnitude those of a ship-canal, several ingenious forms of transporters were used for raising the excavated material from the bottom of the cutting, and depositing it in high heaps on the land along the sides (pp. 40-42).

Dredging is also very largely employed for forming the requisite channel through shallow lakes, or for forming or deepening the channel through low-lying land adjoining the sea or lakes, in which the dredger cuts its own flotation as it advances, or into which water has been admitted after the excavation of the cutting has been partially carried out. In some parts of the Suez Canal, natural depressions, a little height above sea-level, were flooded by water drawn from the fresh-water canal; and dredgers launched upon these artificial lakes, effected the excavation in stages, the water-level being lowered as the work proceeded; whilst the portions of the depressions outside the

[1] *Proceedings Inst. C.E.*, vol. cxli. plate 3.

limits of the canal excavations, were filled with the dredged material deposited from hopper barges. The channel for the Bruges Canal, traversing the flat, low-lying land between Bruges and the North Sea, has been mainly formed by dredging.

Where the land traversed by the canal works is low, the deposit of the dredged materials on the side banks is rapidly and economically effected by long shoots (Fig. 14, p. 52), which were first used in constructing the Suez Canal through the shallow lakes, Menzaleh, Ballah, and Timsah, and the low land near Suez (Figs. 255 and 256, p. 409), and have since been extensively adopted, as, for instance, along the low-lying land of the Panama Canal, and throughout the Bruges Canal. Floating tubes in a long line have sometimes been employed for the conveyance of the dredged materials to form side banks when dredging in lakes, a system extensively used in forming the Amsterdam Canal through Lake Y and Wyker Meer (Fig. 261, p. 410). Travelling bands were used in some places on the Manchester Canal, for conveying the dredged materials from dredgers in the canal to waggons on the bank; and where the banks in places along the Suez Canal were rather too high for deposit from long shoots, elevators projecting with an upward inclination of 1 in 4 over the bank to a height of 45 feet above the water-level, were employed, up which boxes containing 4 cubic yards of dredged materials were drawn by steam-power, and tipped their contents on the land 74 feet back from the edge of the canal slope.

Protection of Slopes against Erosion.—Though the regulations for the traffic oblige large steamers traversing a ship-canal to go quite slowly in proportion to the narrowness of the waterway, the wave generated by their passage breaks upon the side slopes, and tends to erode the banks near the water-level. Berms, or wide benches, have sometimes been formed at the sides a little below the water-level, where

SAND-DAM ON BALTIC CANAL.

Fig. 254.—Section.

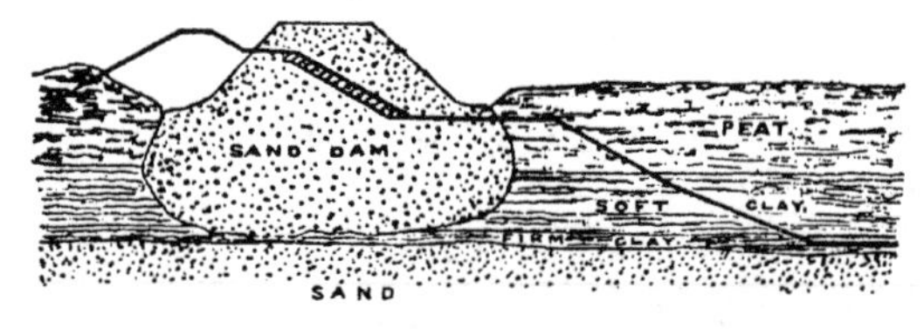

the canal traverses low-lying land or shallow lakes, and where consequently the excavation is not much increased by this arrangement, with the object of stilling the wave in shallow water before it reaches the bank, examples of which are furnished by the Baltic Canal in low ground (Fig. 249, p. 404, and Fig. 254), the Amsterdam Canal through the lakes (Fig. 250, p. 404), and the Suez Canal across Lakes Menzaleh and Ballah (Fig. 257, p. 409). As, however, in this last instance, the berm appears to give an impetus to the breaking wave, it is proposed to do away with the berm when the final widening is carried out. Narrow benches are sometimes introduced to check slips in the slopes, or to limit their extent (Figs. 249, 253, 257, and 258). In carrying the Baltic Canal across marshes for about 6½ miles, where the ground is very soft, an embankment of sand was tipped along each side of the canal, which, settling down on to a firm stratum of sand below, formed a dam retaining the upper very soft stratum outside it from slipping into the canal excavations; and the upper slope of the canal, near the water-level, was excavated in this sand-dam, which also partly supports the side embankment (Fig. 254).

The slopes are generally protected by stone pitching, or concrete, from some feet below the water-level to three or four feet above it, to withstand the wash of the waves raised by passing vessels, a line of stakes being often driven down into the slope, along the bottom of the pitching, to secure its toe; and the pitching on the somewhat steep slopes of the Manchester Canal, has been carried down to the bottom of the canal in soft ground. The growth of suitable plants near the water-line has been found to have a protective influence on the slopes of the deeper cuttings of the Suez Canal; and fascine-work, which has been extensively used to protect the banks in regulating rivers in Holland and the United States, could be advantageously applied to preserving the slopes of ship-canals near the water-level.

Supply of Water to Ship-Canals.—Those ship-canals whose normal water-level is similar to that of the seas they join, except for such fluctuations as are produced in the seas by tides and wind, are amply supplied with water from the sea. Ship-canals, however, whose water-level has to be raised on proceeding inland, by means of locks, have to derive their supply of water from the nearest river. Thus the portions of the Manchester Canal above the tidal reach are fed by the rivers Mersey and Irwell, which are to a large extent absorbed by the canal; and it is proposed to supply the locked portion of the Panama Canal with water from the rivers Chagres, Gatun, and Trinidad, which flow through the area proposed to be converted into the Gatun Lake with a normal water-level of 85 feet above mean ocean level.

Bridges across Ship-Canals.—Bridges carrying roads and railways over ship-canals have either to be made opening, low-level bridges, or high-level fixed bridges, giving a sufficient headway for the passage of masted vessels underneath them. Where the ship-canal is approximately level with the land, low-level swing bridges give no trouble with the approaches, and are much cheaper to construct than high-level fixed bridges with very long, costly approaches; but, on the other hand, opening bridges are liable to cause delays to the traffic, and involve an annual expenditure in working. Swing bridges have been adopted for crossing over the Amsterdam Canal, turning on a pier erected at the foot of one of the slopes; whilst at the Manchester Canal, where the available headway is restricted to 75 feet by the Runcorn railway bridge, built several years before the canal, all the railways cross the canal on four high-level bridges with this headway, and two roads also cross the canal in the same manner; whilst seven roads cross the canal on swing bridges, and minor roads and footpaths are accommodated by ferries. Where ship-canals traverse high ground, the approach to a high-level fixed bridge is comparatively easy; but the bridge has to span a much wider cutting. Thus the two fixed bridges over the Baltic Canal, affording a headway of 137¾ feet above the water-level of the canal, have spans of 513 feet and 536 feet; and the single railway and road bridge crossing the Corinth Canal, with a clear headway of 141 feet above the water-level, has a span of 262 feet, the required span in this case being less owing to the steepness of the slopes of the rock cutting. Two railways and one road cross the Baltic Canal by swing bridges; and the rest of the roads are served by fourteen ferries. The crossing of the Bruges Canal has been provided for by three swing bridges, two passing over the entrance lock, and one ferry; but no bridge of any kind has had to be constructed across the Suez Canal, owing to its position in the desert quite away from frequented routes.

Differences in Ship-Canals.—The cross sections of the waterway of ship-canals are very similar (Figs. 247 to 253, p. 404), with only such differences in the side slopes as the nature of the ground requires, and such differences in bottom width as the prospects of traffic and financial considerations necessitate, from 72 feet in most cases, when passing-places have to be provided at suitable distances apart, up to 120 feet, which may be regarded as the minimum width allowing two vessels of moderate size to pass one another very slowly, and inadequate for the largest class of vessels, which will only be able to pass each other at a fair speed on the Suez Canal when its proposed widening has been fully completed (Figs. 257 and 258, p. 409).

The variable conditions, however, of the sites, such as the width of the intervening land to be cut through, the general character and height of the ground above the proposed trench of the canal, the variations in the level of the sea at the entrances, the exposure of the approaches, and the extent of interference with vested interests, as for instance roads, railways, navigations, and drainage crossed by the canal, produce great differences in the nature, extent, conditions, and cost of the works required for the construction of a ship-canal easy of access and convenient for navigation.

The Suez Canal and the Corinth Canal are both perfectly open channels connecting two seas, formed by cutting a channel across their respective isthmuses; but whilst the Suez Canal, 100 miles long, traverses, for a great portion of its length, low-lying lands very little above sea-level, and several depressions below sea-level which it has converted into lakes, and the highest land on the route is only about 80 feet above the bottom of the Canal (Figs. 255 and 256), the Corinth Canal, only 4 miles long, passes through a neck of land rising in the centre to an elevation of 286 feet above the bottom of the canal. Currents occur in both canals, due in great measure to alterations in the sea-level at the ends, produced chiefly by wind, and also, to some extent, by tidal oscillations, which, though very small in the Mediterranean Sea, attain 4¾ feet at Suez, and cause currents in the southern part of the Suez Canal as far as the Bitter Lakes. The entrances to the Corinth Canal have been both

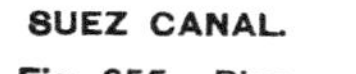

SUEZ CANAL.

Fig. 255.—Plan.

NOTE.—The dimensions shown on Figs. 257, 258, are in process of enlargement.

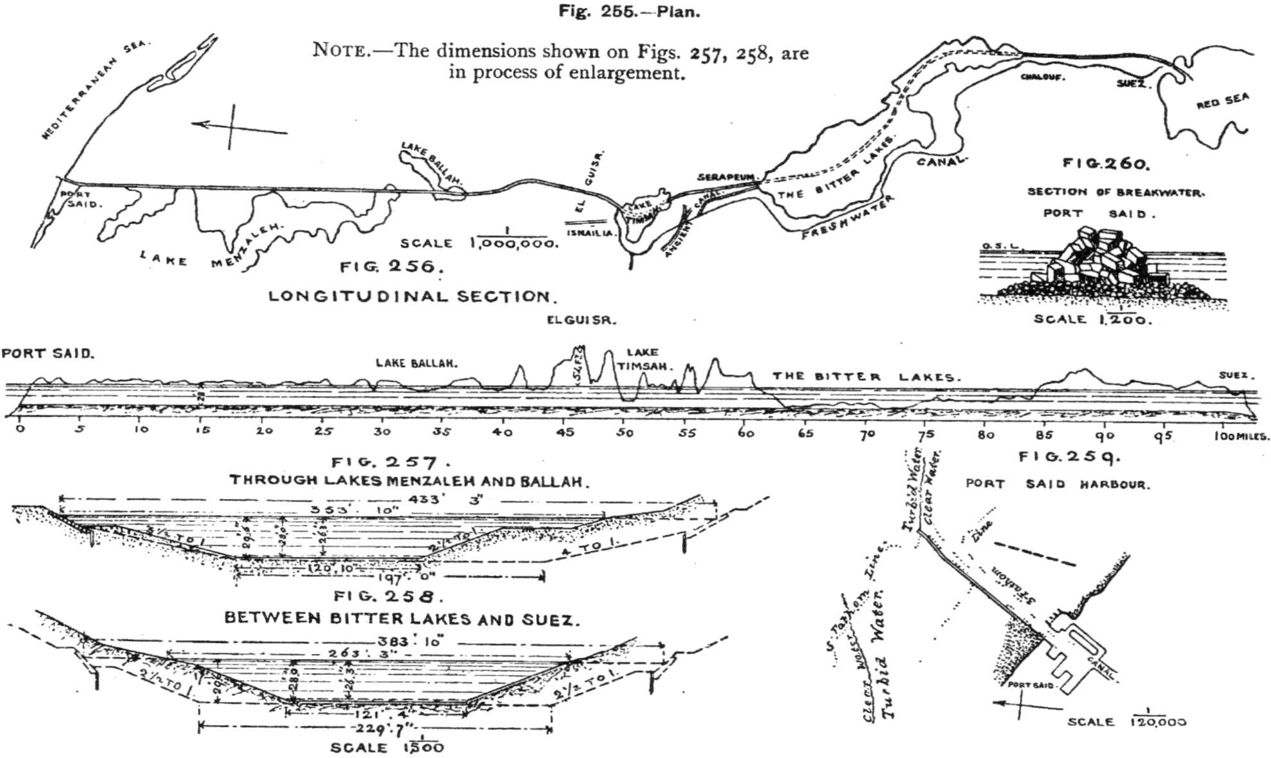

protected by breakwaters, under whose shelter approach channels have been dredged; and the Port Said entrance to the Suez Canal has been protected by large breakwaters, the western breakwater serving also to shelter the dredged approach channel from the easterly littoral drift, which, however, has constantly to be maintained by dredging (Figs. 259 and 260, p. 409). The Suez Canal works, situated in the desert, had to be supplied with every requisite from a long distance; and the fresh-water canal bringing fresh water from the Nile to the works, formed a very necessary preliminary undertaking, not only for supplying fresh water, but also for the conveyance of plant and provisions to the site.

The Suez and Corinth canals are the only instances of open channels; for the Amsterdam and Baltic canals have locks at each end, and the Terneuzen Canal has a lock near its entrance from the River Scheldt, with gates pointing both ways for controlling and regulating the water-level. The rolling caissons closing the lock near the sea end of the Bruges Canal, serve the same purpose; while the Panama Canal as a lock canal is only designed to be open to the sea at each end as far as the first locks; and the tidal reach of the Manchester Canal is only exposed to tidal influences for a short period before high water of spring tides, not exceeding $1\frac{3}{4}$ hours at the highest tides.

The Amsterdam Canal, about 16 miles long, and the Baltic Canal, $61\frac{1}{3}$ miles long, are both kept at a fairly constant level between their controlling locks at either end. The Amsterdam Canal was constructed

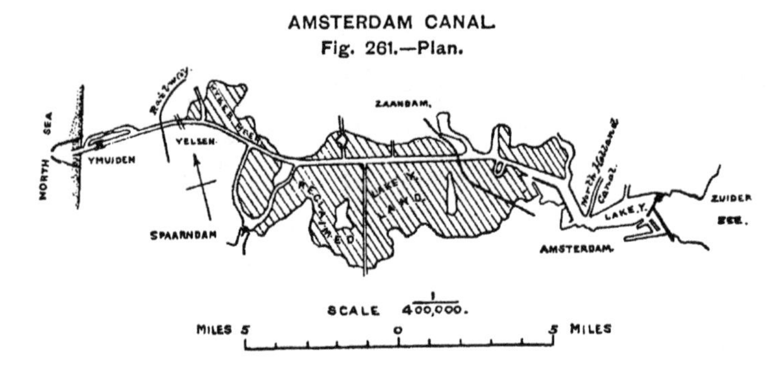

AMSTERDAM CANAL.
Fig. 261.—Plan.

through shallow lakes for about $10\frac{1}{2}$ miles from Amsterdam, and for the rest of the distance across a belt of sand-dunes separating the lakes from the North Sea, and rising to a maximum height of 95 feet above the bottom of the canal (Fig. 261); whilst the Baltic Canal was carried through some small, deep lakes, and across land of very variable height in the Eider basin, and through a long stretch of low, marshy ground lying between the Elbe and the ridge separating the basins of the Elbe and the Eider, in traversing which ridge the deepest cutting, with a maximum depth of 108 feet, had to be excavated. The shallow lakes traversed by the Amsterdam Canal, were reclaimed by aid of the banks formed at the sides of the canal by sand excavated from the cutting at the western end, and subsequently by the material dredged from the lakes; and the lakes along the Baltic Canal had their water-level lowered to that of the canal, and were partially filled up and reclaimed by the excavations for the canal. The western entrance of the Baltic Canal, being situated on the Elbe estuary, through which it is in communication with the North Sea, and the eastern entrance in Kiel Harbour, a sheltered creek of the Baltic Sea, no protective works were needed at either end; but two converging breakwaters had to be constructed to shelter the western approach to the Amsterdam Canal, on the perfectly open North Sea coast, and to protect the dredged channel across the foreshore from the inroad of drifting sand (Fig. 261).

The Terneuzen Canal going from the Scheldt estuary at Terneuzen to Ghent, and the Bruges Canal connecting Bruges with the North Sea, both traverse quite flat, low-lying land throughout their whole lengths of about $20\frac{1}{3}$ miles and $7\frac{1}{2}$ miles respectively, and are therefore exceptionally favourably situated; but whereas the Terneuzen Canal has been gradually enlarged to provide for the growth of traffic and the increasing requirements of Ghent, the Bruges Canal is comparatively a new work. The entrance, however, to the Terneuzen Canal in the Scheldt estuary is well sheltered; whilst the entrance to the Bruges Canal on the North Sea coast has to be protected by a large breakwater, lately under construction, curving round from the shore on the south-west side till it covers the entrance (Fig. 307, p. 474).

The Manchester Canal is the first regular ship-canal which has been constructed with reaches at different levels, surmounted by locks; but the Panama Canal, now under construction, is designed to ascend by locks to a summit-level in the Gatun Lake and Culebra Cutting maintained by a dam at Gatun, thus reducing excavation and lessening

PANAMA CANAL.

Fig. 262.—Longitudinal Section.

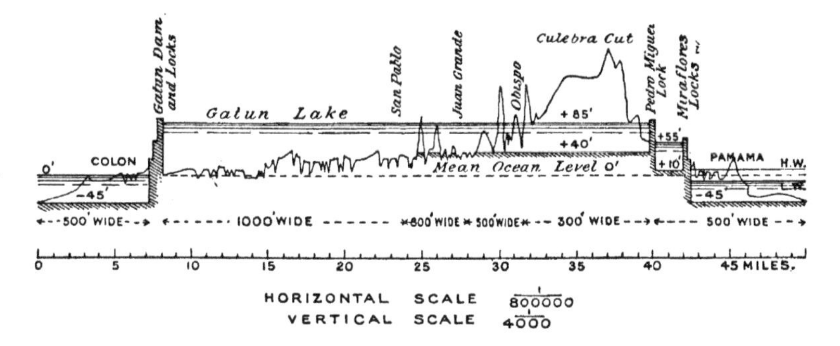

cost (Fig. 262); while the suggested Nicaragua Canal would have risen by locks to the level of Lake Nicaragua. The Manchester Canal, however, 35 miles long, only penetrates this distance inland to serve Manchester, rising with the general upward slope of the land (Fig. 263);

whereas the Panama Canal, 50 miles in length, ascending by locks to the artificial lake to be maintained by the Gatun Dam at a summit-level of 85 feet above mean ocean level, is designed to cut through the isthmus of Panama, together with the central ridge at the Culebra Cut separating the Atlantic and Pacific slopes, and thus, like the Suez Canal,

MANCHESTER CANAL.

Fig. 263.—Longitudinal Section.

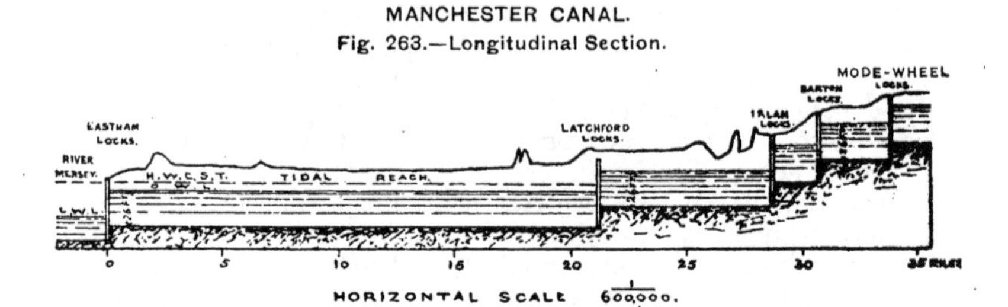

HORIZONTAL SCALE $\frac{1}{600,000}$.
VERTICAL SCALE $\frac{1}{2,000}$.

by doing away with the necessity of a great *détour* to the south, to alter and shorten some of the shipping routes of the world.

Locks on Ship-Canals.—The locks on ship-canals, whether regulating or lifting, should be made adequately wide, and with a sufficient length between the gates, to admit the largest vessels capable of navigating the canal, though in the case of a regulating lock, it would be possible for a longer vessel to pass through when the water outside is level with the canal; whilst the sills of the locks should be at least level with the bottom of the canal, and may with advantage be placed two or three feet lower if there is any prospect of a subsequent deepening of the canal to this extent. Thus a new lock has had to be constructed in a side cut near the North Sea entrance of the Amsterdam Canal, with an available length of 740 feet, a width of 82 feet, and a depth over the sills at low tide of $30\frac{1}{3}$ feet, as the original large North Sea lock, 393 feet long, 59 feet wide, and $23\frac{1}{3}$ feet depth of water over the sills, had become quite inadequate, in the course of about twenty years, for the increased sizes of vessels, and the enlarged and deepened canal; whilst the locks on the Panama Canal at Gatun, Pedro Miguel, and Miraflores, are to be 1000 feet long, 110 feet wide, and 41·3 feet deep on sills at mean tide level, the canal itself being 45 feet deep.

The regulating locks at each end of the Baltic Canal have been made double, to provide for the simultaneous passage of incoming and outgoing vessels; and each lock is furnished with three double pairs of gates pointing in both directions, to admit of locking up and down in reverse directions, according to the relative levels of the water in the canal and outside, and to enable smaller vessels to be passed through with less delay by using the central intermediate pairs of gates (Fig. 264), an arrangement also adopted in the new Amsterdam Canal lock, with the two intermediate pairs of gates placed nearer the lower end to adapt the lock better for accommodating vessels of various lengths. The locks have an available length of 492 feet, a width of 82 feet, and a depth over the sill of $31\frac{1}{3}$ feet with the normal water-level in the canal; and each lock is provided with a longitudinal sluiceway and twelve lateral openings in each side wall, to ensure the rapid filling and emptying of the lock-chamber[1] (Figs. 264 and 265). The length provided is somewhat short, in view of the increasing dimensions given of late years to the larger class of sea-going vessels; but the Holtenau Locks at the Baltic end of the

ENTRANCE LOCKS AT HOLTENAU, BALTIC CANAL.

Fig. 264.—Plan of Locks.

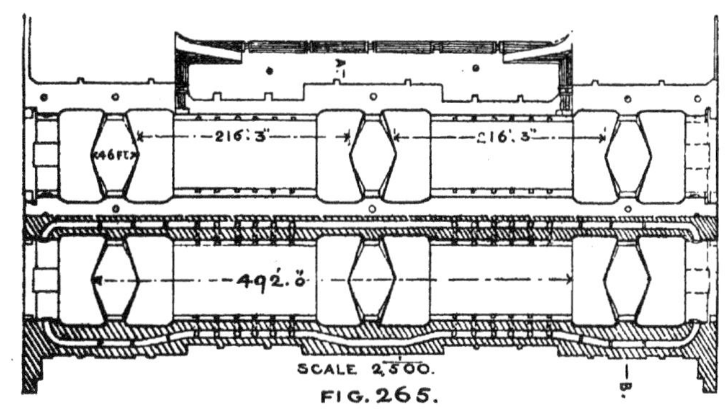

FIG. 265.

SECTION OF LOCK-CHAMBERS ALONG A.B.

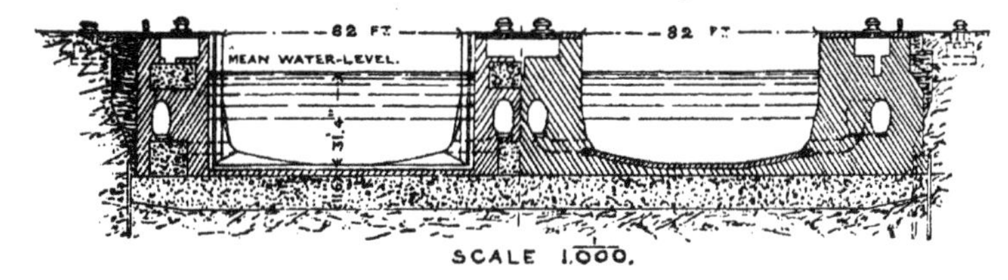

canal are usually kept open, so long as the water-level in the Baltic does not differ from its mean level (which is the normal water-level of the canal) by more than $1\frac{2}{3}$ feet, the main variations being caused by wind; and though the Brunsbüttel Locks are generally closed against the tides in the Elbe, having a maximum range of $27\frac{1}{2}$ feet, they can occasionally be fully opened for a short period when the conditions are favourable.

The entrance lock to the Bruges Canal, with its pair of rolling caissons, each capable of holding up a head of water on either side, like a double pair of ordinary lock-gates pointing in opposite directions, maintains the normal water-level in the canal, and excludes the tidal oscillations of the North Sea; and vessels can be locked into or out of the canal, both up and down, according to the relative levels of the water in the canal and outside (Figs. 266 and 267). The lock has an

[1] "Geschichte des Nord-Ostsee-Kanals," Carl Loewe, Berlin, 1895, plates 2 and 4.

available length between the caissons of 840 feet, an entrance width of $65\frac{2}{3}$ feet, widened out in the chamber, for a length of 518 feet, to a bottom width of $83\frac{2}{3}$ feet, with a slope on the widened side, and a depth

ENTRANCE LOCK, BRUGES CANAL.

Fig. 266.—Plan of Lock.

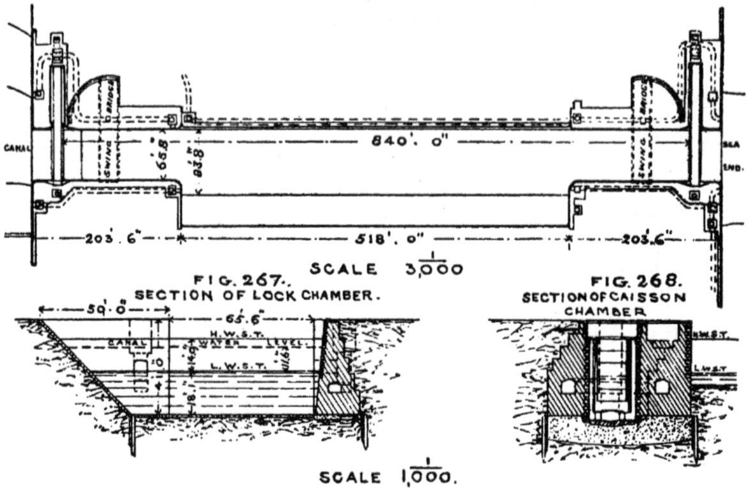

SCALE $\frac{1}{3,000}$

FIG. 267. SECTION OF LOCK CHAMBER.

FIG. 268. SECTION OF CAISSON CHAMBER.

SCALE $\frac{1}{1,000}$.

over the sills of $29\frac{1}{2}$ feet with the normal water-level in the canal, admitting of an increase in the depth of the canal of about three feet when required.[1] The caissons, worked by electricity, can be drawn into or out of their chamber in about two minutes for opening or closing the lock (Figs. 266 and 268, and Fig. 299, p. 467).

Two locks of different dimensions, placed side by side, and each furnished with a pair of intermediate gates, have been constructed at the four changes of level on the Manchester Canal (Fig. 263, p. 412), for facilitating and expediting the passage of vessels of different sizes.[2] The large locks have chambers 600 feet long and 65 feet wide, and the small locks 350 feet by 45 feet; and their sills are placed 28 feet below the normal water-level of the canal; whilst their lifts vary from 13 feet to $16\frac{1}{2}$ feet (Figs. 269 and 270). As the rivers Mersey and Irwell have been taken into the canal, provision has been made for the discharge of their surplus water during floods, through sluice openings built alongside the locks, closed by counterbalanced sluice-gates sliding on free rollers, which serve to regulate the discharge. Three locks of different dimensions, without intermediate gates, admit vessels from the Mersey estuary into the tidal reach, and together with two sluice openings alongside,

[1] *Proceedings Inst. C.E.*, vol. cxxxvi. p. 288, and plate 5, figs. 8 to 12.

[2] *Ibid.*, vol. cxxxi. pp. 18, 20, and 54, and plates 3 and 5; and "Rivers and Canals," vol. ii. p. 585, and plate 12, figs. 7 to 11.

afford passages for the influx of the tide 21 miles up the canal, to the first lifting lock, when it rises above the water-level in the canal. These locks have chambers 600 feet by 80 feet, 350 feet by 50 feet, and 100 feet by 32 feet, whilst the inner sill of the large lock is 28 feet below the normal water-level in the canal, and the outer sill 11 feet lower, to admit vessels at a low state of the tide whenever the approach channel shall have been adequately deepened by dredging. Protection against waves

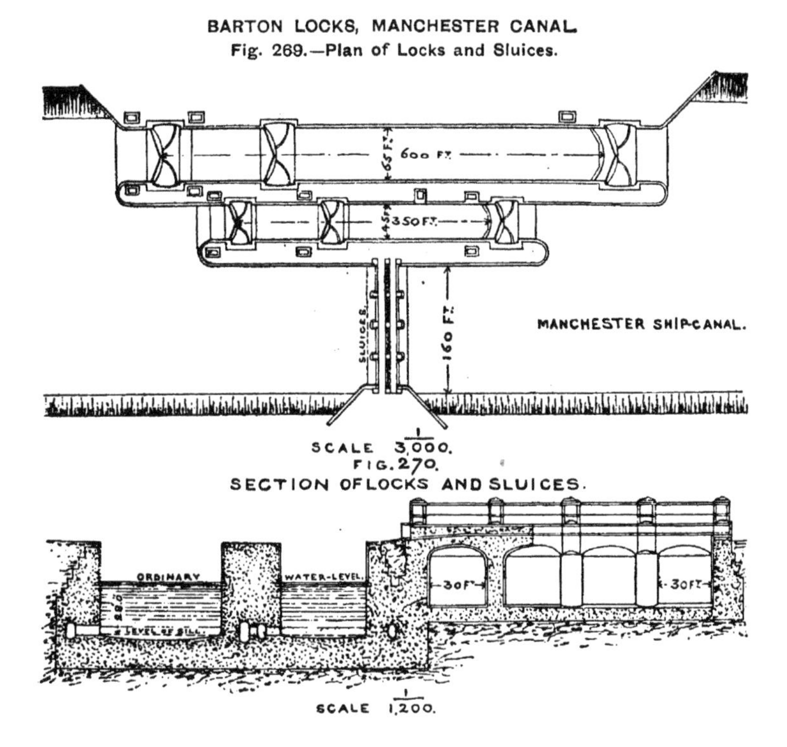

BARTON LOCKS, MANCHESTER CANAL.
Fig. 269.—Plan of Locks and Sluices.

FIG. 270.
SECTION OF LOCKS AND SLUICES.

coming up the estuary has been afforded by a pair of reverse gates near the outer end of each of these three locks; but the closing of these gates against high tides is not allowed.

Special Features of some of the Principal Ship-Canals.—Some of the interesting features of the Suez Canal, such as its isolated position, the necessity of providing a fresh-water canal, and the protection of the banks of the canal, have been already alluded to; whilst the general features of the work are illustrated by Figs. 255 to 260, p. 409. As regards the canal itself, the chief difficulties experienced in its construction were the extent and magnitude of the excavations in a locality quite devoid of resources, and the much larger cost of the works than originally estimated; whilst the widening of the canal has involved the removal of rock under water in the rock cuttings between the Bitter Lakes and Suez, which has been effected by first breaking up the rock by the repeated blows of rock-breaking rams (Fig. 13, p. 51). The low level, for the most part, of the land traversed, and the extensive depressions on the route, together with the circumstance that no locks or bridges were required, constituted very favourable conditions; and the chief engineering difficulty has been the formation and maintenance of an approach channel across the sandy foreshore between deep water in the Mediterranean and the canal at Port Said, in a locality exposed to drift along the coast due to the prevalent westerly winds and the turbid current from the Nile. Though the western breakwater protects the channel dredged under its shelter from the drift, there appeared a danger during the earlier years that the progression of the shoreline produced by the arrest of the drift, might be so rapid as to cause sand to heap up against the breakwater at its upper end, and be carried over into the harbour, and also might eventually lead to the formation of a shoal in front of the channel round the outer end of the breakwater (Fig. 259, p. 409). The rate of advance, however, of the shoreline gradually diminished; and its progress has been checked of late years by forming openings in the upper part of the breakwater, through which some of the drifting sand passes and deposits in the channel, from which it is readily removed by dredging. Nevertheless, shoaling is taking place in front of the harbour, as proved by the receding of the lines of sounding seawards, which, though it has not as yet impeded the access to the dredged approach channel, may in time necessitate an extension of the western breakwater to enable the depth to be maintained by dredging under its shelter.

The canal across the isthmus of Panama is the only undertaking comparable with the Suez Canal in importance, whilst exceeding it in many respects in the magnitude of the works involved. The principal engineering difficulties in the way of carrying out the Panama Canal have been the immense mass of earthwork required for cutting through the central ridge, which would have involved a cutting with a maximum depth of 346 feet for a level canal, in a very unhealthy climate, and exposed to tropical rains, where the upper strata consist of very treacherous clays and schists, and the high, sudden floods of the River Chagres which crosses the line of the canal in several places. The adoption of the design with locks has greatly reduced the amount of excavation, and the difficulty of dealing with the floods of the Chagres, and has rendered the scheme practicable (Fig. 262, p. 411), especially as firm rock, not difficult to excavate, has been reached in the Culebra cutting, so that under present arrangements, with a mean slope of 1 to 1½ with benchings (Fig. 262), a bottom width of 300 feet through the cutting with a depth of 45 feet is obtainable, the water-level being at 85 feet above mean ocean level as stated. The adoption of the lock canal has not, however, been arrived at without strenuous discussion as to the relative merits of locks versus a sea-level canal, while the majority of the International Board of Consulting Engineers, including all the European representatives, reported in 1906 in favour of the latter. Opponents of the lock system have also laid stress upon certain conditions alleged to exist with regard to the foundations of the Gatun Dam, upon the stability of which the lock system depends, while on the other hand the difficulties arising from torrential discharges of the River Chagres are minimised by the creation of the artificial lake, while diminished cost, an estimated shorter period of completion, and certain facilities of navigation are afforded by a lock canal.

The construction of the Baltic Canal necessitated large excavations through the higher ground along the eastern portion of the canal, and in the ridge separating the basin of the Eider from the Elbe;[2] and the formation of the canal through the soft, marshy ground leading to the Elbe, was a troublesome work, involving the consolidation of the slopes by sand-dams (Fig. 254, p. 406); but the sheltered position of both entrances relieved this canal from any necessity for harbour works.

The Amsterdam Canal, with variations of the water-level at its entrances very similar to those at the Suez Canal, as the changes of level in the Zuider Zee are almost wholly caused by wind, and the tidal range at the North Sea entrance is only about $5\frac{1}{2}$ feet, has, nevertheless, had to be provided with regulating locks at both ends, to maintain the canal at the low level of only 14 inches above low water in the North Sea, for the sake of the drainage of the low-lying lands into it, the canal having been shut off from the Zuider Zee by a dam in which the eastern locks are placed (Fig. 261, p. 410). Owing to the reclamation of the lakes, branch canals had to be formed to maintain water communication with Amsterdam for the villages formerly bordering the lakes; and as the drainage waters which used to flow into the lakes, and also those pumped from the reclaimed lands, pass into the canal, provision has been made for their removal by pumps placed alongside the eastern locks, and by opening sluice-gates alongside the locks at each end, for the outflow of the surplus water whenever the level of the water outside is low enough. The breakwaters in the North Sea constituted a large auxiliary work, costing more than one-third of the whole original expenditure, under shelter of which a channel has to be maintained by dredging.

The breakwater now constructed in the North Sea at the entrance to the Bruges Canal, 7340 feet long, forms by far the most important feature of the works, with its open viaduct across the higher foreshore to avoid interference with the littoral currents, and the huge concrete blocks of 2000 to 4400 tons forming the foundations of the harbour and sea walls of the solid breakwater; but this breakwater is intended, not merely to shelter the approach to the canal, but also to form a port of call for passing steamers (Fig. 307, p. 474, and Fig. 315, p. 480).

The Manchester Canal exhibits the unique peculiarity of a reach, 21 miles long, which becomes fully open to the tide in the Mersey estuary, at the Eastham entrance, for the final rise of between 7 and 8 feet by which the highest spring tides surpass the normal water-level of this reach (Fig. 263, p. 412). The canal along this reach borders

[1] "Rivers and Canals," vol. ii. plate 13, figs. 9 and 10.
[2] *Ibid.*, vol. ii. plate 12, figs. 12 and 13.

the estuary, and has for some distance in places been reclaimed from the foreshore of the estuary by lines of embankments across bays, formed by tipping material from the excavations, and protected on the outer 1½ to 1 slope by pitching; and these embankments separate the canal from the estuary. The admission of the tidal water into the canal was, accordingly, arranged to compensate, as far as practicable, for the loss of tidal capacity in the estuary resulting from these reclamations. The upper layer of water thus admitted into this long reach at the highest tides, is let out again higher up the estuary through sluice openings built in the outer embankments enclosing the canal, controlled by sluice-gates sliding on free rollers, and chiefly through the ten sluices each 30 feet wide in front of the River Weaver, which has been shut off from the estuary by the canal embankment; and the gates of these sluices, which can be raised 13 feet above their sills, also enable the water-level of the canal to be regulated during floods of the River Weaver, which flow into the canal, by discharging the surplus water into the estuary. The tidal influx creates strong currents through the narrowed section of the canal, constructed with nearly vertical sides through rock, for a short period before high water of the highest tides, which, though not barring the passage of vessels, is inconvenient, and might have been avoided if the tide could have been excluded, and a second pair of reverse gates placed at the upper end of each of the Eastham locks, to enable vessels to be locked down into the canal near high water during the highest tides. The Bridgewater Canal is carried across the ship-canal in a swing aqueduct, which consists of a trough with lifting gates at the ends, similar to the troughs of canal lifts, with the sole exception that the trough in this case is turned on a central pier in the ship-canal round a quarter of a circle, to open the passage along the ship-canal, or in the reverse direction to enable barges on the Bridgewater Canal to pass over the ship-canal, instead of being raised or lowered vertically from one reach to another. Various other works also, in addition to bridges, had to be constructed to avoid interference with the many vested interests which exist in such a district as that traversed by the Manchester Canal, as for instance locks in the canal embankments to preserve the access of certain places with the estuary, and siphons to maintain the discharge of streams into the estuary.

Methods of increasing the Capacity of Ship-Canals for Traffic.—Since, with the exception of the Manchester Canal, ship-canals have, for the sake of reducing their cost, been constructed in the first instance only of just adequate width for the passage of a single vessel at a time, with passing-places at intervals, their capacity for traffic is necessarily limited, owing to the delays at the passing-places to allow vessels coming in opposite directions to cross, and clear the adjacent sections of the canal, resembling the essential regulation of the traffic on a single line of railway. The simplest expedient for shortening the time of transit, and thus increasing the capacity of a ship-canal when the traffic becomes congested, is the augmentation of the number of passing-places, and the consequent shortening of the intervening sections of the canal which vessels have to traverse in one direction at a time.

Another plan of facilitating the traffic consists in the lighting of a ship-canal with electric lamps and light-giving buoys, to enable vessels to navigate a narrow waterway by night as well as by day. By establishing leading lights and light-giving buoys along the Suez Canal in 1886, and making vessels carry four powerful electric lights, it became possible to navigate the canal safely by night, whereby the capacity of the canal for the passage of vessels has been nearly doubled. The Baltic Canal has been lighted at night from the commencement, with electric lamps on the banks, and by light-giving buoys through the lakes.

The final and most effective means of augmenting the capacity of a ship-canal is by widening the canal sufficiently for two of the largest vessels navigating it to be able to pass one another easily at any place, which is the method gradually being carried out on the Suez Canal (Figs. 257 and 258, p. 409); but in this case, allowing a sufficient space between two of the largest class of vessels passing each other, and a suitable margin on the outside from the line of buoys marking the deep channel, it has been considered advisable to provide ultimately a bottom width of 100 metres, while the radii of curves are to be increased to a minimum of 2500 and 3000 metres to the north and south of the Bitter Lakes respectively. This widening will not only enable vessels to pass one another at a fair speed without danger of collision, thus doing away with all the delays at passing-places; but it will also enable steamers to pass along the canal, when the way is clear, at a much greater speed than at present, without creating such a wash as to injure the slopes, and without experiencing to any extent the retarding influences affecting vessels navigating a shallow and restricted waterway.

Remarks on Ship-Canals.—In addition to the types of ship-canals of which examples have been given, namely, ship-canals at a uniform level traversing low ground, the Manchester Canal rising by locks inland, and the Panama canal now under construction rising on each side with locks to a summit-level, and thus surmounting a dividing ridge, there are instances, on a smaller scale, of lateral canals navigated by sea-going vessels, in place of maritime rivers. Thus a canal has been constructed alongside a portion of the Loire estuary, $9\frac{1}{3}$ miles long, with a bottom width of $78\frac{3}{4}$ feet, and a depth generally of $19\frac{2}{3}$ feet, connected with the tidal estuary at each end by a regulating lock, to enable vessels trading with Nantes to avoid the tortuous, shallow, shifting channel through the central portion of the Loire estuary encumbered with sandbanks. A deep outlet also has been provided for vessels navigating the Rhone, by the St. Louis Canal, which is used in place of the outlet channel of the river, which is impeded by a bar. This canal starts from the left bank of the river, 4 miles above the mouth of the river; it is a little over 2 miles long, and has a bottom width across the land of $98\frac{1}{2}$ feet, rapidly widening out in traversing the foreshore to 656 feet at its termination in the sea, and a depth at the lowest sea-level of $19\frac{2}{3}$ feet; and it is shut off from the turbid river current by a lock at its upper end, pointing down-stream.

The configuration and nature of the ground affect so much the excavations required for ship-canals, the physical conditions at their entrances determine so largely the extent and cost of the harbour works which may be needed, and the incidental works vary so greatly with the special circumstances of the locality traversed by the canal, that it would be impossible to compare the various ship-canals. The works also, as a whole, for ship-canals are of such magnitude that they cannot be profitably undertaken, except where the conditions are specially favourable, or the prospects of traffic very large. Accordingly, though several schemes for further ship-canals have been brought forward from time to time, and some of them may eventually be carried out, such works must necessarily always remain somewhat limited both in number and extent, while the increasing demands made by recent advances in Naval Architecture in the length, beam and draft both of ships of war and of the mercantile marine will probably require in future dimensions of waterway largely exceeding, as in the case of the Panama Canal, those hitherto deemed sufficient.

CHAPTER XXVI.

IRRIGATION WORKS.

Objects of irrigation works—Sources of irrigation water—Tanks collecting rainfall, construction in India, merits and disadvantages—Irrigation reservoir dams, sites and general features of works, earthen and masonry dams, conditions of choice, description of Periyar concrete dam and subsidiary works, general methods of construction of masonry dams—Masonry reservoir dam with flood-discharge sluices, across River Nile at Assuan, description, storage provided for, importance for Egypt, inadequacy of reduced storage, prospect of reduction of reservoir by deposit of silt—Storage of water in lakes for irrigation, advantages, necessary supply for Egypt available, proposed increase of height of Assuan Dam—Water for irrigation drawn from rivers, distributed by canals · Inundation canals, their principles, arrangements with basins in Egypt, in the Punjab and Sind, dimensions, value—Perennial canals, their advantages, two types, important points in design—Formula of discharge for irrigation canals—Upper perennial canals, works required, fall, inclinations given to canals, dimensions and discharges, examples of aqueducts and siphons, lengths of some of these canals with their branches—Head-works of upper perennial canals, construction of weirs and examples, site for head and regulating sluices—Remarks on upper perennial canals—Deltaic perennial canals, arrangement, length of weirs, Nile sluices, Asyut Barrage, Zifta Barrage, Esneh Barrage—Remarks on deltaic canals, merits, objections.

IRRIGATION works are needed in hot, dry regions to supply water to lands where the rainfall is deficient, very variable in amount, or of very short duration, in order to enable crops to be grown, to prevent their failure in years of drought, and to allow agriculture to be carried on during the dry season.

Sources of Irrigation Water.—The water required for irrigation may be obtained from the underground waters, often met with only a moderate depth below the surface, by sinking wells to the water-bearing strata, from which the water is drawn up, either by manual labour or with the aid of oxen, through the medium of buckets, water wheels, or other simple contrivances, for irrigating the adjoining land; for steam-power would be too costly for such a purpose, and the water thus raised cannot be economically conveyed to a distance. Water, indeed, for irrigation differs from water for the supply of towns, in requiring to be obtained at a cheaper rate to be profitably employed on land; whilst silt deposited from irrigation water is beneficial for the land.

Where the rainfall is heavy but of short duration, it may be collected in large tanks; and the flow of streams which fail in the dry season may be stored up in reservoirs, formed by a dam placed across the valley of the stream at a suitable site.

The most abundant supply, however, of water for irrigation is derived from large rivers, by means of canals which either draw off the water in flood-time from rivers having a very variable flow, or procure a constant supply from the upper part of rivers possessing a perennial flow, which they distribute over the land by gravitation to considerable distances, or from the head of the delta of a river for irrigating the low-lying lands constituting the delta.

Tanks collecting Rainfall for Irrigation.—Tanks are readily formed by enclosing a natural depression with an earthwork embankment, in which a heavy rainfall is collected as it falls; or sometimes a series of these embankments, placed across a valley at intervals, draw the flow of rain off the ground from a larger area, forming a succession of small reservoirs. These tanks have been very extensively used in Mysore and Madras, and also to a smaller extent in Bengal, for collecting water for irrigation, varying in area from a few acres up to several square miles. The water thus stored up in the rainy season is drawn off for irrigation in the dry season, through one or more sluice openings built in masonry in the embankment; and a waste weir, or an escape side channel is provided for the discharge of the surplus water, to secure the enclosing embankment from being overtopped by the water, which would result in a breach and the loss of the contents of the tank.

These tanks were formed from remote times by the natives of India, being simple and cheap in construction, and efficient where there is a considerable rainfall in a short period; but they are subject to considerable loss of water from evaporation during hot, dry weather; and they are very liable to have their capacity greatly reduced by the deposit of silt from the water flowing off the land into them, which can only be partially remedied by scouring through the sluices on the first filling of the tank. Sometimes, as a last resort, crops are grown in the fertile alluvium of a silted-up tank.

Reservoir Dams for storing up Irrigation Water.—Large volumes of water can be stored up for irrigation by erecting an earthen or masonry dam across the lower part of the valley of a stream or river, arresting the flow of water during floods till the water has filled the reservoir space above the dam, the volume of water thus stored up depending upon the height of the dam and the configuration of the valley above. A narrow part of the valley is generally selected which widens out considerably higher up, so that a comparatively short dam across a gorge retains a large volume of water; and a waste weir is provided, over which the surplus water flows away harmlessly into the channel below, when the reservoir has been filled. The water in these reservoirs being deep, is much less exposed to loss from evaporation than in shallow tanks; and some of these reservoirs have been purposely formed in high regions on account of the reduced evaporation, which arrangement necessitates a longer conduit to reach the plains to be irrigated, but provides also a greater available fall in proportion to the length, owing to the increased steepness of the upper parts of valleys. When earthen dams are used, it is advisable to carry the discharge conduit in tunnel through the slope of the valley, beyond one extremity of the dam, to secure the dam from the infiltration of water under pressure at its base, which is liable to occur when the discharge culvert is carried across at the lowest point under the bottom of the dam, endangering its stability; but the discharge may be effected through a masonry dam, though even in this case, a discharge at the side, away from the dam, is preferable.

Several reservoirs have been formed by earthen and masonry dams in the hilly districts of the province of Bombay, for irrigating tracts having a small rainfall, as being the only method available on account of the variable flow of the rivers in those regions. Earthen dams are ordinarily used for moderate heights, and in places where compact rock is not reached at a moderate depth below the surface; and masonry dams are adopted for considerable heights, where a solid rock foundation can be obtained. Further details about the construction of reservoir dams, and the pressures on masonry dams and their forms, will be given in Chapter XXXII., on water-supply for towns.

A concrete dam faced with rubble masonry, 178 feet high, and retaining a head of water, with the reservoir full, of 162 feet, was constructed across the bottom part of the valley of the River Periyar in 1888–96 (Fig. 271, p. 424), with the object of diverting the flow of the river from its own valley in the rainy district of Travancore, on the western side of the Ghats, through the narrow dividing ridge, into the valley of the River Vaigai, for irrigating the dry district of Madura, on the eastern side of the mountain range which intercepts the rain coming from the Arabian Sea.[1] The dam has also formed a reservoir with an area, when full, of 6405 acres, from which the water can be drawn down 31 feet to the level of the sill of the diversion, affording a volume of 252 million cubic yards for supplementing the discharge from the river at its low stage, which has a flow varying from 15 million up to 450 million cubic yards in a month. The dam, accordingly, in this case, enabling the flow of the river to be utilized for irrigation, so that the upper layer of water impounded in the reservoir forms only an auxiliary supply, will permanently maintain a depth of water in the reservoir of 131 feet; but though this depth is more than four times the depth of the layer which can be utilized, the volume of water retained in the reservoir is 12 million cubic yards less than that available for drawing off, owing to the notable decrease of every reservoir in area towards the bottom. The water is led from the reservoir in an open cutting, 21 feet wide, to the tunnel, rather over a mile in length, piercing the rocky ridge separating the river basins; and the tunnel, having a sectional area of 90 square feet, has been given a fall of 1 in 75 for discharging the flow of the Periyar into a tributary of the Vaigai, from which river it is directed by a weir into the distributing

[1] *Proceedings Inst. C.E.*, vol. cxxviii. p. 140.

channels for irrigation. A waste weir, 420 feet long, has been formed across a depression on the right bank of the Periyar valley, but separated from the dam by a ridge of rock, with its crest 162 feet above the bottom of the reservoir, and 11 feet below the top of the solid dam, over which the surplus waters are discharged during a high flood, which it is estimated might raise the level of the reservoir 9 feet.

In the construction of this kind of dam, all loose materials have to

IRRIGATION RESERVOIR DAMS.

Fig. 271.—Periyar Dam. Fig. 272.—Assuan Dam.

NOTE.—In connection with the raising of the dam, provision is also made for additional thickness, and for further protection of the apron.

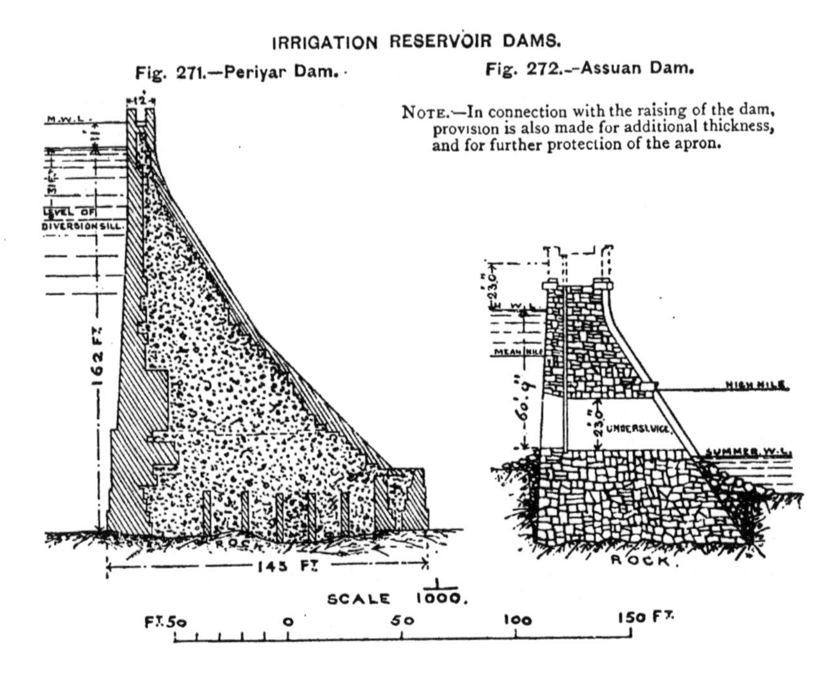

be removed from the base, and the foundations laid upon the solid rock, so as to avoid all chance of infiltration of water, subjected to a high pressure with a full reservoir, under the base of the dam. The work has to be carried on in stages whilst the river is low; and the flow has to be temporarily diverted from the foundations, which are sometimes further protected from the influx of water by enclosing temporary dams; whilst the water coming into the foundations has to be removed by pumping. The foundations may advantageously be laid in sections, provided the sections are eventually thoroughly connected together, so as to avoid having to keep a large extent of excavation free from water. Unless the flow of the river can be discharged through a tunnel at the side, beyond the dam, as the work proceeds, the flow may be passed over the dam at suitable places kept low for a time whilst the other parts are being raised, or through permanent sluices formed in the dam, and the masonry or concrete then gradually carried up. Special care has to be taken with high dams to ensure the construction of a watertight mass, whether of masonry or concrete, for some distance into the dam from the inner face, to avoid the chance of percolation of water through the dam under the high pressure towards the base, which is particularly liable to occur along horizontal joints of masonry, and which is so noticeable on the outer face of the very thick Gileppe masonry dam in Belgium retaining a head of water of 147¾ feet.

Masonry Reservoir Dam with Flood-Discharge Sluices.— Reservoir dams for storing up water for irrigation or water-supply, are commonly constructed across the narrow, mountainous gorge of a small river or stream, whose waters during floods gradually fill the reservoir thus formed above the dam; and the moderate volume of surplus water in wet years, passes over the waste weirs. When, however, a dam is constructed across the channel of a large river, such as the Nile, to store up a portion of the flow towards the close of the flood season, as designed to be accomplished by the masonry dam constructed at Assuan, provision has to be made for the discharge of the maximum flood through numerous openings near the bottom of the high dam, which are only closed when the flood has begun to abate, and the river, having fallen considerably, is carrying along comparatively little silt (Fig. 272). The Assuan dam stretches across the Nile at the first cataract, and now retains a maximum head of water of 65⅔ feet, rising itself 10 feet above the highest water-level of the reservoir. It is pierced by one hundred and forty under sluices 23 feet high and 6½ feet wide, and forty upper sluices of the same width, but only half the height.[1] The sixty-five lowest sluices, situated in the deeper channels, have their sills about 3¼ feet above the lowest water-level on the down-stream side of the dam (Fig. 272); and the other under sluices are built at a uniform level, with their sills about 16 feet higher up. The sluices are arranged in groups of ten, separated by piers 16½ feet wide; whilst there is a thickness of 33 feet of masonry between each group. All the under sluices have been closed by lifting gates sliding on free rollers. The foundations have been carried down into the solid granite rock forming the bed of the river; and the total length of the dam along the top is 6398 feet. The navigation of the river is provided for at the dam, by a channel excavated along the left bank, with locks for surmounting the fall at the dam. The volume of water retained by the Assuan dam, available for increasing the flow of the Nile below the dam during its lowest stage, after deducting the estimated loss from evaporation of a depth of 3¼ feet over the whole surface of the reservoir, is estimated at about 990 million cubic metres, or 1295 million cubic yards.

It is most important that sufficient water should be stored to ensure that the flow of the Nile at Assuan shall never supply less than 500 cubic metres (654 cubic yards) per second at the canal heads for distribution, so that all the lands in Upper Egypt intended to be under perennial irrigation, may receive with certainty the requisite supply of water throughout the summer, even in years when the discharge of the

[1] *Proceedings Inst. C.E.*, vol. clii.

Nile falls very low; for a failure of water ruins the summer crops, causing great loss to the cultivators, and leading to the lands being left uncultivated.

The Nile begins to rise in June at Assuan, and reaches its maximum height in September, and then soon begins to fall somewhat rapidly till the end of the year, and subsequently more slowly, till it drops to its lowest level at the end of May or early in June. The discharge of the river, accordingly, attains a maximum in September, and falls to a minimum in June, the discharge having reached 13,200 cubic metres per second in a maximum year, 1878–79, and fallen to 210 cubic metres per second in a minimum year, 1877–78,[1] in which year the flow was below 500 cubic metres per second for sixteen weeks (Fig. 273).

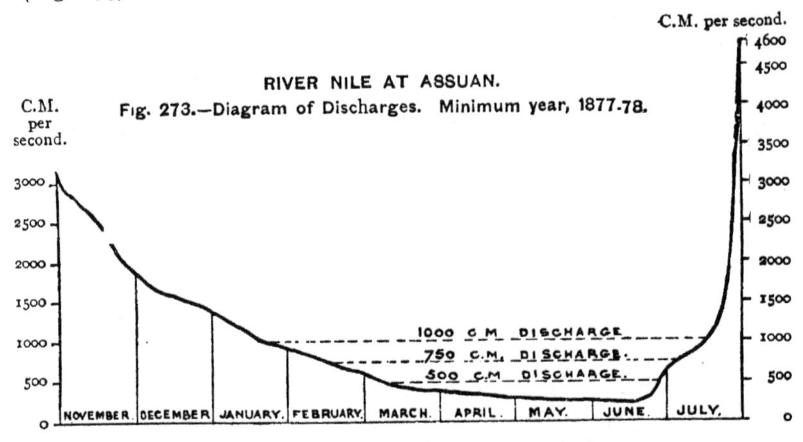

Fig. 273.—Diagram of Discharges. Minimum year, 1877-78.

The first suggestion was that a storage of 4000 million cubic metres should be provided, so as to secure an additional supply of 500 cubic metres per second being always furnished at the canal heads at Assiout for distribution, in addition to the average supply of 250 cubic metres per second from the river during the critical seventy-five days in a bad year at the same place, in order to ensure perennial irrigation for all the lands of Upper Egypt on which summer crops could be advantageously grown. This storage was reduced in the original scheme for a dam at Assuan, to 3700 million cubic metres, with the reservoir level 105 feet above the lowest water-level below the dam; and the Commission appointed to consider the project, by lowering the proposed water-level of the reservoir $13\frac{1}{8}$ feet, reduced the proposed storage to about 2660 million cubic metres. Eventually, in order to avoid submerging the temple of Philæ, the water-level of the reservoir was lowered $26\frac{1}{4}$ feet more, down to the present modified level, reducing the storage to little more than one-fourth the amount at first suggested, and securing an additional flow at the canal heads of only 120 cubic metres per second

1 "Egyptian Irrigation," W. Willcocks, 2nd edition, 1899, pp. 46 and 47.

during the critical period, or 370 cubic metres per second including the river flow in low years, in place of the 500 cubic metres per second essential for securing the lands at present under perennial irrigation from drought, and instead of the 750 cubic metres per second at first proposed to place summer irrigation in Upper Egypt in a satisfactory position. Moreover, there is a probability that the reservoir capacity provided by the Assuan dam, will be gradually reduced by deposit of silt in still water at the sides, beyond the influence of the scouring current through the sluices, which deposit will be further promoted by the checking of the river current near the upper end of the reservoir, when flowing into the inert mass of water already in the reservoir during the latter stage of its filling.

Storage of Water in Lakes for Irrigation.—The inadequacy of the storage at Assuan, in consequence of the reduction in height to which the reservoir has been retained, affecting the widespread upper layers, and consequently diminishing the capacity of the reservoir out of all proportion to the diminution in height, seems to suggest the provision of a supplemental storage elsewhere to satisfy the urgent needs of Egypt. Lakes possess two very important advantages for storing water, over wholly artificial reservoirs formed in a river valley, namely, that the lowest layers of water stored by damming up a lake, amount to a considerable volume spread over a wide area in a large lake, in place of the insignificant amount of water contained in the narrow bottom of an artificial reservoir; and that a lake furnishes a reservoir in which silt can accumulate for a long period without affecting the storage capacity. Moreover, in the case of water for irrigation, the river itself provides a channel for the conveyance of the supply during the dry season, from a reservoir formed in the upper part of its valley. Lakes exist near the heads of both the White and Blue Niles; but though lakes Albert and Victoria Nyanza, at the head of the White Nile, afford a very ample space for storage with their large areas, the White Nile constitutes a very unsuitable channel for the conveyance of the water down to Khartoum, owing to its very small fall in places, and to its passing through extensive swamps where the flow is checked and the water dispersed. Lake Dembea or Tsana, however, near the source of the Blue Nile in the high, rainy district of Abyssinia, though considerably smaller in area than the equatorial lakes, could be converted by the aid of a regulating weir at the outlet into a reservoir, the storage capacity of which in an average year would be three thousand million cubic metres.[1] Political reasons appear to militate against this scheme, and it is now proposed to increase the capacity of the Assuan Reservoir by raising the dam and water-level 7 metres (Fig. 272). Further schemes for the control of the discharge of the Upper Nile have also been suggested.

Water for Irrigation drawn from Rivers.—Water may be drawn from rivers during their floods by means of inundation canals, leading the water at a somewhat high level from the embanked river to the lands at the sides, the influx being sometimes regulated by sluice-

[1] "Report upon the Basin of the Upper Nile, with proposals for the improvement of that river," by Sir W. Garstin. Cairo, 1904.

gates, and ceasing directly the river falls. Water is also obtained for irrigation from the upper part of rivers having a perennial flow, by canals which, being given a suitable fall, convey the water for long distances to irrigate arid plains at a considerable distance from any natural watercourse; and shorter canals are constructed branching off lower down on such rivers, to irrigate lower lands away from the river. Canals, moreover, starting from the head of a delta, where the water of the river is backed up by a weir during the dry season, supply water for irrigation to the low-lying lands situated between the branches of the river traversing the delta. In the first case, the irrigation is intermittent, only taking place when the river is in flood, and depending for its extent on the height of the flood; whilst the last two methods are intended to furnish a perennial irrigation, discharging a constant supply of water throughout the dry season.

Inundation Canals.—As a silt-bearing river flowing through an alluvial plain raises its banks on each side by deposit in flood-time, and also in some cases its bed to a certain extent, its floods rise some feet above the adjacent low lands on each side. Moreover, the river follows a winding course, so that by forming canals in a straight line, branching off from the river at suitable points, it is possible to utilize the available fall to its full extent, and thus by means of longitudinal and branch canals to extend the area of inundation to lands further off from the river. The canal should start from a point where the river-bank is stable, and the flow of the river moderate and in a central channel, so that the entrance to the canal, which should be protected from scour, may be easily maintained. The bottom of the canal is placed well above the bed of the river, so that the heavier detritus brought down may not enter the canal; and the canal should be given as uniform a fall throughout as practicable, so as to ensure a uniform velocity of flow, and thus keep down the deposit of silt in the canal. The size of the canal, and its depth below the lowest flood of the river, must be regulated by the discharge required, and the fall available. The flood-waters conveyed by inundation canals are charged with fertilizing silt, which, if the flow is not checked, deposits eventually on the land; and, consequently, these waters are of much more value for agriculture than the clear waters discharged from reservoirs, or the less turbid waters drawn from rivers held up by weirs, unless the stored-up waters have a subsequent opportunity of picking up again a burden of silt, by passing down a long length of river before reaching the land to be irrigated. One of the principal aims of irrigation works, after arranging for the conveyance of the water in the most economical and efficient manner from the river to the land to be irrigated, is to prevent silt depositing in the canals, and in the case of lands like Egypt, which depend for their fertility on the yearly layer of silt, to cause the current to carry its burden of silt on to the land. This result can only be secured by keeping out the heavier silt brought down by the river, which the weaker current of the canal could not carry along, by avoiding any checking of the current in the canal, and by so laying out the canal as to ensure as uniform a velocity of flow as possible. Numbers of irrigation canals have to be annually cleared, at a considerable cost, by excavation or dredging, from the deposit of silt, which would have been valuable if spread over the irrigated lands.

In Upper Egypt, most of the cultivated land bordering the Nile is divided into extensive basins enclosed by earthen embankments, which are filled by turbid water from inundation canals to a depth of about 3 feet during the height of the flood, which is retained on the land for about forty days, when, having deposited a layer of slimy, red mud, it is allowed to flow back again into the river; and the crops from the seed sown in November are reaped the following March. The extent, however, of the irrigation which can be effected by water drawn fairly direct from the river, varies greatly with the rise of the flood, which in the Nile in Upper Egypt has an extreme range of about 9 feet between the lowest and highest known floods. Moreover, summer crops are grown in some parts of Upper Egypt, particularly on the raised ground bordering the river and the higher lands of the valley; and these lands, in years when the flood rise is deficient, have to be irrigated by raising water from shallow wells tapping the underground flow, or by canals at a higher level, drawing their supply from the river some distance higher up, and enabled by their direct course to deliver the water at a higher level than the river, some distance lower down.

Large tracts of land in the almost rainless districts of the Punjab and Sind, are irrigated by numerous inundation canals drawing their waters from the Indus and its tributaries, more particularly the Sutlej and the Chenab, in flood-time (Fig. 274, p. 430) during which period these rivers bring down large quantities of silt, whereby the land on which it is spread from the canals is yearly fertilized as in Egypt.[1] The melting of the snows on the Himalayas causes the Indus to rise at Sukkur in April; its floods attain their highest level in August; and the river then falls rapidly till October. The area irrigated each year largely depends on the height attained by the flood; though when the supply from the canals is deficient, it is supplemented to some extent by raising water by a chain of buckets turned by a wheel.

The dimensions and fall of inundation canals vary considerably, the canals in Egypt having bottom widths of from 13 to over 300 feet, and falls ranging between 1 in 20,000 and 1 in 33,000, and in the Punjab, lengths of 8 to 60 miles, widths at bottom of 6 to 50 feet, and falls of 1 in 4000 to 1 in 10,000. The beds of these canals are from 1 foot to 10 feet below a low Nile flood in Egypt, and from 5 to 10 feet generally below the floods of the Indus in the Punjab and Sind.

Inundation canals are specially valuable when applied to the irrigation of valleys having a poor soil, and on which hardly any rain falls, such as Egypt and the lower valley of the Indus, and drawing their supplies from silt-laden rivers fed by mountain rains and snows; for the canals furnish water and silt, both of which are essential for the cultivation and fertility of the soil. The outlying and higher districts, however, are dependent with such a system on the height of the flood, and cannot be secured from periodical droughts.

[1] "Irrigation Works in India and Egypt," R. B. Buckley, Map of India, and p. 9.

Perennial Canals.—By drawing water from rivers possessing a fairly good flow throughout the year, irrigation canals can be given a

IRRIGATION CANALS OF INDIA.

Fig. 274.—Map.

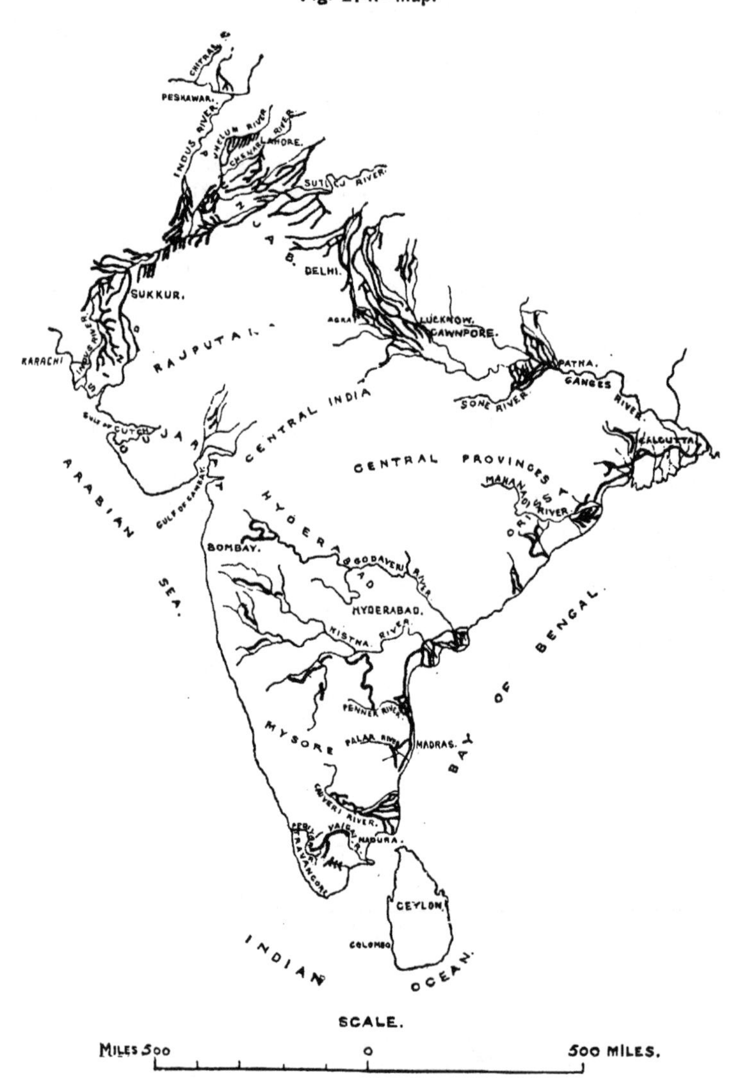

perennial discharge so as to supply arid lands with water whenever necessary, and not merely during the height of the river floods, enabling the districts served to be more fully cultivated, and with greater certainty, than by the intermittent system of inundation canals dependent on the flood rise. There are two types of perennial canals, namely, those which draw their supplies from rivers in the higher part of their course, and convey the water to the lower parts of their valleys over long distances, with the good fall generally available, away from the course of the rivers; and, secondly, those which start from a deltaic river at the head of its delta, and irrigate the low lands lying between the diverging branches of the delta (Fig. 274).

Perennial canals form offshoots from rivers, providing artificial watercourses in regions where natural ones do not exist; but instead of receiving tributaries in their course like rivers, they are eventually separated into several minor branches, which are again divided into a number of distributing channels to spread the supply of water over the land (Fig. 274).

Great care has to be taken in designing these canals, to adjust their cross section and their fall, so that they may discharge the full volume of water required to irrigate the district they command, and that their current may not be rapid enough to erode their bed, or so sluggish as to favour the growth of weeds and the deposit of silt. Where the available fall is large, the channel when traversing rock may be reduced in section; but in soft soil it becomes necessary to introduce vertical falls at suitable points to avoid creating an undue current, as adopted on the Ganges Canal.

Formula of Discharge for Irrigation Canals.—The most reliable formula for calculating the discharge in an open channel is in the form, $D = AV = AF\sqrt{RS}$, where D is the discharge in cubic feet per second, A the area of the cross section of the water in square feet, V the mean velocity of flow in feet per second, F a factor deduced from experimental investigations, R the hydraulic radius, or hydraulic mean depth, in feet, and S the slope, or the fall in a given distance divided by that distance. The hydraulic radius of any channel is obtained by dividing the cross section A of the stream flowing in the channel by the wetted perimeter of the channel, or, in other words, is the depth of a rectangle which, with a length equal to the wetted bed across the stream, has a sectional area equal to that of the stream. Various values have been assigned to F by different hydraulicians, according to the size and conditions of the streams whose velocities have been gauged; but the only general formula which embraces the flow of small streams and large rivers, assigns the following value to F in the case of channels in earth in good order,

$$F = \frac{72{\cdot}44 + 41{\cdot}6 + \frac{0{\cdot}00281}{S}}{1 + \left(41{\cdot}6 + \frac{0{\cdot}00281}{S}\right)\frac{0{\cdot}025}{\sqrt{R}}},$$ in which, for small channels with a good fall, the factor $\frac{0{\cdot}00281}{S}$ might be omitted.

Upper Perennial Canals.—The works for these canals consist, firstly, of the main canal conveying the water from the upper part of a large river to the district to be irrigated, together with its branches and minor channels for distribution in this district; and, secondly, the head-works at the intake of the canal, comprising a weir across the river a little below the entrance of the canal, to raise the water in the river during its low stage in front of the intake, so as to ensure an adequate flow into the canal, and regulating sluice-gates across the intake to control the influx into the canal, and to exclude high floods passing down the river, which, if admitted into the canal, would injure, and might break its banks, and introduce large quantities of silt which would be deposited in the canal as the initial velocity of influx decreased.

The works for the main canal resemble those for forming the channel of a navigation canal, with the exception that the bed of the canal must be given a fall to provide for the flow of its waters, and that siphons are sometimes used, in place of embankments or aqueducts, for crossing valleys, and also for passing under streams. Though upper perennial canals have been sometimes adapted for navigation, especially those passing through flat districts, the requirements of irrigation differ from those of navigation. Thus in irrigation canals, it is expedient to make as much use as practicable of the available fall, since for a given discharge the cross section of the channel varies inversely with the fall; it is advisable to give the current a velocity of not less than 1½ to 3 feet per second, according to the nature of the soil in which the channel is excavated, so as to keep the canal free from weeds and the deposit of silt; and the canal should be reduced in section as the minor channels branch off from it. The average fall given to these canals, except in tunnels and aqueducts where the fall is often increased to reduce the size of the conduit, ranges from 1 in 10,560, or 6 inches in a mile, on the Lower Ganges, Sone, and Agra canals, 1 in 8000 on the Sirhind Canal, or 1 in 5280 on the Bear River Canal, Utah, and the Turlock Canal, California, 1 in 4000 on the Cavour Canal, 1 in 3333 on the Marseilles Canal, up to 1 in 2640 on the Arizona and Idaho canals, and 1 in 2112 on the Del Norte Canal, in the United States. The Ganges Canal, with a bottom width of 172 feet, discharges a maximum volume of 6800 cubic feet per second; the Sirhind Canal, with a width of 200 feet, and the Sone Canals with a width of 180 feet, discharge 6000 cubic feet; the Lower Ganges Canal, with a width of 216 feet, and a depth of water of 8 feet, discharges 5100 cubic feet; the Cavour Canal, with a width of 65 feet, discharges 3585 cubic feet; the Del Norte Canal, with a width of 60 feet, and a depth of water of 5½ feet, discharges 2000 cubic feet; the Turlock Canal, with a width of 70 feet, and a depth of 7½ feet, discharges 1500 cubic feet; and the Arizona and Bear River canals, with bottom widths of 36 and 50 feet, and depths of water of 7½ and 7 feet respectively, discharge 1000 cubic feet per second.

Instances of masonry aqueducts carrying irrigation canals across river valleys are furnished by the Solani aqueduct on the Ganges Canal,

920 feet long and 24 feet high, with fifteen arches of 50 feet span; the Kali Nadi aqueduct on the Lower Ganges Canal, with fifteen arches of 60 feet span; the Dora Baltea aqueduct on the Cavour Canal, 635 feet long, with nine arches of $52\frac{1}{2}$ feet span; and the Roquefavour aqueduct over the Arc on the Marseilles Canal, 1253 feet long and 271 feet maximum height, with three tiers of arches of $52\frac{1}{2}$ feet span. The Cavour Canal is also carried under some rivers in siphon culverts; whilst siphons were adopted for conveying the discharge of the Verdon Canal across several valleys, being for the most part tunnelled through the rock underlying the valley, and lined with masonry; but in two instances, where the rock was unsound, wrought-iron tubes were used. In one case, a tube, $7\frac{1}{2}$ feet in diameter, was adopted for the central portion only of the siphon across the bottom of the valley; and in crossing the valley of St. Paul, with a dip of 118 feet below the canal level, the siphon is composed of two tubes, each $5\frac{3}{4}$ feet in diameter, having a hydraulic gradient between their extremities of 1 in 990, and a length of 851 feet. Wooden troughs, or flumes as they are termed, supported on trestles, are very commonly employed for conveying the water of irrigation canals in the United States across valleys. Tunnels also are often employed in the rugged country sometimes met with in the upper valley of a river, for carrying the irrigation water through ridges and spurs; whilst such hilly country is traversed by the Marseilles Canal in its course from the River Durance to Marseilles, that it is in tunnel for nearly one-fifth of its length of $51\frac{1}{2}$ miles.

The largest examples of these canals are the Ganges Canal, with 440 miles of main and branch canals, and 2500 miles of distributing channels, irrigating 1,600,000 acres (Fig. 274, p. 430); the Lower Ganges Canal, with 560 miles of main and branch canals, and 2100 miles of distributing channels, commanding an area of about 1,190,000 acres, only partially irrigated at present; the Sirhind Canal, drawing its supply from the Sutlej, with 544 miles of main and branch canals, and 4400 miles of distributaries, irrigating 800,000 acres in the Punjab; and the Sone Canals, 370 miles long, commanding 1,016,000 acres, but only irrigating 370,000 acres, with 1200 miles of distributaries. The Cavour Canal, 53 miles long, drawing its supply from the Po, can irrigate 490,000 acres; and the Idaho Canal in the United States can irrigate 350,000 acres with water drawn from the Boise River.

Head-works of Upper Perennial Canals.—A solid weir of masonry, rubble stone, or even gravel, according to the nature of the bed and materials available, protected in the two latter cases by masonry or pitching on the top, and secured from a flow of water underneath, in a permeable river-bed, by sinking two or more rows of wells into the bed across the river, or by the weight and width of the weir, is constructed across the river just below the entrance to the canal, to raise the water-level during the low stage of the river for the influx of water into the canal; whilst floods pass over this obstruction with only a moderate elevation of their level, as indicated by a small drop of the surface of the river on passing the crest of the weir, indicated on the section of the weir at Dehree across the River Sone at the intake of the

Sone Canals (Fig. 275). This weir is $2\frac{1}{3}$ miles long across the river, and retains a head of water of about 10 feet; but there is a row of movable iron shutters all along the crest of the weir, each 18 feet long and $2\frac{1}{4}$ feet high, raising the water-level an additional 2 feet.

As a solid weir across a river bringing down detritus causes an accumulation of deposit in the river-bed above the weir, which would

RIVER WEIR FOR IRRIGATION CANALS.

Fig. 275.—Weir across River Sone at Dehree.

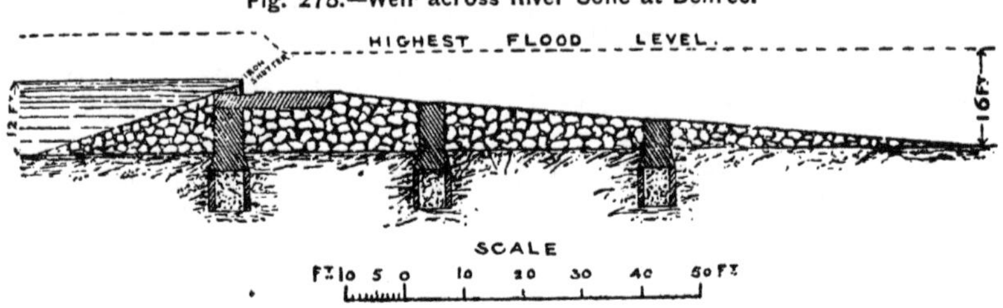

be liable to block up the entrance to the canal, sluice openings are formed through the weir near the bank, closed by draw-doors, needles, or movable shutters, the scour through which when opened keeps the river-bed in front of the entrance of the canal free from deposit. In some instances, central sluice openings have been formed with a view to keep down the deposit; but in a long weir, such openings of moderate width have only a local influence; and it has been found preferable to make the solid weir low, and raise the water-level to the requisite height by movable shutters on the top. The Sone weir is pierced by twenty openings on each bank just below the intakes of the canals, and sixteen openings in the centre of the river, each $20\frac{1}{2}$ feet wide;[1] but whilst the side openings have kept the channel clear in front of the canal head on each bank, the central openings have not prevented the silting up of the river-bed. The openings in the Sone weir are closed by hinged shutters which, having to be raised with the current flowing through the openings, are controlled by hydraulic brakes, consisting of a piston attached to the under side of each main shutter, and drawn along a cylinder filled with water and pierced with a series of holes; and as the shutter rises, the number of holes through which the water pressed upon by the piston can escape, becomes less as the piston travels along the cylinder; and, consequently, the retarding influence of the piston on the shutter increases in proportion as the shutter rises and becomes more exposed to the action of the current (Fig. 276).

The diversion weirs across rivers for supplying irrigation canals, are commonly constructed in the United States of timber cribwork filled with rubble stone; but the weir across a rocky gorge of the San Joaquin River consists of a masonry dam, 103 feet high, founded on solid rock, raising the water sufficiently to pass it through a tunnel in

[1] "Irrigation Works in India and Egypt," R. B. Buckley, p. 143.

rock, 500 feet long, for supplying the Turlock Canal in California; and masonry weirs have generally been constructed to raise rivers in Spain for supplying irrigation canals.

The head of a perennial canal should be placed where the river-bank is stable, the flow of the river uniform, and its course fairly straight, so that the entrance to the canal may not be liable, either to be eroded by the current, or silted up by a change in the river-bed. The supply of water to the canal is regulated by a set of sluices across

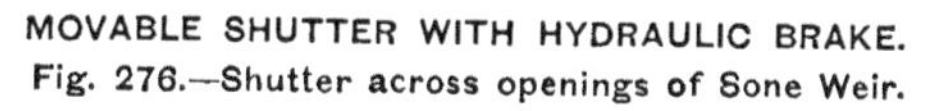

MOVABLE SHUTTER WITH HYDRAULIC BRAKE.

Fig. 276.—Shutter across openings of Sone Weir.

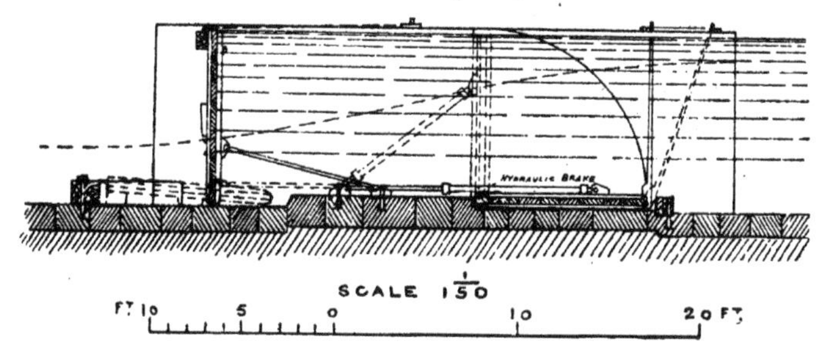

its head, divided by masonry piers supporting a bridge, from which the sluice-gates are raised or lowered, consisting generally of draw-doors sliding between grooves in the piers, but sometimes formed of movable shutters, rolling-up curtains, planks, or needles. The sills of the sluices are raised above the river-bed to exclude the heavy detritus brought down; whilst the sluice-gates, or the bridge sheltering them, are carried up above the highest flood-level to prevent the floods from pouring into the canal.

Remarks on Upper Perennial Canals.—When these canals start in the higher part of a river valley, the fall obtainable is generally ample, and the weir across the river of very moderate dimensions; but the upper part of the canal is liable to have to traverse rugged country involving costly works, the flow of the river is generally very variable, and carries down large detritus which must be excluded from the canal, and the distance to the lands to be irrigated is long. When, however, the canal begins in a lower part of the valley, like the Sone Canals, or even the Lower Ganges Canal, the considerable width of the river necessitates a long weir, the works across the head of the canal may have to be founded on alluvial soil, and the available fall is moderate; but, on the other hand, the river has a larger and less variable flow than higher up, the sediment carried along is finer and more suitable for spreading on the land, the canal works pass through a fairly flat country, and the lands to be irrigated are comparatively near.

Deltaic Perennial Canals.—The flat, alluvial lands between the branches of a river delta, only raised slightly above sea-level by the gradual deposit of the sediment of the river in the sea, are for the most part lower than the land bordering the branches, owing to the rapid deposit of silt on the banks directly the river begins to overflow; and, accordingly, there is no difficulty in irrigating these low-lying lands by raising the river at its low stage by a weir across the head of the delta, from which water is supplied to deltaic canals following the higher ground bordering the branches; and there is a sufficient fall from these canals for distributing channels to convey the water on to the lands of the delta. As a river before it reaches its delta has received the flow of all its tributaries, which, in an extensive river basin, may derive their supplies from regions subjected to a very varied rainfall, the river generally has a fair discharge at the head of its delta; and as a large river has usually attained a considerable width towards the termination of its course, a long weir is required across the head of its delta, reaching $2\frac{1}{2}$ miles at the Godaveri delta, $1\frac{1}{4}$ miles at the Mahanadi delta, and 3290 feet across the two main branches of the Nile delta, or, with the connecting wall, nearly $1\frac{1}{4}$ miles.

The Godaveri and Mahanadi weirs are similar in type to the Sone weir (Fig. 275, p. 434), being solid weirs over which the floods pass with a slight drop below the crest, percolation being guarded against by two rows of masonry wells sunk down into the bed, supporting walls of masonry with rubble between, and protected on each slope, by pitching in the Mahanadi weir, but with a thick masonry floor all along the top in the Godaveri weir. The river is held up about 16 feet above its low stage by the Mahanadi weir, with hinged shutters, 3 feet high, on its crest. The Nile weirs, or sluices, consist of a high wall, or bridge, pierced by a number of openings, $16\frac{1}{2}$ feet wide, separated by piers, $6\frac{1}{2}$ feet in width, standing upon a platform of considerable width along the river (Fig. 277). There are sixty-one openings across the

IRRIGATION WEIR AT HEAD OF DELTA.

Fig. 277.—Barrages at Head of Nile Delta.

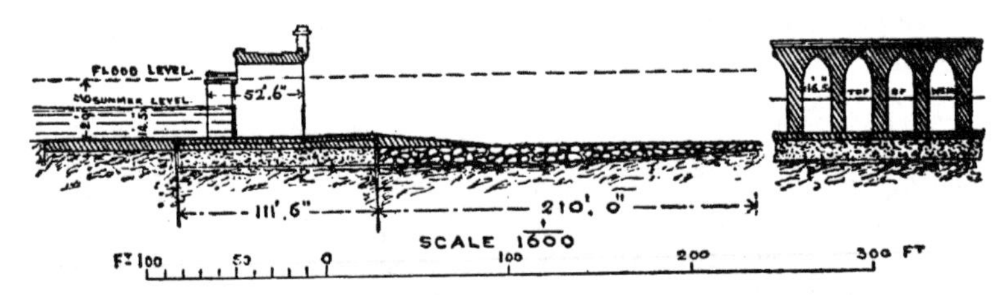

Rosetta Branch, and seventy-one across the Damietta Branch, through which the floods of the Nile pass. As the water at first escaped under the floor, and threatened to undermine the structure when an attempt was made to lower the revolving, cylindrical sluice-gates originally provided, the floor and apron of the weir have been considerably thickened and extended, and the shaken work consolidated; and the openings are now closed with lifting gates sliding on free rollers, while subsidiary weirs down stream of the Barrage have been constructed in order to reduce the head of water against the Barrage to a maximum of 3 metres. The Barrage across the Nile on Asyut[1] is situated about 1½ mile north of that town, and a short distance downstream of the entrance to the Ibrahimia Canal, one of the chief irrigation canals in Egypt, and which supplies a large portion of Middle Egypt and the Fayum with water. The Barrage is designed to hold up a head of water of about 2½ metres at summer level, while passing the full flow of the river in flood time. The total length of the barrage is 2691 feet, and the waterway consists of 111 openings of 5 metres width, the total superficial area of flood way being about 64,000 square feet. The foundations consist of two rows of cast-iron piles 87 feet apart, driven to about 23 feet below floor surface, between which a solid floor of concrete and rubble masonry about 10 feet thick is laid. The river bed is further protected for a width of 65½ feet on each side beyond the piles, making a total width of 218 feet of protected bed.

The sluice gates are of riveted girder work with rollers, working in cast-iron grooves, and are raised or lowered by travelling winches. A lock is provided on the left bank of the river for traffic.

The Zifta Barrage[2] on the Damietta Branch of the Nile is, in its main features, of similar design to that at Asyut, and serves to regulate the distribution of water to the lands in the north of the Delta.

The Esneh Barrage, situated 110 miles to the north of the Assuan dam, and between the latter and the Asyut Barrage, is also of a similar type to that at Asyut. This barrage, 2868 feet long, with 120 sluice openings, and carrying a roadway 20 feet in width across the Nile, was completed and opened in 1909.

The regulating sluices at the head of deltaic canals are similar to those constructed for upper perennial canals; and the canals themselves present no difficulty in construction, being formed throughout in flat land. Owing to the small fall available, these canals are not liable to be eroded by the current; but with the small velocity of their flow, they are exposed to accumulations of silt. Perennial irrigation of deltaic lands necessitates their protection from floods, by embanking the branches of the delta; but this involves the loss of the fertilizing silt carried by the river over the flooded lands, which cannot be compensated for by the comparatively clear water carried on to the lands by the canals drawing their supply from the dammed-up river during its low stage.

The canals of the Cauveri delta (Fig. 274, p. 430) can irrigate 919,500 acres; those of the Godaveri delta, with a maximum discharge of 8500 cubic feet per second, 612,000 acres; those of the Kistna delta, with a discharge of 8100 cubic feet per second, 475,000 acres; and the Orissa Canals of the Mahanadi delta, with a discharge of 6060 cubic feet per second, can irrigate about 400,000 acres. The canals of the Nile delta, which, since the repairs of the floors of the weirs, have been enabled to discharge 13,420 cubic feet per second in an ordinary year,

[1] *Proceedings, Inst. C.E.*, vol. clviii.
[2] *Ibid.*, vol. clvi.

but whose discharge might fall to 7060 cubic feet per second in a minimum year, can irrigate an area for summer crops of 1,520,000 acres, the total cultivated land in the delta reaching 3,430,000 acres, and the area capable of cultivation by inundation canals and basins amounting to 3,930,000 acres.

Remarks on Deltaic Perennial Canals.—The long weir at the head of a delta, and the regulating works for controlling the flow in the canals, all of which have to be founded on alluvial soil, are the only works which present any difficulty in providing for the perennial irrigation of deltas. Efficient drainage, however, of the irrigated land is also required for certain crops, and to prevent the accumulation of salt deposits on the surface of the land; and it is not always easy to find a good outlet for this drainage from the low-lying lands.

Deltaic irrigation works have yielded very large returns in some cases in India, as exemplified by the revenues derived from the Cauveri, Kistna, and Godaveri deltaic canals. Nevertheless, there remains the serious objection to deltaic perennial irrigation, where the land, as in Egypt, requires periodical renovation by the deposit of the fertilizing mud brought down by the river, that a considerable portion of this silt is excluded by the embankments for protecting the summer crops; whilst the land is more rapidly exhausted by raising summer as well as winter crops. Moreover, the silt excluded from the formerly flooded area by the embankments, is confined within the embanked channel, and either deposits in the channel, raising its bed and contracting its section, or being carried forward to the outlet, obstructs the outflow of the river, and hastens the advance of the delta in front of the branches, thereby further impeding the drainage of the protected low-lying lands. In view of the gradual deterioration of the land of Lower Egypt, and the loss that must be entailed in the cultivation of summer crops, from the failure of an adequate supply in years when the discharge of the Nile falls to a minimum, it would have been better to retain the old system of basin irrigation in the Nile delta, which was so successful in former times with its yearly renewal of the fertile layer of silt, and to have restricted summer irrigation to the higher lands rarely reached by the yearly inundation.

PART IV.

DOCK WORKS; AND MARITIME ENGINEERING.

CHAPTER XXVII.

RIVER QUAYS, BASINS, AND DOCKS.

Suitable sites for docks; jetty ports; seacoast ports; quays—River quays, instances—Foundations for quay walls alongside rivers, well-foundations on Clyde, compressed-air foundations at Antwerp, piles for Rouen and New York quays, large blocks for Dublin quays, detached piers connected by arches—Basins and quays on seacoast, concrete blocks in courses, in sloping rows—Remarks on river and maritime quays, relative merits of different systems of quay walls, precautions in backing up, depth obtained for vessels; Liverpool landing-stage, compared with quays of docks—General arrangements of basins and docks, definition of terms "basin" and "dock;" objects of the arrangements, at Southampton, Marseilles, Albert and Victoria Docks, London, at old ports; main and branch docks—Approaches to basins and docks, requirements, instances, naturally sheltered, sheltered by breakwaters, jetty channel—Depth of basins and docks, dependent on depth of approach channel, fixed by sill of entrance and height of neap tides—Excavation of basins and docks, arrangements—Foundations for dock walls, borings, concrete, piling, wells—Pressures against dock walls, defined, influence on walls, counteracting resistances—Provisions against movement of dock walls described—Forms of basin and dock walls, batter on face, widening out at back downwards in steps, height, proportion of widths to heights—Construction of dock walls, materials, various requirements—Pitched slopes and jetties for docks, objects, suitable for coal-tips.

DOCKS are most conveniently constructed on low-lying land, or high foreshores, bordering a tidal river or estuary, near the deep-water channel affording access to the sea; for by this arrangement the construction is effected on land and on the reclaimed foreshore, the excavations for the dock serve to raise the land alongside for forming quays some feet above the highest tide, the entrance is in a protected situation, and a natural approach channel is provided for vessels trading with the port. Moreover, many of the principal seaport towns requiring docks for the accommodation of their traffic, are situated upon tidal rivers or estuaries, alongside which low-lying land in the neighbourhood of the port is generally found. In the absence, however, of large rivers, ports have sometimes been long ago established on the seacoast at the mouth of small rivers or creeks, affording a natural protection for small craft, the tidal flow and ebb from which maintained an outlet channel to the sea across the foreshore; and to meet the exigencies of the increased draught of vessels, these channels have been guided by jetties and deepened, at first by the scour at low tide of the discharge from sluicing basins formed in very low land and filled at high water, and subsequently in a more effectual manner by sand-pump dredgers, as illustrated by the history of such ports as Calais, Dunkirk, and Ostend, which have been furnished with docks communicating with the improved channels (Fig. 290, p. 452).[1]

In tideless seas, the delta-forming rivers flowing into them are liable to be obstructed by a bar at their mouths; and consequently ports have had to be constructed on the open seacoast, with breakwaters protecting the basins formed along the shore for the accommodation of vessels, of which Marseilles, the first port of France, furnishes a notable example (Fig. 283, p. 445). Occasionally also, where tidal rivers are shallow or exposed to accretion, a harbour is constructed on the seacoast to provide a deep, sheltered approach to the port, as adopted at Sunderland, and at Havre (Fig. 286, p. 449).

River ports are sometimes merely provided with quays along the banks, especially where the rise of tide is small, and the river perfectly sheltered; and the river in this case serves as the basin in which vessels lie, and its banks serve as the quays for the unloading and shipping of goods. Ports of this kind, with a large trade, are provided in addition with open basins alongside the river where the rise of tide is small, to increase the accommodation for vessels and the quay space for merchandise. Where, however, the range of tide at river ports is large, docks are constructed, approached through entrances or locks, whose gates retain the water in the docks during the fall of the tide in the river, so that the vessels in them may always remain afloat, and not experience great variations in level, which inconvenience shipping operations.

River Quays.—The construction of river quays is generally combined with a regulation and deepening of the river channel; and quays are usually the first works undertaken for a river port where the tidal rise is moderate. Thus quays were erected along the banks of the Clyde below Glasgow simultaneously with the improvement of the river, and long before basins were constructed; and the long line of quays carried out on the right bank of the Scheldt in front of Antwerp and above, and the quays along both banks of the Seine at Rouen, have been laid out so as to regulate the course and width of these rivers; whilst dredging has been resorted to for increasing the depth of water along their face where required.

Foundations of Quay Walls alongside Rivers.—The foundations for river quay walls in soft alluvial soil, liable to extend some depth down, where the water cannot be excluded, and the walls have occasionally to be laid right in the bed of the river, as has been the case for part of the Antwerp quays where the river had to be narrowed, necessitate special methods of construction, to ensure the stability of the walls when the river in front is deepened and loads are placed on

[1] "Harbours and Docks," L. F. Vernon-Harcourt, pp. 148, 152, and 155, and plate 1, figs. 5, 7, and 10.

the quays at the back, exposed as these walls are to a variable supporting water-pressure at their face according to the state of the tide. Well-foundations were adopted for some of the quay walls on the Clyde (Fig. 22, p. 67); and foundations excavated and laid inside caissons by aid of compressed air (Fig. 31, p. 75), have been employed for the Antwerp quay walls, built very substantially of masonry and brickwork (Fig. 278). A much lighter and more economical construction has been used at Rouen, resting upon piles driven down through the alluvial bed

RIVER QUAY WALLS.

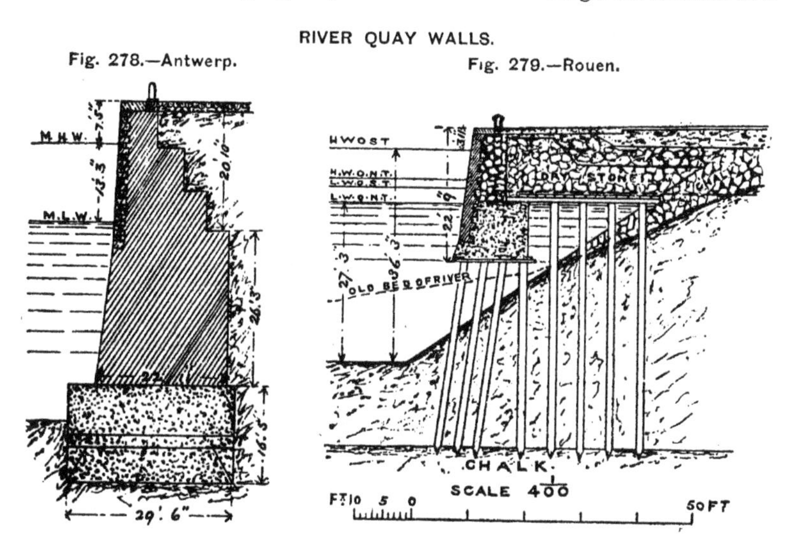

Fig. 278.—Antwerp. Fig. 279.—Rouen.

of the Seine to a chalk stratum, with a small rubble masonry quay wall on the outer face down to low water, built upon blocks of concrete, faced with brickwork like the wall, deposited within watertight timber caissons sunk on to piles cut off at the level of $9\frac{3}{4}$ feet below low water (Fig. 279). These quay walls, costing £36 6*s.* per lineal foot, inclusive of the making up of the quays for a width of 200 feet from the face, have proved perfectly stable, though the quays are liable to be loaded with goods imposing a weight of 7 tons on the square yard. The New York quays bordering the Hudson, or North River, which were founded upon a stratum of silt of such a thickness that no firm bottom could be reached, resemble the Rouen quays in general construction, with the exception that the ground round the upper part of the piles, and under the quay wall, was consolidated at the outset with small stones, deposited in a dredged trench to a depth of 30 feet below low water, through which layer the piles were driven, and were further supported both behind and in front by mounds of rubble stone down to the same depth, raised at the back above the top of the piles, and in front to the base of the quay wall;[1] but in spite of all precautions, some settlement

[1] "Harbours and Docks," p. 428, and plate 14, fig. 18.

could not be avoided on this very soft foundation. The cost of this quay wall was £49 16*s.* per lineal foot.

The lower portion of the Dublin quay wall alongside the River Liffey, up to a little above low water, has been formed by a single row of rubble concrete blocks weighing about 360 tons. These large blocks were built on a staging projecting into the water, from which, after having become thoroughly set, they were lifted one by one by a very powerful floating derrick, which transported the blocks, and deposited them in their proper position in the river at low water, on a bottom previously excavated and levelled by men in a diving-bell, in a depth of about 28 feet at low tide.[1] On the top of this row of blocks, resting on the bottom, and yet high enough to emerge out of water at low tide, a continuous masonry wall has been built up to quay-level (Fig. 280).

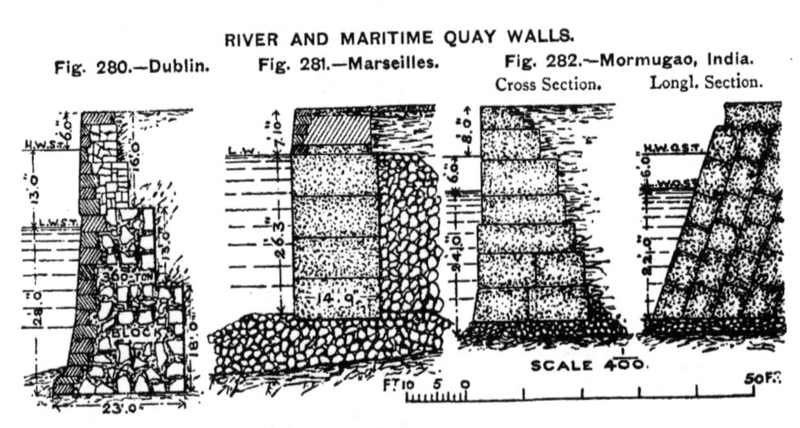

RIVER AND MARITIME QUAY WALLS.

Fig. 280.—Dublin. Fig. 281.—Marseilles. Fig. 282.—Mormugao, India. Cross Section. Longl. Section.

Occasionally, instead of making the lower part of a river quay wall continuous, piers are sunk through the alluvial bed of the river to a solid stratum at intervals of from about 26 to 40 feet, to reduce the cost of the foundations; and the intervening spaces are spanned, near low-water level, by iron girders or arches supported by the piers, upon which a continuous quay wall is constructed, the filling at the back for the quay being retained behind the apertures by a rubble mound, which, in the case of arches, forms the base for the dwarf wall at the back of the arch. Quay walls along the River Tagus at Lisbon have been constructed on this principle, with iron girders cased in masonry spanning the intervals of 26 to 33 feet between the piers, on which a quay wall, in a line with the front of the piers, has been built; whilst at the quay wall along the river Garonne at Bordeaux, masonry arches, with a rise of 6½ feet, have been adopted for bridging the intervals between the piers sunk 39⅓ feet apart, built on centering floated into position.[2] In both these cases, compressed air was employed for

[1] *Proceedings Inst. C.E.*, vol. xxxvii. p. 342, and plates 15 and 16.
[2] "Entreprises de Travaux Publics et Maritimes," H. Hersent et ses Fils, Paris, 1900, pp. 29 and 35.

carrying down the caissons, on which the piers were erected, to a suitable foundation, the caisson with its enclosure above being floated out between pontoons to its proper position, according to the system used for the Antwerp quays (p. 75); but it is evident that large masonry wells might also be applied to the same purpose under favourable conditions, though the sinking vertically of the structure cannot be effected with the same precision by this method as with compressed air.

Basins and Quays for Ports on the Seacoast.—In ports on the open seacoast, protected by an outlying breakwater parallel to the shore, such as Marseilles and Trieste, basins are formed within the shelter of the breakwater, by carrying out wide, solid jetties at right angles to the shore, lined with quay walls alongside which vessels can lie; whilst the jetties and embanked shore furnish the requisite quay space for the handling of goods, the erection of sheds, and the laying down of lines of way (Fig. 283). The quay walls surrounding these

BASINS ON SEACOAST.

Fig. 283.—Port of Marseilles.

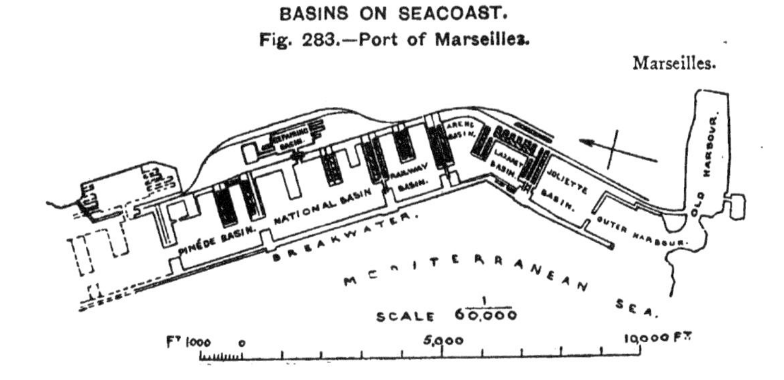

basins, like river quay walls, have to be constructed in the water; and those at Marseilles and Trieste consist for the most part of a foundation mound of rubble stone laid on the soft sea-bottom, surmounted by a wall of concrete blocks, from 11 to $14\frac{3}{4}$ feet in width, brought up to sea-level, and capped by a small masonry and concrete wall raised about 8 to 10 feet out of water up to quay-level (Fig. 281, p. 444). The most recent quay walls, however, at Marseilles, having reached a site where the bottom is rocky, have been constructed of masonry throughout by the help of compressed air, with a batter on the face and stepping out at the back, so that the width increases with the depth,[1] as in ordinary quay walls.

At the port of Mormugao, on the western coast of India, a quay wall under shelter of the breakwater protecting the harbour, has been formed with concrete blocks laid in sloping rows, on a thin layer of rubble stone, by a steam-crane travelling along the finished wall, with its jib overhanging the outer end (Fig. 282, p. 444).[2] The two quay walls shown

[1] "Ports Maritimes de la France, Marseille," Paris, 1899, vol. vii. pp. 118 and 120.
[2] *Proceedings Inst. C.E.*, vol. xcvii. plate 7.

in Figs. 281 and 282, with blocks deposited in courses in the sea on a rubble mound, resemble some breakwaters in their method of construction, but have a slighter section owing to their sheltered position.

Remarks on River and Maritime Quays.—The walls of river quays and of quays in the sea, having to be founded under water and constructed in the water up to the lowest water-level, are necessarily costly works if built in a solid manner, whether founded on wells, or by aid of compressed air, or raised above water by large concrete blocks or courses of blocks; but such structures are very durable, and well suited for ports with a large traffic, and accommodating the biggest class of ocean-going steamers. Dwarf quay walls resting upon piles and timber-work always immersed in the water, provide a quay of a more economical type, and suitable for ports where the traffic is not very large, and the steamers frequenting the port are of moderate size, or where the badness of the foundations prohibits the construction of a heavier wall. Timber wharfing on piles, and iron lattice-work jetties carried on cast-iron piles or cylinders, provide the cheapest forms of construction for accommodating river traffic, and are convenient for the initial development of a port, and for passenger traffic; but timber wharfing is subject to decay; and both kinds of structures are liable to injury from shocks, and do not furnish such suitable quays for the transhipment of goods as the arrangements previously described. At Rouen, the timber wharves originally erected alongside the river and the basins, are being gradually replaced by the dwarf quay walls supported on piles, adopted for several years past in extending the quays of the port (Fig. 279, p. 443).

The driest and most durable materials available should be put at the back of quay walls, such as rubble stone, slag, chalk, gravel, and sand, to secure them from the pressure caused by clay or silt, swelling and rendered almost fluid by the accumulation of water behind the wall, tending to push the wall forward. In spite of the weight of the Antwerp quay wall, and the stability of foundations carried out under compressed air, a portion of the recent extension above Antwerp was pushed forward a little, owing to a too rapid backing up for forming the quay behind. In consequence, the foundations of the remainder of the wall, in course of construction, have been carried down lower at the back, giving the base an upward slope towards the river; and the foreshore in front of the toe has been weighted with leaden mats, with the object of preventing the wall sliding forward into the river; whilst the backing of the wall in the river has been formed of layers of light fascines up to low-water level, overlaid with rubble stone next the wall, followed by gravel, and sand at the back, to reduce the pressure behind the wall to a minimum.[1]

Maritime quay walls must be extended into deep enough water for vessels of the largest draught frequenting the port to come alongside them; and depths of from $26\frac{1}{4}$ to 30 feet have been obtained in front of the most recently constructed quay walls at Marseilles. The depth also

[1] "Entreprises de Travaux Publics et Maritimes," H. Hersent et ses Fils, Paris, p. 58.

in front of tidal river quays should, if possible, be made sufficient for vessels to be able to come alongside and remain afloat at the lowest tides, which has been accomplished at Antwerp, Rouen, and Dublin (Figs. 278, 279, p. 443, and 280, p. 444); whilst the newest quays at Southampton, along the rivers Itchen and Test near their confluence give depths of 28 and 32 feet at L.W.O.S.T., obtained by dredging, and are therefore always accessible to large vessels (Fig. 284). At Liverpool,

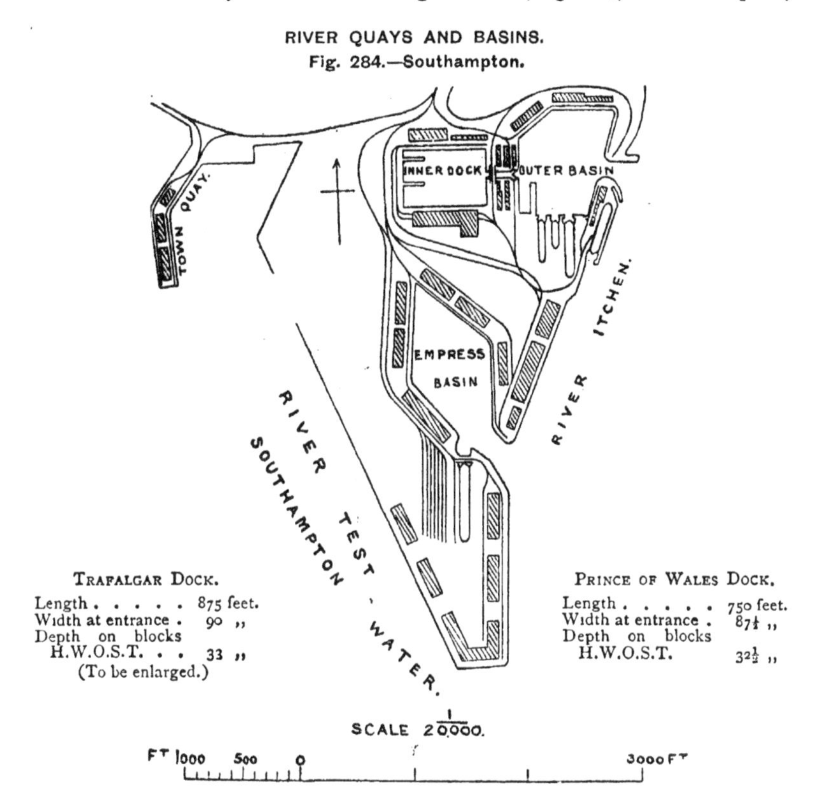

RIVER QUAYS AND BASINS.
Fig. 284.—Southampton.

where the tidal rise of 31 feet at the highest springs is too large to admit of river quays being used, the large Atlantic liners have been enabled of late years to embark and disembark their passengers and luggage in the River Mersey by means of a floating landing-stage.

Atlantic Liners have also recently found a convenient port of call at Fishguard Harbour on the north coast of Pembrokeshire, where works have been carried out which include a breakwater some 2000 feet long, sheltering an area of 500 acres, and supplemented by quays, railway accommodation, and electrical equipment.

River quays possess one important superiority over inner basins and docks, namely, the facility they afford for vessels to come alongside or to depart without having to pass through entrance channels or locks;

but this advantage is somewhat counterbalanced by the vessels being exposed to the river currents, and also to tidal oscillations which are for the most part excluded from docks. Nevertheless, very extensive discharging and loading operations are carried on with large ships along the Antwerp quays, in spite of an average tidal range of $13\frac{1}{4}$ feet. The construction, however, of quay walls alongside a river, or in the sea, involves considerably more difficult operations than the building of basin and dock walls within the shelter of the land and cofferdams, where the foundations are kept dry by pumping.

General Arrangements of Basins and Docks.—Though basins with open entrances to the sea or river are often termed docks, it would

ARRANGEMENT OF DOCKS AT A COALING PORT.

Fig. 285.—Barry Docks.

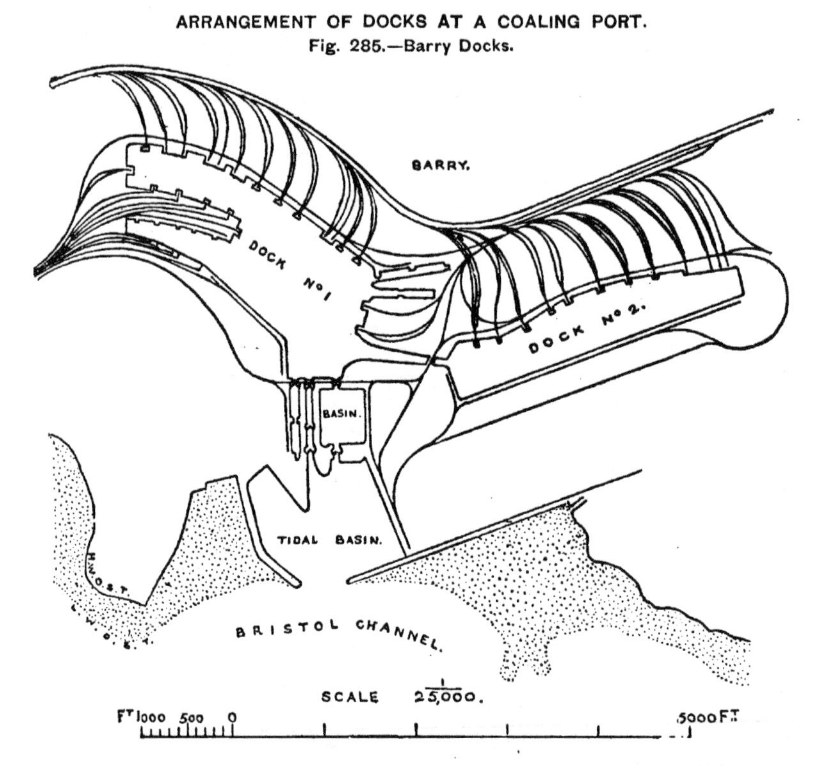

be more convenient to confine the term "dock" to those basins which are shut off by entrances or locks to maintain them at a fairly uniform level, and to employ the term "basin" for those partially enclosed areas of water which are approached by open entrances, and are subject to whatever variations in the water-level may occur outside, as well as for those outer basins, called sometimes half-tide basins, which are often interposed between a dock and an outer entrance or lock, whose water-level is generally drawn down during a rising or falling tide, in order that they may serve as large locks to facilitate the exit or entrance of vessels some time before or after high water (Fig. 285).

The exact arrangement and forms of the basins and docks of a port, must depend upon the available site; but the great object to be aimed at in their design, is to obtain as long a length of quay in proportion to the water area as is compatible with the convenient access of vessels to their berths. At Southampton, this result was partly effected by the diamond shape given to the Empress Basin opening from the River Itchen; but it is more fully accomplished by the extension of deep-water quays in Southampton Water, utilizing the existing water space, and by the proposed basin designed to open out of the River Test (Fig. 284, p. 447). At Marseilles, the length of quays depends merely upon the number and length of the solid jetties carried out from the shore; whilst the open basins thus formed are made readily accessible by the open waterway left between the breakwater and the ends of the jetties, which can be reached round either extremity of the breakwater (Fig. 283, p. 445). A good length of quay can be attained by forming a long, somewhat narrow dock, along the centre of which there is sufficient room for vessels to pass between the rows of vessels lying along the

PROTECTED APPROACH TO PORT ON SEACOAST

Fig. 286.—Havre Docks and Harbour.

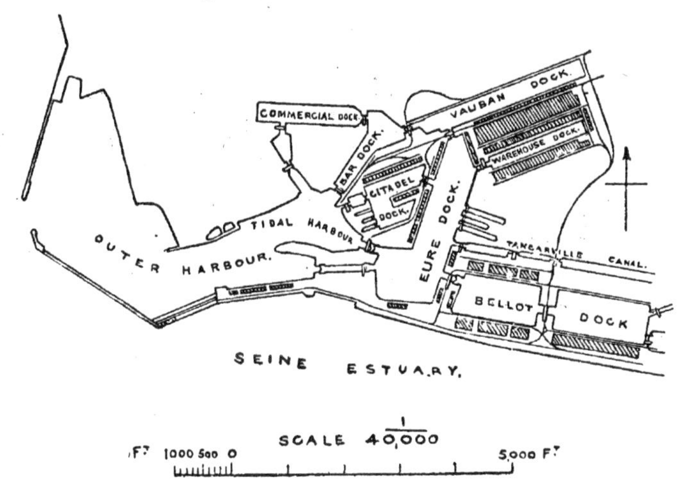

quays on each side, as exemplified by the Albert Dock, London; or by putting out a jetty into a large basin, as adopted at Keyham Dockyard Extension; both these systems are illustrated by the Barry Docks in the Bristol Channel (Fig. 285).

In old ports, where docks have been gradually extended with the growth of trade, the older docks are generally small and connected together by passages; and the newer docks are made larger on the nearest suitable site, and in some cases on land reclaimed from the foreshore of the adjacent estuary, of which Liverpool, Hull, and Havre (Fig. 286, p. 449) furnish typical instances. A very convenient form of docks consists of an outer, main dock leading to a series of narrow

ENTRANCES, HALF-TIDE BASIN, AND BRANCH DOCKS.

Fig. 287.—Liverpool Docks, Northern Portion.

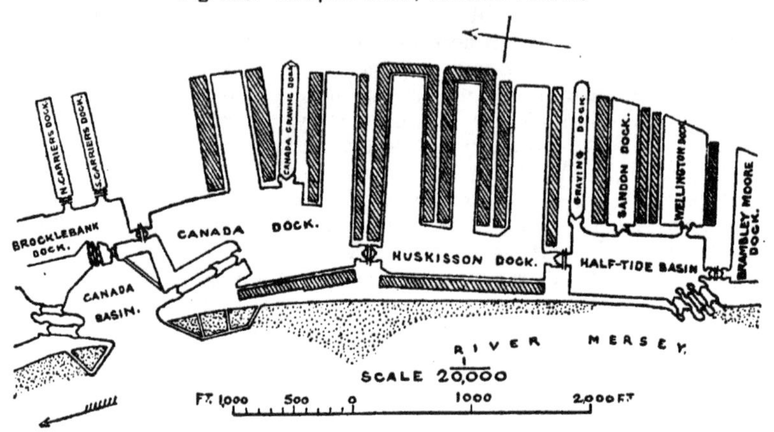

branch docks, a system adopted in several of the more recently constructed docks, such as the northern Liverpool Docks (Fig. 287), the Tilbury Docks (Fig. 288), and the Dunkirk Docks (Fig. 290, p. 452), as well as the Manchester and Salford Docks, and the basins of the Prince's Dock on the Clyde near Glasgow. This arrangement combines easy access for vessels with a long length of quay in a more perfect manner than the other systems.

Approaches to Basins and Docks.—The two requirements for an approach to a port are good shelter, and an adequate depth. Inland ports, such as Rouen, Rotterdam, Antwerp, Montreal, Glasgow, and Manchester, possess perfectly sheltered approaches to their quays, basins, and docks; but their approach channels need generally considerable improvement, and in the case of Glasgow and Manchester have had practically to be created, in the first instance by very extensive dredging, and in the second case by the construction of a ship-canal. Few ports, moreover, combine, like Southampton, excellent shelter, and a natural deep-water approach from the sea, needing comparatively little improvement to render the Southampton quays accessible at any state of the tide; whilst the channels to many ports can only be navigated near the time of high water. Thus Liverpool, though close to the seacoast, has only been rendered accessible at all times by removing large quantities of sand from the channel across the bar by suction dredgers, and by resorting to a landing-stage in the river; and in spite of the general good condition of the Thames estuary, a considerable amount of dredging is needed to render the navigable channel to the

Port of London accessible at all times to vessels of large draught; whereas ports under the conditions of Cardiff and Newport, where, with a large tidal range, the approach channels become very shallow at low water of spring tides, must always remain tidal ports, though possessing an ample depth for the largest vessels at high water.

In some ports on the open seacoast, an outer harbour provides

MAIN AND BRANCH DOCKS, TILBURY DOCKS.

Fig. 288.—Plan. Fig. 289.—Basin Wall.

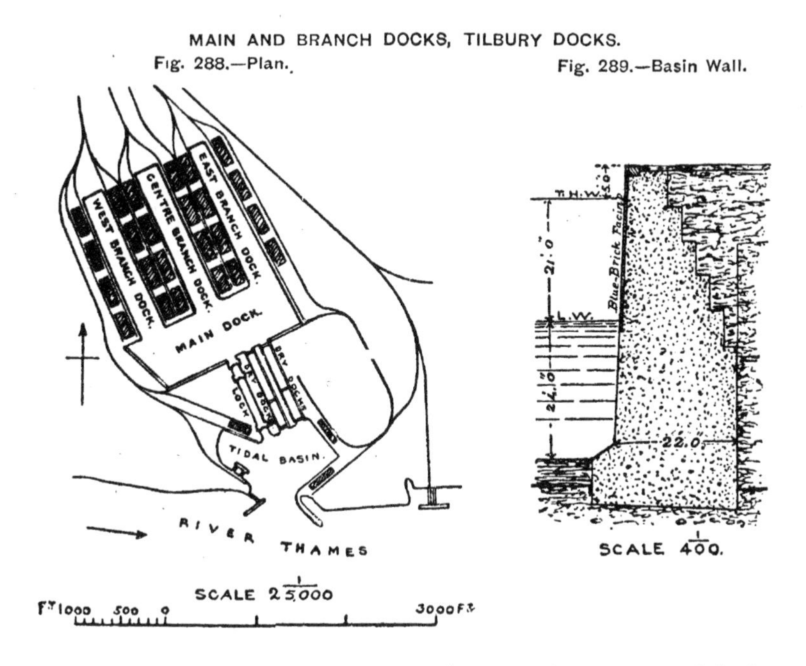

shelter for vessels, as well as a protected approach to quays and docks, of which Dover, Colombo (Fig. 305, p. 474); and Table Bay afford examples; whereas in some instances, breakwaters merely facilitate the access and improvement of the entrance to a river forming the approach channel to a port, such as the works at the mouths of the Tyne, the Tees, the Wear, and the Nervion. The Port of Havre exhibits the peculiarity of having its entrance changed from a somewhat sheltered position near the mouth of the Seine estuary, to the open seacoast between converging breakwaters, to secure it from reduction in depth by the accretion taking place in the estuary (Fig. 231, p. 382, and Fig. 286, p. 449).

Several French and Belgian ports on the Channel and North Sea coasts, have an entrance channel which is guided across the beach and foreshore by parallel jetties, generally built solid up to a little above low water, and with open timberwork above, and has been considerably deepened by suction dredging, and maintained by the same means (Fig. 290, p. 452). This approach channel opens into a tidal harbour, through which access is obtained to the well-sheltered locks leading to the inner docks. As regards the docks, these ports are practically tidal; but where a passenger and mail service exists between these ports and England, the entrance channel and berths of the mail-steamers have been gradually deepened sufficiently to enable these vessels, with their moderate draught, to enter and leave the harbour at any state of the tide. As the entrance channels to these ports are exposed to the introduction of sand drifting along the shore, or stirred up from outlying banks by waves in storms on these sandy coasts, the amount of maintenance required increases with the extension of the depth; but the outlying sandbanks serve to shelter the entrances to some extent from the worst waves, which, otherwise, being narrow and close to the shore, would be difficult of access during storms.

PORT WITH JETTY ENTRANCE.

Fig. 290.—Dunkirk Docks.

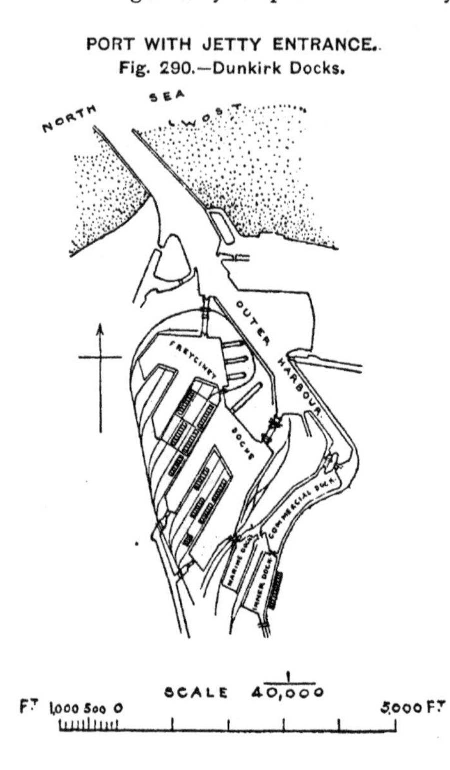

Depth of Basins and Docks.—The depth to be provided in basins and docks must depend upon the available depth in the approach channel, or the depth which there may be a prospect of obtaining by improvements. Basins may be subsequently deepened by dredging, to such an extent as may not imperil the foundations of the quay walls round them; but the level of the sill at the entrance to a dock determines the available depth of the dock. Generally, both sills of the lock to a dock are placed at the same level; but where the rise of tide is large, or the depth of the approach channel is good, the outer sill of the entrance lock may be placed at a lower level than the upper sill, to enable vessels to enter or leave the dock some time before or after high water, by being locked up or down. The depth of water on the upper sill at high water of the lowest neap tides, fixes the limit of draught for vessels using the dock; and a margin should always be allowed between the keel of the vessel of greatest draught and the sill, especially when chains are used for working the gates, as they lie across each other on the sill with the lock open, and thus reduce somewhat the available depth.

Excavation of Basins and Docks.—Where the site for basins or docks is on land, the excavation for forming them is carried out in the ordinary manner by aid of barrows, waggons, and locomotives, and often by the use of excavators, of the type best suited to the local conditions. When the site is partially covered by the tide, an embankment is sometimes made by tipping suitable excavated material from waggons, so as to enclose the site and exclude the tide, the outer toe of the embankment being often protected from scour, and from undue spreading out and settlement in soft silt, by a mound of rubble stone or chalk; and the outer face of the embankment has to be protected by pitching in exposed situations. The final closing of such an embankment, when the enclosed area is large and the tidal range considerable, is somewhat difficult, and often fails when the aperture through which the tide flows and ebbs is narrowed by a high embankment on each side, owing to the scour of the current rushing through the restricted opening. The best plan is to bring up the embankment in successive layers over a long length, gradually penning up the water inside as it is raised, so that the tidal influx and efflux takes place over a wide opening, and is gradually restricted in depth and volume as the embankment is brought up and the water-level in the enclosure is correspondingly raised.

A cofferdam of the type shown in Fig. 18 is frequently constructed in front of the site of the entrance or lock to enable the foundations of these works to be put in, but in many cases a timber dam of this type has been used to enclose the entire site of the excavations for basins or docks with their foundations, as at H. M. Dockyard Extensions at Gibraltar, Keyham, Simons Bay, Hong Kong, and in commercial dock works elsewhere.

Pumps of various kinds are employed, with sumps which drain the lowest level of the foundations and discharge over the cofferdam such sea or land water as may be encountered in the progress of the work, and the cost of such pumping may be an important item.

The excavated material may either be removed by waggons and tipped to make up the levels of the spaces between docks or basins, or if there be a surplus may be deposited by means of staging transporters, etc., in hopper barges and carried out to sea, or run by waggons to spoil in some convenient situation.

Foundations for Dock Walls.—Borings should be taken along the lines of the walls of docks and their locks, in order to ascertain the nature of the strata on which these walls have to be founded, as the design of the walls must depend upon the nature of the foundations, which very largely affects the cost of such works; and in a dock surrounded by quay walls, this cost constitutes a very important portion of the total expenditure. In most cases, the ground at some depth below the surface is sufficiently firm and compact to sustain the weight of a dock wall resting on a broader bed of concrete; but sometimes a stratum of quicksand or silt is met with at the required depth, which is too thick to excavate, or is too charged with water, or whose removal might lead to settlement and slips beyond. Bearing piles, about 4 or 5 feet apart, may then be driven into the stratum, being capped and connected at the top by walings and planking, or by a layer of concrete, upon which the wall is built; or circular brick or concrete wells have been sunk to a solid stratum, as adopted for founding the walls of the most recent basins on the left bank of the Clyde below Glasgow. Sometimes the foundations for the walls have been enclosed within a double row of sheet-piling, confining and consolidating the soft stratum, which can then be partially excavated within the enclosure, and is capable when thus confined of bearing a greater load; whilst the sheet-piling at the back relieves the wall to some extent from the pressure behind, and in front forms a support for the toe of the wall against a forward movement.

In a few instances, a masonry well forms part of the base of the dock wall as well as the foundations, especially the masonry wells of the Bellot Dock at Havre, sunk in the alluvial foreshore of the Seine estuary before the completion of the reclamation bank, in order to expedite the work (Figs. 286 and 293, pp. 449 and 455), and to a minor extent the masonry wells forming the piers of the upper, continuous wall, sunk to a considerable depth in silty alluvium for the St. Nazaire and Bordeaux docks.

Pressures against Dock Walls.—The weight of half the prism of earth, at the back of a dock wall, comprised between the back of the wall and a line from the base of the wall at dock-bottom to quay-level, representing the natural slope of the ground at the back, has to be borne by the dock wall; and under the action of water accumulating at the back of the wall, the material may swell, and also assume the nature of fluid pressure corresponding to the density of the mixture of silt and water. These pressures tend to make the wall slide forward, to sink at its toe, and to go over at the top. Till the water is let into the dock on the completion of the works, the only counteracting resistances are the weight of the wall, the opposition of the mass of earth in front of its toe to any motion forward, and the firmness of the foundation at its toe.

Provisions against Movement of Dock Walls.—In order to reduce the pressure behind the wall, the backing should be formed of materials unaffected by water, and deposited in thin horizontal layers well consolidated, so as to press as little as possible against the wall, and not be liable to slip forward; and where water is liable to accumulate at the back, pipes should be laid through the wall a little above dock-bottom to drain off the water, which are closed just before the admission of the water into the dock. Moreover, by constructing the lower part of the wall in a timbered trench, the amount of loose filling at the back can be greatly reduced. The sliding forward of a dock wall depends upon the nature of the stratum on which it rests, as well as on the pressure at the back; and the sliding has sometimes occurred below the base of the wall, between two smooth, horizontal surfaces of clay quite unconnected, or from which an intervening vein of sand has been washed away. Accordingly, where the wall is founded on clay or other slippery, though hard material, especially where horizontal dislocations may be apprehended, the resistance at the toe must be strengthened by carrying down the foundations some feet below dock-bottom, or by giving the base a downward slope towards the back (Figs. 291 and 292), so that the wall in sliding forward would either have to rise, or push a large mass of earth in front of it, or by driving a row of sheet-piling along the toe of the wall. The liability of a dock wall to sink at the toe, and to come forward at the top, can only be counteracted by taking the foundations down to a solid stratum by excavating and filling in with concrete, by driving bearing piles or sinking wells, or by special care in the selection and filling in of the backing, or, in very treacherous ground,

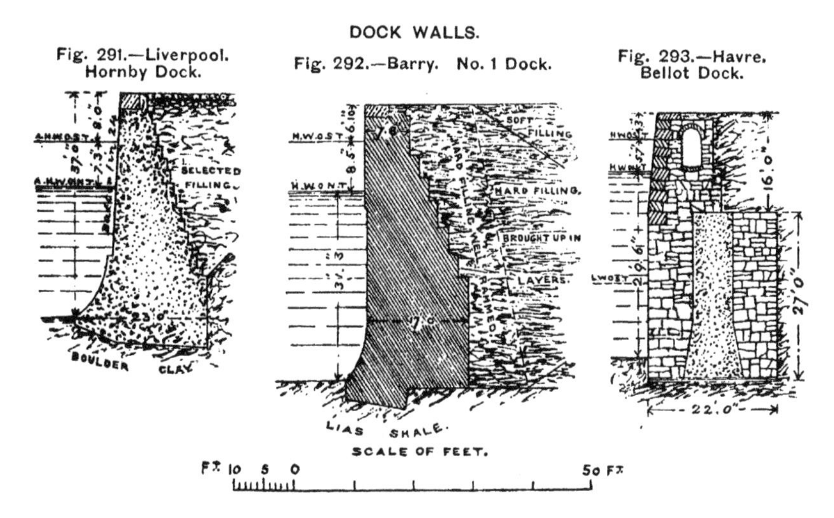

DOCK WALLS.

Fig. 291.—Liverpool. Hornby Dock.

Fig. 292.—Barry. No. 1 Dock.

Fig. 293.—Havre. Bellot Dock.

by deferring the completion of the backing up behind the wall till the water has been admitted, thereby providing a compensating pressure against the face of the wall.

Forms of Basin and Dock Walls.—The only difference between ordinary basin walls, open to the tide, and dock walls, built generally in both cases in excavations from which water is excluded, is that basin walls, like the walls of the tidal basin at Tilbury (Fig. 289, p. 451), are exposed to variations in the supporting pressure of the water in front, to the extent of the tidal range, and should therefore be given a somewhat greater stability; though both dock and basin walls, if fully backed up before the water is let in, are subjected for a short period to a greater pressure at the back than they are subsequently called upon to resist. Owing also to the greater depth to which basin walls are generally excavated, to allow for the fall of the tide, basin walls have to be given a correspondingly greater height.

As the thickness of a retaining wall should increase from the top downwards, similarly to the pressure at the back, in proportion to the height, dock and basin walls, and also river quay walls, are generally designed with a moderate thickness at the top to enable them to resist shocks of vessels, and with an increasing thickness downwards by a batter on the face and steppings out at the back (Fig. 289, p. 451, and Figs. 291, 292, and

293). A batter on the face has the advantage of allowing of a slight coming forward of the wall at the top, without presenting the unsightly appearance of an overhanging wall; but the straight sides given to modern vessels has necessitated the adoption of a vertical face, except near the base, to enable vessels to lie close alongside the quay (Fig. 292, p. 455). The height of a wall above dock-bottom varies with the depth provided below the lowest water-level in front of the wall, and the greatest tidal range, between low water and high water for a basin wall quite open to the tide, and between high water of neaps and high water of springs for a dock wall. Though the depth to which a dock wall is carried depends mainly upon the nature of the ground, the stability of the wall against sliding forward increases with the depth of its base below dock-bottom. On the average, the thickness of a dock wall should be about one-third of its height above dock-bottom halfway above this level; and the thickness of the wall at dock-bottom should be equal to about half the height of the wall above this level.

Construction of Dock Walls.—Dock walls are constructed of masonry, brickwork, or concrete, or of a combination of concrete with one of the others. Masonry may be employed where suitable stone is easily obtainable and cheap, while brickwork is often used under like conditions, but concrete either in mass or in blockwork has been largely adopted of late years. In the latter form a type of interlocked blockwork has been used to a large extent in recent works as at Gibraltar, Simons Bay and Hong Kong, where basin or wharf walls have been constructed in the sea, without cofferdams, the blockwork being brought up to above the level of low water, and the upper portion of the walls constructed in mass concrete with masonry facing and coping.

Concrete walls are liable to abrasion by vessels and hawsers, and require protection by fendering or facework in a harder material, as at Tilbury, Fig. 289, p. 451, where the face has been protected by a lining of blue bricks, a material also used, together with granolithic concrete, for the vertical faces of altars or walls of graving docks. The design of dock walls has also to include the requirements of crane foundations, the anchorage of bollards and fairleads, and the provision of subways for hydraulic, compressed air, or electrical mains.

Pitched Slopes and Jetties for Docks.—Sometimes, for the sake of economy in construction, pitched slopes are formed along the sides of a dock, in place of dock walls; and jetties are run out at intervals across the slope, at the ends of which vessels can lie, instead of alongside a continuous quay wall. This arrangement is convenient at ports where the growth of traffic is not very rapid, as the jetties can be constructed as required; and a continuous wharfing can be eventually erected along the foot of the slope to serve as a quay. The system, however, is especially suitable for the portions of the docks at coal-shipping ports from which vessels are loaded with coal, where coal-tips are established at definite points, to which waggons of coal are brought along sidings to be tipped, and where, therefore, jetties projecting across the slope, and provided with a coal-tip at their extremities, fulfil precisely the object required (Fig. 285, p. 448).

CHAPTER XXVIII.

DOCK ENTRANCES, AND LOCKS; AND GRAVING DOCKS.

Arrangements for approaches to docks—Tidal basins, depth, deposits, maintenance—Dock entrances and locks compared, relative advantages, in conjunction with a half-tide basin—Entrances to docks, arrangements, storm gates—Locks leading to docks, description, arrangements—Dimensions of entrances and locks; increased sizes of vessels; instances of largest entrances and locks—Construction of locks, description of methods, details, materials used, forms of sill, sluiceways—Construction of dock-gates, wood, iron, relative advantages—Forms of dock-gates, straight, gothic-arched, segmental; rise, pressures—Support of dock, gates, methods; roller, object, disadvantages—Working of dock-gates, with chains, with piston—Caissons for Dock Entrances, types, sliding, rolling, combined sliding and rolling, ship, descriptions, method of working, comparison between caissons and gates—Graving docks, object, dimensions of large examples, general construction, peculiar arrangements at Tilbury and Barry—Machinery for Graving Docks; pumps, main and drainage, capstans, caisson engines; hydraulic, electric or compressed air power; cranes—Maintenance of depth in docks, deposit, dredging, floating docks.

THE facilities for access which have to be provided for docks depend mainly on the shelter furnished by the approach channel. Where the approach is a well-sheltered river, like the higher part of the Thames below London Bridge, the Scheldt at Antwerp, and the Usk at Newport, or a minor channel branching off from an estuary, like the entrance channel to the Cardiff Docks, or a jetty channel like Dunkirk (Fig. 290, p. 452), and Leith outer harbour, the entrances or locks leading to the docks are readily reached through a dredged channel with guiding timber jetties. In less sheltered approaches, tidal basins are provided with a wide entrance, through which vessels pass to enter the sheltered entrances or locks, of which arrangement the Liverpool, Tilbury, and Barry docks afford instances (Figs. 287, 288, and 285, pp. 450, 451, and 448). At exposed sites, entrance harbours have to be formed by breakwaters for sheltering the access, as exemplified by Sunderland and Havre (Fig. 286, p. 449).

Tidal Basins leading to Docks.—The tidal basins should be given the same available depth as the approach channel, or even a somewhat greater depth if delay is liable to occur in entering the docks; and these basins, if opening into a river or estuary whose waters are charged with sediment, are subject to a reduction in depth by an accumulation of deposit from the influx of the river water at every tide. The tidal basin at Tilbury has been given a depth of 24 feet at low water of spring tides, so as to render the docks accessible at any state of the tide; but as it receives a layer of muddy water from the Thames at every flood tide, having a range of 21 feet at springs, considerable deposit takes place, which has to be removed during the ebb by stirring up the mud with powerful jets of water. The approaches to the Liverpool entrances and locks have to be maintained by sluicing regularly at low tide; and the tidal Canada Basin, with its floor two feet below the lowest low water, has its depth maintained by the flow from a series of sluicing pipes connected with the docks, having their outlets spread over the floor, as well as by the current from sluices opening out of the wing walls of the locks. The tidal basin leading to the Barry Docks, has had a channel dredged through it to a depth of 16 feet below low water of spring tides.

The period during which vessels can enter docks at each tide depends upon the depth which can be maintained in the tidal basin, or in the outer harbour, as well as upon the available depth of the approach channel.

Dock Entrances and Locks compared.—An entrance to a dock is a passage closed by a single pair of gates; whereas a lock is closed by two pairs of gates with a lock-chamber between them, like the locks on rivers and canals previously described. An entrance takes up much less space, and is much less costly than a lock, and is therefore convenient when there is little space available behind it for docks, as for instance in the case of the older central, and the southern docks of Liverpool. An entrance, however, necessitates restricting the admission and exit of vessels to the time of high water, or for a certain time previous to high water, on the condition of drawing down the water-level of the dock to that of the tide outside; whereas a lock enables vessels to leave or enter docks so long before or after high water as there is sufficient depth in the approach channel, tidal basin, outer harbour, or entrance channel, and over the lower sill of the lock. Nevertheless, when entrances are used in conjunction with a half-tide basin having an adequate depth for its level to be lowered, and provided with an entrance at its inner end communicating with docks, as well as at its outer end, the half-tide basin serves as a large lock for passing a number of vessels into or out of the docks before high water, a system largely adopted at Liverpool, one of these basins being shown in Fig. 287, p. 450, and of which the Barry Docks furnish another example, where a tidal basin with entrances was originally the only means of access to the first dock (Fig. 285, p. 448). Basins have also been adopted at the neighbouring South Wales ports of Penarth and Cardiff, for several of the docks of the Port of London, and at Hull, Bristol, and other ports; but at the Port of London, and in most other cases, a lock leads to the basin; and a single pair of gates, or a lock, or a passage with two pairs of gates pointing opposite ways, intervenes between the basin and the dock. With an outer lock, the basin chiefly serves for the collection of the outgoing vessels before high water, which are readily passed out by drawing down the basin to the water-level outside shortly before high tide, and opening the lock, and for the admission of incoming vessels on a level, which, after the closing of the gates soon after high water, are passed at leisure into the docks. Where the rise of tide is considerable, the outer gates have to be closed very soon after the turn of the tide, as the closing of the gates with a strong current running out of the opening, would be attended with danger to the gates; and gates pointing both ways are often put in the passages between docks, especially where, as at Liverpool, several docks are in communication (Fig. 287, p. 450), in order that the water-level in the various docks and basins may be regulated independently of the others.

Entrances to Docks.—Entrances consist of a single pair of gates

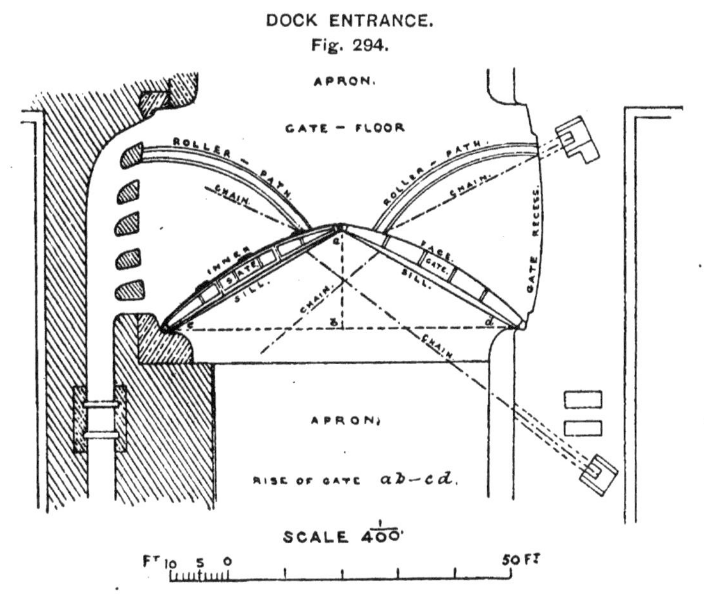

DOCK ENTRANCE.
Fig. 294.

closing against a raised sill at the bottom, with sluices in the side walls, these walls being long enough on the inner side to shelter the open gates in a gate recess on each side, and on the outer side to provide buttresses for the gates when closed; and there is a gate-floor on the inner side of the sill, between the side walls, over which the gates turn, and on which the roller-paths are laid when the gates are supported on a roller, with an apron beyond to the inner ends of the side walls, and an apron on the outer side of the sill, extending to the outer ends of the side walls (Fig. 294). Sometimes at very important entrances, provision is made against accidents and for repairs of the gates, by placing two pairs of gates across the entrance, a plan resorted to at some of the Liverpool entrances (Fig. 287, p. 450), and also at the main 100-feet entrance to the Eure Dock at Havre (Fig. 286, p. 449). Where outer entrances or locks are exposed to waves or swell, tending to force open the gates, which would be injured by closing violently again, the gates are protected by an outer pair of reverse or storm gates which close against the waves, a precaution adopted at Liverpool where the entrances face the unprotected openings of the tidal basins, or themselves open directly on the exposed river, and also at Penarth, Sunderland, and other somewhat exposed entrances.

Locks leading to Docks.—Locks giving access to docks resemble in principle locks on rivers, and at the entrances to ship-canals, already described. They differ from entrances in having two pairs of

ENTRANCE LOCK TO DOCKS.

Fig. 295.—Plan.

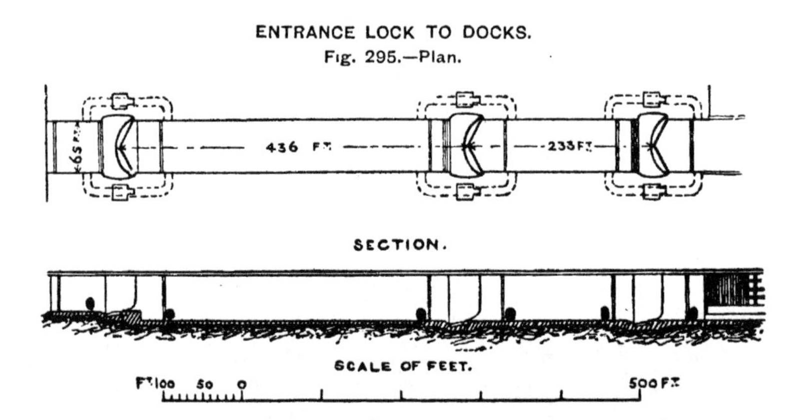

gates separated by a lock-chamber, which should be sufficiently long and wide to receive the largest vessels ordinarily frequenting the port, though longer vessels can be admitted at high water through the open lock, so that an adequate width is more important than the length, and applies equally to entrances, which admit vessels of any length that can be dealt with in the inner docks. The chamber is enclosed by quay walls on each side, and is paved at the bottom by an inverted arch, or invert as it is called, which resists the upward water-pressure under the chamber when the water in it is low, and abuts at each side against the quay walls, effectually securing them against sliding forward, in spite of the frequent removal of the supporting water-pressure in front by the lowering of the water-level in the process of locking. The arrangement of gates, sill, floor, and sluiceways and recesses in the side walls at each end of the locks, is similar to those described above for an entrance. Sometimes an intermediate pair of gates is introduced to expedite the locking of the smaller class of vessels, and also to save water, which cannot be renewed from the sea at the lowest neap tides, and is occasionally derived from an inland source; whilst in places where the rise of tide is large, and a sufficient depth can be obtained in the dock below high water of neap tides, without carrying the excavations down to the level of the bed of the entrance channel, the inner sill of the lock can be laid at about the level of the dock-bottom, and the outer sill carried down to the available depth in the entrance channel, in order to extend to the utmost the tidal period during which the port is accessible (Fig. 295).

Dimensions of Entrances and Locks for Docks.—A measure of the accessibility of a port is the depth of water over the sill of its entrance or lock at high water neap tides, provided this depth is not greater than the available depth of the approach channel, but in docks for naval purposes recently constructed provision has in many cases been made for the entry of deep draught vessels at low water spring tides, in view of the strategic advantages to be obtained. The progress of naval architecture of late years has led to a continual increase in the length beam and draught of ships for commercial purposes, nor is it certain that this increase has yet attained its final limit. The largest vessels now afloat are the Mauretania and the Lusitania, nearly 800 feet in length, 88 feet beam, and 34 feet draught, displacement nearly 39,000 tons, while vessels of still larger dimensions are under construction.

The depth of the Suez Canal, which in former years limited the draught of ships using this route, is being increased to 34⅓ feet, while that of the Panama Canal will be not less than 41 feet.

A few entrances and locks were made 100 feet wide many years ago, apparently to accommodate large paddle-wheel steamers, of which Liverpool, Birkenhead, Barrow, and Havre furnish instances; but the largest locks on the Thames have a width of 80 feet, and this width is still ample for the larger vessels. Deeper sills, accordingly, and longer lock-chambers have been the chief improvements effected in recent works. At Liverpool, a new entrance has been constructed leading into a new large half-tide basin, with a width of 90 feet, and a depth of 10½ feet below the lowest low water over the sill, 8½ feet deeper than the lowest previous sills, and affording a depth of 29 feet at high water of the lowest neap tides (Fig. 287, p. 450). The most recent entrance lock in the Port of London, leading to the Tilbury Docks, with the standard width of 80 feet, has a lock-chamber 700 feet long, and a depth over its outer sill of 23 feet at low water of spring tides, and 40½ feet at high water of neap tides, an increase in length of 150 feet, and in depth of 8 feet, over the previous largest and deepest lock on the Thames; and the inner sill of this lock is 6 feet higher than the other two (Fig. 288, p. 451). The newest entrance lock at Hull has a width of 85 feet, a length of chamber of 550 feet, and a depth over its outer sill of 29½ feet at high water of neap tides, an increase of 5 feet in width, 250 feet in length, and 6½ feet in depth over the principal previous lock on the Humber; and its inner sill is 1½ feet higher than the two others. The lock at the Barry Docks is 65 feet wide, has a chamber 647 feet long, and a depth on the outer and intermediate sills of 41⅓ feet at high water

[1] "The most recent Works at some of the principal British Seaports and Harbours, Paris Navigation Congress, 1900," L. F. Vernon-Harcourt, pp. 1 to 2.

of neap tides, and on the inner sill 9 feet less (Fig. 295, p. 460); whilst the entrance to the Barry tidal basin, 80 feet wide, has a depth of 29⅓ feet over its sill at high water neaps (Fig. 285, p. 448). The lock of the Dockyard Extension works at Keyham leading from the Hamoaze to the Closed Basin has a chamber 730 feet in length between caissons, with entrances 95 feet wide at coping level, and a depth of 32 feet over the sills at low water spring tides, and the lock of the Eure Dock at Havre has a length of 870 feet between gates, a width of 98½ feet, with 35 feet over the outer sill at high water neap tides, or 3½ feet lower than the sill of the adjacent 100 feet entrance (Fig. 286).

Construction of Locks.—The foundations of locks have to be carried out with special care, so as to secure the lock against settlement and the percolation of water under its sills. Locks have been sometimes founded on bearing piles, with a row of sheet-piling across the lock under each sill, and occasionally along the end of the apron, to arrest any flow of water; and settlement and percolation are still more effectually guarded against by enclosing the whole site of the lock within sheet-piling, or so much of it as may have to be built upon a bad foundation. It is, however, preferable, wherever practicable, to carry down the foundations under the sills to a solid, impervious stratum. This is commonly effected by excavations conducted under the shelter of a cofferdam, and kept dry by pumping, and in timbered trenches and short sections where necessary, the trenches being filled in with concrete directly a suitable stratum is reached. In alluvial ground, however, charged with water, where the foundations have to be carried down a considerable depth, compressed air may be required. Thus the two heads of the new entrance lock at Havre, with their side walls, sills, gate-floors, and aprons, have been founded with large caissons by the aid of compressed air. The caisson for the upper head of the lock, constructed in position, is 213 feet across the lock, and 105 feet long in the line of the lock, and has been sunk by means of compressed air to a depth of 31 feet below the lowest low water; whilst the caisson for the lower head, constructed in an adjacent hollow and floated into position, 207 feet by 118 feet, had to be sunk to a depth of 55¾ feet below the lowest low water, before reaching a stratum sufficiently firm to support the foundations of the work.

The side walls at each end of the lock, alongside the gates and sill, have to be founded at the same depth, and with the same precautions as the sill, and have to be given a greater thickness than the dock walls, to contain the sluiceways, sluice-gates, and the machinery for working the gates, together, in some cases, with the support for a swing bridge across the lock. The side walls of the lock-chamber are constructed like the dock walls, for though the supporting water-pressure in front of them may be withdrawn down to the level of low water of spring tides with the outer gates left open, they possess the support of the invert at their toe. The thickness of the invert, with its foundation, is varied with the nature of the ground and the upward water-pressure to which it may be subjected; but its arched form, with a reasonable thickness, and its foundations generally made level with the foundations of the side walls, afford it an ample strength.

The sill-stones, the heel-post stones on which the gates rest, and the hollow quoins in which they turn, are commonly made of granite on account of the wear to which they are exposed; the sluiceways are lined with dressed masonry or blue bricks; the invert is formed of masonry or brickwork, supported by stone skew-backs, or springings, built into the side walls, or occasionally of concrete; and the coping of both the lock and dock walls is made of large dressed stones, usually connected to each other and to the wall underneath, by slate or concrete dowels, so that a shock to a single coping stone may be borne also by the adjoining coping and wall.

The sill, which is generally raised about 2 feet above the gate-floor, may be either made flat, or curved like the invert; but the gate-floor must be made flat for the gates to turn round on their heel-posts over it; and the apron at either end of the lock is generally made flat, especially the outer apron, where the width between the side walls is gradually enlarged seawards. The sill, on plan, consists of two lines meeting in the centre at an angle, these lines being usually straight, but sometimes curved (Figs. 294 and 295, pp. 459 and 460). The sluiceways formed in the side walls behind the gates, serve, not merely for filling and emptying the lock-chamber for the passage of vessels, but also provide, by the discharge from them at low water, a scouring current for removing deposit from the chamber and outer apron of the lock; whilst by putting the inlet sluices in the recesses at the sides of the gate-floor, and level with it, the inflowing current scours the floor (Fig. 294, p. 459). Sluice-gates are placed centrally in the sluiceways for controlling the flow, which are raised and lowered from the top of the wall, and are often in duplicate to allow for repairs. Longitudinal sluiceways have been formed throughout the side walls of the Dunkirk entrance lock, and are designed for the large lock in construction at Havre; but this is an unusual provision for locks to docks, where the difference between the level of the tide outside, when the depth is sufficient for vessels to enter or leave, and the level of the water in the dock or basin, is not generally large.

Construction of Dock-Gates.—The gates closing the entrances and locks of docks are constructed of wood or iron, consisting in the first case of a series of horizontal, framed beams, made thicker and placed closer together towards the bottom to resist the increasing water-pressure, and joined together by the heel-post and meeting-post at the two ends, and by intermediate uprights; and these beams support a close, caulked sheeting of planks on the inner side, forming the skin of the gate[1] (Fig. 296, p. 464). Iron gates comprise an outer and inner skin of iron plates braced horizontally and vertically by plate-iron ribs, the horizontal ribs being placed closer together towards the bottom, and the plates increased in thickness, to allow for the increase of pressure with the depth[2] (Figs. 297 and 298). The heel-post or shutting strip, the meeting-post, and the sill-piece of iron gates are usually constructed of

[1] *Proceedings Inst. C.E.*, vol. xcii. plate 2, figs. 12 and 14.
[2] *Ibid.*, vol. xviii. plate 6.

greenheart, to ensure a fairly watertight closure of these faces. Wooden dock-gates, if exposed to sea-water, are usually constructed of greenheart to resist the attacks of the teredo; and they possess the advantage over iron gates that, owing to their solidity, they are less subject to serious injury if run into by a vessel; but, on the other hand, greenheart gates with their fastenings are considerably heavier than water, and bear

DOCK-GATES.

Fig. 296.—Alexandra Dock, Hull. Sections. Fig. 297.—Victoria Dock, London. Sections. Fig. 298.—Eure Dock, Havre. Sections.

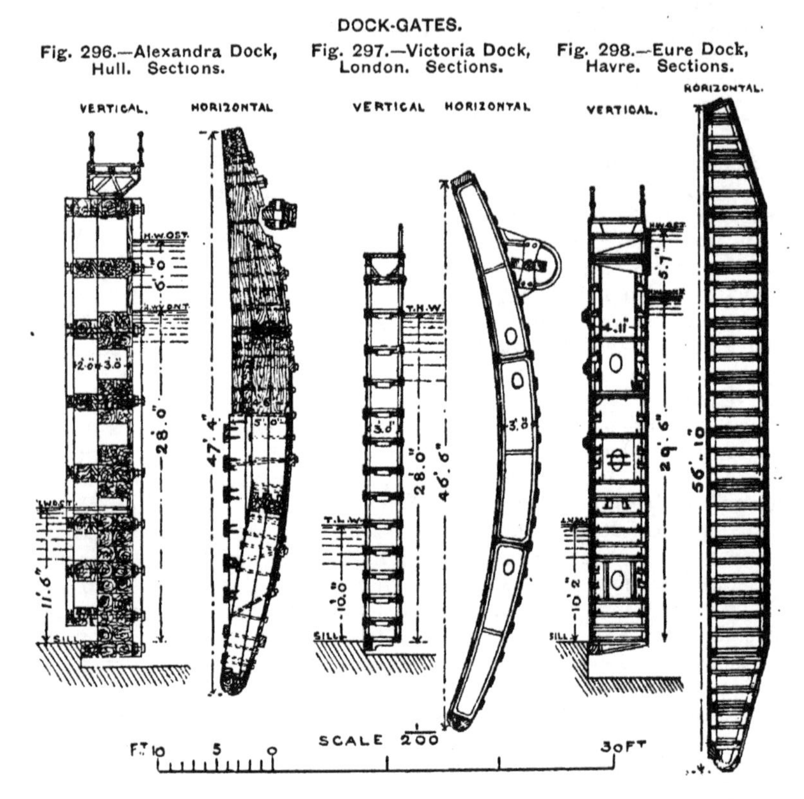

heavily on their pivot and roller when moved with the water somewhat low in the lock, and therefore only very partially immersed. Iron gates, on the contrary, being made cellular and watertight, have to be partially filled with water to resist flotation when immersed; but the water-ballast required can be considerably lessened by leaving the upper part of the gate open to the water on the outside, so that the rise of water above the deck of this open part does not practically increase the flotation of the gate (Fig. 298). No decided preference has been established for wooden or iron dock-gates, for all the gates at Liverpool are constructed of greenheart, and all those in the Port of London of iron; whilst both wooden and iron gates have been adopted at Hull, St. Nazaire, and Antwerp, wooden gates at Avonmouth on the Bristol Channel, and iron gates at Barry a little lower down. Iron gates, however, have been substituted at the 100-feet entrance to the Eure Dock at Havre, in place of the original wooden gates[1] (Fig. 298); and iron has been preferred for the more recent dock-gates at Dunkirk and St. Nazaire. Wooden gates are cheaper than iron gates, and for small gates wood is a very suitable material; but the adaptability of iron to the considerable curvature of most large gates, and the power of adjusting the ballast in cellular gates to the flotation, as well as the greater durability of iron gates, appear to render iron, or steel, preferable to wood for large gates.

Forms of Dock-Gates.—Three forms of dock-gates have been adopted, namely, straight gates, closing against a straight sill and meeting in the centre at an angle (Fig. 298); secondly, curved gates, meeting at an angle so as to form a gothic arch on plan, though often closing against a straight sill by means of a projecting sill-piece at the bottom, made straight on the outer face (Figs. 294 and 296, pp. 459 and 464); and, thirdly, segmental gates, curved so as to form a perfect circular arc along their inner faces when closed, and thus constituting an ordinary arch, though the outer faces coming against the sill are generally given a somewhat flatter curvature than the inner faces, thereby reducing the curvature of the sill, and affording a convenient increase in thickness towards the centre of the gate, near which the manholes for inspection and inside painting are commonly placed (Fig. 297).

The pressures imposed on dock-gates by a head of water against their inner faces, depend upon the form and rise of the gates. The rise of dock-gates is the proportion which the perpendicular distance of the point of the sill from a line joining the centres of the heel-posts, bears to the width between these centres, or the ratio of the projection of the gates to their span (Fig. 294, p. 459); and the rises ordinarily adopted range between one-third and one-sixth of the span. The stresses which a dock-gate has to bear are, firstly, a transverse stress due to the water-pressure against the inner face of the gate, increasing with the head of water and the length of the gate; and, secondly, a compressive stress along the gate due to the pressure of the other gate against its meeting-post, which amounts to half the water-pressure on the gate multiplied by the tangent of half the angle between the closed gates, and varies inversely as the rise with straight gates. Accordingly, a large rise reduces this stress; but at the same time it increases the length of the gate, and the transverse stress upon it, and also the length of the lock. Consequently, though the least amount of material is required for a gate which has a rise of one-third of the span, a rise of one-fourth to one-fifth is generally preferred. By curving the gates, their strength against a transverse stress is augmented and the longitudinal compressive stress is increased, till at last when the gates form a complete circular arc on their inner face, the transverse stress disappears, and the stresses produced by the water-pressure and the opposite gate become wholly compressive and uniform throughout each horizontal

[1] *Annales des Ponts et Chaussées*, 1887 (2), plates 36 to 38.

section of the gates, increasing with the head; and the total stress is equal to the pressure on a unit of surface multiplied by the radius of curvature which, with a given span, varies inversely as the rise.

The dock-gates forming a circular arc are theoretically the best type of gate, with their uniform compressive stress; but these gates are longer than straighter gates with the same span and rise, and their large curvature necessitates deep gate-recesses, cutting unduly back into the line of quay of the lock or entrance, and involves a curved sill more difficult to dress and fit accurately than a straight one. Accordingly, gothic-arched gates, with less curvature and a straight sill, have been more commonly adopted (Fig. 294, p. 459).

Support of Dock-Gates.—The heel-post of a dock-gate is fitted at the bottom with a cap, which fits over a steel pivot embedded in the heel-post stone, on which the gate rests and revolves; and at the top, it is kept against the hollow quoins by an anchor strap encircling its head, tied back by anchor bolts to the wall. A heavy, long, wooden gate, or a long iron gate, without an arrangement for adjusting the ballast to the variable flotation, when being moved in a small depth of water, would throw a considerable strain on the anchorage at the top, and be liable to dip somewhat at its outer extremity; and, consequently, a roller has often been inserted under the gate near its end, running along a cast-iron roller-path laid on the gate-floor, and forming a second support for the gate (Figs. 296 and 297, p. 464). This roller, however, with its adjusting rod and accessories adds materially to the weight to be moved, and, with its roller-path, to the cost of the gate; and, occasionally, through want of adjustment, or obstructions on the roller-path, it has impeded the working of the gates. Accordingly, a roller has sometimes been dispensed with in iron gates, even of large span, where the flotation has been so adjusted as not to increase with the depth of water beyond a certain limit, and the permanent water-ballast counterbalancing the flotation for the lowest water-level at which the gates are moved, is concentrated in the compartments of the gate nearest the heel-post, of which the gates at Havre closing the Eure entrance, 100 feet in width, furnish a remarkable example (Fig. 298, p. 464).

Working of Dock-Gates.—The two dock-gates across an entrance, or at each end of a lock, are usually opened or closed simultaneously by two opening and two closing chains, generally fastened to each gate towards its outer end, at about one-third of its height on both sides, and passing through chain passages with guiding rollers to the drums of the gate machines which actuate them (Fig. 294, p. 459). A hinged hydraulic piston, with its cylinder placed in the side wall of the gate recess, and its outer end fastened to the inner side of the gate, pulling the gate back, or pushing it forward for closing, has been sometimes adopted within recent years instead of chains, on account of its direct action and relative simplicity.

Caissons for Dock Entrances.—Various forms of caissons for closing the entrances to Basins, Locks or Graving Docks are frequently employed. These are sliding caissons; rolling caissons; combined sliding-and-rolling caissons, and ship caissons. Sliding caissons are of a box shape of mild steel plating, stiffened with internal decks and bracing with an air chamber of adequate dimensions to float the caisson, but ballasted sufficiently to maintain such surplus of weight over buoyancy as will ensure stability and smoothness of working. These caissons are provided with steel keels sliding upon smooth granite ways, and numerous examples are to be found in naval dockyards at home and abroad. Rolling caissons are similar to sliding caissons as regards the construction of the body portion, but their keels rest upon rollers as at Bruges (Fig. 299), and at Glasgow, in order to reduce the frictional resistance in hauling. Combined sliding-and-rolling caissons have been successfully introduced into H. M. Dockyards, and combine the features of ordinary sliders with the addition of a pair of cast-steel wheels, one forward and one aft, running upon a cast-steel roller path placed centrally, and actuated by a pair of hydraulic rams arranged to bring pressure on either wheel alternately, as the caisson is hauled in or out. In each of these three types the entrance is opened by hauling the caisson into a recess provided in the side of the entrance. This recess or "camber" is covered by a deck laid flush with the upper deck of the caisson, when road or railway communications are carried across the entrance, and in order that the caisson may enter the camber, the camber deck is caused to lift, or the caisson deck to lower. Hauling engines driven by hydraulic, compressed air, or electrical power, are placed at the farther end of the camber and attached to the caisson by chains.

ROLLING CAISSON.
BRUGES CANAL LOCK.
Fig. 299.—Cross Section.

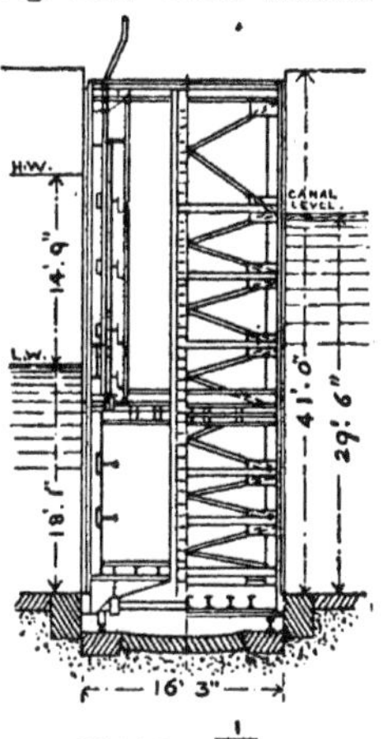

SCALE $\frac{1}{300}$.

Ship caissons, so called from some resemblance of outline to a ship (Fig. 303, p. 468), are floated into position, and sunk, by the admission of water, into grooves in the dock sides and sills, and raised again for removal by emptying water-ballast tanks. This type requires no recess or machinery, and is of less cost than the sliders, but occupies berthing space when removed from the entrance, and requires longer time in handling. The water-tightness of sliding and ship caissons is maintained by the water seal which exists between the caisson itself and the masonry of the sides and sills of the entrance. This is made by the contact of planed greenheart with fine-axed granite, the depth of the surfaces in contact being from 11 to 13 inches. This, under pressure of a head of water, suffices to maintain a degree of water-tightness which is with careful workmanship practically perfect.[1]

An important point in the comparison between caissons and gates is that the former act as bridges capable of carrying the heaviest traffic

[1] For further information on the subject of Dock Caissons see "Notes on Construction in Mild Steel." H. Fidler. Longmans' Civil Engineering Series.

across dock entrances, whereas the latter must be supplemented by a costly swing bridge to meet similar requirements. Recent caissons have been constructed to carry a rolling load on the railway carried by them amounting to 36 tons per axle.

Graving Docks.—Graving, or dry docks form part of the necessary equipment of important seaports, for the execution of any repairs that may be required by vessels frequenting the port. The cleaning of a ship's bottom after a long voyage, so necessary for the maintenance of her speed, and small repairs, may, indeed, be performed in a river or sheltered creek with a good tidal rise, by letting the ship settle down on a timber gridiron with the fall of the tide; but graving docks are used for cleaning, painting, and repairing the larger class of vessels, and for large repairs to smaller vessels, the ship being admitted at high tide

GRAVING DOCK.

Fig. 300.—Plan of Alexandra Graving Dock, Belfast.

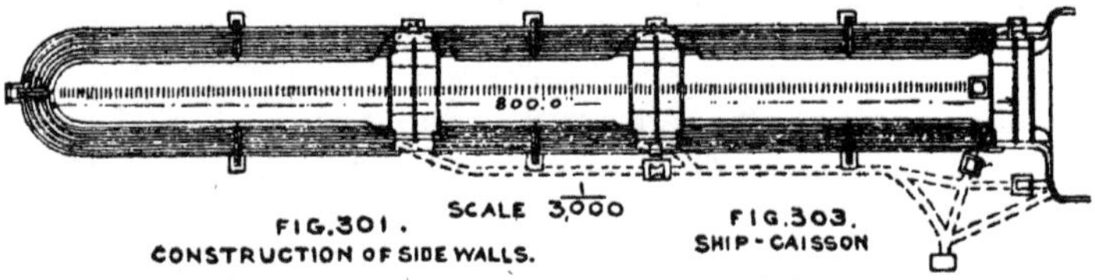

SCALE $\frac{1}{3,000}$

FIG. 301. CONSTRUCTION OF SIDE WALLS.

FIG. 302. CROSS SECTION. DOCK ENTRANCE.

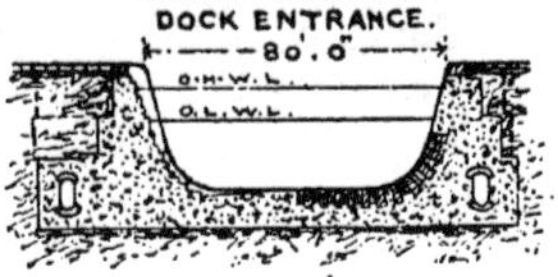

SCALE $\frac{1}{1,200}$

FIG. 303. SHIP-CAISSON

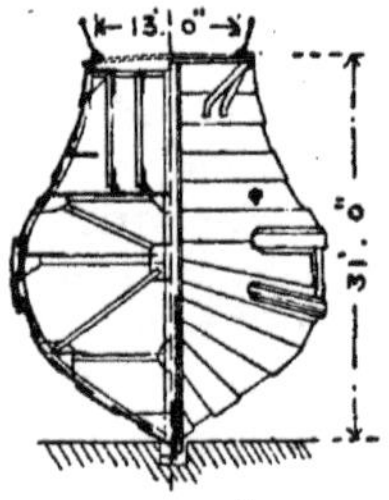

SCALE $\frac{1}{400}$.

from a river, or at any time from a dock, and the water pumped out as soon as the entrance has been closed by a caisson, or a pair of dock-gates pointing outwards. One graving dock at least in a port should be long enough, and have a sufficient width and depth at the entrance, to admit the largest vessel likely to come to the port; and a long graving dock is generally constructed so that it can be divided by an intermediate caisson into two graving docks of different lengths, for accommodating smaller vessels, when not required for a large one. As examples of some of the largest graving docks, may be cited, the graving dock adjoining the Canada Dock at Liverpool, 925½ feet available length, 94 feet width at entrance, and 28 feet minimum depth over blocks and sill, which can be increased to 34 feet; Tilbury, 875 feet by 70 feet, and 31½ feet depth at high water of neap tides; Glasgow, 880 feet, by 83 feet, by 26½ feet at H.W.S.T.; Gibraltar, 850 feet by 95 feet at entrance by $35\frac{1}{2}$ feet over sill at L.W.S.T.; Southampton (Trafalgar), 875 feet, by 90 feet, by 33 feet; Keyham, 745 feet, by 95 feet, by $20\frac{1}{2}$ feet at L.W.S.T., and 741 feet, by 95 feet, by 32 feet at L.W.S.T.; Belfast, 800 feet, by 80 feet, by 25 feet at high water, has two intermediate caisson grooves (Figs. 300, 302), and the mode of construction of side walls in trenches, and the form of ship caisson adopted, are shown in Figs. 301, 303.

Graving docks have their side walls formed with altars, against which the shores supporting the vessel in the dry dock rest; and the floor slopes slightly from the centre, along which the keel blocks are laid carrying the keel of the vessel, down to each side, so that the water may drain off by side channels to the main culverts, from which the water is drawn off by pumps (Fig. 302). Graving docks are constructed of masonry, brickwork, or concrete; and in North America, timber is largely employed, as being cheaper and less subject to injury from frost than masonry or concrete. The altars are sometimes faced with granite or granolithic concrete, the thickness of the walls and floor of the dock being designed to resist the maximum hydrostatic pressure, and the composition of the concrete so proportioned as to render it as watertight as possible, precautions which may fall short of their purpose should there be any leakage under pressure tending to crack or uplift the floor lining; the remedy usually being the provision of drains or boring of holes to relieve the pressure.

The plans of ports in Figs. 283 to 288 and 290, exhibit the positions and general arrangements of the graving docks. At Tilbury, the graving docks are placed parallel to the lock, and extend right across the quay intervening between the docks and the tidal basin, so that in the event of any accident to the lock, these graving docks would afford a way of access to the docks (Fig. 288, p. 451). In two cases at the Barry Docks, two graving docks placed end to end, reached by a common entrance and separated by an intermediate entrance, have been given a width of 100 feet on the floor within the entrances, so as each to receive two vessels side by side.[1]

Machinery in connection with Graving Docks.—Brief reference must here be made to the nature of the mechanical power with which a modern first-class graving dock is usually equipped. This includes the main and drainage pumps, the former for emptying the dock, the latter for freeing the floor of the dock from any accidental leakage which may take place through the gates, caissons or culverts, or in the body of the dock. The main pumps perform a duty which may sometimes amount to upwards of 100,000 tons pumped out of the dock in four to five hours. These pumps are usually located in a pumping-engine house near the dock and connected therewith by suction culverts, with discharge culverts into the sea, these being controlled by penstocks or valves, constructed sometimes of timber, or more commonly of cast-iron faced with gun-metal, and raised or lowered by machinery. Capstans for the control of vessels entering or leaving the dock form

[1] *Proceedings Inst. C.E.*, vol. cxi., and "Barry Docks and Railways, Description of the Undertaking, 1898," plan.

also an important item and are usually machine driven. The gates or caissons at the dock entrance are actuated by the machinery already described, while for the handling of boilers, turbine casings, or other heavy machinery in ships under repair, the graving dock is frequently equipped with locomotive cranes, running on rails alongside, which sometimes attain very large dimensions, such as the electrical crane at Southampton Docks lifting 50 tons at 87 feet radius, the height of jib head at this radius being 60 feet above ground. Another form of crane frequently located on wharf or basin walls for the use of vessels lying alongside is the so-called Giant or Hammer-headed crane, an example of which is found at Clydebank, having a lifting capacity of 150 tons at 85 feet radius, or 80 tons at 133 feet radius, on a tower 125 feet high.

The main and drainage pumps are most commonly driven by steam, but hydraulic, compressed air, or electrical power is used for the other machinery required, and the generating machinery for one or the other of these powers is sometimes placed in the same building as the pumps.

Maintenance of Depth in Docks.—Deposits arising from the replenishing of docks with silt-laden water may lead to a reduction in the depth unless periodically dredged. To avoid the inconvenience involved in dredging operations in docks crowded with shipping, various devices have been employed to ensure a supply of clear water, as in some of the South Wales ports, from inland sources, while partial exclusion of muddy tidal water is effected at Antwerp and St. Nazaire. Fresh water is supplied to the Alexandra Dock, Hull, and a canal supplies the Kidderpur Docks, Calcutta.

Floating Docks.—The modern steel floating dock, lifting ships by pontoon displacement, has superseded earlier forms of lifting-dock worked by hydraulic power. The need for periodical examination of submerged steelwork has led to the adoption of various arrangements of self-docking pontoons, although designs in this respect usually reveal a compromise between self-docking facilities and structural rigidity. Floating docks may be classified as one-sided docks of an |_ shape, either "off-shore" or "depositing," and two-sided docks of a |_| shape, either "sectional," "bolted-sectional," or of the "Havana" or "Hansson" type. "Off-shore" docks are found at Hamburg, North Shields, Sunderland, Cardiff, etc.; "depositing" docks at Barcelona, Zarate, Sebastopol, Vladivostock, etc., while the Bermuda Dock (No. 2) of 16,500 tons,[1] the Stettin of 11,000, the New Orleans of 18,000, and the Pola Dock of 15,000 tons lifting capacity are examples of the "Havana" and "bolted-sectional" types, to which latter class belong docks at Trinidad, Callao, and Hamburg. Comparisons as to relative efficiency, safety, durability, and cost of maintenance between floating and graving docks cannot here be adequately discussed.

[1] *Proceedings Inst. C.E.*, vol clxi.

CHAPTER XXIX.

HARBOUR WORKS.

Varied conditions of site and exposure of harbours, examples—Winds, periods—Waves, motion, transmit force imparted by wind, height and length—Forms of harbours, influence of local conditions, detached breakwater, instances; single breakwater from shore, instances; converging breakwaters, with intermediate breakwater, instances—Closed harbours on open sandy seacoasts, examples at Madras, Ymuiden, Port Said, Boulogne, progression and erosion of shore due to breakwaters—Open viaducts and outer breakwater on sandy coasts, to obviate objections to closed harbours, at Zeebrugge described, scheme suitable for Madras—Entrances to harbours, round breakwater, at Madras, advantage of two entrances for large harbours, widths recently adopted—Types of breakwaters, three types defined—Rubble or concrete-block mound, nature, conditions, zone of disturbance of mound, protected by paving at Plymouth, objects of concrete blocks for mound, examples—Mound with superstructure founded at low water, provisions against damage, instances, defects of system—Mound and superstructure founded below low water, increased depth adopted, instances, defects, sloping-block system, wave-breaker, objections—Upright-wall breakwaters, example at Dover with levelled foundation; concrete-bag foundation, instances; concrete-in-mass within framing under water, precautions; concrete blocks in caissons at Zeebrugge, description, advantages, difficulties—Construction of superstructures and upright walls, staging with gantries; block-setting Titans, revolving Titans—Remarks on breakwaters, three types in construction, causes of injuries to superstructure, remedies, considerations affecting superstructures and upright-wall breakwaters.

HARBOURS have to be formed under very varied conditions as regards site and exposure: in some places partially sheltered bays are available, as for instance at Plymouth, Cherbourg, Genoa, and Barcelona; and sometimes harbours have to be constructed on a perfectly open seacoast, as exemplified by Madras, Ymuiden, and Port Said harbours; whilst in the majority of instances, some sort of natural shelter, or the curve of the coastline, indicates the most favourable position for a harbour, as illustrated by Holyhead, Portland, Dover and Colombo. The exposure of a site, or the distance of the nearest coast, known as the "fetch," in the direction of the strongest and most prevalent winds, determines the direction from which shelter is most needed, and also the strength to be given to the breakwater to resist the impact of the waves, whose size depends upon the distance along which the wind can act in one direction on the sea, together with its force and duration, and also upon the depth of water in the neighbourhood of the coast.

Winds.—In designing a harbour, it is important to know the direction, prevalence, and period of the strongest winds in the locality, from the quarters in which the proposed harbour is open to the sea. In the parts of the world where there are periodical winds, such as the monsoons, the direction and force of the winds vary with great regularity according to the seasons; and places in the track of hurricanes or cyclones are liable to be visited by these storms at definite periods of the year, according to the locality. Even in Europe, where the winds are very variable, strong gales are much more liable to occur on the western and northern coasts in the winter months than at any other period of the year; whilst the calmest weather may be anticipated between May and August.

Waves.—The undulation of waves is due to a sort of cycloidal motion of the particles of water, under the influence of the wind acting on the surface of the sea, which makes the undulation travel in the direction in which the wind is blowing, without any transference of the body of water beyond a slight shifting forward of the upper layer, resulting in a small raising of the level of the sea on the coast facing the wind, in proportion to the violence of the gale and the distance it has travelled in approximately the same direction. The undulation, in fact, exhibits the transmission of the force imparted to the sea by the wind, through the medium of the successive revolving particles of water, and continues its onward course till, on encountering an obstacle, the transmitted force manifests itself in the form of a blow against the obstacle, or on reaching a shoaling shore, the undulation resolves itself into a breaking wave rushing up the beach till the transmitted force is spent. Waves begin to break as soon as the diminishing depth, on approaching the shore, becomes equal to their height; and therefore waves sometimes break on passing over outlying sandbanks, before reaching the shore. The height and length of waves increase with the fetch, and the duration and force of the gale; and in the Pacific Ocean off the Cape of Good Hope, where the exposure is greatest, waves 50 feet high, and from 600 to 1000 feet long, have been observed.

The force of the blow with which waves may strike a breakwater depends upon the size of the largest wave which can reach it, in view of the exposure and the unimpeded depth in front, and also upon the directness of the course of the waves against the face of the breakwater; and where breakwaters sheltering a harbour face different ways, one is often exposed to much heavier blows than another, owing to the greater fetch and force of the wind in a particular direction, though the directions of the greatest fetch and the strongest gales may not coincide.

Forms of Harbours.—The form given to a harbour depends on the configuration of the coastline, the extent of sheltered area required, the quarter from which shelter is needed in relation to the coast, and the position where the requisite depth exists, or can be obtained by dredging.

The site also suited for an entrance exercises an important influence on the choice of form. Thus with a recessed bay of adequate size, a detached breakwater across the wide outlet completes the necessary shelter, whilst leaving an entrance on each side between the extremities of the breakwater and the shore, of which Plymouth (Fig. 304, p. 474) and Cherbourg harbours afford types; and where the sea is shallow near the shore on each side of the outlet, a breakwater may be carried out from each shore, leaving a central entrance, as adopted at Barcelona, or with two entrances between a central, detached breakwater and the ends of the two breakwaters extending out from the shore, as carried out at St. Jean de Luz. In other cases, a single breakwater carried out from the shore on one side, stretches almost across the outlet of the bay, with an entrance left round its outer end, according to the design in progress for converting Peterhead Bay into a refuge harbour; or the breakwater may shelter a portion only of a very large bay, as illustrated by Holyhead and Table Bay breakwaters, with an open approach beyond the extremity of the breakwater.

In a few instances, where the exposure is mainly from a particular quarter, a single breakwater running out from the shore affords an adequate shelter, as for example at Newhaven, Alexandria, and Mormugao. Generally, however, a certain amount of natural shelter is improved, and the sheltered area extended, by breakwaters carried out from projecting points and converging towards each other, so as to enclose an adequate area of water, with the intervention sometimes of a detached breakwater for providing two entrances to the harbour, which should, if possible, face in different directions, in order to give vessels a choice of approach according to the direction of the wind. This latter arrangement is illustrated by Colombo Harbour (Fig. 305, p. 474), and similar arrangements occur at Portland, Dover and Gibraltar; whilst Algiers and Oran harbours furnish examples of converging breakwaters improving and extending some natural shelter, a system very commonly adopted for small harbours. A peculiar arrangement, resorted to in rare instances, for sheltering the entrance between two converging breakwaters facing the open sea, consists of an outer, detached, curved breakwater, constructed seawards of the entrance, and overlapping it for some distance on each side, with its concave face towards the shore, so as not to protect merely the entrance, but also an outer area to some extent, and providing a wide approach on each side, as exemplified by the harbours of Leghorn, Civita Vecchia, and Cette.[1]

Closed Harbours on Open Sandy Seacoasts.—Harbours have sometimes to be constructed on the open seacoast where no natural shelter exists, and which is often exposed to a considerable drift of sand along the shore, carried along by the waves under the action of the prevailing winds. Thus at Madras, where formerly vessels had to lie off the straight, sandy coast, whilst their passengers and cargo were landed

[1] "The most recent Works at some of the principal British Seaports and Harbours, Paris Navigation Congress, 1900," L. F. Vernon-Harcourt, plate 2, fig. 3.

HARBOURS.

Fig. 304.—Plymouth.

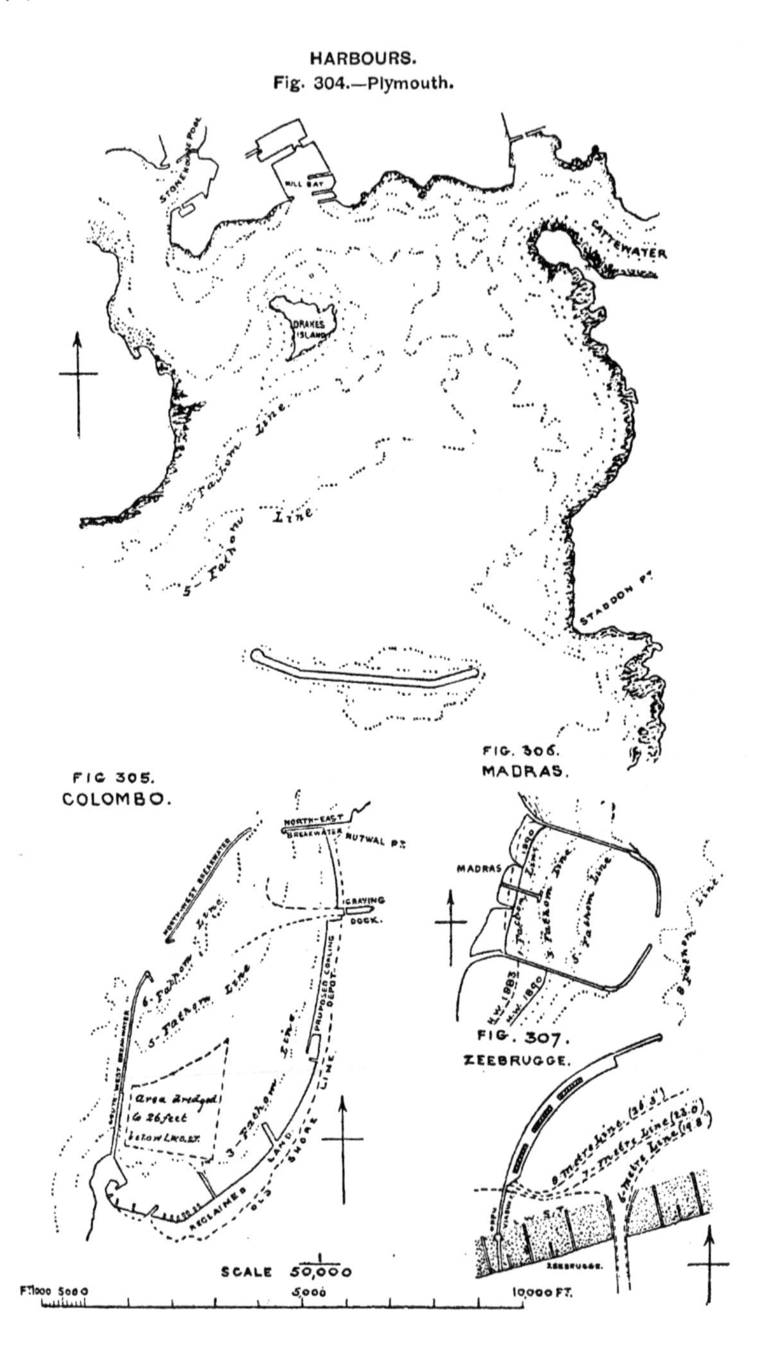

in surf boats, a closed harbour has been formed by two breakwaters, 3000 feet apart, going out straight from the shore for about 3000 feet, and then converging so as to leave an entrance of only 550 feet between their extremities, in a depth originally of 48 feet at low water (Fig. 306). Ymuiden Harbour, on the straight, flat, sandy shore of the North Sea, sheltering the approach to the Amsterdam Canal, is a somewhat similar closed harbour on a larger scale (Fig. 261, p. 410), formed by two breakwaters starting 3750 feet apart at the shore, and converging to an entrance 850 feet wide, in a depth of 28 feet at low water, at a distance of about 4600 feet from high-water mark. At Port Said Harbour, sheltering the approach channel to the Suez Canal, also formed by two converging breakwaters about 4250 feet apart near the shore, the western breakwater, extending into a depth of about 29 feet, has been carried about 3500 feet further seawards than the eastern one, so as to overlap it to this extent on the western side, as the worst winds and all the drift come from that quarter (Fig. 259, p. 409). The harbour at Boulogne, of which only the south-western half has been carried out, is designed to have a fairly square form like Madras Harbour, but on a much larger scale, though in shallower water and far less exposed, with an average width and length of about 5600 feet. The scheme includes a detached outer breakwater in a line with the completed western outer arm, with entrances at each extremity facing west-north-west and north-north-west, having widths of 820 feet and 492 feet, and depths of about 28 feet and 26 feet respectively, and a north-eastern breakwater alongside the dredged approach channel to the port.

In all the above examples, the harbours have been enclosed by breakwaters, with entrances carried into the deepest water available at the outer end, so as to be as far removed as possible from the littoral drift of sand along the coast, and consequently facing the open sea, except in the case of Port Said Harbour, where the entrance is protected from the west. The breakwaters, however, act like groynes, and arrest a portion of the sand drifting along the shore, which accumulates in the angle between the breakwater and the coastline; and where the travel of the drift is almost wholly in one direction, owing to a great predominance of the wind from one quarter, the foreshore advances considerably on the side of the harbour facing the strongest winds, and is eroded on the opposite side, on account of the diversion of the drift by the windward breakwater preventing the replenishing of the losses from erosion on the opposite side. Thus the foreshore has progressed somewhat on both sides of Ymuiden Harbour, which is exposed to both easterly and westerly winds; whereas at Boulogne, where the coast is sheltered from the east, a progression of the shore has occurred on the exposed south-western side. The drift from the west at Port Said is increased by the alluvium discharged by the Nile; and the foreshore exhibits a notable advance since the construction of the western breakwater, extending out into depths of 36 feet some distance seawards of the end of the breakwater; whilst the shoreline has advanced on the western side of the harbour, and has been eroded to a similar extent on the eastern side.[1] The accumulation, however, of alluvium against the western breakwater, has been checked to some extent by allowing a portion to pass into the navigable channel along the inner face, through apertures formed across the crest of the breakwater, whence the deposit is readily removed by dredging; and the advance of the shore becomes slower as it progresses towards deeper water, and also because the onward motion of the drift-bearing current is facilitated by the curved line given to the shore by the filling up of the angle between the old shoreline and the breakwater. Madras Harbour manifests a similar, but much more rapid advance on its southern side, and a corresponding erosion on its northern side, owing to the south-west monsoon blowing with much greater force than the north-east monsoon (Fig. 306, p. 474); and the drift of sand along the shore from the south is exceptionally large, owing to the heavy surf which beats upon that open coast. In most cases, a sort of equilibrium is reached after the shore has advanced a certain amount against a breakwater projecting from a sandy coast; but at Madras, it appears probable that the shore will gradually creep out right to the end of the south breakwater; and the advance of the foreshore on the south side has already reduced the depth at the entrance some feet.

On a shore exposed to considerable littoral drift, it appears preferable to give a breakwater a direction sloping away somewhat seawards from the line of drift, as carried out at Ymuiden and Port Said, rather than to place the breakwater at right angles to the line of drift, as arranged at Madras, in order to prevent a less direct obstacle to the travel of the drift.

Open Viaduct and Outer Breakwater on Sandy Coasts.— Two objections have been raised against closed harbours on sandy coasts, with an entrance facing the open sea, namely, the advance of the shore against the projecting breakwater facing the direction from which the drift comes, and the deficiency of shelter inside the harbour with waves rolling in through the exposed entrance during a storm. Though neither of these objections are generally as serious as has been sometimes anticipated, they are both experienced at Madras Harbour. To obtain better shelter from onshore gales, and at the same time to avoid the favouring of accretion in the sheltered area, proposals have been made from time to time to form an outer harbour beyond the zone of littoral drift, with the sheltering breakwater connected with the shore by an open viaduct. This arrangement was proposed for Port Elizabeth, and partially carried out, on a much smaller scale, on the east coast of Ireland near Wexford, for a steamboat passenger service from Fishguard in Wales;[2] and the system has now been adopted for the harbour constructed at Zeebrugge, for sheltering the approach to the Bruges Canal, and also to serve as a port of call for passing steamers (Fig. 307, p. 474). At Zeebrugge, an embankment, protected on the sea side by a wall, and on the harbour side by a pitched slope, carries two

[1] *Proceedings Inst. C.E.*, vol. cxli. plate 2, fig. 5.
[2] "Harbours and Docks," p. 359.

lines of railway out to low-water mark, from which they run on to a metal viaduct supported on piles, extending out into a depth of about 20 feet at low water, and leading to a breakwater of concrete-blocks, which, curving round till it becomes parallel with the shore, forming the harbour, and sheltering the entrance to the canal, provides also for a quay with sheds and sidings protected by a high parapet on the sea side[1] (Fig. 315, p. 480). There is an open entrance to the north-east, as the site is fairly sheltered by the curving coastline from that quarter; and it is hoped that the open viaduct will admit of the free passage of the littoral current from the south-west with its burden of drift, and that thus the sheltered area will be preserved free from deposit. This work will afford an interesting experience as to how far it may be practicable in this manner to form a sheltered harbour on a sandy coast, capable of maintaining its depth in the presence of littoral drift. A certain amount of deposit can hardly fail to take place within the sheltered area, from the silty sea-water introduced at each tide, and from the fringe of the littoral current passing through the harbour; but the depth could be maintained by a suction dredger, specially designed to draw up a stream of silt containing so little water as to be readily deposited in the hopper of the dredger, such as is at present employed for maintaining the entrance channel to the Bruges Canal lock, where the rate of deposit amounts to a layer 3½ feet thick in a year.

If the drift of sand along the Madras shore ceases at a moderate distance from high-water mark, it would have been possible to cross the littoral current with two open viaducts at the sides of a wide harbour, leading to an outer breakwater parallel to the coast along its central portion, and curving round towards each end to join the viaducts, with an entrance placed conveniently for vessels, and facing the least-exposed quarter fulfilling this condition. By this arrangement, the large advance of the shore to the south, and the threatened invasion of the harbour by the drift might have been avoided, and a better sheltered harbour secured.

Entrances to Harbours.—It has already been pointed out that with single breakwaters protecting partially sheltered bays or recesses in the coastline, a wide approach is generally available round the end of the breakwater, of which Portland Harbour, previously to its further enclosure, furnished an instance. Where, however, harbours are enclosed by breakwaters, the entrances are restricted in width to what is necessary for the safe entrance of vessels; for a wide entrance facing an exposed quarter, especially if the harbour is small and the water deep at the entrance, admit waves large enough to disturb the tranquillity of the harbour. Thus Madras Harbour, with an entrance of the moderate width of 550 feet, is inconveniently disturbed during storms by waves rolling in from the open sea, owing to the exposure of the entrance, the depth of about 40 feet in which it is situated, and the

[1] *Proceedings Inst. C.E.*, vol. cxxxvi. plate 5, figs. 2 and 4; and "Construction du Môle du Port d'Escale de Zeebrugge, Congrès de Navigation, Paris, 1900," J. Nyssens-Hart and C. Piens, plates 1 and 2.

comparatively small area for the waves to expand over in entering the harbour. Vessels necessarily require a wider entrance where its position is exposed than where it is sheltered; whereas a wide entrance affects the tranquillity of a harbour in proportion to its exposure. These conflicting interests can only be reconciled by so placing the entrance that, without interfering with the access of vessels, it may not directly face the most exposed quarter. Two entrances facing in different directions may with great advantage be provided at large enclosed harbours; for this provision is very convenient for vessels; and two openings produce little additional disturbance in the harbour, on account of their different exposure.

The widths adopted recently for the entrances to important harbours range between about 500 and 800 feet; for the entrance to the new harbour of moderate size at the mouth of the River Wear, is 480 feet wide; the entrances between the new breakwaters enclosing the north side of Portland Harbour, to protect it from night attacks by torpedo vessels, are each 700 feet in width; the entrances to Colombo Harbour are 700 and 800 feet wide (Fig. 305, p. 474); and the widths designed for the entrances to Dover and Boulogne harbours are 600 and 800 feet, and 492 and 820 feet, respectively. The entrance, moreover, to the new harbour constructed at Havre, has been given a width of 656 feet.

Types of Breakwaters.—There are three distinct types of breakwaters, namely, a rubble or concrete-block mound; a mound surmounted on the top by a solid masonry or concrete superstructure, constituting a mixed type of breakwater; and a vertical wall, built up solid from the bottom to the top without any mound. The choice of type depends on the materials available at the site where the breakwater is to be constructed, the depth of the sea at the site, and the nature of the sea-bottom. In the first type, there is no quay or means of access along the breakwater; the second type, which, though variable in form, is the most common, has generally a quay of some kind, usually on the superstructure protected by a parapet wall, but sometimes formed on the inner side of the mound under the shelter of the superstructure, as at Marseilles and Trieste; whilst the third type has a readily accessible quay on the top of the wall, sheltered to some extent by a promenade or parapet wall (Figs. 308 to 315, p. 480).

Rubble or Concrete-block Mound Breakwaters.—The rubble mound is the simplest form of breakwater, consisting merely of rubble stone tipped into the sea from barges, or from waggons running along staging, in a definite line, till it emerges out of water, the mound being consolidated and its slopes regulated by the action of the waves. This system of construction is only applicable where large quantities of stone can be readily obtained from neighbouring quarries, as for instance at Plymouth (Fig. 308, p. 480), Portland, and Table Bay; and the amount of stone required increases rapidly with the depth, the exposure, and the rise of tide. At an exposed site, the large waves draw down the mound to a very flat slope on the sea side; and as the waves act on the mound from several feet below low water up to a few feet above high water, the scope of their action is greater in proportion to the tidal range. In order to reduce the effect of this disturbing action, the largest stones are laid on the sea slope of the mound; but during storms, the waves in recoiling are liable to draw down the stones on the exposed face, or, dashing over the breakwater at high tide, to carry over the stones near the top on to the inner slope of the mound, so that very exposed rubble mounds often need replenishing, after a severe storm, with fresh stone on the top and upper sea slope. The cross section of the Plymouth breakwater shows what a flat slope is formed by the waves on a mound exposed to the Atlantic (Fig. 308, p. 480); and the upper part of the mound has been protected by a granite paving set in cement, to reduce the erosion of the mound by the breaking waves.

As a rubble mound uses up a large quantity of stone, especially in deep water and an exposed position, occupies a wide space on the sea-bed, and is liable to disturbance, large concrete blocks have been substituted for stone, or employed on the sea slope and top of the mound, where stone is deficient, as at Port Said (Fig. 260, p. 409), or where the depth is considerable, as at Algiers, where the steep slope of the large concrete blocks on the sea face effects a large saving of material; or where increased stability is secured, and greater rapidity in construction attained by adopting concrete blocks for the outer half of the mound, as carried out at Alexandria (Fig. 309, p. 480). The concrete blocks are constructed within timber frames on shore, and after being left for a sufficient time to set, are carried out to their site on the deck of barges; and the lower blocks of the mound are tipped into the sea from an inclined plane on the barge; whilst the blocks near, and above the water-level, are lifted from the barge and deposited on the mound by a floating, steam derrick. Large blocks of concrete are usually lifted by means of lewis bolts passing through holes formed in the blocks when manufactured; but at Alexandria, the blocks were slung from angular pieces of iron embracing the corners of the block, which could be readily released by the pull of a rope, enabling forty to fifty 20-ton blocks to be deposited by the derrick in a day, so that the breakwater, nearly two miles long, was constructed at the very rapid rate of almost a mile in a year.

Rubble mound breakwaters abroad have been frequently formed of sorted materials placed in the mound according to their sizes, the smallest materials constituting the core of the mound, and increasing in size on approaching the outside, a method which involves sorting the materials, but adjusts the sizes of the rubble to their exposure in the mound, the largest rubble being placed on the outer slope. Concrete blocks, besides being stable at a steeper slope than ordinary rubble, and consequently economizing both materials and space, possess the advantage of being able to be increased in size in proportion to the exposure of the site.

Mound with Superstructure founded at Low Water.—A solid superstructure placed on a mound protects the top of the mound from the action of the waves, serves to form a quay or shelter, and reduces the mass required for the mound in proportion to the depth at which it is commenced. In the earlier examples of this mixed type of breakwater, the superstructure was commonly founded on a rubble mound at low water of spring tides; and this level has been adopted for founding the superstructures of the breakwaters in progress at Havre

BREAKWATERS. CROSS SECTIONS.

Fig. 308.—Plymouth.

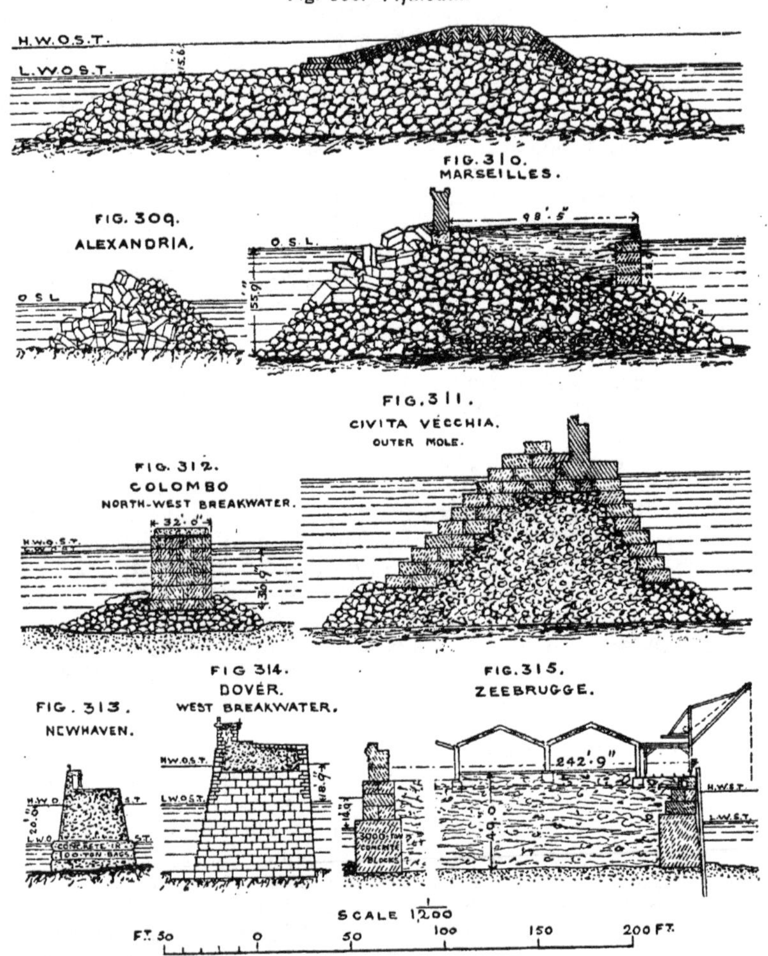

on a rubble mound; whilst the Boulogne breakwater furnishes a fairly recent example of a superstructure founded some feet above low water. In exposed places, however, the experience of severe storms soon proved that the heavy waves breaking on the rubble mound and dashing against the superstructure, lowered the rubble mound in front of the sea face in their recoil, and threatened to undermine the superstructure. This result was guarded against, either by raising the rubble mound against the face of the superstructure, reaching up to high water at the Holyhead and inner Portland breakwaters, and making it up again with fresh deposits of stone as it was lowered by the waves; or more effectually, by laying an apron of large concrete blocks on the top of the mound along the sea face of the superstructure, as carried out at Cherbourg and Genoa, or a covering of concrete blocks deposited at random over the sea slope of the mound, as for instance adopted at Marseilles (Fig. 310), and Oran breakwaters; or a combination of the two, as exemplified by the Boulogne and Havre breakwaters. In some of the more recent breakwaters on the Italian coasts, the concrete blocks covering the rubble mound down to the limit of the possible disturbing action of the waves, have been laid in regular courses stepping back from the face of the slope, of which the outer breakwater at Civita Vecchia, in deep water, is a good example (Fig. 311); for this arrangement has proved more stable than concrete blocks thrown pell mell on the slope, owing to the blocks resting on a flat bed and offering less surface to the impact of the waves.

Marseilles breakwater provides an instance of a quay formed on the inner part of the mound, bounded by a quay wall on the harbour side, and sheltered by the superstructure founded at sea-level (Fig. 310); and the quays on the Holyhead and inner Portland breakwaters are similar in principle. At Cherbourg, St. Jean de Luz, Leghorn, Boulogne, and Havre, the quay is formed on the top of the wide, solid superstructure, and sheltered by a parapet on the sea side,[1] resembling the form commonly given to the quay of upright-wall breakwaters (Figs. 313 and 314).

The defects of this form of the mixed type of breakwater are similar in two respects to those of mound breakwaters, which it much resembles, namely, that it requires a large amount of materials, and that its sea slope is subject to disturbance by breaking waves, unless covered by large blocks. Moreover, though the superstructure protects the top of the mound on which it rests, and prevents the materials on the sea slope and at the top from being carried over by the waves into the harbour, it increases the force of the recoil of the waves, and is liable to be undermined unless the mound in front of its toe is effectually protected by an apron of large blocks. Accordingly, though the amount of materials required is somewhat reduced in this form of breakwater by the adoption of a superstructure, as compared with a simple mound, the sea slope has to be more strongly protected in proportion to the protection afforded to the quay; and therefore the main advantage of this form, over a mound breakwater, consists in the provision of a quay and means of access to the end of the breakwater.

Mound and Superstructure founded below Low Water.— A more efficient method of providing against the disturbance of the mound and the undermining of the superstructure, consists in stopping

[1] "Harbours and Docks," plate 7, figs. 1, 2, 6, 8, 11, 21, and 22.

the mound at a depth at which it is not liable to be disturbed by wave-action, and founding the superstructure on the mound at this level. The depth to which the wave disturbance extends, increases with the exposure of the site; and the low level of the mound necessitates the construction of a considerable portion of the superstructure below low water, where the ordinary method of connecting the blocks together by mortar cannot be resorted to.

When this form of breakwater was first introduced, with the object of avoiding the erosion of the rubble mound and the undermining of the superstructure, and also of reducing the amount of materials, it was supposed that wave-action ceased at about 12 feet below low water; but extended experience has proved that the action is felt, in exposed situations, at considerably greater depths. Thus the breakwater at Alderney, exposed to the Atlantic, was commenced with its superstructure founded at low water of spring tides, and the rubble raised against its sea face up to about half-tide level; the foundation of the superstructure, as the breakwater progressed, had soon to be carried down to 12 feet below low water, which towards the outer part settled on the mound to about 17 feet below low water, and had to be protected by frequent deposits of stone; and the head, in a depth of 130 feet at low tide, was founded 24 feet below low water, which was increased to 30 feet by settlement.[1] The solid, concrete-block superstructure of the outer portion of the south-west breakwater at Colombo, exposed to the south-west monsoon, founded for about 1100 feet at 16 feet below low water, and protected by rubble and two rows of concrete in bags on the sea slope, was eventually carried down to 20 feet below low water along the outer half of the breakwater, and the pier-head to 23¾ feet; and the superstructure of the north-west breakwater, subsequently constructed, has been taken down to 30¾ feet, and protected near its base, on the sea face, by rubble overlaid with concrete in bags (Fig. 312, p. 480).[2] Lastly, the breakwater at Peterhead, which, commencing as an upright wall on a rocky bottom, is, in deeper water, founded on a rubble base, was designed to be founded at a depth of 30 feet below low water, but as during a storm in 1898, disturbance was found to take place at 36½ feet, the superstructure was founded at 43 feet below low water. The Detached Mole at Gibraltar in a less exposed situation is founded on a rubble mound at 36 feet below low water.

This type of breakwater is adopted where the depth is considerable, or the sea-bottom is soft or easily eroded. In securing, however, the mound from disturbance, and the superstructure from undermining, the superstructure becomes the weak point of the system; for besides sharing with other superstructures the liability to considerable settlement on a high rubble mound, which its weight due to its greater height increases, its blocks below low water, being unconnected by mortar, are liable to be displaced in detail. The unequal settlement on a high,

[1] *Proceedings Inst. C.E.*, vol. xxxvii. p. 61, and plates 3 and 4.
[2] *Journal of the Society of Arts*, December 15, 1899, pp. 85 and 86; and "Recent Works at British Seaports and Harbours, Paris Navigation Congress, 1900," plate 2, figs. 8, 9, and 10.

yielding mound, which occurs in building out the superstructure in stepped, bonded courses, as the weight is increased at the inner end of the length in progress by raising the wall to its full height, tends to dislocate the wall more or less below low water, and to produce cracks in the connected masonry above low water, which are readily enlarged by the waves during storms.

In constructing several breakwaters of the mixed type, as for instance at Gibraltar, Simons Bay, Karachi, Mormugao, and Colombo, settlement on a rubble mound has been provided against by depositing the blocks for the superstructure on a slope, in a succession of transverse, unconnected sections inclined outwards towards the bottom, as shown in Fig. 316, with the rows of blocks sloping sufficiently back on the

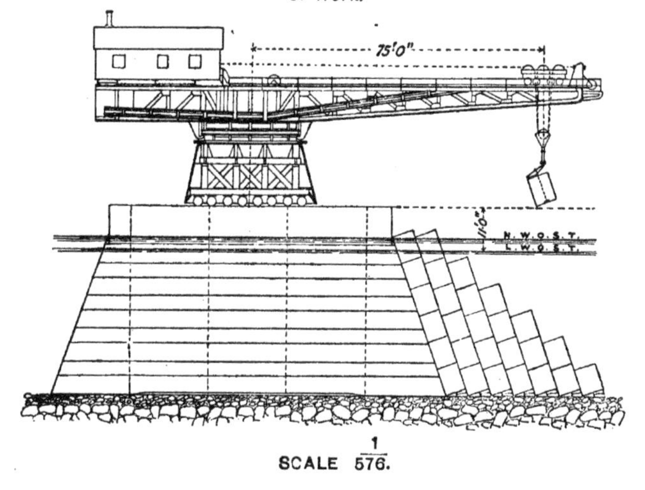

REVOLVING TITAN LAYING SLOPING BLOCKS.

Fig. 316 — Superstructure on Rubble Mound, with Caisson for commencement of Work.

SCALE $\frac{1}{576}$.

preceding rows as not to be liable to tip over forwards, and yet not so much as to prevent each section settling independently. In the earlier designs of this type, as at Madras and Karachi, two unconnected rows of concrete blocks were used with a vertical joint between them. This was seen to be liable to displacement and has been superseded in later examples by a system of break of bond throughout, combined, as at Gibraltar, with the use of concrete joggles whereby the blocks are mortised and tenoned to each other, the superstructure being capped, after settlement has ceased, by concrete in mass, with or without a parapet wall, joggled and tied by steel rails to the blockwork below. At Colombo the sections are connected together from top to bottom by vertical grooves formed in the sides of the blocks and filled with concrete in bags serving as dowels, but in other examples the sections are independent, and free to take up any unequal settlement, movement in a transverse direction being prevented by the capping. At Gibraltar the sloping block construction gives place, at the heads, to a horizontal system forming an abutment to the sloping blocks.

The superstructures of the outer arms of the Madras breakwaters, facing the open sea, when reconstructed after their overthrow, on the eve of completion, during a cyclone in 1881, were protected on the sea side by a mound of concrete blocks, termed a wave-breaker, deposited at random on the rubble mound. The original superstructure was undermined in places by the waves, though founded at a depth of 22 feet below low water, so that the rubble mound needed some protection against the recoil of the waves near the sea face of the superstructure; but the wave-breaker is mainly intended to break up the waves before they reach the superstructure, and thus preserve the superstructure. A similar wave-breaker was adopted at Mormugao, for protecting the superstructure constructed after the damage to the Madras breakwaters had occurred, and was introduced at Ymuiden to secure the upright-wall breakwaters there against a similar injury. A mound of concrete blocks, however, is a costly and indirect method of using a large quantity of additional material for strengthening the superstructure; and it would be preferable to add directly to the strength of the superstructure, by connecting the blocks as previously described, and widening it if necessary, and to increase its stability by carrying the foundations lower down, or by protecting the sea slope of the mound near its face with concrete blocks or bags, according to the system developed at Colombo.

Upright-wall Breakwaters.—Where the depth is not too great, and the bottom is rocky, or firm enough to be readily protected from scour by a layer of rubble, a breakwater may advantageously be built up from the bottom in the form of a thick, fairly upright wall. This arrangement reduces the amount of material required to a minimum; it avoids all danger of unequal settlement upon a yielding mound, and the erosion of the mound by waves breaking on it; but it involves, like the deep superstructures of the mixed type, the building of a considerable height of wall under water, necessitating special care and contrivances for rendering the wall compact, and for preventing unconnected blocks being forced out by the sea; for when once a hole is formed in a structure in the sea, it is rapidly enlarged by waves in storms.

One of the earliest applications of this system was at the west breakwater at Dover, known as the Admiralty Pier, extending out from the shore into a depth of about 35 feet at low water of spring tides, where the chalk bottom was levelled by men in a diving-bell, and the wall was built of bonded courses of concrete blocks laid by the same means up to low water, where granite ashlar facing was introduced, and eventually concrete-in-mass for the upper part of the backing (Fig. 314, p. 480). The same system of construction has been used for the large extensions which now constitute the enclosed naval harbour at Dover, with the exception that concrete blocks faced with granite rubble masonry have been used in place of granite ashlar; and the weight of the concrete blocks has been increased from about 8 tons used at the commencement, up to 20 tons in the East or Prince of Wales Pier, and a maximum of 42 tons in the extensions; whilst in addition to being bonded, the blocks are connected together by dowels at their sides. An apron of concrete blocks on the chalk bed, along the face of the wall, protects the toe of the breakwater from being undermined by the scour of the currents.

The levelling of the foundation under water with the requisite exactness to lay the blocks uniformly throughout, is a costly and slow process even in chalk; and, accordingly, the irregular granite bottom on which a portion of the south breakwater at Aberdeen was built, was levelled with concrete in bags to receive the foundation course of blocks, and was extended, in constructing the north breakwater, to a series of layers of 50-ton concrete bags carried up to a little above low water, in place of blocks, upon which the upper part of the breakwater, formed of concrete-in-mass deposited within framing, was constructed. The latter system, dispensing with the laying of any blocks under water, was followed for the Newhaven breakwater, where 100-ton concrete bags were deposited from a specially-constructed hopper barge, extending across the full width of the breakwater, raising the base above low water, on which the concrete-in-mass upper portion was erected (Fig. 313, p. 480). The jute cloth enclosing the concrete, protects it from wash during its passage through the water; and sufficient cement oozes out from the bags, under pressure, to join the bags together in a solid mass.

In the absence of funds for providing an expensive plant, small breakwaters for sheltering fishery harbours have sometimes been constructed in a solid mass, on a firm bottom, by depositing concrete within timber framing, raised up to low water by lowering the concrete in a closed skip with a movable bottom, through the water, to the sea-bed, or to the top of the concrete previously deposited, and then opening the flaps of the skip and releasing the concrete. By this means, the concrete is shielded as far as practicable from the wash of the water in the process of being deposited; and the timber framing, lined with jute sacking, protects the mass of concrete from currents and waves till it has set thoroughly, when the framing is removed for the construction of another length of breakwater. The breakwater is raised to its full height by concrete-in-mass brought up in layers, within framing, on the top of the subaqueous concrete, thereby forming the breakwater in a solid block not liable to be disturbed by the sea; but as a long, continuous length of concrete is liable to crack at intervals when exposed to changes of temperature, it is advisable to form the concrete-in-mass above low water with vertical joints, from about 15 to 30 feet apart, so as to provide for expansion and contraction, and to avoid the formation of unsightly irregular cracks across the breakwater. The concrete below low water must be made with a larger proportion of Portland cement than the concrete above, to allow for loss from wash, and with small stones to ensure the compactness of the mass; but the concrete deposited out of water may with advantage have large stones embedded in the central mass, away from the face where a rendering of strong Portland cement mortar should be introduced. The bottom along the sea face and end of the breakwater, except where it consists of hard rock, should be protected by an apron of concrete in bags against possible erosion by currents or recoiling waves.

The breakwater at Zeebrugge, for forming a harbour at the entrance to the Bruges Canal, and intended to serve also as a quay, has been formed by a sea wall and harbour wall with intermediate filling. Each wall is brought up from the bed of the sea to above low water, by a single row of concrete blocks of 2500 to 3000 tons, which were constructed within iron caissons erected, and partially filled with concrete, in the dry bed of the canal,[1] landwards of the lock, before the canal was filled with water, and have now been floated out one by one in calm weather, and sunk in position by the admission of water; after which, the block is completed by filling up the caisson with concrete (Fig. 315, p. 480). The firm sea-bottom, on which the caissons are stranded, is levelled where necessary beforehand by small rubble; and a layer of rubble is deposited alongside the sea face of the blocks of the sea wall, to protect them from being undermined. The narrow, outer portion of the breakwater beyond the quay (Fig. 307, p. 474) is constructed from the bottom, up to 3¼ feet above low water, by a single row of still larger blocks, formed in caissons, 82 feet long, 29½ feet wide, and 28¾ feet high, and weighing about 4400 tons when completed; and upon these foundation blocks, 55-ton blocks are lowered and bedded in mortar to form the upper part of the breakwater.[2] This is a large extension of the system first successfully employed for forming the lower portion of the superstructure of the western breakwater protecting the mouth of the Nervion, founded on a rubble base, 16¼ feet below low water, with blocks of 1400 tons towed out in caissons when partially formed (Fig. 227, p. 377), after the attempt to construct a superstructure founded at low-water level on a mound of large concrete blocks laid upon a rubble base, had failed in two successive winters at that very exposed site. The value of the system consists in the formation of the foundations of a breakwater with very large blocks capable of resisting a very heavy sea, constructed under more favourable conditions than can be attained by other systems of laying foundations under water in the open sea; but, on the other hand, there is some difficulty in placing such large blocks in the exact line and perfectly level; and very calm weather is required for the operation of towing out and sinking the blocks in position, which, moreover, are not perfectly secured till filled with concrete.

Construction of Superstructures and Upright Walls for Breakwaters.—Formerly, the superstructures of the mixed type of breakwater, and also upright-wall breakwaters, were constructed by means of staging carried out into the sea along the line of the proposed structure. In moderate depths, the staging was serviceable in the tipping of the mound of a composite breakwater, as well as in the construction of the superstructure by carrying travelling gantries for handling the stones or blocks, as exemplified by the Holyhead and inner Portland breakwaters, where trucks conveyed the stone direct from the quarries. Massive staging, consisting of steel girders on Tasmanian blue-gum piles, and carrying powerful gantries, was also used at Dover Harbour Extension

[1] *Proceedings Inst. C.E.*, vol. cxxxvi. plate 5, figs. 5 to 7.

[2] "Construction du Môle du Port d'Escale de Zeebrugge, Congrès de Navigation, Paris, 1900," J. Nyssens-Hart and C. Piens, plate 2.

Works. Such staging is costly and like other Harbour Plant exposed to loss or damage by storms.

Where the depth is considerable, or where local conditions require it, rubble mounds are sometimes deposited from hopper barges towed to the site, as at Alderney and at the Detached Breakwater at Portland. In the latter case the rubble was tipped direct from the truck lowered down the inclines from the quarries into specially designed hopper barges, by powerful steam cranes.

In order to expedite the progress of the works, and to avoid the injuries to which staging is subject in exposed situations, overhanging block-setting cranes, called "Titans," were introduced, which, travelling along the completed breakwater, can set blocks from the outer end of the finished work, and can be run back into shelter on the approach of a storm [1] (Fig. 317, p. 488). Simultaneously with the introduction of Titans, the sloping-block system of construction for superstructures on a rubble mound was adopted, reducing the amount of overhang necessary, as compared with horizontal courses which have to be carried out in steps in advance. The earlier Titans could only travel forwards and backwards on the breakwater, though admitting a sufficient lateral movement of the trolley, from which the block was suspended at its proper slope, to enable two rows of blocks to be laid side by side. Later Titans, however, have been constructed which can swing round on a ring of rollers placed on the top of the travelling truck which carries the Titan, and, consequently, have a greatly increased range of work, being able not only to lay blocks in advance of the breakwater, but also to deposit large blocks or concrete bags for forming an apron along the sea face. Titans of this type were used in the construction of Gibraltar Detached Mole, the commencement of this work in the open sea, and without staging, being made by the sinking of a steel caisson filled with concrete,[2] of sufficient dimensions to enable the erection of the first Titan to be completed, when it proceeded to make its own road by the deposit of sloping blockwork, Fidler's Patent Block-Tilting Gear being used for this purpose (Fig. 316). By this means room was made for the erection of a second Titan working in the opposite direction. Each Titan lifted a proof load of 45 tons at 75 feet radius. The Peterhead Titan (Fig. 317), laying 50 ton blocks in horizontal courses, lifts a proof load of $62\frac{1}{2}$ tons at 100 feet radius. Titans at Colombo and South Shields lift proof loads of 42 tons at 60 feet, and 45 tons at 75 feet respectively. All the examples cited are worked by steam power and are provided with motions for lifting, slewing, racking in and out, and self-propelling, together with hydraulic brakes for lowering heavy loads. Electrical power is occasionally employed, as in the Titans used at the mouth of the Nervion and at Zeebrugge.

Remarks on Breakwaters.—All the three types of breakwaters

[1] *Proceedings Inst. C.E.*, vol. lxxxvii. plate 2, figs. 10 and 11.

[2] For detailed information as regards this caisson, see "Notes on Construction in Mild Steel." Longmans' Civil Engineering Series.

[3] *Mémoires de la Société des Ingénieurs Civils de France*, 1900 (2), p. 87; and "Harbours and Docks," plate 8, fig. 8.

described above are still carried out for sheltering harbours; for the rubble mound has been quite recently used for enclosing the northern end of Portland Harbour against night attacks by torpedo vessels, and is the system adopted for the northern breakwater constructed at Colombo for sheltering the harbour from the north-east monsoon; breakwaters of the mixed type, of mound and superstructure, have been used at Havre, Gibraltar Detached Breakwater, Colombo north-west breakwater, and Peterhead, with their superstructures founded at low-water, and 36, 36, and 43 feet below low water of spring tides, respectively; and

REVOLVING TITAN.

Fig. 317.—Laying Blocks in Horizontal Courses.

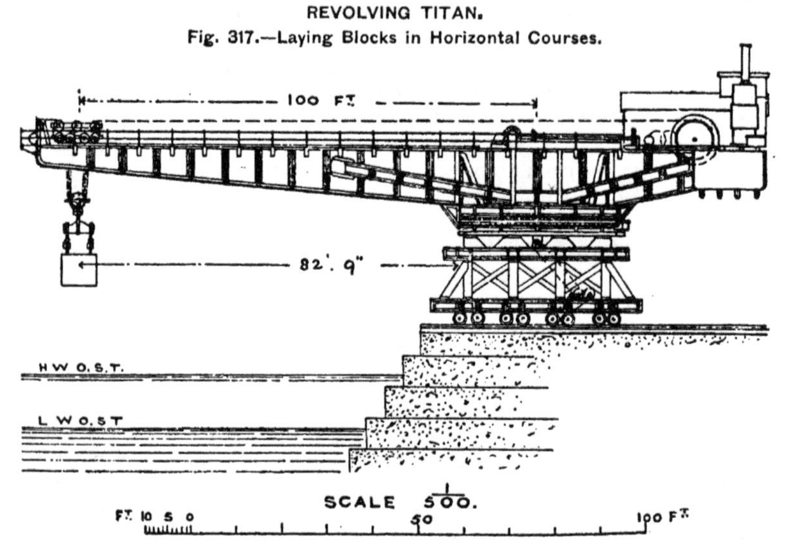

upright-wall breakwaters have been employed at Dover and Zeebrugge. In all breakwaters, the portion exposed to the direct stroke and recoil of the waves is the most subject to injury; and the zone of this action is increased in proportion to the tidal rise, and its influence in proportion to the exposure. Mound breakwaters require their exposed sea slope to be protected by large blocks, or smooth pitching in cement.

The most vulnerable part of a superstructure, or of an upright-wall breakwater, is at low-water level, and for a few feet below, especially where settlement occurs, or a comparatively narrow sea wall is adopted, with rubble filling between it and the harbour wall,—a somewhat common practice in former days, and even followed along the first 1326 feet of the south-west breakwater at Colombo. The blocks at, and below low water, have open joints, into which air penetrates on the recoil of a wave, and also fills any cavities behind due to settlement or other causes; and then the succeeding wave, compressing the air inside, leads to the gradual forcing out of a block by the pressure from behind on the retreat of each wave during a storm, unless the weight of the blocks above, or its connection with adjacent blocks by joggles or dowels, prevents this movement. When once a face-stone has been drawn out, the enlargement of the hole by the sea, and, in the case of a narrow sea wall, the piercing of the wall and the scooping out of the rubble filling, resulting in a breach through the breakwater, merely depends upon the duration of the gale. These injuries may be avoided, by making the superstructure or upright wall solid throughout, with blocks laid as close together as practicable; by laying the blocks in sloping rows to prevent irregular settlement, and consequent dislocation of the superstructure on a high rubble mound; and by connecting the blocks under water firmly together by joggles and dowels, and giving the wall only a slight batter so as to throw the weight of the upper part of the wall on to the lower face-blocks. The stroke also of the waves against the superstructure or upright wall may be somewhat reduced, when the provision of a quay is not required, by raising the wall only just sufficiently high to shelter the harbour, and dispensing with a parapet, so that the upper part of the waves may pass harmlessly over the wall into the harbour, instead of dashing against the breakwater; and the recoil of the waves is also thereby reduced. The most certain way, however, of securing, in a moderate depth of water, the portion of a superstructure or upright wall below low water from injury, is by forming the whole of the subaqueous part of the work of very large concrete blocks, which, with the upper part laid in mortar, constitute masses too heavy to be disturbed by the sea, as effected for a composite breakwater at the mouth of the Nervion, and for an upright-wall breakwater at Zeebrugge.

Though superstructures founded at low water upon a protected rubble mound may prove stable in fairly sheltered situations, such as Havre and Boulogne, it is evident, from the experience at the western breakwater at the mouth of the Nervion, and the breakwater at St. Jean de Luz, that this arrangement is not suited for very exposed sites, even when the rubble mound is capped with large concrete blocks. It appears, accordingly, expedient, in the case of the mixed type of breakwater, to found the superstructure at a sufficient depth for the mound, protected by an apron of concrete bags, not to be disturbed by the sea, which can readily be carried out by modern block-setting Titans, or in moderate depths, and under suitable conditions, by large caisson blocks towed out to the site, and also to adjust the width of the superstructure to its exposure.

Undoubtedly the upright wall provides the most satisfactory and stable type of breakwater on a firm bottom, and in a moderate depth of water; but a breakwater built up in courses on a levelled bed, like the breakwaters at Dover, is costly; concrete-bag foundations up to low water require to be very carefully laid, and necessitate special plant; whilst subaqueous foundations formed with concrete deposited by skips within framing, though economical for moderate depths, demand the greatest care in execution to avoid injury to the concrete by the wash of the sea. The large increase, however, in the size of blocks laid by Titans, combined with their much more efficient connection together under water, the formation of large monoliths of concrete deposited in a heap of bags, and in mass within framing, and the construction and placing in position of large blocks within steel caissons, have added greatly to the stability of breakwaters in exposed situations.

CHAPTER XXX.

LIGHTING COASTS AND CHANNELS.

Sites of lighthouses, on land, on isolated rocks, materials employed—Foundations of rock lighthouses, arrangements, connections of rock and stones of tower, provisions at New Eddystone for expediting work—Arrangements for building rock lighthouses—Forms of rock lighthouses, batter, cylindrical base at New Eddystone and Bishop Rock—Internal arrangements of rock lighthouses, cost—Lights exhibited from lighthouses, electric light, gas, oil, distinguished by flashes of concentrated rays, increased rate of rotation, flashes at Cape Béar, Eddystone, and Bishop Rock, intensity of electric arc lights—Fog signals at lighthouses, bells, explosions; sirens, range of sound—Screw-pile lighthouses, instances—Beacons, materials employed, methods of construction, instances founded under water—Lights on beacons, with compressed oil gas, provided with flashing lights, power—Light-ships, means of increasing stability—Lights on light-ships, suspended lamp with improved stability, illuminant and power, flashing light—Light-giving buoys, objects, instances of use, stable type for exposed sites, power of light, details.

LIGHTHOUSES are erected on headlands, islands, and rocks in the sea, to guide the mariner as to his position and course at night, and to warn him of his approach to the coast and the neighbourhood of outlying reefs; and they are also placed on pierheads, and at the mouths of navigable rivers, to mark the entrance to harbours, and to direct vessels into river channels, two leading lights being often so placed as to appear in line when a vessel is following its proper course towards the entrance channel. The erection of lighthouses on land, or to indicate the entrance to ports, presents no difficulty; and the chief points to be determined in such cases are, the height and power of the light, which settle its range of visibility, its distinguishing characteristic, and with leading lights, their most suitable relative positions and arrangements to guide and warn the pilot; whilst with very high lighthouse towers in very open, and consequently exposed positions, such as the tower recently constructed on Vierge Island off the north-western part of the French coast, with a height of about 270 feet and a base of 52½ feet diameter, their stability under very vehement blasts of wind has to be secured, by giving them an ample base on a solid foundation, and building them of the strongest masonry set in cement mortar. Lighthouses, however, which have to be built on isolated rocks in the sea at some distance from the coast, both to guide and warn vessels, offer considerable difficulties in their foundations and lower courses, in proportion to the lowness in level of the rock, and the exposure and depth of the sea at the site.

Rock lighthouses are generally constructed of granite, as being the hardest available stone; and lighthouses on land are usually built of masonry or iron; whilst concrete-in-mass was employed for the Corbière Lighthouse on a high rock projecting from the coast of Jersey, and also for the Stannard's Rock Lighthouse in Lake Superior.

Foundations of Rock Lighthouses.—The outlying rock on which it is proposed to erect a lighthouse in the sea, has to be carefully

ROCK LIGHTHOUSES.

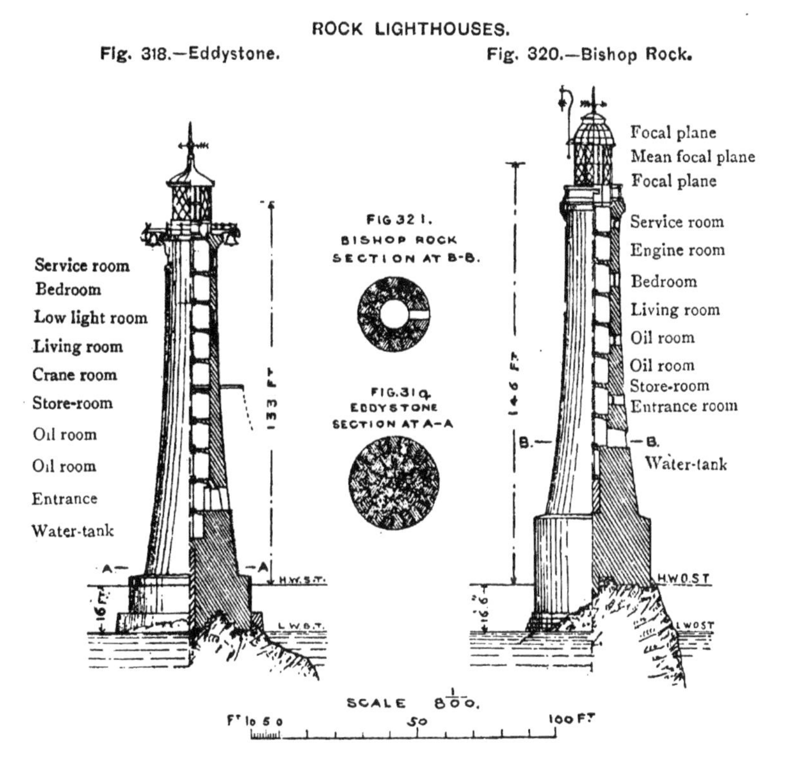

surveyed, so that the area available for the foundations, the general configuration of the rock, and the levels of its various parts, with reference to low water of spring tides, may be accurately known; and these data enable a design for the lighthouse tower to be prepared. The rock has then to be dressed in steps and levelled to receive the foundation stones, the lowest stones being first laid, and carried up in courses stepping up in the rock over an increasing portion of the area, till at last, on rising above the highest level of the rock, complete circular courses can be laid. Though some lighthouses in the Great Lakes of North America have been founded on reefs under water, by enclosing the area within a cofferdam or a wrought-iron casing, and pumping out the water, this system would not be feasible on irregular rocks, very limited in area, and in an exposed situation in the open sea: and therefore generally the whole of the area of the isolated rock on which a lighthouse tower has to be founded, should be above the lowest low water, to enable the foundation stones to be cemented to the rock. Fortunately, rocky shoals in the sea, having resisted the action of the surf for an indefinite period, are composed of the hardest strata; and, accordingly, though troublesome for preparing the foundations, they afford a very firm base for lighthouse towers.

Whilst the rock is being prepared to receive the foundation courses, the stones are dressed in the work-yard to the exact sizes for their position in the tower, and numbered. The stones in the lower courses, besides being set in cement, are dovetailed together, both horizontally and vertically (Figs. 319 and 321, p. 491); and the lowest courses are often connected with the rock by vertical bolts fixed firmly into holes bored in the rock, and passing through the foundation stones, which are further secured by their beds being sunk below the level of the surrounding rock. The benched dressing of the lowest part of the rock foundation, for which blasting cannot be employed for fear of shaking the adjacent rock, and the laying and fixing of the lowest stones, necessarily constitute the most tedious part of the work, especially where the foundations approach low-water level and the site is exposed; for the possibility of working at all on the rock, depends on the conjunction of calm weather with low water of spring tides. To facilitate this work, the foundations of the new Eddystone Lighthouse were enclosed by a brick cofferdam set in quick-setting Roman cement, 7 feet high (Fig. 318, p. 491); and a granite ashlar platform was built up in the centre to 10 feet above low water of spring tides. The enclosure, moreover, was divided into three sections by three radial walls, out of each of which the water was pumped successively through two hoses in connection with pumps on the attendant steamer, as soon as the tide had fallen to the level of the top of the brick cofferdam. The central platform afforded a landing-place for the men on a falling tide, where they were in readiness to commence work on the rock at the earliest possible time; the first section pumped dry afforded an earlier opportunity of starting work than would have been otherwise possible; and the shelter of the cofferdam enabled the work to be continued till the tide rose again above the dam, or to be carried on in the most sheltered section when the wash of the sea prevented work on the windward side.

Arrangements for building Rock Lighthouses.—As the work-yard where the stones are dressed is on land, and often many miles from the site of the lighthouse, the stones have to be conveyed by vessels to the locality. In some recent cases, the steamer carrying the stone has served also for accommodating the workmen, and for landing the stones. In the case of the Beachy Head New Lighthouse, a granite tower was founded on the chalk foreshore about 700 feet from the cliff and surmounted by a first order lantern, the focal plane being 103 feet above high water. The depôt, work-yard, and workmen's barracks were situated on the summit of the cliff, about 433 feet above low water, and communication was established by means of a cable-way from the depôt to temporary staging 23 feet above high water of spring tides. By these means workmen and materials were conveyed up and down.

Forms of Rock Lighthouses.—Lighthouse towers have generally been given a curved batter on the face, so as to spread out more rapidly towards the base, and thus secure as large an area at the foundations as the rock conveniently affords; though occasionally a straight batter has been adopted in lighthouses of moderate height, to suit special conditions. The curved batter, however, though affording the most stable form of tower, has been found to favour the upward run of the waves, which eventually rising still higher as spray over the lantern, have occasionally hidden the light for several seconds at the Bell Rock Lighthouse, as well as at the old Eddystone and Bishop Rock lighthouses. Accordingly, in rebuilding the two latter lighthouses, the lower parts of the towers have been made cylindrical, up to 2½ feet above high water of spring tides at the new Eddystone Lighthouse,[1] and 23 feet at Bishop Rock Lighthouse,[2] which is exposed to a still heavier sea than the Eddystone (Figs. 318 and 320, p. 491); and this cylindrical base causes the waves to divide and break, and thus, though receiving heavier strokes itself, checks the waves from running to the same extent up the tower. The cylindrical base is 44 feet in diameter at the Eddystone, and 41 feet at the Bishop Rock, and its top forms a very convenient platform for landing stores; and in each case, the lighthouse tower, with a curved batter, starts from the top of the cylindrical base, having a bottom diameter of 36 feet at the Eddystone, and 32 feet at the Bishop Rock.

Internal Arrangements of Rock Lighthouses.—To strengthen the tower where most exposed to the stroke of the waves, it is brought up solid for some distance, according to the exposure of the site and the height of the tower. Thus, with the exception of the small hollow required for the water-tank, the Eddystone Lighthouse has been carried up solid from its foundation to a height of 25½ feet above high water of spring tides, and the Bishop Rock Lighthouse 47 feet above high water. The entrance to the lighthouse is placed on the top of the solid portion, and is reached from below by a metal ladder fixed to the side of the tower; and the several rooms for the stores, and the accommodation of the lightkeepers, are arranged in floors above, the service room for the lamps being placed just below the lantern (Figs. 318 and 320, p. 491). The rooms are generally made of the same size, the thickness of the walls being gradually reduced upwards by the batter of the tower; and they are commonly reached by a light iron staircase at the side of each room.

[1] *Proceedings Inst. C.E.*, vol. lxxv. plate 2.
[2] *Ibid.*, vol. cviii. plate 4, fig. 3.

A cornice at the summit of the tower provides a gallery round outside the lantern, which, at the Eddystone Lighthouse, has two fog bells suspended from its under side; and a copper lightning-conductor, extending from the top down to below low water, secures the structure from being damaged by lightning.

The cost of rock lighthouses ranges from about 18*s*. to £2 per cubic foot of contents.

Lights exhibited from Lighthouses.—The intensity given to a light placed in the lantern at the top of the lighthouse, is determined by the importance of the light for shipping, and the facilities for its production. Electricity, gas, mineral oil, and colza oil are all used as illuminants for lighthouse lamps; but the electric arc light is restricted to very important headlands, such as St. Catherine's, the southernmost point of the Isle of Wight, and Cape Gris-Nez on the French coast, where a very powerful light is so important that its cost is a secondary consideration, and ample space is available for generating the electricity. Gas has also been adopted for several lights on land; and mineral oil is used for a great number of lighthouses on shore. Colza oil, on the contrary, is ordinarily employed for rock lighthouses, owing to its being less readily inflammable than mineral oil, and the importance of avoiding any chance of fire in such isolated buildings as rock lighthouses.

The power of the light is greatly intensified by concentrating the rays emitted from the lamp by means of panels of lenses, so constructed and placed as to direct the light in one line (Fig. 322); and, by causing the optical apparatus to revolve, the concentrated ray sweeps successively the different points of the horizon. Moreover, by varying the rate of rotation, and the times of the duration of the light and of the intervening obscuration, a distinctive character can be given to each light, so that the sailor does not merely see the light, and has to judge from the position of his ship what light it is, but the light exhibited, by its succeeding periods of light and darkness peculiar to itself, informs him which the lighthouse is as indicated on the charts, and prevents the possibility of one lighthouse being mistaken for another, which in former days was occasionally the cause of shipwrecks. Formerly, the rate of rotation was slow, and the character of the light was wholly determined by the length of duration of the light and dark periods; and in some cases, coloured lights were introduced to differentiate more clearly between the various lights. Eventually, however, as coloured lights have much less intensity than white light, shorter flashes in groups were adopted, with a longer interval of obscuration between each group, affording greater scope for varying the character in order to distinguish each light. Finally, the rotation of the optical apparatus has been greatly facilitated by making its circular outer rim float in a corresponding annular bath of mercury, enabling the rotation to be effected with little power, and at a speed ranging from 25 seconds down even to 5 seconds. This notable augmentation of the speed, by diminishing the duration of the lightning flashes, in France, towards the necessary limit for perception, estimated at $\frac{1}{10}$ second, has reduced the number of panels of lenses required for each flash, and enabled lenses of large surface, and, consequently, increased power, to be introduced, materially increasing the intensity of the flashes.

The optical apparatus of Cape Béar Lighthouse is shown in Fig. 322, which completes a revolution in 20 seconds, and emits a group of three flashes, each lasting 0·1 second, with intervals between each flash of 3·9 seconds, and an interval of 11·9 seconds between each group; whilst the power of the flash amounts to 237,500 candles, with an

LIGHTHOUSE LENSES.

Fig. 322.—Three-flash Light at Cape Béar.

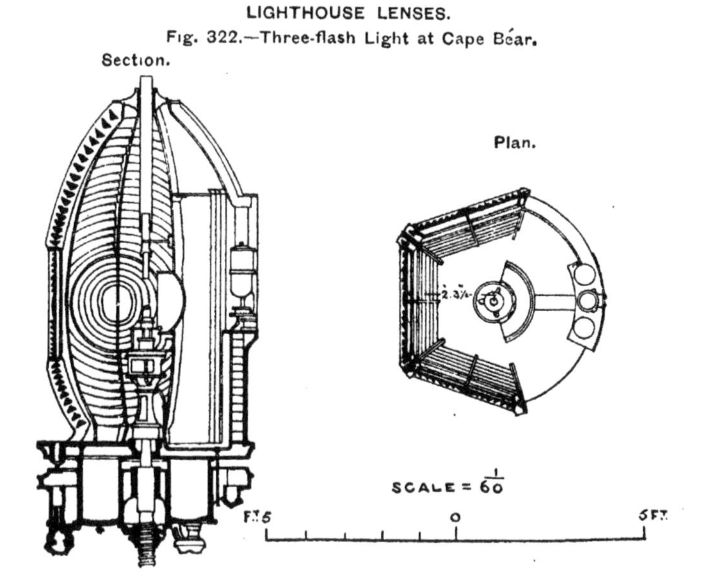

incandescent light derived from petroleum vapour.[1] The Eddystone Lighthouse exhibits flashes in groups of two, at intervals of 21 seconds, each flash lasting $2\frac{1}{2}$ seconds with an interval of 4 seconds between the two; and a revolution occupies 3 minutes, with six groups of flashes. The light in this case consists of two superposed burners, fed with colza oil, having a combined intensity of 1900 candles, which is increased by the optical apparatus to 79,000 candles for the flashes.[2] The Bishop Rock Lighthouse exhibits five similar groups of double flashes, of $4\frac{3}{5}$ seconds' duration, with $4\frac{3}{5}$ seconds' interval between the flashes, in a revolution lasting 5 minutes; and the light produced by two superposed burners fed by heavy mineral oil, gives out flashes having an intensity of 170,500 candles. The advantage of these superposed burners is that they enable the intensity of the light to be doubled in foggy weather; whilst in clear weather, only one burner is used, and is worked considerably

[1] "Notices sur les Appareils d'Éclairage, Modèles et Dessins, exposés par le Service des Phares," Paris, 1900, pp. 33, 36, and 37.

[2] "The Admiralty List of Lights, Part I., The British Islands," 1900, p. 10.

below its maximum intensity, thereby emitting about one-third of the power of the light available in foggy weather. In rebuilding the Eddystone and Bishop Rock lighthouses, the range, as well as the power of these important guiding lights was increased, by making the focal plane of the new Eddystone 61 feet higher than that of Smeaton's lighthouse, and by raising the focal plane of the Bishop Rock light 36 feet.

Far greater intensity can be attained for shore lights by the adoption of the electric arc light, than could possibly be exhibited from rock lighthouses, as exemplified by the two most powerful electric lights on the French coast, established at Cape Gris-Nez and Eckmühl, with a maximum intensity in each case of 28,500,000 candles.

Fog Signals at Lighthouses.—Mist and fog greatly diminish, in proportion to their thickness, the penetration through them of the most powerful lights; and the electric light is more affected in this respect, in relation to its intensity, than other lights, owing to the large proportion of rays towards the violet end of the spectrum in its composition, which are more rapidly cut off by fog than the red rays. Sounding signals have, accordingly, to be provided to replace as far as practicable the more or less obscured lights, and for use during the day in very thick weather. Bells have been hung for this purpose on several rock lighthouses, as, for instance, at the new Eddystone (Fig. 318, p. 491); but the explosion of charges of gun-cotton, suspended from the top of a curved steel rod attached to the lantern, was adopted as the fog signal at the rebuilt Bishop Rock Lighthouse, in place of the former bells, as well as at some other rock lighthouses, and has proved a more efficient warning than bells.

The most powerful fog signal consists of the blasts of a siren trumpet, actuated by compressed air at a pressure of from 20 to 28 lbs. per square inch; and the deep note produced by 326 vibrations per second has been found very suitable, and is heard further off than a high note with the same expenditure of air. With a consumption of about 14 cubic feet of air per second, a siren with a drum of 6 inches in diameter has been found to give the best results.[1] The most powerful sirens cease sometimes to be audible beyond 2 sea-miles; but their range extends generally to at least 4 sea-miles, and under favourable conditions may be much greater. There is not generally sufficient space available at rock lighthouses for the establishment of a siren with its necessary machinery, though the rock lighthouse of Ar-men, on the west coast of France, has been provided with a siren on an adjacent rocky projection; but sirens are important adjuncts to the principal lighthouse stations on seacoasts.

By grouping the strokes on a bell, the explosion of charges of gun-cotton, and the blasts of a siren, with definite intervals, corresponding to the flashes of the lights, lighthouses are characterized by sounds in the same manner as by flashes.

[1] "Notices sur les Appareils d'Éclairage, Modèles, et Dessins," Paris, 1900, pp. 154 and 157.

Screw-Pile Lighthouses.—Where a sandy shoal lies near the track of vessels, a light structure has sometimes been erected on iron screw-piles carried well down into the sand, from the top of which a light is exhibited to mark the shoal. A structure of this kind was erected many years ago on the Maplin Sands in the Thames estuary; whilst the Walde Lighthouse (Fig. 323) was similarly constructed on a sandy beach to the north-east of Calais; and numerous lighthouses of this type mark shoals along the coasts of the United States. The round, slender piles offer little opposition to waves; and the system is convenient for marking soft shoals close to the shore in somewhat sheltered positions, but would not be suitable for sandbanks in deeper water in the open sea.

SCREW-PILE LIGHTHOUSE.

Fig. 323.—Walde, near Calais.

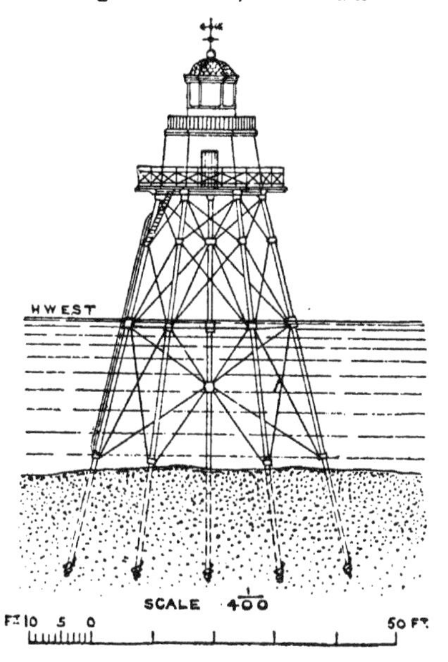

Beacons. — Minor shoals, somewhat out of the proper track of vessels, are often indicated by beacons, constructed of open ironwork, or solid in masonry, brickwork, concrete, and even occasionally neat cement. Beacons are generally built on detached reefs which emerge out of water at low tide, but whose position requires to be indicated at high water. The construction of these beacons is readily accomplished where the rock is not near, or below low water, nor very exposed. In France formerly, these beacon towers were built of masonry; but about twenty years ago Portland cement concrete, deposited within framing, was adopted, as being quicker in building, less liable to injury in construction, and enabling the work to be carried out by sailors alone. Subsequently, the circular form of tower was abandoned, as the framing for it was not suited for resisting the waves, and was otherwise inconvenient; and the concrete was deposited within octagonal timber framing (Fig. 324, p. 498), fastened at the corners by vertical, cast-iron corner pieces, with a groove on each side in which the boards of the framing slide and are kept firmly in place.[1] The lower and more exposed portion of these concrete beacons was made with a large proportion of cement, to augment its strength; and subsequently, in erecting beacons for the first time on rocks below low water, cement mortar has been employed,

[1] "Notices sur les Appareils d'Éclairage, Modèles et Dessins," Paris, 1900, pp. 283, 285, and 301.

and, finally, neat Portland cement for the portion under water, deposited within a dam of cement bags, and more recently within a framing of stiffened wire gauze, by skips provided with holes at the top for admitting water to the cement during its descent. The foundation of the Grande-Vinotière beacon, on a reef off the coast of Finistère, shown in Fig. 324, was built, in 1896, of neat Portland cement deposited under the shelter of a dam of sacks of quick-setting cement; whilst the mass of the beacon was constructed of cement mortar, with pieces of granite embedded in it, deposited within octagonal framing. Still more recently, the cylindrical foundation of a beacon, on the very exposed Rochebonne reef, has been constructed in sections, by depositing neat Portland cement within an enclosure of wire gauze, divided by partitions, in a depth of $26\frac{1}{4}$ feet below the lowest low water; and in this manner, a bottom layer has been formed, $3\frac{3}{4}$ feet high and 23 feet in diameter, which is surmounted by a second layer, 5 feet high and $19\frac{2}{3}$ feet in diameter. The large expenditure on material involved in using neat cement, is compensated for by the saving effected by rapidity of execution; whilst a considerably stronger beacon is obtained by constructing the most exposed portion of neat cement.

CEMENT AND CONCRETE BEACON.
Fig. 324.—Grande-Vinotière.

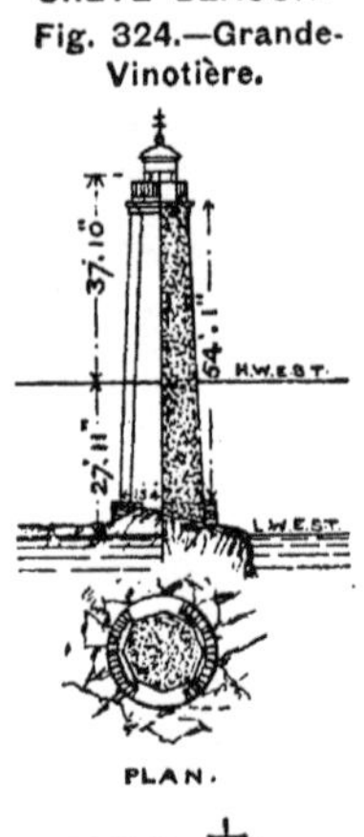

Lights on Beacons.—By the introduction of compressed oil gas as an illuminant, it became possible to establish permanent lights on beacons and buoys, which only need to have their reservoirs of gas replenished at intervals of about three months, and, consequently, are well suited for exposed sites inaccessible in stormy weather. By this means, beacons serve now for marking shoals by night as well as by day, which was to some extent effected by bell and whistling buoys before light-giving buoys were introduced.

Owing to the ease of rotation attained by floating the optical apparatus in an annular mercury bath, it has proved possible to advance a step further, and besides exhibiting a permanent light on beacons, concentrated by lenses into a much more powerful ray in a definite direction, to give the lights distinguishing characters by flashing lights in groups, on the same principle, though on a smaller scale, as adopted for lighthouses. An electric battery, which can continue in action for over four months, drives a Gramme motor, which effects the rotation of the apparatus; and the rate of rotation, which averages 10 seconds for one revolution, is regulated by an electro-magnetic brake. The reservoir of compressed oil gas, and the receptacle for the rotating machinery, with its battery, are contained in the body of the beacon, which thus with its flashing light fulfils the functions of a small lighthouse, needing only occasional visits. One of these beacon lights, recently established in France, has a power of 950 candles.

Light-Ships.—Where a shoal out at sea, near the track of vessels, is always covered with water, and especially where the shoal is sandy, precluding the erection of a lighthouse or a light-giving beacon, a light-ship is moored close to, or over the shoal (Fig. 325). Light-ships, accordingly, are of great importance for the safety of navigation; but being placed at very exposed sites, the visibility of their light is liable to be materially impaired by their rolling. The aim, consequently, of recent improvements has been to augment their stability during severe storms, and to increase the power of the light. The period of oscillation of waves in the worst storms, though varying considerably in different

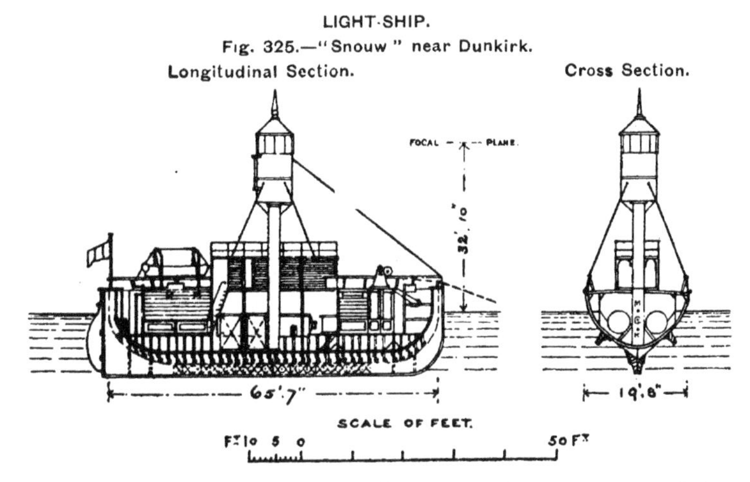

LIGHT-SHIP.
Fig. 325.—"Snouw" near Dunkirk.

localities, is approximately constant in any particular place; and therefore, having ascertained the period for the site where the light-ship has to be moored, the vessel must be so designed, with a low centre of gravity and adjusted ballast, that the period of its roll shall differ materially from that of the wave oscillations. Moreover, the rolling of a light-ship may be checked by giving the vessel a good draught, and also deep bilge keels (Fig. 325).

Lights on Light-Ships.—The visibility of the light is improved by the increased stability given to the vessel; but the stability of the light itself has been still further augmented by supporting the light, with its optical apparatus, on a vertical rod pivoted below the light on knife-edges on a bracket, and carried down below its point of suspension, with a counterpoise at the bottom of the rod, and also above the optical apparatus, so as to form a sort of compound pendulum arranged so as to have a period of oscillation differing widely from the roll of the vessel.[1] The intensity, moreover, of the light has been greatly increased by the adoption of an incandescent lamp fed with compressed oil gas; and the power of the light exhibited by this form of lamp has been brought up to 33,250 candles, or more than double the intensity of the lights previously employed on light-ships. The rotation of the apparatus

[1] "Notices sur les Appareils d'Éclairage, Modèles et Dessins," Paris, 1900, p. 204.

is effected by means of a clockwork arrangement actuated by a weight, which turns the vertical rod; and a flashing light is thus exhibited, which gives a distinguishing character to the light-ship.

Light-giving Buoys.—Buoys furnish a convenient means of marking minor shoals, and the limits on each side of the navigable channel in estuaries and ship-canals; and by supplying them with a lamp fed with compressed oil gas from a reservoir inside the buoy, they effect their object by night as well as by day. Thus the navigable channel along the Suez Canal is indicated by pairs of light-giving buoys moored at intervals of 820 feet; and the channel through the estuary of the Seine is similarly marked at intervals of about 4900 feet. The type of buoy adopted in the Seine estuary is shown in Fig. 326, having the shape of a boat, which is a convenient form in a channel with a current running alternately up and down; and the light is exhibited on the top of a short mast, with its reservoir of oil gas placed inside the boat.

Where a buoy has to be moored to mark a shoal out at sea, it is important to raise its light, and to give it a stable form, so that it may

LIGHT-GIVING BUOYS.

Fig. 326.—Boat Buoy. Fig. 327.—Large Buoy.

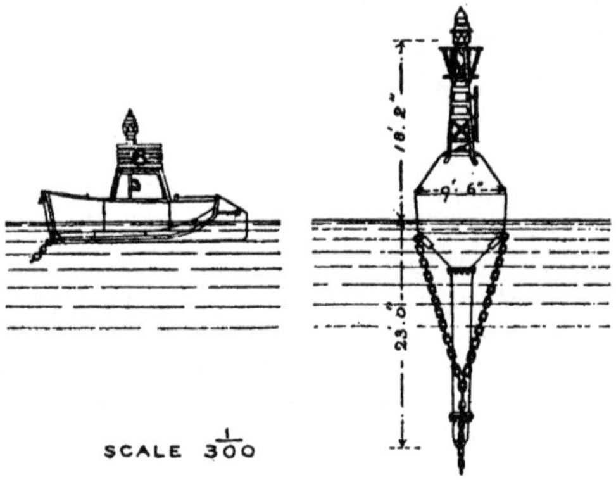

be seen some distance off, may not be obscured by spray, and may remain fairly vertical when exposed to high waves. A type of large buoy which fulfils the above conditions is shown in Fig. 327, its stability being secured by a long, weighted tube dipping down below the plate-iron body of the buoy, which, in the largest of these buoys, extends to depths of from 23 to 29 feet; whilst the light is exhibited at heights of from 18 to 26$\frac{1}{3}$ feet above the water-level, and is supplied with compressed oil gas from a reservoir with a capacity ranging from 512 up to 635 cubic feet. The light exhibited from the largest of these buoys has a power of 237 candles. Smaller buoys of the same type often satisfy the local conditions; whilst the largest buoys can be moored in deep water, and have advantageously replaced light-ships in some instances, owing to their being able to withstand a rougher surf without breaking from their moorings, and as, by increasing their number, an extensive shoal can be more thoroughly indicated.

CHAPTER XXXI.

LAND RECLAMATION; AND COAST PROTECTION.

Accretion in estuaries and bays, and erosion of seacoasts—**Land Reclamation**: advantages—Reclamation from estuaries, facilitated by fixing channel, accretion, enclosed by embankments, instances, conditions requisite — Reclamation of land adjoining seacoast, by protected embankments, examples, conditions affecting enclosure; closing embankments, methods adopted—Drainage of reclaimed lands, arrangements for draining by gravitation, pumping, instances in Holland—**Coast Protection**: by groynes, and sea walls—Groynes, littoral drift, causes of sea encroachment; slanting groynes on south coast of England, description of groynes at Blankenberghe, denudation of beach leeward of Dungeness, and raising it by groynes, best arrangement for groynes—Sea walls, protected sea bank, examples at Ostend; forms of upright sea walls, relative merits and disadvantages, best forms, protection required at base.

CHANGES are continually occurring along seacoasts: in some places the sea is gradually receding, especially in sheltered bays and estuaries, owing to the accumulation of deposit resulting from littoral drift, or from the settlement of sediment brought into estuaries by the flood tide aided by wave-action, or carried down by the rivers themselves; whilst in other parts, the coasts are being eroded by the sea during storms, this action being assisted by the disintegrating influences of rain and frost on cliffs composed of soft strata. These changes have led to the execution of two fairly distinct classes of work: in the one case, endeavours are made to accelerate the natural process of accretion, and as soon as the land is sufficiently raised, it is permanently reclaimed from the sea by enclosing it within embankments; and in the other case, protective works are undertaken to arrest the ravages of the encroaching sea.

LAND RECLAMATION.[1]

Works for reclaiming land from the sea have been carried out from very early times, since the gain to be derived from the cultivation of fresh alluvial land, generally of a very fertile nature, was obvious enough, and as land which has been raised nearly to high-water level

[1] The subject of Coast Erosion is (1909) under the consideration of the Royal Commission on Coast Erosion, the Reclamation of Tidal Lands, and Afforestation. See Reports and Evidence.

by accretion, can be shut off from the sea by works of a simple character.

Reclamation of Land from Estuaries.—When sediment is brought into an estuary by the flood tide, or carried down into it by the river, accretion, under natural conditions, takes place very slowly along the sheltered sides and recesses of the estuary, as any general accretion is prevented by the periodical erosion of deposits produced by the frequent shifting of the channel. Directly, however, the channel is permanently fixed by longitudinal embankments or training walls, especially if these works are raised to the level of high water, the scour of the tides and fresh-water discharge is restricted for the most part to this fixed channel; and accretion progresses rapidly from the deposit of sediment in the slack water behind the embankments, and at the sides of the estuary, being no longer displaced by the wanderings of the channel. The rate of accretion depends on the proportion of sediment introduced by the tidal or river waters, or both combined, and upon the shelter of the site; but by degrees the foreshores in the upper part of the embanked estuary, and at the sides, are raised sufficiently for samphire to make its appearance, and, later on, a coarse grass; and, eventually, the tide is excluded by the construction of enclosing embankments raised above the highest tide, with a flat slope on the exposed side, protected by clay, sods, fascines, or stone pitching, according to the exposure and the depth of water at the highest tides against its face. The growth of accretion, or warping as it is termed, is promoted by anything which checks the flow of the water off the foreshores during the ebbing tide, till it has deposited the material it holds in suspension, as for instance placing rows of faggots or sods across the line of flow; and the enclosure of the highest parts of the foreshore by banks, accelerates the accretion on the adjacent foreshore below, by increasing the period of stagnation of the silt-bearing water near high tide, and reducing the flow over the foreshore on the ebb, and consequently the scour.

The embanking of the Thames below London, whereby the adjacent low-lying lands were reclaimed from its estuary, was carried out at a very remote period; the Romans effected reclamations in the Fen districts; and large reclamations were accomplished in the Dee estuary, below Chester, in the eighteenth century. In recent times also, reclamations have been gradually carried out in many estuaries, as, for example, in the estuaries of the Seine (Fig. 231, p. 382), the Ribble, and the Tees[1] (Fig. 234, p. 383, and Fig. 328), where they have been assisted by the training of these rivers; and the method in which the extension of training works along the outfalls of the Fen rivers into the gradually accreting Wash, is followed up by reclamation banks,[2] is illustrated by Fig. 329. As the light, fertilizing alluvium only deposits in shallow water at high tide, beyond the influence of tidal currents, the enclosure should not be effected till the foreshore has become sufficiently raised for this deposit to have taken place, which raised level, moreover, enables the reclamation to be accomplished with low embankments, renders the closing of the

[1] *Proceedings Inst. C.E.*, vol. xc. plate 8, fig. 6.
[2] *Ibid.*, vol. xlvi. plate 8.

banks comparatively easy, and ensures a more efficient drainage of the reclaimed land. Whilst the enclosing is in progress, the increased shelter favours the deposit of alluvium, except near the outlet; and sometimes, after the final enclosure of the land, a rapid deposit of warp can be effected by admitting the turbid water through sluices in the bank

RECLAMATION EMBANKMENTS IN ESTUARIES.

Fig. 328.—Bank in Tees Estuary.

Fig. 329.—Bank at back of Training Wall. Outlets of Fen Rivers in the Wash.

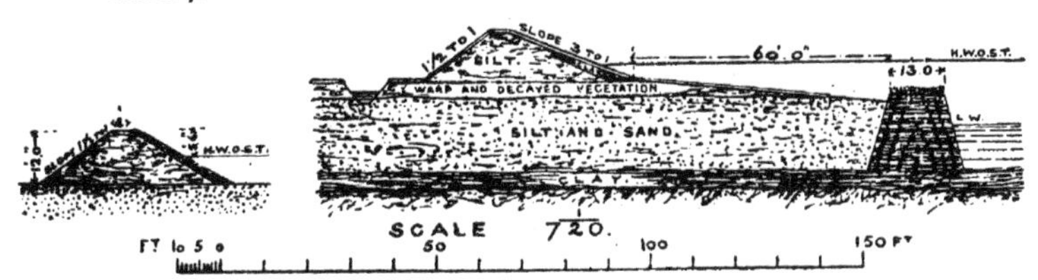

near high tide, and retaining it till the whole of the silt has been deposited, after which the clear water is let out slowly towards low tide.

Reclamation of Land adjoining the Seacoast.—Marsh lands adjoining the seacoast, and more or less subject to inundation at high tides, can be permanently reclaimed by embankments; but the bank facing the sea, unless protected to some extent by sand-dunes or a shingle beach, must be made much stronger, be given a flatter and better protected slope, and raised higher above the highest tides, than the sheltered reclamation banks in an estuary. The width of the bank at the base will generally secure it from the percolation of water underneath; but where suitable impervious material cannot be obtained for forming the bank, it may be necessary, in addition to the protective facing on the sea slope, to introduce a central core of puddled clay, or a row of sheet-piling, to prevent infiltration; whilst a line of sheet-piling along the outer toe of the bank in an exposed site, preserves the bank from being undermined by the recoil of the waves on the beach at its base. The embankment must be raised high enough to be secure from being overtopped by the waves during the severest onshore gales coinciding with spring tides; for the waves in passing over the top, would erode the inner slope, and soon form a breach in the bank, which would rapidly be enlarged by the rushing in and out of the tide.

Romney Marsh was reclaimed long ago by the Dymchurch Wall[1] (Fig. 330, p. 504); and a large portion of Holland has been reclaimed from the sea by embankments[2] (Fig. 331, p. 504). Such an embankment should not generally be placed further down on the foreshore than half-tide level, as the cost of the construction and maintenance of the bank would be increased quite out of proportion to the additional area of land gained; and the proper site for these reclamations is where accretion is taking place, and the sea, consequently, gradually receding.

[1] *Transactions of the Society of Engineers*, Nov. 1, 1897, p. 146, and plate 2, fig. 9.
[2] *Proceedings Inst. C.E.*, vol. xxi. plate 19, fig. 5.

The cost of reclaiming land depends upon the length of embankment required; and, consequently, it is more economical to reclaim a large area at a time, instead of enclosing it gradually in sections. On the other hand, however, it is more difficult to close an embankment reclaiming a large area, on account of the large volume of tidal water, flowing in and out of the contracted opening. The closing of a reclamation embankment is best effected by gradually raising a fairly long length of bank across the final opening, and either leaving the enclosure full of water to the height of the unfinished bank, as in closing an embankment for a dock, or letting it out through sluices in

RECLAMATION EMBANKMENTS ON SEACOAST.

Fig 330.—Dymchurch Sea Wall.

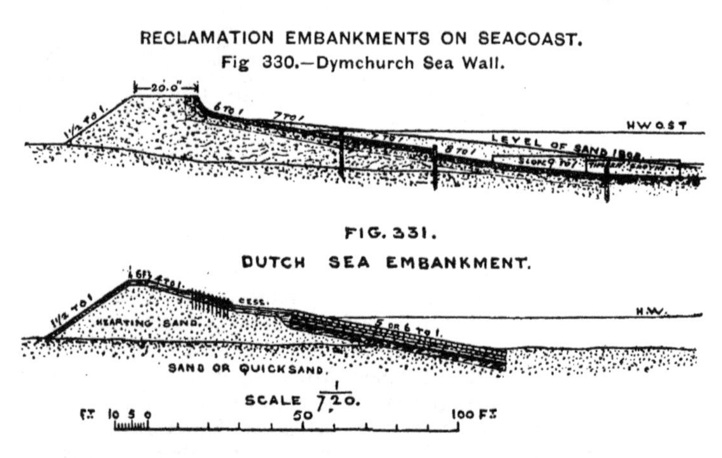

FIG. 331.

DUTCH SEA EMBANKMENT.

SCALE $\frac{1}{720}$.

the finished bank, and admitting it again in the same way on the return of the tide. The embankments in Holland are closed by sinking long fascine mattresses across the opening, and up the side slopes at the ends of the finished banks, which are weighted with clay and stones, and very effectually withstand the scour through the gap. The closing of a breach through a reclamation embankment is more difficult than that of the gap in completing the work, on account of the depth to which the scour of the current extends on the first opening of the narrow breach; and the Dutch generally, on the occurrence of a breach which cannot be closed in a single tide, enlarge the aperture to prevent the formation of a deep gulley at the bottom of the breach, which is difficult to close again. Owing to the deep channel thus eroded in a breach across the line of the embankment, the original line has sometimes been abandoned in repairing a breach, and a longer embankment formed on the land side of the breach, curving round and joining the portions of the original embankment, on each side of the breach, a short distance from their ends. By this arrangement, the new connecting bank is formed on solid ground, and affords a considerably wider opening than across the breach, for the influx and efflux of the tide over the bank as it is gradually raised. Gaps or breaches in embankments have also been closed with the aid of rows of planks fastened at the back of piles; by sliding panels down grooves between piles driven at intervals across the opening; or by sinking barges across the base of the gap, on the top of which the bank is raised; whilst bags filled with sand or clay serve to protect the embankment against the scour of the current in the process of raising.

The gradual drying of reclaimed land causes the surface of the land to settle down some two or three feet, so that if a breach occurs some years after the completion of the reclamation, the land is more exposed to inundation than when it was reclaimed. Accordingly, it is very important to prevent the occurrence of breaches by promptly repairing any damage to sea banks produced by storms; and in the event of a breach being formed, it should be closed at the earliest possible opportunity, as the breach is liable to be rapidly enlarged and deepened, and the land is damaged, by the rapid influx and efflux of the tide.

Drainage of Reclaimed Lands.—Special attention has to be devoted to the drainage of the low-lying reclaimed lands, to prevent these lands, after being relieved from inundation by the sea, from being flooded by land waters. Where higher ground adjoins these lands, catch-water drains formed at a higher elevation, and utilizing the good fall of the upper ground, intercept the rainfall of the higher lands, and prevent it from flowing straight down the slopes on to the low-lying reclamations. Straight drains are also formed through the reclaimed lands, to convey the rain falling on these lands, together with any flow from springs, and discharge these waters near low tide into the river draining the basin, or direct into the sea. The drains have to be given a sufficient capacity to store up the water flowing into them, till the tide outside has fallen sufficiently for it to be discharged. The flood tide, moreover, must be prevented from flowing in through the outlets of these drains, either by vertical sluice-gates for closing the sluiceways as soon as the tide has risen to the level of the water in the drains; or by self-acting gates pointing outwards, which close across the channel directly the water-level on the sea side becomes the highest, as adopted for the outlet sluices across the Fen rivers; or by placing a hinged tidal flap against the outlet, which is shut by the rising tide.

Where the land is so low that the drainage cannot be effected by gravitation, or the available fall at low water is so slight that the discharge of the land waters can be only partially accomplished during the short period that the tide is low enough, the water has to be raised by pumps, and discharged over the bank into a channel formed at a higher level to provide an adequate fall, or direct into the sea or adjacent estuary. This system has been largely resorted to in Holland, where the extensive tracts of reclaimed land, known as polders, are in the lowest places as much as 16 to 19 feet below ordinary high-water level in the Zuider Zee; and the low polders have to be wholly drained by pumps, formerly worked by windmills, but now for the most part by steam. The lands reclaimed from Lake Y, in the construction of the Amsterdam Canal, by embankments bordering the canal on each side through the lake, are drained by steam pumps discharging the water straight into the canal, or into branch canals joining the ship-canal; and this drainage water is discharged from the canal, partly by gravitation at low tide through sluices adjoining the North Sea and Zuider Zee locks, and partly by centrifugal pumps which raise the water from the canal over the embankment at the eastern end, and discharge it into the Zuider Zee (Fig. 261, p. 410). The inland lake known as Haarlem Meer, about 44,100 acres in area, which was formerly in communication with the Zuider Zee through Lake Y, and occasionally inundated the adjacent low lands, was reclaimed in the middle of the nineteenth century by enclosing it by an embankment, with an encircling canal formed at a high enough level to drain into Lake Y, and pumping out the water from the lake; and the drainage waters of this reclamation are lifted by powerful steam pumps into this canal, from which it now flows through a branch canal into the Amsterdam Canal.

COAST PROTECTION.

Where the sea is tending to encroach on the land, and especially where important seaside towns have been established near the shore, it becomes necessary to arrest the inroads of the waves. The protection afforded to the coast is of two kinds, namely, groynes, which projecting from the coastline down the beach, though to some extent breaking the waves, are mainly designed to collect and heap up the drift on the beach; and sea walls or pitched slopes, which, in forming a high-water barrier to the sea, prevent the erosion of the coastline and cliffs, and at the same time are often built out sufficiently on the beach to form behind them a promenade or drive alongside the sea.

Groynes.—Along most coasts, there is a littoral drift of sand or shingle following the direction of the strongest prevailing winds in the locality, owing to the action of the waves on the beach and foreshore. Under normal conditions along a straight coast, the beach, though constantly changing as regards the actual materials of which it is composed, retains its general form, since the materials removed by littoral drift are replaced by fresh materials from the same cause. When, however, a portion of the coast is specially exposed to the run of waves, or to the current from a river, or is deprived of its due renewal of drift from windward by a natural or artificial projection in that quarter arresting its travel, more material is removed from the beach than is brought to it, and the sea consequently gains on the land.

The south coast of England, between Bognor and Brighton, is exposed to the south-west gales from the Atlantic, nearly in the direction followed by the coastline; and, consequently, the sea has eroded the shore. Groynes of timberwork, however, were carried out some years ago across the beach between Lancing and Shoreham (Fig. 332), about 270 to 300 feet long, and 400 to 500 feet apart, which, impeding the travel of the drift towards the east, have raised the shingly beach sufficiently to produce an advance of the high-water mark 85 feet seawards.[1] Groynes are generally placed at right angles to the line of coast; but in this case the groynes have been given an easterly slant, so as to slope somewhat

[1] *Surveyors' Institution, Transactions*, vol. xxii. p. 346.

more away from the run of the waves under the influence of the prevalent south-westerly winds, which, owing to the great predominance of the exposure from this quarter, has favoured the conveyance of some of the drift over the groynes to the leeward side, and sheltered it in that position close under the groynes, so that the accumulation of drift on the windward side of the groynes, and its scarcity on the leeward side, are not so marked as is usual, under such conditions, with groynes at right angles to the coast.

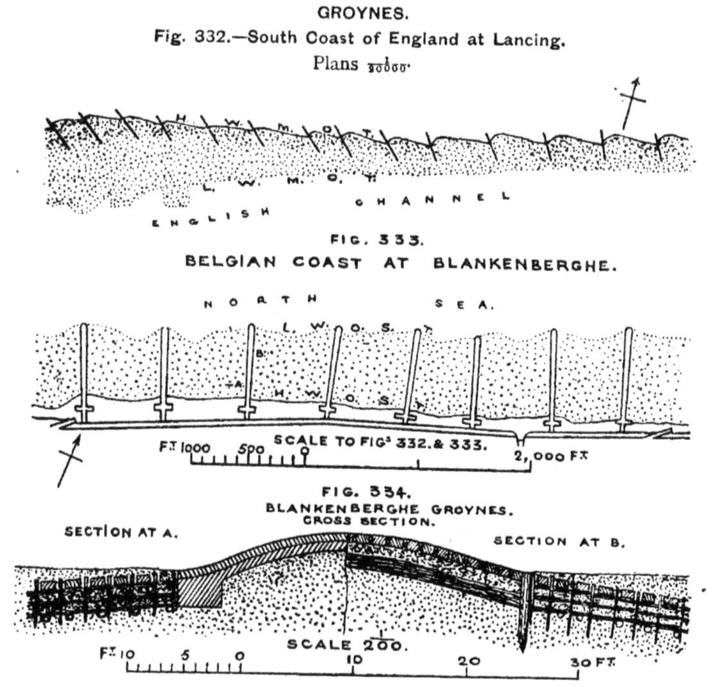

GROYNES.

Fig. 332.—South Coast of England at Lancing.

Plans $\frac{1}{30000}$.

FIG. 333.
BELGIAN COAST AT BLANKENBERGHE.

FIG. 334.
BLANKENBERGHE GROYNES.
CROSS SECTION.

Owing to the eroding action of the outflowing current from the River Scheldt, deep water approaches the coast in front of Blankenberghe, and the sea slope is steep; and, consequently, the waves in storms tend to encroach upon the shore at this part. This encroachment has been arrested by carrying out groynes at right angles to the coastline, down to beyond low water, about 820 feet long and 680 feet apart on the average (Fig. 333), which have caused the sand, drifting on this coast in both directions according to the wind, to accumulate on the beach, and checked the erosive currents, so that a gently-sloping beach has been formed, which reduces the wave-action on the shore, and has replenished the loss of beach occasioned by waves during storms.[1] The groynes, in fact, in the present and preceding instances,

[1] "VIIme. Congrès International de Navigation, Bruxelles, 1898, Guide-Programme," p. 289, and plate 20, figs. 1 to 6, and plate 21.

compensate for the loss of beach resulting from erosive currents along the shore and a considerable exposure to the sea, by promoting the accumulation of drift. The groynes in front of Blankenberghe have been made wide, to secure the sides from scour, and curved on the top, to present as little obstruction as possible to the waves, and are raised only slightly above the beach, so as to minimize the erosive action of waves dashing over them in storms; and they are constructed with a foundation of fascines and concrete protected by a facing of brickwork or stone pitching (Fig. 334, p. 507). In some places, as for instance at Zeebrugge, the upper part of the beach is further protected by shorter intermediate groynes (Fig. 307, p. 474).

The influence of breakwaters projecting from the coast, in leading to erosion of the shore on their leeward side, owing to the arrest of the travel of drift on their windward side, has been already pointed out, in the cases of the Madras and Port Said breakwaters, in the chapter on "Harbour Works;" whilst Dungeness furnishes an example of a natural projection from the general line of coast, which, by completely arresting the easterly drift of shingle, is continually growing out seawards, and has led to the denudation of the beach under its shelter to the east in front of the Dymchurch Wall; for this beach, though fairly protected from westerly storms by the point, receives no fresh supply of shingle to make good the losses due to the action of the waves. As the safety of the reclamation embankment was becoming endangered by the gradual lowering of the beach in front of it, numerous groynes have been extended down the beach since 1894, as far as, and at right angles to, low-water mark, from 300 to 1000 feet in length. These groynes consist of rows of planks on edge, fastened horizontally to piles fixed firmly in the beach, 7½ feet apart, the planks following down the inclination of the beach in steps, being only slightly raised above the level of the sands; and additional planks are put on as the beach becomes raised by the fine, drifting sand, which is retained by the groynes. By this means, the beach has been raised about 7 feet in some parts; and the cutting off of the shingly drift resulting from the advance of Dungeness, has been compensated for by the accumulation of the fine sand arrested by the projecting groynes.

It is evident that the efficiency of groynes in collecting drift is proportionate to the distance to which they can be carried out seawards, and that they should always be extended to low-water mark; whilst by raising them only slightly above the beach, the accumulation of drift is equalized, owing to the ease with which the drift is carried over and deposited on the further side of the groyne, and the erosive action of waves passing over the groynes on to the leeward side is reduced. Moreover, the gradual raising of the groynes as the drift accumulates, causes a gradual general elevation of the beach, instead of the heaping up on the windward side, and the depression on the leeward side of the groynes, resulting from the erection of groynes considerably above the level of the beach. Drift generally travels in both directions along a coast, according to the quarters from which the wind blows; but the prevailing wind determines the preponderating travel of the drift; and whereas groynes collect the drift from both directions, the cutting off of the main drift by a natural or artificial projection from the coast, in the absence of groynes, produces a denudation of the beach on the leeward side.

Sea Walls.—Where the coast is somewhat sheltered by outlying sandbanks, and the beach protected from erosion by a regular series of

SEA BANKS AND WALLS.

Fig. 335.—Ostend Sea Bank. Beyond High Water.

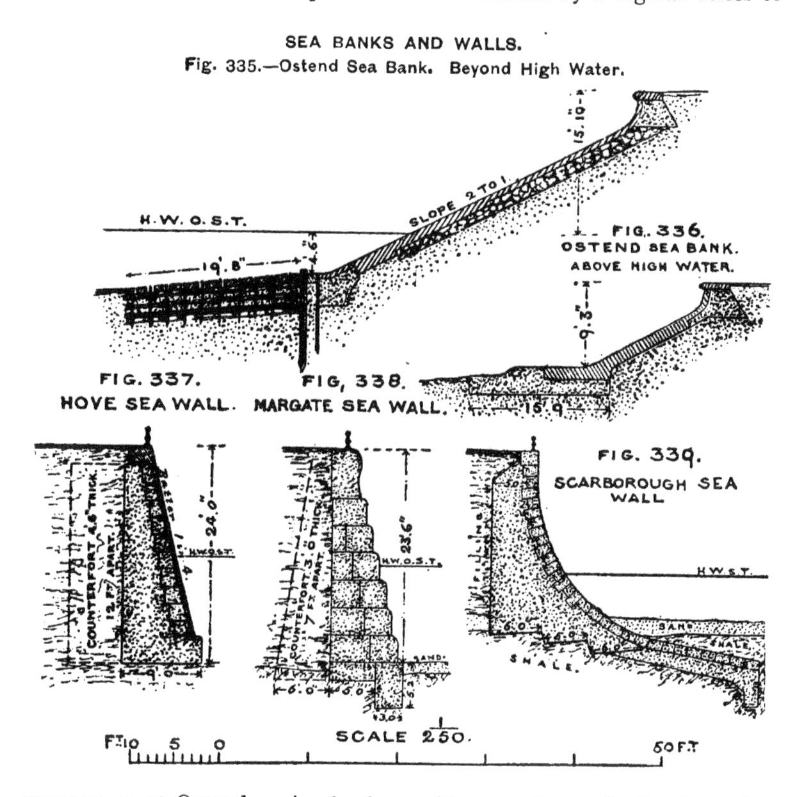

groynes, as at Ostend, a simple slope with a maximum inclination of 2 to 1, paved with brickwork or masonry laid on a bed of clay, rubble, or concrete, suffices to protect the face of the sand-dunes, or an embankment for a promenade in front of them [1] (Figs. 335 and 336). Along the parts where this sea bank has been carried out beyond high-water mark, to gain a strip of land from the sea for an esplanade, the toe of the slope has to be protected from undermining by the recoiling waves, which has been effected with piles and planks, and an apron of pitching laid on fascines, or of concrete, extending down the foreshore in front of the slope (Fig. 335). Where, however, the paved slope merely has to

[1] "VIIme. Congrès International de Navigation, Bruxelles, 1898, Guide-Programme," p. 297, and plate 20, figs. 7–15.

protect the slope at the foot of the dunes, above high-water mark, from waves in exceptionally severe onshore gales, a short slope with a moderate protection at its toe is sufficient to secure the promenade at the back (Fig. 336). The top of these slopes is ended off with a curving-back face, to throw off the surf rushing up the slope in storms, away from the esplanade.

Upright sea walls with some batter on the face have been constructed in front of several seaside towns, with the object of forming a promenade or drive, and at the same time protecting the foot of a cliff at the back, or the slope of the shore, from erosion by waves during storms. The concrete sea walls erected in front of Hove,[1] Margate,[2] and the North Cliff at Scarborough[3] (Figs. 337, 338, and 339, p. 509), exhibit straight, stepped, and curved forms of batter, having some importance from their affecting the erosion at the toe of these walls, which in the absence of the protection of a beach collected by groynes in front, or an apron, might occasion the destruction of the wall. A very sloping, and also a curved batter facilitates the rising of a wave up the face of the wall, and consequently the force of its recoil; and therefore the advantages afforded by such forms, of reducing the stroke of the wave against the wall, and directing the blow of the wave on the foreshore at its foot away from the vertical, are practically neutralized by the increase in the recoil. A wall with a vertical face offers most direct opposition to a wave, but is the least favourable form for the wave to rise up against it, and therefore minimizes the recoil; whilst the stepped face tends to break up both the ascending and recoiling wave, in proportion to the recession of the steps; though, on the other hand, the greater the projection of the blocks, the greater is their liability to displacement by the action of the waves. The curved form of the Scarborough sea wall, though diverting the recoil at the base to a direction approaching the horizontal, did not prevent the erosion of the shale bed on which the wall is founded; and an apron had to be subsequently added in front of the toe, to secure the wall from injury. The wall in this instance, besides providing a marine drive, and preventing the further encroachment of the sea on the base of the cliff, serves also, with the filling at the back, as a buttress to the undercliff, and assists, in conjunction with other special works, in rendering the cliff less subject to slips.

Walls with almost a vertical face, or formed in steps, appear to be the best forms for sea walls; and owing to the erosion at the foot, which is necessarily produced by waves on encountering such an obstacle, unless the foreshore consists of hard rock, the beach in front of the wall should be protected by an apron, or the beach made up sufficiently by means of groynes to compensate for erosion during storms, and thus prevent the undermining of the wall, which would eventually result in its destruction.

[1] *Proceedings of the Association of Municipal Engineers*, vol. x. p. 146.
[2] A section of the Margate sea wall was supplied me by Mr. A. Latham.
[3] *Proceedings Inst. C.E.*, vol. cv. plate 8, fig. 2.

PART V.

SANITARY ENGINEERING

CHAPTER XXXII.

SOURCES AND STORAGE OF WATER-SUPPLY.

Importance of water-supply, derived from rain—**Sources of Water-Supply:** Tanks for collecting rain-water, in tropical countries, in temperate regions, merits, disadvantages—Springs, origin, occasionally intermittent, value, salts in solution, instances—Wells, shallow and deep, tap underground flow, caution needed with shallow wells, advantages of deep wells, sun- to water-bearing strata, instances, adits, limits to supply, artesian wells, salts in well water—Water-supplies from Rivers and Streams, advantages, objections, mountain streams, examples of supplies to cities from rivers, salts in solution—Lakes for water-supply—**Storage of Water-Supply:** in lakes by raising level, facilities for storage, amount dependent on drainage-area and available rainfall, examples of Loch Katrine and Thirlmere—Impounding Reservoirs for storage, formed by dam across river valley, preliminary investigations, amount of storage to be provided, compensation water—Reservoir Dams, earthen and masonry dams, conditions affecting choice; reservoirs in steps, examples in Longdendale, Furens, and Elan and Claerwen valleys; earthen dams, form, puddle wall, construction, precautions, central masonry wall in Croton embankment, selected material in Indian dams; masonry dams, pressures, lines of resultant pressures with reservoir empty and full, distribution of pressures, form, varied by maximum pressure allowed on masonry, height dependent on level of rock below surface; bye-channel and waste weir, for discharge of floods, construction, waste weir in masonry dams, examples—Outlets from reservoirs, objections to culvert under embankment, culvert at side of valley, instances through masonry dam, and also beyond dam.

An adequate supply of pure water is essential to the healthiness and well-being of every community; and it is more particularly of vital importance in the case of the dense population crowded together, within a restricted area, in large cities. All water collected on land is derived in the first instance from rain, due to evaporation by the sun's heat of water and moisture from the sea and land, forming clouds which restore the water to the earth as rain. The water, accordingly, descending as rain, being distilled by the sun, is in its purest form; but rain-water, besides being insipid to the taste when unaërated, would be troublesome, in most cases, to collect directly in sufficient quantities for a large supply; and therefore sources are generally sought where the naturally collected rainfall yields a more abundant supply of water.

Sources of Water-Supply.

In addition to the collection of rain-water as it falls, water may be readily obtained in considerable quantities from streams and rivers fed by the rainfall flowing off the basins which they drain, or from springs which constitute the outlet of rain which has percolated from the surface through permeable strata; and it can also be raised from wells sunk down so as to tap underground supplies contained in water-bearing strata resting on impermeable beds. The volume of water which can be derived from such sources, depends upon the available rainfall of the district, or the actual rainfall less losses from evaporation, absorption, and percolation, and the area from which the source draws its supply.

Tanks for collecting Rain-water.—In tropical countries, where the rainfall is practically confined to a certain period of the year, and is generally considerable in volume during that period, it is very commonly collected, during the rainy season, in large tanks, excavated to some depth in the ground, to provide a supply of water for the rest of the year. These storage tanks are an absolute necessity at a distance from rivers, and where an underground supply of water from wells is either not available, or cannot be depended upon. In towns, however, these open tanks are very liable to pollution; and a considerable loss of water from evaporation during the dry season has to be allowed for.

In temperate regions also, the rain flowing off the roofs of houses and sheds in rural districts, can be collected with advantage in underground brick or concrete tanks, arched over at the top, and rendered watertight with an inside lining of cement, where other sources of supply are deficient or unsatisfactory. A considerable quantity of water can be stored up in this way, for almost all the rain falling on the area occupied by the buildings can be led into the tanks by the gutters and pipes, owing to the steep slope and impermeability of the roofs; whilst loss from evaporation is also to a great extent prevented by the roofing over of the tanks. Dead leaves and other *débris* are liable to be brought into the tanks by the rain-water, and necessitate the periodical cleaning out of such tanks; but even if the water from this source, after filtration, and boiling when necessary, has not to be relied upon for drinking purposes, it affords a very serviceable supply of soft water for washing and other objects. In large towns, however, the rain-water coming off the roofs is subject to contamination by soot, dust, and other impurities, and therefore ceases to be a suitable source of supply.

Springs.—A very valuable source of water for villages, or other small, or supplemental supplies, is provided by springs issuing out of the ground. Springs are derived from rain which, falling on a permeable stratum, or flowing off an impermeable stratum at a higher level, percolates through the permeable stratum till it encounters an impermeable bed, over which it flows to the lower outcrop of these strata, where it issues at the lowest point of the permeable stratum, on the top of the impervious bed, as a spring. The volume of water discharged by the spring depends upon the rainfall, and the surface area drained by the spring. Where the plane of separation of the two strata dips, at some underground part, below the lowest level of the dividing line at the outcrop, sufficiently for the level of saturation to fall occasionally in dry weather below the outlet of the spring, the water ceases to flow, and the spring is intermittent, only commencing its discharge again when the rain has raised the plane of saturation once more above the outlet. Such a spring usually derives its supply from a small area with a moderate rainfall, and is unsuited for supplying water, unless its flow can be stored. Most springs, however, in temperate regions have a continuous flow, though variable in quantity according to the rainfall and the season of the year.

Water from springs is very clear and free from organic impurities, having been filtered by the strata through which it has passed; and owing to its passage underground, it issues with a very uniform temperature; and, accordingly, in these respects, spring water is very suitable for a domestic supply. The water, however, in its underground flow, dissolves any soluble gases and salts contained in the strata through which it percolates; and, consequently, spring water often contains some inorganic substances in solution, depending on the nature of the strata it has traversed and the duration of its flow. Some spring waters, indeed, are so impregnated with mineral salts, such, for instance, as salts of iron, sodium, and sulphur, as found, for example, at Tunbridge Wells, Homburg, and Harrogate, that they are only suited for medicinal purposes; but most springs are either very fairly pure, as those on the Malvern Hills, or coming for instance from the chalk, contain calcium salts which are quite innocuous. Though springs generally yield only a moderate quantity of water with a variable flow, some towns situated at the foot of hills, on whose slopes numerous springs flow out, are wholly supplied with water from them; and London has now for nearly three centuries drawn a portion of its water-supply from the springs in the chalk at Amwell and Chadwell in Hertfordshire, by the New River conduit, 40 miles in length, constructed early in the seventeenth century.

Wells.—The wells sunk for the purpose of procuring a supply from the underground waters contained in permeable strata, are of two kinds, namely, shallow wells sunk nearly to the bottom of a superficial, permeable stratum, and deep wells sunk through an upper, impermeable stratum, some distance down into an underlying, water-bearing, permeable stratum. In each case, the wells tap the underground flow, or subterranean reservoirs of water derived from the percolation of rainfall, which find a natural outlet in the form of springs at the outcrop of the strata, or, near the seacoast, through the cliffs direct into the sea, or sometimes are permanently retained in the lower portions of the permeable stratum, from the absence of any outlet at a low enough level to drain off the waters during dry weather.

Shallow wells afford access to the rain-water percolating through a permeable, surface stratum, whose downward flow is arrested by an underlying, impermeable stratum, such as the sand or gravel overlying the London Clay in the Thames basin. Moderate supplies of water are readily obtained in this manner at a small cost; but great caution is needed to avoid contamination, from surface impurities being carried down to the well through the shallow stratum; and the establishment of cesspools in the neighbourhood must be absolutely prohibited, as fatal to the use of water from such a source. Accordingly, this source of supply is only admissible in country districts, where the houses are far apart, where the surface is not exposed to pollution, and stringent regulations are enforced as regards sewage disposal.

Deep wells are not open to the same objections; for the upper, impervious strata shut off, for the most part, the surface impurities; and the thickness of the strata which have to be traversed before the water drawn into the well is reached, protects the water to a great extent by filtration from organic impurities. These wells, when very deep, are usually constructed by sinking an ordinary brick-lined well for the upper portion, and then carrying down a steel tube considerably further by boring to the requisite depth; but in many instances, wells have been bored and lined with tubes throughout, which is a more economical method of construction. The Chalk, Oolite, New Red Sandstone, and Lower Greensand formations are the strata in England from which supplies by sinking deep wells can be most readily obtained. Thus the water supplied to parts of London by the Kent Waterworks, is obtained by pumping from deep wells in the chalk underlying the London Clay; and the supplies for Brighton and Southampton are wholly derived from the same source; whilst Liverpool was formerly supplied with water from wells in the New Red Sandstone.

The depth at which water is found, and the volume that can be regularly obtained by deep wells, vary considerably even under apparently similar conditions, owing to local differences in the state of the stratum, and sometimes to the existence of faults, which are liable to arrest the underground flows. Thus the flow in the chalk takes place mainly along fissures, which a well may not happen to intersect; and, accordingly, in some cases, as at Brighton, the volume of water is increased by driving adits branching off horizontally from the bottom of the well to tap fresh fissures. Moreover, two wells have sometimes been sunk near together, one yielding an abundant supply, and the other proving almost dry, on account of a barrier caused by a fault intervening between the wells. Occasionally, also, when a well has been deepened to increase the supply, the only result has been the loss of the existing supply, owing to the lowering of the well merely giving the water an outlet into lower unsaturated strata. Though deep wells sunk under favourable conditions in water-bearing strata, often yield an abundant supply of water, and frequently merely intercept a portion of the underground flow, which would otherwise have discharged into the nearest stream or river, or occasionally, near the coast, straight into the sea, there is a limit to the extension of this source of supply, depending as it does on the amount of rain falling on the area draining into the permeable stratum, which serves to replenish this underground reservoir. Accordingly, the sinking of several fresh wells within a limited area, tends to divert some of the flow from the older wells, and by lowering the level of saturation of the permeable stratum by the increased drain on its supply, augments the depth from which the water has to be raised.

Artesian wells are those wells in which, when carried down by a boring in a valley to a water-bearing stratum enclosed between two impermeable strata, the water rises above the surface of the ground, the name being derived from the French province of Artois, in which the earlier artesian wells in Europe were bored in the twelfth century, though traces of more ancient borings for water have been found in Asia and Africa. The overflow of the water from artesian wells is due to the confined permeable stratum rising at the side so much higher than its level where the boring is made, that the plane of saturation is higher than the top of the well; and, consequently, when the incumbent, impervious stratum is pierced at a lower level, the water is forced up the aperture by the hydrostatic pressure. The well at Grenelle, near Paris, bored to a depth of 1798 feet in 1833–41, discharges water to a height of 32 feet above the surface; and several deeper artesian wells have been formed in the United States.

Well water, like spring water, generally contains some inorganic salts in solution which it has collected in its course, though free from organic impurities owing to the filtration it has undergone in traversing the considerable thickness of the strata intervening between the surface and deep wells. The salts most commonly found in solution in well water are calcium bicarbonate and sulphate, constituting what is termed "hardness" in water, which, owing to its action on soap, is unsuitable for washing and certain manufactures, though harmless for drinking; these salts being collected by the water in its passage through strata containing calcium deposits, such as the Oolites and New Red Sandstone, and more especially chalk, the wells in which yield the hardest water, quite unsuitable for manufacturing purposes.

Water-supplies from Rivers and Streams.—Rivers and water-courses afford a very convenient and accessible source of supply; and one of the principal reasons for towns in olden times having been established by the banks of rivers, is supposed to have been the facility with which, in such a situation, an ample supply of water was secured. Rivers, however, become turbid in flood-time; and they are subject to contamination from the sewage of towns situated on their banks, which tends to augment with the increase of population. The suspended matter, nevertheless, can be removed by passing the water pumped from rivers through filter-beds; whilst stringent legislation against the direct discharge of sewage into rivers, which lower down may be utilized for water-supply, minimizes the liability to pollution; and organic impurities are gradually oxidized as they are carried down by a river, so that the river is by degrees cleansed by natural changes, provided the source of pollution is at an adequate distance above the place where the water has to be drawn off, for these changes to be accomplished.

The ideal source of supply is found in streams draining hilly, uninhabited, and uncultivated districts, where the sole impurity, besides a certain turbidity during floods due to the conveyance of some of the loose soil into the stream by the rapid flow off the ground, consists of a slight brown discoloration when the water has flowed through peat, often met with on moorlands. The turbidity is readily removed by deposit when the water is at rest in a reservoir or settling tank; and the peaty discoloration is bleached by exposure to light. Such streams, however, being situated in the upper part of a river basin, are generally at a considerable distance from large towns, which are usually established on the lower parts of rivers, where the volume of the river is greater and its flow more uniform; and, moreover, the flow of these mountain streams is too irregular, under natural conditions, to furnish a sufficient, reliable supply. Cities and towns, accordingly, situated on rivers possessing a sufficiently large constant flow, very commonly draw their principal supply from the river at a suitable point higher up. Thus, for instance, London obtains its chief supply from the Thames at different pumping stations above the tidal limit at Teddington. Calcutta derives its water-supply from the River Húgli about 14 miles above the city, but within the tidal portion of the river; and Paris formerly drew its supply from the Seine, which is still used for general purposes, such as watering the streets, flushing the sewers, and extinguishing fires, though the water for domestic use is taken from the Vanne, the Dhuis, and the Avre, supplying a good potable water derived in great measure from springs in the chalk. Very often in tropical countries, where rivers with a sandy bed appear to dry up during the dry season, there is still an underground flow which can be reached by sinking wells, and utilized for water-supply; and during floods in the rainy season, the water drawn from these wells is obtained in a clearer state than if it was taken direct from the turbid river.

As the flow of rivers is to some extent derived from springs, their waters contain certain inorganic substances in solution in very variable quantities, mainly, in general, calcium carbonate and sulphate, and magnesium carbonate, depending upon the stratum constituting their basins, as well as some organic matter; whilst water drawn from a river within the tidal limit contains a varying quantity of sodium chloride, or common salt, from the sea-water mingling with the fresh water when the discharge of river water is small.

Lakes as Sources of Water-supply.—Lakes in hilly districts, away from habitations, fed by streams and rain flowing off steep mountain slopes, furnish natural reservoirs providing an excellent source of water-supply, from which the water can be conveyed by gravitation to a town lower down the valley, or can be carried by tunnelling through the dividing ridge, to a town situated in an adjoining river basin. A lake is formed by a natural barrier across the valley at its outlet, so that the enclosed hollow above the barrier has been filled at some remote period with the water draining into the valley, which had to rise above the top of the barrier before the drainage of the valley could commence to be discharged.

STORAGE OF WATER-SUPPLY.

Lakes used for Storage of Water.—In order to interfere as little as possible with vested interests in a settled country, instead of draining off the lake water and thus lowering it, the customary course is to raise its level by artificially heightening the natural barrier by a dam, to provide the storage required in the driest year to secure the necessary supply, so that the lake may never have to be drawn down much below its original average level. The extent to which the lake has to be raised depends upon the area of the surface of the lake, as well as upon the volume of water required for the supply; and in this respect, a lake furnishes a very convenient and capacious reservoir for storage, for from the very outset, any raising of the water-level provides a large volume when spread over the area of an extensive lake, as pointed out in the chapter on "Irrigation Works;" whereas the lower layers of water in a reservoir formed in a valley, are small in area, and consequently in volume, they often cannot be entirely drawn off, and they are subject to reduction by the accumulation of deposit. The lake is raised by impounding the flood-waters of the streams flowing into it, till it has reached the desired level. The extent to which such a source can be drawn upon for water-supply, depends upon the area of land draining into the lake, or the gathering ground as it is called in the case of a reservoir in a mountain valley, the rainfall in the driest years, and the proportion of the rainfall which reaches the lake; for the storage water should be either fully replenished in flood-time each year, or the amount of water stored up should at least be made sufficient that, in spite of an exceptional drain upon it during a year of minimum rainfall, it should afford the requisite supply during the three driest consecutive years, without the original water-level of the lake being unduly infringed upon.

Loch Katrine and Thirlmere furnish notable examples of the use of lakes as reservoirs for water-supply, with the requisite storage provided by raising their level. As Loch Katrine has an area of 3119 acres, it was possible, by merely raising the level of the lake 4 feet, and allowing for a drawing down of the lake 3 feet below its original level, to obtain a storage of 5687 million gallons, and provide a supply for Glasgow of 50 million gallons per day of the purest water; and by recent works, including a second conduit, it has been possible to double the available supply. Though the lake is about 34 miles distant from Glasgow, the water can be very easily conveyed by gravitation, as the lake is 367 feet above sea-level. Thirlmere is about 100 miles from Manchester; but as the lowest level of the lake is 182 feet above the water-level of the service reservoir, situated 4 miles short of Manchester, there is an available fall of nearly 2 feet per mile, enabling the water to be readily conveyed by gravitation. Thirlmere, however, has an area at its original water-level of only 328 acres; and as this water-level is not allowed to be lowered, the lake will eventually be raised 50 feet to provide a storage in this case of 8130 million gallons, for securing a future total supply of 50 million gallons of water per day delivered to Manchester, and also 5½ million gallons a day of compensation water for the stream flowing out of the lake. This raising of the level of the lake has been provided for by a concrete dam across the outlet, faced with masonry, carried down to the solid rock and built up to a height of 57 feet above the former level of the lake, which, when the water is impounded to the full height of 50 feet, will have an area of 783 acres.[1] It is specially important in this instance to provide an ample storage, as the total drainage-area from which the rainfall will be collected, even when some streams outside the basin are diverted into it, only amounts to 11,000 acres; but fortunately the rainfall in this hilly district has been found to average 71 inches a year in three consecutive dry years, whilst the steep, rocky nature of the ground makes the proportion of rainfall reaching the lake exceptionally large; and the storage contemplated would be obtained with an available rainfall of 32⅔ inches from the whole drainage-area.

The existence of lakes proves that the basin which they fill is enclosed by fairly watertight strata, and, consequently, that the site is suitable for storage; and the lakes serve as settling reservoirs, in which the streams flowing into them deposit their silt, as so strikingly illustrated by the turbid Rhone flowing into the Lake of Geneva, and also provide a space, in proportion to their depth, in which the silt can accumulate for an indefinite period without affecting their storage capacity.

Impounding Reservoirs for storing up Water.—Where no lake is available, water is stored up in the wet season, for use during the dry months, by forming an artificial lake by erecting a dam across a narrow part of the valley of a mountain stream, and impounding the water of the stream above the dam in flood-time. It is important to select a site in the valley where the strata are impervious, and where a considerable volume of water can be impounded by a dam of moderate length and height. This can be best effected where a tolerably wide valley contracts abruptly to a narrow, deep gorge; and it must be borne in mind, that an increase in the height of the dam augments the capacity of the reservoir out of all proportion to the ratio of the addition to the height, on the same principle as the raising of lakes, especially with an extensive reservoir. Accordingly, if practicable, a stable, high dam is preferable to two or three dams in steps down the valley making up altogether a similar height.

Before designing a storage reservoir, the drainage-area of the stream to be impounded, or gathering ground, must be ascertained, and the available rainfall over this area must be estimated. The proportion which the available rainfall bears to the actual rainfall, may be found approximately by gauging the flow of the stream and observing the corresponding rainfall; but the rainfall of the three driest consecutive years can only be obtained with precision from a series of rainfall observations, at several stations in the district, extending over a period of at least twenty years. If no such record exists, the observations made must be compared with those of similarly situated stations where

[1] *Proceedings Inst. C.E.*, vol. cxxvi. p. 2, plate 1, and plate 2, fig 13.

a long record has been kept, and an estimate of the required rainfall deduced from it. As a general rule in England, it has been found that the rainfall of the wettest year is one-half greater than the mean, of the driest year one-third smaller than the mean, and of the three driest consecutive years one-fifth, on the average, less than the mean; and the quantity of water Q, in gallons per day, which can be collected from a gathering ground having an area A in acres, with a mean annual rainfall R in inches, and loss by evaporation E in inches, which in England amounts to from 10 to 18 inches in the year, is expressed by the following formula [1]:—

$$Q = 62{\cdot}15A(\tfrac{4}{5}R - E).$$

The storage which has to be provided for a given daily supply must be sufficient, after allowing for loss by evaporation, to enable the whole of the supply to be drawn from the reservoir, for the number of days in succession during which it is possible that there may not be an adequate rainfall over the gathering ground, to have any practical influence in replenishing the reservoir. This period varies according to the local conditions, and is less in wet districts, where the variations from the mean rainfall are less than in dry districts. At Loch Katrine, in a wet district, the storage originally provided was adequate for 113 days, without allowing for loss from evaporation; and at Thirlmere there is storage for 156 days, where, owing to the restricted drainage-area, the fluctuations of the rainfall are liable to be greater than over a more extensive gathering ground. After making due allowance for loss from evaporation, which has been estimated at a depth of 3 feet in a year over the surface of reservoirs in England, and varies with the climate, a storage sufficient for a supply for 100 days may be regarded as a minimum, only suitable for a very wet district, and a supply for 250 days in very dry districts; whilst in the parts of tropical countries where the rains are liable to fail sometimes in the rainy season, it may be necessary to provide storage for two successive dry seasons.[2]

In addition to the water-supply stored up in reservoirs, it is necessary, in damming up streams on which vested interests lower down exist, to provide for compensation water which, in consideration of the abstraction of the flood-waters of the stream, has to be discharged from the reservoir into the stream below, so as to ensure that the flow of the stream shall never be less than a stipulated amount. As this compensating volume has to be discharged during the dry weather, when no surplus water is flowing over the waste weir, it has to be stored up in the reservoir in flood-time, together with the necessary storage for the water-supply; and where manufactories are established on the stream below, compensation water equivalent to one-third of the total volume impounded may have to be allowed. Storage reservoirs, accordingly, in addition to their primary object, reduce the floods in the river below in proportion as they impound the flood discharge: and they also

[1] "The Theory and Practice of Hydro-Mechanics," Inst. C.E., p. 44.

[2] "Water-Supply: Rainfall, Reservoirs, Conduits, and Distribution," 2nd edition, A. R. Binnie, Chatham, 1887, p. 36.

ensure, by the enforced discharge of compensation water, that the flow in the driest seasons shall never fall below a definite volume.

Reservoir Dams.—As previously stated with reference to reservoirs for irrigation in Chapter XXVI., dams for storing up water in river valleys are constructed, either as earthen embankments, or in the form of masonry or concrete walls. Earthen embankments are well suited and economical for dams of moderate height, up to about 80 feet; and for many years, storage-reservoir dams in England were exclusively constructed of earthwork. Masonry dams were, indeed, constructed long ago in Spain for reservoirs, though, for the most part, with a very incorrect section; but the earliest high masonry dam with its section correctly adapted to withstand the pressures increasing with the depth, was erected in France across the River Furens,[1] a tributary of the Loire, in 1859–66 (Fig. 344, p. 528). The amount of materials required for an earthen dam, with its flat slopes on each side, increases rapidly with the height; and the danger of percolation through the dam with the increased water-pressure, and of slips on the outer slope, is also augmented, so that though earthen dams exceeding 100 feet in height have been constructed, they become rather costly, and are less reliable than solidly-founded masonry dams. Masonry dams, however, must be founded on solid rock; for any settlement of a masonry dam would result in the formation of cracks in the dam, endangering its stability. Accordingly, whilst both types of dams must rest on an impervious stratum, to guard against infiltration of water under pressure underneath the dam, tending to undermine it, a masonry dam can only be constructed with safety where a foundation of compact, hard rock can be reached at a reasonable depth below the surface.

In a valley where a large volume of water has to be stored, and earthen embankments are rendered necessary from the absence of a rock foundation, or are preferred to masonry, reservoirs can be formed in a series of steps down the valley by dams of moderate height, as illustrated by the six reservoirs constructed down the Longdendale valley by earthen dams from 70 to 100 feet high, for supplying Manchester with water, the total area of these reservoirs amounting to 497 acres, and their combined capacity to 4160 million gallons. The same course, moreover, has sometimes to be adopted even with high masonry dams in a narrow valley; for in spite of the great height of the Furens dam of 175 feet above the bed of the river, retaining a maximum head of water of 164 feet, the reservoir formed by it has a capacity of only 353 million gallons; and a second masonry dam, a little higher up the valley, supplements the storage. The works (opened in 1904) for the supply of Birmingham in the future, are designed to include six reservoirs, in the somewhat narrow valleys of the Elan and Claerwen in South Wales, formed by means of concrete dams from 98 to 128 feet in height, for storing 18,000 million gallons of water, from a watershed of 45,560 acres, to secure a daily supply of 75 million gallons, and compensation water reaching 27 million gallons a day, the storage provided being equivalent to the requirements for about 180 days.[2]

[1] "Water-Supply," Encyclopædia Britannica, 9th edition, vol. xxiv. p. 407.
[2] *Proceedings Inst. C.E.*, vol. cxxvi. pp. 71–73.

The advantage, in respect of storage, of a single high masonry dam, over several dams of moderate height, is exemplified by the Gileppe masonry dam near Verviers in Belgium, which retains a head of water of $147\frac{2}{3}$ feet, and has formed a reservoir 198 acres in area, with a capacity of 2650 million gallons, which was constructed in 1869–75, instead of four dams, 95 feet high, originally proposed for storing up a similar quantity of water.

Earthen Dams for Reservoirs.—An earthwork embankment for a reservoir dam should have a width at the top of not less than 8 to 12 feet, with a 3 to 1 slope on the upper side facing the reservoir, protected by stone pitching or a layer of concrete against the wash of waves raised in the reservoir by gales; and a 2 to 1 slope at least should be formed on the lower side, or flatter towards the base if the dam is high,

RESERVOIR EMBANKMENT WITH PUDDLE WALL.

Fig. 340.—Cross Section.

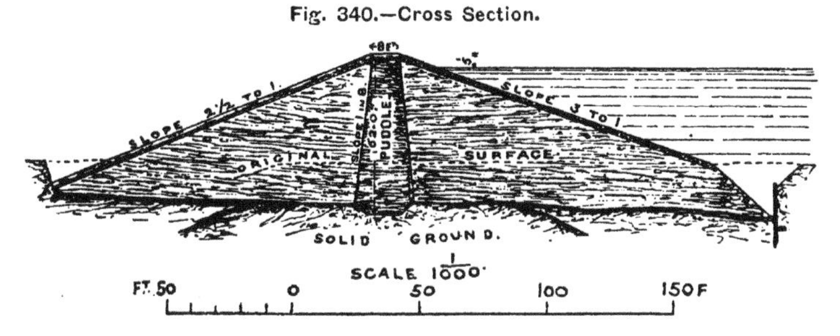

or the available material of inferior quality. A wall of puddled clay in the centre of the embankment, carried down in a trench to a solid, impervious stratum, or occasionally a narrow concrete wall, is the method adopted for forming a watertight barrier to prevent the infiltration of water from the reservoir through the embankment (Fig. 340), as it would be difficult to make the inner slope watertight with a layer of puddled clay or concrete, owing to the liability of the embankment to unequal settlement; and, moreover, such a layer would be more costly than the central wall, owing to the large area of the flat inner slope of a high dam. Provided suitable materials can be obtained within the site of the reservoir, the excavation of these materials for the construction of the dam enlarges at the same time the capacity of the reservoir. The most tenacious, clayey materials should be deposited near the puddle trench, so as to avoid any abrupt change in the nature of the adjoining earthwork, tending to irregular settlement; and the harder and looser materials should be reserved for the outer portions. The central puddle wall is kept moist, and preserved from cracking, by the materials close against it on each side; and the construction of the outer portion of a dam with comparatively dry, loose material well drained, renders the outer slope far less liable to slips than it would be if the dam was constructed entirely of clay, which would crack on the exposed face in hot weather, and be disintegrated by rain. The embankment must be raised high enough to be in no danger of being overtopped by waves in the reservoir, and more especially by any rise of the water in the reservoir; for if once water begins to flow over an earthen bank, its destruction rapidly follows; and therefore any undue settlement must be most carefully guarded against.

Before the dam is commenced, all loose material should be removed from its site, and the excavation for the puddle trench taken down to a sound and impervious stratum; and after the trench has been filled in with puddled clay, the embankment must be slowly carried up by spreading the materials in a succession of thin, horizontal layers, consolidated by punning and rolling, with the aid of water in dry weather.

RESERVOIR EMBANKMENT WITH MASONRY WALL.

Fig. 341.

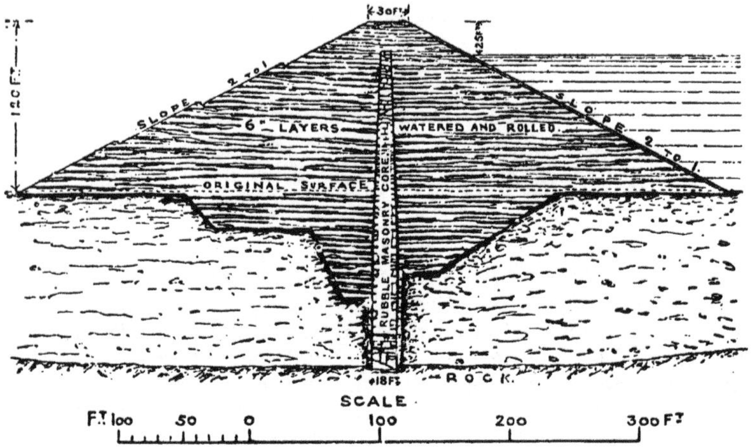

Earthen dams with a puddle wall, having the outer portion on the downstream side of the wall well drained, and in some cases provided with berms on the outer slope as an additional safeguard against slips, have been constructed to a height of about 125 feet. The southern portion of the New Croton dam, forming a reservoir for extending the water-supply of New York, was originally intended to be of the type shown in Fig. 341,[1] having a watertight wall along the centre line of the embankment founded on rock and carried up to the highest water-level. A desire for greater security led, however, to the substitution of a solid masonry wall, and the entire dam was ultimately constructed of the type shown in Fig. 345, having a total length of 1168 feet exclusive of the spillway. The dam is 297 feet high from the lowest foundations, 206 feet wide at base, 18 feet wide at summit, which is 16 feet above water-level. The spillway, 1000 feet long, is constructed with granite

[1] "The Water-Supply of the City of New York," E. Wegmann, plate 132.

steps, crossed by a steel arch 200 feet span. The face of the dam is of granite ashlar with rubble masonry backing, there being about 855,000 cubic yards of masonry in the structure. Similar dams at Olive Bridge, forming with the Beaver Kill dykes, the proposed Ashokan reservoir are under construction, and will with the Catskill aqueduct further augment the New York supply.

In places where clay for puddle cannot be obtained, as in parts of India, a larger mass of the most retentive material available is placed in the central part of the dam, with the looser material for the outer portion on each side[1] (Fig. 342). The Gatun earthen dam, which, with its spill-ways, diversion channels, and adjacent locks, forms so important a feature of the Panama Lock Canal, is now in course of construction, and will retain, when completed, the waters of the artificial Gatun Lake at the level of 85 feet above mean sea level (Fig. 262).

RESERVOIR EMBANKMENT WITH SELECTED MATERIAL.

Fig. 342.—Culvert with Valve Tower.

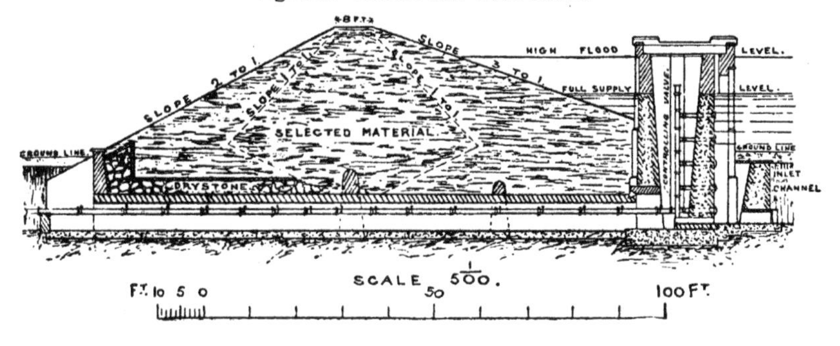

Masonry Dams for Reservoirs.—In earthen dams, the mass of the embankment is always ample for withstanding the pressure of the water in the reservoir; and these dams only fail owing to infiltration of water, settlement, and slips. Masonry dams, on the contrary, can be exactly proportioned in thickness at the different levels to the pressures they have to support, which vary with the depth below the surface of the water; and such dams, if well and solidly built, are only liable to fail on account of standing on yielding or fissured foundations, or from being given an inadequate thickness at any part, producing tension at the inner face, or an excessive pressure on the masonry at the outer toe. So long, however, as the dam is made adequately thick to resist the pressures, any further addition to its width merely increases the weight on the masonry below, and on the foundations, to no purpose.

A reservoir dam is subjected to two distinct pressures, namely, a vertical pressure due to the weight of the dam, with its resultant passing through

[1] *Proceedings Inst. C.E.*, vol. cxxxii. plate 3, fig. 25.

the centre of gravity of the mass, and a horizontal pressure, or a pressure at right angles to the inner face of the dam, due to the pressure of the water in the reservoir against the inner face. When the reservoir is empty, the only pressures on the dam are those due to its own weight; and the line of resultant pressures, with the reservoir empty, may be obtained by dividing up the cross section of the dam by a series of horizontal lines, finding the centre of gravity of each portion comprised between each line and the top, and then drawing vertical lines from the centres of gravity to the corresponding base lines, and joining the several points of intersection of these lines by a line, which represents the resultant pressures (Fig. 344, p. 528). The total pressures on each base line correspond to the weight of the portion of the dam above it; and the pressures are necessarily equal on each side of the resultant line, since it represents the mean position of all the pressures on any horizontal line across the dam.

The pressures on the dam with the reservoir full, are compounded of the weight of the dam and the pressure of the water on the inner face, which latter is equal to the weight of a triangular prism of water having

DIAGRAM OF PRESSURES IN MASONRY DAM.

Fig. 343.—Reservoir Empty and Full.

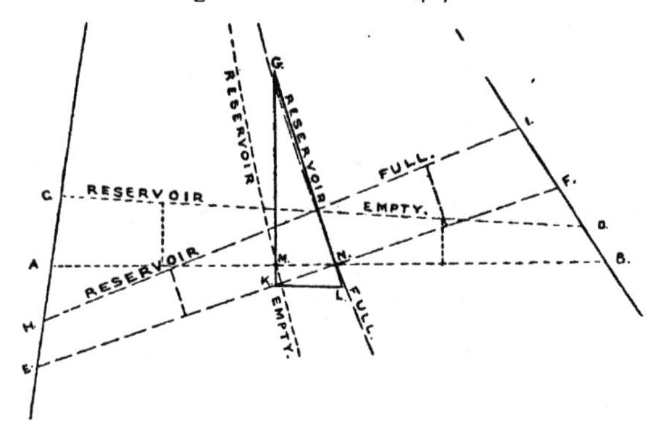

a height equal to the depth of the dam below the water, a base of the same length, and a unit of width along the dam, usually taken as a foot; and the resultant of the pressures over the inner face may be regarded as acting at the centre of pressure, which is two-thirds of the depth below the water-level. The resultant of these pressures can be obtained graphically, on the principle of the parallelogram of forces, by drawing a line vertically through the centre of gravity **G**, of the portion of the cross section of the dam situated above the horizontal line **AB** (Fig. 343), making the length of the line **GK** represent the weight of the dam above **AB** in tons on the square foot to a given small scale, drawing the line **KL**[1] at right angles to the inner face, representing approximately the water-pressure in

[1] The water-pressure is assumed, for the sake of simplicity, to act also at G, though, owing to the form given to these masonry dams, the centre of gravity is a little higher up than the centre of pressure.

tons to the same scale, and joining **GL**. The line **GL** represents the resultant pressure in tons per square foot to the given scale, and also its direction. The line of resultant pressures with the reservoir full, can be obtained by joining the intersections of the lines corresponding to **GL** in each portion of the section of the dam, with their respective base lines, when the line joining these intersections will be the required line. In this case also, the pressures are equally divided on each side of the line of resultant pressures, in regard to the line at right angles to the direction of the resultant with the reservoir full. Accordingly, in the diagram (Fig. 343), the pressure on **EN** with the reservoir full, is equal to the pressure on **NF**; and with the reservoir empty, the weight on **AM** is equal to the weight on **MB**. The actual pressure, however, per unit of surface at any point, depends on the length of these lines, and also on its position owing to the pressure varying along the lines. The maximum and minimum pressures on any base lines with the reservoir empty and full, such as **AB** and **EF**, can be obtained graphically in the following manner. Let W and P represent the total pressures in tons on **AB** and **EF**, with the reservoir empty and full respectively; and dividing $\frac{W}{2}$ by the lengths in feet of **AM** and **MB**, and $\frac{P}{2}$ by the lengths in feet of **EN** and **NF**, the average pressures on these portions of the dam are obtained in tons per square foot, with the reservoir empty for the first two, and full for the second two. Then selecting a suitably large scale to represent tons on the square foot, perpendicular lines are drawn from the centre of each line **AM**, **MD**, **EN**, and **NF**, having lengths corresponding to the tons per square foot obtained above on the selected scale; and the ends of these pairs of lines, representing by their lengths the respective average pressures, are joined by the straight lines **CD** and **HI** prolonged to the faces of the dam, forming two bands whose widths indicate the variations in pressure across the dam along the lines **AB** and **EF**, with the reservoir empty and full respectively, so far as the rigidity of the masonry may not modify this distribution (Fig. 343). The actual pressure in tons per square foot can be measured at any point along **AB** and **EF**, by the width of the band to the given scale; and the diagram shows that with the reservoir empty, the maximum pressure is at the inner face and the minimum at the outer face, and the reverse with the reservoir full, and that there is no tension so long as the lines of resultant pressures are within the middle third of the cross section of the dam, which tension would be indicated by the crossing of the pairs of lines forming the bands, namely **AB**, **CD**, and **EF**, **HI**, respectively.

As the total water-pressure on the inner face of the dam increases with the head of water retained, in proportion to half the square of the depth below the surface, this pressure augments rapidly with the depth; and whereas the pressures are small in the upper part of a masonry dam, and it is only necessary to provide a sufficient width to prevent the lines of resultant pressures passing outside the middle third, the pressures in the lower part of a high dam become so great, that the dam has to be widened out rapidly towards its base by a curved or very sloping batter on its outer face, to avoid excessive pressures on this face with the reservoir full (Fig. 344). The width, indeed, of the upper part of a high masonry dam has to be regulated by the condition of keeping the resultant pressures within the middle third, though the top width is generally enlarged sufficiently for a road; whilst the width of the lower portion is determined by the pressure which it is considered may be safely imposed on the masonry.[1]

The rational form for high masonry dams, with widths varying in proportion to the pressures which have to be borne at different depths, having an inner face deviating only slightly from the vertical, and an outer face sloping out rapidly towards the base, is indicated by the section of the Furens dam, retaining a maximum head of water of 164 feet, with the moderate maximum pressure on the masonry of 6 tons on the square foot (Fig. 344); and the general correctness of this form for a given pressure, has been proved by the test of experience. By allowing, however, a greater pressure on the masonry, it has been possible to make the width towards the base somewhat less in proportion to the depth than at Furens, as exemplified by the slighter Ban dam erected subsequently in France, in which the maximum pressure on the masonry was increased to 6·6 tons, and also by the New Croton masonry dam in the United States (Fig. 345).

FORM OF HIGH MASONRY DAM.

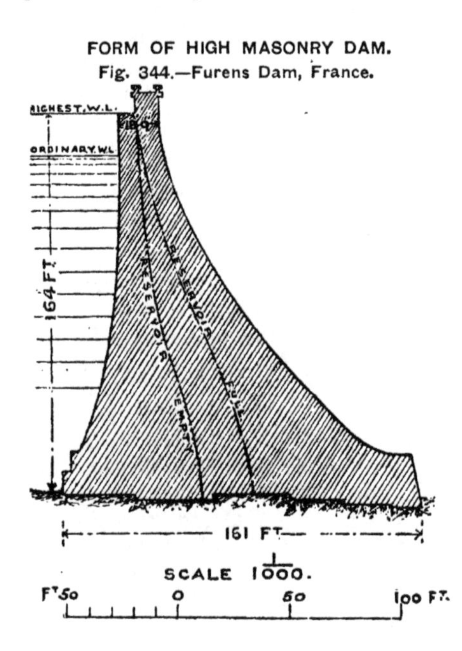

Fig. 344.—Furens Dam, France.

The Furens dam, though in a very narrow valley, gains at any rate the full benefit of its height, as a solid rock foundation has been met with close to the surface (Fig. 344); whereas both the New Croton and Vyrnwy dams, though more favourably situated for impounding water, have had to be carried to considerable depths below the surface to reach a rock foundation (Figs. 345 and 346). Thus the New Croton dam has a maximum height to the surface of the reservoir of 281 feet, though the depth of the reservoir alongside it, when full, is only 157 feet; and the Vyrnwy dam,[2] with a maximum height to the water-level of the full reservoir of 144 feet, retains a maximum depth of water of only 84 feet in the

[1] The student should refer to the papers on the experimental and mathematical determinations of the stresses in dams contained in vol. clxxii. of the *Proceedings Inst. C.E.*

[2] *Proceedings Inst. C.E.*, vol. cxxvi. p. 29.

reservoir; whilst the masonry and concrete dam at Thirlmere has been carried $58\frac{1}{2}$ feet down below the surface of the rock.[1]

The reservoir formed by the New Croton dam has a capacity of

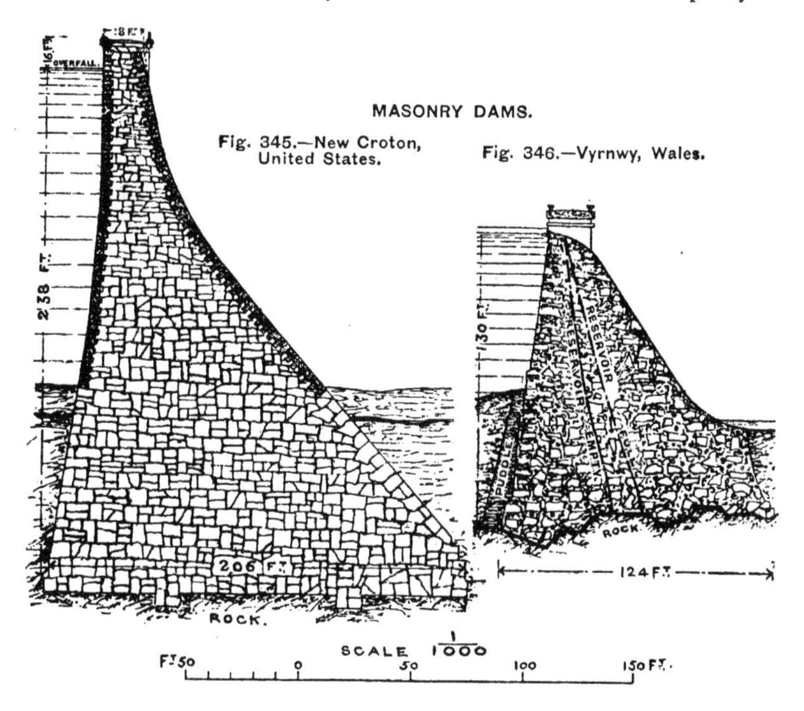

Fig. 345.—New Croton, United States.

Fig. 346.—Vyrnwy, Wales.

32,000 million gallons, and is supplied from a drainage-area of 240,840 acres, with an average rainfall of about $48\frac{1}{3}$ inches in the year, half of which reaches the Croton River; and the Vyrnwy reservoir, with an area of 1121 acres and an available capacity of 12,130 million gallons, is fed from a drainage-area of 23,500 acres, with an average annual rainfall of about 68 inches. The Vyrnwy reservoir is capable of providing Liverpool eventually with a water-supply of 40 million gallons a day, as well as compensation water to the River Vyrnwy, equivalent during the drier months to an average daily discharge of $15\frac{1}{4}$ million gallons; and the reservoir, accordingly, will afford sufficient storage for about 220 days.

Bye-channel and Waste Weir for Reservoirs.—The formation of a bye-channel in cutting, round the side of a reservoir, for conveying away the flood discharge of the stream supplying the reservoir, is very advantageous during the construction of an earthen dam, and also very useful for diverting the turbid and discoloured flood-water away from a reservoir which has been already filled. These bye-channels, controlled by gates or sluices, and leading the flood discharge

[1] *Proceedings Inst. C.E.*, vol. cxxvi. plate 2, fig. 13.

from the stream above the reservoir into the stream below the dam, have to be protected along their bed and sides with puddle, pitching, or a lining of masonry or concrete, according to the velocity of flow through them, and the erosive nature of the stratum through which they are constructed. A small settling reservoir is also sometimes provided above the storage reservoir, in which the sediment brought down by a flood is deposited before the water passes on to the bye-channel or into the storage reservoir, and from which the deposit can be readily removed at intervals.

A waste weir is formed at a suitable place at the side of a reservoir, with its sill level with the surface of the water of the full reservoir, so that the surplus water discharging into a full reservoir in flood-time, may escape over the weir; and the width of the weir is made sufficient for the maximum influx into a full reservoir to be discharged over the weir, without raising the water-level above the limit allowed for in designing the dam. Where the reservoir is formed by an earthen dam, the waste weir should be formed in cutting at one side, beyond the end of the dam; and the channel leading from the weir into the stream below, with a fall equivalent to the head of water retained by the dam, must be effectually protected against the scour of the overflow from the reservoir. The velocity of this discharge is best checked by forming the escape channel, below the waste weir, with a succession of very long, shallow steps constructed so as to retain a pool of water on the top of each step, resembling a series of diminutive weirs separated by pools which deaden the impact of the falling water.

When the water is retained in the reservoir by a masonry dam, a portion of the dam itself may serve as the waste weir by having its crest terminated at the requisite level. Thus for instance at the Vyrnwy dam, the spaces under the nineteen central arches between the piers carrying a roadway across the top of the dam, form the main waste weir, having lower sills than the spaces under the side arches, and affording a total length of weir of 456 feet; and as the surplus water flows down the outer face of the dam, a sharp curve has been given to this face near the level of the stream below the dam, to receive the shock of the falling water (Fig. 346, p. 529). At the New Croton dam, the waste weir is provided by a special masonry dam, 1000 feet long, extending from the high central masonry dam to one side of the valley, with its top level with the surface of the full reservoir, and its outer face constructed with a series of steps from top to bottom to break the fall of the overflowing water.

Outlets for Reservoirs.—Provision has to be made for drawing off the water stored in the reservoir for water-supply, and also for compensation water. Formerly, the outlet culvert for drawing off the supply was laid across the base of earthen dams, and usually at their lowest point, as by this means the culvert could be readily constructed, or the line of pipes laid, before the embankment was carried up, and served to convey the water away during the progress of the dam; and by placing it at the lowest point, the drainage was most effectually provided for during construction, and the reservoir most fully drawn down when in regular use. Culverts and pipes, however, so placed are liable to be injured by the unequal settlement of the embankment over them, and particularly where they pass through the puddle wall and rest on the puddle trench; and the leakage of water under pressure through the fractures thus formed, may lead to the subsidence, and eventual destruction of the dam, especially when the valves controlling the discharge are placed at the outer end, or in the centre of the embankment, precluding access to the culvert for repairs. A culvert or pipe under an embankment must be commanded throughout its whole length by a valve tower at its inner end in the reservoir, thereby rendering the conduit accessible for repairs (Fig. 342, p. 525). Even if this essential precaution is adopted, and the culvert is firmly bedded in a trench excavated in the solid ground, a leakage of water under pressure is liable to take place along the line of the culvert or pipes, impairing the stability of the embankment. Accordingly, where a reservoir is formed by an earthen dam, the outlet culvert should be carried to one side of the valley beyond the end of the dam; and by controlling the flow into the conduit by a valve tower erected in the reservoir, with inlets at different levels, the water can be drawn from the reservoir, with its variable water-level, where it is least turbid or discoloured, which is generally close to the surface.

The outlet culvert from a reservoir closed by a masonry dam, is sometimes carried through the dam at a low level, with the valve chamber formed in the dam, of which arrangement the Villar dam, storing up the flow of the River Lozoya for the supply of Madrid, and the New Croton dam furnish examples. It is, however, preferable to avoid the construction of a large culvert through a masonry dam, by carrying it in tunnel at one side of the valley beyond the dam, as resorted to at the Vyrnwy reservoir, where the outlet supply culvert, carried from the side into a tunnel, has its flow controlled, and the inflowing water strained through copper wire gauze, in a masonry tower erected in the lake; but the culverts for the compensation water have been carried through the dam, as the most direct course to the river channel below the dam. At Thirlmere, it was possible to provide a direct course for the compensation water outside the dam, owing to the existence of a projecting rock in the centre of the line of the dam, through which the discharge culvert has been constructed; whilst, as the upper end of the lake is nearest to Manchester, the outlet for the supply is situated at the opposite end of the lake to the dam, the flow into the conduit being controlled by valves in a shaft, and strained in a well, sunk 65 feet into the rock.

CHAPTER XXXIII.

CONVEYANCE, PURIFICATION, AND DISTRIBUTION OF WATER-SUPPLY.

Aqueduct required from storage reservoir—**Conveyance of Water-supply:** extended meaning of aqueduct; hydraulic gradient for aqueducts, definition, uniform or variable; aqueducts from impounding reservoirs in tunnel, cutting, and pipes laid as inverted siphons, following hydraulic gradient, or mainly in pipes under pressure, relative merits of the two systems, instances; arrangement and sizes of aqueducts, Thirlmere and Elan aqueducts described, sections of Loch Katrine and New Croton aqueducts, Vyrnwy aqueduct described; service reservoirs, arrangements, construction, objects—**Purification of Water:** methods, impurities; separating weirs, description; settling reservoirs, settlement in impounding reservoirs, for clearing river water by subsidence; filtration, object, description of filter bed, suitable rate of flow for purification, reduction of organic matter; purification by contact with iron, through spongy-iron filters, by iron borings, resulting chemical reaction; softening of hard water, definition of term "hard," disadvantages of hard water, methods of softening, scale of hardness, instances of hard river waters—**Distribution of Water:** methods of distribution; supply per head of population, measure of requirements, average consumption; intermittent and constant supply, methods described and contrasted; prevention of waste, requirements, means of detection of waste, importance; water-meters, charge by rental, charge by meter the fair system, objections urged, methods of obviating them, positive and inferential meters described; different supplies for special purposes, increasing difficulty of obtaining pure water, impure supply could be used for municipal purposes, purest water for domestic use, adopted in Paris, might be resorted to in London.

WATER stored up in a lake, or in a river valley, by a dam across the outlet, is usually a long distance away from the town for which the supply is required; and an aqueduct has to be constructed from the storage reservoir, for the conveyance of the water to a service reservoir, in the neighbourhood of the town, formed at a sufficient elevation for the water collected in it to be readily distributed throughout the town at an adequate pressure. Fortunately, an impounding reservoir is always situated in a hilly district, towards the upper end of a river basin, and therefore at a sufficient height above the town which it is formed to supply, for the water to flow down an aqueduct from it into the service reservoir under the action of gravity.

CONVEYANCE OF WATER-SUPPLY.

The term "aqueduct," though formerly applied to the magnificent masonry bridges erected by the Romans for the conveyance of water across deep valleys, and consequently to all bridges carrying a channel containing water, whether for navigation, irrigation, or water-supply, properly denotes any form of channel serving for the conveyance of water. An aqueduct in this more extended sense may, in addition to bridges, comprise open and covered channels or conduits, tunnels, and metal pipes.

Hydraulic Gradient for Aqueducts.—The hydraulic gradient is the vertical fall between two points on an aqueduct divided by the length, and is generally expressed in feet or inches per mile; and this gradient determines the size to be given to a conduit for the conveyance of a definite volume of water. The hydraulic gradient can be maintained uniform throughout, provided the aqueduct is constructed of the same form and dimensions along its whole length, or if its dimensions are merely varied so as to compensate for an increased resistance to flow due to a modification in its form, as for instance in changing from a channel to pipes. On the other hand, it may be sometimes expedient to vary the hydraulic gradient at different parts of the same aqueduct, in order to adjust the level of the aqueduct more nearly to the configuration of the ground, or so as to modify the size of the tunnel, conduit, or pipes at certain places, for the sake of economy in construction.

Aqueducts from Impounding Reservoirs. — An aqueduct either consists of a channel formed to the inclination of the hydraulic gradient, carried through ridges in tunnel, and traversing fairly level ground, or contouring the slopes, in cutting, in which the water flows freely; or where the ground in a valley dips below the hydraulic gradient, it is formed by one or more lines of pipes, constituting inverted siphons, through which the water is conveyed under pressure, instead of continuing the channel along the hydraulic gradient by carrying it on the top of a high, arched bridge across the valley, like the Roman works. Long aqueducts leading the water from impounding reservoirs in hilly districts, having generally to traverse irregular ground and some river valleys, are usually formed in part by tunnels and cuttings, and partly by pipes; but this combination of the two systems of aqueducts is effected on two different principles, according to the local conditions. Where the configuration of the ground is fairly favourable, the hydraulic gradient is followed for considerable distances by contouring the slopes in a somewhat circuitous course, as well as by tunnelling through high ridges; and pipes are only resorted to where the land along the line selected for the aqueduct, dips considerably in descending the valleys of the streams and rivers which have to be traversed. When, on the contrary, the land along the route of the aqueduct lies for the most part below the hydraulic gradient, the aqueduct is mainly formed by pipes following closely the irregularities of the ground, and taking a fairly straight course; whilst the hydraulic gradient is only followed in the tunnels, and is reached elsewhere at points where balancing reservoirs can be introduced for reducing the pressure of the water on the lowest part of the pipes, by dividing the aqueduct into sections.

The first system, following the hydraulic gradient as far as practicable, involves a greater amount of cutting for constructing the conduit with a uniform gradient, and also a more circuitous course in contouring the slopes to avoid abrupt changes in the level of the ground; but it dispenses with balancing reservoirs, as each inverted siphon constitutes an independent section of the aqueduct, with the pressure at the lowest point proportionate to the depth of the dip below the upper end of the siphon. The second system, on the other hand, affords the advantages of a straighter, and therefore shorter course, increasing the available gradient, and a uniform depth of cutting merely sufficient to protect the pipes from injury and frost; but a pipe aqueduct requires a steeper gradient, or a larger water section, than a free channel, to pass the same volume of water; and the pressure at the lowest point of a long line of pipes would be unduly great during any stoppage of the flow, if not relieved at intervals by balancing reservoirs. Long lengths of channel with a free flow at the hydraulic gradient, have been adopted for the Loch Katrine, Thirlmere (Fig. 347),[1] and Elan aqueducts: whilst the Longdendale and Vyrnwy[2] aqueducts (Fig. 348) are formed with lines of pipes and tunnels.

Arrangements and Sizes of Aqueducts.—The Thirlmere aqueduct has a length of nearly 96 miles, of which 45 miles consist of 48-inch, 40-inch, and 36-inch cast-iron pipes laid as inverted siphons down the valleys, $36\frac{3}{4}$ miles of covered conduits, and $14\frac{1}{8}$ miles of tunnels. The conduits formed of concrete on the cut-and-cover system, and crossing small streams on masonry bridges, and the tunnels, lined for the most part with concrete, are constructed of the same internal section (Fig. 349, p. 536),[3] which, with a uniform hydraulic gradient of 20 inches in a mile, are capable of conveying the maximum proposed supply of 50 million gallons a day; one line of cast-iron pipes as at first laid, followed by a second line which has been laid subsequently, affording the supplemental supply at present required for Manchester. Only three lines of pipes, 48 inches in diameter, are to be laid in the Lake district for the full supply, to reduce the disfigurement of the country in laying them; but five lines of the smaller pipes will be required elsewhere. A uniform hydraulic gradient of 20 inches in a mile is maintained throughout for about $83\frac{1}{6}$ miles from Thirlmere; but in the last $13\frac{2}{3}$ miles, there is an available fall for the continuous line of pipes, of about 32 inches in a

[1] *Proceedings Inst. C.E.*, vol. cxxvi. plate 1, fig. 4.
[2] *Ibid.*, vol. cxxvi. plate 4, fig. 2.
[3] *Ibid.*, vols. cxxvi., clxvii.

mile, which enabled the diameter of this line of pipes to be reduced to 36 inches. Each siphon has a rectangular chamber or well at either

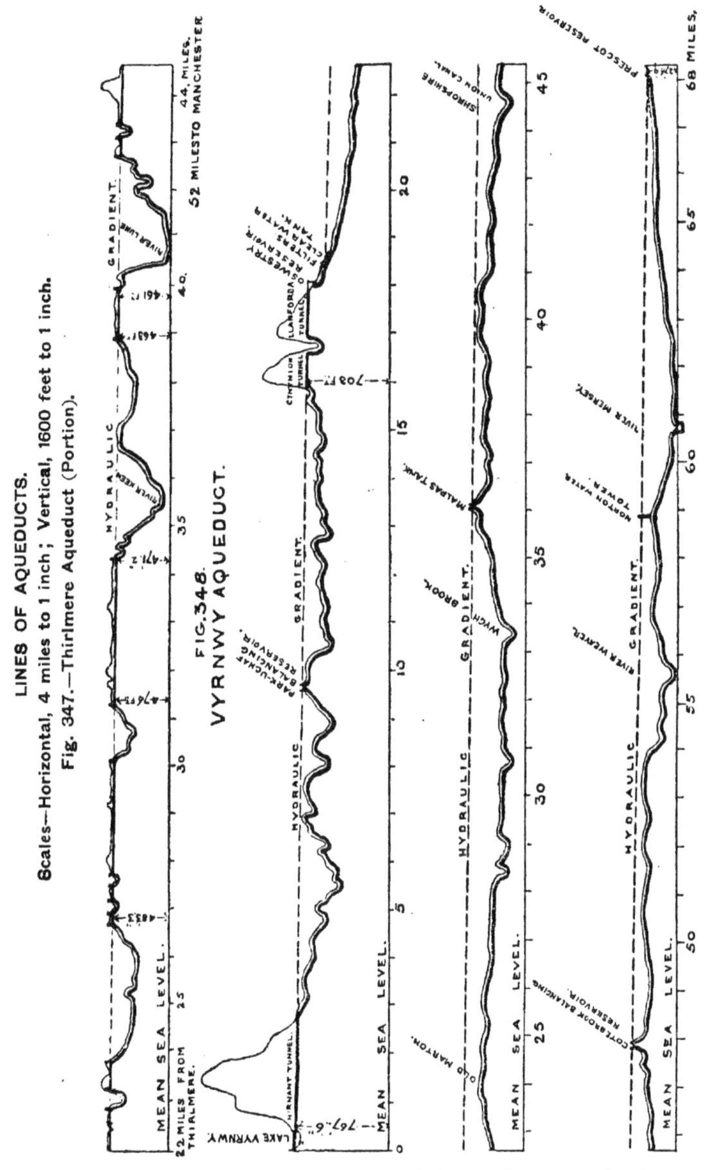

LINES OF AQUEDUCTS.
Scales—Horizontal, 4 miles to 1 inch; Vertical, 1600 feet to 1 inch.
Fig. 347.—Thirlmere Aqueduct (Portion).
FIG. 348. VYRNWY AQUEDUCT.

end, to connect it with the adjacent conduit; and a valve in the upper well closes automatically on the bursting of a pipe in the siphon, from the fall of a float in the well, owing to the resulting lowering of the water-level. The most important siphon, 9½ miles long, crosses the valley of the Ribble; but the siphon across the Lune valley has the greatest dip, in which the head of water reaches a maximum of 427 feet (Fig. 347, p. 535). The pipes are in most cases carried over the rivers, at the bottom of the valleys, on bridges, thereby reducing the amount of dip, and rendering them readily accessible for repairs.

The Elan aqueduct is very similar in general character to the Thirlmere aqueduct; but the conduits and tunnels, with a hydraulic gradient of only 15·84 inches per mile, and having to provide for a larger eventual discharge, have had to be given a greater section (Fig. 350);[1] whilst by allowing a gradient of 36 inches in a mile

SECTIONS OF AQUEDUCTS.

Fig. 349. Thirlmere, for Manchester. Fig. 350. Elan, for Birmingham. Fig. 351. Loch Katrine, for Glasgow. Fig. 352. New Croton, for New York.

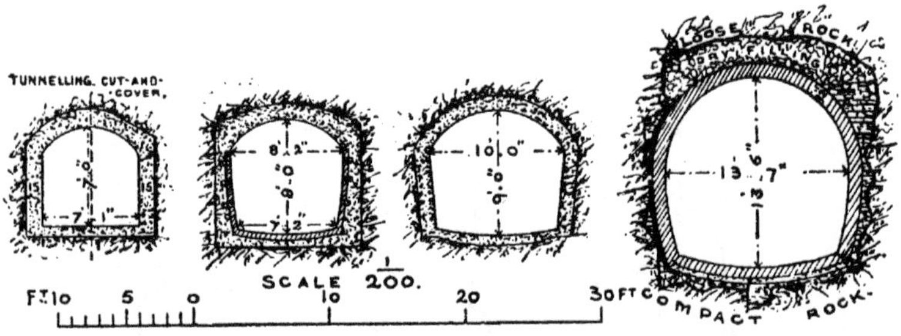

for the siphons, pipes very similar in diameter to those of the Thirlmere aqueduct, possess the requisite increased discharging capacity. The siphons in the Elan aqueduct will eventually consist of six lines of 42-inch iron or steel pipes; but only two lines of pipes have been laid in the first instance, to provide for the present needs of Birmingham. The longest siphon on this aqueduct has a length of 17 miles; and the greatest dip of the pipes of 550 feet below the hydraulic gradient, is where the River Severn is crossed by a bridge near Bewdley.

A section of the new Loch Katrine aqueduct,[2] for augmenting the Glasgow supply, is given in Fig. 351, and of the New Croton aqueduct in Fig. 352. When an aqueduct passes through an unlined tunnel in hard rock, the section provided is slightly increased, to allow for the increased friction of flow in a channel with a rough bed and sides. In all cases, the conduits and tunnels along an aqueduct are alone constructed to the full dimensions at the outset; and the siphons are restricted to the lines of pipes needed to supply the immediate requirements, additional lines of pipes being laid as the demands for water increase.

[1] *Proceedings Inst. C.E.*, vol. cxl. p. 242. [2] *Ibid.*, vol. cxxiii. p. 410.

The Vyrnwy aqueduct, nearly $68\frac{1}{2}$ miles long, is divided into six distinct sections by balancing reservoirs, four of these reservoirs having been formed at places where the land attains the level of the hydraulic gradient; but as there was no land high enough along the last $20\frac{1}{2}$ miles of the course of the aqueduct to reach the hydraulic gradient, it was necessary, in order to make a break in this long line of pipes under pressure, to construct the fifth balancing reservoir on the top of a high tower built on the summit of Norton Hill, so as to raise the reservoir to the hydraulic gradient, 110 feet above the surface of the ground (Fig. 348, p. 535). The water for this reservoir is contained in a circular basin or tank, 80 feet in diameter, formed of steel plates affording a central depth of 31 feet. The sections into which the aqueduct is thus divided up, vary in length from $6\frac{1}{2}$ miles up to $17\frac{5}{8}$ miles, in which latter length the maximum head of water of 480 feet occurs in passing under the Wych Brook; and the hydraulic gradient ranges in general between 2 feet per mile in the three tunnels, 7 feet in diameter, and having a total length of four miles, up to 6·87 feet in a mile for the long siphon between Oswestry and Malpas, the ordinary gradient for the other siphons being 4·5 to 4·8 feet per mile, with sharper gradients where steel pipes are introduced for deep crossings difficult of access, to reduce the size of the pipes. Only one line of pipes has hitherto been laid; but eventually the siphons will consist of three lines of pipes, 42 inches in diameter for the normal gradients of 4·5 to 4·8 feet per mile, 39 inches for the long siphon with the gradient of 6·87 feet per mile, and reduced to 36 inches, and 32 inches diameter where steel pipes are used for passing under the Manchester Ship-Canal, and the River Mersey respectively, with a sharper gradient. Automatic valves, in suitable positions, begin to gradually close the pipes directly the velocity of flow is considerably increased by the bursting of a pipe; and air-valves at all the summits of the siphons, and on some of the falling lengths, provide for the escape of the air which, if allowed to accumulate in the pipes, would arrest the flow of the water.

Water pumped up from a river, or a deep well, is conveyed in a line of cast-iron pipes to the service reservoir for distribution; whilst water collected from springs issuing at a good elevation, is carried by an aqueduct similarly to the supply from a storage reservoir.

Service Reservoirs.—The supply of water conveyed by an aqueduct from an impounding reservoir, or pumped up from a river or wells, or collected from the flow from springs, is delivered into one or more service reservoirs, situated at a sufficient height above the town to be supplied, to be carried by gravitation through cast-iron mains to the top of the houses in every part of the town, and at an adequate distance if possible from the town, to be free from contamination by smoke, fumes from manufactories, and other sources of pollution. Where a town is built on land varying considerably in level, and the water has to be pumped up into the service reservoirs, the town is advantageously divided into two or more high-level and low-level districts served by reservoirs at different elevations, to avoid lifting the supply higher than necessary for securing the requisite head for distributing the water at the different levels. When these reservoirs have to be placed close to a town, they must be covered, to secure them from the introduction of soot, dust, and dirt; and the covering over of the reservoirs, commonly effected by brick arches, possesses the further advantages, of keeping the water at a more uniform temperature, cooler in summer, and less exposed to frost in winter, and, together with a depth of water of not less than 15 feet, of impeding the growth of aquatic plants, the germs of which exist in some waters.

Service reservoirs are formed by excavating them in the ground, or by surrounding them with watertight embankments or retaining walls; and they are lined, if necessary, with puddle, concrete, or brickwork coated with cement. Their chief objects are to equalize the pressure, which cannot be secured by pumping directly into the mains; to provide for the variable consumption of water at different periods in the 24 hours, without varying the influx or the rate of pumping; and to supply adequate storage to meet sudden emergencies, such as a fire, the bursting of a main, an accident to the aqueduct, or a breakdown of the pumps. The consumption is largest in the morning hours from about 7 o'clock, reaching a maximum between 9 and 10; and it falls to a minimum in the middle of the night. The storage available should evidently not be less than about a couple of days' ordinary consumption, and might have to be extended to a week's supply, or even more, if the conveyance of the water to the reservoir is liable to occasional interruptions, owing to injuries of the aqueduct or other causes.

The reservoirs recently formed by embankments near the Thames at Hampton and Staines, resemble service reservoirs in their construction, and in being filled by water pumped up from the river; but in reality they fulfil the functions of storage reservoirs, by being filled whilst the river has a large discharge, in order to store up water for supplementing the supply of London when the flow of the river becomes very low during a dry summer.

PURIFICATION OF WATER.

Water is very liable to contain impurities, both in suspension and in solution, composed of inorganic and organic matters. The heavier particles carried along by a current readily settle when brought to rest in a settling tank, or in an impounding reservoir; whilst the finer and floating matter, with the exception, in some cases, of organic matter in an extremely fine state of division, can be removed from the water by straining it through fine wire gauze, and by filtration. Calcium bicarbonate, generally found in solution in spring and river waters, is easily removed; and other inorganic salts usually existing in solution in much smaller quantities, which cannot be separated by any practicable process, are harmless; but organic matter in a very fine state, or in solution, is difficult to deal with, and being very objectionable even in minute quantities, has led to special care being exercised in the selection of the sources of supply, to avoid its introduction into the water as far as possible. Water collected in a reservoir from a hilly, uninhabited watershed, spring water, and water from deep wells are usually adequately free from organic impurities; but river water, which sometimes constitutes the only available source of supply for towns situated on flat, alluvial plains, generally contains some organic matter.

Separating Weirs.—When water from mountain streams crossing an aqueduct, is admitted into the aqueduct, provision has to be made to prevent the influx of these streams in flood-time, when liable to be turbid, and discoloured by peat. This is effected by erecting a separating, or leaping weir across the stream where it crosses the aqueduct; and the clear, tranquil stream, falling gently over this weir, drops through an aperture at the foot of the weir, into the aqueduct; but when in flood and turbid, and consequently flowing rapidly over the weir, the stream leaps over the opening and is discharged beyond the aqueduct.

Settling Reservoirs. — The water drawn from impounding reservoirs is free from the heavier particles in suspension which, if not previously intercepted in a subsiding reservoir, or carried away along a bye-channel when the river supplying the reservoir is in flood, are deposited in the reservoir itself before reaching the intake of the aqueduct; and the floating *débris*, and any light, flocculent matter, are arrested by the strainers at the entrance to the aqueduct. River water, however, even when drawn from the upper layers of the stream, which always contain the smallest amount of suspended matter, and especially when the river is in flood, contains a considerable amount of materials in suspension, which, though too fine to be caught on strainers, deposit to a considerable extent when left perfectly still for a time in a settling reservoir. The amount of purification thus effected from the heavier and coarser matter, depends upon the general nature of the sediment brought down by the river; but generally a certain proportion of the finer and lighter impurities are not removed by this process of subsidence, as well as the substances in solution.

Filtration.—It is the general practice in this country to pass water destined for domestic supply, through filter beds before discharging it into the service reservoirs, thereby removing the fine, light particles which do not separate by subsidence. This process of purification, which is essential for rendering most river waters suitable for supplying towns, more especially in very unfavourable instances, such for example as the turbid River Húgli, affording the only available supply for Calcutta, is usually advisable even for the comparatively pure waters collected in impounding reservoirs, owing to the possibility of their contamination at some part of their course, and the greater liability of the purest waters to be affected by pollution than waters already impregnated with mineral salts. Thus the water from the Vyrnwy reservoir passes from the aqueduct through filter beds at Oswestry (Fig. 348, p. 535), before being conveyed to the Prescot service reservoirs, 3½ miles from the Town Hall of Liverpool; whilst straining wells have been relied upon for clearing the water from Thirlmere, on issuing from the Prestwich service reservoir of 4½ acres, on its way to Manchester, 4 miles distant.

Filter beds are placed in shallow reservoirs, usually lined with brickwork, from the bottom of which drains lead the filtered water into the pure-water tank, from which the water is conveyed to the service reservoir. The filtering materials are laid in successive horizontal layers over the floor of the reservoir; and the top layer consists of 6 inches or more of sharp sand, which constitutes the actual filtering medium, arresting the impurities contained in the water, and the surface portion of which has to be periodically scraped off, and washed by a jet of water before being replaced. Below this sand comes a layer of coarse sand or fine gravel, then a layer of coarse gravel, and, lastly, about a foot of rubble at the bottom, or sometimes bricks laid loose and dry, or tile drains, the whole of the layers occupying a thickness of about 4 feet, through which the water is passed gradually, to ensure which, the depth of water over the filter bed described should not exceed about one foot. If, however, a thicker layer of sharp sand is adopted, the head of water may be augmented, owing to the slower passage of the water through a somewhat fine, thick layer. A flow of 2½ to 3 gallons of water an hour per square foot of filter bed, has proved efficient for the purification of river water such as that of the Thames, necessitating a provision of about 16,500 to 14,000 square feet of filter bed for each million gallons per day of supply. To this, however, must be added an extra area to allow for some filter beds being out of use for cleansing or repairs, depending mainly on the amount of impurities in the water supplied.

Filtration appears not merely to remove the suspended impurities in the water, but also to reduce the organic pollution. This may be traced to the aëration, and consequent oxidation of the nitrogenous compounds contained in the water, just as storage for some time in a reservoir, or a prolonged flow in a river, has been found to reduce the organic impurities. Moreover, the slower the rate of filtration, and the thicker the top layer of sand, the more perfect is the purification effected. The aëration which results from the ordinary process of filtration, is sometimes increased by causing the water flowing into the reservoir for filtration, to fall down a series of steps formed at the side, or to be discharged, like a fountain, from a pipe rising some feet above the water-level in the centre of the reservoir.

Purification of Water by Contact with Iron.—Very impure river waters have been efficiently purified by using spongy iron mixed with gravel, or in some instances carbide of iron, as the filtering medium. The iron, however, is costly, and becomes rapidly clogged with deposit, necessitating frequent breaking up of the hard top crust and renewals. A more economical and rapid, as well as an equally efficient system, subsequently introduced, consists in causing the water to flow through long, horizontal, revolving cylinders, provided along the inside with projecting, curved plates fastened at intervals round the cylinder, which, as they revolve, scoop up the cast-iron borings and turnings introduced into the cylinder, which have fallen to the bottom, and discharge them into the water on reaching the top, so that these particles of iron are constantly falling through the flowing water.[1]

[1] *Proceedings Inst. C.E.*, vol. lxxxi. p. 279, and plate 12.

The iron, being brought in contact with the water, is first converted into ferrous oxide, FeO, which is partially dissolved as bicarbonate, and then by a further oxidation into ferric oxide, Fe_2O_3, which is readily precipitated, and can be easily removed by filtration through sand, only a small portion of the oxide remaining in solution. This combination of the iron with the oxygen of the water purifies the water from organic matter, partly in consequence of the presence of iron being prejudicial to the growth, or even existence of bacteria, and partly owing to the precipitate of ferric oxide, in settling, dragging down with it the finely divided organic matter in suspension.

Softening of Hard Water.—Water derived from limestone formations, or from springs or deep wells in the chalk, is termed *hard*, owing to its property of decomposing or curdling soap, which is due to inorganic salts in solution, mainly salts of lime, and especially calcium bicarbonate. This *hardness*, though not apparently at all prejudicial to health, is very disadvantageous for washing, and for manufactures necessitating the use of soap; and boilers, hot-water pipes, and kettles become encrusted with lime deposits, owing to the soluble calcium bicarbonate being converted into the insoluble calcium carbonate by the driving off of carbonic acid gas in boiling. Hard water, accordingly, is to some extent softened by boiling; but the most efficient softening process consists in adding lime-water or caustic lime, CaO, in a suitable proportion to the water, so as to combine with half of the carbonic acid, CO_2, in the calcium bicarbonate, $CaO2CO_2$, in solution in the water, thereby forming a double portion of insoluble calcium carbonate, $CaOCO_2$, which consequently is precipitated, and can be removed. The actual reaction which takes place in this softening process is represented by the formula $CaO + CaO2CO_2 = 2CaOCO_2$. The fine precipitate of calcium carbonate requires some time to settle and leave the softened water clear; and, consequently, in certain softening powders, some alumina is mixed with the lime to hasten the settling of the precipitate. Water after being softened has been found to contain less organic matter than before, which must be due to the precipitated calcium carbonate carrying down with it the very finely divided organic matter in suspension in the water.

In order to compare the relative hardness of different waters, a scale has been adopted in which each grain of calcium salts contained in a gallon of water, is represented by 1° of hardness; and all waters containing more than 5° on this scale, or over five grains of calcium salts in a gallon, are denoted as hard. The waters of some rivers from which supplies are obtained, are decidedly hard, such for instance as the Thames with 15°, or even more, of hardness, and the Trent with 27°; whilst the water from chalk wells usually exceeds 20°, and is occasionally much harder. Where the hardness is mainly due to calcium bicarbonate, as in Thames water and wells from the chalk, it can be for the most part removed by the softening process, by which water of 22° has been reduced to 5°; but so far as the hardness consists of calcium sulphate, it cannot be reduced, so that the Trent water, with 21° of its hardness due to this salt, must remain permanently hard.

The composition of river water varies with the discharge of the river, being mainly derived from springs in dry weather, and from the flow of rainfall off the ground in flood-time; whilst the matter carried down in suspension increases with the velocity of the current, and the flow off the ground.

Distribution of Water.

The distribution of the water from the service reservoirs is effected by cast-iron mains, which, on reaching the town, branch off along each street, from which the service pipes to the several houses diverge. The mains and service pipes should be laid at least 3 feet below the surface of the ground, to be secure from injury by the passing traffic, and to be protected from frost, especially in the event of any stoppage of the flow, which, in very severe weather, may result in the freezing of the water in shallow-laid mains, thereby cracking them, and causing the cutting off of the water-supply from a whole district for a long period; and each service pipe should be controlled by a stop-cock before entering the house, so that the water may be shut off in the event of an accident in the house, or for repairs, without interfering with the general supply. The discharge through pipes can be calculated with accuracy by the simplified formula, $D = AV = \dfrac{\dfrac{1{\cdot}811}{n} + 41{\cdot}6}{1 + \dfrac{41{\cdot}6n}{\sqrt{R}}} A\sqrt{RS}$ where D is the discharge in cubic feet per second, A the sectional area of the flow in square feet, V the velocity in feet per second, n the coefficient of roughness, varying between 0·009 and 0·04, R the hydraulic radius in feet, and S the slope. Tables, moreover, have been calculated from this formula for the most ordinary sizes of pipes and falls, which greatly facilitate the determination of the requisite sizes of pipes and culverts under the conditions usually met with in practice.[1]

Supply of Water required per Head of Population.—The volume of water required for supplying any town or district is reckoned in gallons per head per day, so that, having determined the quantity to be supplied per head, it is only necessary to multiply it by the population to be supplied to obtain the daily volume which has to be provided; whilst the future probable increase in the demand which has to be taken into consideration in selecting a source of supply, depends upon the estimated growth of population during the period to be provided for in the scheme. The estimated supply per head, which necessarily varies with the habits of the population, does not merely deal with domestic supply, which depends largely on the frequency of water-closets and baths, but also includes watering the streets and roads, flushing the sewers, water for stables and gardens, and special supplies for trade and manufacturing purposes, fountains, and extinguishing fires, mainly determined by the conditions of each locality.

[1] "Canal and Culvert Tables," 2nd edition, L. D'A. Jackson, pp. 129-179.

The actual consumption in different towns exhibits a very wide range, being very small in the absence of water-closets and baths, and with a careful administration, and large where baths are very common, the demand for public purposes or trade requirements is considerable, and where waste is not kept under control. An average consumption with a plentiful supply kept under supervision as regards waste, and under normal conditions, may be estimated at from 25 to 35 gallons per head per day. The consumption is lowest during the cold winter months, and reaches a maximum in the hottest summer months.

Intermittent and Constant Supply.—Formerly, with the object of preventing waste, a cistern in each house adapted to its size was filled with water from the main, by turning on the water for a short period once or twice a day; and by this means the consumption was limited to the contents of the cistern between the periods of turning on the water; and the supply to be provided could be very closely estimated, and was only required at fixed periods. These open cisterns, however, were exposed to various sources of contamination, and were rarely cleaned out; water might not be available in an emergency; and in the event of a fire, perilous delay was occasioned by having to summon the turncock before water could be procured. Accordingly, though the old system is still in existence in several places, the constant system of supply is being generally adopted.

A constant supply is obtained by drawing the water through the service pipe direct from the main, under the pressure due to the head provided by the elevation of the service reservoir, so that the water is always obtainable fresh from the main, without being exposed to pollution, and to any extent required. The increased pressure, however, involved in changing from an intermittent to a constant service, necessitates careful overhauling and modification of the house fittings, to prevent leakage, which, in a constant service, if unchecked, may greatly raise the general consumption, and impose extravagant demands upon the water-supply.

Prevention of Waste.—Waste can only be prevented with a constant supply, by insisting upon the provision of suitable screw-down taps, and proper ball-cocks for the flushing cisterns, with the overflow pipe so placed that any escape of water cannot fail to be observed; and the sources of waste and the need for repairs can be detected at night, when the flow is mainly due to waste, by measuring the flow with a waste-water meter temporarily attached at a suitable point on the main where waste is suspected, and by further locating the escape by a sounding bar, which, when put in contact with a pipe, renders any flow audible in applying the ear to the upper end of the bar.[1] Having thus localized the origin of the waste, an inspection of the houses to which the waste has been traced, enables the cause to be detected, and the waste to be stopped by repairs or other measures. Leakage and careless waste of water, if unattended to, are liable to become so great in the aggregate, that their prevention serves to reduce notably the annual cost of the supply,

[1] *Proceedings Inst. C.E.*, vol. xlii. p. 143; and vol. cxvii. p. 147.

and, moreover, by enabling the supply to be extended to a larger area, defers the period when additional capital has to be expended in providing an increase in the supply.

Water-Meters.—Large supplies of water for trade purposes are commonly measured by meter, and charged for according to the quantity supplied; but domestic supplies in England are usually paid for in proportion to the rateable value of the house, and are charged for even when a breakdown, or a fracture of the mains prevents the proper supply being delivered. This convenient arrangement for the water companies may be one of the reasons why the system of payment by meter has been so little extended in this country; but payment according to the actual quantity of water supplied, is evidently the fair system, and would powerfully conduce to the prevention of waste. Two objections have been raised against the use of meters for domestic water-supply, namely, the cost of supplying meters to all the small houses, and the inexpediency of encouraging the poorer population in large towns to restrict their use of water. Water-meters, however, can now be supplied at a moderate cost, and have been successfully introduced into some large cities, as for instance in Berlin in 1878; and it would be easy to get over the objections brought against the system of payment by meter, either by retaining the existing system for houses below a certain rental, or by making a fixed minimum charge for the quantity of water considered requisite in small tenements, and only beginning to charge by meter, in such cases, for any excess of water beyond the sanitary allowance provided for.

There are two kinds of water-meters, namely, positive or piston meters, and inferential or turbine meters. Positive meters measure the actual quantity of water supplied, by admitting the water into a cylinder of a known capacity, at the bottom and top alternately, thereby moving a piston up and down the cylinder, the strokes of which are recorded by suitable wheel-work. Inferential meters contain a small turbine which is turned by the flow of the current passing through the meter; and the rate of rotation, varying with the velocity of influx, affords a measure of the quantity of water supplied, as recorded on an index. Positive meters are the most accurate, and record correctly very small flows of water; whereas turbine meters are cheaper, and though sometimes liable not to revolve under a small flow turned on very slowly, they are sufficiently correct in recording supplies under ordinary conditions, if periodically inspected.

Different Supplies for Special Purposes.—Whilst a higher standard of purity is being required for the water supplied for domestic use, water of unquestionable purity is becoming year by year more difficult to obtain in very populous countries; and therefore pure water has often to be procured at a large cost from long distances; or a considerable expenditure has to be incurred in the adequate purification of river water, generally much nearer at hand, but gradually becoming more liable to pollution, owing to the constant increase of population along the valley of the river. Much of the water thus supplied at a large cost, is used for purposes for which pure water is not at all needed;

and the problem of procuring an adequate supply of pure water for a rapidly increasing population, would be greatly simplified if any unpurified sources of supply in the neighbourhood could be utilized for watering the streets, flushing the sewers, extinguishing fires, and manufacturing purposes, reserving the more costly pure supply of water for domestic use. This system has already been to some extent resorted to in Paris, where the streets are watered and the sewers flushed with water drawn direct from the Seine; whilst water for supplying the houses is obtained from the purest tributaries in the higher part of the Seine basin, deriving a considerable part of their flow from springs in the chalk. In the event of London obtaining the proposed supply of pure water from the Welsh hills, the water drawn from the Thames and the Lea might be used unfiltered for ordinary municipal purposes, and the extinction of fires; whilst the pure Welsh supply would very advantageously replace the river water for domestic consumption. A duplicate system, indeed, necessitates two lines of mains; but a large extension of the water-supply, to meet the demands of a rapidly increasing population, might be secured by laying down a second line of mains for the municipal supply, and restricting the old mains to the increasing domestic consumption, instead of providing a line of larger mains to convey the extended supply.

CHAPTER XXXIV.

SEWERAGE WORKS.

Enhanced importance and difficulty of sewage disposal—Primitive methods, objections, in rural districts—Precautions to be adopted in houses, traps, ventilation, disconnection, flushing—Sewers, conveniently carried along roads, materials used ; forms, object of egg-shape, circular form suitable for very small and large sewers, ordinary typical sections, Paris sections with bench, importance of uniformity of current in sewers, inadequate fall rectified by pumping—Outfall sewers, object, large size necessary ; arrangement of London northern outfall sewer ; Clichy outfall sewer, section, object, methods of construction with shield and centerings ; Clichy to Herblay outfall, construction, description, pumping stations, siphons, capacity—Removal of deposit from sewers, methods of temporary flushing, reservoirs for flushing—Storm overflows, to relieve sewers, arrangement—Ventilation of sewers, importance, openings in roadway, shafts, manholes, inlets for fresh air, artificial methods—Hydro-pneumatic ejector for lifting sewage, description, advantages—Separation of rain-water from sewage, advantages, interception of foul water from streets, objection to cost of double system of sewers, conditions for adoption of the separate system.

THE increase of population and its concentration in large towns, the growing demands for improved sanitary conditions, and the urgent necessity of preventing the pollution of rivers and streams which may be required for water-supply, have greatly enhanced the importance and augmented the difficulties of providing efficient systems of sewage disposal.

Primitive Methods of Sewage Disposal.—The midden system, in which the crude sewage was collected in a heap, sometimes for considerable periods, and only occasionally removed and placed on the land, is now universally condemned. Earth, or dry-ash closets provide a less offensive system, and one in which the sewage is more conveniently disposed of on the land, and furnish a suitable method of disposal in rural districts. The pail system has also been extensively employed in some towns in the north of England, notably in Rochdale, and supplies a mode of readily removing undiluted sewage. The objection, however, to these systems in towns consists in their not dealing with the whole of the sewage; and the polluted water from sinks, sculleries, and stables, has to be collected in sewers and dealt with.

Where water-closets are used, the whole of the liquid refuse from each house and stables is often, in country districts, led by drains into a brick-lined underground cesspool, where the solids settle at the bottom, and the overflow is occasionally discharged into a ditch, or is more commonly led into an underground tank, from which it is periodically pumped up and led by troughs to different parts of a garden or field; and when the solids have nearly filled the cesspool, they are removed at night to manure land in the neighbourhood. This system, however, though suitable and economical for the disposal of the sewage of houses standing in fair-sized gardens, which can be thus irrigated and manured by the liquid refuse, is not expedient, even in country villages, for cottages with little or no gardens.

Precautions to be adopted in Houses.—Every pipe in a house leading to a drain, should be efficiently trapped by an adequate seal of water provided by a suitable bend in the pipe, so that there may be no danger of sewer gas penetrating into the house. Each closet should be connected as directly as possible with the soil pipe on the outside of the house; and the soil pipe should be ventilated by being itself carried up about 4 feet above the roof and left open at the top, or by a separate smaller pipe, protected from being blocked up with leaves by a wire covering over its outlet; and the scullery sink outlet pipe, and all other outlet pipes, should discharge freely outside into an open gully trap near the ground-level, connected with the drain. Flushing cisterns of from 2 to 3 gallons capacity should be provided for the water-closets, and a larger flushing cistern in the scullery, for periodically scouring out the drain where the sink gully discharges its contents, which is liable to be blocked by accumulations of grease and become very foul.

Sewers.—In cities and towns, and also in large villages where a general system of drainage is adopted, sewers have to be constructed for collecting the flow from the house drains, and conveying the sewage to the place of discharge, whatever method of disposal may be resorted to. The sewers are generally carried along under the streets and roads, where they receive the sewage from the houses on each side, can be readily inspected by providing manholes in convenient places, can be efficiently ventilated, and can be repaired or enlarged by merely taking up the road. The 4-inch house drains, and small 6-inch to 18-inch sewers, are made of glazed stoneware pipes, except where having to be carried near the surface, and therefore subjected to the shocks of heavy vehicles, iron pipes have to be adopted; but masonry, brickwork, and concrete are used for the larger sewers, iron pipes being only substituted in exposed positions, and for siphons and river and other crossings.

The forms commonly adopted for the principal sewers, are the egg-shaped section for the branch sewers, and a circular section for the larger main, collecting, and outfall sewers. The flow through sewers is very variable, the sewage from houses being largest in the morning, decreasing towards midday, increasing again somewhat in the evening, and falling very low during the night; the influx of refuse from manufactories is more regular, but introduces great variations in the chemical composition of the sewage; whilst the greatest fluctuations in the flow occur when the rainfall is conveyed away by the sewers. The resulting irregularities in the discharge, combined with the small amount of fall obtainable in flat, low-lying districts, render it very important to provide a channel which shall offer the least possible hindrance to a small flow, and thereby avoid an undue checking of the current leading to the deposit of the sediment. This is most effectually accomplished by an egg-shaped sewer, which, with its narrow channel at the bottom (Fig. 353), offers less frictional resistance to the flow than a circular sewer having the same sectional area, owing to the smaller extent of the sides wetted by the current, resulting in a superior hydraulic mean depth, so that the velocity of the current is better maintained. In house drains and small sewers, the small radius renders the curvature adequate at the bottom, and the fall can generally be made larger for the short distances traversed; whilst in the larger main and intercepting sewers, and particularly in large outfall sewers, the larger and more uniform

BRANCH AND MAIN SEWERS.

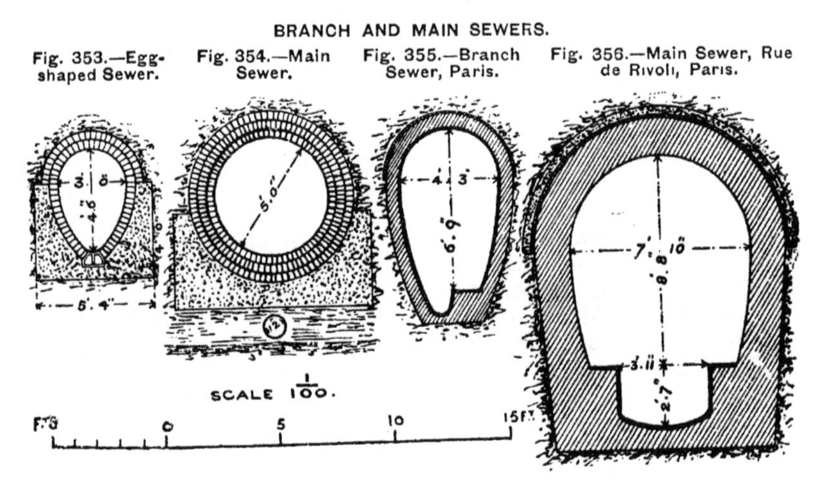

Fig. 353.—Egg-shaped Sewer. Fig. 354.—Main Sewer. Fig. 355.—Branch Sewer, Paris. Fig. 356.—Main Sewer, Rue de Rivoli, Paris.

flow makes the simpler and more economical circular form quite satisfactory.

Typical sections of branch and main sewers are given in Figs. 353 to 356. The first section shows the form of egg-shaped sewer very commonly adopted at the present day, with a bottom block or invert of glazed stoneware or blue brick, having a flat bed, which simplifies the construction, and renders the bottom more stable and easier to lay. In soft or treacherous ground, a bedding of concrete, enclosing the lower half of the sewer, secures the work from disturbance and strengthens it (Fig. 353). The ordinary form for the larger main sewers is circular, with a thickness proportional to the size of the sewer and the pressure of earth on it; and the sewer is bedded in concrete where the ground is soft, and the foundation unreliable[1] (Fig. 354). In Paris,

[1] *Proceedings Inst. C.E.*, vol. xxiv. plate 20.

provision has long been made for facilitating the inspection of the large sewers, by constructing flat benches at the side along which men can walk; and this plan has been more recently extended to the egg-shaped branch sewers, by forming a bench along the bottom at one side, and the channel suitable for a small flow on the other side adjacent to the bench[1] (Fig. 355). The usual form of sewer laid under the main streets in Paris, is shown in Fig. 356, where between the benches a comparatively narrow channel is provided, with a well-curved invert at the bottom, to facilitate small discharges. The Paris sewers are made of very ample section, as they are utilized as subways for the water-pipes placed on brackets high up at the sides, and also for underground telegraph and telephone wires.

The size of the sewers must be proportional to the maximum estimated discharge with the available fall. To prevent the accumulation of sediment in the sewers, the current should be as uniform as practicable, which necessitates a larger fall being given to the branch sewers with their very variable flow; whilst the main sewers, collecting the discharge of the branch sewers leading into them, as well as that of the house drains of the street along which they pass, can be given a smaller fall on account of their greater and more uniform discharge. The available fall, and consequently the size of the sewers for a given maximum discharge, depends upon the conditions of the locality drained, and the distance and relative level of the outfall sewer into which the sewage is eventually discharged. In low-lying districts, where the fall is inadequate in relation to the distance to which the sewage has to be conveyed, an artificial fall has to be created by raising the sewage at a suitable place, by pumping or an ejector, from its low-level sewer to another constructed at a higher level. Where a town is built at different elevations, it is divided into districts, which are served by sewers laid at suitable levels.

Outfall Sewers.—Where the sewage of a town has to be conveyed some distance off to the place of discharge, an intercepting and outfall sewer has to be constructed to receive the collected sewage of the town and carry it to its destination. These outfall sewers have necessarily to be made of large capacity, in proportion to the maximum flow of sewage they may have to receive; but owing to the large area from which they collect the sewage, the volume of liquid passing through them is generally sufficient to keep them tolerably free from deposit with a moderate fall, which is often all that can be provided.

The section of the outfall sewer conveying the London sewage, on the north side of the Thames, from the Abbey Mills Pumping Station to the sewage reservoir at Barking, whence, after precipitation of the solids and purification, it is discharged into the river during the earlier part of the ebb-tide, is shown in Fig. 357, p. 550.[2] The sewage of London along the north side of the Thames is collected into five lines of sewers,

[1] "Salubrité urbaine, Distributions d'Eau, Assainissement," G. Bechmann, Paris, 1888, p. 572; and *Annales des Ponts et Chaussées*, 1895 (1), plate 3, fig. 3.
[2] "Metropolitan Main Drainage and Intercepting Works. Contract Drawings," Sir J. W. Bazalgette, London, 1859–73.

a high-level, middle-level, and low-level sewers, converging together to Abbey Mills, where the sewage of the low-level sewers and the Isle of Dogs branch, is raised 36 feet by pumps to the level of the three other sewers; and the whole of the sewage is conveyed to Barking by the outfall sewer, formed of five similar culverts running side by side. These culverts are built within a single structure of brickwork laid with a fall of 2 feet per mile, founded for the most part on a thick bed of concrete, taken down through the peat of the marshes to a solid stratum of gravel (Fig. 357); but for about $1\frac{1}{2}$ miles near Barking, where the layer of peat has a considerable thickness, it is carried on brick arches supported on concrete piers, 21 feet apart, founded upon the gravel; and the whole structure is encased in an earthen embankment with side slopes, being raised above the low-lying marshes, and carries a roadway on the top. The outfall sewer near Abbey Mills is conveyed over two railways, a road, and a watercourse, by cast-iron culverts similar in dimensions to

OUTFALL SEWERS.

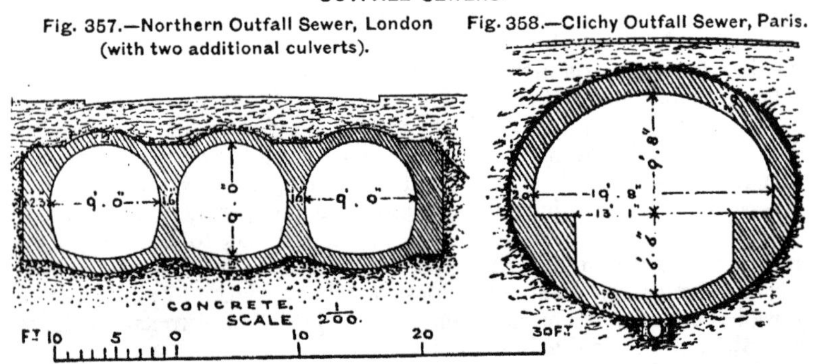

Fig. 357.—Northern Outfall Sewer, London (with two additional culverts).

Fig. 358.—Clichy Outfall Sewer, Paris.

the brick culverts; but further on, the roads are crossed by the brick culverts on bridges.

The new Clichy collecting and outfall sewer, of which a cross section is given in Fig. 358, is built, like most of the Paris sewers, of rubble masonry with an inside lining of Portland cement, and with a protecting lining on the upper outside part of a thin layer of cement concrete or of Portland cement, and with side benches after the usual Paris type, to facilitate inspection and clearance from deposits;[1] and it has been laid with a fall of 1 in 2000, equivalent to $2\frac{2}{3}$ feet per mile. This sewer has been rendered necessary by the increased discharge involved in the decision of 1894 to suppress all cesspools in Paris, and remove everything by the sewers. The Clichy sewer derives its supply from some of the large collecting sewers on the right bank of the Seine, and thus has relieved the Asnières collecting and outfall sewer on the right bank, and enabled it, by a siphon under the Seine a little above the Pont de la Concorde, in its turn to relieve the Alma Bridge siphon, and the

[1] *Annales des Ponts et Chaussées*, 1895 (1), p. 305, and plate 3.

Marceau outfall sewer going from this siphon to Clichy, which was previously overcharged (Fig. 360, p. 552). These three large outfalls unite their discharges at Clichy, whence the sewage is lifted by pumps and conveyed across the Seine for irrigating the permeable, alluvial lands bordered by bends of the Seine at Gennevilliers and Achères. A portion of the Clichy sewer was constructed under the Boulevard National by the process of tunnelling with a shield, where the cover overhead was only from 10 feet down to 2½ feet, in order to avoid the great inconvenience involved in the complete stoppage of a large traffic converging into Paris, in constructing such a large culvert in open cutting in a comparatively narrow roadway.[1] Thirty-eight cast-iron centres, bolted together 3¼ feet apart, were erected immediately behind the shield, and not merely enabled the masonry to be carried forward almost directly after the completion of a length of excavation under the shelter of the shield, but also served as a support for the hydraulic presses in pushing forward the shield, and kept the superincumbent earth and roadway in place after the withdrawal of the shield from under them in its advance. This portion of the sewer, near the surface, has a length of 5752 feet; whilst the remainder, lying within the Paris fortifications, constructed under a second contract and 8450 feet long, was similarly executed by tunnelling, under shelter of a shield, through higher ground, so that the invert of the sewer attains a maximum depth of 128⅔ feet below the surface. In the first length, the arch was built first, and the lower half of the sewer was subsequently constructed by underpinning as the work proceeded; whereas in the second length, the cast-iron frames were made annular to the full elliptical size of the excavation for the sewer, braced together only 2 feet apart, by aid of which the whole ring of the sewer was simultaneously constructed.

The land at Gennevilliers serves only for the purification by irrigation of a portion of the sewage formerly delivered at Clichy and by the Northern outfall; and, consequently, in order to cease polluting the Seine below Clichy, and to provide for the increased discharge resulting from the decree of 1894, it was necessary to extend the irrigated area to lands lower down the Seine valley. This has been provided for by the construction of a new outfall sewer from Clichy to Herblay, about 9 miles long, from which a branch passing under the Seine, leads the sewage on to the lands of Achères; whilst a prolongation of this outfall in the future, with branches, will enable further lands to be brought under irrigation for the disposal of an increased volume of sewage[2] (Fig. 360, p. 552). A longitudinal section of this outfall sewer to its present termination at Herblay, and of its branch to Achères, is shown in Fig. 359, p. 552. This sewer is constructed for the greater part of its length, namely from Asnières, on the left bank of the Seine opposite Clichy, to the Seine at Colombes, and from near Argenteuil to Herblay, as a circular culvert, 10 feet in diameter, formed of masonry, or of concrete or cement mortar strengthened by interlaced iron bars or bands embedded in it, laid with

[1] *Annales des Ponts et Chaussées*, 1897 (1, i.), p. 270, and plate 7.
[2] *Ibid.*, 1897 (1, ii.), p. 14, and plate 9.

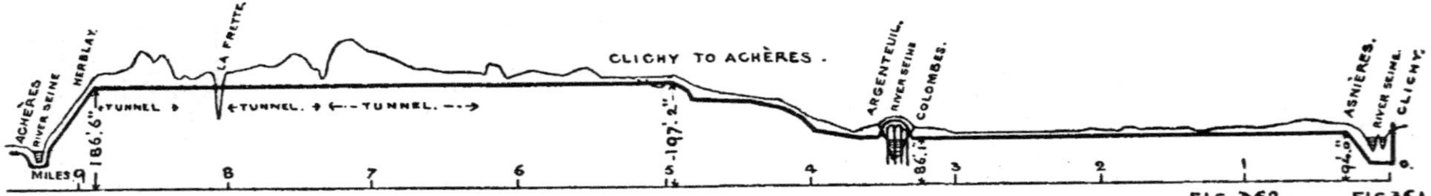

CLICHY TO ACHÈRES.

FIG. 362.  FIG. 361.

MANHOLES.

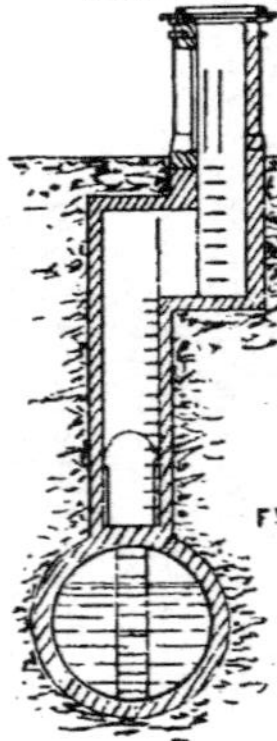

FIG. 364. DEEP SEWER.

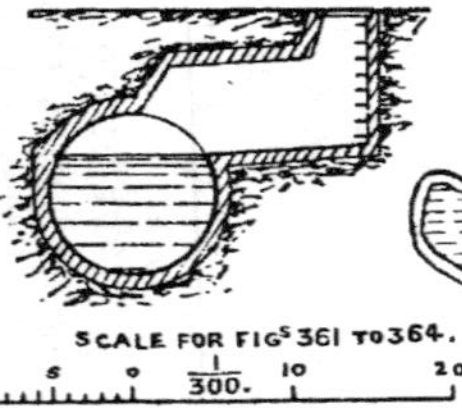

FIG. 363. SHALLOW SEWER.

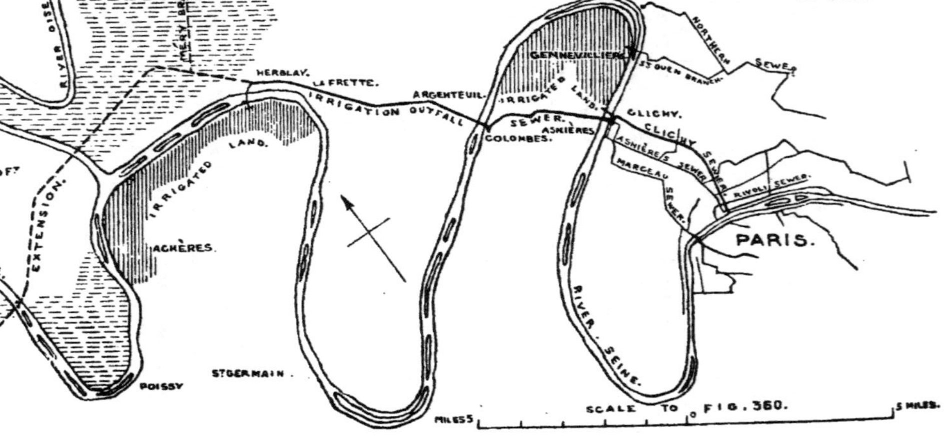

SCALE FOR FIGS 361 TO 364.

FIG. 360.

IRRIGATED AND IRRIGABLE LANDS BY PARIS SEWAGE.

SCALE TO FIG. 360.

a fall of $2\frac{2}{3}$ feet per mile, in which the sewage flows freely under the action of gravity. The culvert was constructed in open cutting between Asnières and Colombes, of an annular form in the deeper cutting (Fig. 362), and with a flat base in lower, very soft ground (Fig. 361); and between Argenteuil and Herblay, it was constructed for rather over half the distance in tunnel, and the remainder in cutting and embankment, on arches, and over a bridge. The sewage is raised $19\frac{1}{2}$ feet at Clichy by pumping it into a tank, from which it passes under pressure through a siphon, 1520 feet long, formed by a cast-iron pipe, $7\frac{1}{2}$ feet in diameter, laid in a tunnel constructed under the Seine by means of a shield and compressed air, and discharged into the culvert, through which it flows freely from Asnières to Colombes (Fig. 359). At Colombes there is a second pumping station, where the sewage is raised under pressure a height of 111 feet, so as to discharge into the culvert conveying it with a free flow to Herblay; and the sewage is carried from Colombes over the Seine, under the roadway of a bridge, in four steel pipes, $3\frac{3}{5}$ feet in diameter, and thence by two pipes, 5 feet 11 inches in diameter, formed for part of their length of steel, and the remainder of cement strengthened by iron framing. The branch from the end of the outfall sewer at Herblay to the Achères domain, consists of two cast-iron pipes, $3\frac{1}{4}$ feet in diameter, dipping down to the Seine, and conveying the sewage under pressure to the siphon crossing under the Seine, 671 feet long, composed of two wrought-iron pipes of the same diameter, which were weighted with old rails and lowered into a trench dredged in the bed of the river, and then protected by a layer of cement concrete; and the sewage is led from the siphon right on to the land at Achères, about 2470 acres in extent, by two pipes of cement strengthened by iron, 1755 feet long. This outfall sewer would be capable of conveying a flow of 344 cubic feet per second; whereas the present discharge from the Paris sewers is only about half that amount, so that a good provision has been made for future increase. Access to the sewer has been provided for by about six manholes, on the average, per mile, two types of which, where the sewer is near the surface, and at some depth below, respectively, are shown in Figs. 363 and 364.

Removal of Deposit from Sewers.—The heavier matters carried into the sewers, especially from the streets, are liable to deposit when the flow is small, and impede the discharge. The sand and crushed stone from the streets is to some extent kept from reaching the sewers by providing catchpits under the gullies; and the amount of this heavy matter which finds its way into the sewers, has been considerably reduced by the substitution of asphalt and wood paving for macadam. Occasionally, in the small sewers, the sediment has to be removed by hand; but generally flushing is resorted to, sometimes by placing a temporary barrier across the sewer, and suddenly removing it, or allowing the pent-up water to rush out underneath it, which barrier is moved by hand, by a trolley, or in the large sewers by a boat.[1] The siphon crossing the Seine at the Alma Bridge in Paris, is scoured out by intro-

[1] "Salubrité urbaine, Distributions d'Eau, Assainissement," G. Bechmann, p. 576

ducing at its upper end a wooden ball a little smaller in diameter than the pipe, so that as the ball is carried through the siphon by the flow of sewage, a strong sluicing current is produced underneath the floating ball, forcing forward the sediment as it advances. The clearance, however, is most conveniently effected, especially in small sewers, by providing flushing tanks at suitable places, filled with water, which by being rapidly emptied create a powerful scouring current. The requisite sluicing is also sometimes carried out by pouring down, through an opening into the sewer, a stream of water from a large hose connected with the hydrant of a water-main.

Storm Overflows.—Where the rainfall in a town, as well as the sewage, passes down the sewers, the flow in the sewers may become very large on the occurrence of a very heavy fall of rain during a brief period, which runs rapidly off roofs, pavements, and roads, into the sewers. When a sewer becomes overcharged, the water rises in the manholes or any other exits it may find, and is liable to flood the streets and basements of the houses; but excessive falls of rain occur so rarely, that it would be unduly onerous to have to construct larger sewers throughout, in order to provide a sufficient capacity to cope with these exceptional cases. Accordingly, these unusual discharges are afforded a vent through storm overflows, which furnish a direct channel for the surplus discharge into the sea, estuary, river, or neighbouring water-course, reached by an aperture at the side of the sewer as soon as the water in the sewer rises above the height at which the sill of the aperture is placed. By this means, the sewers are preserved from being overcharged during the short periods of excessive rainfall at long intervals apart; whilst the chance of pollution by the influx of some of the discharge from the sewers direct into the river, is minimized by the large volume of rain-water with which the sewage is diluted, as well as by the circumstance that this connection only occurs during a period when the river is likely to be in flood, and therefore for a time unfitted by its condition to serve as a source from which to draw water for the supply of towns.

Ventilation of Sewers.—The fluctuations in the flow in a sewer, causing variations in its liquid contents, would produce corresponding variations in the pressure on the air in a closed sewer, so that the air inside would be liable to force a passage through the traps, and introduce sewer gas into the houses. Moreover, it is very important to get rid of the noxious emanations from decomposing sewage as rapidly as possible, so that they may not collect, and eventually find a vent into places where they might prove very deleterious to health. The necessary change of air, accordingly, in sewers has to be accomplished by providing openings for the escape of the foul gases at frequent intervals, depending on the volume and nature of the sewage, and the conditions of the locality. These openings are sometimes made at the surface in the centre of the street, in which case it is expedient to make the issuing gas come in contact with some deodorizing material below the roadway; but it is preferable to provide special ventilating shafts emitting the gases above the traffic in the streets, and away from houses, as for instance at shelters or lamps in the centre of the streets, and in open spaces at the side of the roads in suburban districts. Manholes also are sometimes utilized as ventilators (Fig. 364, p. 552). Inlets for the admission of fresh air have to be provided at a lower level than the outlet shafts; but as, under the variable conditions in a sewer, and alterations in atmospheric influences, an inlet is liable occasionally to become an outlet, if it is unadvisable for this change to occur at any special place, a flap-valve must be provided to close the inlet against an outgoing current of air.

Artificial ventilation can be obtained by producing a continuous current of air through a sewer, by drawing air out from the outlet at its upper end by the draught of a furnace with a high chimney; or an upward current can be created in a high shaft by the revolution of a fan, as in the ventilation of tunnels. A simple expedient for promoting ventilation up small shafts in streets, is by making the shaft furnish the necessary supply of air to a gas lamp placed above its outlet.

Hydro-pneumatic Ejector for lifting Sewage.—In place of centrifugal and lift pumps, an ejector, worked by compressed air, is sometimes used for raising sewage from a low-level sewer to another at a higher elevation, in order to compensate for a deficiency in fall in low-lying districts. The sewage flows into the cast-iron ejector **E**, through the inlet pipe **A** (Fig. 365), and filling the ejector up to the underside of the bell **B** at the top, encloses the air in this bell; and in continuing to rise outside the bell, it finally compresses this air sufficiently to lift the bell and spindle, so as to open the valve **V**, admitting compressed air into the ejector. The compressed air rushing in, drives the sewage out at the bottom of the ejector, into the rising outlet pipe **D**, through which it passes to the higher level required.[1]

HYDRO-PNEUMATIC EJECTOR.
Fig. 365.—Cross Section.

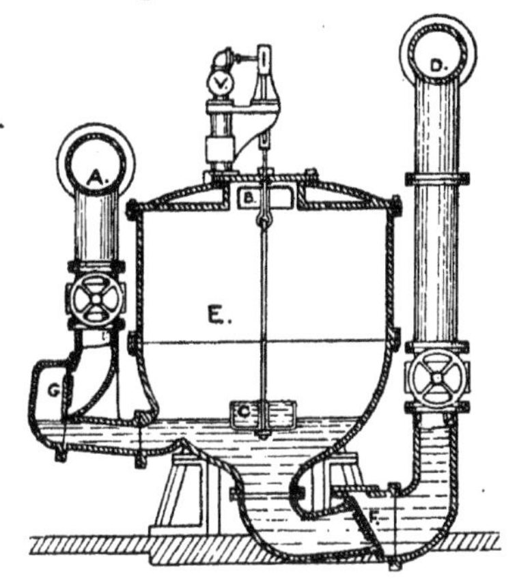

Directly the sewage in the ejector has fallen below the full cup **C** sufficiently to leave it unsupported, the cup descending by its own weight, pulls down the bell and spindle to their original position, closing the compressed-air valve, and enabling the compressed air in the ejector to escape. As soon as the pressure in the ejector is removed, the valve **F** at the bottom of the outlet pipe closes, and the valve **G** at the bottom of the inlet pipe opens, and admits a fresh charge of sewage, to be raised similarly to a higher level by a repetition of the cycle of operations just described. The main advantages claimed for

[1] "The Shone Hydro-pneumatic Ejector," Illustrated description, Chester, 1892.

the ejector are, that there are no parts liable to be injured by the action of the sewage or of grit upon them, or to get out of order; there is a free passage provided throughout for the solids in the sewage; the heavier solids, falling to the bottom of the receptacle, are first ejected into the outlet pipe by the compressed air; a powerful flush of liquid is discharged into the upper sewer at each emptying of the ejector; and the ejector forms a complete severance between the lower and upper sewers. Several ejectors can be worked from one central compressing-air station, enabling a town to be readily divided into several minor, independent drainage districts.

Separation of Rain-water from Sewage.—The addition of rain-water to the sewage produces great irregularities in the flow, necessitates considerably larger sewers and the provision of storm overflows, increases the difficulties of sewage disposal by the augmented volume to be dealt with, and though diluting the effluent, renders any process of purification more costly and tedious. Accordingly, it is very advantageous to keep the rainfall and subsoil waters out of the sewers; but in many cases, especially in large towns where the volume to be provided for is the greatest, as for instance in London and Paris, the separation of the rainfall from the sewage has been considered impracticable. In the smaller towns, however, and in suburban districts, the separation has often been satisfactorily effected; and the problem of dealing with the sewage has been thereby much simplified.

The objection has been raised against the separate system, that it involves the discharge of foul water from the streets direct into the nearest river or watercourse; but this difficulty has been obviated by allowing the dirty water from the first washing of a street by a shower, to drop straight through an opening into the sewer, which, as the flow increases in volume and velocity, the current leaps over when the water has become much purer after the initial cleansing. Moreover, the rainfall falling off the cleansed street, and passing down the gullies direct to the river, is not liable to contain impurities nearly so dangerous to health, as certain infected matters often carried down in the sewage, which, though much diluted, may find an exit into the river during heavy rainfalls, through the storm overflows.

In large towns, the cost and inconvenience of laying a second system of drains and sewers throughout, so as to separate the rainfall from the sewage, offer a very serious obstacle to the introduction of this method; but in smaller towns, especially when a large extension or reconstruction of the sewers has to be effected, it may be possible to utilize the old sewers for the rainfall, and to construct a complete network of new sewers for the sewage, on the most approved modern principles. In suburban and rural districts, moreover, particularly when a system of sewerage is being introduced for the first time, the rain-water may be left to its former means of efflux, and be excluded from the sewers, which, under these conditions, need only be made of adequate size for the conveyance of the sewage; and thus the Drainage Boards are relieved from the unnecessary burden of having to provide for the disposal of the rainfall.

CHAPTER XXXV.

DISPOSAL OF SEWAGE.

Methods of disposal, direct into a tidal estuary or the sea, irrigation, precipitation of solids, bacterial purification—Discharge of sewage into a tidal estuary or the sea, arrangements of outfall, discharge aided by pumping; outfall sewers for discharge of sewage at a distance, for Torquay and Brighton, inadequate for London, clarification adopted; object of system—Irrigation of land with sewage, difficulties, requirements; two systems—Broad irrigation, conditions, methods of distribution, volumes of sewage used for Berlin and Paris sewage farms, rest essential, crops —Intermittent irrigation, with limited areas of land, ridge and furrow system of distribution, rest and digging up required—Chemical processes for the treatment of sewage, clarification by precipitation of solids, sludge and effluent; lime process, merits, objections; ferrous sulphate and lime used at Barking and Crossness, ferric salts preferable; the A.B.C. process; ferrozone and polarite process, mode of action; no system universally applicable, owing to great differences in sewage and local conditions—Electrolytic processes, similarity to chemical processes; sewage decomposed by electrical currents, mode of action; decomposition of sea-water by electricity, used as disinfectant—Settling and precipitation tanks, ordinary open tanks, construction, method of working; cylindrical tanks, description, method of working; modified cylindrical tank with series of chambers, mode of working—Disposal of sewage sludge, pumped into vessels for deposit in the sea; drying of sludge by evaporation, absorption of moisture, or heat; filter press forming solid cakes of sludge, value of sludge, difficulty of disposal, occasionally burnt in destructor—Bacterial purification of sewage, varieties of bacteria, anaërobic and aërobic bacteria in sewage decomposition, changes effected in three stages, nature of changes; best conditions not complied with in irrigation; septic tank and filtration, upward filtration—Concluding remarks.

WITH the increase of population and the growth of towns, the simple plan of discharging sewage direct into the nearest river, stream, or watercourse, so that it may be carried seawards by the current, has had to be abandoned in many parts. In very populous countries, only towns situated on a tidal estuary or the seacoast can, without danger to health or of polluting a possible source of water-supply, discharge their sewage direct into the adjacent estuary or the sea; and even towns so situated have generally been obliged, for their own sakes, to prolong their outfall sewers to low water or beyond, at points where the ebbing tide carries the sewage well away from their neighbourhood, and in some cases, with this object, to construct an intercepting and outfall sewer conveying the sewage to a distant outlet.

Inland towns and large villages in the centre of populous districts, being debarred from any direct discharge of their sewage, have been obliged to resort to preliminary methods of purification, and, more particularly in the neighbourhood of rivers, to some system furnishing an innocuous effluent. Three principal methods, varying largely in details, have been applied to the purification of sewage discharged from sewers, namely, (1) Irrigation of land, or sewage farms; (2) Precipitation of organic matters by various chemical processes, thereby separating the sewage into deposited sludge and a purified effluent; and (3) Bacterial purification of sewage by septic tanks, or by upward filtration.

Discharge of Sewage into a Tidal Estuary or the Sea.—This system merely consists in carrying the sewage by one or more outfalls, at suitable points, to an estuary or the sea, and discharging it if possible during the earlier portion of the ebb, so that it may not be brought back by the returning flood tide; and a hinged flap at the extremity of the outlet, closes the opening as soon as the outside pressure of the water, with a rising tide, exceeds that of the issuing current, and thus prevents the tidal water flowing up the sewer. In a large estuary or the sea, the discharge of the sewage is often allowed to proceed unchecked so long as the level of the tide outside permits, trusting to the mixture of the sewage with the tidal water, or the distance of the outlet, to prevent the creation of a nuisance. This plan has the advantage of prolonging considerably the period of discharge, and thus reducing the reservoir space needed to receive the flow when the outlet is closed by the tide. A reservoir has to be provided, or the sewer has to be enlarged towards its outlet, to afford sufficient capacity for the collection of the sewage during the closure of the outlet, depending upon the level of the outlet, the rise of the tide, and the volume of sewage brought down. Where the outfall sewer is at a low level, the direct discharge of the sewage may be impracticable, or the available tidal period of too short duration to complete the discharge; and pumping must then be employed for effecting or completing the discharge, as exemplified by the London Southern Outfall sewer at Crossness, 11½ feet in diameter and having a fall of 2 feet per mile, where, owing to the requirement of discharging the sewage during the first two hours of the ebb, the sewage has to be pumped up into a reservoir, 6½ acres in area, with a lift of from 10 to 30 feet, from which it is discharged into the Thames within the requisite period.

Where a seaside town is situated in a bay, the conditions are unfavourable for a direct outfall, as the sewage is not effectually carried away by the sea in a sheltered position, and therefore is liable to pollute the beach. Accordingly, the sewage of Torquay has been carried by an outfall sewer to an outlet into the open sea beyond Torbay, within the shelter of which bay Torquay is situated.[1] Moreover, where the

[1] *Proceedings Inst. C.E.*, vol. lxi. p. 144, and plates 4 and 5.

sea bordering a large town is much frequented by pleasure boats, direct outfalls, even if quite unobjectionable as regards the town, may lead to unpleasant appearances in the sea near the shore on calm days. On this account, the eight Brighton outfall sewers which formerly discharged into the sea in front of the town, were intercepted some years ago by a main outfall sewer, $7\frac{1}{2}$ miles long, which conveys the sewage of Brighton and Hove to Rottingdean, about three miles to the east of the town, where the run of the tides and the prevailing westerly winds prevent its return in front of the town.[1]

The main drainage of London, which in 1864 and 1865 diverted the sewage of the metropolis from the Thames within the metropolitan area, to northern and southern outfalls in the river, about 11 and 13 miles respectively below London Bridge, effected a great purification of the river in the neighbourhood of London, and appeared for a time to have solved the problem of the disposal of the sewage of London. Before long, however, it became evident that the pollution had been, to a considerable extent, only transferred to a more remote and wider part of the river nearer the sea. Eventually, in 1887, works were undertaken for the clarification of the sewage by precipitation with chemicals, so that only the liquid products are discharged into the river; whilst the sludge, which settles in tanks, is pumped into special vessels which convey it out to sea for deposit. Consequently, the disposal of the London sewage is no longer a simple discharge of sewage direct into a tidal river, as originally arranged; but it furnishes an instance of the treatment of sewage by chemical means on a very large scale.

The only aim of the system of discharging sewage into an estuary or the sea, is to get rid of an insanitary nuisance in the cheapest possible way consistent with health; and this forms the sole justification for such a system. The system often involves large expenses to render it perfectly safe; it requires a favourable, open situation, with tidal currents and prevalent winds tending to promote the removal of the discharge; and the opportunities for its application are comparatively limited.

Irrigation of Land with Sewage.—The restoration to the land of the constituents which have in a large measure been derived from it, appears to conform to natural laws; but various obstacles present themselves at the outset. Land near a town for a sewage farm is liable to be costly; and the price is enhanced by the opposition generally offered by adjoining landowners, for fear that land used for such a purpose would depreciate the value of the contiguous property. Land at some distance might be cheaper; but it involves more extensive works for bringing the sewage to the land, and renders the produce of the farm more difficult to dispose of. Such land, moreover, should be permeable and somewhat loamy, for soils of this nature oxidize and purify the effluent more readily, and can bear a greater dose of sewage without becoming swampy and sour, than impermeable soils like clay,

[1] *Proceedings Inst. C.E.*, vol. xliii. p. 193, and plate 8.

which is liable to pass sewage through its cracks without purification, and requires considerable preparation to fit it for irrigation. The land also should, if possible, be at such a level in relation to the town, that the sewage can be conveyed by gravitation, as pumping necessarily adds considerably to the yearly cost.

Sewage irrigation would be very valuable for certain crops if the sewage could be applied in a somewhat concentrated form, in the exact quantities, and at the particular periods required; but this is not the way in which a sewage farm of limited extent can be conducted. The sewage must be disposed of over the land with regularity, and, if combined with the rainfall, in larger quantities in wet weather, when the land could dispense with it, than in fine weather. By the adoption, however, of the separate system, the supply of sewage corresponds approximately to the consumption of water, and is consequently larger in hot, dry weather than in the winter, and corresponds, therefore, more nearly to irrigation requirements; but the difficulty of disposal is not thereby entirely removed, owing to the unsuitability of surface irrigation for certain plants at particular stages of their growth.

Two systems of irrigation have been commonly adopted, namely, Broad Irrigation, and Intermittent Irrigation, or Intermittent Downward Filtration.

Broad Irrigation with Sewage.—Broad irrigation is adopted where an ample area of sloping land is available for surface irrigation; and the sewage is gradually distributed over the land. On very sloping ground, a series of contour drains are formed, into which the flow from the land above is successively received and distributed by overflow to the land below; and on flat land, the main distributing channels, formed of brickwork, concrete, or stoneware, are carried at intervals of about 40 feet, on somewhat raised parallel ridges, from which the liquid is distributed down the gently sloping ground on each side. The amount of sewage that can be advantageously distributed over the land necessarily depends on the nature of the soil. On the Berlin sewage farms, where there is a moderate depth of light soil, the volume of sewage that can be disposed of, consistently with due purification, is equivalent, on the average, to between 2000 and 3000 gallons per acre per day, though larger volumes have been often applied; and on the Paris sewage farms, the regulation quantity is about 1000 gallons per acre per day, applied at intervals of three to five days according to the crops. Unless the soil consists of a thick layer of permeable soil, under-draining is necessary to prevent the land becoming waterlogged. It is important, moreover, to avoid the formation of a layer of slime on the surface, resulting from excessive, and too constant sewage irrigation; and the land is preserved in the best condition, and the most satisfactory results are obtained, when the waterings are alternated with good intervals of rest, allowing time for the production of the requisite chemical changes, and the final aëration of the sewage. Various kinds of market-garden produce, small fruit such as strawberries, and certain grasses and root crops have been very successfully cultivated on this system.

Intermittent Irrigation with Sewage.—Often the amount of suitable land obtainable is limited, or the cost of an adequate area for broad irrigation is too great, or the land is too flat to allow of a regular surface flow over it; and in such cases, intermittent irrigation, with thorough under-drainage, is adopted. This is practically a process of filtration through land, the chief object being to pass as much sewage on to the restricted area of land as possible, consistently with an adequate purification of the effluent, the benefit of the crops being quite a secondary consideration in this case; though certain special crops are cultivated successfully under this system. The sewage is sometimes spread over the land as in a tank, which thus serves as a simple filter; but as this course is liable to injure the leaves of the crops at certain periods, a preferable system consists in laying out the land in broad, flat ridges directly over the lines of drains, with wide furrows between. The sewage is then admitted to the furrows, which resemble wide, shallow ditches, till they are nearly filled, causing the liquid to percolate laterally to the roots of the vegetables and root crops grown on the ridges, on its way to the under-drains, and thus irrigates the crops without injuriously affecting them during their growth. This irrigation must be conducted intermittently, so that ample periods of rest may prevent the land becoming choked with sewage; and the ground should be frequently dug up or ploughed, so as to break up the slimy layer which forms on the top as a result of the filtration of the liquid, and to enable the soil to recover from the effects of its large dose of sewage by being well aërated. When the soil is somewhat ill-adapted for irrigation, it may be necessary to resort to some treatment of the sewage with chemicals before applying it to the land; but in such a case, the purification is really a chemical process, with merely a further improvement of the effluent by irrigation.

Chemical Processes for the Treatment of Sewage.—A great variety of processes have been tried for purifying sewage by mixture with one or more chemical reagents.[1] The results attained by the most successful processes, consist in a more or less efficient clarification of the sewage by the precipitation of the suspended solids, and a certain reduction in the organic matters contained in the effluent, mainly owing, no doubt, to the effect of the precipitation (as in the process of softening water) in drawing down the very fine organic particles in suspension, and in some cases perhaps on account of a slight chemical action on the liquid. In all the processes, the sewage is separated into two portions, namely, the solid precipitate known as sludge, which settles to the bottom of the tank in which the process is applied, and has generally to be dried and pressed before it can be disposed of, and the effluent, which is drawn off from the tank in a purified condition in proportion to the efficacy of the process employed, and is discharged into the nearest watercourse, or in some cases is further purified by irrigation on land. As chemicals can do little more than clarify the sewage, and they increase the bulk to be dealt with, the quantities used

[1] "Sewage Disposal Works," 2nd Edition, W. Santo Crimp, pp. 67 to 104; and "Sanitaιy Engineering," 3rd Edition, Mocre and Silcock.

for purification should be kept down as low as is consistent with the requisite precipitation.

Lime has been extensively used for the precipitation of sewage, either alone as milk of lime well stirred up with the liquid, or preferably as lime water, or in conjunction with other inorganic substances. Lime is efficient as a precipitant, and cheap; but unless the sewage is very acid from the refuse products of certain manufactures, the lime renders the effluent alkaline, favouring decomposition of organic matters when discharged into rivers, and injurious to fish. Accordingly, though the alkalinity may be reduced by ample aëration, in being made to flow down a series of steps in its course to the river, the effluent from the lime process is liable to prove a nuisance in rivers during hot weather; and generally lime should only be used in conjunction with some other salts. The addition of a small proportion of choride of lime, or bleaching powder, has been found beneficial in the lime process; and salts of aluminium and iron have often been used in conjunction with lime. Ferrous sulphate, or copperas, and lime are used for precipitating the solids in the London sewage at Barking and Crossness, before discharging the effluent into the Thames.[1] Ferric salts are superior to ferrous salts, both as precipitants and in oxidizing; but copperas possesses the advantage of cheapness. The A.B.C. process was at one time much advocated, its title being derived from the first letters of its original constituents, namely, alum, blood, and clay, though the blood has been abandoned and charcoal substituted; and in this process, the alum acts as the clarifier, and the charcoal and clay as deodorizers. In a somewhat recently introduced process, ferozone, consisting mainly of ferrous sulphate, magnetic oxide of iron, and some silica, is the precipitant, and the resulting effluent is filtered through polarite, composed chiefly of magnetic oxide of iron with silica; and good results are obtained by the ferrous sulphate precipitant being aided in producing rapid subsidence by the heavy magnetic oxide, which oxide of iron, moreover, forming the main constituent of the filter, acts as a powerful oxidizer of the organic impurities in the effluent.

Amongst the numerous chemical processes which have been tried for the purification of sewage, some of them with satisfactory results in particular instances, not one of them has been found universally applicable, or so superior to the rest as to be undoubtedly the best. This may be traced to the variable conditions and dilution of sewage, the differences introduced by manufacturing refuse and other extraneous sources of pollution, the relative cost of the clarifying substances at different places and their influence on the effluent, the standard of purity required for the effluent, depending on the special circumstances of the locality, the volume of sewage requiring treatment, the plant and method of working adopted, and the care and skill exercised in conducting the process.

Electrolytic Processes.—Chemical changes produced by electrical action have been used in two ways for the purification of sewage; and

[1] *Proceedings Inst. C.E.*, vol. cxxix. p. 80.

these processes differ only from chemical processes in the purifying agents being produced by electrolysis, instead of being applied directly, with the advantage generally attributed to salts or gases in a nascent state, of entering more readily into combination with substances they may come in contact with in this state, than under ordinary conditions.

In one electrolytic process, the sewage is made to flow between iron or aluminium electrodes. The electric current decomposes the liquid sewage, causing chlorine and oxygen to be liberated at the positive pole, thereby deodorizing and purifying the sewage; and when iron plates are used, the iron salts formed by the electrolysis precipitate the solids. When, however, the much more costly aluminium plates are adopted, the aluminium hydrate produced acts as the precipitant. An objection has been raised against this system, that some of the sewage is liable to pass the electrodes without being brought under the action of the electric current.[1]

In the second process, sea-water, or a solution of magnesium and sodium chlorides, is decomposed by electricity, the magnesium chloride being converted into magnesium hydrate, which is deposited, and hypochlorous acid, which acts as a disinfectant. The liquid thus produced is either added to the sewage at the head of the sewer, or is used to flush the water-closets and drains.

Settling and Precipitation Tanks.—The sewage to be clarified

SETTLING AND PRECIPITATING TANK.

Fig. 366.—Longitudinal Section.

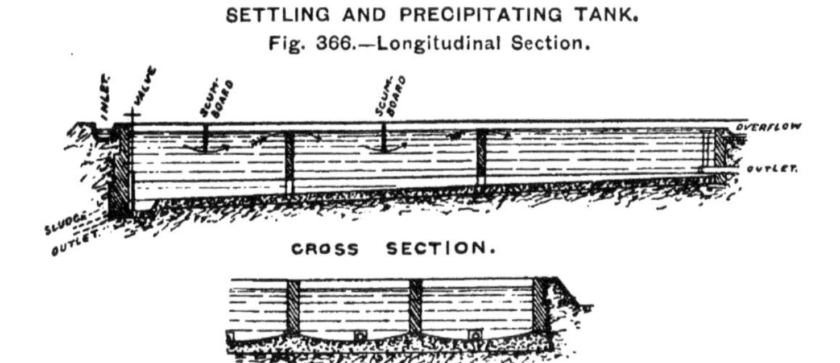

by some chemical process, is discharged from the sewer into a tank either by gravitation, or, if necessary, by the aid of pumps, though it appears expedient, if practicable, to dispense with pumping, as it has been found to retard the precipitation by subdividing and introducing air-bubbles into the solids. The tanks are usually rectangular and oblong in shape, and a few feet in depth, open at the top, and with the bottom dipping down towards the inlet, lined with brickwork faced with a coating of cement or with concrete (Fig. 366); and they are generally constructed above ground, so that they may be emptied by gravitation, and the outer walls protected by an embankment against them. At some

[1] "Sewage and the Bacterial Purification of Sewage," S. Rideal, p. 150.

places, however, as for instance at Frankfort on the Main, they are placed underground and covered over, enabling the sewage to flow into the tanks direct, but necessitating the pumping up of the effluent for discharging it into the river. By means of cross walls, over which the liquid has to flow, and intermediate, floating scum boards, under which it has to pass, a due circulation of the liquid in its flow along the tank is ensured; and the heavier parts of the sewage, and the precipitate formed from the other solids by chemical reagents, settle gradually to the bottom of the tank. During the influx, the excess of clarified liquid flows out at the far end; and at intervals, when an adequate quantity of sludge has been collected, the liquid above it is drawn off at the far end, through a hinged, sloping pipe in connection with the outlet, which is so attached to a float at its upper end, that its orifice is always kept a little below the surface as the liquid is drawn off, thereby avoiding the influx of the scum floating on the top. When the liquid has been removed, the sloping bottom enables the sludge

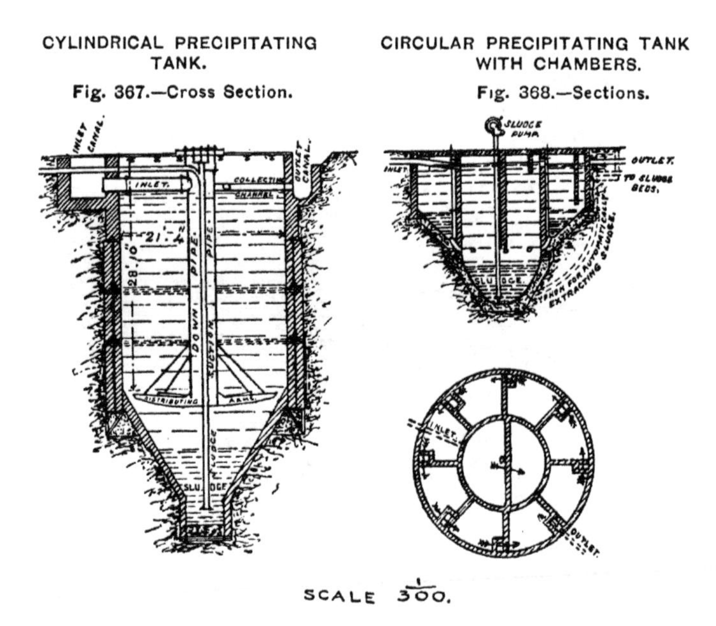

CYLINDRICAL PRECIPITATING TANK.

Fig. 367.—Cross Section.

CIRCULAR PRECIPITATING TANK WITH CHAMBERS.

Fig. 368.—Sections.

SCALE $\frac{1}{300}$.

to be readily drawn off through a pipe at the inlet end of the tank. Sufficient tanks must be provided to receive the sewage during the drawing off of the liquid, and removal of the sludge, from one or more tanks, and to allow for occasional cleaning out.

Where the space available for tanks is very limited, cylindrical tanks sunk down into the ground, with an inverted conical bottom for receiving and concentrating the sludge, have been adopted (Fig. 367). This system was first resorted to at Dortmund, where the sewage rapidly becomes very foul, owing to the fermenting of large quantities of the refuse from numerous breweries; and the ordinary precipitating tanks proved unsatisfactory from the nuisance caused by the putrefying sludge, and are now only used as settling tanks previously to the treatment of the sewage. After being mixed with the precipitating chemicals, the sewage is passed down a pipe in the centre, to the bottom of the cylinder, where it is distributed through the whole area of the cylinder by wooden, radial arms, and, gradually rising through the suspended flocculent matter, effects a thorough admixture of the chemicals with the solids of the sewage, which finally accumulate at the bottom of the inverted cone. The purified effluent, rising to the top, passes off through overflow channels to its destination; and the sludge, whose settlement is promoted by the conical form of the lower part of the tank, is periodically drawn off by suction into exhausted receivers, through a central pipe dipping down nearly to the bottom of the tank, and communicating with the receivers at the top, from which the sludge is readily forced out by merely placing them in communication with the open air. Each cylinder at Dortmund is about 30 feet deep and $21\frac{1}{3}$ feet in diameter; and the cone gives an additional depth to the tank of about 18 feet. The sludge is said to tend to adhere to the sides of the cone, and diminish the purity of the effluent; but it is evident that, provided these very sloping sides are made smooth and of material unaffected by sewage, the sludge must be much less likely to remain on them, than on the comparatively flat bottom of an ordinary precipitation tank.

Another form of cylindrical tank with an inverted conical bottom, is shown in Fig. 368. This arrangement, which was first adopted at Nuneaton,[1] consists of a series of chambers, into which the treated sewage passes in succession, so arranged at each outlet that the sewage has to pass under one wall and over the next, thereby leaving a portion of its suspended solids in each chamber, so that finally it passes out of the last chamber to the outlet as a duly purified effluent. The solids left behind by the sewage in each side chamber, settle to the bottom, and are periodically discharged through sluiceways into the central conical receptacle, in which the sludge accumulates, and from which it is eventually removed by pumping or other means.

Disposal of Sewage Sludge.—The sludge, as drawn off from the precipitating tanks, is too liquid to be dealt with, unless it merely has to be pumped into vessels and taken out to sea for deposit, as resorted to at Barking, Crossness, Manchester, and Salford, or deposited on land and buried, as adopted at Birmingham. In other methods of disposal, the drying of the sludge is a necessary preliminary process. This may be effected in dry countries, partly by absorption, and partly by evaporation, by discharging the liquid sludge into a wide trench excavated in porous soil, or wholly by evaporation in tanks. In moist countries, the evaporation may be accomplished by artificial heat; but the method is costly, and liable to create a nuisance by the fumes driven

[1] "Sanitary Engineering," 3rd Edition, Moore and Silcock.

off. The most convenient means, however, of making the sludge portable, and capable of being utilized as manure, is by pressing out the liquid, and forming the sludge into solid cakes, by putting the sludge between a series of flat plates enclosed within jute canvas, or other filtering medium, to prevent any solids passing out with the liquid, and then applying pressure, which is gradually increased, till the moisture has been forced out, and only the solid cakes of sludge remain. These cakes can then be employed as manure on the land; and in some cases, the solid sludge has a manurial value equivalent to ordinary farmyard manure, and is readily disposed of. The value of the sludge, however, as manure depends upon the original nature of the sewage, and the chemical process used for its precipitation; and very often it is difficult to get rid of the cakes of sludge, sometimes even if given away. Accordingly, in some instances, the dried sludge has been burnt in the furnace of a destructor, together with the other town refuse; and under such conditions, the sludge is merely a waste product formed in the purification of the effluent, at a considerable expense.

Bacterial Purification of Sewage.—Certain occult chemical changes, such as fermentation, have been somewhat lately found to be due to the action of innumerable living micro-organisms, known under the general name of bacteria; and the decomposition of sewage has been recently discovered to result from a similar cause. Bacteria are now known to exist in a variety of forms; and whereas certain bacteria, generated by disease, furnish the most insidious forms of infection, and must be kept out of any water-supply by every possible means, many kinds of bacteria are not merely innocuous, but actually produce beneficial changes in decomposing matter, such as sewage; and their action should therefore be promoted, by placing them under the most favourable conditions for their growth.

Two distinct kinds of bacteria are concerned in the purifying changes produced in sewage, namely, *anaërobic* bacteria, which flourish best in the absence of air, and *aerobic* bacteria, which need air to perform their functions. The first decomposition of the sewage as it is discharged from the sewers, after having undergone a slight oxidation in its passage through the sewers, should be effected by the anaërobic bacteria, and therefore, as far as possible, out of contact with air. In this first stage, albuminous matters, cellulose, and fats are acted upon by the anaërobic bacteria, with the production of soluble nitrogenous compounds, phenol derivatives, gases, and ammonia, resulting in anaërobic liquefaction of the solids of the sewage. There is then a second stage, in which aërobic bacteria begin partially to act in conjunction with a partial continuation of the anaërobic action, so that the presence of a moderate amount of air is necessary; and ammonia, nitrites, and gases are produced. In the third and final stage, thorough aëration is essential, so that the aërobic bacteria may act with their full effect in the nitrification of the sewage; and the ammonia and carbonaceous residues are converted into carbonic acid, water, and nitrates,[1]

[1] "Sewage and the Bacterial Purification of Sewage," S. Rideal, pp. 75 and 207.

thereby completing the purification of the liquefied sewage without the deposition of sludge.

As the bacteria in the sewage are competent, when placed under suitable conditions, to effect the necessary purifying changes, it is unadvisable to attempt to sterilize sewage by chemical reagents, for this arrests the beneficial action of the bacteria; and in order to ensure the fullest efficiency of bacterial treatment, the air should be excluded from the sewage at the outset, then partially admitted, and complete aeration provided towards the end. This order of sequence, as regards aëration, is unfortunately not complied with in the more ordinary processes of irrigation on land; for in broad irrigation, the action of the aërobic bacteria is promoted at the outset, and rest is necessary to enable the anaërobic bacteria to perform their functions; whilst in downward filtration, the proper order of action is also somewhat reversed, preventing the two classes of bacteria from exercising their full influence, and necessitating intermittent action to give the anaërobic bacteria an opportunity of intervening.

One method by which the proper sequence of bacterial changes is sought to be obtained, is by means of a septic tank, and subsequent filtration. The sewage is introduced gently near the bottom of the septic tank, from which air and light are as far as possible excluded to enable the anaërobic bacteria to produce the initial changes undisturbed, under conditions favourable to their action; whilst the succeeding filtration affords the aëration of the sewage requisite for promoting the action of the aërobic bacteria.

A more perfect system, however, for carrying out the cycle of changes under the best conditions, appears to be upward filtration through suitably prepared filters, where the sewage on entering at the bottom is not exposed to the air, leaving the anaërobic bacteria free to perform their functions; next the sewage, on rising in the filter, begins to be aërated; and at length it undergoes perfect aëration by suitable arrangements, such, for instance, as being passed over a series of perforated trays containing coke.[1]

Concluding Remarks on Sewage Disposal.—None of the various methods tried for the disposal of sewage have hitherto resulted, as was at one time hoped, in a commercial success; and the variable nature of the sewage, its dilution by the water-closet system, and the heterogeneous ingredients mingled with it from various sources, present serious obstacles to its being converted into a profitable manure. Irrigation on land, however, though hampered by many onerous conditions, and not complying with the proper cycle for bacterial purification, has proved a satisfactory means of utilizing, and disposing effectually of sewage, with very decided benefit to the irrigated land. Sewage must be regarded as a waste product, like ordinary town refuse, which has to be removed in the most innocuous, cheapest, and, if possible, in a useful way, depending on the circumstance of each locality.

[1] "Sewage and the Bacterial Purification of Sewage," S. Rideal, pp. 209 and 222–223.

Bacterial treatment appears likely to furnish the simplest, most rational, and most efficient means of sewage disposal; and its highly nitrified effluent seems likely to prove very valuable for agriculture, and a means of returning the nitrogen to the soil which it so much needs, and which is entirely lost in some methods of sewage disposal.

INDEX

B

C

E

F

G

H

I

J

K

M

N

O

Q

R

T.

V

W

Y

THE END.

PRINTED BY WILLIAM CLOWES AND SONS, LIMITED, LONDON AND BECCLES.

CPSIA information can be obtained
at www.ICGtesting.com
Printed in the USA
LVHW052303030119
602732LV00010B/108/P

9 781330 199923